PROCESS MINERALOGY VI:

Applications to Precious Metals Deposits, Industrial Minerals, Coal, Liberation, Mineral Processing, Agglomeration, Metallurgical Products, and Refractories, with Special Emphasis on Cathodoluminescence Microscopy.

PROCESS MINERALOGY VI:

Applications to Precious Metals Deposits, Industrial Minerals, Coal, Liberation, Mineral Processing, Agglomeration, Metallurgical Products, and Refractories, with Special Emphasis on Cathodoluminescence Microscopy.

Proceedings of a symposium presented in six sessions on Process Mineralogy held during The Metallurgical Society Annual Meeting, New Orleans, Louisiana, March 2-6, 1986.

Edited by

Richard D. Hagni
University of Missouri-Rolla
Department of Geology and Geophysics
Rolla, Missouri

A Publication of The Metallurgical Society, Inc.

Library of Congress Cataloging-in-Publication Data

Process Mineralogy VI.

"Applications to precious metals deposits, industrial minerals, coal, liberation, mineral processing, agglomeration, metallurgical products, and refractories, with special emphasis on catholdoluminescence microscopy."

Sessions sponsored by the AIME Process Mineralogy Committee.

Includes index.

1. Ores-Congresses. **2.** Mines and minerals resources-Congresses. **3.** Metallurgy-Congresses. **4.** Oredressing-Congresses. **5.** Optical mineralogy-Congresses. **I.** Hagni, Richard D. **II.** TMS-AIME Process Mineralogy Committee. **III.** Metallurgical Society of AIME. Meeting (1986: New Orleans, La.)

QE390.P75 1987 553 86-23576
ISBN 0-87339-054-7

A Publication of The Metallurgical Society, Inc.
420 Commonwealth Drive
Warrendale, Pennsylvania 15086
(412) 776-9000

Printed in the United States of America.
Library of Congress Catalogue Number 86-23576
ISBN NUMBER 0-87339-054-7

Process Mineralogy Committee

EXECUTIVE

CHAIRMAN

Richard D. Hagni
University of Missouri-Rolla
Rolla, Missouri

HONORARY CHAIRMAN

Frank F. Aplan
Pennsylvania State University
University Park, Pennsylvania

VICE CHAIRMAN

Andreas H. Vassiliou
Rutgers University
Newark, New Jersey

PAST CHAIRMAN

James N. Harley
Battelle, Pacific Northwest Laboratories
Richland, Washington

SECRETARY

Susanne Pignolet-Brandom
University of Missouri-Rolla
Rolla, Missouri

SECTION CHAIRPERSONS

APPLICATIONS OF PROCESS MINERALOGY TO:

Mineral Processing
Tsu-Ming Han
Cleveland-Cliffs Iron Company
Ishpeming, Michigan

Metallurgy
Daniel C. McLean
Sacremento, California

Exploration & Evaluation
David J. Carson
Noranda Exploration Co.
Toronto, Ontario

Refractories Minerals
Richard D. Hagni
University of Missouri-Rolla
Rolla, MO 65401
Past chairman, 1981-82

Energy (coal, uranium, oil shales)
Charles A. Wert
University of Illinois
Urbana, Illinois

Industrial Minerals and Aggregates
James A. Soles
CANMET
Ottawa, Ontario

Precious Metals
Donald M. Hausen
Newmont Exploration
Danbury, Connecticut
Past chairman, 1980-81 and 1983-84

Brent M. Hamil
University of Alaska
Fairbanks, Alaska

Liberation (Methodology)
William Petruk
CANMET, Dept. of Energy, Mines, and Resources
Ottawa, Ontario
Past chairman, 1982-83

ASSOCIATE CHAIRPERSONS

Refractories
Larry A. Harris
Oak Ridge National Laboratory
Oak Ridge, Tennessee

Energy (coal, uranium, oil shales)
Beverly A. Chom,yn
Atomic Energy of Canada Limited
Ottawa, Ontario

Precious Metals
Martha B. Sizgoric
INCO Metals Company
Mississauga, Ontario

Liberation (Methodology)
Paul Mainwaring
CANMET
Ottawa, Ontario

COMMITTEE MEMBERS

Won C. Park
Dowell Schlumberger Inc.
Tulsa, Oklahoma
Past chairman, 1979-80

William J. Mallio
Resource Engineering Inc.
Waltham, MA
Past Chairman 1978-79

Jerry R. Odekirk
AMAX Laboratory
Golden, Colorado

James R. Craig
Virginia Polytechnic Institute and State University
Blacksburg, Virginia

J. D. Lowell
Pillar, Lowell and Associates
Tucson, Arizona

FOREIGN REPRESENTATIVES

West Germany
G. C. Amstutz
Der Universitat Heidelberg
Heidelberg, West Germany.

Mexico
Frank L. Marley
Mexico City, Mexico

Australia
Keith L. Henley
AMDEL
Eastwood, South Australia

South Africa
Dieter K. Hallbauer
Chamber of Mines
Johannesburg, South Africa

Preface

This volume contains the proceedings of 39 papers presented in six sessions on Process Mineralogy held during the AIME Annual Meeting in New Orleans, Louisiana, March 3-5, 1986. These sessions marked the eighth consecutive year in which the Process Mineralogy Committee has sponsored sessions at the AIME Annual Meeting since the inception of the committee in 1978. Although the AIME Process Mineralogy Committee was initiated under the sponsorship of the Metallurgical Society (TMS), in 1982 it became a joint committee of TMS and the Society of Mining Engineers (SME) upon the mutual agreement of the two constituent AIME societies. As a result of its joint status, the principal responsibility for scheduling and publishing the Process Mineralogy Committee papers is alternated between the two societies and the present volume has resulted from the efforts of The Metallurgical Society.

This volume is the sixth in a series of Process Mineralogy Committee-sponsored proceedings volumes that have been published by AIME. The initial volume was titled "Process Mineralogy: Extractive Metallurgy, Mineral Exploration, Energy Resources" and was published by TMS. The second volume, also published by TMS, was titled "Process Mineralogy II: Applications in Metallurgy, Ceramics, and Geology". SME published the third volume titled "Process Mineralogy III: Applications in Metallurgy, Coal, Concrete, Smelting and Exploration. Process Mineralogy IV was titled "Applied Mineralogy", published by TMS, and contained the proceedings of the second International Congress on Applied Mineralogy in the Minerals Industry that was held as a pre-meeting symposium in conjunction with the annual AIME meeting in Los Angeles, California in 1984. Process Mineralogy V was titled "Mineralogy-Applications to the Minerals Industry", published by SME, and contained the proceedings of the Paul F. Kerr Memorial Symposium that was held as a post-meeting symposium in conjunction with the annual AIME meeting in New York City in 1985.

The interdisciplinary character of Process Mineralogy requires the involvement of engineers and scientists with diverse backgrounds, expertise, and interests. These diverse areas of applications of process mineralogy were reviewed during the 1986 AIME Process Mineralogy Committee-sponsored sessions and include the following seven areas of application or topics: 1) precious metal deposits, 2) predictive metallurgy in exploration, 3) industrial minerals, 4) coal, 5) liberation, 6) mineral processing, 7) agglomeration 8) metallugical products, 9) refractories, and 10) cathodoluminescence microscopy. Each of these formed a subject of presentation and discussion at the meeting and constitutes a chapter in the present proceedings volume. This volume, Process Mineralogy VI, places an emphasis upon the process mineralogy applications to precious metal deposits, deals especially with applications to mineral beneficiation and liberation, and discusses a wide variety of applications of cathodoluminescence microscopy to process mineralogy problems, a technique that has not been treated in previous Process Mineralogy volumes. The Committee also co-sponsored a session with the Society of Economic Geologists (SEG) that was part of a three-session symposium on nature, distribution and reduction of sulfur in coal, the papers from which will by published separately by SEG.

Dr. Peter T. Luckie, Associate Dean for Research in the College of Earth and Mineral Sciences at The Pennsylvania State University, presented the keynote address, "An Assessment of the Problems of Quantifying Mineral Liberation." While working for the McNally Group from 1972 to 1979 as Corporate Director of Research and Development and for Penn State since 1979 as Professor of Mineral Engineering, his research interest have centered on the analysis of comminution circuitry and gravity separation circuits. It is therefore not surprising that Dr. Luckie became interested in liberation. Working with Dr. L. G. Austin, he continues to analyze liberation phenomenon in order to integrate liberation and comminution knowledge into a predictive model that can be coupled with predictive separation models, thereby producing mineral processing CAD flowsheets.

The recognition of the importance of process mineralogy has grown markedly in the past eight years, and its future appears to be bright. Two Process Mineralogy-sponsored sessions are planned for the fall 1986 AIME meeting in St. Louis, Missouri. A total of six sessions are planned for the next annual meeting to be held in Denver, Colorado. It is intended that the papers from both of those SME meetings be published in the projected Process Mineralogy VII Volume.

Richard D. Hagni
University of Missouri-Rolla
Department of Geology and Geophysics
Rolla, Missouri

Acknowledgements

The following individuals have gratiously served as reviewers for the papers contained in this volume.

F. F. Aplan
G. Barbery
R. B. Berg
R. S. Boorman
M. L. Boucher
T. Bundtzen
Y. P. Chugh
W. L. Cornell
W. G. Davenport
W. M. Dressel
H. M. Erten
A. W. Fletcher
O. L. Forchheimer
R. N. Guillemette
S. K. Grant
T. M. Han
H. Harlan
D. M. Hausen
O. C. Kopp
A. N. Mariano
E. Martinez
D. J. Marshall
D. G. McLean
R. E. Moore
S. W. Nabbs
J. R. Odekirk
W. C. Park
W. Petruk
J. H. Puffer
D. G. C. Robertson
F. H. Sharp
R. P. Stevens
W. G. Staley
G. V. Sullivan
R. L. Voss
J. L. Watson
R. L. Wiegel
C. A. Wert

Table of Contents

PROCESS MINERALOGY APPLICATIONS TO REFRACTORIES

Keynote Address

AN ASSESSMENT OF THE PROBLEMS OF QUANTIFYING MINERAL LIBERATION

L.G. Austin and P.T. Luckie

Mineral Processing Section
The Pennsylvania State University
University Park, PA 16802

A review is given of the key recent publications concerning the theory of liberation in grinding of ores, with a view to combining models of liberation with models of size reduction. Although a number of approaches are promising, it is concluded that there is no adequate model of liberation of binary systems suitable for incorporation into a mill model. A simplified partial model is derived and future development is suggested.

Introduction

Mineral Processing engineers convert worthless ores into marketable concentrates by selectively separating valuable constitutent particles from gangue constitutents. The desired particles are produced by comminution, the unit operation that dominates the capital and operating costs of a concentrator. It is for these reasons that there has been extensive development of predictive comminution engineering models. Although there has been an impressive evolution of these types of models with their ability to predict the product capacity and size consist of a size reduction circuit, it is the liberation process - the creation of valuable constitutent particles - that is of interest to mineral processing engineers. Thus, predicting the product size consist is necessary, but not sufficient. Within each narrow size interval, we would like to know the volume percent of valuable constitutent in each particle after some period of grinding.

The importance of such a prediction was recognized by Gaudin (1), who described the degree of liberation as the percentage of that mineral or phase occurring as free particles (consisting of a single mineral) in relation to the total of that mineral occurring in all particles. Of course, such free particles are rare birds, which means that a certain amount of associated gangue is acceptable. However, this in turn makes the definition of liberation dependent upon the separation process - what is liberated for gravity concentration may not be liberated for separation by froth flotation. Gaudin derived a simple geometric liberation model for a binary mixture with a lean phase which allowed him to predict free particles created during random fracture. He stated that liberation by "detachment", i.e., selective fracture along grain boundaries, must be determined by size consist analysis. Such detachment should produce a bimodal distribution. Unfortunately, since a bimodal distribution can be produced by other mechanisms, its existence is not unique. Gaudin also derived a "locking factor" since he realized that a microscopic examination of the products of liberation would "overstate the liberation of the material under consideration".

Because of the unrealistic nature of the geometry of Gaudin's model, other researchers have continued to derive appropriate liberation models. However, Gaudin set the stage for things to come -

- the model will have to incorporate theoretical and empirical aspects
- the model will have to handle both random and non-random brittle fracture
- the evaluation and parameter estimation techniques will also require further development.

Indeed, because of the need for model evaluation and parameter estimation, other researchers have continued to derive techniques for examining the host rock and the products of liberation. As important as this area may be, this paper gives only a brief review of this field, primarily to show how it interacts with liberation modeling.

This paper primarily reviews the current state-of-the-art in formulating models of liberation, with particular emphasis on the problems involved in joining such models to a realistic size reduction scheme. It is not intended to be a comprehensive review of the rapidly-growing literature on all aspects of liberation, but rather it selects key papers for illustration of the various approaches.

Definition of the Problem

The term "A model for liberation" has several distinct meanings current in the literature, ranging from prediction of the size distribution of B grains occluded in a host rock A from image analysis measurements on a polished section, to prediction of the entire liberation matrix of material leaving a grinding circuit. In this paper a "binary liberation model" is defined in a more specific sense, (2): if a feed material AB containing a known proportion of B is reduced in size to a measured product size distribution with a density function of p(x), a liberation model predicts the function f(r x) such that the fraction of product material which is in size range x to x+dx and has a composition range of r to r+dr is

$$p(x,r)dxdr=f(r\ x)p(x)dxdr \tag{1}$$

where r is most conveniently expressed as the volume fraction of B in AB. Particles are pure A for r=0 and pure B for r=1.0. Clearly, Eq. 1 enables the ready calculation of the mass distribution of B in AB (knowing the specific gravities of A and B), the fractional distribution of B in each size-composition range, etc. However, since the object of a liberation model is to provide the information necessary to model a following separation process, the function f(r x) is not generally a sufficient description and the concept used by Schaap (3) and Klimpel (4) must be introduced; where f(r x) is the sum of the fraction of the material in size range x to x+dx which has a composition range of r to r+dr, but where the B grains are Wholly- Contained (encapsulated) in the AB particle, and the equivalent fraction where the B grains are Surface-Exposed, or

$$f(r\ x) = f_{WC}(r\ x)+f_{SE}(r\ x) \tag{2}$$

It is self-evident that these two types of material will behave differently in a flotation or leaching operation even though they may appear to be identical to a separation process based on specific gravity.

The simple splitting of f(r x) according to Eq. 2 can only apply under the restrictive condition that each particle contains only zero or one B grain or B fragment, since in principle a particle can contain some B in surface-exposed form and some in wholly-contained form. The more general formulation then is the density function f(r,o x), where f(r,o x)drdo represents the fraction of material of size x to x+dr which is of composition class r to r+dr and contains o to o+do fraction of r as surface-exposed (o=0 is all B is surface-exposed, o=1 is all B is wholly-contained). However, as discussed by Schaap and Klimpel, the common case of a small volume fraction of B in AB, with size reduction to the level where B is being liberated as pure B, gives the conditions necessary for the probability of more than one B grain or B fragment per particle to be negligible. Under actual conditions it is more convenient to express Eqs. 1 and 2 as shown in Tables 1 and 2 (5).

The approaches to providing the functions of Eqs. 1 and 2 or the matrices of Tables 1 and 2 can be generally grouped into three major classifications: (1) image analysis of polished sections; (2) geometrical treatments; and (3) back-calculation of empirical functions, as discussed below.

Image Analysis: Measurement of Liberation

There is a large collection of literature on the preparation and optical examination of polished sections for mineralogical analysis, but here we will give only representative selections based on the use of automated scanning

TABLE 1: The mass distribution matrix: surface-exposed; % distribution of volume in surface-exposed (flotable) classes: grinding time, τ=5 minutes, B grain size 420-150μm; mean volume content of B=1%.

VOLUME FRACTION x10^4

Int. no.	Top Size μm	0-10%	10-20%	20-30%	30-40%	40-50%	50-60%	60-70%	70-80%	80-90%	90-100%	Pure B	Sum
5		-	-	-	-	-	-	-	-	-	-	-	-
6		9.00	4.27	2.51	1.40	0.62	0.08	-	-	-	-	-	17.88
7	420	4.95	3.01	2.06	1.74	1.31	1.07	0.92	0.68	0.58	0.47	0.50	17.29
8		2.03	1.49	1.22	1.12	0.82	0.73	0.68	0.60	0.56	0.53	2.90	12.68
9		0.74	0.59	0.52	0.49	0.41	0.39	0.37	0.34	0.33	0.32	4.55	9.05
10		0.26	0.21	0.19	0.18	0.17	0.16	0.15	0.15	0.14	0.14	5.08	6.83
11	105	0.08	0.07	0.07	0.06	0.06	0.06	0.06	0.05	0.05	0.05	4.80	5.41
12		0.03	0.02	0.02	0.02	0.02	0.02	0.02	0.02	0.02	0.02	4.20	4.41
13		0.01	0.01	0.01	0.01	0.01	0.01	0.01	0.01	0.01	0.01	3.57	3.67
14		-	-	-	-	-	-	-	-	-	-	2.98	2.98
15		-	-	-	-	-	-	-	-	-	-	2.48	2.48
16		-	-	-	-	-	-	-	-	-	-	2.04	2.04
17		-	-	-	-	-	-	-	-	-	-	1.69	1.69
18		-	-	-	-	-	-	-	-	-	-	1.39	1.39
19		-	-	-	-	-	-	-	-	-	-	1.15	1.15
20	<5	-	-	-	-	-	-	-	-	-	-	5.49	5.49
Sum		17.1	9.7	6.6	5.0	3.4	2.5	2.2	1.9	1.7	1.5	42.8	94.4

TABLE 2: The mass distribution matrix; wholly-contained; non-flotable classes (see Table 1).

VOLUME FRACTION x10^4

Int. no.	Top Size μm	0-10%	10-20%	20-30%	30-40%	40-50%	50-60%	60-70%	70-80%	80-90%	90-100%	Pure A	Sum
1	3350	241.6	-	-	-	-	-	-	-	-	-		241.6
2		484.2	-	-	-	-	-	-	-	-	-		484.2
3		724.0	-	-	-	-	-	-	-	-	-		724.0
4		885.0	-	-	-	-	-	-	-	-	-		885.0
5		973.4	-	-	-	-	-	-	-	-	-		973.4
6		39.1	9.55	4.78	2.85	1.82	0.69	-	-	-	-	889.2	948.0
7	420	5.9	4.96	1.29	0.78	0.89	0.33	0.21	0.23	0.07	0.02	868.1	882.8
8		0.4	1.02	0.37	0.22	0.59	0.09	0.06	0.06	0.02	0.01	780.1	783.0
9		-	0.02	0.11	0.07	0.10	0.03	0.02	0.02	0.01	-	669.4	669.8
10		-	-	-	-	-	-	0.01	-	-	-	565.7	565.7
11	105	-	-	-	-	-	-	-	-	-	-	469.1	469.1
12		-	-	-	-	-	-	-	-	-	-	389.5	389.5
13		-	-	-	-	-	-	-	-	-	-	320.9	320.9
14		-	-	-	-	-	-	-	-	-	-	265.3	265.3
15		-	-	-	-	-	-	-	-	-	-	219.7	219.7
16		-	-	-	-	-	-	-	-	-	-	181.7	181.7
17		-	-	-	-	-	-	-	-	-	-	150.6	150.6
18		-	-	-	-	-	-	-	-	-	-	125.0	125.0
19		-	-	-	-	-	-	-	-	-	-	103.7	103.7
20	<5	-	-	-	-	-	-	-	-	-	-	522.5	522.5
Sum		3354	15.6	6.6	3.9	3.4	1.1	0.3	0.3	0.1	0.03	6520	9905

instruments, either optical or SEM-electron microprobe, since only such instruments give information which can be quantified with sufficient accuracy for the purposes of mathematical model building. Such instruments, whether they use optical reflectance, back-scatter electron energy density or X-ray fluorescence for discrimination, are all based on measuring chord lengths or exposed section areas or perimeters, as shown in Figure 1. Data can be taken from random lines (beam traverses) or can be taken from systematic parallel traverses covering a whole field of view. The techniques have been applied to polished sections either of rock or of particles mixed in a matrix (e.g. mounting resin).

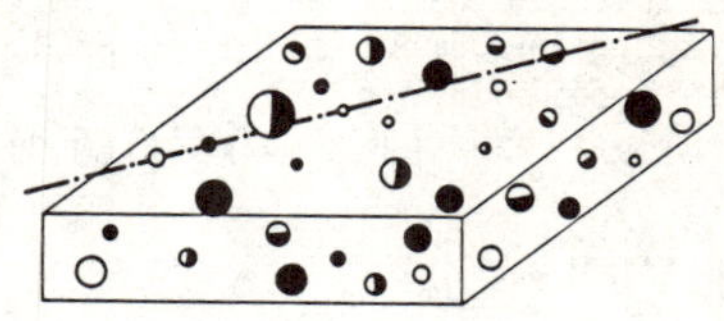

Figure 1. Illustration of problem of measuring compositions from a random polished section of solid block containing randomly oriented spheres of a single diameter consisting of half white and half black: complete ranges of sections from white to black, from zero to D in diameter are seen (14).

Amstutz and Giger (6) have discussed the difficulty of making general quantitative statements about the complex mineralogical morphologies (texture) seen in ore deposits. However, almost all workers in liberation modeling have explicitly or implicitly assumed that as material is reduced in size it behaves as if the original morphology was a size distribution of (unbroken) B grains spaced randomly throughout the volume of a host matrix. We will return to this problem later, but at the moment the discussion is restricted to cases where there is physically a distribution of grain sizes in the host rock. It is clear from Figure 1 that the mean diameter of exposed cross-sections is much less than the diameter of grains in a rock or particles in matrix. It is simple to derive the equations for the distributions of measured diameters (or chord lengths) resulting from any size distribution of spheres, assuming perfectly random cuts through the sphere volumes. Oosthuyzen (7) has given a good discussion of the practical difficulties of achieving valid experimental results for particles mounted in a matrix. In principle, then, it is possible to reverse-calculate the size distribution of spheres leading to the measured areal or chord distributions, thus performing stereological transformation of the measured data.

However, Dinger (8) and Jones and Horton (9), among others, have shown by Monte Carlo simulation or geometrical analysis that random cuts through non-spherical geometries give quite different distributions for different geometries. In addition, the size distributions of, for example, intercept lengths produced from a grain size distribution are not very sensitive to changes in the shape of the grain size distribution, which automatically makes the reverse calculation of the true grain size distribution from measured intercept lengths sensitive even to the smallest experimental errors. It is normally necessary to postulate functional forms (such as a Beta Function) to reduce the complexity of the calculation (9).

King (10) has given a formalism for determining the size distribution of original particles from measurements of distributions of intercept lengths, section areas, etc. based on inserting an experimental measurement of the distributions produced by mounting, polishing and scanning a narrow screen size range of particles (see also Petruck (1), with extrapolation to a single sieve size). This assumes that the distribution of particles shapes (all of the same sieve size) leading to the overall intercept length distribution, or area distribution, etc., is independent of the particle size, which seems a reasonable assumption. He finds typical particles to give

$$P(L/D) = 1-(1-\frac{L}{D})\exp[\frac{-R^2L}{D}] \qquad (3)$$

where P(L/D) is the fraction of accumulated random intercept lengths which is in intercepts of length 0 to L, D is maximum sieve size, 1.2 and R = 2 or 2 as appropriate. Thus, the maximum length of an intercept is about 1.2 times the maximum screen size. However, the use of this equation to describe the sectioned products of grains in a host rock is much more open to question since such grains are not the natural products of mechanical size reduction.

In practice, because of the mathematical complexity and difficulty of calibrating stereological transformations, it is common to find workers neglecting the corrections entirely. Actually, a number of <u>mean</u> parameters can be obtained from image analysis without stereological transformation. For example, the total cross-sectional area of B exposed by cuts divided by the cross-sectional area of AB exposed by cuts is the <u>mean volume fraction</u> of B in AB, if a sufficiently large amount of data is collected over many surfaces or particles. However, if a particle which is mounted and sectioned exposes only B, it does not mean that the particle is pure B, or if it exposes pure A it does not mean it is pure A. It is common to examine single sieve size fractions mounted and sectioned and to calculate "apparent" liberation values defined by

pure A	= particles which show A only in section	
pure B	= particles of which show B only in section	
0-10% B in AB	= particles showing 0 to 10% B in AB in the exposed section, etc.	(4)
Surface-exposed B	= particles where B touches D matrix in the exposed section	
Wholly-contained B	= particles where B is surrounded by A in the exposed section.	

Without some reasonably accurate way of applying the statistics of stereological transforms these values in untransformed form are substantially in error. Always they tend to over-estimate the proportions of pure A and pure B, under-estimate the locked fractions, and over-estimate the wholly-contained B even when averaged over all particles showing exposed A and B together. There is no way of telling how inaccurate the splits of material are in the various composition classes as far as surface-exposed versus wholly-contained is concerned.

Petruk (11,12,13) and Petruk and Hughson (14) have used apparent liberation values with discretion, drawing conclusions by comparitive trends only. Berube and Marchand (15) also use untransformed values with the knowledge that they are incorrect but in the belief that the errors are small; we do not agree, since there will always be consistent errors which do not cancel by fortuitous compensations. Results from the QEM-SEM system are quoted (16,17), in untransformed form as if they were accurate and fully meaningful, which is highly mis-leading. Reproducibility of the analytical data by the instrument is not the question: it is the accuracy of assigning material to the various blocks in Tables 1 and 2 which is at question.

Barbery, Huyet and Gateau (18) discussed the problem of using untransformed values and they showed the errors involved by analysis of a sample prepared by Stewart and Jones (19). This consisted of closely sized particles (495x417 m) of iron oxide-quartz in a narrow specific gravity fraction (3.57x3.68), so that the material was essentially all AB material of almost constant composition. Figure 2 shows the result obtained by Barbery, Huyet and Gateau. They also gave a method of stereological transformation which led to a result much closer to the correct one. The paper did not give sufficient details for the method to be clear, but it has been described again in an excellent recent review by Barbery (20). Unfortunately, the method does not at present distinguish surface-exposed from wholly-contained B material; in addition, Jones and Horton (9) state that since the method involves calculation of the fourth moment of an intercept distribution it is subject to the usual errors involved in taking high order moments. More recently, Moore and Jones (21) and Barbery et al. (22) state that the treatment is mathematically incorrect and Barbery, et al (23) have also used apparent liberation values without transformation.

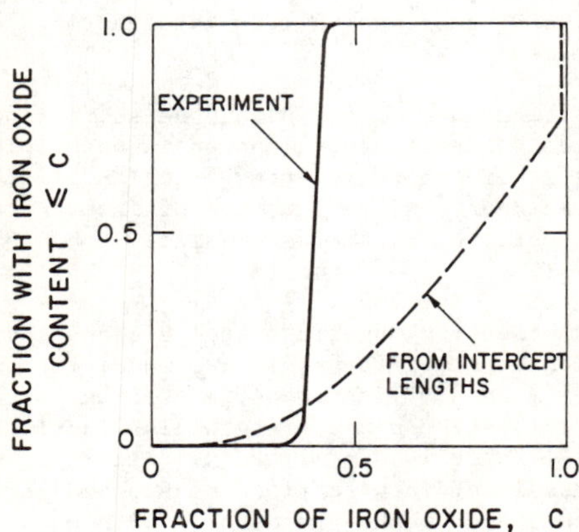

Figure 2. Material of almost uniform composition (∿40% iron oxide shows ∿25% material of 100% iron oxide and ∿10% of material of 0% iron oxide, by intercept analysis (after Barbery, Huyet and Gateau, ref. 18).

Steiner (24), as quoted by King (25,26), intuitively defines the fraction of B still locked with some A by the expression

$$\text{Fraction of locked B in particles of size D} = S_B^i / S_B^c(D) \qquad (5)$$

where S_B^i is the interfacial surface area existing between A and B per unit volume of B and S_B^c is the exposed surface area of B per unit volume of B in particles of size D. If there is no fracture along the interfaces between A and B then S_B^i is conserved and does not vary with D. However, the treatment cannot predict the distribution of B in AB or the fraction which is surface-exposed versus wholly-contained without making further assumptions which are incorrect. It is not correct to assume that any B which appears as wholly-surrounded by A in a cross-section can be assigned to wholly-contained B, since it may be part of a B grain which extends to the particle surface. On the other hand any B section which borders on mounting medium is properly assigned to surface-exposed B. However, the surface area ratio in Eq. 5 can be obtained by the ratio of appropriate perimeters in the exposed sections.

King points out that Eq. 5 has no theoretical basis and he proposes an alternative method for deducing the volume fraction of B which is completely liberated, using measurements of intercept scans on mounted particles of mesh size D. In essence, it is accepted that particles of size D which are of pure B will give some fractional number of B intercepts which are αD long. It is assumed that this fraction is the same as that measured from AB intercepts from mounted AB particles of size D. Then the fraction of liberated (pure) B is calculated from the average length of B intercepts, the average length of AB intercepts, the measured fraction of αD intercepts which are entirely B and the standard fraction of intercepts which are expected to be αD in length. Of course, if particles of size D report no αD long intercepts of only B, there is no pure B liberation at mesh size D.

Again, at present the technique is not developed sufficiently to measure the distribution of B in locked particles, although in principle it would be possible to back-calculate it if the calibrating function $g(\ell_B/D_B|D)$ were known. This function is the mean density function of intercepts of B of length ℓ_B/D_B derived from sectioned particles of size D containing grains of size D_B, $D_B<D$.

King (27) has also proposed linking image analysis and the concepts of stereological transformation to create a predictive model for liberation. The formulation is as follows. A random line probe is used on polished sections of the rock to obtain the cumulative number-length distributions of intercepts of rock (A) and valuable (B), that is, $F_A(x)$, $F_B(x)$. It is assumed that these intercepts are the total result of intersections of exposed sections resulting from a volume size distribution of whole grains of B within the rock. A closely sized screen fraction of the ground rock is mounted, sectioned, polished and intercepts measured for a random line probe across these particles, to give $n(\ell/D)$, the number density function of intercept length ℓ from particles of mesh size D. It is assumed that all particles sizes will give this same density function in terms of the dimensionless intercept length, ℓ/D.

The random line probe through the rock is now imagined to exist on the rock containing a maze of cracks (as if it were the product size distribution of particles reassembled as a three-dimensional jig-saw puzzle) and that this would give the same answer in terms of intercept lengths of AB and B between the cracks occuring along the line as if the particles were mounted in a matrix and examined. It is then assumed that, on the average, if one could examine intercepts originating only from particles of mesh size D, then the number fraction of particles of size D which contain a fractional volume < r of B, N(r|D) say, is given by the weighted sum of intercepts which have contained B intercepts lengths less than or equal to rℓ, where ℓ is the intercept length A+B across a particle section, that is,

$$N(r|D) = \int_{\ell=0}^{D_u} \phi(r|\ell)\, n(\ell/D)\, d\ell \tag{6}$$

where $\phi(r|\ell)$ is the probability that the test line will have a total length of intersection across B grains < rℓ as it traverses a particle of total intersection A+B of ℓ, and D_u is the maximum intercept length from size D. The value of the function $\phi(r|\ell)$ is then replaced by equations derived by random statistical placement of B intercepts with respect to A+B intercepts, in terms of the previously measured functions $F_A(x)$, $F_B(x)$. The extension to a size distribution of particles is clearly

$$p(r) = \int_0^{\infty} N(r|D)\, p(D)\, dD$$

Unfortunately, we see no reason why Eq. 6 should be valid since it suffers from the same basic problem as the definitions of apparent liberation used in Eq. 4. A host rock containing B grains of a constant size would still lead to a distribution of B/(A+B) intercept lengths, which would still then be interpreted as a distribution of r for all statistically possible cases of B/(A+B) intercept lengths. Similarly, the extensions of this treatment given by Finch and Petruk (28) and Schapp (3) are open to the same objections.

Geometrical Treatments

The essence of such treatments is to imagine B grains distributed in host rock with superimposed three-dimensional fracture patterns, and then perform geometrical calculations of the probability that fracture which creates particles of size D will give some fraction containing r fraction of B in AB, wholly-contained or surface-exposed. The fundamental problem, of course, is that choosing unrealistic geometries to make the system amenable to analysis leads to solutions which are radically in error. For example, the probability calculations for spherical grains lying within spherical particles are relatively easy to perform, but unfortunately, fragments of spheres do not form spheres. The difficulty in performing three-dimensional geometrical calculations is that the geometry of a grain (or particle) is not preserved on fracture. Any assemblage of particles produced by fracture will generally have a variety of shapes in order to fit together to make the original particle. The distribution of shapes for a large number of particles of volume size V is probably the same as that for, but the steps to go from a large particle of one shape to a small particle of the same shape must involve passage through other shapes.

The pioneering attempts by Gaudin (1) and Wiegel (29,30,31) and Wiegel and Li (32) are well known and will not be discussed here except to point out that King (27) has found them to be in error when applied to experimental data. Klimpel and Austin (2) have attempted to avoid geometrical complexity and incorrect assumptions concerning particle geometry by treating the problem as one-dimensional in volume space. The treatment clearly owes a great deal conceptually to the work of King, but it tries to eliminate the need for image analysis experiments. Their treatment is not correct since it treats a three-dimensional problem as one-dimensional, but it at least served to emphasize the importance of being able to describe locked material as surface-exposed or wholly-contained.

If it is assumed that the system can be envisaged as shown in Figure 3, then simple geometrical treatments with restrictive assumptions can be generally formulated as follows. Consider grains of B randomly dispersed in a host matrix of A. The material is broken into a size distribution which is determined experimentally by sieving or other means. It is assumed that the fractures do not occur preferentially along A-B grain boundaries, because a fracture surface can pass through a B grain due to the high velocity and energy of propagating cracks. Thus B grains lie at random with respect to the surfaces of the broken particles. The assumption of randomness means that all particle sizes will contain the same overall mean fraction of B in AB. Liberation will not occur until the size distribution is fine enough to approach the size of the B grains, and if the volume fraction of B in AB (denoted by ϕ) is small there is only a small probability of a particle containing more than one B grain or fragment of a B grain. In fact, under these circumstances, most particles will contain no B at all.

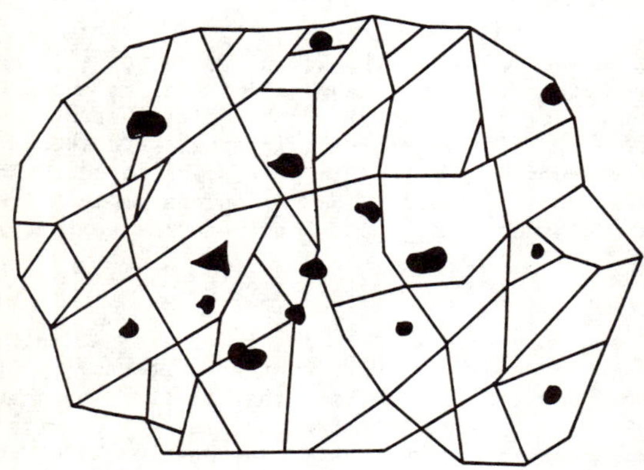

Figure 3. Illustration of B grains (black) occurring at random with respect to the breakage of host matrix (white) into particles.

The simplest sub-system to formulate is that where it can be assumed that the B material is present as a small volume fraction, so that a realistic degree of size reduction gives particles containing only zero or one B grain or B grain fragment per particle. Under this condition, the statistics derived from a single grain volume size V_B and a single particle volume size V are readily extended to distributions of grain and particle sizes, because of the statistical independence implicit in this situation. The fractional distribution of mass of particles of size V in the various classes are defined by

feed to the mill to split it into the various classes, which assumes random fracture and little liberation at this stage. Third, the breakage of a class containing a high volume fraction of B required a new set of liberation equations based on random placement of V_i volumes on particles of volume V_j containing a volume $V_B=rV_i$ as a single inclusion. Clearly, a particle containing a surface-exposed inclusion cannot generate a product particle with a wholly-contained B fragment.

A Partial Liberation Model

A treatment is given in the Appendix which is an attempt to combine the best features of the Klimpel-Austin treatment with that of the Meloy-Gotoh treatment. The results are summarized as the equation set:

$$H_A = \begin{cases} [1-(aV_B/V)^{1/3}]^3[(N_1-1)(\phi V/V_B)]+(1-N_1\phi V/V_B) & a \le V/V_B \le 1/N_1\phi \\ [1-(aV_B/V)^{1/3}]^3(1-\phi V/V_B) & 1/\phi \ge V/V_B \ge \max\{a,1/N_1\phi\} \\ 0 & a \le V/V_B \le 1/\phi \\ 1-\phi[1+(V/bV_B)^{1/3}]^3 & 0 \le V/V_B \le \min\{b,b[(1/\phi)^{1/3}-1]^3 \\ 0 & b[(1/\phi)^{1/3}-1]^3 \le V/V_B \le b \end{cases}$$

with grade of zero;

$$H_B = \begin{cases} 0 & a \le V/V_B \\ \phi[1-(V/cV_B)^{1/3}]^3 & 0 \le V/V_B \le c \end{cases}$$

with grade of 1;

$$H_W = \begin{cases} [1-(aV_B/V)^{1/3}]^3(\phi V/V_B) & a \le V/V_B \le 1/\phi \\ 1 & V/V_B \ge \max\{a,1/\phi\} \\ 0 & 0 \le V/V_B \le a \end{cases}$$

with grade of V_B/V for $H_W \le 1$ and ϕ for $H_W=1$;

$$
H_S = \begin{cases}
\{1-[1-(aV_B/V)^{1/3}]^3\}(N_1\phi V/V_B) & a \le V/V_B \le 1/N_1\phi \\
1-[1-(aV_B/V)^{1/3}]^3 & 1/\phi \ge V/V_B \ge \max\{a, 1/N_1\phi\} \\
0 & V/V_B \ge \max\{a, 1/\phi\} \\
\phi\{[1+(V/bV_B)^{1/3}]^3-[1-(V/cV_B)^{1/3}]^3 & 0 \le V/V_B \le \min\{c, b[(1/\phi)^{1/3}-1]^3\} \\
1-\phi[1-(V/cV_B)^{1/3}]^3 & b[(1/\phi)^{1/3}-1]^3 \le V/V_B \le c
\end{cases}
$$

with mean grades of

$$
\begin{cases}
V_B/N_1V & a \le V/V_B \le 1/N_1\phi \\
\phi & V/V_B \ge \max\{a, 1/N_1\phi\}
\end{cases}
$$

$$
\bar{r}_s = \begin{cases}
\dfrac{\{1-[1-(V/cV_B)^{1/3}]^3\}}{\{[1+(V/bV_B)^{1/3}]^3-[1-(V/cV_B)^{1/3}]^3\}} & 0 \le V/V_B \le \min\{c, b[(1/\phi)^{1/3}-1]^3\} \\
\phi & b[(1/\phi)^{1/3}-1]^3 \le V/V_B \le c
\end{cases}
\tag{9}
$$

where a, b, c are numbers of the magnitude of unity and $N_1 \ge 2$, becoming ~ 8 for $V/V_B = 1$. These equations apply for a single volume size of grain V_B in particles of volume size V.

Providing that the stipulation of only zero or one B grain (or B fragment) per particle is met, the extension to a volume size distribution of B grains with a density function of $g(V_B)$ is simply

$$\bar{H}_K = \int_{V_B} H_K g(V_B) dV_B \tag{10}$$

where $\bar{H}_K$ is the H_A, H_B, H_W or H_S of Eq. 9; $g(V_B)dV_B$ is the volume fraction of ϕ which is of grain volume size V_B to V_B+dV_B; and $\bar{H}_K$ is the overall volume fraction of material in one of the four classes for particle size V. For a range of particle size from V_J to V_{J+1}

$$h_{K,J} = \int_{V_{J+1}}^{V_J} \bar{H}_K p(V) dV \tag{11}$$

where $h_{K,J}$ is the volume fraction in the appropriate K class for material between V_J and V_{J+1} in size and $p(V)dV$ is the volume fraction of the product particle size distribution in the range V to V+dV. The integrations must conserve the constraints of Eq. 9. The mean grades are similarly weighted sums or integrals.

Table 3 gives the results for a single grain size at a volume fraction 1% and Table 4 shows the result for a volume fraction of 5%. In these tables, it has been assumed that both pure B and surface-exposed B will ideally float:

the volume grade and recovery are those for the material of that particle size.

TABLE 3: The simple Liberation Model: 1% by volume B in AB (ϕ=0.01)

Rel. Size V/V_B	x/x_B	% Pure A H_A	% Pure B H_B	% Locked B WC H_W	SE H_S	Ideal Grade of Floats, %	Recovery of B, %
10	2.1	81.54	0	1.54	16.92	5.0	84.6
5	1.7	90.36	0	0.36	9.28	10.0	92.8
1	1	92.00	0	0	8.00	12.5	100
0.5	0.8	94.23	0.01	0	5.76	17.3	100
0.25	0.6	95.67	0.05	0	4.28	23.0	100
0.125	0.5	96.63	0.13	0	3.25	29.7	100
0.06125	0.4	97.29	0.22	0	2.49	36.8	100

TABLE 4: The Simple Liberation Model: 5% by volume B in AB (ϕ=.05)

Rel. Size V/V_B	x/x_B	% Pure A H_A	% Pure B H_B	% Locked B WC H_W	SE H_S	Ideal Grade of Floats, %	Recovery of B, %
10	2.1	7.69	0	7.69	84.61	5.0	84.6
5	1.7	51.79	0	1.79	46.42	10.0	92.8
1	1	60.0	0	0	40.0	12.5	100
0.5	0.8	71.15	0.04	0	28.81	17.3	100
0.25	0.6	78.35	0.25	0	21.40	23.0	100
0.125	0.5	88.13	0.63	0	16.25	29.6	100
0.06125	0.4	86.45	1.11	0	12.44	36.9	100

It is clear that this model is incomplete within our original definition of a liberation model, since it does not split the locked surface-exposed H_S term into its component parts, that is, the fractions with r within a certain range, as in Table 1. Logic states that the $H_S(r)$ term will depend on the texture of B inclusions in the host matrix, and hence, on the shape of the resultant apparent B grains as well as on N_1. It is readily possible to derive an expression for $H_S(r)$ in terms of a distribution function which is a definition of how grains of size V_B randomly placed to be intersected by fracture surfaces of particles of size V each will split into two or more fragment. It seems likely that this function will have to be determined experimentally, much as King (10) has determined the distribution functions of chord lengths of polished sections of mounted grains of known sieve size.

Discussion and Conclusions

It is clear that the fragmentation of an AB material into particles, with no restrictions on the relative volumes of A and B, is a problem in three-dimensional stochastic geometry (38), complicated by the need to define the geometry of the various shapes of grains or texture of A-B interactions. Davy (39) has studied the problem from this point of view, but the paper is written in mathematical terminology and does not appear to give explicit solutions which can be calculated. Barbery (40) has also followed this approach by assuming that the texture of the material is described by A and B shapes formed by Poisson planes in three dimensions. In this treatment, any chosen mean spacing gives rise to B grains of Poisson polyhedra of varying shapes and volumes, so that it is not possible to consider a system of a single size of B grains disposed in a matrix. No distinction was made between wholly-contained or surface-exposed locked B. Such advanced mathematical treatments are surely the future path to the solution of the problem, but they must be cast into a form where is is clear how unknown parameters or functions can be deduced from experimental values.

Similarly, the theorems of conservation of AB interfacial area, etc., during size reduction, as proposed by Meloy (41), do not as yet appear to lead to sets of equations from which the general liberation function f(r,o x) or the more restricted forms of Eq. 2 or Tables 1-2 can be calculated.

It can be argued that distributed grain models can never be valid for many types of ore structure. However, we agree with the concept proposed by Schapp (3) that it may not be necessary to consider the complex geometrical forms catalogued by Amstatz and Giger (6). If data is taken on material at a degree of size reduction corresponding, for example, to a rod mill product, and this data is used to back-calculate an apparent original grain size distribution which would forward-generate the data according to some model, then this apparent original grain size distribution may be sufficient for further purposes.

We feel that the equations leading to Tables 3 and 4 are properly formulated within the assumptions of the model. However, the predicted results do not look reasonable when compared to experimental experience, so it seems probably that it will be essential to allow for non-random breakage as volume fractions of B become significant compared to the grain size.

Since some of the best intellects in mineral processing are working on those advanced formulations, including Barbery, Davy, Moore and Jones, King and Meloy, among others, it can be expected that equations for precise calculation of the liberation distributions will become available in the next few years.

Acknowledgements

We gratefully acknowledge financial support for this work from The National Science Foundation under Grant No. BCT-8303154.

References

1. A.M. Gaudin, Principles of Mineral Dressing, McGraw-Hill, New York; 1939, 70-91.

2. R.R. Klimpel and L.G. Austin, "A Preliminary Model of Liberation from a Binary System", Powder Technology, 34; 1983, 121-130.

3. W. Schaap, "Illustrated Liberation - Flotation Recovery Model for a Disseminated Mineral in a Low-Grade Ore", Trans. IMM London, Section C88; 1979, C220-228.

4. R.R. Klimpel, "Some Practical Approaches to Analyzing Liberation from a Binary System", Process Mineralogy II, Ed. W. Petruk, SME-AIME; 1984, 65-81.

5. L.G. Austin and R.R. Klimpel, "A First Study of Liberation Combined with Breakage Kinetics in a Mill", SME-AIME Annual Meeting, New Orleans, 1986.

6. G.C. Amstutz and H. Giger, "Stereological Methods Applied to Mineralogy, Petrology, Mineral Deposits and Ceramics", J. of Microscopy, 95; 1972, 145-164.

7. E.J. Oosthuyzen, "Application of Automatic Image Analysis to Mineralogy and Extractive Metallurgy", Spec. Publ. No. 7, Geological Society of South Africa; 1983, 451-464.

8. D. Dinger, "New Stereological Method for Obtaining 3-Dimensional Grain Size Distributions and Shape Factors from Planar Sections, Application to Characterization of Mineral Matter in Coal" (Ph.D. Thesis, The Pennsylvania State University, University Park, PA., USA).

9. M.P. Jones and R. Horton, "Recent Developments in the Sterological Assessment of Composite (middling) Particles by Linear Measurements", Proc. 11th Commonwealth Mining and Metallurgical Congree, Hong Kong, Ed. M.J. Jones, IMM (London); 1978, 113-122.

10. R.P. King, "Determination of the Distribution of Size of Irregularly Shaped Particles from Measurements on Sections or Projected Areas", Powder Technology, 32; 1982, 87-100.

11. W. Petruk, "Application of Quantitative Mineralogical Analysis of Ores to Ore Dressing", C.I.M. Bulletin, 69; 1976, 146-153.

12. W. Petruk, "Correlation Between Grain Sizes in Polished Sections with Sieving Data and Investigation of Mineral Liberation Measurements from Polished Sections", Trans. I.M.M., Section C87, 1978, C272-277.

13. W. Petruk, "Image Analysis Measurements and Data Presentation", Applied Mineralogy, Eds. W.C. Park, D.M. Hausen, D. Hagni, SME-AIME; 1984, 127-140.

14. W. Petruk and M. R. Hughson, "Image Analysis Evaluation of the Effect of Grinding Media or Selective Flotation of Two Zinc-Lead-Copper Ores", C.I.M. Bulletin, 70; 1977, 128-135.

15. M.A. Berube and J.G. Marchand, "Evaluation of the Mineral Liberation Characteristics of an Iron Ore Undergoing Grinding", Int. J. Min. Proc., 13; 1984, 223-237.

16. A.F. Reid, P. Gottlieb K.J. MacDonald and P.R. Miller, "QEM-SEM Image Analysis of Ore Minerals, Volume Fraction, Liberation", Applied Mineralogy, Eds. W.C. Park, D.M. Hausen and R.D. Hagni, SME-AIME; 1984, 191-204.

17. A. Lynch "Adelantos y Problemas en la Simulacion de Circuitos de Reduccion de Tamano", Proc. IV Simposium Sobre Molienda, Armco Chile; 1984, 259.

18. G. Barbery, G. Huyet and C. Gateau, "Liberation Analysis with the Help of Image Analysis Theory and Applications", Proc. 13th IMPC, Warsaw Poland; 1981, 568-594.

19. P.S.B. Stewart and M.P. Jones, "Determining the Amounts and Compositions of Composite (middling) Particles", Proc. 12th IMPC, Sao Paulo, Brazil; 1977, paper 4.

20. G. Barbery, "Mineral Liberation Analysis Using Stereological Methods; a Review of Concepts and Problems", Applied Mineralogy; 1984, 171-190

21. S.W. Moore and M.P. Jones, "Mathematical Methods for Estimating Particle Compositions from the Results of Image Analysis", Proc. 15th IMPC, Cannes, France; 1985, 3-11.

22. G. Barbery, et al., "Measurements of Mineral Liberation of Comminuted Ores, Archiwum Gornictwa (Poland) 30, 1985, 297-310.

23. G. Barbery, R. Bloise, C. Gateau and C. Reinhart, "Mineral Liberation Measurements and Their Interpretation", Proc. ICAM 81, ed. J.P.R. de Villiers and P.A. Hawthorn,Spec. Pub. No. 7, Geological Society of South Africa; 1983, 469-473.

24. H.J. Steiner, "Liberation Kinetics in Grinding Operations, Proc. XI IMPC, Cagliari, Paper No. 2 (1975) 33-58.

25. R.P. King, "Stereological Methods for the Prediction and Measurement of Mineral Liberation", Spec. Publ. No. 7, Geological Society of South Africa; 1983, 443-447.

26. R.P. King, "Measurement of Particle Size Distribution by Image Analyzer", Powder Technology, 39; 1984, 279-289.

27. R.P. King, "A Model for the Quantitative Estimation of Mineral Liberation of Grinding", Int. J. Min. Proc., 6; 1979, 207-220.

28. J.A. Finch and W. Petruk, "Testing a Solution to the King Liberation Model", Int. J. Min. Proc., 12; 1984, 305-311.

29. R.L. Wiegel, "A Quantitative Approach to Mineral Liberation", Proc. 7th Int. Miner. Proc. Cong., New York; 1964, 3-9.

30. R.L. Wiegel, "Liberation in Magnetic Iron Formation", Trans. AIME, 258; 1975, 247-256.

31. R.L. Wiegel, "Integrated Size Reduction Mineral Liberation Effects", Trans. AIME, 260; 1976, 147-152.

32. R.L. Wiegel and K. Li, "A Random Model for Mineral Liberation by Size Reduction", Trans. AIME, 238; 1967, 179-189.

33. T.P. Meloy and K. Gotoh, "Liberation in an Homogenous Two Phase Ore", Int. J. Min. Proc., 14; 1985, 45-55.

34. J.R.G. Andrews and T.S. Mika, "Comminution of Heterogenous Material: Development of a Model for Liberation Phenomena", Proc. 11th IMPC, Cagliari, Italy; 1976, 59-88.

35. R.J. Finlayson and D.G. Hulbert, "The Simulation of the Behavior of Individual Minerals in Closed Circuit Grinding", Proc. 3rd IFAC Symp., Montreal, Canada; 1980, 323-332.

36. J.A. Herbst, K. Rajamani, J.D. Miller and C.L. Lin, "Development of a Multicomponent-Multisize Mineral Liberation Model", presented at SME-AIME Annual Meeting, New York, 1985.

37. R.R. Klimpel and L.G. Austin, "The Back-Calculation of Specific Rates of Breakage from Continuous Mill Data", Power Technology, 38; 1984, 255-266.

38. J. Serra, Image Analysis and Mathematical Morphology, Academic Press, London, 1982,

39. P.J. Davy, "Probability Models for Liberation", J. Applied Probability, 21; 1984, 260-269.

40. G. Barbery, "Random Sets and Integral Geometry in Comminution and Liberation of Minerals", presented at SME-AIME Annual Meeting, New York, 1984.

41. T.P. Meloy, "Liberation Theory: Eight, Modern, Usable Theorems", Int. J. Min. Proc., 13; 1984, 315-324.

Appendix

Derivation of a Simplified Partial Liberation Model.

It is assumed that (a) the degree of size reduction is taken to the stage where there is zero or one B grain or B fragment per particle and (b) there is random occurence of B with respect to fracture planes.

Consider first the case of $V/V_B > 1$, as illustrated in Figure A1, and let mean particle (D) and grain (D_B) diameters be defined by $V=k_1D^3$, $V_B = k_3 D^{*3}_B$.

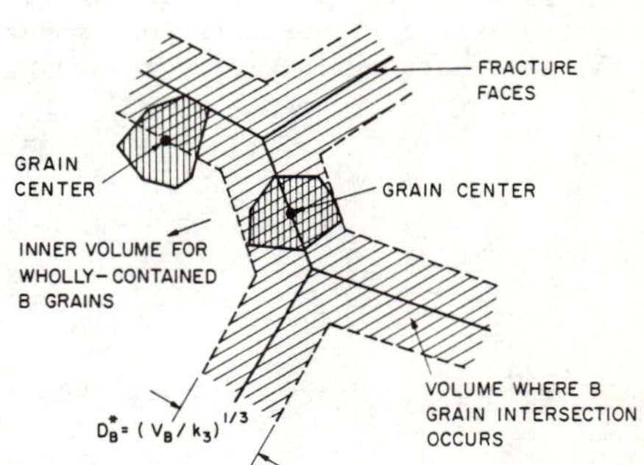

Figure A1. Illustration of B grains of volume V_B lying at fracture faces of particles of volume V.

Define a mean spacing in from the surface of the particles by $V_B = k_3(D_B)$ such that all grains whose centers lie closer in than this spacing are wholly-contained and not intersected by the fractures forming the particle surface. Since it is assumed that B grains lie at random, the fraction of V_B volume which is wholly-contained is then

$$= \frac{\text{volume within inner spaces}}{\text{total volume}}$$

$$= k_1(D-D_B^*)^3/V$$

$$= [1-(aV_B/V)^{1/3}]^3 \quad , \quad a=k_1/k_3$$

The fraction of V_B volume which is cut and surface-exposed is then $1-[1-(aV_B/V)^{1/3}]^3$. Every B grain is associated with a volume V_B/ϕ of AB , by the definition of ϕ, so a single wholly-contained B grain represents a volume $(V_B/\phi)-V$ of pure A and a volume V containing $r=V_B/V$ of B. Thus, the associated volume fraction of pure A is

$$H_{A1} = \begin{cases} [1-(aV_B/V)^{1/3}]^3(1-\phi V/V_B) & V/aV_B \geq 1,\ \phi V/aV_B \leq 1 \\ 0 & \phi V/V_B \geq 1 \end{cases}$$

The stipulation $\phi V/aV_B<1$ is necessary to meet the requirement that there is only zero or one B grain per AB particle.

On the other hand, an intersected B grain will sometimes lie in two particles, sometimes in three, in four, etc., depending on its relative size and the mean shape of the particles in terms of numbers of adjoining particles. Let the fraction of pair intersections be n_2, of triple intersections be n_3, etc., where $\Sigma\, n_i=1$. Let the mean number of fragments formed per intersected grain be N_1, where $N_1=(2n_2+3n_3+\cdots)$. N_1 will be a function of relative size, $N_1(V_B/V)$, but will always be > 2. The mean number of particles containing surface-exposed fragments per intersected B grain is also N_1, so that the associated volume of pure A particles is $V_B/\phi - N_1V$, giving

$$H_{A2} = \begin{cases} \{1-[1-(aV_B/V)^{1/3}]^3\}(1-N_1\phi V/V_B) & V/aV_B \geq 1, N_1\phi V/V_B \leq 1 \\ 0 & N_1\phi V/V_B \geq 1 \end{cases}$$

Thus

$$H_A = \begin{cases} [1-(aV_B/V)^{1/3}]^3[(N_1-1)\phi V/V_B]+(1-N_1\phi V/V_B) & V/aV_B \geq 1,\ N_1\phi V/V_B \leq 1 \\ [1-(aV_B/V)^{1/3}]^3(1-\phi V/V_B) & \phi V/V_B \leq 1,\ N_1\phi V/V_B \geq 1 \\ 0 & \phi V/V_B \geq 1 \end{cases}$$

$$
H_W = \begin{cases} [1-(aV_B/V)^{1/3}]^3(\phi V/V_B) & V/aV_B \geq 1,\ \phi V/V_B \leq 1 \\ 1 & \phi V/V_B \geq 1 \end{cases}
$$

The fraction of volume containing surface-exposed B is thus

$$
H_S = \begin{cases} \{1-[1-(aV_B/V^{1/3}]^3\}(N_1\phi V/V_B) & V/aV_B \geq 1,\ N_1\phi V/V_B \leq 1 \\ 1-[1-(aV_B/V)^{1/3}]^3 & N_1\phi V/V_B \geq 1,\ \phi V/V_B \leq 1 \\ 0 & \phi V/V_B \geq 1 \end{cases}
$$

with a <u>mean</u> grade of

$$
\bar{r}_s = \begin{cases} (V_B/N_1 V) & N_1 \geq 2,\ N_1\phi V/V_B \leq 1 \\ \phi & N_1\phi V/V_B \geq 1,\ \phi V/V_B \leq 1 \end{cases}
$$

For $\phi V/V_B \geq 1$ there can be no pure A or surface-exposed material because of the of the assumption of only zero or one B grain or fragment per particle, all grains are wholly-contained, and the grade is ϕ. N_1 will tend to 2 as V/V_B becomes larger.

The case of $V/V_B < 1$ is treated similarly. In this case, the particles of volume V which lie within the inner volume $k_2[(V_B/k_2)^{1/3\ 3}-(V/k_4)^{1/3}]^3$ are pure (see Figure A2), the particles of volume V which lie between the inner

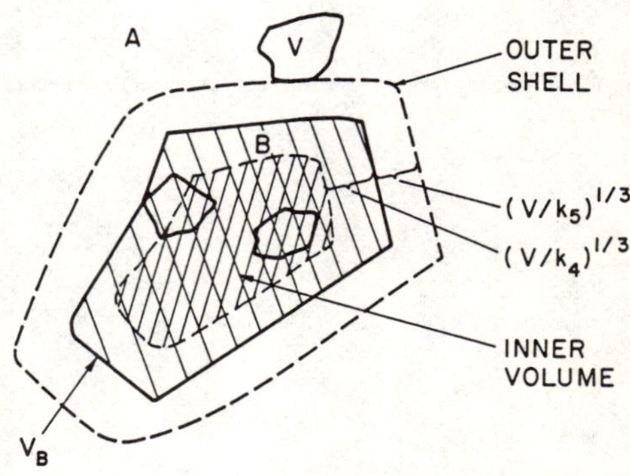

Figure A2. Grain of volume V_B with fracture particles of volume V.

volume and the outer shell are intersected AB (surface-exposed B), and particles outside the shell are pure A; obviously there is no wholly-contained B of size V_B. The volume associated with each B grain is V_B/ϕ so the volume fraction of pure B is clearly $\phi[1-(V/cV_B)^{1/3}]^3$, where $c=k_4/k_2$ and $V/cV_B<1$. When V is small, particles of V lying outside the outer shell still lie within the associated volume, so the volume fraction of pure A is $1-\phi[1+(V/bV_B)^{1/3}]^3$ for $\phi[1+(V/bV_B)^{1/3}]^3<1$, where $b=k_5/k_2$. However, as the particles of V approach size V_B the volumes intersecting with the B grain may exceed the associated volume, so that there is no pure A. This stage is reached when (and if) $\phi[1+(V/bV_B)^{1/3}]^3=1$. Thus

$$H_A = \begin{cases} 1-\phi[1+(V/bV_B)^{1/3}]^3 & V/bV_B\le 1,\ \phi[1+(V/bV_B)^{1/3}]^3\le 1 \\ 0 & \phi[1+(V/bV_B)^{1/3}]^3\ge 1 \end{cases}$$

$$H_B = \phi[1-(V/cV_B)^{1/3}]^3 \qquad V/cV_B\le 1$$

The fractional volume of material with surface-exposed B is $\phi\{[1+(V/bV_B)^{1/3}]^3 - [1-(V/cV_B)^{1/3}]^3\}$ providing some pure A exists, but becomes $1-H_B$ when $H_A=0$, giving

$$H_S = \begin{cases} \phi\{[1+(V/bV_B)^{1/3}]^3-[1-(V/cV_B)^{1/3}]^3\} & V/cV_B\le 1,\ \phi[1+(VbV_B)^{1/3}]^3\le 1 \\ 1-\phi[1-(V/cV_B)^{1/3}]^3 & \phi[1+(V/bV_B)^{1/3}]^3\ge 1 \end{cases}$$

with a <u>mean</u> grade of

$$\bar{r}_s = \begin{cases} \{1-[1-(V/cV_B)^{1/3}]^3\}/\{[1+(V/bV_B)^{1/3}]^3-[1-(V/cV_B)^{1/3}]^3\} & \phi[1+(V/bV_B)^{1/3}]^3] \le 1 \\ \phi & \phi[1+(V/bV_B)^{1/3}]^3\ge 1 \end{cases}$$

Grouping the constraints in a form convenient for computation gives Eq. 9 as quoted in the text.

Introduction

RECENT PROGRESS IN THE FIELD OF PROCESS MINERALOGY

Richard D. Hagni

Department of Geology and Geophysics
University of Missouri - Rolla
Rolla, Missouri 65401

The recognition of process mineralogy and its varied industrial applications has undergone marked progress during the past eight years. Since the AIME Process Mineralogy Committee was first established in 1978, great strides have been made in the recognizion of the value of process mineralogy, the utilization of process mineralogy in an ever widening sphere of applications, and in publication of new techniques and the results of the industrial applications of process mineralogy. The creation of additional national and international organizations whose principal interest is directed toward process mineralogy has contributed significantly to the advance of process mineralogy. All of these organizations have provided forums for the presentation and discussion of the results of process mineralogy investigations, new techniques, and the publication of relevant papers. The plans for the future by these organizations bodes well for the future of process mineralogy.

Introduction

For many years, individuals interested in the application of mineralogy to industrial processes were without an organization to sponsor the presentation, discussion, and publication of the results of their efforts. Although significant industrial research was conducted by geologists, metallurgists, ceramists, and others, the results of that research remained largely in the files of individual companies. The reason for this was partly due to the fact that many journals prefer to publish the results of fundamental research rather than that of applied research. The confidentiality of much of the applied research prevented its release and also contributed to the lack of literature in this area. Perhaps the most important factor, however, was the lack of a scientific organization or group specifically devoted to applied mineralogy. Although single sessions were sometimes devoted to applied mineralogy, such as that on "Ore Microscopy Applied to Beneficiation Problems" at the 1976 International Geological Congress meetings in Sydney, Australia, it has been largely due to the inception of the Process Mineralogy Committee of AIME in 1978 that process mineralogy has flourished.

History of Process Mineralogy

Although isolated papers on the subject matter of process mineralogy were given at various meetings (1,2), the formal organization of a group to specifically sponsor sessions on process mineralogy began with the organization of the Process Mineralogy Committee of AIME. The group was initiated in 1978 as an ad hoc committee on process petrography in The Metallurgy Society (TMS) of AIME and organized its first session at the AIME annual meeting in 1978. The following year the committee was fully recognized as the Process Mineralogy Committee of TMS, and held three sessions at the AIME annual meeting. The Committee has found that there exists a great need for information in the area of process mineralogy, that there is a wealth of unpublished information available for distribution, and that there is a significant desire by industrial and academic researchers to share the results of their process mineralogy studies. Therefore, the Committee has remained very active through the sponsorship of six or seven sessions for the mutural discussion of process mineralogy applications to a wide variety of problems in the minerals industry, which have been held at each of the subsequent annual meetings of AIME and the publication of those papers in proceedings volumes. The Process Mineralogy Committee was asked by the directors of the Society of Mining Engineers (SME) of AIME during 1981 to become a joint TMS-SME committee. The approval for joint status in 1982 resulted in the principal responsibility for scheduaing and publishing Process Mineralogy papers to subsequently be alternated between the two societies.

The efforts of the TMS-SME Process Mineralogy Committee of AIME have resulted in the publication of five important volumes that contain papers that deal with a wide range of applications of process mineralogy. The first volume, titled "Process Mineralogy: Extractive Mineralogy, Mineral Exploration, Energy Resources" was published by TMS (3). It contains the papers presented at the Committee-sponsored sessions held during the first three years, 1978, 1979, and 1980. The papers are organized into eight chapters: 1) instumentation, methods and concepts, 2) applications to mineral explorations, 3) applications to mineral dressing, 4) applications to extractive metallurgy, 5) applications to mine and plant operations, 6) applications to energy resources: coals, 7) applications to energy resources: oil shales, uranium and geothermal systems, and 8) other applications: refractories, zeolites, volcanic ash, bauxite and kaolin deposts. The volume also contains a keynote address by Paul F. Kerr on "Reminiscences in Applied Mineralogy".

The second volume, also published by TMS, is titled "Process Mineralogy II: Applications in Metallurgy, Ceramics, and Geology"(4). The volume is comprised of eight chapters: 1) techniques of automation, 2) applications to beneficiation, 3) applications to extractive metallurgy, 4) applications to refractories, 5) applications to cement, 6) applications to coal and coke, 7) applications to environment, and 8) applications to exploration. Two keynote addresses are included: 1) "Recollections of Industrial Alpplications of Mineralogy", by G. C. Amstutz, and 2) "Use of the Petrographic Microscope in Refractories Technology" by C. Burton Clark.

The third volume was published by SME and is titled "Process Mineralogy III: Applications in Metallurgy, Coal, Concrete, Smelting and Exploration" (5). Its six sections deal with: 1) applications to beneficiation, 2) applications to energy, 3) applications to building materials, 4) applications to mineral exploration, 5) applications to metallurgy, and 6) process methodologies.

The fourth and fifth volumes were special ones and are dealt with in the following section of this article. The fourth volume, published by TMS, comprised the papers delivered at the second International Congress on Applied Mineralogy in the Minerals Industry held in conjuction with the 1984 annual meeting of AIME in Los Angeles, California (6). The fifth volume, published by SME, contains the papers delivered at the Paul F. Kerr Memorial Symposium held in conjunction with the 1985 annual meeting of AIME (7).

Recent Developments in Process Mineralogy

A number of important symposia on process mineralogy have been held in in the past few years that have provided significant contributions to the understanding of process mineralogy technique and applications.

The principal thrust for the promotion of process mineralogy came initially from and continues to be provided by the TMS-SME-AIME Process Mineralogy Committee that has sponsored symposia and sessions at every annual AIME meeting since 1978. A great debt of gratitude is due to that committee for their diligent work.

The first International Congress on Applied Mineralogy in the Mineral Industry (ICAM) was organized and convened in Johannesburg, South Africa in June, 1981. During that ICAM meeting, the International Committee (later changed to Council) on Applied Mineralogy was formed to guide the future of ICAM. The congress proceedings volume, ICAM 81" was published in 1983 as Special Publication No. 7 of the Geological Society of South Africa, edited by J. P. R. De Villiers and P. A. Cawthorn (8). The volume contains 58 papers arranged into five sections: 1) exploration, 2) coal and carbonaceous material, 3) applications in mineral dressing, 4) high-temperature techniques, and 5) mineralogical methods. Also contained are opening, closing, keynote, and four plenary addresses.

Recogition of the importance of process mineralogy in beneficiation was shown by the devotion of a session with 12 papers centered on "Mineralogy

Applied to Ore Dressing" at the XIV International Mineral Processing Congress meeting on "Worldwide Industrial Application of Mineral Processing Technology" held in Toronto, Ontario in October, 1982. Most of the papers dealt with new mineralogical techniques and several emphasized the use of automated systems of image analysis for rapid assessment of liberation results during beneficiation. The papers are included in the Congress proceedings volume for session VIII (9). Abstracts for the 12 papers are in French, German, and Russian as well as English.

In 1981, the Mineralogical Society of France organized a discussion group that meets once a month. For two years, 1981-82, one of the discussion groups dealt with process mineralogy, and this group continues to meet occasionally for continued discussions.

In 1983, the Mineralogical Society of South America established a section on process mineralogy. The section first met in 1983, and held a second meeting in May, 1984.

The second triennial International Congress on Applied Mineralogy in the Minerals Industry (ICAM) was held in February, 1984 in Los Angeles, California, and the proceedings volume, "Applied Mineralogy", was published in 1985 (6). The large volume is nearly 1200 pages and includes 71 papers that are organized into eight sections on process mineralogy: 1) general methodology, 2) image analysis, 3) beneficiation, 4,5) high and low temperature processes, 6) gold, silver and platinum group metals, 7) exploration and ore genesis, and 8) medical sciences.

Recognizing the importance of process mineralogy, the International Mineralogical Association organized the Commission on Applied Mineralogy (IMA-CAM) in 1982. During the 1984 ICAM meeting in Los Angeles, representatives from IMA-CAM and the International Council for Applied Mineralogy met to discuss areas of mutual interest in applied mineralogy and possible means of collaboration. Those discussions resulted in the agreement that both groups would continue to exist and hold separate meetings, and that the two organizations would co-sponsor symposia or sessions on topics of mutural interest at subsequent IMA and ICAM meetings. It was agreed that the initial co-sponsored symposia would be held at the Stanford, California meeting of the IMA.

The Paul F. Kerr Memorial Symposium, sponsored by the Process Mineralogy Committee of AIME was held in New York in 1985 as a special post-meeting symposium prior to the regular AIME annual meeting. The proceeding volume is titled "Mineralogy - Applications to the Minerals Industry", was published by AIME in 1985 (7), and contains 17 papers organized into four chapters dealing with: 1) innovations in methodology, 2) uranium investigations, 3) clay mineral investigations, and 4) geochemistry, alteration, and ore microscopy. The volume also includes a keynote address by Leo J. Miller on Dr. Kerr's contribution to ore discovery and memorials and testimonials to Paul Kerr, one of the early process mineralogists.

The importance of process mineralogy was recognized in the organization of four special sessions with 27 papers on "Applied Mineralogy in Science and Technology" at the joint annual meetings of the Geological Association of Canada (GAC) and the Mineralogical Association of Canada (MAC) held at Fredericton, New Brunswick in May, 1985 (10). Areas of applied mineralogy treated included coal technology, science and technology, occupational health, metals industry, and smelting. A special issue of the Canadian Mineralogist is devoted to Applied Mineralogy to include papers from this symposium. At the 1986 GAC-MAC meeting in Ottawa, two sessions were devoted to Applied Clay Science.

The Metallurgical Society (METSOC) of the Canadian Institute of Mining and Metallurgy has organized a section on Mineral Sciences that includes process mineralogy as a major contributor. Sessions were conducted at the METSOC meetings in 1984 and 1985 and a three-day symposium in 1986.

The role of process mineralogy was further recognized by the inclusion of a session on mineralogical studies of ores and concentrates in the International Complex Sulfide Symposium held in San Diego, California in November, 1985 (11). The sessions treated especially volcanogenic massive sulfide and disseminated shale-hosted deposits. The proceedings volume, "Complex Sulfides: Processing of Ores, Concentrates and By-Products" has been published by AIME and includes a chapter on "mineralogy" with three papers and three abstracts.

Process mineralogists working in the cement industry have recognized the importance of cement microscopy. After initial meetings in 1978 and 1980, the group organized the International Cement Microscopy Association (ICMA) in 1981 that has met annually and published the papers in six conference proceedings volumes (12,13,14,15,16,17). The papers deal with the application of reflected light, SEM, electron microprobe, X-ray, and other techniques to the study of cement, clinker, concrete, and other building materials. Emphasis has been given to microscopic studies of structures, hydration, effects of additives, quality control, and the relationship between microstructure and mineralogical composition. Recent conferences have been held in Nashville, Tennessee (1983), Albuquerque, New Mexico (1984), Fort Worth, Texas (1985), and Orlando, Florida (1986).

Other international groups that have involved cement microscopy in recent meetings include the International Congress of Cement Chemistry that has met every four years for a total of seven times. The second International Conference on the Use of Fly Ash, Silica Fume, Slag, and Natural Pozzolans in Concrete that met in Madrid, Spain in 1986 included all techniques for studying the products and the primary phases in a reacting concrete system. An international group that met in Montreal in 1985 for a conference on Alkalie-Aggregate Reactivity and the Determination of Parking Garages" considered minerals that are important to those reactions.

Some of the first books on process mineralogy were published during the past two years. "Mineralogy for Metallurgist: An Illustrated Guide", by H. W. Fander was published in 1985 by the Institution of Mining and Metallurgy, London, England (18). The small book consists largely (60%) of color photomicrographs that deal with the ore microscopy of ores and concentrates. "Process Mineralogy of Ceramic Materials" by W. Baumgart, A. C. Dunham, and G. C. Amstutz was published by Ferdinand Enke Verlag Stuttgart (19).

Future Developments in Process Mineralogy

Two symposia on process mineralogy will be held at the International Mineralogical Association meeting at Stanford, California in July, 1986. The symposia will deal with: 1) applied mineralogy, for which ICAM is the lead organization, and 2) surface analytical methods, for which IMA-CAM is the lead organization.

Two sessions, sponsored by the Process Mineralogy Committee, are scheduled for the fall meeting of the Society of Mining Engineers of AIME to be held in St. Louis, Missouri in September, 1986. One session will deal with applications to exploration, the second session will treat applications to minerals, beneficiation, and pyrometallurgy.

The eighth International Congress of Cement Chemistry, which will include microscope application, will be held in Rio de Janeiro, Brazil in September, 1986. The 7th International Conference on Alkali-Aggregate Reaction will be held at Carleton University in Ottawa in August, 1986.

Seven sessions, sponsored by the Process Mineralogy Committee, are scheduled for the annual AIME meeting to be held in Denver, Colorado in February, 1987. Three sessions will deal with process mineralogy applications to precious metals, two of which will concentrate on shale-hosted, finely disseminated gold deposits. The other sessions will encompass applications to 1) mineral processing, 2) pyrometallurgy and refractories, 3) liberation, 4) exploration and evaluation.

The third International Congress on Applied Mineralogy (ICAM) will be held in Orleans, France in July, 1987. The Congress is being organized by the International Council for Applied Mineralogy with sponsorship of the Group of European Mineralogists, Societe francaise de Mineralogie et de Cristallographie, Societe de l'Industrie Minerale, Universite d'Orleans, Bureau de Recherches Geologiques et Minieres, and Centre National de la Recherche Scientifique.

References

1. Richard D. Hagni, "Ore Microscopy Applied to Beneficiation", Mining Engineering, 30 (1978) pp. 1437-1447.

2. Richard D. Hagni, "Process Mineralogy: Past, Present, and Future", pp. 29-36 in Process Mineralogy II: Applications in Metallurgy, Ceramics, and Geology, Richard D. Hagni, Ed.; The Metallurgical Society of AIME, New York, NY, 1982.

3. Donald M. Hausen and Won C. Park, Eds., Process Mineralogy: Extractive Metallurgy, Mineral Exploration, Energy Resources, The Metallurgical Society of AIME, New York, NY, 1981, 713 pp.

4. Richard D. Hagni, Ed., Process Mineralogy II: Applications in Metallurgy, Ceramics, and Geology, The Metallurgical Society of AIME, New York, NY, 1982, 503 pp.

5. William Petruk, Ed., Process Mineralogy III: Applications in Metallurgy, Coal, Concrete, Smelting and Exploration, Society of Mining Engineers, New York, NY, 1984, 322 pp.

6. Won C. Park, Donald M. Hausen, and Richard D. Hagni, Eds., Applied Mineralogy, The Metallurgical Society of AIME, New York, NY, 1985, 1194 pp.

7. Donald M. Hausen and Otto C. Kopp, Eds., Mineralogy-Applications to the Minerals Industry, Society of Mining Engineers, New York, NY, 1985, 285 pp.

8. Johan P. R. de Villiers and P. A. Cawthorn, Eds., ICAM 81: Proceedings of the First International Congress on Applied Mineralogy, Special Publication No. 7, The Geological Society of South Africa, Kelvin House, Johannesburg, South Africa, 1983, 509 pp.

9. "Mineralogy Applied to Ore Dressing", Session VIII, Worldwide Industrial Application of Mineral Processing Technology, Preprints and Abstracts in English, French, German, and Russian, XIV International Mineral Processing Congress, Toronto, Ontario, 1982, 191 pp.

10. "Applied Mineralogy in Science and Technology", Joint Annual Meeting of the Geological Association of Canada and the Mineralogical Association of Canada, Program with Abstracts, University of New Brunswick, Fredericton, New Brunswick, 1985, 77 pp.

11. Alan D. Zunkel, Roy S. Boorman, Arthur E. Morris, and Rolf J. Wesely, Eds., Complex Sulfides: Processing of Ores, Concentrates and By-Products, The Metallurgical Society of AIME, New York, NY, 1985, 938 pp.

12. George R. Gouda, Ed., Proceedings of the Third International Conference on Cement Microscopy, Intl. Cement Microscopy Assoc., Duncanville, TX, 1981, 395 pp.

13. George R. Gouda, James Bayles, and Andrew Kraft, Eds., Proceedings of the Fourth International Conference on Cement Microscopy, Intl. Cement Microscopy Assoc., Duncanville, TX, 1982, 313 pp.

14. George R. Gouda and James Bayles, Eds., Proceedings of the Fifth International Conference on Cement Microscopy, Intl. Cement Microscopy Assoc., Duncanville, TX, 1983, 278 pp.

15. James Bayles, Ed., Proceedings of the Sixth International Conference on Cement Microscopy, Intl. Cement Microscopy Assoc., Duncanville, TX, 1984, 486 pp.

16. George R. Gouda, James Bayles, and Arturo Nisperos, Eds., Proceedings of the Seventh International Conference on Cement Microscopy, Intl. Cement Microscopy Assoc., Duncanville, TX, 1985, 434 pp.

17. James Bayles, George R. Gouda, and Arturo Nisperos, Eds., Proceedings of the Eighth International Conference on Cement Microscopy, Intl. Cement Microscopy Assoc., Duncanville, TX, 1986, 344 pp.

18. H. W. Fander, Mineralogy for Metallurgist: An Illustrated Guide, Institute of Mining and Metallurgy, London, England, 1985, 77 pp.

19. W. Baumgart, A. C. Dunham, and G. C. Amstutz, Process Mineralogy of Ceramic Materials, Ferdinand Enke Verlag Stuttgart, 1986, 230 pp.

Two sessions, sponsored by the Process Mineralogy Committee, are scheduled for the fall meeting of the Society of Mining Engineers of AIME to be held in St. Louis, Missouri in September, 1986. One session will deal with applications to exploration, the second session will treat applications to minerals, beneficiation, and pyrometallurgy.

The eighth International Congress of Cement Chemistry, which will include microscope application, will be held in Rio de Janeiro, Brazil in September, 1986. The 7th International Conference on Alkali-Aggregate Reaction will be held at Carleton University in Ottawa in August, 1986.

Seven sessions, sponsored by the Process Mineralogy Committee, are scheduled for the annual AIME meeting to be held in Denver, Colorado in February, 1987. Three sessions will deal with process mineralogy applications to precious metals, two of which will concentrate on shale-hosted, finely disseminated gold deposits. The other sessions will encompass applications to 1) mineral processing, 2) pyrometallurgy and refractories, 3) liberation, 4) exploration and evaluation.

The third International Congress on Applied Mineralogy (ICAM) will be held in Orleans, France in July, 1987. The Congress is being organized by the International Council for Applied Mineralogy with sponsorship of the Group of European Mineralogists, Société française de Minéralogie et de Cristallographie, Société de l'Industrie Minérale, Université d'Orléans, Bureau de Recherches Géologiques et Minières, and Centre National de la Recherche Scientifique.

References

1. Richard D. Hagni, "Ore Microscopy Applied to Beneficiation", Mining Engineering, 30 (1978) pp. 1437-1447.

2. Richard D. Hagni, "Process Mineralogy: Past, Present, and Future", pp. 29-36 in Process Mineralogy II: Applications in Metallurgy, Ceramics, and Geology, Richard D. Hagni, Ed.; The Metallurgical Society of AIME, New York, NY, 1982.

3. Donald M. Hausen and Won C. Park, Eds., Process Mineralogy: Extractive Metallurgy, Mineral Exploration, Energy Resources, The Metallurgical Society of AIME, New York, NY, 1981, 713 pp.

4. Richard D. Hagni, Ed., Process Mineralogy II: Applications in Metallurgy, Ceramics, and Geology, The Metallurgical Society of AIME, New York, NY, 1982, 503 pp.

5. William Petruk, Ed., Process Mineralogy III: Applications in Metallurgy, Coal, Concrete, Smelting and Exploration, Society of Mining Engineers, New York, NY, 1984, 322 pp.

6. Won C. Park, Donald M. Hausen, and Richard D. Hagni, Eds., Applied Mineralogy, The Metallurgical Society of AIME, New York, NY, 1985, 1194 pp.

7. Donald M. Hausen and Otto C. Kopp, Eds., Mineralogy-Applications to the Minerals Industry, Society of Mining Engineers, New York, NY, 1985, 285 pp.

Cathodoluminescence Microscopy Applications in Process Mineralogy

INDUSTRIAL APPLICATIONS OF CATHODOLUMINESCENCE MICROSCOPY

Richard D. Hagni

Department of Geology and Geophysics
University of Missouri-Rolla
Rolla, Missouri 65401

Cathodoluminescence microscopy should be more fully utilized in applications to a wide range of industrial products. The study of mineralogy and textures or intergrowths of minerals can be greatly enhanced by cathodoluminescence microscopy used in conjunction with transmitted light, reflected light, and other microscopic techniques. Important applications of cathodoluminescence microscopy include the detection of valuable minerals, recognition of ore guides and hydrothermal alteration assemblages, and the study of beneficiation, refractories, and pyrometallurgical products. Cathodoluminescence microscopy can provide means to detect phases that have been previously overlooked, forms a rapid means for distinguishing cathodoluminescent phases from non-cathodoluminescent minerals and also between phases that show different colors of cathodoluminescence, aids in the recognition of subtle textures within single grains, and provides an excellent technique for better determining the distributions of those phases that cathodoluminesce. Cathodoluminescence microscopy should become a standard item of equipment available in all process mineralogy laboratories.

Introduction

Although cathodoluminescence microscopy has found important application to the petroleum industry, the technique has been inadequately applied or neglected in the study of process mineralogy problems in other areas of geology and in the mining, metallurgical, and ceramic industries.

For petroleum exploration cathodoluminescence microscopy has been utilized as an important aid in the distinction of the primary calcitic constitutents from those calcite crystals that were subsequently deposited as pore-filling cements. The ratio of these constituents provide a measure of the degree of original porosity in carbonate rocks, which forms one means of indicating the potential for a given rock to have acted as a petroleum conduit or reservoir rock. Cathodoluminescence also has been utilized to study the cements and other features of sandstones, to distinguish authigenic from clastic feldspar, and especially to detect certain accessory minerals, such as apatite and zircon, in sedimentary, igneous, and metamorphic rocks. References to these and other applications of cathodoluminescence microscopy have been given in Hagni (1) and need not be repeated here. An extensive bibliography on cathodoluminescence applications also is given by Barker and Wood (2).

Because cathodoluminescence microscopy provides a means for readily detecting and studying the character and distribution of many gangue phases and some ore minerals, the technique should be applied to areas of mineralogical study beyond those noted in the preceding paragraph. Its applications to process mineralogy can include the following: 1) detection of valuable ore minerals, 2) recognition of mineralogical guides to the presence of ore deposits, 3) resolution of beneficiation problems, 4) study of refractories, and 5) applications to pyrometallurgy. The purpose of this paper is to briefly outline some of the applications of cathodoluminescence microscopy to these areas of process mineralogy, provide some specific case histories from the writer's experience, and especially to emphasize to the value of cathodoluminescence microscopy in the study of refractories both before and after their use in pyrometallurgical and other applications.

Instrumentation and Operation

The writer conducts his cathodoluminescence microscopy with a Nuclide Corporation model ELM 2A Luminoscope. A second instrument, the Technosyn, has recently become available. Both instruments are mounted on standard petrographic microscopes, and form a beam of electrons that are directed upon the specimen, placed within a vacuum chamber, to produce cathodoluminescence (Figure 1). Because the ability of a mineral to cathodoluminesce is due principally to the presence of certain minor elements, called activators, it is generally not possible to predict in advance whether the phases present in a given ore or product will cathodoluminesce and be ameanable to study by this method. Some minerals, however, exhibit intrinsic cathodoluminescence and can be expected to consistently respond to the technique.

The Luminoscope was operated at an excitation beam energy of 14 to 18 keV and a beam current of 0.7 to 0.9 ma, with a beam area of about 1 mm. Five objectives, 3.2X, U-20 (12.7X magnification without the universal stage hemisphere), U-32 (20.2X), UD-40 (25.2X), and U-50 (31.5X), and four ocular (5X, 6.3X, 8X, and 10X) provided total magnifications from 16X to 315X. Cathodoluminescence photomicrography presents special problems due to the

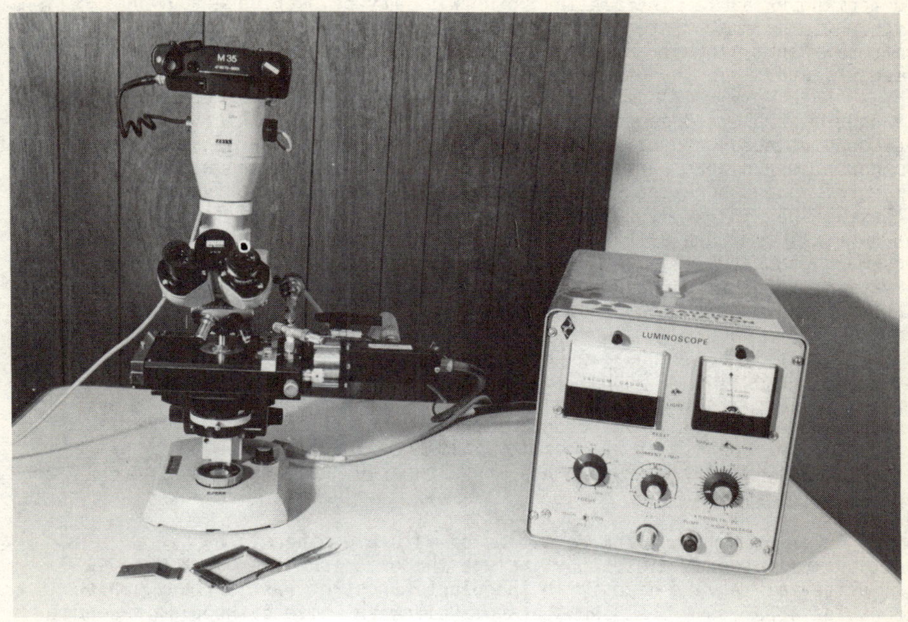

Fig. 1 - Nuclide Luminoscope with vacuum chamber mounted on a Zeiss petrographic microscope with a Zeiss M35 automatic camera.

low level of cathodoluminescence shown by some phases, the desire to photograph phases that are present only in minute amounts, and the need to minimize burning of the specimen by the electron beam. For these reasons, exposure times as long as 20 minutes (included 30-second periods with the beam off between a series of 12 one-minute exposures) have been required obtain a single photomicrograph of materials such as the small phosphate grains present in the Birmingham, Alabama iron ores and concentrates. Exposure times for those materials recently have been reduced to less than one minute through several instrument and equipment modifications. These modifications involve the use of a microscope that allows the passage of more light, a camera system that provides a higher percentage of light to the film, and a film with a faster speed. Specifically, we have: 1) replaced a previously used, less expensive microscope with a Zeiss standard petrographic microscope, 2) use a Zeiss M35 camera with a Zeiss MC 63 automatic exposure meter, and 3) employ 3M film with 1000 ASA.

Product samples are best prepared into thin sections or polished thin sections for cathodoluminescence microscopy. Although the Technosyn chamber will accept a thicker specimen than that of the Luminoscope, neither instrument will accept a polished section as thick as 1½-1 3/4 cm, a common size range prepared in the industry. Therefore, previously prepared normal polished sections must be thinned to about 3/4 cm for cathodoluminescence study. A high degree of polish is desirable for the study of either polished sections or polished thin sections.

Detection of Valuable Minerals

Cathodoluminescence microscopy is an excellent ancillary technique for detecting valuable minerals that otherwise may be overlooked during conventional transmitted and reflected light microscopic studies. Those minerals that exhibit cathodoluminesence may stand out like "beacons of light", and draws the attention of the investigator even when they are present in very small amounts in a matrix that consists predominantly of gangue minerals. Because the ability of a mineral to exhibit cathodoluminesence usually is a function of the presence of small amounts of minor elements that serve to activate the cathodoluminesence, the valuable minerals noted here are ones that commonly contain the activator elements or they exhibit intrinsic cathodoluminesence. Valuable metallic and non-metallic minerals that may show cathodoluminesence include emerald, diamond, cassiterite, sphalerite, fluorite, barite, scheelite, wollastonite, kyanite, andalusite, corundum, willemite, and certain uranium minerals, such as schroeckingerite and uranopilite.

Hagni (1) has emphasized and illustrated the cathodoluminesence as a means of detection of scheelite, wollastonite, tremolite, andalusite, fluorite, and zircon. Kyle (3) has noted that the greenish yellow and reddish orange cathodoluminesence shown by two varieties of sphalerite in the sulfide deposits at the Hockley Salt Dome in Texas are the result of their cadmium and silver contents (and low iron contents) that serve as the activator elements.

Other examples of the detection of valuable minerals by cathodoluminescence microscopy include the ores from the Clara mine, a fluorite-barite vein deposit in the Black Forest of West Germany, which contains sphalerite that shows bright yellowish white cathodoluminesence, fluorite that exhibits bright blue cathodoluminesence, and barite with dark green cathodoluminesence (Figure 2). Emeralds from the Gravelotte mine in South Africa exhibit strong crimson red cathodoluminesence (Figure 3). Kyanite from the Graves Mountain mine in Georgia exhibits dark purple cathodoluminesence that serves to distinguish it from the more abundant potash feldspar that shows light blue cathodoluminesence (Figure 4).

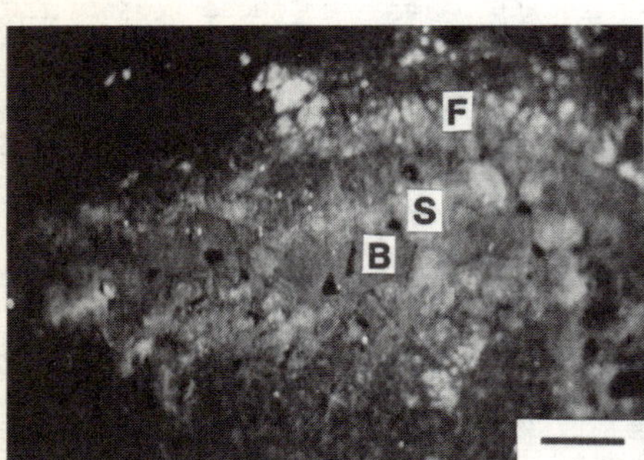

Fig. 2 - Micrograph showing sphalerite (S) (bright yellowish white cathodoluminescence), fluorite (F) (bright blue cathodoluminescence), and barite (B) (dark green cathodoluminescence) in vein ores from the Clara mine in the Black Forest of West Germany. Cathodoluminescence. Bar = 0.2 mm.

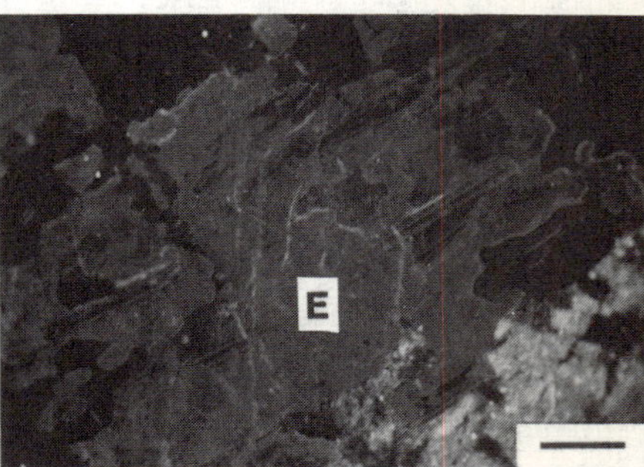

Fig. 3 - Micrograph showing emerald (E) crystals (strong crimson red cathodoluminescence) in ore from the Gravelotte mine, South Africa. Cathodoluminescence. Scale bar = 0.2 mm.

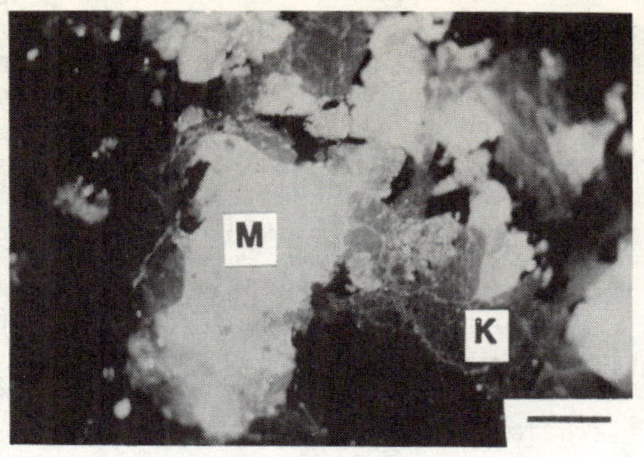

Fig. 4 - Micrograph showing kyanite (K) (dark purple cathodoluminescence) and microcline (M) (bright light blue cathodoluminescence) in ore from the Graves Mountain mine in Georgia. Cathodoluminescence. Scale bar = 0.2 mm.

Recognition of Ore Guides

Cathodoluminescence microscopy forms an outstanding technique for determining the presence of ores and utilizing certain types of guides to ore deposits. The study of hydrothermal alteration products in the host rocks within and near ore deposits commonly is greatly enhanced by cathodoluminescence microscopy. Mariano (4) has emphasized the fact that the typical blue cathodoluminescence of potash feldspar changes to a trivalent iron-activated characteristic red cathodoluminescence near the intrusions of carbonatite bodies and that this property may form an excellent guide to ore deposits associated with those igneous intrusions. Strong blue quadravalent titanium-activated cathodoluminescence has been described and illustrated for secondary potash feldspar in the potassic alteration zones at the Henderson mine in Colorado by Hagni (1) and noted for other porphyry copper and molybdenum deposits by Mariano (4). The fact that sillimanite exhibits blue cathodoluminescence where it is associated with the Benson iron ores in the Adirondack Mountains of northern New York but shows red cathodoluminescence where it is distal from ore has been suggested as a potential guide to Adirondack-type iron ore deposits by Hagni (1).

The application of cathodoluminescence microscopy to the study of gangue dolomite in Mississippi Valley-type lead-zinc deposits has revealed the presence of growth zoning, the character of which constitutes a potential guide to these ore deposits. Ebers and Kopp (5) studied the character of growth zoning in dolomite crystals from the East Tennessee ores and found that most of the zones could be traced throughout district. Kopp et al. (6) have further found that many of those growth bands may correlate with similar growth zones in dolomite from the Central Tennessee Zinc District, about 125 miles to the west of the East Tennessee district. Voss and Hagni (7) were able to recognize four growth zones in gangue dolomite deposited with the lead-zinc ores in the Viburnum Trend in southeastern Missouri, and they found that all four zones could be recognized and correlated in sparry dolomite crystals throughout the entire 45 mile long Trend. They also noted that cathodoluminescence revealed periods of dolomite dissolution that commonly occurred after the deposition of the first two bands of dolomite and before the deposition of zones three and four. Because the extent of occurrence of the dissolution increased toward the ore deposits, they suggested that the study of that feature by cathodoluminescence may be useful as an guide to ore deposits in the Viburnum Trend. Gregg (8), Rowan (9), and Farr (10) have shown that the four zones of Voss and Hagni (7) can be extended outside the Viburnum Trend throughout much of Missouri. Gregg and Hagni (11) have detected by cathodoluminescence the presence of ghost crystal faces within sparry dolomite crystals from the Viburnum Trend and concluded that their presence is an indication of their deposition from metal-bearing brines and may consititute a potential ore guide.

Hydrothermal alteration of the host limestones in association with the zinc-lead ore deposits in the Tri-State district has been previously described by Hagni and Saadallah (12). Recent study of those carbonate rocks shows that cathodoluminescence microscopy readily detects the influence of the ore fluids far beyond the ore deposits themselves and reveals the character of that alteration process. Distal from ore, the unaltered Mississippian limestone consists largely of calcitic crinoid fragments that show a deep red cathodoluminescence (Figure 5A). Certain unaltered stratigraphic units distal from ore consist of oolitic limestone (Figure 5B). Within about 200 feet of the ore deposits, the initial replacement of the crinoid fragments by sparry calcite that shows a brighter light red cathodoluminescence tends to clearly outline the darker areas of crinoidal calcite (Figure 5C). The margins of the partly replaced crinoidal calcite become ragged and impart a patchy appearance to the partly altered rock under cathodoluminescence microscopy. Toward the ore deposits, the host carbonate rock becomes progressively replaced by sparry calcite (Figure 5D). Close to the ore deposits, the crinoidal limestone has been almost entirely altered to sparry calcite, and its original crinoidal character is revealed only by the presence of circular to ovate shapes of the former crinoid fragments and locally by irregularly shaped replacement remnants of the original limestone that has a much duller cathodoluminescence in contrast to that of the sparry calcite (Figure 5E). Within the ore deposits themselves the carbonate rock has been replaced by coarsely crystallized dolomite (Figure 5F), but which locally contains unreplaced remnants of the sparry calcite that attests to the replacement process. Locally, the dolomite crystals have suffered subsequent partial dedolomitization to calcite.

Applications to Beneficiation Problems

Cathodoluminescence microscopy has only rarely been applied to beneficiation problems. However, it can provide an important tool for the recognition of small amounts of valuable and deleterious constituents in the beneficiation products, including the heads, concentrates, and tailings produced from a wide variety of ores. The detection with cathodoluminescence microscopy of small amounts of the deleterious phosphate mineral, carbonate-fluorapatite, which occurs in the Birmingham iron ores, the recognition of 15 varieties of phosphate grain sizes and shapes, and the determination of the extent of their intergrowths with the iron minerals in various beneficiation products has been discussed in some detail elsewhere by Hagni (1,14,15).

Cathodoluminescence microscopy has also been used to detect large amounts of collophane in the Lorraine limonitic iron ores of France (1). The phosphorous has commonly been thought to be present in chemical combination with the iron rather than present as a separate phosphorous

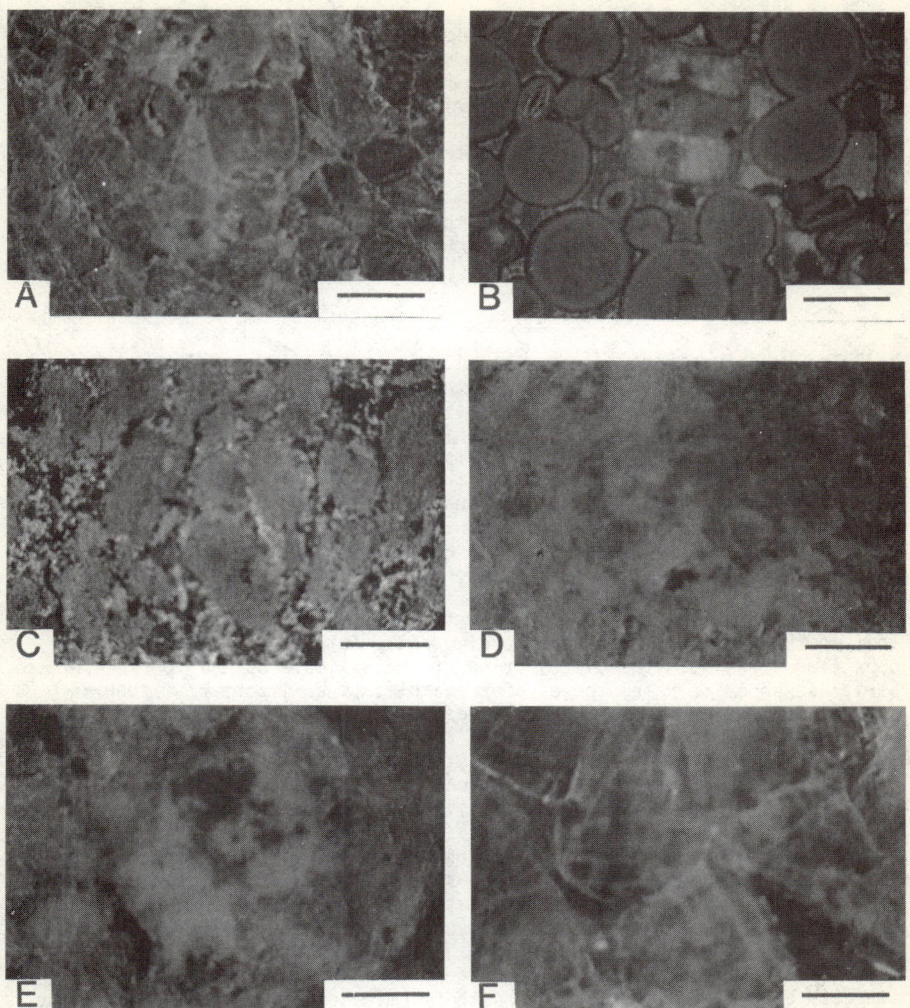

Fig. 5 - Cathodoluminescence micrographs showing an alteration sequence from unaltered Boone Formation (Mississippian) crinoidal limestone to sparry calcite rock and dolomite associated with ore deposits in the Tri-State Zinc-Lead District. (A) Crudely ovate shapes of abundant crinoid stem fragments (dark reddish orange cathodoluminescence) in unaltered limestone distal from ore deposits. (B) Abundant ooids in unaltered Short Creek limestone distal from ore deposits. (C) Crinoidal limestone (dark reddish orange cathodoluminescence) showing initial replaced by sparry calcite (brighter orange cathodoluminescence) along the grain boundaries of the crinoid fragments. (D) Crinoidal limestone (upper right; dark) about 65% replaced by sparry calcite (lower left; light). (E) Sparry calcite rock close to ore deposits with about 5% remnants of the original crinoidal limestone (dark). (F) Coarsely crystallized dolomite (red cathodoluminescence) within the core of the ore deposits. Scale bars = 0.2 mm.

phase in these ores.

The extent of free vs. locked dolomite-collophane particles has been assessed by cathodoluminescence microscopy for experimental phosphorous beneficiation products from the South Florida phosphate district that contained excessive amounts of deleterious magnesia (1).

At Mittersill, Austria the tungsten concentrate from the newer underground mine contains larger objectionable amounts of calcite that dilute the concentrate grade. Because both the scheelite and calcite from the Mittersill ores exhibit strong cathodoluminence (1), cathodoluminescence microscopy would be a good means to investigate this beneficiation problem.

Applications to Refractories

The use of cathodoluminescence microscopy in the general microscopic study of refractories has been neglected. Literature reports that apply cathodoluminescence to ceramic materials are those which deal with an assessment of the optical and electronic properties of semiconductors (15) and with the detection of monoclinic material in zirconia ceramics (16). Refractories are especially well adapted to examination by cathodoluminescence microscopy because many of the constituent phases show strong cathodoluminescence. The cathodoluminencent refractory phases are more readily detected, their distributions become very apparent, and the extent of cation diffusion into some refractory grains can be uniquely observed.

One of the important problems in the manufacture of refractories is that of the distribution of the bonding phase between the principal constituent grains. The distribution of the silicate bond in periclase and periclase-chromite refractories, for example, is an important factor in determining the hot strength of the refractory. It is desirable to have a uniform distribution of silicate bond throughout the magnesite refractory to avoid localized concentrations of deleterious glassy phase and thereby reduce the rate at which slags will react with the vulnerable silicate phases. That is, if large areas of silicate pockets or continuous veinlets of silicate are present, the slag attack reactions degrade the brick much more rapidly and the product gives shorter service life. Under ordinary light the silicate bond tends to be obscured by the dark area associated with the diffusion of light at periclase grain boundaries (Figure 6A). Cathodoluminescence microscopy provides an excellent means by which to detect the silicate bond because the silicate phases commonly exhibit a bright yellowish white cathodoluminescence (Figure 6B). Locally, the silicate bond exhibits bright red cathodoluminescence, probably due to the presence of ferric iron, and rarely it shows a bluish grey cathodoluminescence. The distribution of the silicate bond is especially well shown by its cathodoluminescence. The magnesite refractory shown in Figure 6B, for example, is revealed by cathodoluminescence to have a very continuous distribution of the silicate bond. The distribution of silicate bond also can be studied in greater detail at higher magnifications under cathodoluminescence microscopy (Figure 6A). Particles that originally were natural magnesite may be inferred by the presence of very abundant silicate bonding phases within periclase particles in some magnesite refractories (Figure 6B).

Cathodoluminescence microscopy provides a very useful means for detecting minor phases that may be present in refractories. For example, the presence of ferrite spinel may be easily overlooked in a magnesite refractory under ordinary light (Figure 7A), but with cathodoluminescence it is readily detected by its red color (Figure 7B). In addition, the presence

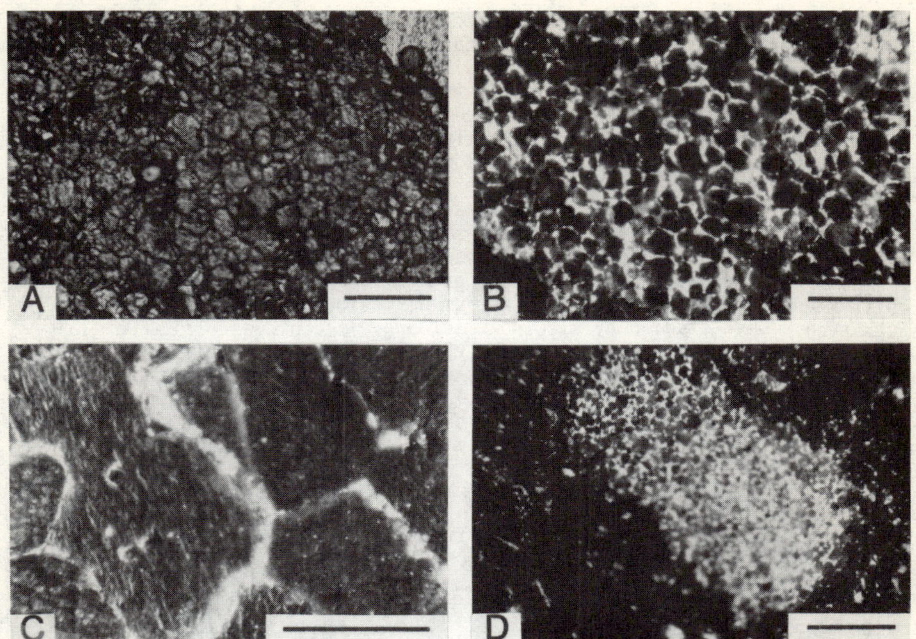

Fig. 6 - Micrographs showing the abundance and distribution of silicate bond in an unused magnesite refractory. (A) Silicate bond phases are largely hidden in the periclase grain boundaries due to the diffusion of light. Transmitted light. (B) Silicate bond distinctly shown by its bright yellowish white cathodoluminescence. Cathodoluminescence. (C) Magnified silicate bond in Figure 6B to show greater detail of its distribution. Cathodoluminescence. (D) Particle containing very abundant silicate bond phase, which may have originally been natural magnesite. Cathodoluminescence. Scale bars = 50 μm.

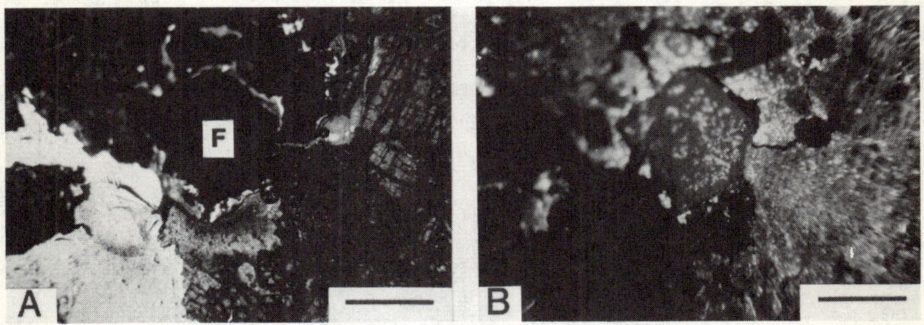

Fig. 7 - Micrographs illustrating the detection of the presence of the minor phase, ferrite spinel in an unused magnesite refractory. (A) Transmitted light. Ferrite (F) spinel crystal is dark and nearly indistinguishable from other refractory phases. (B) Cathodoluminescence. Ferrite spinel crystal is readily detected by its red cathodoluminescence which contrasts to the orange cathodoluminescence of the surrounding periclase. Scale bars = 0.2 mm.

of abundant silicate inclusions that cathodoluminesce yellowish green, and which were not detectable under ordinary light, are readily recognized and studied.

The constituent phases in chrome-magnesite bricks are not readily distinguished under ordinary transmitted light microscopy because chromite is opaque and the diffusion of chromium into the periclase grains during firing may render them translucent to nearly opaque (Figure 8A). With cathodoluminescence microscopy the orange cathodoluminescence of the periclase grains readily serves to distinguish them from the non-cathodoluminescent chromite grains (Figure 8B). The deep orange cathodoluminescence of the periclase grains is a result of ferric iron diffusion from the chromite. Magnesium diffusion into chromite grains has been studied by electron microprobe techniques (17). Partial diffusion of magnesium into the margins of chromite grains also can be detected by cathodoluminescence microscopy through the recognition of the local presence of bluish grey cathodoluminescent rims (Figure 8B). The distribution of rims of fine-grained periclase about chromite grains are shown especially well under cathodoluminescence microscopy (Figure 9).

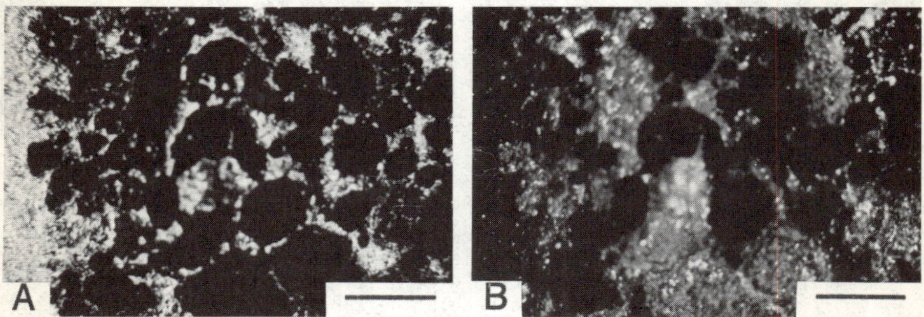

Fig. 8 - Micrographs showing periclase and chromite in an unused chrome-magnesite refractory. (A) Transmitted light. Periclase grains are nearly as dark as and difficult to distinguish from chromite grains. (B) Cathodoluminescence. A deep orange cathodoluminescence serves to distinguish periclase grains from non-cathodoluminescent the chromite grains. The rims of the chromite grains show a gray cathodoluminescence due to magnesium diffusion from the periclase. Scale bars = 0.2 mm.

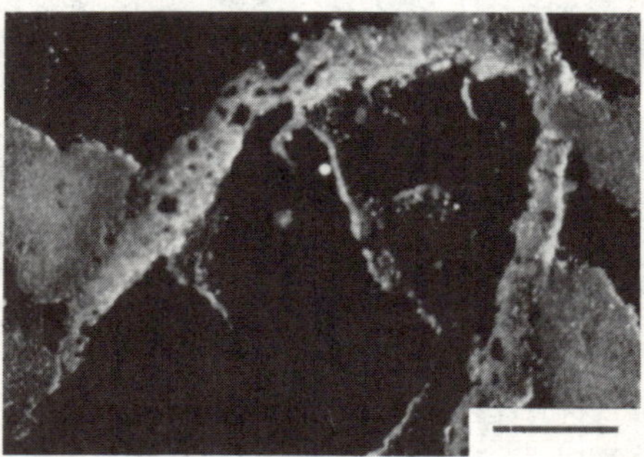

Fig. 9 - Micrograph showing fine-grained periclase (bright orange cathodoluminescence) that forms a rim around a large non-cathodoluminescent chromite grain in an unused chrome-magnesite refractory. Cathodoluminescence. Scale bar = 0.2 mm.

Zircon characteristically shows an intense intrinsic cathodoluminescence and it may exhibit various colors that result from the presence of activator elements, such as trivalent dysprosium. Because of these properties of cathodoluminescence, zircon refractories are especially well adapted to study by cathodoluminescence microscopy. For example, zircon sand grains in a zircon refractory may cathodoluminesce dark blue (some grains cathodoluminesce bright white), and the sand grains are readily distinguishable from the light bluish grey cathodoluminescence shown by the fine milled zircon where they are sintered together in the refractory (Figure 10B). Such detail cannot be distinguished under ordinary transmitted light microscopy (Figure 10A).

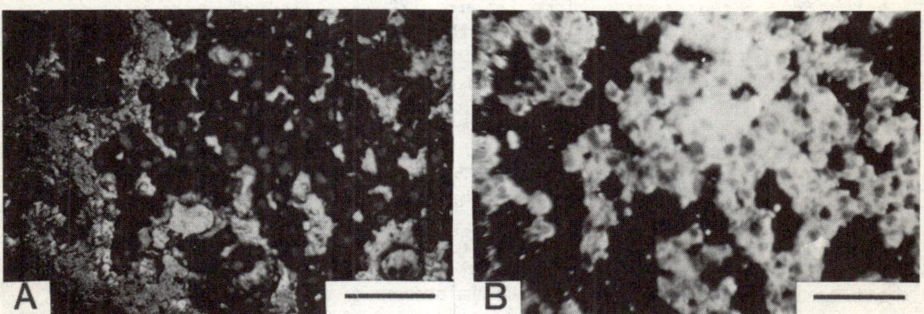

Fig. 10 - Micrographs showing zircon sand grains and fine milled zircon sintered together in an unused zircon refractory. (A) Transmitted light. Zircon sand grains and fine milled zircon have a similar appearance. (B) Cathodoluminescence. Zircon sand grains (dark blue cathodoluminescence) are easily distinguished from the fine milled zircon (light blue cathodoluminescence). Scale bars = 0.2 mm.

Mullite examined by the writer shows a dull violet cathodoluminescence in mullite refractories and dull blue cathodoluminescence in fireclay refractories. The presence of smaller amounts of glass intergrown with mullite in fireclay refractories are readily detected by its light bluish gray cathodoluminescence. Corundum, which is formed as the principal phase in high alumina refractories, exhibits dull bluish green cathodoluminescence. Although most crystals of silicon carbide in silicon carbide refractories do not cathodoluminesce, local crystals exhibit a distinct reddish brown cathodoluminescence (Figure 11).

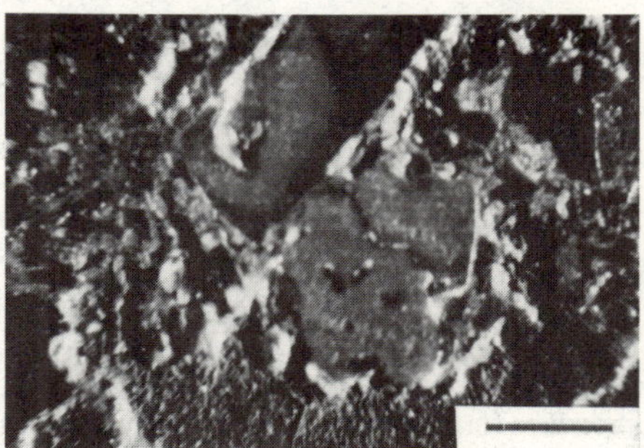

Fig. 11 - Micrograph showing two silicon carbide (reddish brown cathodoluminescence) crystals in an unused silicon carbide refractory. Cathodoluminescence. Scale bar = 0.2 mm.

Applications to Pyrometallurgy

Cathodoluminescence microscopy also can be effectively utilized in the study of refractories after they have been put to use as smelter linings. The technique provides an excellent method by which zones of attack within a refractory or even within single grains can be detected and studied. An example is the invasion of iron into periclase at and near the hot face of a magnesite refractory taken from an experimental lead furnace. The periclase grains in most unused magnesite refractories and those that are present toward the cold face of used refractories exhibit a strong, dark blue cathodoluminescence. In contrast, periclase grains at the hot face exhibit a bright red cathodoluminescence (Figure 12A). Their red cathodoluminescence is a function of the diffusion of ferrous iron from the matte into the periclase grains, evenually forming magnesiouwustite (Mg,Fe)O. Upon cooling of the brick, in air, the iron oxidizes and MgO Fe_2O_3 and/or (Mg,Fe)O Fe_2O_3 form depending on the oxygen partial pressure. It is well known that the phase magnesioferrite eventually forms due to the continued reaction of ferric iron with magnesium oxide in magnesite bricks during use in open hearths, electric furnaces, and basic oxygen furnaces. Within the reaction zone between the hot and cold faces some periclase grains exhibit red cathodoluminescence whereas others have blue cathodoluminescence. The periclase grains that cathodoluminesce red are those into which the activator element, ferric iron has diffused, whereas those grains that show blue cathodoluminescence are free from iron diffusion. The cause for iron diffusion into certain grains, in preference to other grains, may be readily apparent under cathodoluminescence microscopy. Those grains preferentially selected for the initial iron diffusion commonly exhibit cracks that have promoted the iron diffusion (Figure 12B). Rapid movement of iron took place along the crack that served as a channel into the periclase grain, and then iron diffused into the grain from the crack. Uncracked grains in the reaction zone have a greater tendency to resist iron diffusion. The character of the subgrain structure is excellently revealed under cathodoluminescence microscopy as shown in Figures 12A,B. The extent of iron diffusion within single periclase grains also can be closely assessed by observing the distributions of their colors of cathodoluminescence. Those portions into which iron has diffused show red cathodoluminescence, whereas the remnant portions of the periclase grain that are free from iron diffusion show blue cathodoluminescence (Figure 12C). Because the finer grained matrix periclase has assimilated greater amounts of iron at the hot face due to its larger surface area, it exhibits a brighter, more intense red cathodoluminescence that contrasts to that of the duller red cathodoluminescence of the particles of coarser grained periclase (Figure 12D).

A second example of the application of cathodoluminescence microscopy to pyrometallurgy illustrates the diffusion of copper into a firebrick refractory that was used in a copper reverberatory smelter. Under ordinary transmitted light the reaction of the refractory with copper matte can be detected only by the presence of very minor local darkening of the periclase (Figure 13A). With cathodoluminescence microscopy it can be observed that there is significant diffusion of copper into the periclase grains from their grain boundaries (Figure 13B). This is shown by a blue cathodoluminescence where copper has diffused into and activated the margins of the periclase grains; the blue margins gradually give way to a yellow cathodoluminescence in the central portions of the grains where the copper ions have not penetrated. The presence of introduced slag also is readily detected by its green cathodoluminescence in this refractory (Figure 13C).

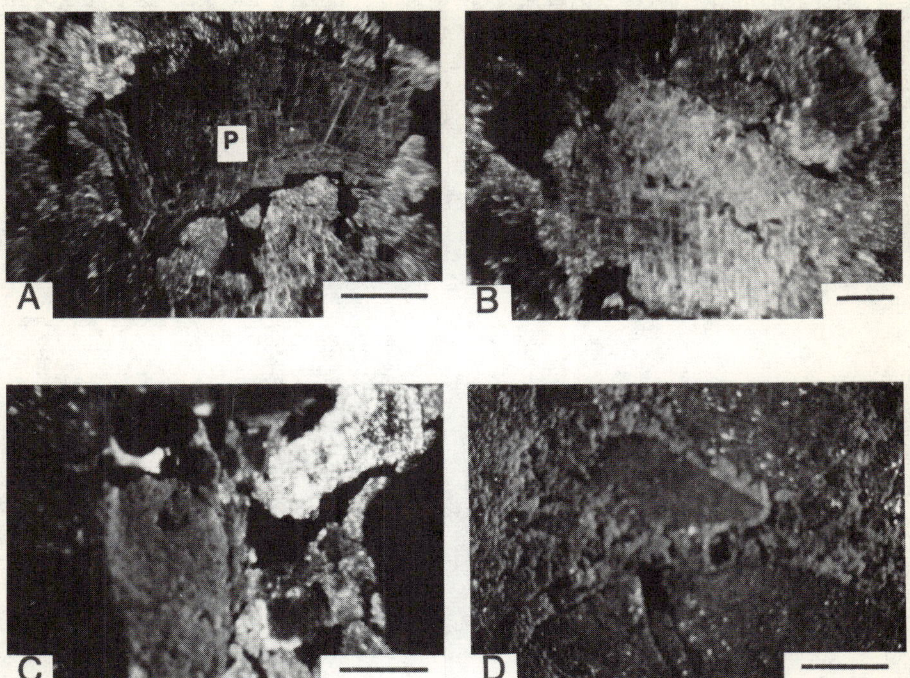

Fig. 12 - Cathodoluminescence micrographs showing periclase in a magnesite refractory after use in an experimental lead furnace. (A) Large periclase (P) grain at the hot face that shows bright red cathodoluminescence that has is the result of ferric iron diffusion from the matte. Periclase grains at the cold face show bright blue cathodoluminescence. (B) Large red cathodoluminescent periclase (P) grain within the reaction zone that consist principally of blue cathodoluminescent periclase. A crack in the large periclase grain has promoted iron diffusion into that grain causing red cathodoluminescence. Note that the character of the subgrain structure also is revealed in micrographs 12A,B. (C) Single periclase grain from the reaction zone that shows a remnant core free from iron diffusion that shows red cathodoluminescence surrounded by a rim into which ferric iron has diffused and that shows blue cathodoluminescence. (D) Fine-grained matrix periclase at the hot face shows a brighter orange cathodoluminescence than that of the coarse periclase grains due to their greater surface area that promotes the uptake of ferric iron during use in the furnace.

A third example of the application of cathodoluminescence microscopy to pyrometallurgy involved a study of the extent of corrosion of a high-silica (35% SiO_2) slag from an experimental lead furnace on a magnesite refractory. Under ordinary light the extent of corrosion appears to be nearly total, leaving only a single rounded particle of periclase that is unaffected except for a single transgressive veinlet (Figure 14A). Cathodoluminescence microscopic examination of the same area revealed the presence of abundant remnants of periclase within the area of corrosion (Figure 14B). Scale bars = 0.2 mm.

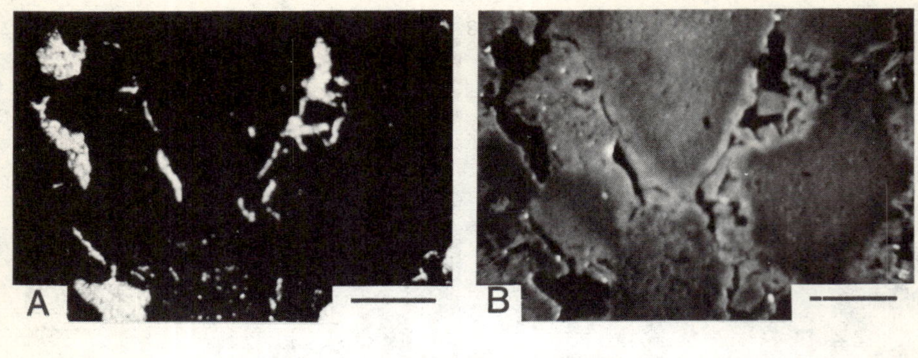

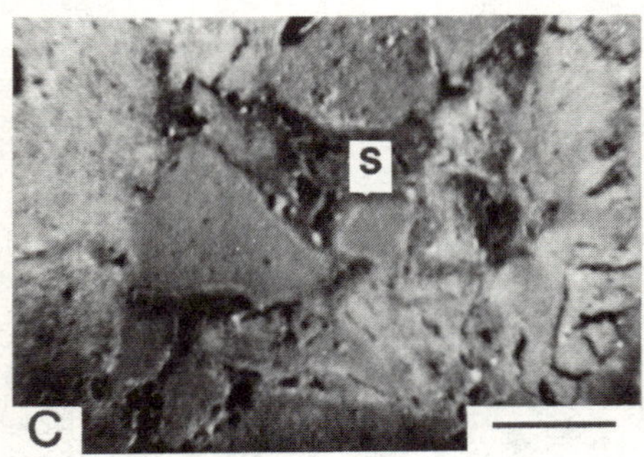

Fig. 13 - Micrographs showing copper diffusion into a firebrick refractory after use in copper reverberatory smelter. (A) Transmitted light gives little indication a the reaction with the copper matte. (B) Cathodoluminescence. The extent of copper diffusion into the refractory is revealed by blue cathodoluminescence (light) that contrasts with the yellow cathodoluminescence of the copper-free portion of the refractory. (C) The local presence of introduced slag (S) is detected by its green cathodoluminescence that contrasts with the yellow cathodoluminescence of the refractory and the blue cathodoluminescence where copper has diffused. Scale bars = 0.2 mm.

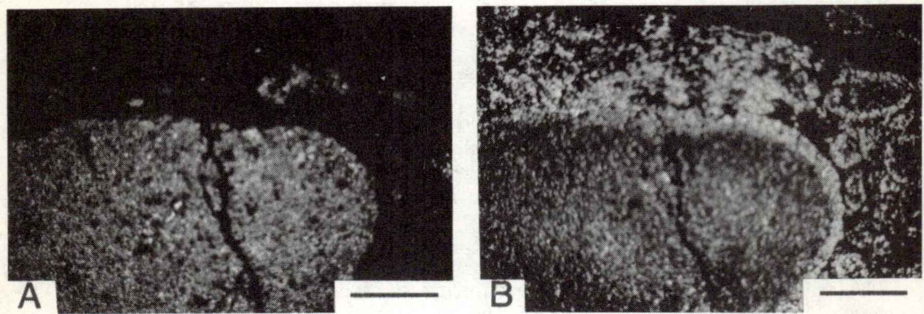

Fig. 14 - Micrographs showing the extent of corrosion of a magnesite refractory taken from an experimental lead furnace. (A) Transmitted light. In the upper and right portions of the micrograph, the magnesite refractory (medium gray) appears to be almost completely replaced by the high-silica slag (black). The remaining area of periclase also is veined and permeated by the slag. (B) Cathodoluminescence reveals that abundant (locally up to 85%) periclase remains as replacement remnants in the area that appeared under transmitted to light to be completely replaced. Scale bars = 0.2 mm.

Summary and Conclusions

Cathodoluminescence microscopy is a valuable technique that can be applied to a wide variety of mineralogic problems in mineral recognition, minerals exploration, minerals beneficiation problems, refractories, and pyrometallurgy. The technique should be more widely applied to process mineralogy problems in those areas than has previously been the case. The method should become a standard tool in all process mineralogy laboratories.

Acknowledgements

The writer wishes to thank Michael G. Schroer for providing some of the refractory samples. The manuscript was reviewed by Robert Moore and Robert Stevens.

References

1. Richard D. Hagni, "Cathodoluminescence Microscopy Applied to Mineral Exploration and Beneficiation", pp. 41-66 in Applied Mineralogy, Proceedings of the Second International Congress on Applied Mineralogy in the Minerals Industry, Won C. Park, Donald M. Hausen, and Richard D. Hagni (editors), AIME, New York, 1985.

2. Charles E. Barker and Teresa Wood, "Bibliography of Applied Cathodoluminescence", in this volume.

3. J. Richard Kyle, "Mineralogical Investigation of Sulfide Concentrations in the Hockley Salt-Dome Cap-Rock, Texas", pp. 1065-1082 in Applied Mineralogy, Proceedings of the Second International Congress on Applied Mineralogy in the Minerals Industry, Won C. Park, Donald M. Hausen, and Richard D. Hagni (editors), AIME, New York, 1985.

4. Anthony N. Mariano, "The Application of Cathodoluminescence for Carbonatite Exploration and Characterization", Proceedings of the International Symposium on Carbonatites, Pocos de Caldas, Minas Gerais, Brazil, 1978.

5. Michael L. Ebers and Otto C. Kopp, "Cathodoluminescent Microstratigraphy in Gangue Dolomite, the Mascot-Jerrerson City District, Tennessee", Econ. Geol., 74 (1979) pp. 908-918.

6. Otto C. Kopp, M. L. Ebers, L. B. Cobb, T. B. Crattie, T. L. Ferguson, R. M. Larsen, R. A. Potosky and R. T. Steinberg, "Application of Cathodoluminescence Microscopy to the Study of Gangue Carbonates in Mississippi Valley-Type Deposits in Tennessee: The Search for a Tennessee Trend", in this volume.

7. Robert L. Voss and Richard D. Hagni, "The Application of Cathodoluminescence Microscopy to the Study of Sparry Dolomite from the Viburnum Tend, Southeast Missouri", Chapter 5, pp. 51-68 in Mineralogy-Applications to the Minerals Industry, Proceedings of the Paul F. Kerr Memorial Symposium, New York, NY, 1985.

8. Jay M. Gregg, "Regional Epigenetic Dolomitization in the Bonneterre Dolomite (Cambrian), Southeastern Missouri", Geology, 13 (1985) pp. 503-506.

9. Lanier Rowan, "A Study of Fluid Inclusions within Cathodoluminescent Zones in Gangue Dolomites, Viburnum Trend Pb-Zn District, Southeast Missouri", in this volume.

10. M. R. Farr, "Regional Isotopic Variation in Bonneterre Formation Cements: Implications for Brine Migration Pathways and Sources" (abs.), p. 8 in Jay M. Gregg and Richard D. Hagni (editors) Symposium on the Bonneterre Formation (Cambrian), Southeastern Missouri: Stratigraphy, Sedimentology, Diagenesis, Geochemistry, and Economic Geology, University of Missouri-Rolla, Rolla, Missouri, 1986.

11. Jay M. Gregg and Richard D. Hagni, "The Use of Cathodoluminescence Microscopy to Reveal Hidden Crystal Faces in Gangue Dolomite Cements, Viburnum Trend, Southeastern Missouri", in this volume.

12. Richard D. Hagni and Adnan A. Saadallah, "Alteration of Host Rock Limestone Adjacent to Zinc-Lead Ore Deposits in the Tri-State District, Missouri, Kansas, Oklahoma", Econ. Geol., 60 (1965) pp. 1607-1619.

13. Richard D. Hagni and Michelle Cooper, Characterization of Accessory Minerals in the Birmingham Red Iron Ores and Eufaula Ferruginous Bauxite Deposits in Alabama, U.S. Bureau of Mines, NTIS microfiche, 1981, 119 pp.

14. Richard D. Hagni and Michelle Cooper, "The Nature of Phosphorous-bearing Mineral Grains in the Birmingham, Alabama Sedimentary Iron Ores and an Assessment of Their Potential Liberation by Beneficiation", pp. 95-117 in Richard D. Hagni (editor) Process Mineralogy II: Applications in Metallurgy, Ceramics, and Geology, AIME, New York, NY, 1982.

15. B. G. Yacobi and D. B. Holt, "Cathodoluminescence Scanning Electron Microscopy of Semiconductors", Journal of Applied Physics, 59 (1986) pp. R1-R24.

16. J. T. Czernuszka and T. F. Page, "Cathodoluminescence: A Microstructural Technique for Exploring Phase Distributions and Deformation Structures in Zirconia Ceramics", J. Am. Ceram. Soc., 68 (1985) pp. C-196-C-199.

17. Robert P. Stevens, "The Petrographic Examination of Refractory Service Samples", pp. 633-644 in Donald M. Hausen and Won C. Park (editors) Process Mineralogy: Extractive Metallurgy, Mineral Exploration, Energy Resources, AIME, New York, NY, 1981.

APPLICATION OF CATHODOLUMINESCENCE MICROSCOPY TO THE STUDY OF GANGUE CARBONATES IN MISSISSIPPI VALLEY-TYPE DEPOSITS IN TENNESSEE: THE SEARCH FOR A "TENNESSEE TREND"

O. C. Kopp, M. L. Ebers, L. B. Cobb, T. B. Crattie, T. L. Ferguson, R. M. Larsen, R. A. Potosky and R. T. Steinberg

Department of Geological Sciences, University of Tennessee, Knoxville, Tennessee 37996-1410

Abstract

Cathodoluminescence microscopy has been used to study zoned gangue carbonates in Mississippi Valley-type sphalerite deposits in east and central Tennessee. In east Tennessee, the same microstratigraphic pattern can be traced from the the Mascot-Jefferson City district to the Copper Ridge district over a distance of nearly 50 km. In the Central Tennessee district, a microstratigraphic zonation pattern in the gangue carbonates can be traced over a distance of 40 km eastward from the Elmwood mine to the mineralized Right Fork area in Jackson and Overton Counties. In east Tennessee, the gangue carbonates are dominated by dolomite with minor, late calcite. At Right Fork, zoned dolomite and calcite occur in subequal amounts. At Elmwood, zoned calcite dominates with only minor dolomite. The apparent mineralogical trend from dominant dolomite to dominant calcite is consistent with a relatively hotter source of fluid(s) to the southeast.

When adjacent Paleozoic basins (Appalachian, Illinois, and Michigan) are considered as potential sources of basinal brines which might have introduced the ore minerals, several lines of evidence suggest that the Appalachian Basin was the most likely source for those fluid(s), including the distribution of known sphalerite occurrences in Tennessee with respect to the Nashville dome (Cincinnati arch) and trace element studies in central Kentucky which suggest that the Cincinnati arch acted as a barrier to fluid migration.

If further study confirms that the present districts in east and central Tennessee had a common source of ore-bearing fluid(s) and a common distribution system (perhaps the extensive paleoaquifer developed at the top of the Knox Group carbonates), then the districts in east and central Tennessee would all belong to a single district extending well over 200 km. We propose that this district be called the "Tennessee Trend" by analogy with the well-known, but less extensive Viburnam Trend in Missouri.

Introduction

Previous Studies

In 1965 (1) it was suggested that electron-excited luminescence (cathodoluminescence) could be applied to studies in petrology. Because no commercial unit was available, we began construction of our own instrument in early 1966 (2,3). Among our early conclusions (2) were, "This investigation has shown that cathodo-luminescence may reveal structures and textures unobservable by other methods..." and "Sphalerite mineralization appears to be closely related to this zoned dolomite, and perhaps pertinent information pertaining to the origin of the mineralization can be obtained through the study of additional specimens". The untimely demise of our home-made device was followed by a period of waiting for funds to purchase a newly available LUMINOSCOPE (R) (Nuclide Corporation, Acton, MA). During the past fifteen years, several theses in which cathodoluminescence (CL) was used to study gangue carbonates in Mississippi Valley-type (MVT) deposits in Tennessee have been completed.

There have been numerous other studies of MVT ore deposits and much attention has been devoted to deposits in Tennessee. Their mineralogy is relatively simple, comprising dolomite, calcite, various forms of silica, clays, and sphalerite, with minor pyrite, marcasite, galena, chalcopyrite, barite, fluorite, etc. Deposits typically occur in association with breccia bodies in Upper Knox Group carbonates, but they are also found in bedded structures in certain limestone and medium- to coarse-grained dolostone beds. In breccia bodies, carbonate gangue and ore minerals occupy the space between breccia fragments. It should be noted that many of these mineralogical and textural features are quite similar to those in MVT deposits hosted in the Knox (and Knox equivalents) in other locations.

Although there is no consensus, many present-day workers seem to agree that brines derived from nearby basins were responsible for transporting the ore minerals (4). Support for this view comes from the locations of many MVT deposits near the margins of basins and from the common association of MVT mineralization and hydrocarbons. Still, there are many unresolved questions about the fluid(s) responsible for the emplacement of MVT ore deposits, concerning such aspects as: 1) the nature of the fluid(s)- (were there one or two? or more?): 2) the source(s) of the fluid(s)- (how much was introduced and how much was derived locally?); and 3) the timing of mineralization- (when did it begin and did it occur more than once?); etc. In the present paper, we will not consider alternative hypotheses proposed to explain the origin of these deposits. The interested reader is directed to several review articles (5,6,7,8,9,10) on this subject.

Purpose

The purposes of this paper are: 1) to review the status of knowledge concerning gangue carbonates which are ubiquitous to MVT deposits in Tennessee and, 2) in a preliminary way, to consider adjacent basins which might have contributed fluid(s) that not only introduced ore, but gangue minerals as well. Although the conclusions reached must be considered speculative and in need of much additional research, we believe that information obtained through the study of gangue carbonates (using CL and trace element geochemical studies) is essential to our understanding of the complex history of these deposits in Tennessee (and elsewhere). We realize that other techniques (fluid inclusions, carbonate petrology, stratigraphic analyses, isotopic analyses, etc.) are essential; however, through the study of gangue carbonates (which can be traced over many kilometers using CL) we have the potential to link these deposits together.

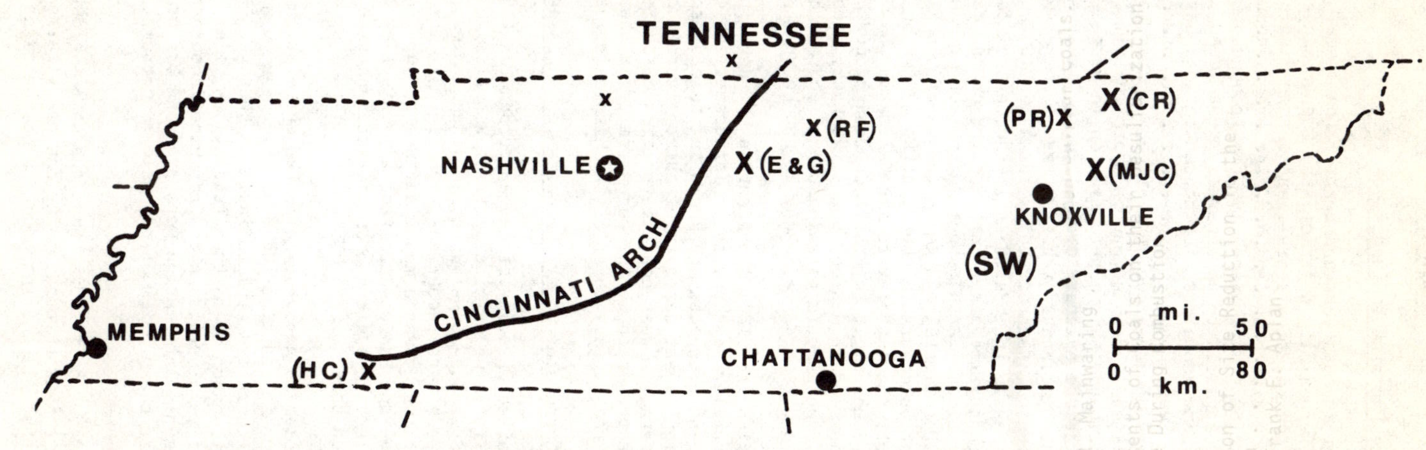

Figure 1 - Mississippi Valley-type ore deposits in Tennessee. (CR = Copper Ridge district, MJC = Mascot-Jefferson City district, PR = Powell River district, SW = Sweetwater district, RF = Right Fork area, E & G = Elmwood and Gordonsville mines, HC = Hardin County.)

Discussion

Mississippi Valley-type Deposits in Tennessee

MVT deposits occur in both east and central Tennessee, see Fig. 1, including the Mascot-Jefferson City district, the Copper Ridge district, the Sweetwater district, (the relatively small and rather distinctive Powell River district which contains both lead and zinc ores), and the Central (or Middle) Tennessee district. Until recently, occurrences of sphalerite (excluding trace amounts) appeared to be generally restricted to east and central Tennessee (11). Some anomalous zinc concentrations have been discovered in water wells in Henry County, Tennessee (12). More recently, approximately seven meters of ore-grade sphalerite was discovered in an exploratory oil well drilled in Hardin County (approximately 150 km southwest of Nashville) (13). Although this mineralization in west-central Tennessee seems to occur west of the Cincinnati arch, it more likely occurs along (slightly south of) the Cincinnati arch, see Fig. 1, based on maps of the district (14) and tectonic maps produced by the AAPG (15) and USGS (16). Presently mined major sphalerite orebodies in the Central Tennessee district (Elmwood and Gordonsville mines) are located on the eastern flank of the Nashville dome (Cincinnati arch) about 70 km east of Nashville. Still further to the east (approximately 40 km northeast of the Elmwood mine) significant mineralization occurs in the Right Fork area in Jackson and Overton Counties. {Note that sphalerite mineralization has been discovered west of the Nashville dome, e.g., in Robertson County, north of Nashville near the Kentucky border and in Monroe County, Kentucky, near the Tennessee border (13); however, the genetic relationship of these occurrences to major districts in east and central Tennessee is not known.}

Barite (17) and fluorite (18) also occur in MVT deposits, sometimes in association with the sphalerite deposits in Tennessee. While many barite and fluorite deposits and prospects are known in the eastern and central parts of the state, significant deposits of these minerals have not been identified in the western (or west-central) portion of the state. In part this is due to the greater depth to the unconformity at the top of the Knox Group and also to the sedimentary cover of the Mississippi Embayment.

CL Studies of Gangue Carbonates in MVT Deposits in Tennessee

In our first detailed CL study of gangue carbonates (19,20), the gangue microstratigraphy for a major zinc district was first elucidated. Six major zones (plus several sub-zones where individual zones were relatively thick and well developed) were recognized, see Fig. 2. The recognition of this gangue microstratigraphy was a major advance in our knowledge of MVT deposits in general and the Mascot-Jefferson City district in particular. Only a few years earlier, in a discussion of East Tennessee zinc deposits (21) it had been stated that:

> "The main problem in paragenesis is the correlation of the mineral sequences from district to district and even from mine to mine within a district...recent more detailed studies indicate that sequences from mines only a few miles apart cannot be matched readily...similar sphalerites as well as other distinctive minerals in the paragenetic sequences may not be time correlative...sizable information gaps produced by thrust faulting and subsequent erosion make the problem of mineralogic correlation even more difficult."

Our study (19,20) not only showed that the same microstratigraphic (CL) zonation pattern exists throughout the Mascot-Jefferson City district, but it also revealed that: 1) the introduction of the major phase of sphal-

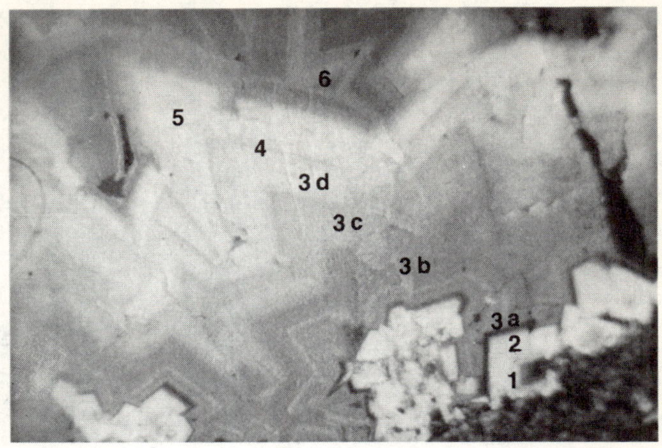

Figure 2 - Zoned dolomite gangue, New Market mine, Mascot-Jefferson City district. Zones 1 through 6 present. (Note that Figures 2, 3, 5, 6 and 7 were taken using CL. The width of each photomicrograph equals approximately 2.8 mm on the specimen.)

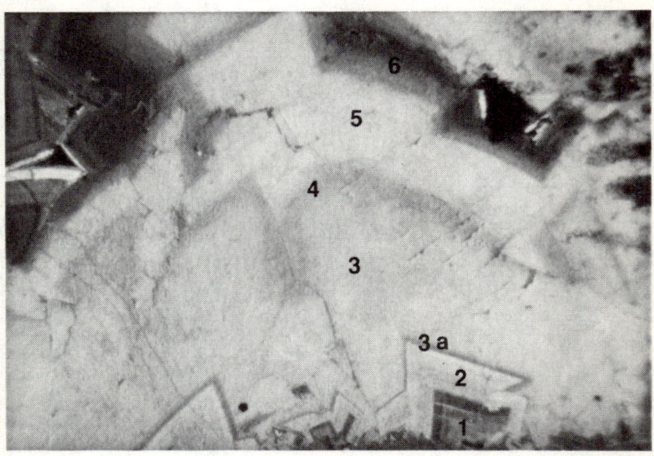

Figure 3 - Zoned dolomite gangue, Flat Gap mine, Copper Ridge district. Zones 1 through 6 present. In addition, some bright, post zone 6 zones can be seen lining small voids in this sample.

Mascot-Jefferson City Zone	Mascot-Jefferson City Color	Copper Ridge (Flat Gap mine) Color
Post zone 6	(Calcite)	Alternating bright and dark red
6D	Very dark red	Dark red (6D–6A)
6C	Medium dark red	
6B	Dark medium red	
6A	Medium dark red	
5	Bright red	Medium bright red
4	Dark red	Dark red
3D	Bright medium red	Medium dark red (3D–3B)
3C	Dark medium red	
3B	Medium dark red	
3A	Dark red	Dark red
2	Very bright red	Medium bright red
1	Very dark red	Dark red

Figure 4 - Comparison of cathodoluminescent zones in carbonate gangue in the Mascot-Jefferson City district and the Copper Ridge district (Flat Gap mine). All the zones are in dolomite except Post zone 6 which is calcite. Note that the relative thicknesses of zones were also considered when establishing the microstratigraphy. The color terms used here and throughout the text are after Ebers (19).

erite (which began at the end of zone 3c and ended before the initiation of zone 6) was essentially time-correlative throughout the district, and 2) essentially the same microstratigraphic pattern, see Figs. 3 and 4, is present in the Copper Ridge district (Flat Gap mine) now located ca. 48 km northeast of the Mascot-Jefferson City district. Prior to deformation, the Copper Ridge district may have been twice as distant. Preliminary evidence suggests that mineralization at Flat Gap took place a little later than in the Mascot-Jefferson City district (19). {Note that during this same time period, CL microscopy was used to trace carbonate cement stratigraphy in New Mexico over a distance of 16 km (22) and more recently, to trace CL zones in the Viburnum Trend over a distance of nearly 30 km (23).} Zoned dolomite from the Mascot-Jefferson City district was analyzed using an electron microprobe (24). Although the Mn contents could not be measured with confidence (because of the detection limits for that element) the alternating brighter and darker CL zones could be correlated with changing iron and manganese concentrations (and Fe/Mn ratios). Zones 6a through 6d, which developed after the main period of mineralization, usually luminesce dark red since neither pyrite nor sphalerite were precipitating at that time; hence, more iron went into the dolomite lattice, effectively quenching its luminescence.

Additional information concerning the role of fluids in MVT deposits comes from a recent study of wall rock alteration adjacent to mineralized zones in the Coy mine (Mascot-Jefferson City district) and core from the Right Fork Area in the eastern part of the Central Tennessee district (25). Several wallrock specimens were taken in immediate contact with ore and gangue. CL zones like those observed in gangue (19) appear to be restricted to carbonates deposited in void spaces between breccia fragments and near their edges. Within breccia fragments, distinctive CL patterns, see Fig. 5, appear to be related to dolomitization that occurred during early diagenesis and later burial diagenesis (25).

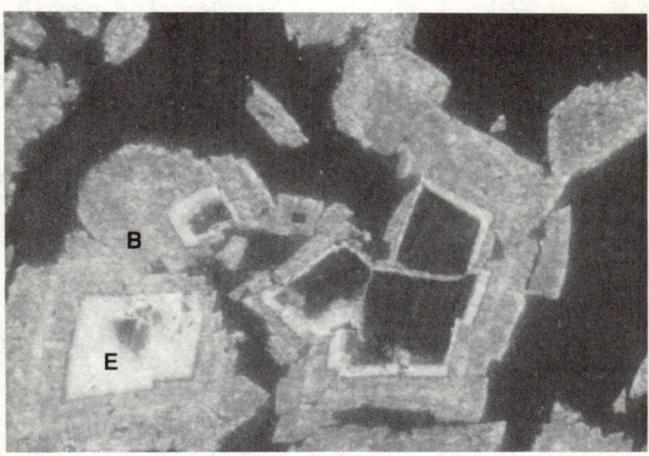

Figure 5 - Early diagenetic dolomite rhombohedra (E) and later, burial diagenetic dolomite (B) which surrounds the first generation dolomite. A similar pattern can be traced vertically for over thirty meters in the Coy mine, Mascot-Jefferson City district.

The timing of mineralization in Tennessee is uncertain. In East Tennessee, some workers (5) postulate an early Middle Ordovician age while others (9) suggest a Lower Ordovician age for at least some of the mineralization based on the occurrence of sphalerite in interlaminated limestones and dolostones of the Kingsport Formation in the Copper Ridge district. Still others (14) suggest a Late Paleozoic age for mineralization in Central Tennessee or multiple stages of sphalerite introduction in the Mascot-Jefferson City district (26). Barite mineralization (with minor late pyrite, marcasite, galena, chalcopyrite and sphalerite) has been studied at the Lost Creek mine (27) located on the southwestern flank of the Powell River district, see Fig. 1. Although a relatively small orebody, it is unique because it is the only MVT deposit in East Tennessee in which brecciation and mineralization are not confined to Lower Ordovician rocks, but extend through the Middle Ordovician ("Knox") unconformity. At Lost Creek, textural evidence suggests that mineralization took place epigenetically at depth while, at the same time, it was taking place syngenetically at or near the surface (the sea floor) on which new carbonate sediments were being deposited. The timing of sphalerite mineralization in Tennessee is in need of much further investigation!

In the Central Tennessee district, we have made two studies (28,29) using CL microscopy. The gangue carbonates at the Elmwood mine, see Fig. 1 (28), are dominated by calcite (with only minor dolomite) in contrast to East Tennessee districts which are dominated by dolomite gangue with only minor, late (perhaps much later than mineralization) calcite. At least three generations of calcite can be recognized with CL. The first generation is zoned, see Fig. 6, and associated with sphalerite mineralization. The same zonation pattern is present in very minor gangue dolomite found in association with the first generation of calcite, see Fig. 6. Sphalerite precipitation began near the end of deposition of the first zone in the calcite. Two later, unzoned generations of calcite appear to be younger and post-sphalerite deposition. The gangue microstratigraphy can be traced for tens of kilometers in the vicinity of the ELmwood mine.

Examination of core from the Right Fork area, see Fig. 1, also reveals three generations of calcite, but in addition, there is a significant amount of gangue dolomite present (29), see Fig. 7. Five CL zones are present in the first generation of dolomite and calcite. Sphalerite deposition appears to have commenced during the deposition of the third zone. The fifth zone is non-luminescent and has a high (ca. 12) Fe/Mn ratio. Zones 3, 4 and 5 at Right Fork appear to be equivalent, respectively, to zones 1, 2 and 3 at Elmwood.

Potential Sources of MVT Fluids in Tennessee

In the discussion which follows, we make the assumption that the mineralizing fluid(s) which were primarily responsible for MVT sphalerite deposits in Tennessee were derived from some nearby basin(s). Other potential sources of mineralizing fluids (e.g., deep-seated intrusions, submarine volcanic vents, downward-percolating groundwater, etc.) have received attention in the past, but are not considered here. Nor do we consider here the role of fluids (such as pore fluids) already present in the rocks, even though these may have played an important role in the localization of ore.

Three major basins (Appalachian, Illinois and Michigan), see Fig. 8 (30), were active during Paleozoic time and are potential sources for MVT fluid(s). Based on geographic location, it seems unlikely that the Michigan Basin was a source of fluids for MVT deposits in east and central Tennessee; hence, only the Appalachian and Illinois Basins will be consi-

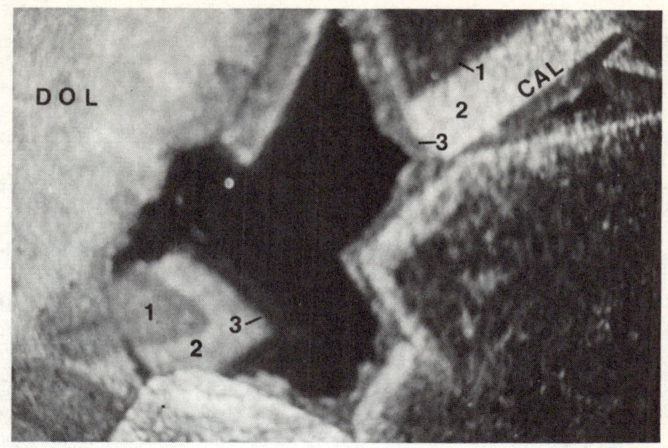

Figure 6 - Zoned calcite and minor zoned dolomite gangue in the Elmwood mine, Central Tennessee district. Zones 1 through 3 present. In many places, zone 1 can be divided into two sub-zones (1a and 1b).

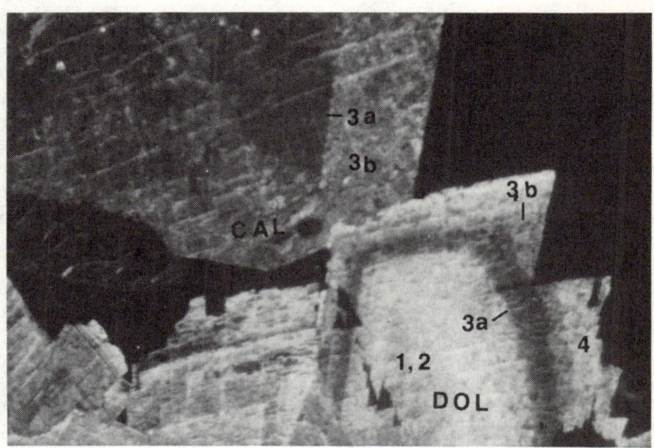

Figure 7 - Zoned calcite and dolomite gangue in the Right Fork area, on the eastern flank of the Central Tennessee district. Zones 1 through 4 present in the dolomite gangue. Zones 3a and 3b correspond to zones 1a and 1b in the Elmwood mine.

dered further. The areal extent of each basin fluctuated as a result of relative rise and fall of sea level and as deformation progressed in the southeast. During part of its existence, the Illinois basin may have extended as far to the south and east as central Kentucky and north-central Tennessee, see Fig. 8. If basin-derived fluid(s) were responsible for the development of MVT deposits in Tenessee, both of these major sedimentary basins should be considered viable sources for such fluid(s).

The Appalachian and Illinois Basins as Potential Sources of MVT Sphalerite Deposits in Tennessee

Comparison of several aspects of the Appalachian and Illinois Basins suggests that the Appalachian Basin would have been the most likely source for MVT fluid(s) in east and central Tennessee. Consider the following:

1. Sediments in the Appalachian Basin are much thicker than those in the Illinois Basin and were subject not only to compaction due to gravity, but also to several pulses of deformation beginning in Ordovician time. Large masses of Cambrian shales (such as shales in the Rome Formation and the Pumpkin Valley, Rogersville and Nolichucky Shales) were derived from the craton and deposited in the Appalachian basin as were Ordovician shales (such as the Sevier and Athens Shales) derived from sources to the southeast. Shales are considered to be the best potential sources for both fluids and metal ions during basinal compaction.

2. Fluids driven out of the Appalachian Basin were probably hotter than those of the relatively passive Illinois Basin (because of the higher geothermal gradient in the Appalachian region where there was extensive volcanic activity, metamorphism, etc.), contributing to their role as "hydrothermal fluids".

3. Study of trace element distributions at the top of the Knox Dolomite in central Kentucky (31) suggests that the unconformity at the top of the Knox controlled the up-dip migration of ore fluids and that fluids were blocked by the structural high of the Cincinnati arch. Anomalies for Pb, Zn, and Ba fall mainly on the southeast side of the Cincinnati arch.

4. Studies of basin characteristics, lithologic units present, etc., (4) suggest that, "the Friedensville deposit, the East Tennessee district, and possibly the Central Tennessee district are associated with the Appalachian basin".

5. The locations of MVT sphalerite occurrences in Tennessee suggest a general southeastward source. Most deposits lie along, or east and/or south of the Nashville dome (Cincinnati arch) which might have served as an impediment to fluid migration as has been suggested for deposits in Kentucky (31). The widespread Knox unconformity, which extends thoughout Tennessee, contains numerous solution channels, paleo-sinks, etc., that could have served as passageways for fluids migrating from a southeasterly source (32).

6. Our CL studies of zoned, mineralization-stage gangue carbonates suggest a trend compatible with fluid(s) moving away from a (relatively) hotter source area to the southeast: in East Tennessee, zoned dolomite gangue is dominant with minor, late calcite; further to the west (on the eastern margin of the Central Tennessee district) at Right Fork, gangue carbonates comprise zoned dolomite and calcite in sub-equal amounts {the last CL zone is quite dark, much like the zone (6) in East Tennessee}; finally, at Elmwood, gangue carbonates are dominated by zoned calcite (with only minor dolomite) which can be correlated with the gangue at Right Fork.

ORNL DWG 82-542

Figure 8 - Location map of the Appalachian, Illinois and Michigan Basins and adjacent structural elements (30). Reprinted with kind permission of T. F. Lomenick, Chemical Technology Division, Oak Ridge National Laboratory, Oak Ridge, Tennessee.

Conclusions

The application of CL microscopy has been, and will continue to be, a most important source of information concerning the paragenesis of individual MVT deposits and their relationship to other deposits and districts.

Based on several considerations, it appears quite possible that the fluid(s) which deposited sphalerite in east and central Tennessee had a common, southeasterly source in the maturing Appalachian Basin. If so, and if our tentative conclusions about the gradual changes in zoned carbonate gangue from east to central Tennessee can be confirmed, then what appear to be isolated districts (Mascot-Jefferson City, Copper Ridge, Central Tennessee) may all belong to a single "district". (Preliminary study of samples from the Powell River district and Austinville, VA, suggest they are different from the rest.) CL microstratigraphic correlation of the present Mascot-Jefferson City and Copper Ridge districts strongly suggests that they belong to a single district. If it can be shown that the Knox breccia-hosted sphalerite deposits in east and central Tennessee had a common source of fluid(s) and common distribution system, this new district would be one of the largest zinc districts in the world, with mines and prospects extending over a linear distance well in excess of 200 km. If the existence of such a district can be confirmed, we propose that it be called the "Tennessee Trend" (by analogy with the well-known, although much less extensive Viburnum Trend).

The search has already begun.

Epilogue

Shortly after this paper was completed, Jack Oliver published an interesting paper in GEOLOGY (33) presenting a "speculative hypothesis" concerning the expulsion of fluids from sediments in continental margins buried beneath thrust sheets. Oliver suggests that the MVT deposits of the eastern and central United States had their source in the Appalachian belt. He recommends that this "tectonic fluids" hypothesis should be tested on certain mineral deposits in the western U. S. and elsewhere.

The MVT deposits in Tennessee (as well as related deposits in Kentucky) will provide an excellent natural laboratory in which to study and to test the "basinal brine" and "tectonic fluid" hypotheses.

Acknowledgements

We appreciate the many forms of assistance provided by ASARCO geologists (especially Del Harper and Ed McCormick), the staff of the Tennessee Division of Geology, and the constructive comments offered over the years by Stuart Maher (Kenwill, Inc.), Kula Misra and Ken Walker (University of Tennessee) during various aspects of this research. We wish to acknowledge Sigma Xi, the Appalachian Basin Industrial Associates, and the Discretionary Funds, Department of Geological Sciences, University of Tennessee, for partial support of this research.

References

1. J. F. Smith and R. C. Stenstrom, "Electron-Excited Luminescence as a Petrologic Tool," Journal of Geology, 73 (1965) 627-635.

2. R. A. Potosky, "The Application of Cathodo-Luminescence in the Field of Petrology" (Master's thesis, University of Tennessee, 1967).

3. R. A. Potosky and O. C. Kopp, "Application of Cathodoluminescence to Petrographic Studies," The Compass, 47 (1970) 63-69.

4. L. M. Cathles and A. T. Smith, "Thermal Constraints on the Formation of Mississippi Valley-type Lead-zinc Deposits and Their Implications for Episodic Basin Dewatering and Deposit Genesis," Economic Geology, 78 (1983) 983-1002.

5. A. D. Hoagland, W. T. Hill, and R. W. Fulweiler, R. W., "Genesis of the Ordovician Zinc Deposits in East Tennessee," Economic Geology, 60 (1965) 693-714.

6. J. Crawford and A. D. Hoagland, "The Mascot-Jefferson City Zinc District, Tennessee," (Ore Deposits of the United States, Graton Sales Volume, New York, American Institute of Mining and Metallurgical Engineers, 1968) 242-256.

7. J. E. McCormick et al., "Environment of the Zinc Deposits of the Mascot-Jefferson City District, Tennessee," Economic Geology, 66 (1971) 757-762.

8. A. D. Hoagland, "Appalachian Zinc-Lead Deposits", Handbook of Strata-Bound and Stratiform Ore Deposits, (K. H. Wolfe, ed., Amsterdam, Elsevier Science Publishers, 1976), 495-534.

9. H. G. Churnet and K. C. Misra, "Sphalerite Mineralization and Its Relationship to Carbonate Facies Boundaries, Copper Ridge District, East Tennessee", (Proceedings, International Conference on Mississippi Valley-Type Lead-Zinc Deposits, University of Missouri-Rolla Press, 1983), 360-372.

10. J. E. McCormick et al., 1985, "Zinc Deposits of the Mascot-Jefferson City and Copper Ridge Districts, East Tennessee" (Field Trips in the Southern Appalachians, 1985 SE GSA, University of Tennessee Studies in Geology, 9, 1985) 1-33.

11. S. W. Maher, "The Zinc Industry of Tennessee" (Information Circular No. 6, State of Tennessee, Division of Geology, 1958).

12. R. E. Hershey and J. M. Wilson, "Anomalous Zinc in Water Wells, Northeastern Henry County, Tennessee" (Report of Investigations 27, State of Tennessee, Division of Geology, 1970).

13. Delbert D. Harper, private communication with author, ASARCO, 24 February 1986.

14. J. R. Kyle, "Brecciation, Alteration, and Mineralization in the Central Tennessee Zinc District," Economic Geology, 71 (1976) 892-903.

15. P. B. King, "Tectonic Map of the United States," (Committee on Tectonics, Division of Geology and Geography, National Research Council, published by the American Association of Petroleum Geologists, 1949).

16. K. C. Bayer, "Generalized Structural, Lithologic and Physiographic Provinces in the Fold and Thrust Belts of the United States," (Department of the Interior, U. S. Geological Survey, 1983).

17. S. W. Maher, "Barite Resources of Tennessee" (Report of Investigations 28, State of Tennessee, Division of Geology, 1978).

18. S. W. Maher, "Fluorspar in Tennessee" (Report of Investigations 42, State of Tennessee, Division of Geology, 1983).

19. M. L. Ebers, "A Study of Gangue Dolomite Mineralization in the Mascot-Jefferson City Zinc District, Using Cathodoluminescence" (Master's thesis, University of Tennessee, 1976).

20. M. L. Ebers and O. C. Kopp, "Cathodoluminescent Microstratigraphy in Gangue Dolomite, the Mascot-Jefferson City District, Tennessee," Economic Geology, 74 (1979) 908-919.

21. W. T. Hill, E. J. McCormick, and H. Wedow, Jr., "Problems on the Origin of Ore Deposits in the Lower Ordovician Formations of East Tennessee," Economic Geology, 66 (1971) 799-804.

22. W. J. Meyers, "Carbonate Cement Stratigraphy of the Lake Valley Formation (Mississippian) Sacramento Mountains, New Mexico," Journal of Sedimentary Petrology, 44 (1974) 837-861.

23. R. L. Voss, and R. D. Hagni, "The Application of Cathodoluminescence Microscopy to the Study of Sparry Dolomite from the Viburnum Trend, Southeast Missouri," (Proceedings of the Paul F. Kerr Memorial Symposium: Mineralogy- Applications to the Minerals Industry, New York, American Institute of Mining and Metallurgical Engineers, 1985) 51-68.

24. L. B. Cobb, "A Study of the Distribution of Trace Elements and Cathodoluminescent Zonation in Mineralized Fracture Filling" (Masters thesis, University of Tennessee, 1974).

25. T. B. Crattie, "A Petrographic and Geochemical Study of Coexisting Limestones and Dolostones from the Mascot-Jefferson City District, East Tennessee and the Right Fork Area, Central Tennessee" (Master's thesis, University of Tennessee, 1985).

26. M. Taylor et al., "Relationships of Zinc Mineralization in East Tennessee to Appalachian Orogenic Events" (Proceedings, International Conference on Mississippi Valley-type Lead-Zinc Deposits, University of Missouri-Rolla Press, 1983) 271-278.

27. R. T. Steinberg, "Geology and Petrology of the Lost Creek Barite Mine, Union County, Tennessee" (Master's thesis, University of Tennessee, 1981).

28. R. M. Larsen, "A Study of the Cathodoluminescent Properties and the Trace Element Distribution of the Gangue Calcite (and Other Phases) at the Elmwood Mine in Smith County, Tennessee" (Master's thesis, University of Tennessee, 1978).

29. T. L. Ferguson, "Petrographic and Trace-Element Characterization of Coarse Crystallization Carbonate Minerals in Fractures and Breccia Bodies and Associated Mineralization in the Right Fork Area, Central Tennessee" (Master's thesis, University of Tennessee, 1981).

30. T. F. Lomenick et al., "Regional Geological Assessment of the Devonian-Mississippian Shale Sequence of the Appalachian, Illinois, and Michigan Basins Relative to Potential Storage/Disposal of Radioactive Wastes" (Report ORNL/TM-5703, Oak Ridge National Laboratory, 1983).

31. R. A. Baird and W. H. Dennen, "A Geochemical Survey of the Top of the Knox Dolomite: Implications for Brine Movement and Mineralization in Central Kentucky," Economic Geology, 80 (1985) 688-695.

32. "A Paleoaquifer and Its Relation to Economic Mineral Deposits: The Lower Ordovician Kingsport Formation and Mascot Dolomite. A Symposium." Economic Geology, 66 (1971) 695-810.

33. J. Oliver, "Fluids Expelled Tectonically From Orogenic Belts: Their Role in Hydrocarbon Migration and Other Geologic Phenomena," Geology, 14 (1986) 99-102.

CATHODOLUMINESCENT ZONATION IN HYDROTHERMAL DOLOMITE CEMENTS: RELATIONSHIP TO MISSISSIPPI VALLEY-TYPE PB-ZN MINERALIZATION IN SOUTHERN MISSOURI AND NORTHERN ARKANSAS

E. Lanier Rowan

U.S. Geological Survey, P.O. Box 25046, Mail Stop 912, Denver, Colorado 80225

Hydrothermal, vug-lining dolomites from mines and drill core in the Viburnum Trend lead district, southeast Missouri, and in drill core from southern Missouri and northern Arkansas, have been examined using cathodoluminescence. A cathodoluminescent (CL) microstratigraphy in the dolomites has provided a time framework to which fluid inclusion measurements have been tied. Equivalent CL zones within microstratigraphies from different locations are interpreted to be time correlative. Local dissolution, much of which is visible only with cathodoluminescence, is associated with sulfide precipitation.

Regionally, the CL microstratigraphy defined in the Viburnum Trend may be traced in dolomite cements for a distance of up to 350 km to the southwest. This observation is consistent with a model in which Mississippi Valley-type lead-zinc mineralization in Missouri and Arkansas is related to regional fluid migration northward from the Arkoma basin.

Introduction

Cathodoluminescence refers to the emission of light in response to electron beam excitation of specific trace elements (activators) within the host mineral; other trace elements quench luminescence. In dolomite Mn^{+2} is the primary activator, although other elements such as Pb^{+2}, if present, may participate (1); Fe^{+2} is primarily responsible for quenching luminescence (2,3). Crystal growth zones which vary in their trace element chemistry are often distinguishable by cathodoluminescence.

Observations made with cathodoluminescence have contributed to an understanding of both regional and detailed aspects of Mississippi Valley-type (MVT) lead-zinc mineralization in southern Missouri and northern Arkansas (Fig. 1). Much of the work presented here draws on a 1985 study by Voss and Hagni (4) which defined and correlated a cathodoluminescent (CL) microstratigraphy in hydrothermal gangue dolomite throughout the Viburnum Trend district (Fig. 2). They showed that a sequence of four zones spanned mineralization. Two of these zones were contemporaneous with the main stages of ore deposition.

Within the microstratigraphy, correlative bands from different locations are considered to have formed essentially synchronously. The CL banding is therefore of great value in placing relative time

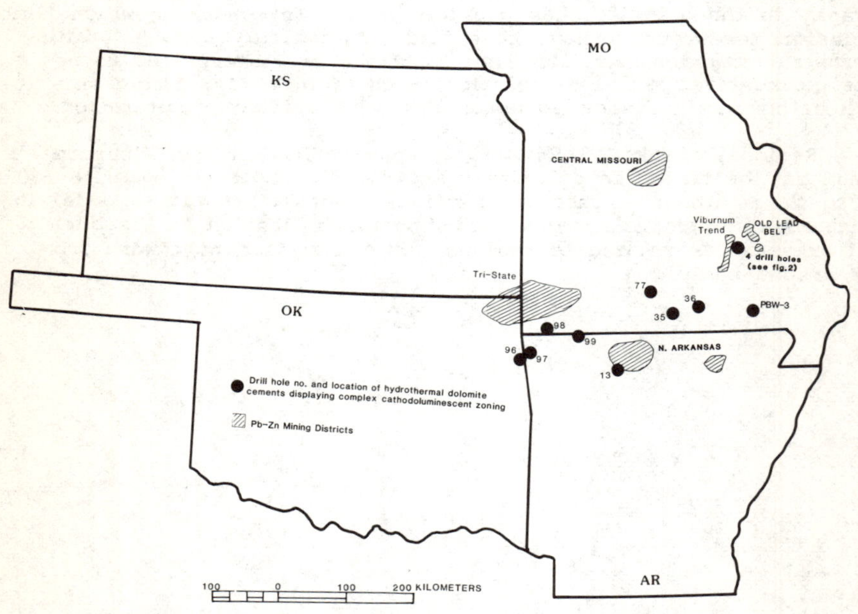

Figure 1 - Map showing locations of the Viburnum Trend and adjacent Mississipppi Valley-type Pb-Zn districts, and drill core examined in this study.

constraints on fluid inclusions or microsamples for isotope analysis. In addition, the timing relative to mineralization, of small scale dissolution textures visible only by cathodoluminescence may be used to infer changes in fluid chemistry in the Viburnum Trend.

On a larger scale, the CL zonation defined in the Viburnum Trend extends to dolomite cements from barren drill core outside the district. Gregg (5) correlated the Viburnum Trend microstratigraphy with the CL zonation in cements 80 km west of the district. Evidence presented here suggests that the microstratigraphy can be correlated over a distance of approximately 350 km to the southwest, near the margins of the Tri-State and Northern Arkansas Zinc districts (Fig. 1). The microstratigraphy therefore represents a tool for defining the flowpaths of the fluids responsible for mineralization.

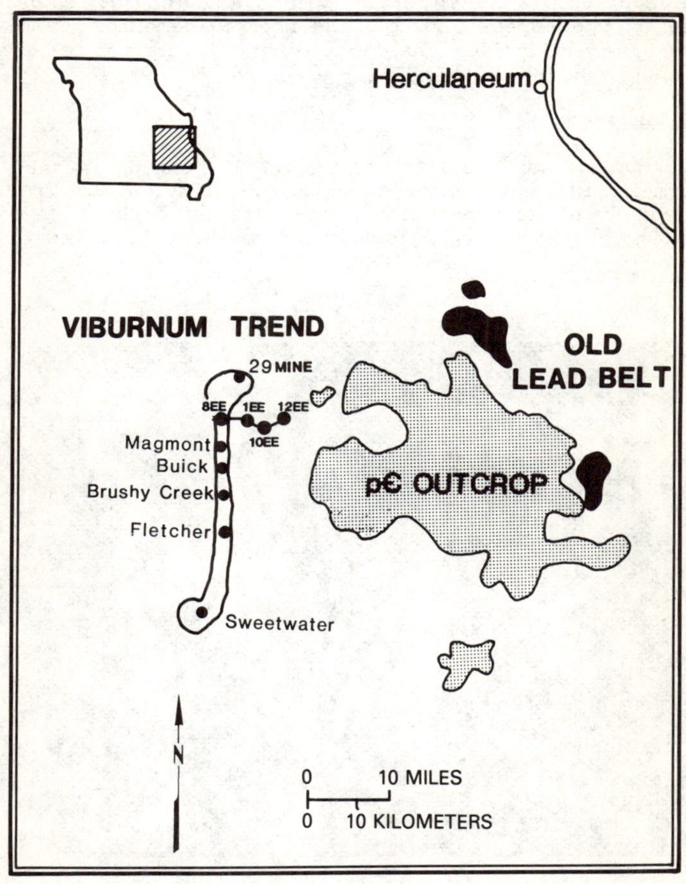

Figure 2 - Map of the Viburnum Trend Pb-Zn district showing location of mines and four drill holes (1EE, 8EE, 10EE, and 12EE) from which gangue dolomites were examined.

Viburnum Trend

General Geology

The Viburnum Trend, southeast Missouri (Fig. 2), is one of the world's largest lead mining districts. Ore is hosted by the Bonneterre Dolomite, the lowest in a sequence of Upper Cambrian carbonates. Most of the mineralization occurs near the reef/back reef transition of an algal stromatolite build-up which developed in the shallow water around a Precambrian basement high. Immediately above the Bonneterre is the shale-dominated Davis Formation (Upper Cambrian). Beneath the Bonneterre Dolomite is the Lamotte Sandstone (Upper Cambrian), which lies directly on Precambrian crystalline basement. The Lamotte Sandstone and the Bonneterre Dolomite served as aquifers for the mineralizing fluids.

The principal ore minerals, galena and sphalerite, occur as replacements, open-space fillings, and breccia cements. Gangue and minor ore minerals include chalcopyrite, pyrite, marcasite, quartz, calcite, and abundant dolomite (6).

Cathodoluminescence

Rickman (7) was the first to report CL banding in the hydrothermal, gangue dolomites (cements) of the Viburnum Trend (Fig. 3). Voss and Hagni (4) then defined and correlated the microstratigraphy over approximately 30 km throughout the district.

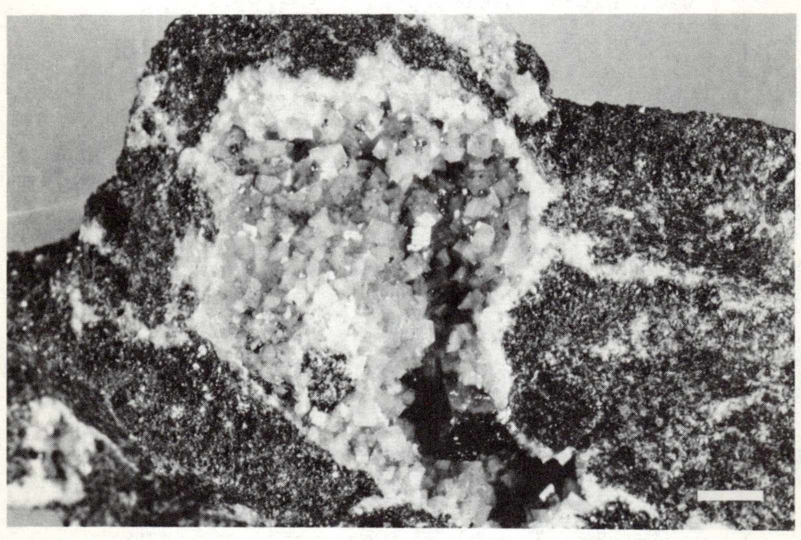

Figure 3 - Sample from the Bonneterre Dolomite (Magmont Mine) showing coarse, vug-lining, hydrothermal gangue dolomite (dolomite cement) typical of the Viburnum Trend. Scale bar=15 mm.

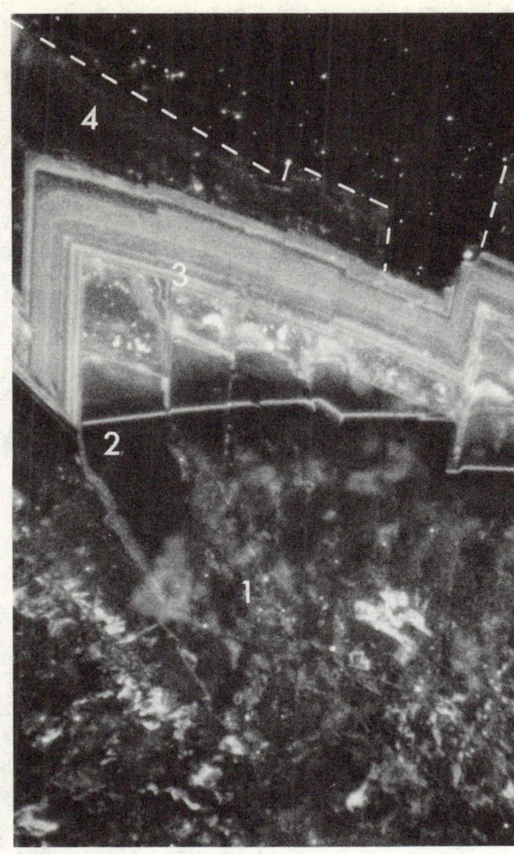

Figure 4 - Photomicrograph of cathodoluminescent zonation in dolomite cements from the Bonneterre Dolomite, Buick Mine, Viburnum Trend. All four of the zones defined by Voss and Hagni (4) are present. Patchy luminescence in zone 3 indicates dissolution. Field of view=1.3 mm.

Their microstratigraphy consists of four major zones within which there may be numerous sub-zones ranging from several microns to fractions of a micron in width (Fig. 4). The four major zones span mineralization from an early pre-ore stage (zone 1) to a late post-ore stage (zone 4). Based on limited petrographic observations made by Voss and Hagni (4) and by the author, zones 2 and 3 can be shown to coincide with main stages of mineralization (Fig. 5). Sulfides appear to replace or precipitate alternately with dolomite, rather than coprecipitating.

Cathodoluminescence of the host rock (Bonneterre Dolomite) ranges from essentially non-luminescent to mottled dull red. Low levels of luminescent light may be scattered off the surfaces within the crystal such as fracture planes and the abundant fluid inclusions, giving varying degrees of 'apparent' luminescence to dull or non-luminescent crystals. Host rock displays no CL banding as do the younger cements.

The host rock is interpreted to have been a limestone dolomitized early in its diagenetic history and subsequently recrystallized to varying extents by the mineralizing fluid (8).

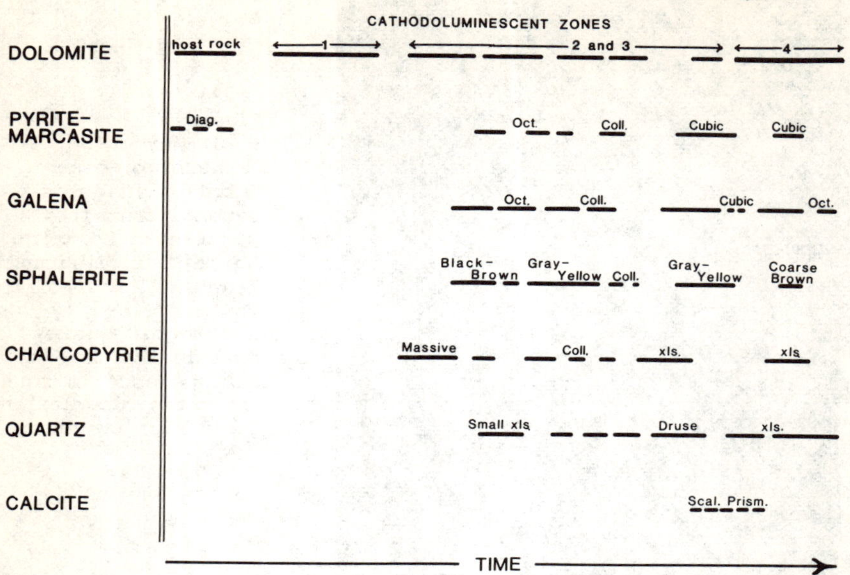

Figure 5 - Simplified paragenetic sequence for the Viburnum Trend showing the relationship of the four cathodoluminescent zones in dolomite to the ore minerals (6). Diag.= diagenetic; Oct.=octahedral; Coll.=colloform; Scal.=scalenohedral; Prism.=prismatic; Xls.=crystals.

Cathodoluminescence evidence for this recrystallization, presented by Braunsdorf and Lohmann (9), consists of small (20-50 micron), bright, yellow-orange patches within a grain luminescing a more dull red color. The bright patches represent replacement and infilling of microporosity in the host rock dolomite by the younger dolomite cements. When observed at high power (e.g. 400X) using universal stage objectives, the bright patches can be seen to consist of very compressed sequences of CL bands. This evidence for recrystallization is corroborated by a shift in host rock oxygen isotope values toward the values of the cements; the magnitude of this isotopic shift decreases away from the heavily mineralized areas (8).

Application to Fluid Inclusion Studies

Fluid inclusions suitable for ice melting and homogenization temperature measurements are relatively uncommon in dolomite. A thorough study requires searching many crystals to locate sometimes not more than one useable inclusion in a given dolomite rhomb. In this situation, fluid inclusion measurements may represent blind sampling of fluids from different ages and stages of crystal growth. CL provides the means to unravel the evolution of the fluid by making it possible to identify the different growth zones hosting individual inclusions. Within the Viburnum Trend, the well documented CL microstratigraphy in dolomite has proved useful in placing constraints on the relative ages of individual fluid inclusions with respect to main stages of mineralization. Without CL, the early pre-ore dolomite cement is often difficult to distinguish from the dolomitic host rock, and the later zonations cannot be detected. Other techniques for identifying growth zones, such as potassium ferricyanide staining for iron and use of UV fluorescence, did not reveal the detail necessary to correlate microstratigraphies between samples.

The ability to correlate inclusion measurements with the CL zones, which in turn can be correlated to the Viburnum Trend ore mineral paragenesis, has made it possible to evaluate changes in the mineralizing fluid through time. Fluid inclusions in the dolomite host rock have variable salinities ranging from 5 to 20 wt.% NaCl equivalent. The lower salinity inclusions may be relicts of an earlier diagenetic event (such as dolomitization) in rock partially recrystallized by higher salinity fluids. In contrast, inclusions in the younger cements have uniformly high salinities (approximately 20 wt.% NaCl equiv.)(10). No significant salinity fluctuations have been recognized between CL zones 1 through 4. Homogenization temperatures in the inclusions fall predominantly between 110° and 120°C throughout the four CL zones, and also show no trend of increase or decrease with time (10). Conclusions drawn thus far from these observations are that sulfide precipitation is not due either to temperature decrease or mixing of the metal-bearing fluid with a lower salinity, meteoric fluid, given the consistent temperatures and high salinities of inclusions in the cements.

Fluid Chemistry

The presence of CL banding indicates changes in activator and/or quencher element concentrations in the cement-forming fluid. Previous work (11) has assumed Mn and Fe to be the dominant controls on luminescence, and has inferred fluctuations in the Eh and pH of the fluid to explain the variation in Fe and Mn activities. However, as discussed by Machel (1), caution should be used in this type of analysis since trace amounts (as low as 10 ppm) of elements such as Pb, Ni, Co, Zn, and Ce have also been shown to influence or control luminescence. In the Viburnum Trend, galena, sphalerite and minor amounts of chalcopyrite, bornite, millerite and seigenite (6, 12) indicate the presence of Pb, Zn, Cu, Ni, and Co in the mineralizing system. Electron microprobe analyses of the Viburnum Trend gangue dolomites confirm that Fe/Mn ratios are higher in the non-luminescent zones than in the moderately to brightly luminescent zones (4). However, the correlation between Fe/Mn and intensity of luminescence is imperfect, suggesting the influence of other trace elements. Detection of elements such as Pb, which may have a significant effect at concentrations below the detection limit of the electron microprobe, may require analysis by other techniques (e.g. atomic absorption or induction-coupled plasma mass spectrometry).

Dissolution textures visible only with CL have been noted within individual dolomite crystals in the Viburnum Trend. It was suggested by Anderson (13) that a decrease in pH resulting from precipitation of sulfides may cause local dissolution of adjacent carbonates. The following reaction, where Me indicates metal such as Pb or Zn, could account for the observed dissolution:

$$Me^{+2} + H_2S = MeS + 2H^+ \qquad (1)$$

This reaction may explain the rarity of clearly coprecipitated dolomite and sulfide. Figure 6 shows replacement and infilling of partially dissolved material within zone 3; the difference in color and intensity of luminescence between the original and infilling material creates a patchy texture.

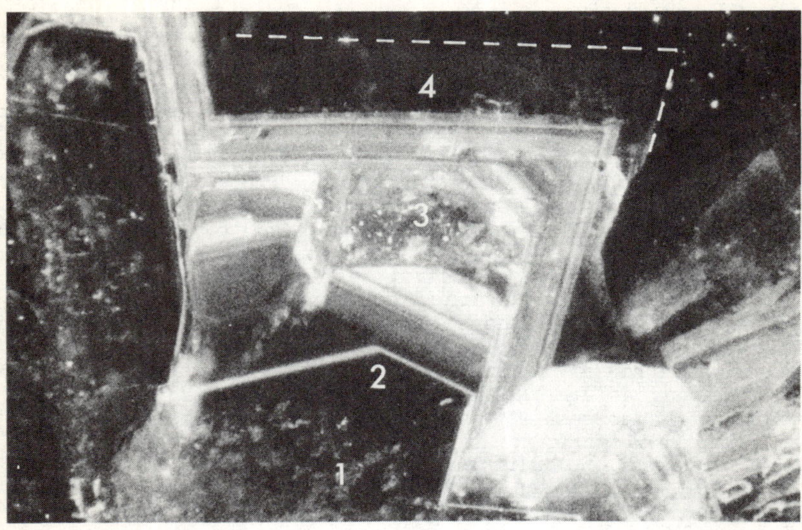

Figure 6 - Photomicrograph of cathodoluminescent zonation in dolomite cements from the Bonneterre Dolomite, Buick Mine, Viburnum Trend. The mottled portion of zone 3 shows dissolution. Incomplete removal of luminescent material was followed by later infilling by dolomite of different luminescence. Field of view=1.0 mm.

In other cases, complete removal of material leaves an uneven, unconformity-like dissolution boundary between zones. In Figure 7 partial replacement of a dolomite rhomb by marcasite has resulted in dissolution and contortion of CL bands which are undisrupted several centimeters away. A reaction similar to equation (1) involving maracasite may be responsible for the dissolution. Rickman (7) also observed strong corrosion textures associated with the presence of marcasite. As pointed out by Voss and Hagni (4) the most abundant and distinct dissolution textures occur in zone 3, supporting the observation that zone 3 coincided with main stages of mineralization. In addition, the abundance of dissolution textures in the dolomites decreases away from heavily mineralized areas, further implicating sulfide precipitation in local carbonate dissolution.

Regional Observations and Discussion

The cathodoluminescence of dolomite cements in drill core (Fig. 8) has been examined for ten locations throughout southern Missouri (Fig. 1). The Viburnum Trend CL microstratigraphy has been recognized in three drill holes up to 7 km to the east of the

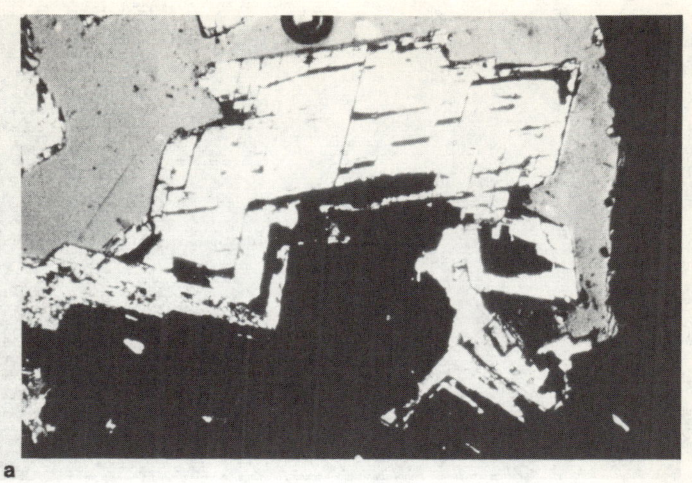

a

b

Figure 7 - a) Transmitted light photomicrograph of a sample from the Buick Mine, Viburnum Trend. A rhomb of dolomite has been incompletely replaced by marcasite. Marcasite precipitation was followed by continued overgrowth of dolomite. b) Replacement of the dolomite was almost complete through the first portion of zone 3. Some of the original CL banding remains, although now highly disrupted, in the unreplaced islands of dolomite. When marcasite precipitation ceased, further growth of dolomite continued uninterrupted through the remainder of zone 3 and zone 4. Field of view = 1.45 mm.

Figure 8 - Core from drill hole 98 (1060 ft.), Eminence Formation. The high porosity is typical of the Cambrian carbonate section of southwest Missouri. The vugs are lined with dolomite cements whose CL zones can be correlated with those of the Viburnum Trend.

district (Fig. 2), as well as in apparently correlative microstratigraphies in drill holes as distant as 350 km to the southwest. Figure 9 shows a comparison of generalized CL microstratigraphies for the Viburnum Trend and southwestern Missouri and northern Arkansas. Correlation is made more difficult by the fact that in any given dolomite crystal, zone thickness is highly variable (even allowing for the variation due to sections cut at angles oblique to the direction of crystal growth), and some zones may be completely absent.

Dolomite cements from the Upper Cambrian Eminence Formation in drill holes 13, 35, 97, 98, and 99 have CL microstratigraphies believed to be correlative with the Viburnum Trend; Figures 10-14 show examples. The presence of CL zones which can be correlated across such distances is interpreted to indicate the regional passage of fluids ultimately resonsible for MVT mineralization in southern Missouri and Arkansas. Cements in drill holes 36, 77, 96, and PBW-3 displayed complex CL zonations, but correlation with the Viburnum microstratigraphy was less certain.

More than one aquifer is believed to have carried the mineralizing fluids. Based on distinct chemical signatures of fluids extracted from inclusions in Viburnum Trend ore and gangue minerals, Viets et al. (14) suggested that both the Lamotte Sandstone (and its facies equivalents to the west), and the Upper Cambrian carbonates higher in the section served as distinct aquifers for the mineralizing fluids. Gregg (5) argued that the Lamotte and the basal six

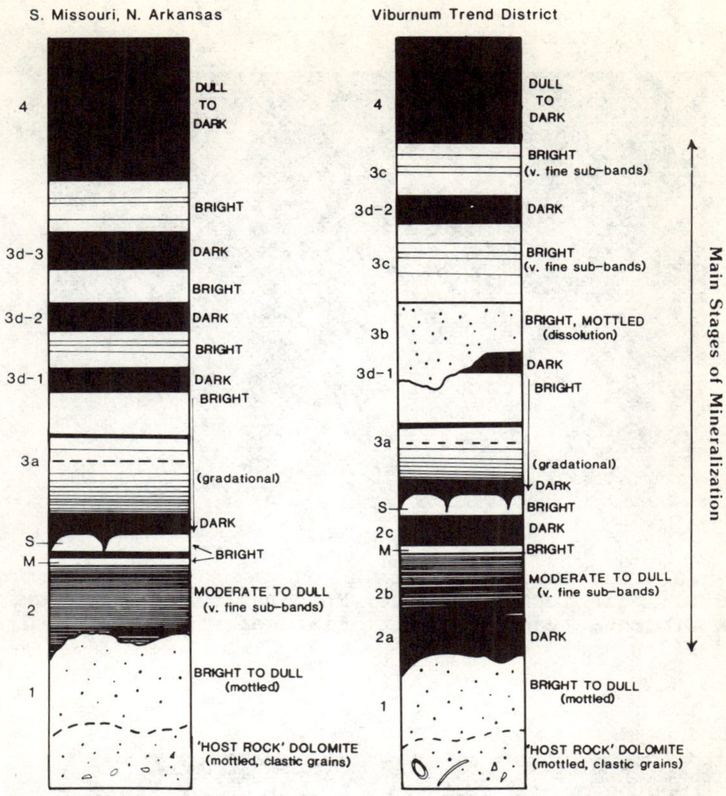

Figure 9 - Generalized CL microstratigraphies for the Viburnum Trend and drill core from southwest Missouri and northern Arkansas. Correlations have been made between most of the distinct bands (or sets of bands) observed in the Viburnum Trend and the zonations observed in distant drill core.

meters of the Bonneterre Dolomite together acted as one aquifer for the mineralizing fluids. Further evidence that Upper Cambrian carbonates, the Eminence Dolomite in particular, served as aquifers is provided by a 1981 study by Erickson et al. (15). A fault system in south central Missouri brings the Eminence Dolomite into contact with the Bonneterre Dolomite (16), very likely establishing hydrologic continuity between the two formations. Erickson et al. analyzed the concentrations of an ore-related suite of metals in insoluble residues from 10 foot intervals within the same drill holes examined in the present study. The examples of CL zoning shown in Figures 11-14 coincide with anomalous concentrations of metals, most notably molybdenum. Erickson et al. interpreted these anomalies to indicate the passage of metal-bearing fluids.

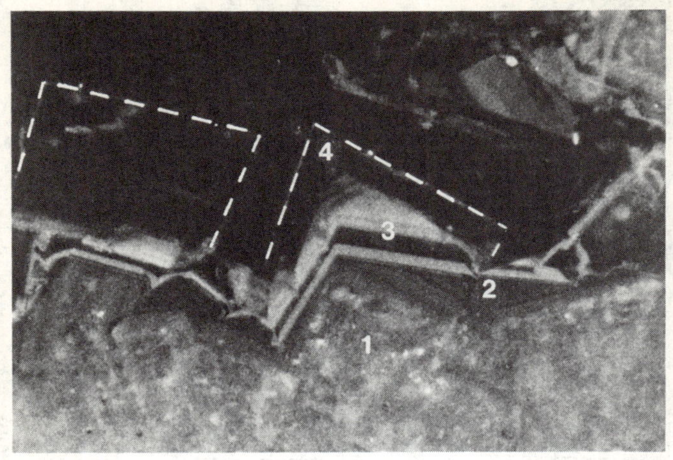

Figure 10 - Photomicrogaph of CL zonation in dolomite cements from drill hole 10EE (513 ft.), Bonneterre Dolomite, east side of the Viburnum Trend district. Field of view = 1.0 mm.

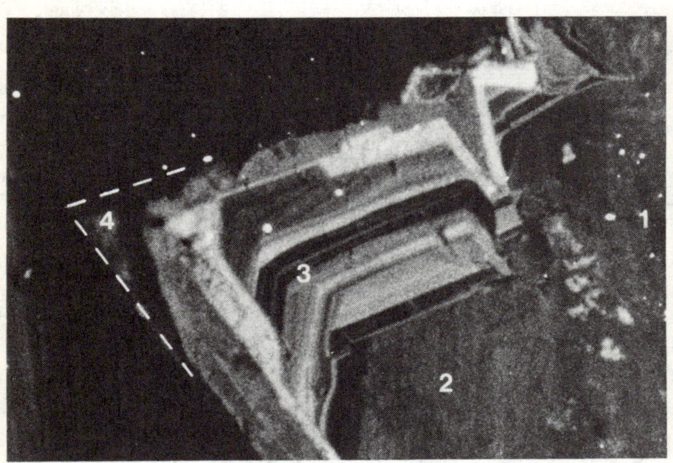

Figure 11 - Photomicrograph of CL zonation in dolomite cements from drill hole 97 (1545 ft.), Eminence Formation, northwest Arkansas. Field of view = 1.0 mm.

Figure 12 - Photomicrograph of CL zonation in dolomite cements from drill hole 98 (1060 ft.), Eminence Formation, southwest Missouri. Field of view = 1.0 mm.

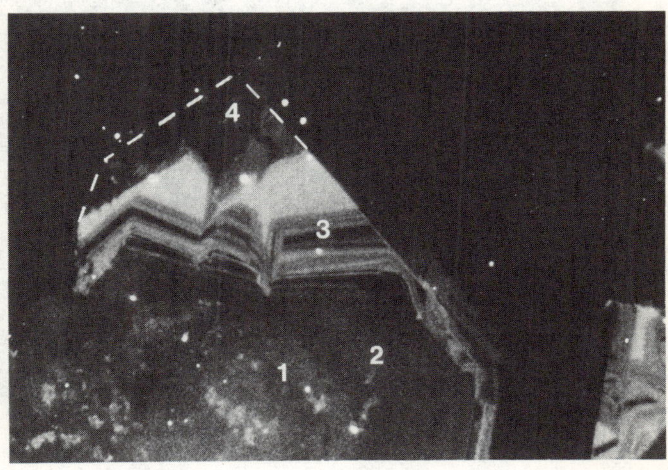

Figure 13 - Photomicrograph of CL zonation in dolomite cements from drill hole 99 (1560 ft.), Eminence Formation, northwest Arkansas. Field of view = 0.6 mm.

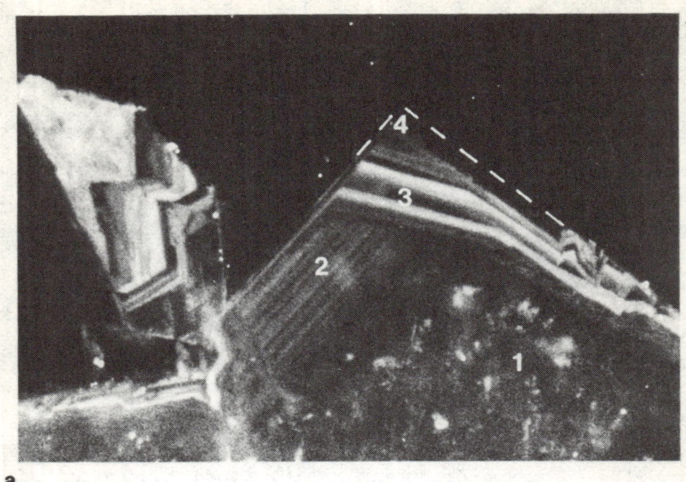

a

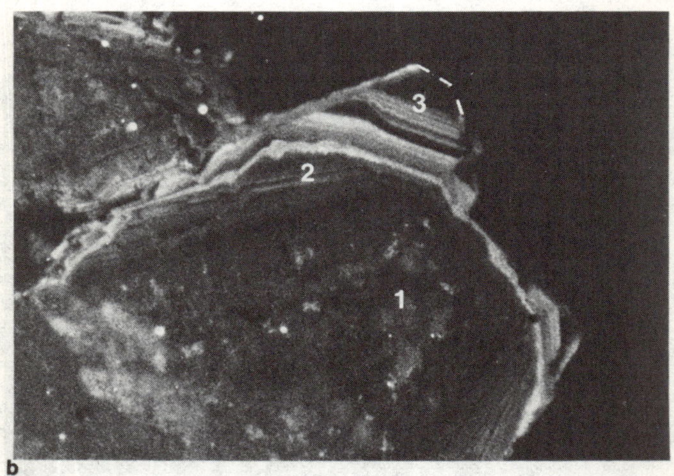

b

Figure 14 - a) and b) Photomicrographs of CL zonation in dolomite cements from drill hole 35 (620 ft.), Eminence Formation, southern Missouri. Field of view = 0.49 mm.

It is reasonable to expect that the fluids responsible for mineralization of the Viburnum Trend flowed through the carbonate section to the south of the district. Numerous lines of evidence discussed in Leach and Rowan (17), including fluid inclusions, secondary chemical remnant paleomagnetism, and thermally reset detrital zircons, indicate a major episode of northward migration of heated, saline fluids. Leach and Rowan (17) develop a model first proposed by Leach (18) which links formation of Mississippi Valley-Type deposits in Missouri, Arkansas, Oklahoma, and Kansas to fluids expelled from the Arkoma basin in response to the Pennsylvanian-Permian Ouachita orogeny (Fig. 15).

A number of previous studies have reported areally extensive correlations of chemical zonations in minerals resulting from fluid flow within confined aquifers. McLimans et al. (19) traced a sphalerite stratigraphy based on Fe banding over 23 km throughout the Upper Mississippi Valley district, Illinois. Ebers and Kopp (20) reported a CL microstratigraphy in gangue dolomite that could be correlated almost perfectly throughout the Mascot-Jefferson City district, Tennessee. They also found good correlation between this district and the CL zonation in gangue dolomites of the Copper Ridge district approximately 48 km to the northeast. Work by Cander et al. (21) reports regional correlation of calcite cements in the Burlington-Keokuk Limestones (Lower Mississippian) based on a CL microstratigraphy.

In all of these studies, equivalent zones from different locations are interpreted to be time correlative and to have formed in response to distinct phases in the chemical/hydrologic evolution of the fluid-flow system. In a hydrologic system such as the one envisioned for southeast Missouri and northern Akansas (17), hydraulic head generated by tectonism in the Ouachita foldbelt, may have governed fluid flow rates, and significant variability would be expected. If the fluids react with the rocks of their aquifers (host rock), then fluid chemistry would vary with the flow rate and residence time of fluids in pore spaces. At present, the extent to which host rock controlled fluid chemistry, and therefore CL banding is not well known.

Evidence for the passage of great volumes of fluid suggests a fluid- rather than rock-dominated system. For example, the broad distribution of hot (90°-120°C) fluid inclusions within and outside of the mining districts implies thermal equilibrium between fluid and host rock (10). Recent work by Viets et al. (14) however, provides evidence for a significant influence of host rock on fluid chemistry. Their study showed a systematic decrease in Na/K ratios northward in fluids extracted from inclusions in sphalerite, galena, and dolomite. It is difficult to explain this trend other than by fluid-host rock interaction. It seems likely that during periods of low fluid flow rates, or relative stagnation, pore fluids would approximate a closed system, reacting to a significant degree with the surrounding rock. During episodes of more rapid fluid flow and shorter fluid residence time in a given pore space, fluid chemistry would be less influenced by reaction with host rock.

The CL observations of this study support a growing body of evidence to link MVT deposits in general to hydrologic systems orders of magnitude greater than once envisioned by economic geologists. The work of Ebers and Kopp (20), McLimans et al. (19), and Voss and

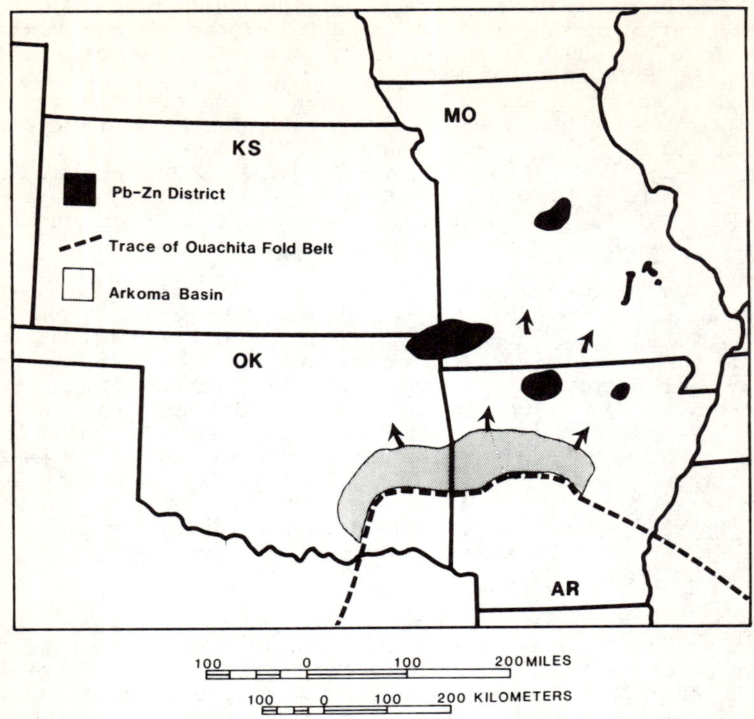

Figure 15 - Schematic representation of northward fluid migration from the Arkoma basin. Arrows show proposed direction of fluid flow. These basinal brines are thought to have been responsible for formation of the Mississippi Valley-type mineralization in Missouri, Arkansas, and eastern Kansas and Oklahoma (17, 18). See text for discussion.

Hagni (4) demonstrated hydrologic continuity within the East Tennessee, Upper Mississippi Valley, and Viburnum Trend districts, respectively. On a larger scale, Garven (22), and Bethke (23), demonstrated through numerical modelling that formation of the Pine Point and Upper Mississippi Valley MVT districts may plausibly be linked to gravity-driven hydrologic systems responsible for fluid flow from the adjacent sedimentary basins.

A gravity-driven hydrologic system with hydraulic head as the primary control on fluid flow rate, might provide possible answers to the enigma of widely correlatable CL zonations. As proposed by Leach (24), the expected fluctuations in hydraulic head at the recharge area would propagate rapidly throughout the hydrologic system causing variations in fluid flow rates. If fluid chemistry were controlled significantly by flow rate, then a mechanism capable of producing virtually simultaneous flow rate changes throughout the hydrologic system might account for the widely correlatable, sharp CL zone boundries.

Summary

The application of cathodoluminescence to the study of MVT deposits in the midcontinent has provided a wealth of information not readily available from other sources. On the scale of a mining district, a CL microstratigraphy such as in the Viburnum Trend provides a precise time framework within which fluid inclusions (or other microsamples) can be placed. Only rarely will a well-defined mineral paragenesis provide a time framework of equivalent precision and detail. On a smaller scale, factors such as pH changes in the fluid may be reflected by dissolution textures visible only in cathodoluminescent light.

The CL banding itself reflects fluctuations in the trace element chemistry of the dolomite and of the cement-forming fluid. Until recently the banding was attributed primarily to Fe^{+2} and Mn^{+2}, and the possible contribution of other elements in concentrations below electron microprobe detection limits was not addressed. A recent review by Machel (1) has emphasized the potential importance of elements such as Pb, Ni, Cu, Co and others in controlling the wavelength and intensity of emitted CL light. Where Mn and Fe can be shown to be the dominant activator and quencher elements respectively, fluctuations in the Eh and pH of the precipitating fluid may be inferred from the CL banding.

Results of the present study indicate that the CL microstratigraphy defined in the Viburnum Trend can be traced as far as 350 km into southwest Missouri, and near the margins of the Tri-State and Northern Arkansas Zinc districts. The regional extent of the CL zonation is evidence for hydrologic continuity within the Upper Cambrian carbonates throughout most of this region. CL observations support the model in which the Viburnum Trend is genetically linked to other Mississippi Valley-type districts of Missouri and Arkansas by an episode of northward regional fluid migration.

Acknowledgements

Drill core used in this study was kindly made available by the Missouri Department of Natural Resources, Division of Geology and Land Survey. Discussions with many people, Otto Kopp, David Leach, and Robert Voss in particular, contributed much to this study.

References

1. Machel, H.G., "Cathodoluminescence in calcite and dolomite and its chemical interpretation", Geoscience Canada, 12 (1985) 139-147.

2. Pierson, B.J., "The control of cathodoluminescence in dolomite by iron and manganese", Sedimentology, 28 (1981) 601-610.

3. Fairchild, I.J., "Chemical controls of cathodoluminescence of natural dolomites and calcites: New data and review", Sedimentology, 30 (1983) 579- 583.

4. Voss, R.L. and Hagni, R.D., "The application of cathodoluminescence microscopy to the study of sparry dolomite from the Viburnum Trend, southeastern Missouri", D.M. Hausen and O. Kopp, eds., Mineralogy-Applications to the Minerals Industry, Proceedings, Paul F. Kerr Memorial Symposium, (New York, AIME, 1985) 51-68.

5. Gregg, J.M., "Regional epigenetic dolomitization in the Bonneterre Dolomite (Cambrian), southeastern Missouri", Geology, 13 (1985) 503-506.

6. Heyl, A.V., "Geologic characteristics of three major Mississippi Valley districts", G. Kisvarsanyi, et al., eds., International Conference on Mississippi Valley Type Lead-Zinc Deposits, Proceedings Volume, (Rolla, University of Missouri, 1983), 27-60.

7. Rickman, D.L., "A thermochemical study of the ore deposits of the Milliken Mine, New Lead Belt, Missouri" (Ph.D. thesis, University of Missouri, Rolla, 1981).

8. Frank, M.H. and Lohmann, K.C., "Cathodoluminescent and Isotopic Examination of Host Rock Dolomite: Bonneterre Formation, Southeast Missouri" (in prep).

9. Braunsdorf, N.R. and Lohmann, K.C., "Isotopic trends in gangue carbonates from the Viburnum Trend: Implications for Mississippi Valley-type mineralization" (submitted to Economic Geology).

10. Rowan, L., Leach, D.L. and Viets, J.G., "Evidence for a Late Pennsylvanian-Early Permian regional thermal event in Missouri, Kansas, Arkansas, and Oklahoma", Geological Society of America, Abstracts with Programs, 16 (1984) 640.

11. Frank, J.R., Carpenter, A.B. and Oglesby, T.W., "Cathodoluminescence and composition of calcite cement in the Taum Sauk Limestone (Upper Cambrian), southeast Missouri", Journal of Sedimentary Petrology, 52 (1982) 631-638.

12. Le Font, M., "Siegenite from the Buick Mine, Bixby, Missouri", The Mineralogical Record, Jan-Feb (1984) 37-39.

13. Anderson, G.M., "Some geochemical aspects of sulfide precipitation in carbonate rocks", G. Kisvarsanyi, et al., eds., International Conference on Mississippi Valley Type Lead-Zinc Deposits, Proceedings Volume, (University of Missouri, Rolla, 1983), 61-76.

14. Viets, J.G., Leach, D.L., Meier, A.L., Rose, S.C. and Rowan, E.L., "Application of induction coupled plasma spectrometry to the analysis of fluids extracted from Mississippi Valley-type deposits of the midcontinent", Geological Society of America, Abstracts with Programs, 17 (1985) 740.

15. Erickson, R.L., Mosier, E.L., Odland, S.K. and Erikson, S.K., "A favorable belt for possible mineral discovery in subsurface Cambrian rocks in southern Missouri", Economic Geology, 76 (1981) 921-932.

16. Kurtz, V.E., Thacker, J.L., Anderson K.H., and Gerdemann, P.E., "Traverse in Late Cambrian strata from the St. Francois Mountains, Missouri to Delaware County, Oklahoma" (Missouri Geological Survey and Department of Natural Resources, Report of Investigations 55, 1975).

17. Leach, D.L. and Rowan, E.L., "Genetic link between Ouachita foldbelt tectonism and the Mississippi Valley-type deposits of the Ozarks" (submitted to Geology).

18. Leach, D.L., "Possible relationship of Pb-Zn mineralization in the Ozarks to the Ouachita Orogeny", Geological Society of America, Abstracts with Programs, 5, (1973) 269.

19. McLimans, R.K., Barnes, H.L. and Ohmoto, H., "Sphalerite stratigraphy of the Upper Mississippi Valley Zinc-Lead district, southern Wisconsin", Economic Geology, 75 (1980) 351-361.

20. Ebers, M.L. and Kopp, O.C., "Cathodoluminescent microstratigraphy in gangue dolomite, the Mascot-Jefferson City district, Tennessee", Economic Geology, 74 (1979) 908-918.

21. Cander, H.S., Kaufman, J., Hanson, G.N. and Meyers, W.J., "Two stage cementation of the Burlington-Keokuk Limestone in Illinois, Missouri, and Iowa", Geological Society of America, Abstracts with Programs, 16 (1984) 631.

22. Garven, G., "The role of regional fluid flow in the genesis of the Pine Point deposit, Western Canada sedimentary basin", Economic Geology, 80 (1985) 307-324.

23. Bethke, C.M., "Hydrologic constraints on genesis of the Upper Mississippi Valley mineral district from Illinois basin brines", Economic Geology, 81 (1986) 233-249.

24. Leach, D.L., private communication with author, U.S. Geological Survey, 1986.

THE USE OF CATHODOLUMINESCENCE MICROSCOPY TO REVEAL HIDDEN CRYSTAL FACES IN GANGUE DOLOMITE CEMENTS, VIBURNUM TREND, SOUTHEAST MISSOURI

Jay M. Gregg[1] and Richard D. Hagni[2]

[1]St. Joe Minerals Corporation
P.O. Box 500
Viburnum, Missouri 65566

[2]Department of Geology and Geophysics
University of Missouri-Rolla
Rolla, Missouri 65401

Discordant cathodoluminescent banding has revealed the existence of complex (non-cleavage rhomb) dolomite crystal faces that developed during the growth of sparry gangue dolomite associated with Mississippi Valley-type sulfide orebodies in the Viburnum Trend, southeastern Missouri. Subsequent stages of crystal growth, however, produced the rhombic crystal habit thus "hiding" the earlier complex faces. Rarely, the complex forms are expressed in the outward crystal morphology where the later stages of dolomite growth did not occur. The hidden crystal faces were indexed using universal stage microscopy. This kind of complex faceting has been observed where simultaneous precipitation of dolomite and halite have occurred under strongly saline conditions. We believe that the presence of trace metals in the mineralizing brines were responsible for the complex faceting of dolomite in the Viburnum Trend.

Introduction

Although cathodoluminescence microscopy has been used to study the paragenesis of dolomites (1,2), ore fluid compositions (3,4), and for correlation of gangue cements throughout mining districts or large regions (1,2, 5), that technique can also be used to study changing crystal morphology of carbonates and thus gain insight into changing conditions in the growth media through time. This paper discusses the application of cathodoluminescence microscopy to the recognition of crystal faces that occur within dolomite crystals associated with the sulfide ores in the Viburnum Trend, Southeast Missouri. These crystal faces usually are not recognizable by other techniques because they are normally overgrown by subsequent generations of dolomite that prevent their expression in the external morphology.

Cathodoluminescence microscopy provides an excellent technique by which to detect the internal hidden faces, referred to as ghost faces in this paper, because variations in manganese and iron content within Viburnum Trend dolomites cause marked variations in cathodoluminescence within the individual crystals. Cathodoluminescence variations reveal intricate growth bands that clearly exhibit the crystal growth morphology. The cathodoluminescence is a result of the presence of trace amounts of manganese that activates the luminescence; increasing ferrous iron tends to depress the cathodoluminescence of dolomite.

Earlier Cathodoluminescence Investigations of Viburnum Trend Sparry Dolomite

Cathodoluminescence microscopy has played an important role in recent investigations of the gangue dolomite that occurs in close association with the lead-zinc ores of the Viburnum Trend in southeastern Missouri (Figure 1A,1B). The first study of Viburnum dolomite by cathodoluminescence was by Rickman (6) who noted that four growth zones could be distinguished within single dolomite crystals at the Milliken mine due to the presence of marked variations in the intensity of cathodoluminescence between the bands (Figure 2A,2B). Subsequent cathodoluminescence microscopy studies of the Viburnum Trend dolomites by Voss and Hagni (1) found that the four growth zones were detectible in sparry dolomite crystals in the ores throughout the 45 mile length of the Trend and established a "dolomite stratigraphy" that is similar to that found in the Tennessee zinc ores by Ebers and Kopp (2) and Kopp et al. (7) and similar to that found for growth bands within sphalerite in the Wisconsin-Illinois zinc-lead district by McLimans et al. (8). Further cathodoluminescence study of sparry dolomite crystals outside the Viburnum Trend, elsewhere in Missouri by Gregg (5), Rowan (4), and Farr (9) has shown that those dolomites are comprised of the same four growth zones. Utilizing the cathodoluminescence character of the four dolomite growth zones, Lohmann (3) and Rowan (4) have examined the variations in isotopic composition, fluid inclusion filling temperatures, and salinities of the dolomite-depositing fluids as a function of time.

During the course of cathodoluminescence microscopic study of the Viburnum Trend dolomites, it was noted by Gregg and Hagni (10) that some dolomite crystals contained crystal faces that were not expressed in their external morphology. The dolomite crystal that first drew our attention to the ghost faces is that shown in Figure 3A and 3B. In that crystal it is readily apparent that although the initial and final crystal morphology was that of the characteristic cleavage rhomb, the external morphology of many of the intermediate dolomite growth layers was that of other dolomite crystal faces. It is important to note that the latter crystal faces, the ghost faces, cannot be detected in this dolomite crystal with ordinary transmitted light microscopy (Figure 3A). Furthermore, because these ghost

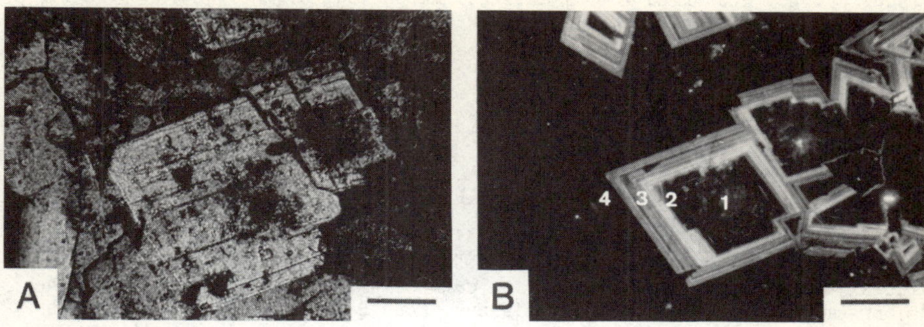

Figure 1 - Micrographs showing typical sparry dolomite that occurs in association with lead-zinc ore deposits in the Viburnum Trend. Bonneterre Formation. (A) Transmitted light. (B) Cathodoluminescence. Four characteristic cathodoluminescent zones are labeled 1 through 4. Scale bars = 0.1 mm.

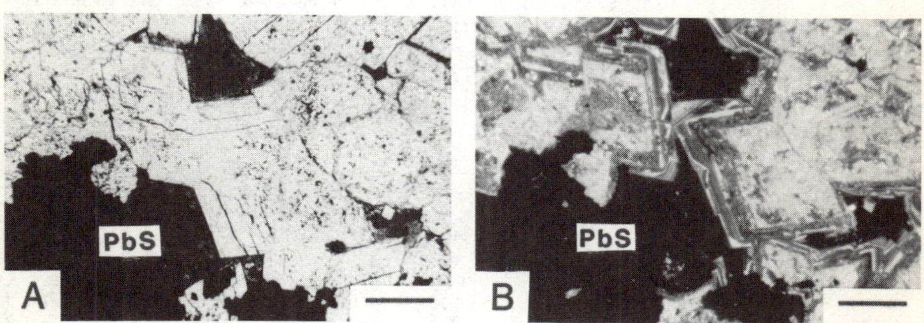

Figure 2 - Micrographs showing sparry dolomite in contact with galena (black) from the Viburnum Trend. Bonneterre Formation. (A) Transmitted light. (B) Cathodoluminescence. Scale bars = 0.1 mm.

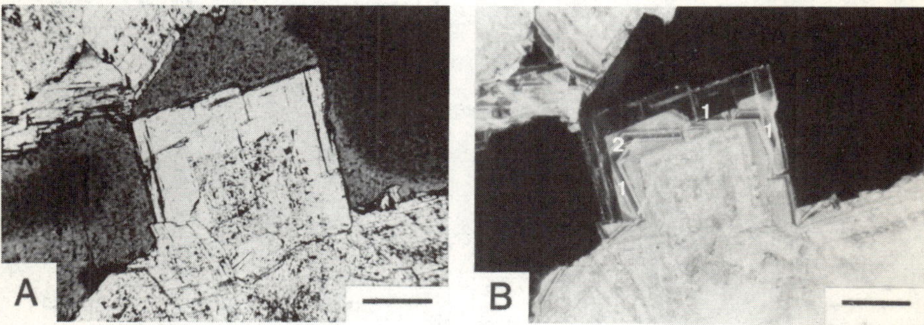

Figure 3 - Micrographs showing a single dolomite rhomb from the Lower Davis Formation in the Viburnum Trend that is a striking example of ghost faces. The crystal has grown into open space that is filled with epoxy. (A) Transmitted light. Most of the ghost faces cannot be detected. Epoxy is dark gray to nearly black. (B) Cathodoluminescence. Four ghost faces were measured, three were found to belong to the (1) acute rhombohedron ($10\bar{1}1$), and the fourth is a (2) scalenohedron ($21\bar{3}4$). Epoxy is non-cathodoluminescent (black). Scale bars = 0.2 mm.

faces are not expressed in the external morphology of the Viburnum Trend sparry dolomite, their faces and interfacial angles cannot be measured by the conventional techniques. For this reason, Gregg and Hagni (10) utilized universal stage techniques to measure and identify those crystal faces. The results of that universal stage study were presented in tabular and stereonet projection form and published elsewhere (10).

Purpose

The purpose of this paper is to discuss the valuable application of cathodoluminescence microscopy to the detection of hidden or ghost crystal faces in sparry dolomite, to provide cathodoluminescence examples from the Viburnum Trend in the Southeast Missouri Lead District, and to suggest the importance of such studies in the exploration for Mississippi Valley-type ore deposits.

Geologic Setting

Most of the samples examined in this study have come from the Viburnum Trend in the Southeast Missouri Lead District. The Vibrunum Trend extends in a northerly direction for about 45 miles, and is located about 40 miles to the east and southeast of Rolla, Missouri (Figure 4).

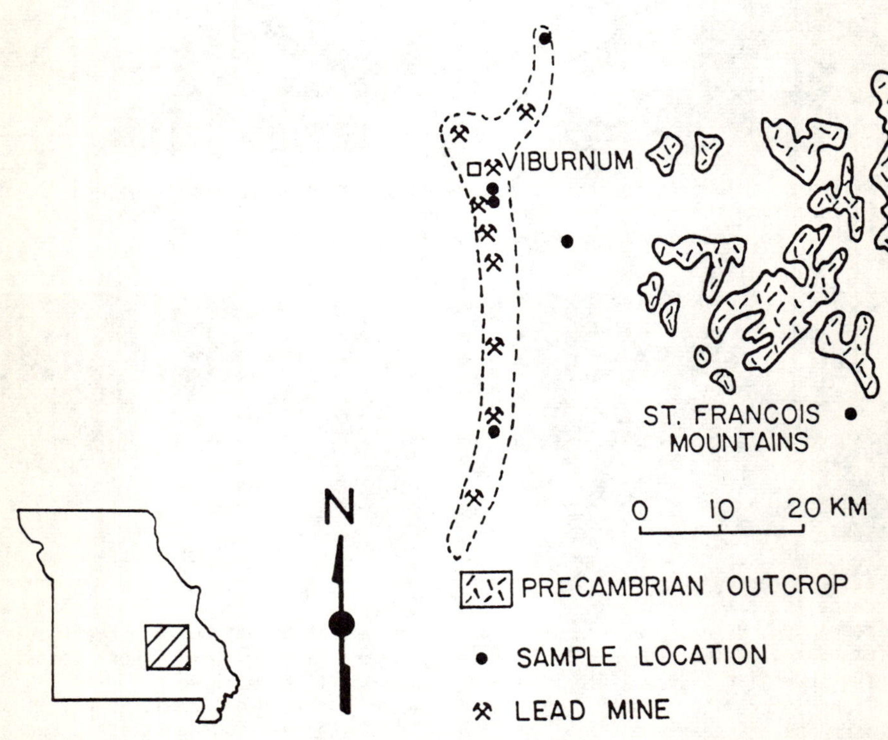

Figure 4 - Map of the Viburnum Trend Lead-Zinc District showing Precambrian outcrops, lead mines, and sample locations.

Many geological aspects of the ore deposits in the Viburnum Trend have been discussed in the literature, particularly in papers that resulted from two symposia held in Rolla (11,12). Therefore, only a brief discussion of the general geologic setting is included here.

The ore deposits of the Viburnum Trend are contained in the Cambrian Bonneterre Formation (Figure 5). The relatively impermeable Cambrian Davis

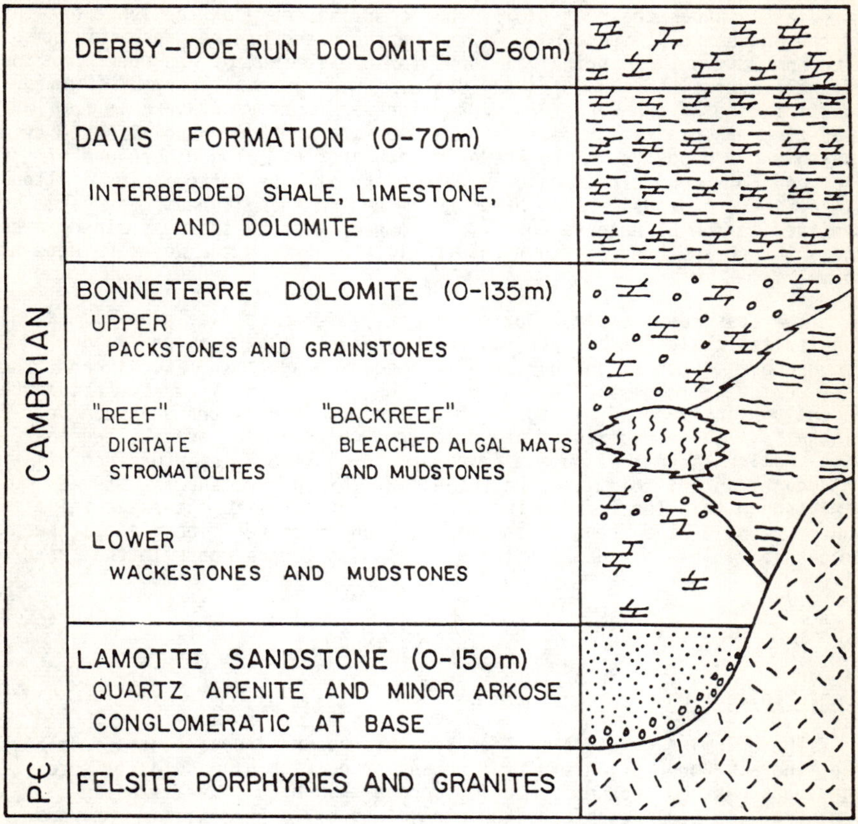

CAMBRIAN	DERBY–DOE RUN DOLOMITE (0-60m)
	DAVIS FORMATION (0-70m) INTERBEDDED SHALE, LIMESTONE, AND DOLOMITE
	BONNETERRE DOLOMITE (0-135m) UPPER PACKSTONES AND GRAINSTONES "REEF" DIGITATE STROMATOLITES "BACKREEF" BLEACHED ALGAL MATS AND MUDSTONES LOWER WACKESTONES AND MUDSTONES
	LAMOTTE SANDSTONE (0-150m) QUARTZ ARENITE AND MINOR ARKOSE CONGLOMERATIC AT BASE
P€	FELSITE PORPHYRIES AND GRANITES

Figure 5 - Columnar section showing Upper Cambrian units of southeastern Missouri and their general facies relationships.

Shale overlies the Bonneterre and its ore deposits. The Cambrian Lamotte Formation, comprised of conglomerate and sandstone, which forms a permeable aquifer throughout much of Missouri, underlies the Bonneterre and the ore deposits. The Precambrian basement, upon which the Cambrian sediments were deposited, consists of ash-flow tuffs, rhyolites, and granites. Surface exposures of the Precambrian rocks are located in the St. Francois Mountains about 20 miles east of the Viburnum Trend.

The distribution of ore deposits are controlled by: 1) the ancient Bonneterre shoreline about the elevated area of Precambrian rocks that formed an area of islands in the Bonneterre shallow sea, 2) Bonneterre sedimentary facies, 3) sedimenary structures, such as submarine slide breccias, 4) local structural highs, and 5) solution collapse breccias that are especially well developed at the Buick, Magmont, Milliken, and Casteel mines.

The Bonneterre sedimentary facies are especially important in controlling the locations of the ore deposits. The ores have been shown by Gerdemann and Myers (13) to occur principally in the dark colored calcarenite facies of the middle one-third of the Bonneterre Formation immediately above the reef facies that comprises much of the lower one-third of the Bonneterre (Figure 5). At the northern end of the Trend, at the Viburnum 28 mine, the principal locus for ore is that of the upper portion of the reef facies. The offshore shaley facies and the backreef ("white rock") facies of the Bonneterre Formation were mostly unfavorable to lead-zinc mineralization, but two ore deposits in the Annapolis subdistrict and some ore deposits in the Fredericktown area at the southern end of the Old Lead Belt subdistrict occur within backreef Bonneterre facies. Although all of the ore deposits that have been mined in the Viburnum Trend are confined to the Bonneterre Formation, some of the ore that was mined in the Indian Creek and Fredericktown subdistricts came from the upper Lamotte Formation.

The ore deposits are comprised largely of the sulfide minerals galena, sphalerite, chalcopyrite, pyrite, and marcasite. The occurrence, ore microscopic textures, and paragenetic sequence for these minerals and other less common ore minerals that occur in the ores of the Viburnum Trend have recently been reviewed by Hagni (14). The principal introduced gangue minerals in the ore deposits are dolomite, calcite, and quartz, together with smaller amounts of potash feldspar and dickite. The abundance of introduced sparry dolomite, the subject of this paper, shows a marked variation within the Trend. The most abundant dolomite occurs at the southern end of the Trend at the Milliken and Fletcher mines, and it becomes progessively less abundant in the ore deposits to the north in the Trend (1).

Method of Study

Sample Collection

The dolomite crystals studied for this paper were collected from St. Joe Minerals Corporation exploration and production diamond drill core. The specimens were collected from drill core within both mineralized and unmineralized portions of the Bonneterre and Davis Formations. Standard thin sections were prepared and polished with 1 μm diamond paste.

Luminoscope Parameters

A Nuclide Luminoscope, model ELM-2 was used for the cathodoluminescence microscopic study. The instrument was operated at 10 to 15 kv and between four and six ma. The model is equipped with a helium atmosphere that reduces the time required for the attainment of a vacuum and improves the cathodoluminescence for photography. Spot sizes were between 0.5 and 1.5 cm.

Cathodoluminescence Microscopy

Cathodoluminescence microscopy provides the most convenient means to study ghost faces in Viburnum Trend dolomite crystals. All of the 20 ghost faces measured and reported in Gregg and Hagni (10) were first detected by

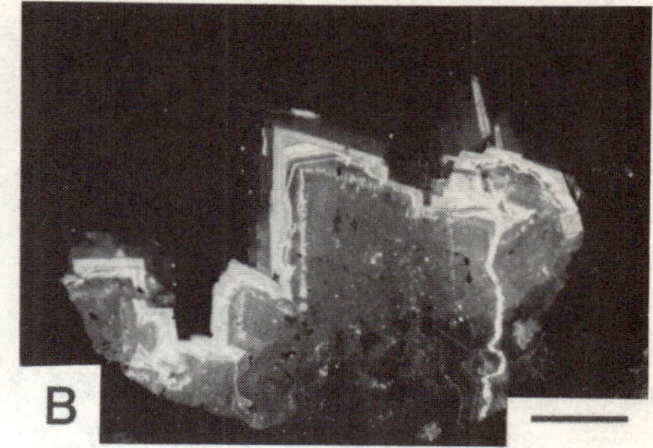

Figure 6 - Micrographs showing several dolomite crystals grown into a vug in lead-zinc ores at the Brushy Creek mine. (A) Transmitted light. Dolomite crystals are light; the mounting epoxy has been burned black by the electron beam. (B) Cathodoluminescence. A prominent ghost face (arrow) is contained in zone three within the large central crystal. Sample provided by R. L. Voss. Scale bars = 0.5 mm.

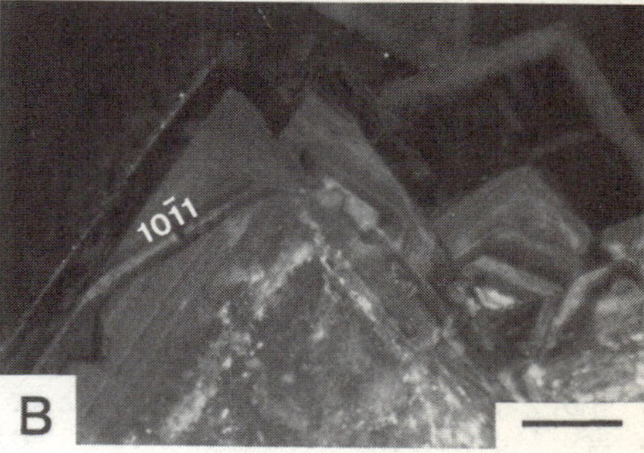

Figure 7 - Micrographs showing a ghost crystal face that is observable as a slight trace (arrow) within a large dolomite rhomb under plane polarized light (A). With cathodoluminescence (B), that crystal face and other ghost faces are distinctly revealed. One measured ghost face belongs to the acute rhomb (1011). Lower Davis Formation, Viburnum Trend. Scale bars = 0.2 mm.

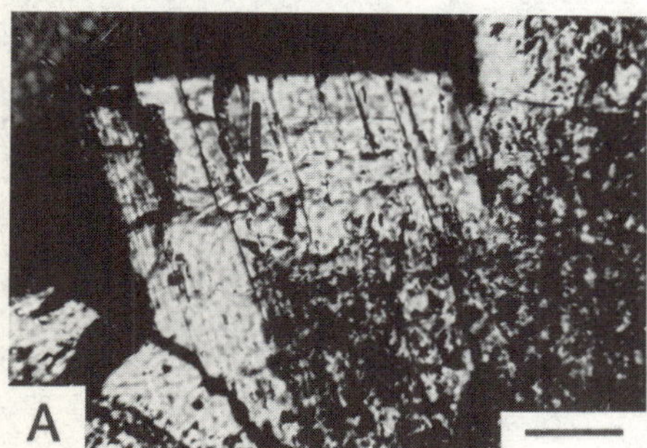

Figure 8 - Micrographs showing ghost crystal faces (arrow) that can be detected under plane polarized transmitted light (A) in a single dolomite rhomb. The ghost faces are clearly revealed with cathodoluminescence (B). One measured ghost face belongs to the unit cell $(10\bar{1}1)$. Lower Davis Formation, Viburnum Trend. Scale bars = 0.1 mm.

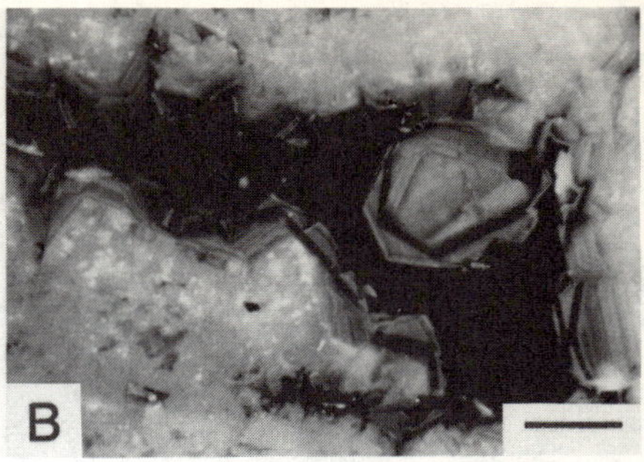

Figure 9 - Micrographs showing a dolomite rhomb from the Bonneterre Formation in the Lake of the Ozarks region about 150 miles to the northwest of the Viburnum Trend, illustrating that such faces are not limited to the Trend. (A) Transmitted light. Ghost faces can not be detected. (B) Cathodoluminescence. Ghost faces are strikingly exhibited. Scale bars = 0.1 mm.

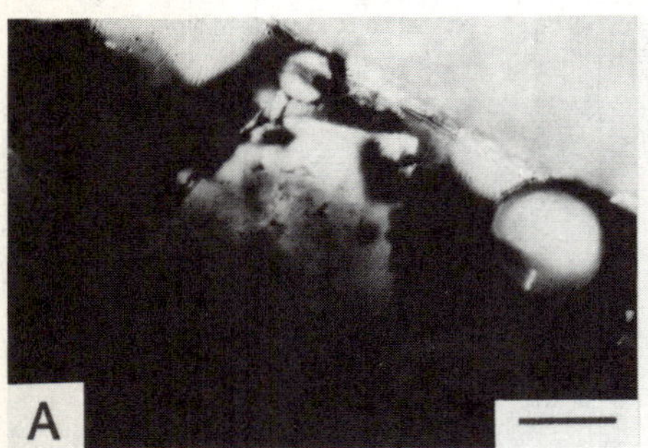

Figure 10 - Micrographs showing a dolomite crystal from the Bonneterre Formation in the Viburnum Trend that exhibits complex crystal faces in its external morphology. Such external morphology is rare in Viburnum Trend gangue dolomite cement. (A) Transmitted light. Dolomite is light; the lower portion of the photomicrograph is black due to burning by the electron beam. (B) Cathodoluminescence. Shows the internal complex crystal faces that also are present in the external morphology. Scale bars = 0.1 mm.

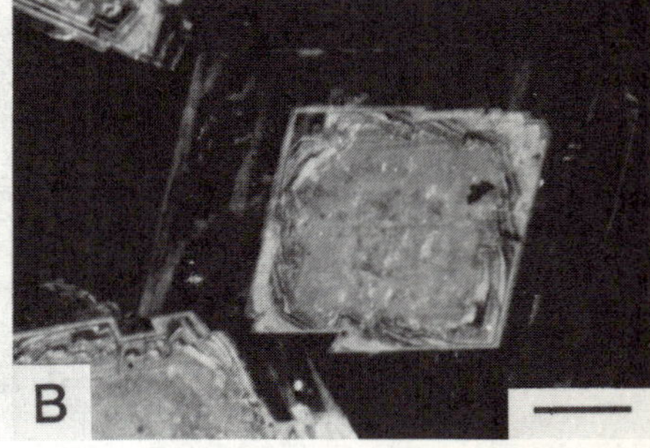

Figure 11 - Micrographs of a single dolomite rhomb from the backreef facies of the Bonneterre Formation which shows dissolution (irregular to rounded, corroded surface at the base of the distinctly banded zone three dolomite) that occurred immediately before the development of the ghost crystal faces. Such an association is common but neither universal nor necessary. (A) Transmitted light. (B) Cathodoluminescence. Scale bars = 0.1 mm.

cathodoluminescence. Viburnum sparry dolomite crystals contain four growth zones: 1) an early cathodoluminescent non-banded zone, 2) a non-cathodoluminescent zone, which may be absent in the northern part of the district, 3) a strongly banded zone comprised of alternating brightly cathodoluminescent and non-cathodoluminescent dolomite, and 4) a dull to non-cathodoluminescent zone. The ghost faces occur within zone three as shown in Figure 6B.

A slight trace of some ghost faces can be observed with ordinary transmitted light, but the ghost faces are best detected and studied with cathodoluminescence microscopy. An example is given in Figure 7A where the trace of a single ghost face can be detected in the right center portion of the photomicrograph, but the cathodoluminescence photomicrograph clearly reveals the full character of the ghost faces within that dolomite crystal. The three ghost crystal faces shown in Figure 7B have been measured with a universal stage, plotted on a stereonet projection, and identified to belong to rhombic (10$\bar{1}$4) (two faces) and acute rhomb (10$\bar{1}$1) forms of dolomite.[1]

A second example is provided in Figure 8A where the traces of two ghost faces can be detected by ordinary transmitted light microscopy, but these faces can be much better observed with cathodoluminescence microscopy. One of the ghost crystal faces shown in Figure 8B was measured with a universal stage, plotted on a stereonet projection, and identified to belong to the negative rhomb (02$\bar{2}$1) form of dolomite.

Cathodoluminescence microscopic study of dolomite crystals elsewhere in Missouri indicates that the presence of ghost faces is not limited to the Viburnum Trend. Figure 9A,B shows ghost faces that are present in sparry dolomite in the Bonneterre Formation in the vicinity of the Lake of the Ozarks about 150 miles to the northwest of the Viburnum Trend.

In rare instances, such as that shown in Figure 10A,B, the internal ghost crystal faces that have been studied by cathodoluminescence microscopy occur in the external morphology. Such dolomite crystals are ones on which zone four is not well represented. Unfortunately these crystals were destroyed during preparation for universal stage study and were not measured.

Cathodoluminescence microscopy also provides an important means to detect the presence of periods of dolomite dissolution that occurred between periods of dolomite deposition. The detection of these periods of dissolution is important because they may serve as potential ore guides. The ghost faces studied for this paper commonly have developed immediately after a period of dolomite dissolution as shown in Figure 11A,B. That association is sufficiently common that we believed that there was a causal relationship during our early studies. The fact that some ghost faces occur in the absence of dissolution features, however, leads us to conclude that the relationship is not a causal one.

[1]In all cases, Bravais indicies within parentheses in this paper refer to the crystal form rather than individual faces; the indicies are based on the hexagonal cell.

Discussion and Conclusions

This paper reports a new application of cathodoluminescence microscopy, that of the recognition and study of ghost crystal faces within dolomite. Although such faces can locally be detected in some sparry dolomite crystals by ordinary transmitted light, cathodoluminescence microscopy provides a much superior technique for their recognition and full study. The association of ghost faces with lead-zinc deposits in southeastern Missouri suggests that their presence may serve as a guide to ore deposits in other Mississippi Valley-type districts.

Ghost crystal faces, such as occur in dolomite crystals in the Viburnum Trend, should not be confused with "pseudoface" development in calcite crystals as described by Walkden and Berry (15). Pseudofaces develop as a result of the coalescing of small facets on a rough surface. The growth plane that develops continuously changes in orientation until it is parallel to a crystal face (15). Pseudofaces are recognizable using cathodoluminescence and may be difficult to distinguish from ghost faces.

Reeder and Prosky (16) reported compositional sector zoning, which is visible using cathodoluminescence microscopy, in dolomite crystals containing the $(10\bar{1}4)$ cleavage rhomb as well as the $(11\bar{2}0)$ 2nd order prism. Electron microprobe analysis showed that the sector under the $(11\bar{2}0)$ face was enriched in Fe, Mn, and Mg relative to the $(10\bar{1}4)$ sector. Complex faceting, therefore, may effect the bulk chemistry of dolomite crystals (16) making identification of ghost faces important when doing chemical analyses on dolomite crystals.

The appearance of ghost crystal faces in sparry dolomite may be the result of one or more factors, including 1) high chloride content, 2) trace metals, and 3) growth rate. A high chloride content of the fluids from which the sparry dolomite was deposited may have been an important factor in their development. The dolomite crystals are known to have been deposited from very saline, chloride-rich fluids because fluid inclusions contained in the crystals have freezing temperatures that indicate an average salinity of about 23% NaCl equivalent (4). Further support for the role of chloride in ghost face development comes from the occurrence of complex polyhedral dolomite crystals (Figure 12),

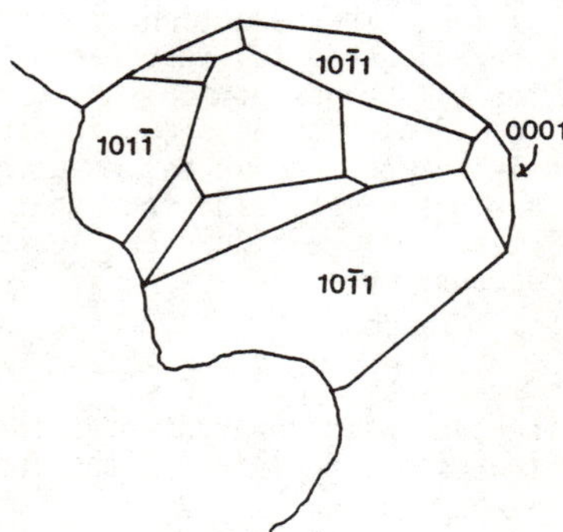

Figure 12 - Line drawing of complex faceted dolomite crystal from the Permian of west Texas (after SEM micrograph from Naimen et al., 17) with the $(10\bar{1}1)$ Acute (Unit Cell) Rhomb, the Basal Pinacoid, and several other forms.

that are associated with halite in the Permian evaporites in west Texas and were precipitated from NaCl-rich brines (17). These polyhedral crystals display many of the same crystal habits observed in Viburnum Trend dolomites (10).

The presence of trace amounts of metals in the brines also may have been an important factor in the formation of ghost faces. The fact that the faces developed at precisely the time of greatest ore deposition (1) supports that idea. Further support comes from the fact that ghost faces are contained in sparry dolomite that was deposited elsewhere in the presence of mineralization. The ghost crystal faces previously illustrated (Figure 9) to occur in sparry dolomite from the Lake of the Ozarks in central Missouri, distal from the Viburnum ore deposits, is associated with traces of zinc-lead-iron sulfide mineralization that are known to occur in this region. Ghost crystal faces were identified in late dolomite cements associated with trace sphalerite and galena on the Central Basin Platform in west Texas (Figure 13A,B) (18). Similarly, ghost faces are present in sparry dolomite associated with trace zinc-lead mineralization in the middle Ordovician Trenton Formation in northwest Ohio (Figure 14A,B) (19).

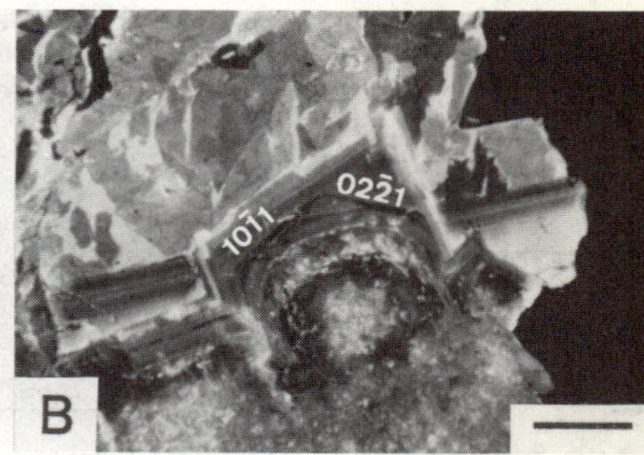

Figure 13 - Micrographs showing ghost crystal faces in dolomite rhombs that illustrate their close association with trace galena and sphalerite (black; right side of micrographs) and calcite (light gray; upper portion of micrographs) from the Central Basin Platform of west Texas. Two measured ghost faces were found to belong to the acute rhomb (10$\bar{1}$1) and the negative acute rhomb (0221). (A) Transmitted light. (B) Cathodoluminescence. Scale bars = 0.1 mm.

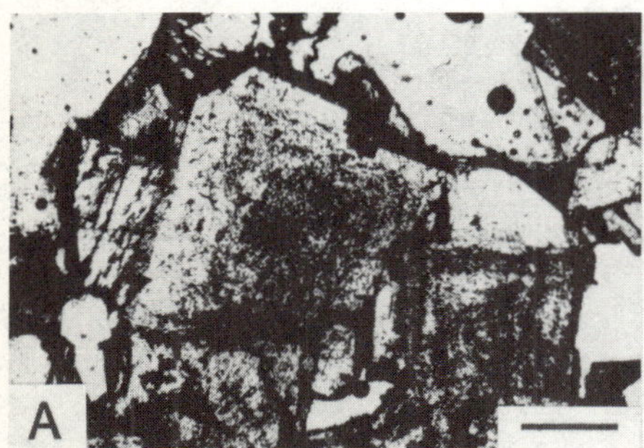

Figure 14 - Micrographs showing ghost crystal faces in dolomite crystals associated with zinc mineralization in the middle Ordovician Trenton Formation in northwestern Ohio, illustrating that the association of ghost faces with base metal mineralization is not limited to the Viburnum Trend. (A) Transmitted light. (B) Cathodoluminescence. Scale bars = 0.5 mm.

Finally, support for this idea comes from the laboratory experimental investigation of Gregg and Sibley (20). Dolomite crystals grown in hydrothermal bombs from 1-2 molar Mg/CaCl solutions at 250-300°C exhibited polyhedral crystal habits. Subsequent chemical analyses of the waters after blank runs have indicated significant trace metal contents of 92 ppm zinc, 2-3 ppm cobalt, nickel, and iron, and 1 ppm copper. These metals have been leached by the saline solution from the stainless steel and copper seals used to construct the bombs. The metals may have been adsorbed upon preferred crystallographic planes, had a poisoning effect to prohibit crystal growth in that direction, leading to the development of other crystal faces, the ghost faces.

A third factor, that of the rate of crystal growth, may also have contributed to the development of the ghost crystal faces. Variations in the rates of deposition have had an important influence on the crystal forms developed by the sulfide minerals in the Viburnum Trend. Hagni and Trancynger (21) noted various morphologies developed especially by pyrite, marcasite, sphalerite, and galena crystals deposited in the ores at the Magmont mine. Clendenin (22) has noted a great variety of galena forms deposited in the ores at the Milliken mine and related these largely to the rate of deposition. The ghost crystal faces contained in Viburnum Trend sparry dolomite may mark a period of rapid dolomite deposition as well as high metal and chloride contents of the fluids that deposited the dolomite.

Irregardless of the cause of the development of the ghost faces, their presence is best recognized by cathodoluminescence microscopy. The observation of ghost faces using cathodoluminescence microscopy may form a useful tool in exploration for Mississippi Valley-type ore deposits.

Acknowledgements

The writers wish to thank Kerry Grant for discussions on crystallographic technique throughout this study, Robert Folk for providing specimens of polyhedral dolomite crystals from West Texas, Duncan F. Sibley and Greg Fox for samples and chemical analyses of waters from hydrothermal bomb runs, and St. Joe Minerals Corporation for permission to the senior author to publish the results of this study.

References

1. Robert L. Voss and Richard D. Hagni, "The Application of Cathodoluminescence Microscopy to the Study of Sparry Dolomite from the Viburnum Tend, Southeast Missouri", Chapter 5, pp. 51-68 in Donald M. Hausen and Otto C. Kopp (editors) Mineralogy-Applications to the Minerals Industry, Proceedings of the Paul F. Kerr Memorial Symposium, AIME, New York, NY, 1985.

2. Michael L. Ebers and Otto C. Kopp, "Cathodoluminescent Microstratigraphy in Gangue Dolomite, the Mascot-Jefferson City District, Tennessee", Econ. Geol., 74 (1979) pp. 908-918.

3. Kyger C. Lohmann, "Integration of Cathodoluminescence and Geochemical Approaches in Studies of Carbonate Diagenesis", in this volume.

4. Lanier Rowan, "A Study of Fluid Inclusions within Cathodoluminescent Zones in Gangue Dolomites, Viburnum Trend Pb-Zn District, Southeast Missouri", this volume.

5. Jay M. Gregg, "Regional Epigenetic Dolomitization in the Bonneterre Dolomite (Cambrian), Southeastern Missouri", Geology, 13 (1985) pp. 503-506.

6. Douglas L. Rickman, "A Thermochemical Study of the Ore Deposits of the Milliken Mine, New Lead Belt, Missouri", Unpublished Ph.D. dissertation, University of Missouri-Rolla, Rolla, Missouri (1981) 295 pp.

7. Otto C. Kopp, M. L. Ebers, L. B. Cobb, T. B. Crattie, T. L. Ferguson, R. M. Larsen, R. A. Potosky and R. T. Steinberg, "Application of Cathodoluminescence Microscopy to the Study of Gangue Carbonates in Mississippi Valley-Type Deposits in Tennessee: The Search for a Tennessee Trend", in this volume.

8. Roger K. McLimans, Hugh L. Barnes, and H. Ohmoto, "Sphalerite Stratigraphy of the Upper Mississippi Valley Zinc-Lead District, Southwest Wisconsin", Econ. Geol., 75 (1980) pp. 351-361.

9. Mark R. Farr, "Regional Isotopic Variation in Bonneterre Formation Cements: Implications for Brine Migration Pathways and Sources" (abs.), p. 8 in Jay M. Gregg and Richard D. Hagni (editors) Symposium on the Bonneterre Formation (Cambrian), Southeastern Missouri: Stratigraphy, Sedimentology, Diagenesis, Geochemistry, and Economic Geology, University of Missouri-Rolla, Rolla, Missouri, 1986.

10. Jay M. Gregg and Richard D. Hagni, "Irregular Cathodoluminescent Banding in Late Dolomite Cements: Evidence for Complex Faceting and Metalliferous Brines", Geological Society of America Bulletin (in press).

11. Society of Economic Geologists, "An Issue Devoted to the Viburnum Trend, Southeast Missouri", Econ. Geol., 72 (1977) 524 pp.

12. Geza Kisvarsanyi, S. Kerry Grant, Waldemar P. Pratt, and John W. Koenig (editors) International Conference on Mississippi Valley Type Lead-Zinc Deposits, Proceedings Volume, University of Missouri-Rolla, Rolla, Missouri 1983, 603 pp.

13. Paul E. Gerdemann, and Harold E. Myers, "Relationships of Carbonate Facies Patterns to Ore Distribution and to Ore Genesis in the Southeast Missouri Lead District", Econ. Geol., 67 (1972) pp. 426-433.

14. Richard D. Hagni, "Mineral Paragenetic Sequence of the Lead-Zinc-Copper-Cobalt-Nickel Ores of the Southeast Missouri Lead District, U.S.A.", pp. 90-132 in James R. Craig, Richard D. Hagni, W. Kiesl, I. M. Lange, N. V. Petrovskaya, Tatanya N. Shadlun, G. Udubasa, and S. S. Augustithis (editors) Mineral Paragenesis, Theophrastus Publications S.A., Athens, Greece, 1986.

15. G. M. Walkden and J. R. Berry, "Natural Calcite in Cathodoluminescence: Crystal Growth during Diagenesis", Nature, 308 (1984) pp. 525-527.

16. Richard J. Reeder and Joan L. Prosky, "Compositional Sector Zoning in Dolomite", Jour. Sed. Petrology, 56 (1986) pp. 237-247.

17. E. R. Naiman, A. Bein, and Robert L. Folk, "Complex Polyhedral Crystals of Limpid Dolomite Associated with Halite, Permian Upper Clear Fork and Glorietta Formations, Texas", Jour. Sed. Petrology, 53 (1983) pp. 549-555.

18. S. J. Mazzullo, "Mississippi Valley-Type Sulfides in Lower Permian Dolomites, Delaware Basin, Texas: Implications for Basin Evolution", Amer. Assoc. Pet. Geol. Bull., in press.

19. Ralph J. Haefner, "Mississippi Valley Type Mineralization and Dolomitization of the Trenton Limestone, Wyandot County, Ohio", Unpub. M.S. Thesis, Bowling Green State University, Bowling Green, Ohio, 1986, 123 pp.

20. Jay M. Gregg and D. F. Sibley, "Epigenetic Dolomitization and the Origin of Xenotopic Dolomite Texture", Jour. Sed. Petrology, 54 (1984) pp. 908-931.

21. Richard D. Hagni and Thomas C. Trancynger, "Sequence of Deposition of the Ore Minerals at the Magmont Mine, Viburnum Trend, Southeast Missouri", Econ. Geol., 72 (1977) pp. 451-464.

22. C. William Clendenin, "Suggestions for Interpreting Viburnum Trend Mineralization based on Field Studies at Ozark Lead Company, Southeast Missouri", Econ. Geol., 72 (1977) pp. 465-473.

TEXTURAL AND CHEMICAL ALTERATION OF DOLOMITE: INTERACTION OF MINERALIZING FLUIDS AND HOST ROCK IN A MISSISSIPPI VALLEY-TYPE DEPOSIT, BONNETERRE FORMATION, VIBURNUM TREND

Marc Hilary Frank and Kyger C Lohmann, Department of Geological Sciences,

The University of Michigan, Ann Arbor, MI 48109-1063

ABSTRACT

Textural and geochemical characteristics of dolomites from the Bonneterre Formation were examined to determine the nature and style of interaction between the host rock and mineralizing fluids, and to clarify the relative importance of pre-ore and syn-ore alteration in developing porosity pathways for distribution of the mineralizing solutions. Core samples, from a traverse extending thirty-five kilometers to the east of the intensely mineralized Ozark Lead Mine, were examined with transmitted light and cathodoluminescence (CL) microscopy. On the basis of Fe^{2+} -content, dolomite crystallinity and CL microfabric, three end member dolomites, Type I, Type II and Type III were identified; two transitional dolomites exhibiting mottled CL fabrics were also observed, Type II*m* and Type III*m*. End member dolomites show excellent separation based on isotopic compositions, with mottled dolomites possessing compositions intermediate along either the Type I-Type II or Type I-Type III trends.

The compositional and textural trends of host rock dolomites are produced through partial to pervasive alteration of a pre-ore, Type I dolomite by mineralizing fluids of either Type II or Type III composition. Because an unambiguous record of the chemical evolution of mineralizing solutions is preserved in the sequentially-zoned dolomite gangue spars, the timing of host rock alteration can be discerned relative to episodes of migration of the mineralizing fluids. Moreover, the development of mottled CL microfabrics both within gangue spars and host dolomites is of particular significance, for these document the corrosive style of fluid-host rock interaction. Although an early porosity network, formed by pre-ore dolomitization, may have guided the initial flow of mineralizing fluids, active corrosion during mineralization was likely dominant in controlling the present distribution of mineralization.

INTRODUCTION

The Viburnum Trend of southeast Missouri is a classic example of a sediment-hosted, Mississippi Valley-Type (MVT) lead and zinc deposit. This regionally extensive ore deposit is hosted dominantly within the Upper Cambrian Bonneterre Formation, a shallow shelf carbonate sequence which formed on the flanks of exposed remnants of the Precambrian St. Francois Mountains (Thacker and Anderson, 1977). This relationship of lead and zinc mineralization in carbonate terranes has long been noted and serves to characterize MVT ore deposits (Beales, 1977; Ohle, 1980; and Sangster, 1983). In the Bonneterre Formation, the carbonate sequence is totally dolomitized in areas of intense mineralization. Thus, it is important to determine the role of host rock dolomitization as a control on the distribution of mineralization. Specifically, in order for mineralization to proceed, fluids must have access for flow , and this is controlled by the porosity system of the host sediment. Was the porosity network intrinsic to primary depostional facies or was it defined by early pre-ore diagenetic events, such as dolomitization, which may have provided an optimum porosity pathway for later migration of mineralizing fluids? Alternatively, mineralizing fluids might actively create the necessary porosity through processes of dissolution and dolomitization. Such questions require analysis of the fabric and chemisty of the host sediment to discern their relationship to patterns of mineralization.

Although extensive research has been completed on the distribution, geochemistry and paragenesis of the ore sulfides (Rickman, 1981; Hagni, 1983; Sverjensky, 1980, 1981b), little has focused on the character of the sediment host which potentially controls the site of mineralization. Few studies have examined the geochemical and diagenetic history of the Bonneterre Formation (Larsen, 1977; Lyle, 1977; Hannah and Stein, 1984; Gregg, 1985). Sedimentologic analysis of the Bonneterre Formation has been largely restricted facies studies to reconstruct the paleoenviromental setting of the carbonate shelf, and in turn link the distribution of carbonate facies to patterns of mineralization (Howe, 1968; Gerdemann and Myers, 1972).

This study focuses on the carbonate components of the Bonneterre Formation to decipher the history of interaction of mineralizing fluids with host rock sediments. Petrographic, cathodoluminescence (CL) and stable isotope geochemical techniques were utilized to characterize both rock fabric and chemistries. Two rock records are present: gangue carbonates, which were precipitated from ore fluids in association with metal sulfides; and the host rock carbonate sequence. Where zoned crystals are present in gangue carbonates, they provide an unambiguous record of the evolving chemistry of the ore-forming fluids. In a similar fashion, host carbonates, comprised of primary sediment grains, may possess chemical and fabric records of diagenetic events which might have constructed the porosity network prior to migration of ore-forming fluids. If significant interaction of mineralizing fluids with host rock has occurred, the record of primary (pre-ore) chemistries and fabrics would be altered. Comparison of chemical trends in gangue carbonates with changes in the fabric and chemistry of the host carbonate, will elucidate the timing of dolomitization and define the style and spatial extent of fluid - host rock interaction.

AREA OF STUDY: BACKGROUND AND SAMPLING SCHEME

In order to contrast possible interaction of ore fluids with host carbonate sediments, a sampling scheme was chosen which traverses areas of intense sulfide mineralization and extends laterally into areas of relative barren host rock (Figure 1). The traverse was initiated in the Ozark Lead Mine utilizing samples collected from in place mine workings. From this mine location, a series of thirteen subsurface drill cores housed at the Missouri State Geological Survey in Rolla was sampled to define a study section extending 35 kilometers eastward of the Viburnum Trend. Rock macrofabrics were described for all cores, and samples were collected for

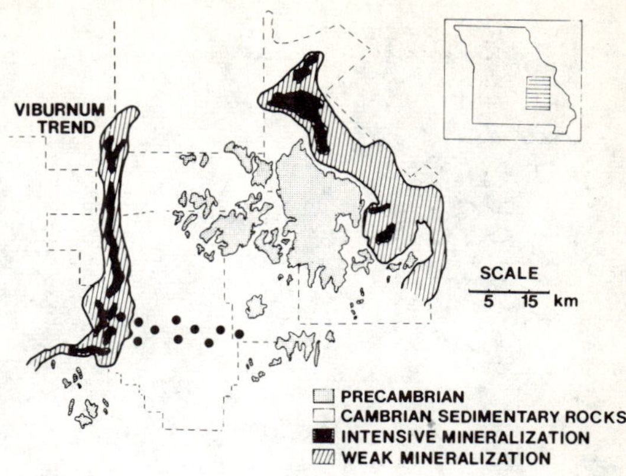

FIGURE 1: Host rock sampling traverse. Location of drill core sites defines position of traverse beginning at the Ozark Lead mine and extending approximately 35 km to the East. This traverse lies normal to the strike of depositional facies and extends from areas of intense to weak mineralization.

---

petrographic and geochemical analysis. The basal contact of the Bonneterre Formation, with the underlying Lamotte Sandstone, was also recorded for each core to be used as a correlation datum.

PETROGRAPHY AND CATHODOLUMINESCENCE

Petrographic analysis of samples was performed on thin sections and polished chips stained with Potassium Ferricyanide for qualitative evaluation of Fe^{2+} variations. Cathodoluminescence of polished samples was undertaken in a Nuclide Luminoscope utilizing a 12 to 14 KV accelerating potential, 0.8 milliampere beam current and 2 millimeter focused beam with air as the ionizing medium.

Ganuge Dolomite Spar

Gangue dolomite spars, chemical precipitates derived from mineralizing fluids, are a volumetrically significant part of the late diagenetic record. These spars, precipitated on substrates of host dolomite or metallic sulfides, are most abundantly developed in vuggy and cavernous porosity in areas of intense mineralization, but they are also present as thin overgrowths in intergranular and intercrystalline porosity of the host dolomite throughout the study area.

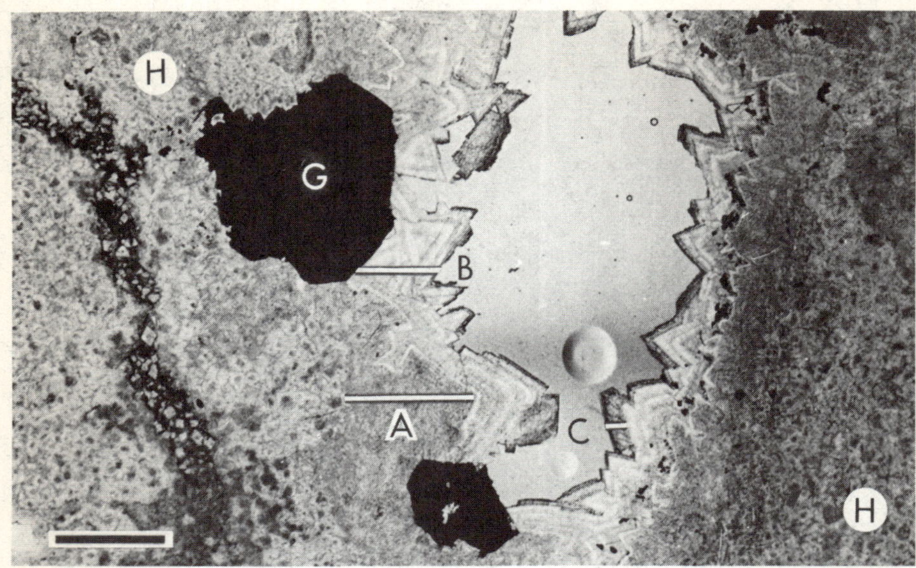

FIGURE 2: Coarse saddle crystals of dolomite gangue spar developed in vuggy porosity of the multicrystalline host rock. Identification of Zones **A**, **B**, and **C** within gangue spar is aided by staining with K-Ferricyanide which expresses variation in Fe^{2+} content. Host rock dolomite **H** in this example contains moderate Fe^{2+} and is lightly stained. Note large galenas **G** precipitated directly on host rock and within Zones A and B; Disseminated sulfides are present in early Zone B (left side of pore). Scale bar is 2 millimeters. Transmitted light, Ozark Lead Mine.

Sparry Gangue dolomites exhibit a subtle growth zonation in transmitted light marked by a variation in the density of inclusions, ranging from clear to translucent. This zonation is best resolved with a combination of K-Ferricyanide staining and cathodoluminescence microscopy (Figures 2 and 3). Three sequential zones are easily separated (see Braunsdorf, 1983) which are similar to zones 2,3 and 4 of Voss and Hagni (1985):

- Zone A, an early, slightly Fe^{2+} -rich dolomite which luminesces dull purple-red to non-luminescent.
- Zone B, an intermediate Fe^{2+} -free dolomite comprised of numerous alternating bright red and non-luminescent subzones
- Zone C, a late, very Fe^{2+} -rich dolomite which is non-luminescent.

Cathodoluminescence at high magnifications of Zones A and B (200x to 400X) reveals complex boundaries among subzones within A and B likely recording alternation of corrosion and precipitation during the precipitation of these zones of gangue spar. As noted by Braunsdorf

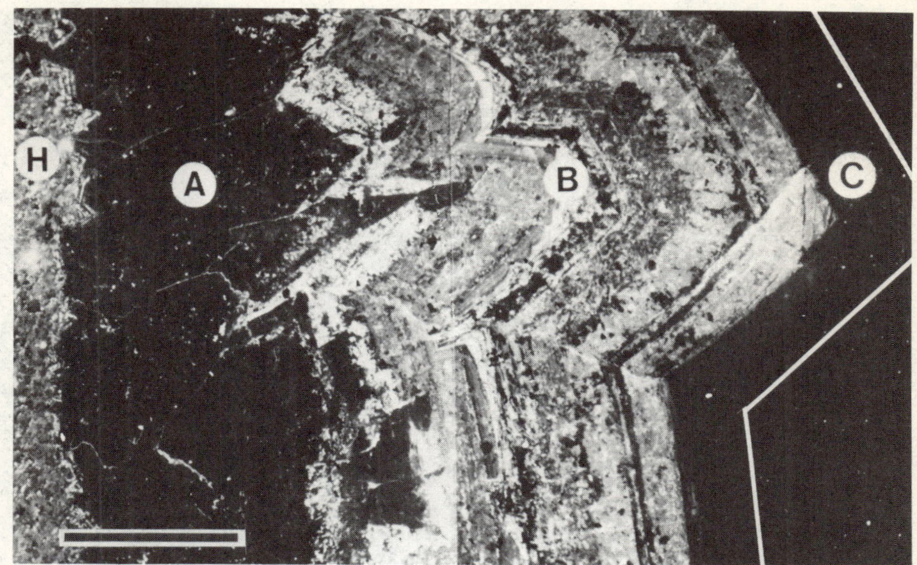

FIGURE 3: Cathodoluminescent Zonation of dolomite gangue spar. Based on cathodoluminescence and microfabric, three consistently identifiable zones comprise the gangue spars. Zone **A**, although in this example appears entirely non-luminescent, locally luminesces dull purple to red. Zone **B** is characterized by subzones comprised of alternating bright to dull luminescing dolomite. Zone **C**, which is highly enriched in Fe^{2+} , is uniformily non-luminescent. The microfabric of the dull luminescent host rock **H** suggests destruction of well-defined subcrystals of dolomite. Note evidence of corrosion among zones within Zone B. Scale bar is 0.5 millimeters.

(1983), emplacement of zones A and B is coincident with the principal episodes of sulfide mineralization (Figure 2); whereas, zone C postdates all but minor marcasite mineralization.

Host Rock Dolomites

Petrographic characteristics of Bonneterre Formation host dolomites have been documented by several workers (Larsen, 1977; Lyle, 1977). One important characteristic of the host rock is the variable preservation of primary fabric. Locally, well developed ghosts of former grains are easily identified; whereas, adjacent samples exhibit total loss of primary grain fabric. More striking, is the variation in rock crystallinity, the character of individual dolomite crystals which comprise the total rock mass. In samples where remnant grain fabric is preserved, individual dolomite crystals are euhedral to subhedral and range from fine 20-100 microns in size. Such a fabric is common in samples of the planar stromatolite facies and in the upper the Bonneterre at all core locations. In samples where grain fabrics have been destroyed, dolomite crystals are coarse, ranging in size up to 2 millimeters, and form an anhedral interlocking dolomite mosaic. Although samples comprised of the coarse anhedral mosaics are most common in the lower third of the Bonneterre Formation, coarsely crystalline larnellae are present locally throughout the section on a regional scale.

Under cathodoluminescence, host dolomites exhibit variable fabric which is difficult to resolve except at high magnifications. For example, dolomite crystals from highly mineralized areas exhibit a homogeneous, uniform luminescence. Euhedral crystals from upper core levels are commonly constructionally zoned while intermediate depth samples exhibit a complex mottling reflecting intermixtures of bright luminescent dolomite intermixed within a "corroded" fabric of individual dolomite rhombs (Figure 4b). When luminescence characteristics are combined with Fe^{2+} -content, estimated from K-ferricyanide staining, and overall dolomite crystallinity, the host dolomite can be divided into three end member dolomite types:

- Type I -- a euhedral, zoned-CL, Fe^{2+} -free dolomite (Figure 4a).

- Type II-- an anhedral, uniform, purple-red CL, Fe^{2+} -rich dolomite (Figure 4c).

- Type III -- an anhedral, uniform, orange-red CL, Fe^{2+} -free dolomite (Figure 4d).

In addition, two dolomites exhibiting fabrics transitional between Type I and either Type II or III were observed. These include:

- Type II*m* -- anhedral, complexly mottled CL, Fe^{2+} -rich.

- Type III*m* -- andhedral,complexly mottled, Fe^{2+} -poor (Type (Figure 4b).

These distinctive dolomite types serve as the petrographic framework on which dolomite isotope chemistry was performed.

CARBON AND OXYGEN ISOTOPES

On the basis of petrographic and cathodoluminescent fabric described above, texturally distinct dolomite components were microsampled for analysis of carbon and oxygen isotope composition. Microsampling was performed on a microscope mounted drill assembly with a motorized X-Y translation stage to provide for accurate removal crystalline materials for analysis. Drilling tools faceted to a 20 micron point were employed where individual subzones of gangue dolomite spar were subsampled. This allowed areas as narrow as 100 microns to be mechanically separated for analysis. Samples of host rock were taken with 300-500 micron diameter drilling tools to obtain .2-.3 milligrams of powdered sample.

After mechanical separation from polished rock chips, all powdered samples were roasted at 380^0 C *in vacuo* for one hour to remove volatile organic contaminants. Samples were then reacted in anhydrous phosphoric acid at 50^0 C in separate reaction vessels for 8-12 hours, and CO_2 was purified in an extraction line connected directly to the inlet of a VG Micromass 602E gas ratio mass spectrometer. Isotope enrichments have been corrected for ^{17}O contributions and are reported relative to the PDB carbonate standard. Precision of analysis was 0.12 (std. dev.) for oxygen and 0.08 for carbon based on daily analysis of NBS-20 analyzed as an unknown carbonate powder.

Gangue dolomite spar was sampled from fine intergranular pores and coarse vuggy porosity. Figure 5 illustrates the distribution of data obtianed in this study relative to data of

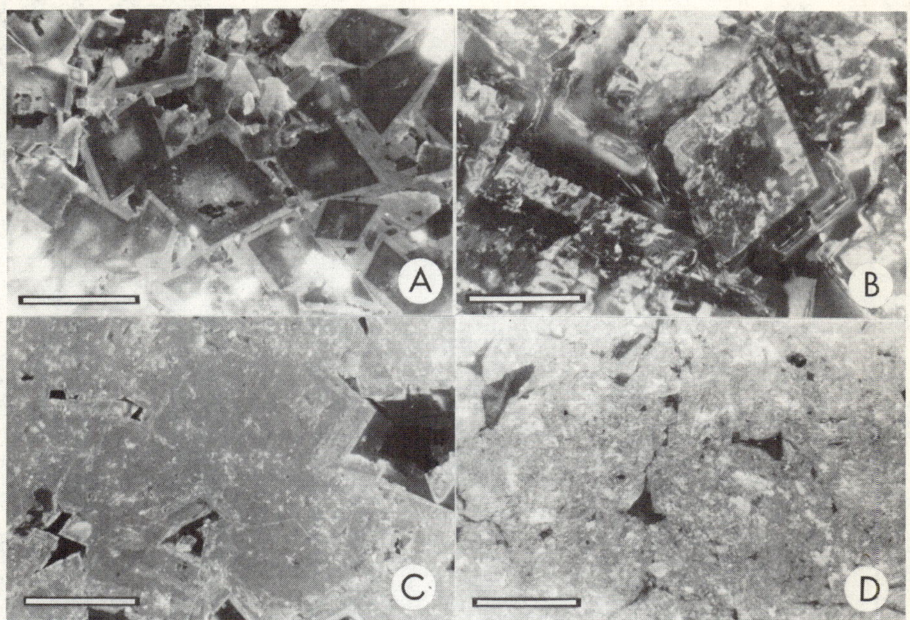

FIGURE 4: Cathodoluminescence of dolomitic host rock.

A--Type I host rock is characterized by constructional CL zoning and an overall euhedral to subhedral crystal form. The non-luminescent cores may exhibit minor replacement by brightly luminescent dolomite. Scale bar is 50 microns.

B-- Mottled fabric present in Type II*m* and Type III*m* dolomites is distinguished by the corrosion microfabric under cathodoluminescence. Outlines of the euhedral crystal cores are retained despite extensive solution and secondary dolomite replacement. Scale bar is 100 microns.

C-- Type II host rock characteristic of mineralized zones is often pervasively luminescent and may result in total loss of primary grain fabrics. This example illustrates a uniformly luminescent host rock which because of variable levels of Fe^{2+} content, varies from moderate to dull. Remaining intergranular porosity is in-filled by Zone C dolomite spar. Scale bar is 200 microns.

D-- Type III host rock exhibits near pervasive loss of primary crystal fabric and unlike Type II dolomite exhibits bright overall luminescence because it is virtually free of Fe^{2+}. Scale bar is 200 microns.

Braunsdorf (1983) who selectively analyzed individual growth zones within gangue dolomite crystals collected from the Ozark Lead Mine. Analyses of samples from this study coincide with the range of compositions documented by Braunsdorf (1983). An important observation of Braunsdorf (1983) was the nearly invariant composition with subzones of the gangue spar, except for late zone B. Zone A has a composition of -8.0 $\delta^{18}O$, -1.0 $\delta^{13}C$ with less than 1.0 per mil total variation. Early zone B is indistinguishable from zone A, whereas, late zone B

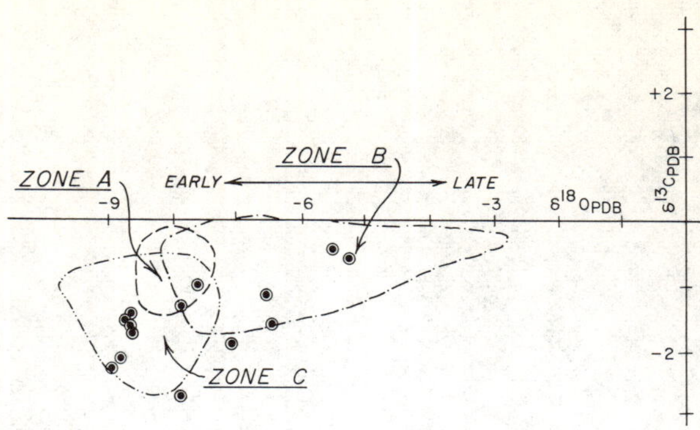

FIGURE 5: Carbon and oxygen isotope composition of dolomite gangue spar. Analyses obtained from this study are plotted relative to the fields defined by Braunsdorf (1983) for individual luminescent zones within the gangue spar. All zones overlap at a composition of -8.0 $\delta^{18}O$ and -1.0 $\delta^{13}C$ and exhibit minor variation,with the exception of late Zone B.

---

exhibits a distinctive enrichment in oxygen and carbon isotopic composition up to -3.0 $\delta^{18}O$ and 0.0 $\delta^{13}C$. Zone C returns to a composition overlapping Zone A spars.

Host rock dolomites exhibit a compositional variation consistent with the separation of dolomite types defined petrographically. Type I euhedral, zoned dolomites possess enriched $\delta^{13}C$ compositions distinct from both gangue spars and all anhedral dolomite types. Anhedral dolomites, define two compositional trends which diverge from euhedral, zoned Type I dolomites and extend to endpoint compositions. Samples exclusively of Type II dolomite lie uniformily at a -8.0 $\delta^{18}O$, -1.0 $\delta^{13}C$ endnpoint. Type III dolomites lie near a -1.0 $\delta^{18}O$, +0.0 $\delta^{13}C$. For example, host rock collected from mineralized mine areas exhibits Type II characteristics and has an invariant composition (Braunsdorf,1983). In contrast, dolomites exhibiting a complexly mottled cathodoluminesce fabric (Type II*m* and Type III*m*) are highly variable in composition and lie intermediate along these trends between Type I and either Type II or Type III endnmember dolomites.

Utilizing the dolomite composition and fabric types, it is possible to evaluate their stratigraphic and regional distributions. As dolomite Types II and II*m* are most abundant, these have been separated and plotted with Type I and Type (III + III*m*) dolomites relative to stratigraphic position for each core location (Figure 7). A consistent distribution is observed for most dolomite types. Type I dolomites always are present in the uppermost part of each core. Similarly, Type I dolomite occurs in the basal Bonneterre Formation in contact with the Lamotte Sandstone. Type II*m* dolomite forms a transitional zone separating upper, Type I, from lower, Type II, dolomite types. Interestingly, in areas of intense mineralization, the distribution of Type II dolomite extends stratigraphically higher in the cores. Type III dolomites exhibit a non-uniform distribution, occurring dominantly to the East and co-occurring laterally with Type II*m* dolomites.

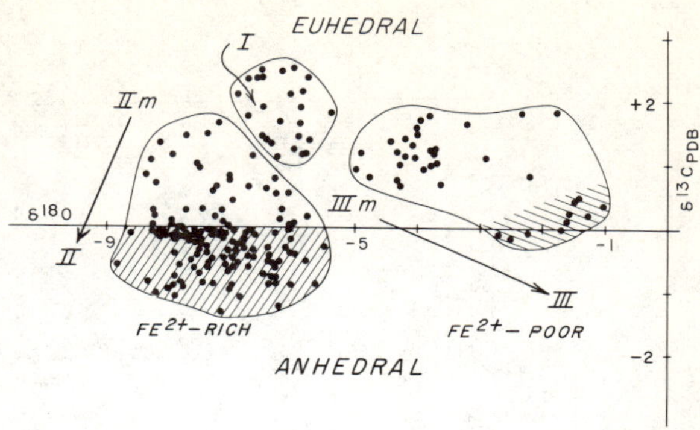

FIGURE 6: Carbon and oxygen isotope composition of host rock dolomite. Based on crystal fabric, luminescence and Fe^{2+} -contents, three broad compositional fields are defined. With Type II and Type III dolomites, trends from mottled (II*m*) to pervasively (II) luminescent dolomites are observed. Stippled areas indicate dolomites exhibiting uniform luminescence and pervasive destruction of primary fabric. All isotope enrichments are expressed relative to the PDB carbonate standard.

DISCUSSION

Sequential zones of gangue dolomite spar provide a decipherable record which constrains possible variations in the chemistry of the mineralizing fluids, or at least the composition of carbonate which they could precipitate. While each zone is petrographically distinct and consistent differences in Fe^{2+} -content are present (varying from moderate in Zone A to low in Zone B to extremely high in Zone C), one might envision derivation of mineralizing fluids from several sources. Alternatively, the sequential phases of gangue spar precipitation might represent multiple episodes of fluid migration from a single, chemically-evolving source. In either case, shifts would be anticipated in the isotopic composition of carbonates precipitated during each stage of fluid migration.

Data of Braunsdorf (1983) demonstrate a profound continuity in composition among all stages of mineralization, with the exception of late Zone B. This constancy of composition requires that either: a) the fluid temperature and isotopic composition were invariant throughout the duration of mineralization; or b), that fluid compositions were in some fashion buffered to maintain variation within narrow limits. For example, based on whole crystal analysis of gangue spar and paired with co-occurring host rock in the Buick Mine, Sverjensky (1981a) suggests that fluid compositions were modified through rock-water interaction until isotopic equilibrium with host rock compositions was achieved. This would require a rock-dominated reaction system in which water-to-rock ratios were low. If such were the case, gangue spar compositions should mimic host rock values. Clearly, data of this study document a large variation present in host rock compositions. Rather than a **rock-dominated** reaction system in which the composition of the host rock controls the composition of mineralizing solutions, trends

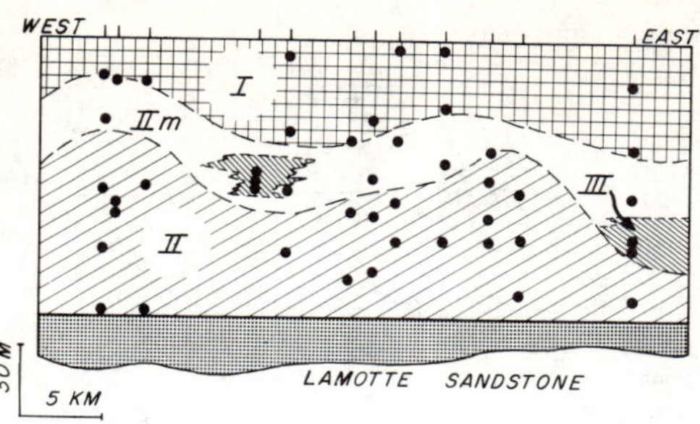

FIGURE 7: Stratigraphic distribution of dolomite host rock. Core locations and sample depths are marked on the West to East traverse. Separation of dolomite types is based on isotopic and petrographic criteria illustrated in Figure 6. Note the uniform occurrence of Type II dolomite in the basal Bonneterre Formation where it contacts the Lamotte Sandstone. Type I occurs at the top of all cores with Type II*m* developed in an intermediate position. Type III dolomite occurs only locally within the area of study.

---

in isotopic composition suggest a **fluid-dominated** system whereby, in areas with free access for fluid flow, host rock is pervasively altered to compositions in equilibrium with the mineralizing fluids.

Textural data of the host rock dolomites support the extensive reaction of mineralizing fluids with host rock dolomite. Dolomite crystallinity varies from euhedral in the upper Bonneterre Formation to anhedral mosaics in areas of intense mineralization and in the basal Bonneterre Formation where it contacts the underlying Lamotte Sandstone, reaffirming that this aquifer served as the principal conduit for regional migration of mineralizing fluids. Recrystallization is supported by changes in the cathodoluminescence microfabric which indicate partial to pervasive replacement of much of the host rock. Constructionally-zoned euhedral crystals present in Type I dolomite from the Upper Bonneterre exhibit increasing evidence of corrosion and replacement by luminescent dolomite (Type II*m*) down section. In areas of intense mineralization, dolomite microfabrics are dominated by uniformily luminescent, anhedral mosaics (Type II) in which primary grain fabrics are destroyed. This is matched by a concomitant increase in the Fe^{2+} -content to levels comparable to Zone A dolomite spar. Thus, changes in microfabric and cathodoluminescence of host rock dolomite indicate progressive reaction of Type I host rock dolomite with mineralizing fluids and variable replacement by dolomites with compositions equivalent to the gangue spars.

Such a process of progressive replacement of an early-formed dolomite, Type I, by dolomite in equilibrium with mineralizing fluids, Type II, is supported by variations in isotopic composition. For example, Type II*m* dolomite, which is distinctively mottled under cathodoluminescence, has isotopic compositions transitional between these two end member dolomites. We suggest that this intermediate composition is evidence of partial replacement of

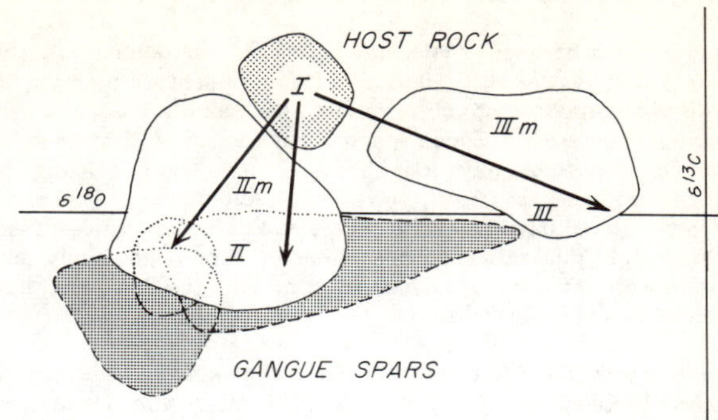

FIGURE 8: Recrystallization trends resulting in petrographic and geochemical variation of host rock dolomite. Recrystallization of host rock, initially of Type I composition (coarsely stippled field), by mineralizing fluids whose composition is reflected in the variation of gangue dolomite spars (finely stippled fields), results in characteristic trends in carbon and oxygen isotope composition. Type II*m* and Type II dolomites underwent replacive reaction with fluids precipitating Zones A and early B gangue spar. Enriched compositions of Type III and III*m* host rock indicates reaction with late Zone B fluids. The absence of host rock containing highly enriched Fe^{2+} -contents precludes significant host rock replacement during the fluid migration event which emplaced Zone C spar.

Type I dolomites by a dolomite with a composition of Zones A and early B gangue spars. Type II dolomite overlaps directly the compositional fields of these gangue spars. The trend in isotopic composition, thus, reflects variation in the degree of replacement (i.e amount of mixing of Type I and Type II dolomite). The coincidence of Type II host rock and gangue spar compositions, therefore requires a water-dominated system in which massive quantities of mineralizing fluid locally reacted with, and replaced, comparatively small volumes of host rock.

The divergence of host rock isotope compositions from Type I toward Type III, coupled with textural criteria for Type III*m*, argues for a similar mechanism of partial to pervasive replacement for the development of this trend. In this case, however, the composition of the replacive dolomite must be approximately -1.0 δ ^{18}O and +0.0 δ ^{13} C in order to produce the Type I to Type III compositional trend.

If we consider that every stage of gangue spar precipitation represents temporally discrete episodes of migration of the mineralizing fluids, it is possible to reconstruct a history of fluid-host rock interaction. Based on the the sequential occurrence of Zones A, B, and C, a temporal framework has been identified which is recorded by changes in isotopic and minor element composition. Therefore, by correlating chemical and fabric modification of the host rock to the paragenetic sequence of the gangue spar, a relative timing of host rock alteration can be determined. For example, through detailed microsampling of subzones in Zone B gangue spar, Braunsdorf (1983) documented a marked shift in composition from early to late Zone B. Because late zone B spar coincides in composition with the replacive dolomite necessary to produce Type

III*m* and Type III host rock, formation of Type III host rock must be coincident with the precipitation of late zone B spar. Similarly, it must post date migration of fluids which precipitated Zones A and early B, and predate emplacement of Zone C spar.

It is significant that host rock containing high Fe^{2+} -contents approaching those of Zone C are absent, thus arguing against significant alteration of host rock by mineralizing fluids during Zone C time. This is compatible with observations that metallic sulfide mineralization is coincident with the precipitation of Zones A and B gangue spars. However, Zone C spar has a very similar isotopic composition, even though it is not accompanied by significant sulfide mineralization. This suggests that Zone C was precipitated from a fluid of similar temperature and composition and derived from the same source as fluids responsible extensive mineralization. The absence of significant mineralization during Zone C time might imply that the controls for sulfide precipitation were intrinsic to the host rock, and not the fluid, and that evolution of the host rock lead to cessation of mineralization.

In addition to elucidating the temporal relationships among host rock alteration and sequential episodes of mineralizing fluid migration, this study provides insight into the nature of host rock dolomitization. Clearly, the mineralizing fluids interacted aggressively with the Bonneterre Formation carbonate sequence. These fluids were active in producing secondary porosity as evidenced by abundant corrosion features both within the host rocks and in zones of precipitated gangue dolomite spar. The saturation of these fluids with respect to dolomite, however, must have varied dramatically. While clear evidence of corrosion is present, these fluid must have alternately precipitated dolomite. Vuggy and intergranular porosity contains abundant dolomite spar. Moreover, dissolution features indicate extensive solution of the host rock accompanying mineralization. Because these solution features have subsequently been filled by dolomite with compositions characteristic of the mineralizing brines, processes of solution-replacement must have occurred almost synchronously.

Implicit in this discussion of host rock - fluid interaction is the existence of dolomite in the Bonneterre Formation prior to the influx of mineralizing solutions. The mottled replacement textures in host rock indicate a process of dolomite-dolomite replacement. The drive for such replacement might include non-stoichiometry of composition (Land, 1980) or disorder of the crystal structure (Reeder, 1981). Although Cathodoluminescence fabrics documented in this study clearly demonstrate corrosion-replacement within most of the host rock dolomites, this does not require an intrinsic difference in solubility or stability between host rock and replacive dolomites. Corrosion features are equally abundant within subzones of the dolomite gangue spars. This argues against an intrinsic drive for recrystallization. Rather, this implies an extrinsic drive related perhaps to a fluctuating dolomite saturation state in the mineralizing fluids in response to sulfide precipitation.

The presence of an early-formed dolomite in the Bonneterre Formation, the remnants of which are preserved as Type I host rock, suggest that a diagenetically-formed porosity system could have controlled the initial distribution of mineralizing fluids as they migrated up dip into the carbonate sequence. This porosity network, while perhaps determining initial flow pathways, was certainly enlarged through the action of corrosion during mineralization. Therefore, the concern of whether mineralizing fluids were capable of dolomitizing the Bonneterre Formation (Gregg, 1985), become secondary. In contrast, understanding the role of these fluids in enlarging the porosity system in the carbonate host is of vital importance in resolving the regional distribution patterns of mineralization.

ACKNOWLEDGEMENTS

Financial support for this project was provided by U.S. Department of Education Grant G008002673 as well as grants from the Geological Society of America, and the Scott Turner Earth Science Fund at the University of Michigan. We wish to thank the Missouri Department of Natural Resources for providing core material, and particularly staff geologist Jim Palmer for additional assistance and early guidance. Thanks to David Dettman who kept the isotope laboratory functioning smoothly and special thanks to Neil Braunsdorf who supplied isotope data from the Ozark Lead Mine as well as stimulating discussion and loyal friendship. The first co-author (MHF) especially wishes to acknowledge Georgia Brown for her tacit yet strongly felt support and encouragement.

REFERENCES

Anderson, G.M., 1975, Precipitation of Mississippi Valley-type ore: Econ. Geol., v. 70, p. 937-942.

Beales, F.W., 1977, Stratigraphic approach to Mississippi Valley-type ore deposits: *in* Ore Deposits Workshop: p. 29-55, Dept. Geology, University of Toronto.

Braunsdorf, N.R., 1983, Isotopic trends in gangue carbonates from the Viburnum Trend: implications for Mississippi Valley-type mineralization: Unpub. M.S. thesis, University of Michigan, Ann Arbor.

Gerdemann, P.E., and Myers, H.E., 1972, Relationships of carbonate facies patterns of ore distribution and to ore genesis in the southeast Missouri lead district: Econ. Geol., v. 67, p. 426-433.

Gregg, J.M., 1985, Regional epigenetic dolomitization in the Bonneterre Dolomite (Cambrian), southeastern Missouri: Geology, v. 13, p. 503-506.

Hagni, R.D., 1983, Ore microscopy, paragenetic sequence, trace element content, and fluid inclusion studies of the copper-lead-zinc deposits of the Southeast Missouri lead district: *in* International Conference on Mississippi Valley Type Lead-Zinc Deposits, G. Kisvarsanyi, S. Grant, W. Pratt, and J. Koenig, eds., p. 243-256, Univ. Missouri, Rolla.

Hannah, J.L., and Stein, H.J., 1984, Evidence for changing ore fluid composition: stable isotope analyses of secondary carbonates, Bonneterre Formation, Missouri: Econ. Geol., v. 79, p. 1930-1935.

Howe, W.B., 1968, Planar stromatolite and burrowed carbonate mud facies in Cambrian strata of the St. Francois Mountain area: Missouri Geol. Survey Water Resources, Rept. Inv. 41, 113 p.

Land, L.S., 1980, The isotope and trace element geochemistry of dolomite: the state of the art: Soc. Econ. Paleon. Miner. Spec. Publ. no. 28, p. 87-200.

Larsen, K.G., 1977, Sedimentology of the Bonneterre Formation, southeast Missouri: Econ. Geol., v. 72, p. 408-419.

Lyle, J.R., 1977, Petrography and carbonate diagenesis of the Bonneterre Formation in the Viburnum Trend area, southeast Missouri: Econ. Geol., v. 72, p. 420-434.

Ohle, E.L., 1980, Some considerations in determining the origin of ore deposits of the Mississippi Valley type, Part 2: Econ. Geol. v. 75, p. 161-172.

Reeder, R.J., 1981, Electron optical investigation of sedimentary dolomites: Contrib. Mineral. Petrol., v. 76, p. 148-157.

Rickman, D.L., 1981, A thermochemical study of the ore deposits of the Milliken Mine, New Lead Belt, Missouri: Unpub. Ph.D. dissertation, Univ. of Missouri, Rolla, 310 p.

Sangster, D.F., 1983, Mississippi Valley-type deposits: a geological melange: *in* International Conference on Mississippi Valley Type Lead-Zinc Deposits, G. Kisvarsanyi, S. Grant, W. Pratt, and J. Koenig, eds., p. 7-19, Univ. Missouri, Rolla.

Sverjensky, D.A., 1980, The origin of a Mississippi Valley-type deposition in the Viburnum Trend, Southeast Missouri: Unpub. Ph.D. dissertation, Yale Univ., 144 p.

Sverjensky, D.A., 1981a, Isotopic alteration of carbonate host rocks as a runction of water to rock ratio -- an example from the Upper Mississippi Valley zinc-lead district: Econ. Geol, v. 76, p. 154-156.

Sverjensky, D.A., 1981b, The origin of a Mississippi Valley-type deposit in the Viburnum Trend, southeast Missouri: Econ. Geol., v. 76, p. 1848-1872.

Thacker, J.L., and Anderson, K.H., 1977, The geologic setting of the southeast Missouri lead district -- Regional geologic history, structure and stratigraphy: Econ. Geol., v. 72, p. 339-348.

Voss, R.L., and Hagni, R.D., 1985, The application of cathololuminescence microscopy to the study of sparry dolomite from the Viburnum Trend, southeastern Missouri: *in* Proceedings, Paul F. Kerr Memorial Symposium, Process Mineralogy V: Hausen, D.M. and Kopp, O., eds, AIME, New York .

COMBINED INSTRUMENTATION FOR EDS ELEMENTAL ANALYSIS AND CATHODOLUMINESCENCE STUDIES OF GEOLOGICAL MATERIALS

Donald J. Marshall,[a] John H. Giles[b] and Anthony Mariano[c]

[a] Nuclide Corporation, Box 315, Acton, MA, 01720
[b] Princeton Gamma-Tech, Inc., Princeton, NJ, 08540
[c] Carlisle, MA, 01741

An accessory has been developed for the Luminoscope (1) which supplements the standard transmitted light, polarized light and cathodoluminescence observations with the acquisition of energy dispersive X-ray spectra. This adds the extra dimension of elemental analysis with a simpler, less expensive, instrument than the electron microprobe or SEM and retains the good quality of optical microscopy observations. Samples of many types, including non-luminescent opaque or conducting mineral samples, can be studied with little or no sample preparation. Examples of spectra obtained from phlogopite, pyrochlore, limestones, soil, and metallurgical standards are presented.

Introduction

The serious use of the cathodoluminescence (CL) technique in the earth sciences began with the electron microprobe (EMP) (,2,3,4,5). Brightly luminescent materials were viewed and photomicrography of relatively brightly luminescing samples was done. The primary purpose of the electron probe is to generate and measure characteristic X-rays and the CL application was a by-product of this. Its promise of useful applications for CL was such, however, that there was a rapid development of simple cathodoluminescence microscope attachments (CMA) which could be mounted on the stage of a conventional optical microscope, retain the full capabilities of the microscope and yet permit the CL observations also (6,7). Several hundred laboratories are presently equipped with this type of instrument, and many others continue to make CL observations on the electron probe or on the SEM.

CL observations in geological samples are extremely useful for such purposes as discerning generations of carbonate cement and distinguishing authigenic from detrital minerals (3,5). In some minerals they provide information on possible trace impurity activators and even their charge states (8). CL observations alone, however, do not provide complete and unequivocal information on composition. There are many instances when the geological observations would be facilitated if some elemental composition data were available to supplement the optical and CL observations. The acquisition of this information has now been facilitated by the development of an X-ray detector accessory for the Luminoscope. This accessory does not interfere with the normal transmitted light (TrL), polarized light (PoL) and CL observations and yet it allows the user, at any time, to acquire an X-ray spectrum from the sample using an EDS X-ray detector. The X-rays are generated, of course, as a byproduct of the electron beam bombardment which is producing the luminescence. We have now come "full circle" from the viewing of CL on the electron probe, whose primary purpose is to generate X-rays, to the acquiring of X-ray energy spectra on the Luminoscope, whose primary purpose is to produce CL.

The Luminoscope, as an electron source for X-ray excitation, provides several advantages and a few disadvantages, as compared with the EMP and the SEM. These include:

Advantages:

1. Price is significantly lower

2. Convenience and simplicity of operation and maintenance

3. Sample preparation is minimal

4. The retention of good optical microscopy capabilities

5. Speed of operation

6. Combined capabilities of TrL, PoL, CL and EDS.

7. Photomicrography of sample is conveniently done.

Disadvantages include:

1. Beam size is larger

2. Beam energy is lower. Also beam energy is often unregulated and has a greater energy spread.

These add up to the general conclusion that the EDS accessory for the Luminoscope provides an excellent tool to supplement CL studies; it is immediately useful for samples with large mineral fragments or for very fine-grained, homogeneous, samples. It provides a convenient, inexpensive screening tool to select those samples which will require the more sophisticated operational features of the EMP or SEM; it provides a qualitative analytical tool for large quantities of samples and the work can be done rapidly because there is little or no sample preparation required; and it is also an excellent teaching tool to prepare individuals for the later use of the EMP or SEM because of its ruggedness, tolerance of operator mistakes, and low maintenance costs.

Finally, there are some situations of great practical importance, where it may provide a better solution than the more expensive instruments because of the wide range of sample sizes it will accept and because of the simple vacuum system it uses. More about all of these points later.

This article will first describe the EDS accessory and its operation and then present several application examples. It is assumed that the reader is familiar with cathodoluminescence microscope attachments (CMA) and with the uses of cathodoluminescence. If not, the references by Kopp (9) and by Nickel (10) are good starting points. It is also assumed that the reader is familiar with the general aspect of EDS detectors and the uses of X-ray data for analysis. A good introductory references to this subject is the publication by Barbi (11).

Description

Figure 1 is a cross sectional view of the Luminoscope showing the relationship between the electron beam, the sample surface, and the optical path for transmitted light and CL. The sample shown is a typical rock thin section. The incident electron beam, of course, generates X-rays from the sample - X-rays which are normally absorbed by the metallic parts of the vacuum chamber or by the lead glass window. These X-rays are emitted over a 4 pi solid angle. Those emitted over the 2 pi hemisphere above the sample are unattenuated by the microscope slide, and therefore their energies are diagnostic of the sample composition. A detector could be located anywhere in this region where it does not interfere with the electron beam or with the optical path in the microscope. In practice, the location opposite the electron beam path is best for flat samples and this is the one normally used. (Figure 1). However, the X-ray analyzer mounting is such that it can be rotated about the optical axis and, for rough chips and some other samples, it may be better to locate it in the quadrant that the electron beam is located in.

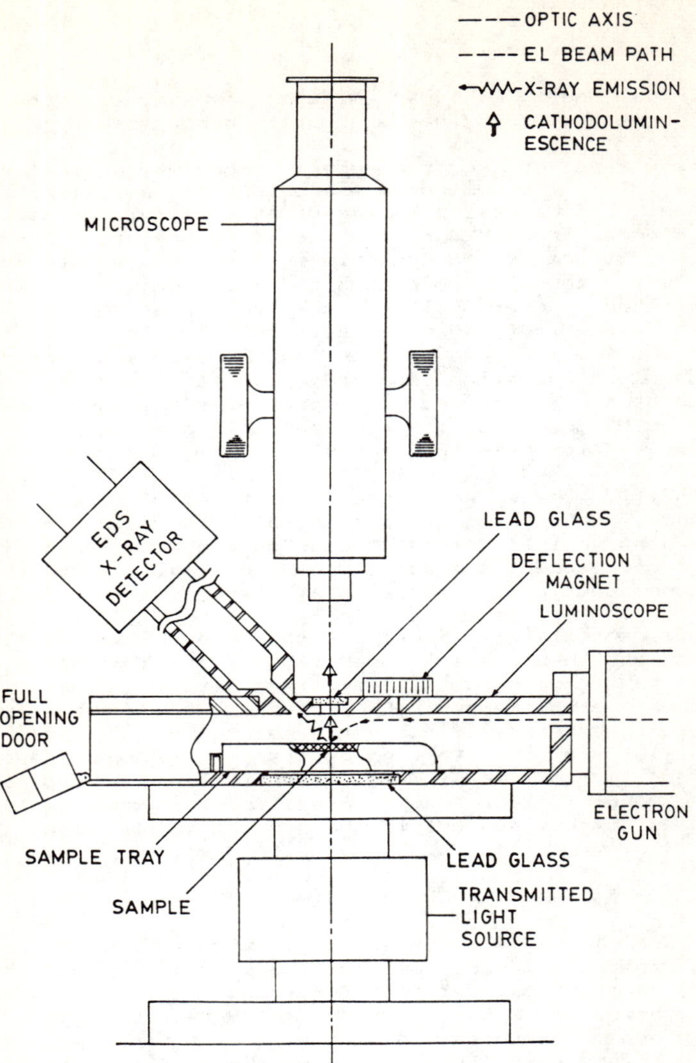

Figure 1. Cross sectional view of the EDS system mounted on the Luminoscope. The path of the electron beam at impingement on the sample is controlled by the deflection magnets which are positioned laterally by the operator, usually for maximum intensity of CL or of X-ray count rate. The X-ray take-off angle is fixed at 45° (for flat specimens) but the collimation of the detected beam can be varied by choosing the size of the collimator and by changing its position with respect to the sample. Note that the windows on both sides of the chamber permit ordinary transmitted light observations as well as CL observations.

Wherever the EDS accessory is located, its presence does not interfere with the loading and unloading of samples in the normal fashion.

The EDS accessory mounts in place of the large viewing window normally provided on the Luminoscope chamber. This large (2") window is replaced with the EDS accessory which contains a 1" viewing window, the latter still being adequate for most normal CL observations.

The EDS accessory is based on the standard system provided by Princeton Gamma-Tech, Inc. of Princeton, New Jersey. It uses a Si(Li) crystal detector. The data is managed by their "System4plus" microanalysis data system (12). The system components are identical to the system that would be provided for other users with SEMs or EMPs. Indeed, if one acquires the EDS accessory for the Luminoscope, all of the power supply, energy analyzer and data acquisition components can also be used on these other, more elaborate, instruments. Only the special detector crystal mount required for the particular sample chamber vacuum system of the SEM or EMP would be needed. Similarly, (and perhaps more importantly), if one already has an EDS system available in the same laboratory space as the Luminoscope, then only the special detector for the Luminoscope needs to be purchased.

The energy resolution for the standard EDS instrument is 155 eV FWHM, measured at Mn-K alpha.

OPERATIONAL PARAMETERS

Beam current 0.1 mA:. Currents up to 1 milliampere are available but it is usually necessary to reduce the current to about 0.1 milliampere to avoid overloading of the detector counting system. Even this current is usually adequate to develop the cathodoluminescence of the sample, if the beam diameter is reduced slightly using the focus coil on the Luminoscope electron gun.

Beam energy 0-25 kV. The standard Luminoscope provides up to 15 - 18 kilovolts electron energy, and a new unit is available which provides up to 25 kV.

Beam size 1 to 2 mm diameter. The beam size can be varied from an unfocused value of about 1 cm diameter to a focused value of between 1 and 2 mm. The beam can be located on individual grains of this size or greater. For some applications with the EDS, where the detector counting system is overloaded, it is also possible to reduce the beam current density by defocusing the beam using the same focusing coil system.

Sampled-area size 1 - 2 mm diameter. The actual size of the area sampled by the EDS detector is determined by the collimation provided in the detector. Of necessity this is highly collimated, since the electron currents used are so much larger than those used in the EMP and the SEM, and the counts must be attenuated. The area sampled is more nearly determined by the collimation of the detector than by the size of the incident electron beam.

The EDS uses a windowed detector so one can break vacuum without concern for the EDS crystal. The operating vacuum in the Luminoscope is usually in the range between 20 millitorr and 100 millitorr, depending on the sample and the desired bombardment energy and current. No apparent degradation of performance of the EDS system over a period of 6 months has been noted, even though the system uses only a mechanical roughing pump. It should be noted, however, that a fore-line trap has been used for the majority of the time. This type of trap removes, or at least reduces, the amount of pump oil that backstreams into the vacuum chamber.

The detector in the EDS system must be cooled during operation and it is best to keep it at liquid nitrogen temperature at all times. (The liquid nitrogen usage is small, averaging about 100-200 cc per day.)

In operation, the System4plus is turned on at the beginning of the session and left on for it's duration. About three minutes after sample introduction and start of pumping, the electron beam is turned on at a low voltage level. During this initial two or three minute pumping period, the necessary initial computer commands can be entered to identify the spectrum being acquired and the acquisition time, etc. The sample is then positioned in the field of the microscope and the desired area of investigation is located by TrL or CL or by a combination of both. At any time when the electron beam is present, one can operate two adjacent keys on the computer keyboard to acquire and display a spectrum. If the spectrum is satisfactory, the operation can be terminated at that point and another spectrum identification entered. However if the preliminary spectrum is not optimum, e.g. the electron energy has not yet reached its final value, or the sample area being looked at is not the best choice, then one can simply strike the same 2 keys and overwrite the previously acquired spectrum with a new one. This continues until the desired spectrum is obtained. Normal spectrum acquisition times are of the order of 20 seconds. Subsequent to the obtaining of the spectrum, one can perform many additional operations including peak identification, background removal, deconvolution of overlapping peaks. etc.

Examples

Phlogopite

The chemical formulas for phlogopite and biotite are both of the form

$$K(Mg,Fe)_3(AlSi_3O_{10})(OH)_2$$

If the iron content is less than about 10%, the mineral is referred to as a phlogopite.

A phlogopite grain with dimensions of approximately 1 cm x 1 cm, in an uncoated thin section of shonkinite was analyzed. The phlogopite was non-luminescent, as expected, and no luminescent impurity inclusions were detected.

A chemical analysis had previously been performed on this sample and yielded the following results (all in weight % oxide).

SiO_2	39.38	K_2O	9.81	Na_2O	0.18
MgO	21.41	FeO	8.48	MnO	0.06
Al_2O_3	12.48	TiO_2	3.93	CaO	0.00

The EDS spectrum, taken at an excitation voltage of 16 kV, reveals the presence of Mg, Al, Si, Fe, K, and Ti. There are minor unidentified peaks in the 5.4 and 5.9 kV range. The original spectrum is shown in Figure 2 and the spectrum with background subtracted is shown in Figure 3.

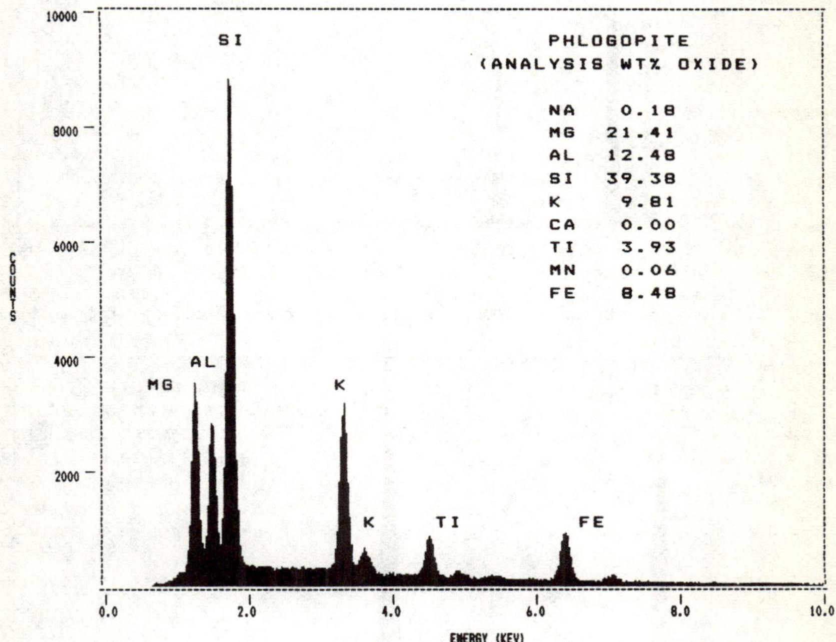

Figure 2. The EDS spectrum of a flake of phlogopite $(K(Mg,Fe)_3(AlSiO_3O_{10})(OH_2))$ taken at 18 kV, 0.01 mA, and 100 seconds.

From the background subtracted spectrum, and the prior chemical analysis, one can make certain estimates of elemental detection limits in this particular sample. We adopt the criterion that a peak is detectable if the maximum count intensity is three times the background intensity -which varies between about 10 counts and 50 counts. This assumption leads to the following results.

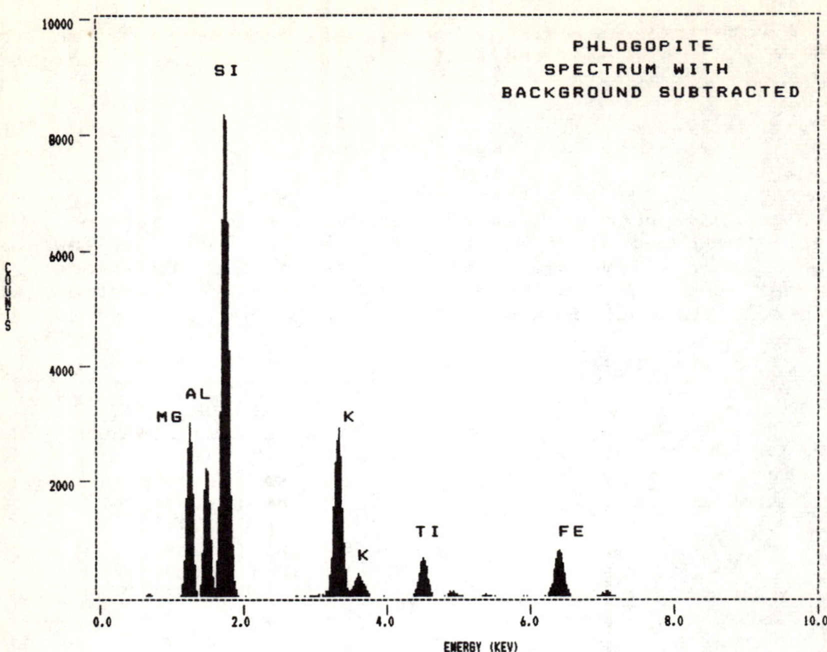

Figure 3. The spectrum of Figure 3, with background subtracted.

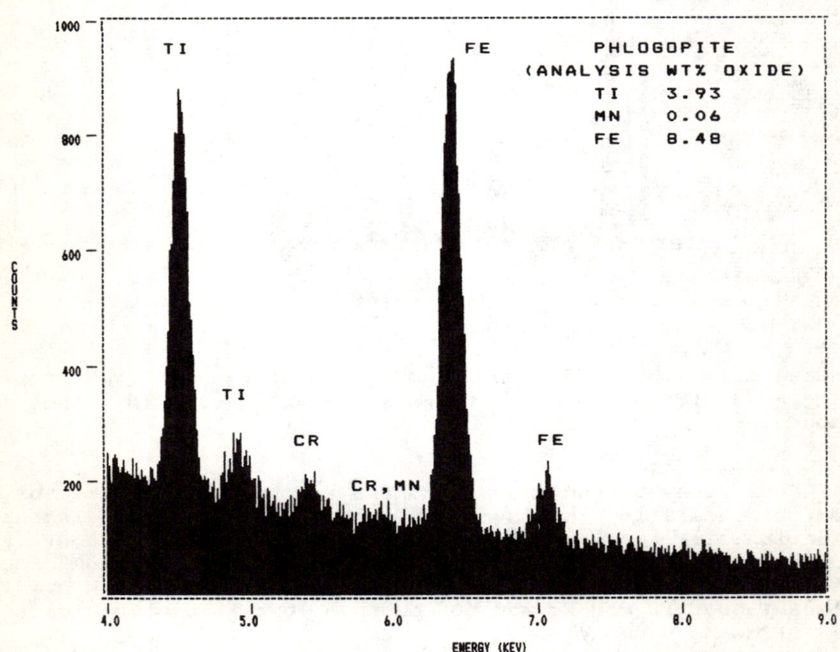

Figure 4. The spectrum of Figure 2, expanded to reveal additional details in the Cr and Mn region.

Element oxide	Concentration (Weight %)	Count Rate	Background Count rate	Calculated Detection Limit
MgO	21.41	3163	50	1.02 %
FeO	8.48	815	15	0.47
TiO_2	3.93	679	10	0.17
K_2O	9.81	2997	40	0.39

Small peaks occur in the EDS spectrum in the 5.4 to 5.9 keV range (Figure 4). These suggest the presence of chromium and possibly manganese. The previous chemical analysis did not reveal chromium and showed only a small amount of manganese.

Pyrochlore

Pyrochlore is an oxide mineral which occurs often in carbonatite localities. Pyrochlore also may occur in pegmatites. The chemical formula is complex and many different substitutions are possible. Generally the formula is stated as

$$NaCaNb_2O_6F - (Na,Ca)_2Ta_2O_6(O,OH,F)$$

Several elements, including U, Th, Ti, & REE can substitute for Ca, Na. Some pyrochlore is mined for its niobium and tantalum content.

The attached EDS spectra (Figures 5,6,7) are for a pyrochlore from the Fen, Norway. All spectra were taken with an electron energy of 16 kV on separated grain(s) mounted on a carbon substrate with a small amount of adhesive. No other sample preparation was necessary.

Pyrochlores do not in general luminesce although some samples did contain finely disseminated luminescing impurities on the surface (probably calcite).

A prior EMP analysis (WDS) of the pyrochlore is available and the result is shown in the table. When a range of numbers are given, these represent the total range for several analyses of different pyrochlores. The value which follows in parentheses is the average value for the locality, but not necessarily for the particular sample.

Fen Pyrochlore

Na_2O	0.00 - 7.60(0.90)
SiO_2	0.00 - 9.26(2.22)
Nb_2O_5	35.39 - 72.76(53.79)
Ta_2O_5	1.32 - 15.11(6.16)
PbO	0.00 - 9.62(3.67)
ThO_2	0.00 - 6.55(2.10)
UO_2	0.00 - 20.48(6.47)
CaO	0.06 - 14.03(5.58)
TiO_2	0.66 - 3.23(1.88)
FeO	0.00 - 19.29(4.99)
Ce_2O_3	0.00 - 2.79
BaO	0.00 - 8.57(3.08)

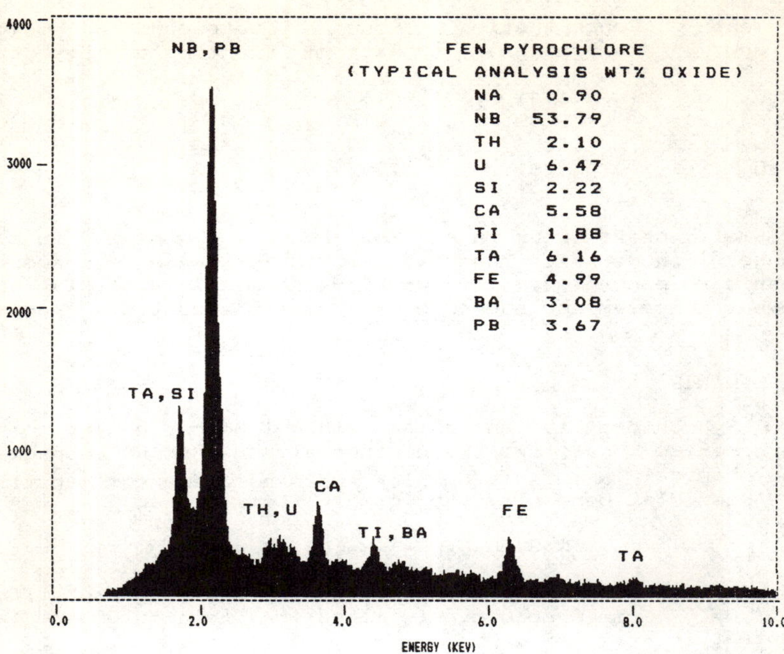

Figure 5. EDS spectrum of a pyrochlore from the Fen, taken at 16 kV, 0.2 mA, and 100 seconds.

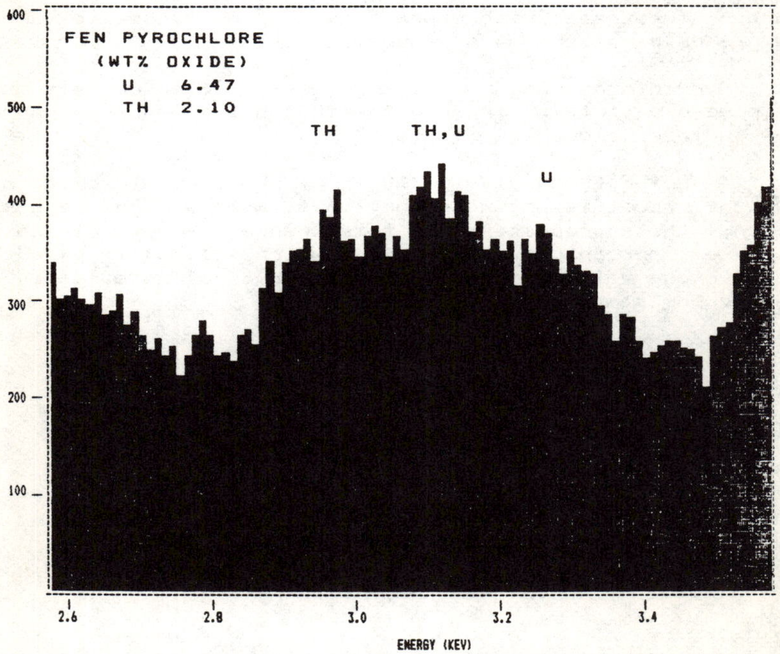

Figure 6. Spectrum of Figure 5, expanded in the U, Th region.

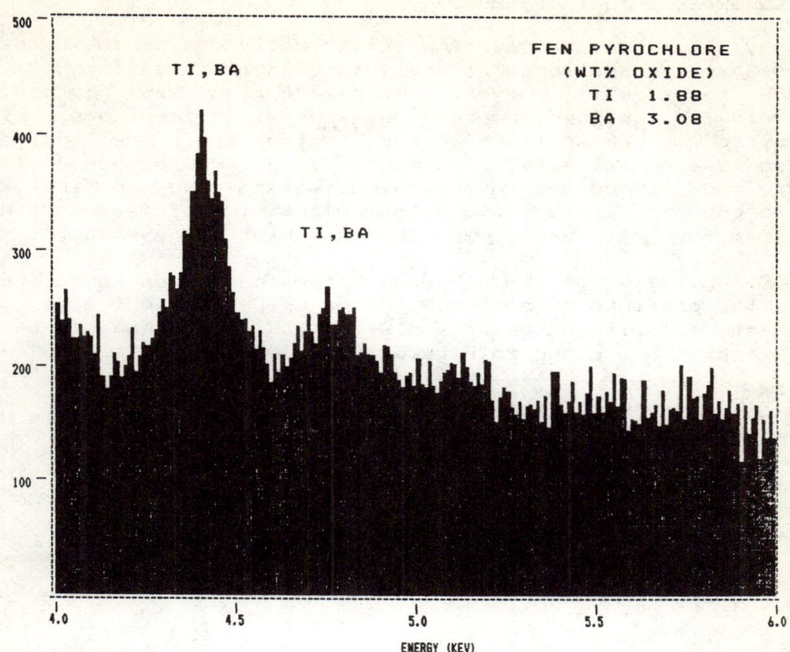

Figure 7. Spectrum of Figure 5, expanded in the Ti, Ba region.

The spectrum of the Fen pyrochlore reveals peaks from most of these elements with three overlapping peak regions, Ta-Si, Pb-Nb, and U-Th. The presence of Ta is confirmed by the higher energy Ta peak at 8.0 kV. Additional work would be needed with fitting of suitable standards or peak stripping to establish the relative contributions of Si and Ta to the peak at 1.8 keV.

The large peak at 2.1 keV indicate the presence of Nb or Pb or both. The relative amounts of these elements could be determined following peak fitting or peak stripping procedures. The much smaller Pb peak at 10.55 kV was not observable.

Uranium and thorium each produce a pair of peaks at about 3.2 kV. The U-Th region for the Fen pyrochlore is shown expanded in Figure 6. There is a peak whose shape and location strongly suggest that both U and Th are present with roughly equal peak intensities.

Finally, the Fen pyrochlore contains barium and titanium. The two most intense L lines of Ba overlap the K lines of Ti. This portion of the spectrum is shown in Figure 7. Although the overlap is severe, the use of peak stripping procedures would allow the separation of the Ba and Ti peaks.

Major impurities in limestones

Many limestone samples are sufficiently medium to coarse-grained that the smaller area analysis capability of the SEM or EMP are necessary. However, there are also many cases of relatively fine-grained, homogeneous appearing limestones, which are suited to EDS analysis on the Luminoscope. The analysis obtained is automatically an average over a large number of fine grains, and may be a more convenient way to make an analysis than to use the SEM to measure the composition of many grains, on a grain-by-grain basis, and then calculate an average.

EDS analysis of homogeneous limestone samples quickly reveals the presence of not only the major Ca peak but also, in some cases, significant peaks for Si, Al, K, and other elements. Three examples are shown in Figures 8, 9, and 10.

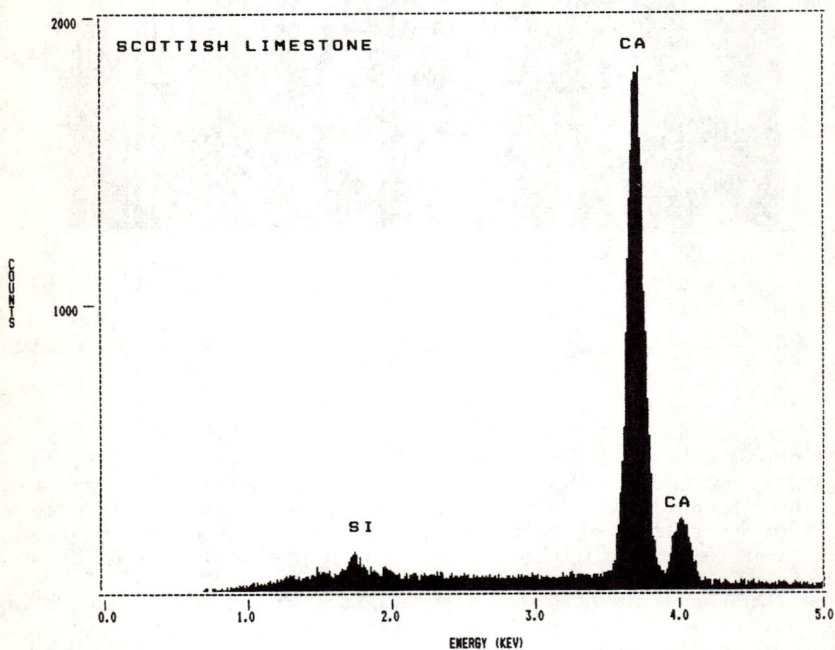

Figure 8. EDS spectrum of a fine-grained, homogeneous limestone, from Scotland, taken at 18 kV, 0.1 mA, and 20 seconds. The spectrum was recorded to higher energies but the only peaks identifiable are the major Ca and a trace of Si. (Sample furnished by Logitech, Ltd.). Polished thin section.

In some cases, the Si peak intensity is quite low and the only other peaks present are for Ca (Figure 8).

In many cases, in addition to the small Si peak and the major Ca peak, there is a small peak at the chlorine position (Figure 9). There is no indication with CL of any mineral phase except carbonate - the sample is uniformly luminescent. This suggests that the silica present is very fine-grained.

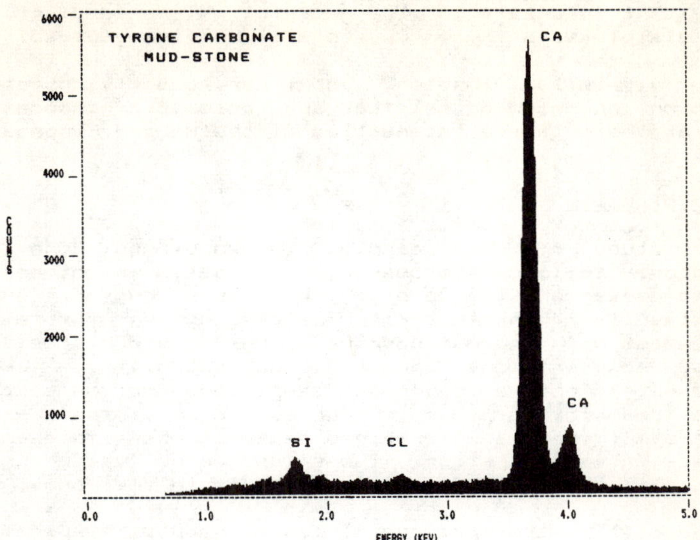

Figure 9. EDS spectrum of a carbonate mud-stone - core section, Tyrone limestone (middle Ordovician) Jessamine County, KY. (13). 18.5 kV, 0.04 mA, 20 seconds. Lightly polished thin section. (Sample furnished by T. Young.)

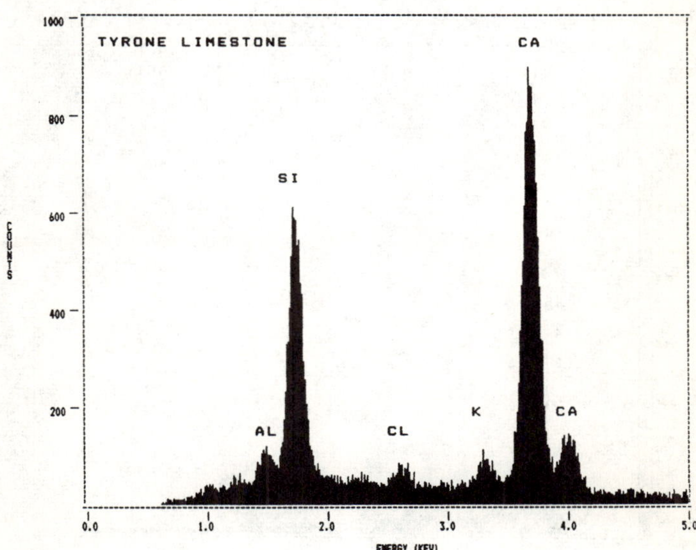

Figure 10. EDS spectrum of an area of pressure solution in the Tyrone limestone (13). Outcrop sample. EDS spectrum recorded at 18.5 kV, 0.02 mA, and 20 seconds. Lightly polished thin section. (Sample furnished by T. Young.)

The presence of additional peaks for Al and K (Figure 10) suggests that feldspars or clay material may be present. These are not visible with TrL or CL and must be fine-grained.

In a similar fashion, one can quickly obtain some information on high-Mg calcites and dolomitic composition by looking at the relative intensities of the Mg and Ca peaks.

Analysis of soils

The study of sand samples is sometimes done in the Luminoscope, the separate quartz, feldspar, and other mineral particles often showing up nicely with CL. However, one does not ordinarily think in terms of analyzing very fine-grained soil samples with the Luminoscope. When the EDS accessory is attached, however, one can study such materials., using the Luminoscope as both a convenient sample preparation system and electron beam column. Sample preparation is minimal - usually one can simply make a water-based paste of the sample and smear this paste on a glass slide. The slide can be 'vacuum-dried' by placing it in the Luminoscope and pumping in the normal manner!

Figure 11 is the spectrum of a soil sample, suspected to be contaminated. A few luminescing grains were noted, probably calcite, K-feldspar, and apatite. The EDS spectrum reveals Mg, Al, Si, P, K, Ca, Fe, a small Ti peak, and a small peak at about 2.3 keV which may represent lead contamination. The presence of sulfur interference cannot be ruled out without a more thorough analysis and application of peak deconvolution procedures. The point is that a qualitative screening of samples can be done quickly before more expensive analytical procedures are used.

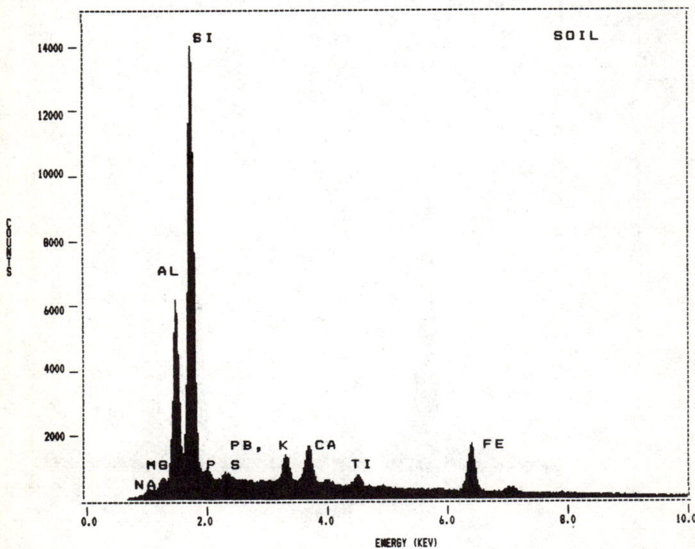

Figure 11. The EDS spectrum of a soil sample indicating the possible presence of small amounts of lead. Sample was in the form of an approximately 1 cm diameter disc of paste, dried on to a microscope slide. EDS spectrum at 18 kV, 0.2 mA, 100 seconds.

Streak plate samples

We have found that a convenient method of sampling some specimens for EDS analysis is to use a mineralogical 'streak plate'. A streak of the specimen is made in the same manner that one might do for mineralogical identification and then the streak is analyzed. The Luminoscope-EDS system will readily accept rather large pieces of insulating materials such as streak plates and the samples do not require a conductive coating - hence this technique for sampling works very well.

Figure 12 is an example of this application. A sample of sphalerite was streaked and the spectra obtained at 18 kV and 0.06 mA are shown. The Zn and S peaks are clearly revealed along with those peak characteristic of the streak plate (Figure 13). A small Fe impurity peak is present in the sphalerite sample. If desired, one can subtract the streak plate spectrum using resident software programs and also obtain some information on the Al and Si content, although not with the same certainty that applies to those peaks which are not in the streak plate spectrum.

Other advantages of the streak plate sampling method, besides rapidity and simplicity, include the fact that it is essentially non-destructive and that the streak plate can be brought to the sample. A streak can often be obtained with essentially no apparent change to the sample. Thus we have been able to make streaks for analysis of such items as jewelry and precious mineral specimens. Streak samples have been made from components of metal vacuum systems while they are in operation! We have also made streak plate analysis of marble blocks. In this case not only can we make the EDS analysis but we can also observe the CL color of the marble (although obviously without any information on fabric). It may be possible to sample precious art objects, e.g. marble statues, using this method.

Quantitative analysis

The question inevitably arises as to the quantitative analytical capabilities of the EDS system installed on the Luminoscope. Some limited work has been done in this area with metallurgical standards. These standards are nickel-based alloys and various other elements such as Ti, Co, Mn, Fe, Cr, etc. are present. The spectrum of such a sample is shown in Figure 14. A number of these standards were analyzed for their Ti content and the ratio of X-ray counts for Ti/Ni versus the concentration ratios of these elements is shown in Figure 15. The working curve anticipated for this type of data is not a straight line but should be a smooth regular curve. The experimental results confirm this over a concentration range for Ti from 0.5% to 8%.

Much more work is required in this area for geological samples but the technique certainly holds promise of providing reasonably quantitative data.

Summary

The EDS attachment for the Luminoscope is a relatively inexpensive, easy to use, attachment which can be installed and put into operation rather rapidly. It facilitates the study of many geological and other specimens because sample preparation is minimal and the pump-down time and data acquisition times are short. Its potential capabilities have already been demonstrated for a variety of different sample types.

ACKNOWLEDGMENTS

The phlogopite sample and the sample of the Fen pyrochlore were furnished by Dr. Anthony N. Mariano of Carlisle, Mass. The polished thin section of Scottish limestone was furnished by Logitech, Ltd. Timothy Young of Eastern Kentucky University furnished the two samples of the Tyrone limestone.

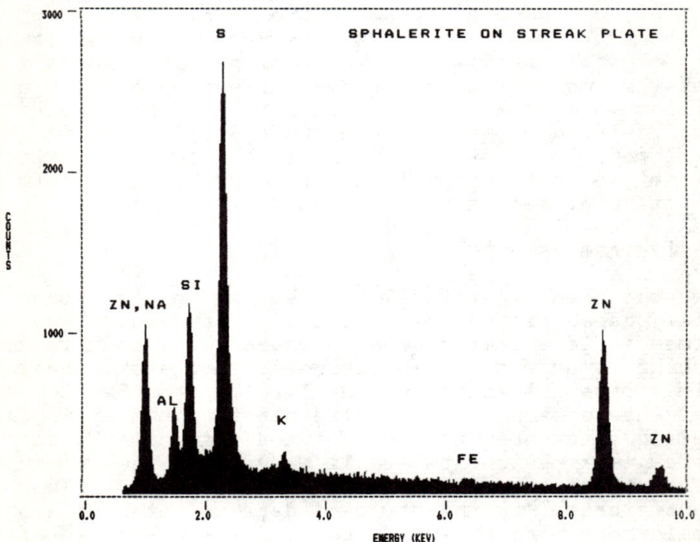

Figure 12. The EDS spectrum obtained from a streak of sphalerite on a streak plate. EDS conditions 18 kV, 0.06 mA, and 20 seconds. The peaks arising from the streak plate (See Figure 13) are Na, Si, Al, and K. The other peaks shown are the major elements, Zn and S, and a small Fe impurity peak.

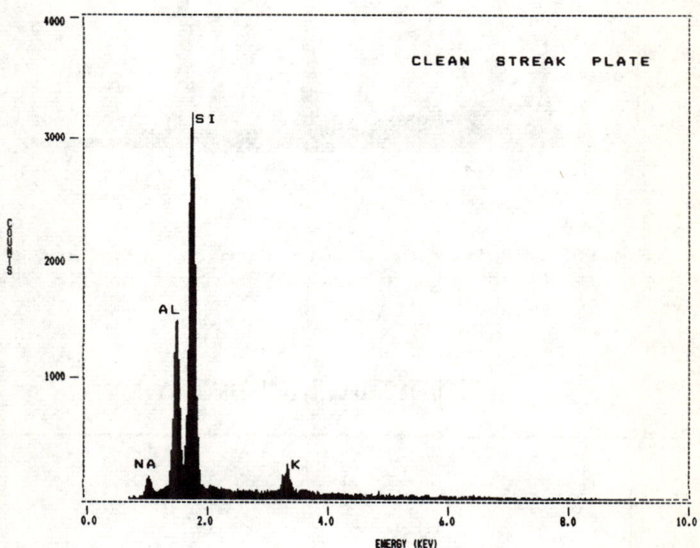

Figure 13. The EDS spectrum of a clean streak plate (Ward's 12E0290) taken at 16.5 kV, 0.2 mA, and 25 seconds.

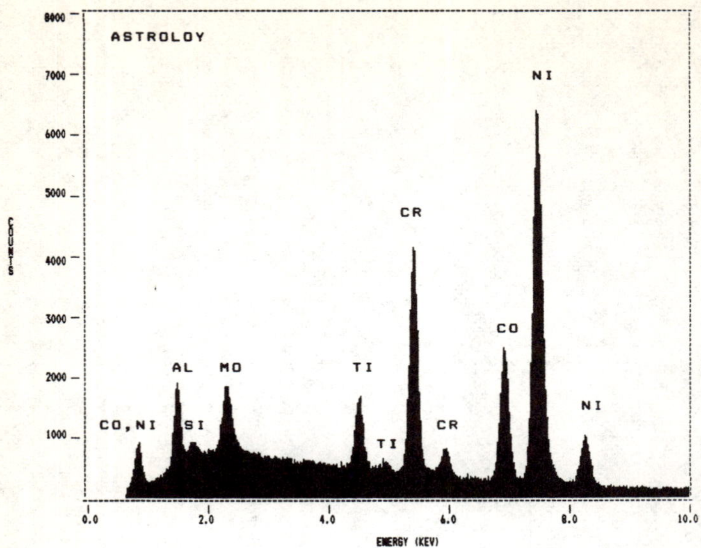

Figure 14. The EDS spectrum of a metallurgical standard taken on the Luminoscope. Sawed metal slab. EDS conditions 18 kV, 0.1 mA, and 100 seconds.

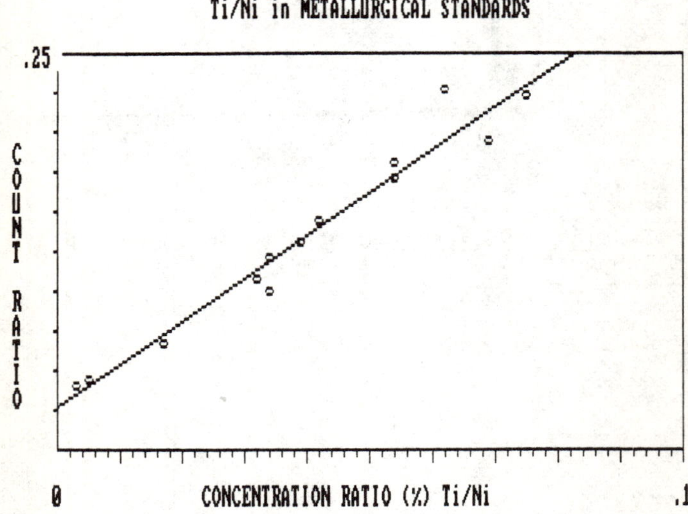

Figure 15. A plot of the count rate ratio versus the concentration ratio for Ti and Ni in 13 metallurgical standards. Ni varies from 41 to 77% in these standards and Ti varies from 0.2 to 4.2%. All spectra were taken at 16.5 keV, 0.1 mA, and 25 seconds.

References

1. Luminoscope is a Registered Trademark of Nuclide Corporation, State College, Pa.

2. W. M. Mueller and M. Fay, eds., Advances in X-Ray Analysis, Vol. 6, Proc. 11th Annual Conference on Applications of X-ray Analysis. August 1962, (New York, NY: Plenum Press, 1963), 276-90. J. V. P. Long

3. J. V. P. Long and S. O. Agrell, "The cathodoluminescence of minerals in thin section," Mineral. Mag. *34* (1965), 318-326.

4. R. C. Stenstrom and J. V. Smith, "Electron-excited luminescence as a petrologic tool," GSA Annual Meeting Abstr. page 39. 1963

5. J. V. Smith and R. C. Stenstrom, "Electron-excited luminescence as a petrologic tool," J. Geol. *73* (1965), 627-635.

6. R. F. Sippel, "Simple device for luminescence petrography," RSI *36* (1965), 1556-1558.

7. L. F. Herzog, D. J. Marshall and R. F. Babione 1970. The Luminoscope - a new instrument for studying the electron-stimulated luminescence of terrestrial, extra-terrestrial and synthetic materials under the microscope. In Space science applications of solid state luminescence phenomena, MRL Pub 70-101, J. N. Weber and E. White (eds.), 79-98. State College, Pa.: Penn. State U.

8. A. N. Mariano and P. J. Ring, "Europium-activated cathodoluminescence in minerals," Geochim. Cosmochim. Acta *39* (1975), 649-660.

9. O. C. Kopp, "Cathodoluminescence petrography a valuable tool for teaching and research," J. Geol. Ed. *29* (1981), 108-113.

10. E. Nickel, "The present status of cathode luminescence as a tool in sedimentology," Minerals Sci. Eng. *10* (1978), 73-100.

11. N. C. Barbi, Electron probe microanalysis using energy dispersive X-ray spectroscopy (Princeton, NJ: Princeton Gamma-Tech, Inc., 1981), 67 pages.

12. SYSTEM4plus is a product of Princeton Gamma-Tech, Inc.

13. T. J. Young, "The examination of fenestral structures and cements in the Tyrone Limestone (middle Ordovician) in central Kentucky" (M.S. thesis, Eastern Kentucky University, 1986).

A REVIEW OF THE TECHNOSYN AND NUCLIDE CATHODOLUMINESCENCE STAGES AND THEIR APPLICATION TO SEDIMENTARY GEOLOGY

Charles E. Barker[1] and Teresa Wood[2]

[1]U.S. Geological Survey, Denver, Colorado 80225 U.S.A. and
[2]S and M Microscopes Inc., Colorado Springs, Colorado 80915 U.S.A.

Abstract

Cathodoluminescence (CL) microscopy is now possible on two commercial stages made by Nuclide (model ELM 2a), and Technosyn (model 8200). The Technosyn uses a coincident electron beam and optical axis design that appears to have some advantage over the Nuclide design which uses manually positioned magnets to deflect the electron beam. The Technosyn vacuum system uses O-ring seals and fewer valves that may make it less susceptible to air leaks than the Nuclide, which uses custom-molded gaskets. Overall, the Technosyn design requires less attention by the microscopist, increasing its utility for cathodoluminescence petrography.

CL microscopy has wide application in geology, especially in sedimentary petrography and geochemistry. Significant applications also exist in economic geology, igneous-metamorphic petrology, and mineralogy. The applications of CL microscopy to sedimentary geology arise from visual information relating to geochemical variations in the pore water during mineral growth. This information on the relative sequence of geochemical changes in the system may not be easily derived by other methods. In sedimentary petrology, CL microscopy shows the fabric and distribution of luminescent minerals, as well as details of cement texture and in some cases, origin of highly altered framework grains. In sedimentary geochemistry, CL microscopy is used both to separate and correlate cement zones for analysis, allowing comparison of data from the same zone over distances up to several kilometers.

Introduction

Cathodoluminescence (CL) microscopy has become an essential tool in petrography and is especially useful in the description of sedimentary rocks (1). The technique is unique because it visually reveals textures and structures in sedimentary minerals and rocks resulting from geochemical variations in detrital grains and cements rather than variation in optical properties. This visual information is not easily obtainable by other methods. Two commercial CL stages, the Nuclide Luminoscope ELM-2A* (2) and the Technosyn model 8200 (3), are now available. We review the design, operation, and problems of CL stages to contrast their potential application in petrography.

Some petrographers are not convinced of the usefulness of CL microscopy. We think this is because most studies use CL microscopy as a supplementary tool and do not utilize its full potential. Our review emphasizes the specific uses of CL microscopy in sedimentary petrography to stress the unusual visual data that can be observed with this method.

Our treatment of CL theory and previous work in sedimentary rocks is not comprehensive and is intended only to supplement earlier reviews (1,4,5,6,7,8; and others). Rock and mineral characteristics under CL, and its use for interpretation of petrogenetic environment are reviewed in several papers (1,2,6,9,10,11,12; and others). A number of color plates of CL textures in rocks and their interpretation have been published (5,6,7,9,11,13,14,15,16). Bibliographies of CL microscopy have been compiled by Broecker and Pfefferkorn (17) and Barker and Wood (3).

Comparision of the Nuclide and Technosyn Stages

The designs of the Technosyn and Nuclide stages include the same basic components: a vacuum chamber with upper and lower windows enclosing an X-Y stage movement with an attached cold-cathode electron source. However, the stages use different methods used to direct the electron beam onto the sample (Fig. 1). The Technosyn stage uses an electron gun aimed at the sample that is centered in the microscope optical axis. The Nuclide stage uses deflection magnets for beam control. The Technosyn and the standard Nuclide stages will accept samples of similiar size (Table 1).

These stages have similar microscope requirements (for the Nuclide stage, see 2,7,18; for the Technosyn, see (3)). Both stages require microscopes with about 5-cm clearance between the objective and microscope stage mount. Most research-grade microscopes that do not have a fixed stage mount are suitable or can be adapted with minor modification. The microscope must be set up to optimize the viewing of the dim CL image through a large free-working-distance lens required by these stages. Light intensity transmitted to the operator during luminescence observations is proportional in part to the degree of magnification. The lower power objectives transmit a brighter image. The lowest power objective available should be used, provided the feature of interest can be seen in sufficient detail. The objective lens must have a minimum free-working distance of 8 mm for the Nuclide stage with the recessed window accessory and 9 mm for the

---

*The use of trade names in this report is for descriptive purposes only and does not constitute endorsement by the U.S. Geological Survey or S and M Microscopes, Incorporated. The complete addresses of the manufacturers cited in this report are listed in Appendix 1.

standard Technosyn stage. Selected objective lenses with the required working distance are listed in Table 2. The optimal microscope oculars are those with the lowest power that will allow effective resolution of details in the sample. For instance, the 8X Nikon ocular transmits 56 percent more light than the Nikon 10X. The highest possible clear magnification through a lens is its numerical aperture multiplied by 1000. The effective resolution of the optical system is determined by the numerical aperture of the objective lens. Therefore, trying to increase the magnification beyond this value by using higher power oculars provides a larger image but decreased resolution. The low-power objective lenses required by these CL stages tend to have small numerical aperatures and low effective-magnification levels. Using high-power oculars (over 15x) is not recommended with low numerical aperature lenses.

A long-working-distance substage condenser is essential to allow Kohler illumination of the sample that produces a high-quality transmitted-light observation and to allow photomicrography when the sample is in the CL stage. Photographs of the same sample view in both CL and transmitted light are very useful for interpretation of sample characteristics.

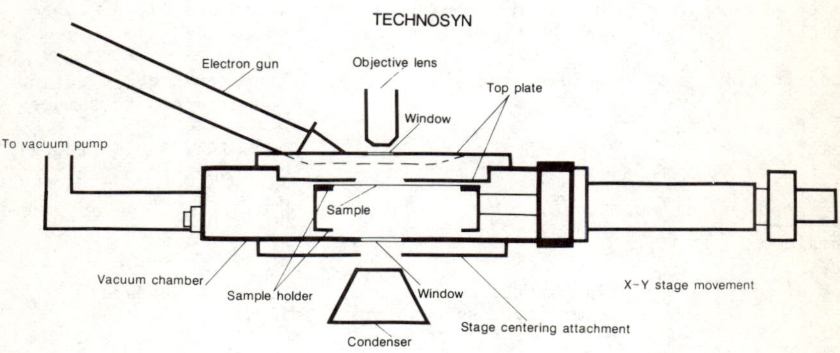

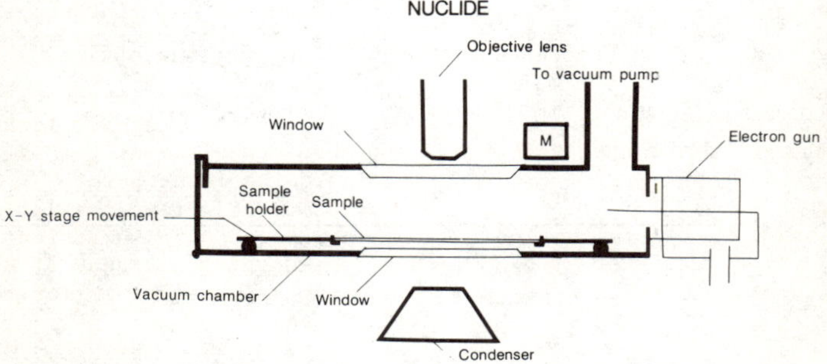

Figure 1. Schematic cross-section views of the cathodoluminescence stages Nuclide model ELM-2a and Technosyn model 8200.

Table I. Specifications of the Nuclide and Technosyn Stages

Specification	Nuclide Stage	Technosyn Stage
Accelerating voltage	3 to 18 kV (early model) to 30 kV (current model)	3 to 30 kV
Operating vacuum	0.025 to 0.050 Torr	0.04 to 0.07 Torr
Current	900 microamps	400 microamps
Current breaker	Manual reset	Automatic reset
System balancing	Manual	Manual or Automatic (optional)
Transmitted light to CL conversion	Push button	Manual
Conversion to high power objective	Replace window and add lead shield	No modification required; Rotate nosepiece.
Electron beam spot on sample	0.5-10 mm subcircular; can be focused	4 x 7 mm elipse; cannot be focused
Beam deflection	Sliding magnets	None required; electron gun directed to sample
Stage chamber Capacity	50 x 75 x 13 mm	50 x 70 x 17 mm (thick) 70 x 80 x 1 mm (thin)
X-Y travel of sample	44 x 70 mm	50 x 70 mm
Window diameter	48 mm (standard) 8 mm (recessed window)	6.5 mm (standard) 12 mm (large window)
Window thickness	3.5 mm 3 mm (recessed window)	3 mm 3 mm (large window)
Top window to sample distance	over 20 mm 8 mm (recessed window)	9 mm 30 mm (wide option)
Bottom window to sample distance	about 15 mm	about 25 mm
Vacuum seals	Custom gaskets	O-rings

Table II. Objective Lenses for Cathodoluminescence Microscopy

(As used by the U.S. Geological Survey)

Source	Lens[1]	Power(n.a.[2]) fwd[3]	Comments
Leitz	EF 4	4X(0.12) 24.0 mm	Inexpensive
Leitz	EF 10	10X(0.25) 8.0 mm	Inexpensive[4]
Leitz	L 25	25X(0.35) 13.5 mm	Long fwd lens
Nikon	M Plan 5	5X(0.10) 20.0 mm	210 mm tube length[5]
Nikon	M Plan 10	10X(0.25) 9.0 mm	
Nikon	M Plan 20	20X(0.40) 10.5 mm	Extra-long fwd
Zeiss	UD16	10X(0.11) 13.5 mm	Universal-stage lens[6]

Notes: (1) Manufacturers' identification code engraved on lens. (2) n.a. = numerical aperture. (3) fwd = free working distance. (4) 8-mm fwd objectives can be used if a glass slide is placed underneath the thin section. (5) All Nikon M-plan lenses are designed for microscopes with a mechanical tube length of 210 mm; and their use on 160 mm tube length microscopes reduces the rated magnification power and n.a. (6) A universal-stage lens, designed for use with glass hemispheres. These lenses have a poor image and lower than rated magnification and n.a. when used without the glass hemispheres (effective magnification and n.a. shown). Any low-power universal-stage objective lens usually can be used, but most lack the high n.a. necessary for good CL observation.

Design Problems

Nuclide stage. Design problems with the Nuclide stage are: (1) the use of custom-molded gaskets on the major ports in the chamber; (2) the use of external valves with associated connectors and seals in the vacuum system; (3) the extreme working distance required by the standard stage (20 mm); (4) the use of manually adjusted sliding magnets (and this is not an easy task) to deflect the electron beam to the sample and (5), an insensitive vacuum control. The use of specialized gasket material makes it necessary to send the parts back to Nuclide to be replaced or repaired. These custom-molded gaskets shrink with time, necessitating periodic rebuilding with new gaskets (cost: $500 in 1985). Further, the extensive use of external valves and associated connectors adds to the potential for vacuum leaks. Leaks often occur when inexperienced users overtighten a valve-packing nut or compression fitting to "eliminate" vacuum leaks. In fact, overtightening often creates a permanent leak. Kopp (19) suggests using the main-vacuum control valve to control stage pressure while the controlled leak valve is closed. He states that this procedure protects the controlled-leak valve against damage and is especially useful for inexperienced users.

The external controls for the X-Y mechanical stage inside the Nuclide chamber have a design flaw that produces large vacuum leaks. If the X or Y control knob is turned too far, the stage reaches a stop and pushes the knob shaft out of its seal causing a rapid vacuum loss in the stage. This knob can easily be turned too far in the darkness during CL microscopy. Our experience is that a common reason the Nuclide stage will not pump down to operating vacuum is because this shaft has popped out during sample changes.

The free-working distance required by the standard Nuclide stage is extremely large (over 20 mm). A 5X objective is the maximum power usable without using the recessed window accessory. This configuration can be modified to a 10X standard objective, or to an even higher power, by using universal stage lenses with an elevated sample mount and extra external deflection magnets (20).

Technosyn stage. Design problems in the Technosyn stage are: (1) a sample tray and X-Y stage that slide out; (2) an adjust kV control that apparently cannot be used to control kV; (3) an insensitive vacuum control; (4) a small upper chamber window; and (5) the excessively long free-working distance substage condenser required for Kohler illumination of the sample.

In practice, almost everyone who uses the stage changes samples by removing the top cover. The sliding tray has little utility and is another source of potential air leaks.

The adjust kV control on the Technosyn stage appears to exert little control over kV levels because the electonic system apparently sets the kV level automatically, according to the vacuum within the stage. Thus, the accelerating potential of the electron gun is actually controlled by the vacuum-control knob.

The Technosyn design requires that the upper chamber window must be small (6.5 mm) to allow the geometry of the stage to accommodate a relatively short working-distance lens (9 mm). The optional wide-window (12 mm diameter) stage cover moves the electron gun away from the optical axis and requires an extremely long free-working-distance lens (3 cm). This free-working distance effectively restricts the use of the wide-window option to stereo microscopes. The small window offers little disadvantage when looking down the microscope because the field of view is limited. The major problem comes when trying to position the beam on a point on a sample, or when trying to move to another sample in the chamber. The view into the chamber is blocked when the objective is in place. The Nuclide stage has an enormous window (48 mm) compared to the Technosyn stage; finding a point on the sample is easier with the Nuclide stage because the view is not blocked as much by the objective.

As described above, a long-working-distance substage condenser allows Kohler illumination of the sample in the CL stage and permits high-quality transmitted-light observation and photomicrography. Kohler illumination requires that the free-working distance of the condenser should be sufficient to focus a condensed beam of light on the upper plane of the sample. The shorter distance between the bottom window of the stage and the upper sample surface in the Nuclide stage makes this possible. The longer free-working distance required by the Technosyn stage prevents available substage condensers from sharply focusing the closed condenser diaphram image in the plane of the sample.

Design Advantages

Helium injection into the controlled leak-valve is now a standard feature on the Nuclide stage. No such accommodation exists on the Technosyn stage. Helium injection into the stage may be important because the electron beam can be stabilized at higher vacuum levels (important if the stage is not sealing properly) and because some operators report increased CL intensity when using helium instead of air.

The operation of the Technosyn stage is streamlined compared to the Nuclide stage because: (1) no electron-beam-deflection magnets are needed to force the beam to strike the sample or to center it in the microscope optical path; (2) the electron beam is wider and brighter, making focusing unnecessary; (3) the sample is close enough to the stage window that a high-power, long-working-distance objective lens can be used without conversion to a recessed window; (4) vacuum seals are O-rings that are easy to change if damaged; and (5) the top cover is easily removable for sample change or routine cleaning of the upper window.

A difference in design between the Nuclide and Technosyn stages is also expressed in maintenance problems. Maintaining a CL system routinely requires that: (1) the gaskets be kept clean and lightly lubricated with vacuum grease; (2) volatile substances be removed from the chamber surface (fingerprints, vaporized plastics, etc.); and (3) the chamber window be cleaned. After each day's use, the chamber window should be cleaned of the material that condenses on its surface with isopropyl or ethyl alcohol. It should then be rinsed with freon solvent (Miller-Stephenson brand MS-180). Cleaning the chamber window is simpler on the Technosyn stage because the top cover is removable. The top cover O-ring seal can also easily be removed or cleaned at this time. The Nuclide stage observation window is easy to remove, but the gasket molded to it is easily damaged during cleaning or reinsertion in the stage. The Nuclide window, once disturbed, also requires longer pumpdown time to reseat the gasket.

Disadvantages common to both stages

These stages have a common disadvantage in that they both have an insensitive vacuum control. This vacuum is set by the "Controlled Leak Valve" on the Nuclide stage, and by the "Vac Control" on the Technosyn. These valves control the entry of gas into the system and allow control of vacuum in the stage (termed "balancing the system"). The valve for this function on both stages is, at best, a crude control. The valve can be turned very far without much change. Then, a sudden rapid change occurs, often causing the system to turn off the electron beam. Another annoying problem is that slightly turning the leak valve when the system is nearly balanced often produces vacuum change out of proportion to the valve movement (due to wear in the valve?). This disproportionate change is often irreversible in that returning the valve to the original position does not restore the system to the original vacuum. The Nuclide controlled-leak valves are also particularly susceptible to damage from overtightening of the valve stem (19).

Many of the design problems with the Nuclide stage have become apparent only after years of use. Time is required to assess whether the apparent design improvements in the Technosyn stage will be advantageous. The decision to purchase a CL stage is a difficult one, centering around the needs of the laboratory and the type of samples available.

In the vacuum chamber of the CL stage, the electron beam is absorbed by the surface it hits. Much of the energy of the incident electron beam is absorbed by the specimen molecules, causing an increase in the energy levels of absorbing atoms. Normally, the excited atoms (also termed cathodoluminescent centers) return to the ground state by transfer of the excess energy to adjacent atoms by inelastic collisions. Under certain circumstances, the absorbed energy is re-emitted as visible light energy before these collisions can take place. Electrons generated in a CL stage are readily absorbed in the sample, penetrating about 1-2 microns (21); consequently little luminescence is emitted from below the surface. The intensity of light emitting from any particular point will be primarily proportional to the surface density of luminescent centers. The conditions for luminescence often occur in impure crystalline substances in which the impurities act as luminescent centers.

Transition metals and the rare earth elements are particularily susceptible to electron-beam excitation. CL can be enhanced (sensitizer element), inhibited (quencher element), or caused (activator element) by their presence in a mineral. Luminescent centers are also caused by radiation damage, crystal defects, and by sometimes still-unidentified sources (22). Coy-Yll (23) found that the luminescent intensity (L) is directly proportional to excitation voltage (V, included in the factor k), current density (i), and factors (k, m) that are dependent on the sample material. For a given material, the Coy-Yll equation ($L = ki^m$) predicts that luminescence is a power function of the voltage and current density ($L = f(Vi)$). Luminescence increases directly with an increase in voltage and exponentially with increase in current density. This equation predicts that the most-effective way to increase CL intensity is by increasing current density. In practice, however, heating of the sample produced by the increased current density makes it more useful to increase CL intensity by increasing voltage at a relatively low current density.

Heating of the sample by the electron beam can decrease CL by thermal quenching. Thermal quenching is produced by increasing the thermal energy in the sample, causing an enhanced environment for transition of excited electrons without the production of luminescence (1). The temperature (T) under the CL spot on the sample is directly proportional to the current density (i) and accelerating voltage (V) and is inversely proportional to the beam radius (r) and thermal conductivity (C) ($T=f(Vi/Cr)$ (Castaing in (23)). Focusing the beam (only possible on the Nuclide stage) and decreasing its radius under constant electron beam energy initially causes an increase in luminescence. However, a point is reached where further decreased of beam radius (or any further increase in the beam energy) causes thermal quenching and a loss of CL intensity (Fig. 2). Carbonate minerals appear to be more susceptible to the apparent effect of thermal quenching than are silicates (Fig. 3). Quartz and Feldspar do not show significant thermal quenching in the beam energies produced within the operating limits of present CL stages. The often-dim CL of quartz needs to be enhanced, and this is probably possible by using higher beam energies than the present 30 kV maximum on commercial CL stages that use a cold-cathode electron gun.

A great disadvantage of CL microscopy is the potential for heating and ultimately destroying the sample. Temperature-sensitive materials or features that are easily altered by heat should not be observed by CL if further analysis is planned. Sample heating by the electron beam can break a thin section and can cause hydrated minerals to lose water and (or) to decrepitate. A more subtle problem with heating is the alteration of low-temperature fluid inclusions in sedimentary rocks. If fluid inclusions are heated beyond their original temperature of formation, they may change (stretch), and the record of the initial conditons under which they formed can be lost. These overheated fluid inclusions all tend to change in a similar way and the difference may not be detected. Another effect of heating by the electron beam is pyrolysis of organic matter. Pyrolysis induced by the electron beam increases the pumpdown time required to reach operating vacuum levels. Materials rich in organic matter are highly altered by the electron beam and should be examined by fluorescence petrography rather than CL to avoid these problems. Pyrolysis byproducts can also coat the vacuum chamber and window degrading both the optical quality and the vacuum system.

Sample Preparation

Sample mounts for CL microscopy should use a cement that is stable at the moderate temperatures generated in the sample. Heating and outgassing from the cement can coat the stage window, delay vacuum stabilization, and extend pumpdown time. Epoxide or epoxy resins (Buehler, Ltd., Epoxy Technology EPO-TEK 301) are satisfactory. A cover glass is not used because it would absorb the electron beam before reaching the sample.

Optimal resolution of sample detail in a CL stage is produced by using high-quality polished surfaces, because the electron beam is absorbed at the surface. A highly polished sample surface allows interpretation of the luminescent pattern resulting from the material itself rather than from its surface roughness. The current trend in polishing techniques is to use a simple method that requires only a few steps, yet produces a low-relief, high-quality polished surface (24,25,26). After sample preparation, lubricants, water, fingerprint oils, and other volatiles must be removed from the sample. An effective sample cleaner and dehydrator is a Freon solvent (Miller-Stephenson).

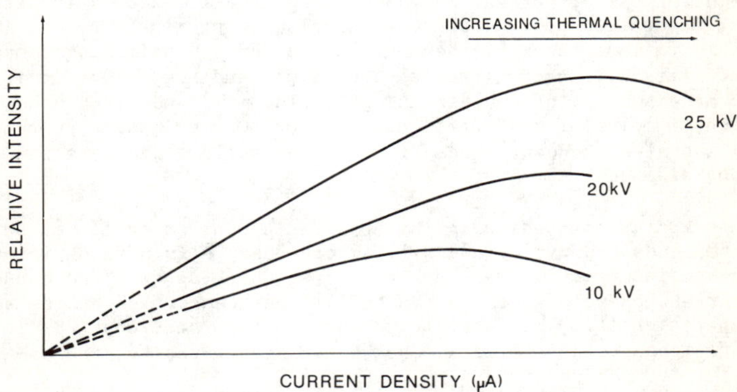

Figure 2. Hypothetical thermal quenching effect on cathodoluminescence intensity with increasing current density at constant accelerating voltages. Modified from Nickel (1979).

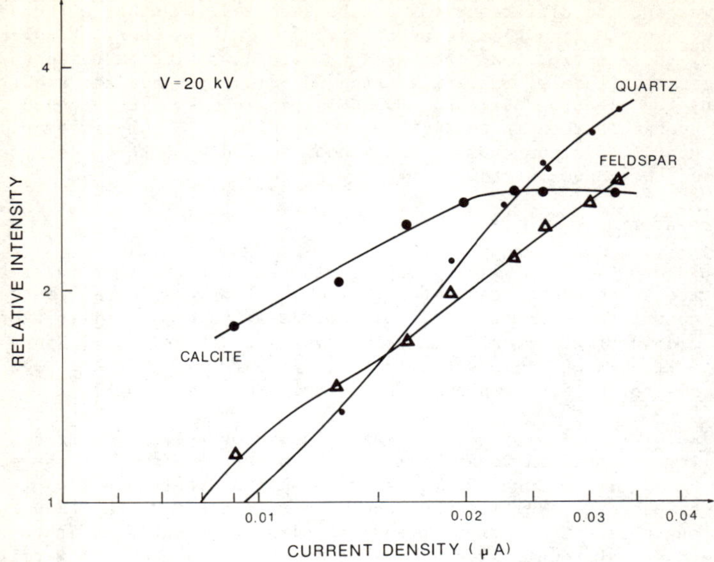

Figure 3. Cathodoluminescence intensity of common carbonate and silicate minerals with increasing current density at constant accelerating voltage. Modified from Amieux (6).

Sources of Spurious Cathdoluminescence

In addition to rocks and minerals that have CL properties, several manufactured compounds and processed natural materials exist that can interfere with intrinsic CL of the sample. A common contaminant in geological samples is the abrasive material used to polish them. Diamond has a bright CL that is often seen as glowing specks imbedded in softer materials, fractures, and voids in the sample. Contamination is recognizable in some cases by the uniform size of the polishing compound (note what sizes were used to polish the sample and compare). It is also useful to make a mount of the diamond polishing compound being used, to observe the CL color of that particular brand. Our polishing compound (Buehler) contains abundant aquamarine, common yellow, and rare red luminescent diamonds.

Some organic compounds also luminesce. Careful selection of sample bonding compounds (epoxy, cyanoacrylate, etc.) can eliminate CL interference from thin-section cement. Many human tissues, such as hair and dander, also have bright CL. Most of these products fall on the surface of the sample after preparation (they may also found under a thin section) and consequently can be identified as such because they rest on the sample surfaces.

Reflection of light from nearby CL sources such as fractures, cleavages, or other planar surfaces can give the impression that the surface itself is luminescing. This can be deceptive, but careful inspection of the sample or moving the suspect area to the edge of the electron beam can be used to detect this source of spurious illumination.

Differences in CL color between stages. Several investigators using both the Nuclide, Technosyn, and (or) a custom stage have noted a difference in CL color between stages when observing the same sample. The color difference between the Nuclide and Technosyn stages may be due to the methods used to direct the beam onto the sample. Because the Technosyn beam is oriented directly on the sample, ions as well as electrons can hit the sample (Technosyn Ltd., 1986, personal communication). The Nuclide stage design apparently causes a large-portion of electrons to strike the sample. Cathode-tube design, efficiency of the apparatus, discoloration of the chamber window, and the light-transmission character of the microscope may also be factors in determining the observed CL color (27).

Photomicrography

Photomicrography is an essential but difficult aspect of CL microscopy. Experience with a particular sample suite is perhaps the best guide. Nickel (1) and Mariano (28) provide basic data on CL photomicrography. Since those publications, new films have been introduced that offer higher speed without significant loss of resolution. We have found these films to be useful:

1) Kodak Ektachrome P800/1600 (for color slides);

2) Kodak Kodacolor 400 (for color prints);

3) Kodak 5294 or equivalent (for both color slides and negatives).

Microscope photometers can yield spurious results when setting exposure for CL photomicrography. Area-averaging meters can set the exposure to long, and "wash out" the bright CL images especially when the luminescing minerals are small relative to the field of view. Spot photometers are superior because they set exposure for the specific area of interest. Correction of area-averaging photometers is possible by setting the ASA on the meter far higher than the rating of the film (or, alternately, by setting the shutter speed much faster or aperature much smaller than indicated). Remember that the beam may not remain stable over long periods of exposure, and a camera that is properly adjusted for the light available at one instant may be wrong the next. Photographs of any crucial image, taken over a range of exposures, is an important safeguard.

Correct color processing can distort the true color as perceived by the microscopist. We have found it useful to send to the processing laboratory a photograph that has color qualities close to the actual CL view. Such a photograph allows the color analyzer to be set properly for realistic color reproduction.

Applied Cathodoluminescence Microscopy

Mineral Identification

CL is useful in petrography because minerals may have diagnostic properties when alternately observed in both transmitted light and CL. CL petrography may allow identification of small mineral grains that are beyond the effective resolution of the optical microscope. However, minerals usually cannot be identified based on CL alone. Mineral CL colors are not unique, and can change with their origin (1,29,27). The lack of distinctive character and overlap of properties for minerals may also result from a paucity of detailed published descriptions of CL properties, and from the fact that descriptions are made at much different operating conditions (30). This problem has led to the proposal of standards for reporting CL data

(31), although such stringent standards may not be practical considering the instability of CL systems (27). Without diagnostic CL data on minerals, the general approach to mineral characterization is to use conventional petrographic procedures to identify the mineral that shows CL, and to assume that other materials with similar CL character are the same mineral *in that sample*.

Modal analysis can be made more accurate and rapid by using CL microscopy, especially in the identification of minerals that have similar optical properties in transmitted light such as potassium feldspar, plagioclase, and quartz (7). A point-counting accessory is available for the Nuclide stage. This technique can be used to evaluate heavy minerals in beach sands (32), to assess the degree of contamination in mineral concentrates produced from separation techniques (33), and to deduce the minerals present in fine-grained matrix that cannot be resolved by transmitted light microscopy (34).

Internal Textures of Minerals

CL microscopy reveals internal textures in minerals based on subtle differences in activator-quencher-sensitizer content. These textures are interpreted as intergrowths of minerals, growth zones that have different CL properties (7; and others), healed fractures (35), and radiation damage halos (36). Thin, almost sub-microscopic, grains of transparent minerals that are luminescent (such as apatite) can be found within enclosing transparent minerals. In carbonate rocks, remnant textures revealed by CL microscopy can be used to suggest the origin of the precusor carbonate in neomorphic spar and sparry cement.

CL microscopy often discloses light and dark banding within a crystal that is interpreted as caused by geochemical changes resulting in a growth-zoned crystal. These zones represent a record of crystal growth that is useful in interpreting the history of diagenesis by analyzing fluid inclusions, isotopes, etc. Each zone that has uniform CL also apparently represents relatively stable geochemical conditions in the system at that time of crystal growth (37). For example, CL has been used to test for homogeneity of a particular cement zone as an aid in interpreting isotope data (38,39). CL microscopy, however, can also show that apparently homogeneous crystals in transmitted light have actually undergone episodes of nondeposition, and (or) corrosion (7).

Distribution of Minerals

The distribution of minerals that show CL in a sample is immediately evident under CL. For instance, heavy minerals concentrated on bedding planes or patchy calcite cementation in siliciclastic rocks are readily observed in the CL stage.

Sediment Source

The distinctive CL characteristics of rocks that may have contributed sediment to a basin may be preserved in the clastic material after deposition. In a fine-grained sediment matrix, minerals that lack distinguishing optical characteristics in transmitted light may be found by CL. The heavy-mineral suites can be readily identified by CL because these minerals may have distinctive CL properties (34). Brightly glowing heavy minerals are easily picked out in the thin section.

Variation of CL by grains of the same mineral in a sedimentary rock suggests that these phases originated from different petrogenetic environments (for example, see 29), or under different conditions over time within the same deposit. These problems requires that sediment source-rock studies generally must establish a regional standard for the CL properties of materials in the area of interest (1).

Discrimination of Authigenic and Detrital Minerals

Authigenic minerals formed at low temperature characteristically do not luminesce (40; and others). Detrital minerals are often luminescent because they are derived from physical weathering of relatively high-temperature igneous, metamorphic, and thermally-mature sedimentary rocks.

Cement Stratigraphy

Meyers (41), using CL microscopy, demonstrated that cement zones in limestone have a uniform stratigraphy that can be traced over considerable distances. Similar results were found in a Mississippi-Valley-type ore deposit (42). This observation can be used to trace a specific cement zone across an area to measure a specific parameter, such as pore fluid composition, temperature, or pressure, within the limited time span represented by a thin cement band.

Fluid Inclusion Petrography

CL can provide evidence on the origin of fluid inclusions. Primary or pseudosecondary fluid inclusions result from fluid trapped in crystal defects during growth. Fluid inclusions related to crystal growth typically are large and occur randomly. However, textural evidence of the processes of primary fluid-inclusion trapping may not be clearly expressed in the rock, and their origin may be obscure. The evidence is complicated by the locally common corrosion of crystals, a process that can introduce fluid inclusions that appear primary but exist in a cement that actually formed earlier. Direct evidence of the primary nature of fluid inclusions is rare in sedimentary rocks, and CL microscopy can aid in their identification. Cathodoluminesence can be used to document the observation that some planes of apparent secondary inclusions are actually primary fluid inclusions occurring along geochemical zones in the cement (36,37). Rowan (46, also see 47) used CL microscopy to compare fluid inclusions that formed in the same cement zone across a wide area. This technique allows selection of fluid inclusions that are related to a specific diagenetic event, such as ore-mineral deposition in limestones.

The use of CL to identify detrital grains that fractured and healed can also aid in determining the time of formation of the fluid inclusion. Sippel (48; also see 33) found that quartz of an apparently uniform origin in transmitted light was shown by CL microscopy to have been actually formed by complex fracturing and healing of secondary quartz. This process produces an intimate association of fluid inclusions of potentially much different origin that may be inseparable in transmitted light.

We emphasize that any combination of CL and fluid inclusion petrography should be used with caution because of the potential for heating and altering the fluid inclusions. The homogenization temperature should be measured first, and then the inclusion can be observed in CL to add evidence as to its origin. Serial thin sections can also be used to identify the general origin of fluid inclusions in a cement and the adjacent thin section analyzed on the heating/cooling stage.

Compaction and Cementation

In petrography, the degree of compaction in a rock is determined by the character and proportion of grain contacts. CL microscopy allows the original grain boundary (often luminescent) to be discriminated after cement (usually nonluminescent) has apparently obscured the boundary when viewed in transmitted light. Thus, the original texture of a orthoquartzite can be reconstructed by CL microscopy (9). Sippel (46; also see 49) was also able to discriminate between pressure solution of grains and secondary quartz cementation of the grain boundary.

Calcite cements in geothermal systems are often complex, because they are formed in several different generations. Barker (50) found a low-temperature (nonluminescent or dim CL) carbonate cement was followed by a high temperature carbonate cement that partially replaced the early cement. Both of these cements were clear in transmitted light. CL microscopy revealed that the later cement often replaced the early cement up to several thicknesses of the fracture away from the vein as indicated by the change of CL characteristics. This widespread replacement of calcite was expressed in isotope measurements where the isotope compostion formed a series between the end members of unaltered rock and the high-temperature calcite of the vein itself.

Structure in Fossils Replaced by Neomorphic Spar

The fine interior septa of some fossils can be observed in the neomorphic spar by CL (and also fluorescence). Growth zones in fossil fish teeth can also be resolved by CL microscopy (51). The different carbonate minerals in a fossil can be observed by their CL color (52). Thus, the aragonitic muscle pad of an unaltered mollusk shells can be easily observed in shells otherwise constructed of calcite.

Mobility of Ions after Chemical Zone Formation

One of the most striking aspects of CL petrography is the often-sharp CL zones that form complex geometric patterns in some CL-active minerals, especially in sparry calcite. The sharpness of these zones in cements that have formed as early as the Paleozoic suggests that ions activating the CL are not significantly mobilized within the crystal at moderate burial temperatures.

Cement Chemical Composition and Pore Water Chemistry

CL variation within a single crystal is an indicator of a systematic variation of activator-quencher-sensitizer content that is not observed routinely in transmitted light. This variation can give useful information even if the activator-quencher-sensitizer identity is unknown. For example, zones of homogeneous CL character have been interpreted as a result of cement precipitation from fluids of relatively uniform composition (43). They also suggest that sharply defined CL zones indicate that the pore water chemistry changed abruptly or, conversely, that a gradual change in CL character across a zone indicated the pore water changed slowly over time. Carpenter (43) postulated that four different environments of calcite deposition can be defined based on pH, Eh, and the total dissolved sulfur content of the pore water (Fig. 4). Each of these environments would produce a particular mineral assemblage and a calcite with either no, dull, or bright CL. Carpenter's hypothesis is that the environment of the pore water can be interpreted from the diagenetic mineral assemblage and from the CL properties of the associated calcite. However, his technique can lead to ambiguous results in some cases (43).

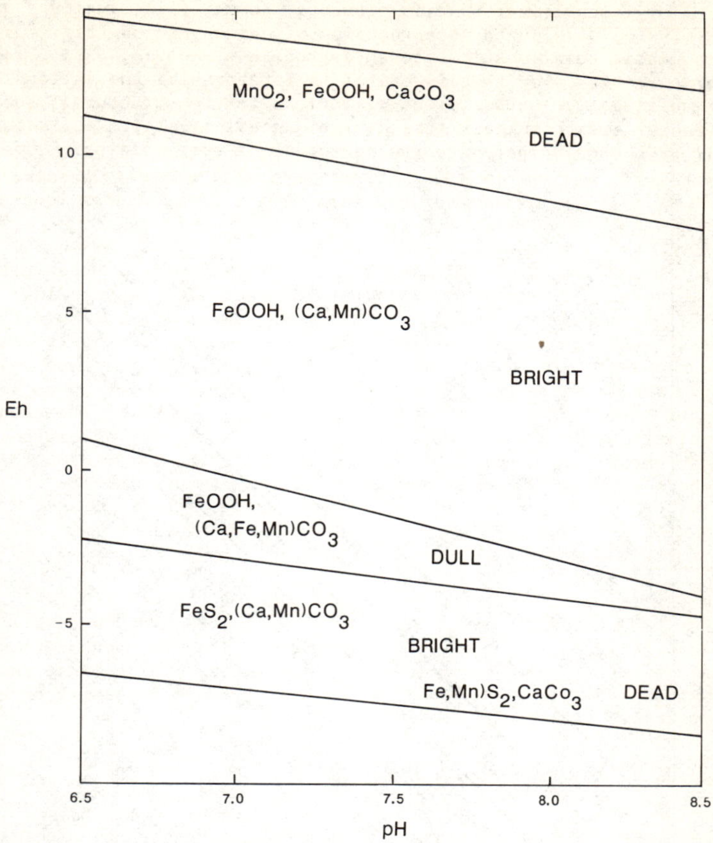

Figure 4. Theoretical mineral assemblage, carbonate composition, and the character of the calcite cathodoluminescence as a function of pore water pH and Eh. Modified from Frank et al. (43).

Sippel and Glover (44) first proposed that growth bands of different CL character result from changes in the pore-water chemistry. Frank et al. (43), using a combination of CL microscopy and microprobe analysis found that the ratio of activator content (Mn) to (Fe) quencher content controls the CL of carbonate cements (Fig. 5). This activator-quencher variation can be traced to variable pore-water-chemistry conditions. However, other factors may be operative in carbonates, and the Mn/Fe ratio may not be a simple, direct function of CL character as Figure 5 suggests (27,45,54).

Geochemistry of Minerals from CL Color

Quartz commonly luminesces various shades of blue and red; Sprunt (12) has shown this color is significantly controlled by impurity content (Fig. 6). Quartz with high titanium content relative to iron content luminesces blue, and quartz with low titanium content relative to iron content luminesces red. Frank et al. (1943) assert that the cathodoluminescence character in calcite is related to the ratio of activator (Mn) to quencher (Fe), rather than to the absolute concentration of either cation (Fig. 5). Thus, extremely iron-rich calcite may still be cathodoluminescent if sufficent Mn or a sensitizer (54) is present. Sommer (53) studied carbonate-emission wavelength as a function of cation radius (Fig. 7). He found a systematic relationship between carbonate structure, cation radius, and the perceived CL color that is useful in interpreting the various shades of color resulting from substitution of cations in the structure. For instance, Sommer's data suggest that a high-magnesium calcite would have an orange-yellow CL when compared to the yellow CL of pure calcite.

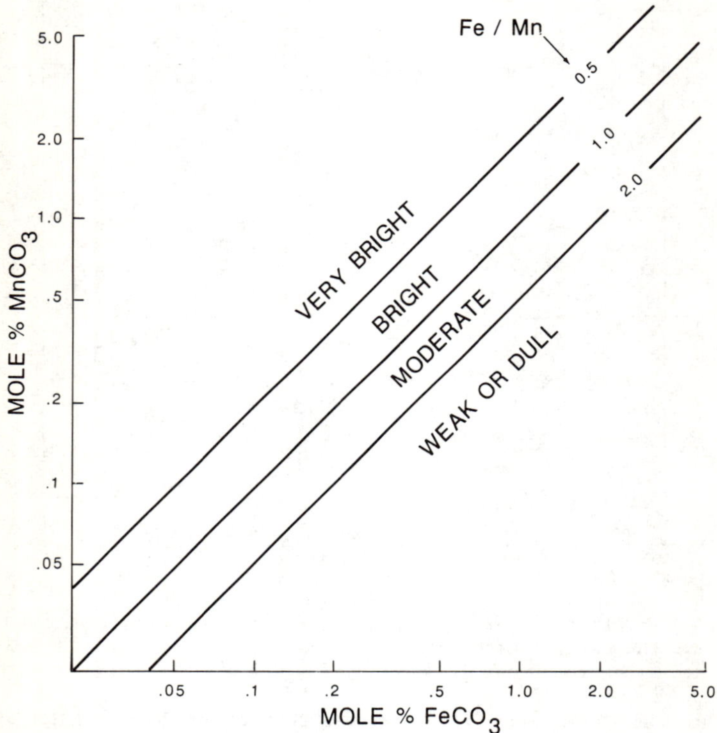

Figure 5. The variation of cathodoluminescence character in calcite with manganese and iron content. Modified from Frank et al. (43).

Discussion

Application of CL microscopy to petrography is useful in the description and interpretation of diagenesis, cementation history, framework grain mineralogy, rock fabric and texture. The utility of this aspect of CL microscopy is that it uses standard thin sections, easily performed, and is nondestructive (except for the possibility of overheating). The greatest problem in CL petrography is the quality of the image transmitted to the microscopist through the stage and the long-working distance lenses, and the dim CL produced by some minerals.

The interpretation of carbonate geochemistry and diagenesis from CL has proven to be more complex than initially thought. Initial attempts at using CL to interpret the geochemistry of pore fluids present during cement precipitation or neomorphism considered that only Mn^{2+} and Fe^{2+} were the only activator and quencher present (54). Gies (55) and others, have shown that a number ot trace elements in carbonate are involved in CL. The main activators in calcite and dolomite are Mn^{2+} and Pb^{2+} and some rare earth elements. The common sensitizers for Mn^{2+} activated CL in calcite and dolomite are Ce^{2+} and Pb^{2+}. The common quenchers are Fe^{2+}, Ni^{2+}, and Co^{2+} in these carbonates (54). The reported ratios of Mn^{2+} to Fe^{2+} necessary for CL activation in clacite and dolomite also vary widely (27,43,45) but these studies did not measure other trace elements that may be involved in CL activation. CL activation still probably depends mainly on Mn^{2+} to Fe^{2+} ratio in carbonates (54) as discussed above.

The possibility of activator-quencher-sensitizer influencing CL requires some assessment of trace elements in carbonate before interpretation of the geochemistry of the ambient fluid present during mineral precipitation. CL spectroscopy analysis is used to identify some activators-quenchers-sensitizers by characteristic emission peaks or bands (54, 56). Also, microprobe analyses for Mn and Fe in light and dark CL zones can suggest whether these elements are controlling CL emissions (57). Dark and light CL zones that do not correspond well with changes in Mn^{2+} to Fe^{2+} ratio suggest control of CL by other elements, and indicates that interpretation of the ambient pore water geochemistry should be approached with caution.

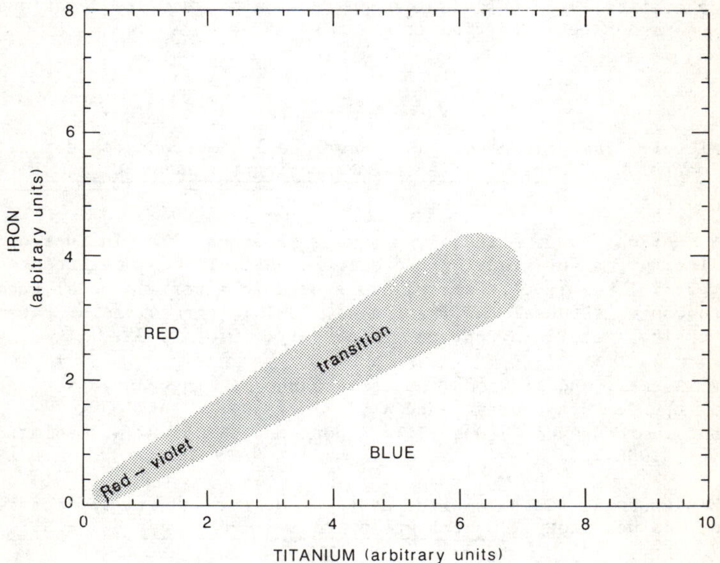

Figure 6. Titanium and iron content in quartz and the resultant cathodoluminescence color. Modified from Sprunt (12).

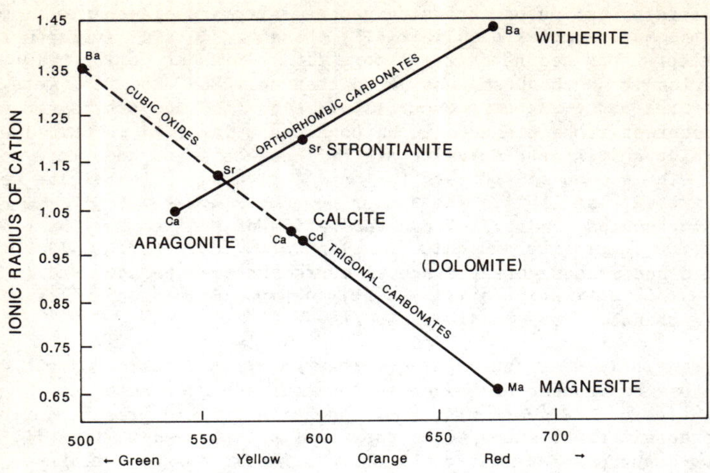

Figure 7. Ionic radius of the cation versus emission wavelength and perceived cathodoluminescence color. Dolomite is a double carbonate of calcium and magnesium that can be modelled as alternating layers of calcite and magnesite. Its cathodoluminescence color results from Mn emissions from both the calcite and magnesite layers. The resultant composite color is plotted schematically on trigonal carbonate line to show its relationship to calcite and magnesite. Modified from Sommer (52).

References

1. E. Nickel, "The present status of cathode luminescence as a tool in sedimentology," Mineral Science Engineering (Johannesburg), 10 (1979) 73-100.

2. L. F. Herzog, D. J. Marshall, and R. F. Babione, "The luminoscope--a new instrument for studying the electron-stimulated luminescence of terrestrial, extra-terrestrial and synthetic materials under the microscope," (Special Publication no. 70-101, Pennsylvania State University, Materials Research Laboratory, 1970) 79-98.

3. C. E. Barker, and T. Wood, "Notes on Cathodoluminescence microscopy using the Technosyn stage, and a bibliography of applied cathodoluminescence," (Open File Report no. 86-85, U. S. Geological Survey, 1986).

4. U. Zinkernagel, "Cathodoluminescence of quartz and its application to sandstone petrology," Contributions to Sedimentology, 8 (1978).

5. D. K. Richter, and U. Zinkernagel, "Use of cathodoluminescence in carbonate petrography," Geology Rundschdau, 70 (1981) 1276-1302. (in German).

6. P. Amieux, "Cathodoluminescence: method of sedimentological study in carbonates," Bulletin des Centres de Recherches Exploration-Production Elf-Aquitaine, 6 (1982) 437-483.

7. O. C. Kopp, "Cathodoluminescence petrography: a valuable tool for teaching and research," Journal of Geological Education, 29 (1981) 108-113.

8. R. D. Hagni, "Cathodoluminescence microscopy applied to mineral exploration and benefication," W. C. Park et al., eds., Applied Mineralogy (New York: A.I.M.E., 1985) 41-66.

9. J. V. Smith, and R. C. Stenstrom, "Electron-excited luminescence as a petrologic tool," Journal of Geology, 73 (1965) 627-635.

10. R. F. Sippel, "Geologic applications of cathodoluminescence," Proceedings, International Conference on Luminescence, (Budapest, Akademiai Kiado, 1966) 2079-2084.

11. G. Remond, "Applications of cathodoluminescence in mineralogy," Journal of Luminescence, 15 (1977) 121-155.

12. E. S. Sprunt, "Causes of quartz cathodoluminescence colors," Scanning Electron Microscopy, pt. 1, (1981) 525-536.

13. M. R. Hall, and P. H. Ribbe, "An electron microprobe study of luminescence centers in cassiterite," American Mineralogist, 56 (1971) 31-45.

14. J. C. Goni, and G. Remond, "Localization and distribution of impurites in blende by cathodoluminescence," Mineralogical Magazine, 37 (1969) 153-155.

15. P. A. Scholle, A color illustrated guide to constituents, textures, cements, and porosites of sandstones and associated rocks. (Tulsa, OK. American Association of Petroleum Geologists, Memoir 28, 1979).

16. P. A. Scholle, and R. B. Halley, "Burial diagenesis, out of sight, out of mind!," Carbonate Cements, N. Schneidermann, and P. M. Harris, eds., (Tulsa, OK., Society of Economic Paleontologists and Mineralogists, Special Publication 36, 1985) 309-334.

17. W. Broecker, and G. Pfefferkorn, "Bibliography on cathodoluminescence," Scanning Electron Microscopy, 9 (1976) 725-737.

18. D. J. Marshall, "The influence of microscope performance on cathodoluminescence observations," (Acton, MA., Nuclide Corporation, publication no. 1019-0281, 1978).

19. O. C. Kopp, "Leak valve wear and tear," Nuclide Spectra, 16(1) (1983) 1.

20. O. C. Kopp, "An inexpensive means for increasing the magnification range of a luminoscope," Journal of Sedimentary Petrology, 51, (1981) 667-668.

21. G. F. J. Garlick, "Cathodo- and radioluminescence," Luminescence of Inorganic Solids, P. Goldbert, ed., (New York: Academic Press, 1966) 685-731.

22. F. A. Kroger, Some aspects of the luminescence of solids (New York: Elsevier, 1948).

23. R. Coy-Yll, "Quelques aspects de la cathodoluminescence des mineraus," Chemical Geology, 5 (1970) 243-254.

24. D. Allen, "A one-stage precision polishing technique for geological specimens," Mineralogical Magazine, 48 (1977) 298-300.

25. C. E. Barker, and T. J. Reynolds, "Preparing doubly polished sections of temperature sensitive rocks," Journal of Sedimentary Petrology, 54, (1984) 635-636.

26. S. J. Mugridge, and H. R. Young, "Rapid preparation of polished thin sections for cathodoluminescence study of carbonate rocks," Canadian Mineralogy, 22 (1984) 513-515.

27. I. J. Fairchild, "Chemical controls of cathodoluminescence of natural dolomites and calcites: new data and review," Sedimentology, 30 (1983) 579-583.

28. A. N. Mariano, "Macro-photography of CL," (Acton, MA., Nuclide Corporation, publication number 1080-1083, 1983).

29. E. S. Sprunt, L. A. Dengler, and D. Sloan, "Effects of metamorphism on quartz cathodoluminescence," Geology, 6 (1978) 305-308.

30. H. Gorz, R. J. R. S. B. Bhalla, and E. W. White, "Detailed cathodoluminescence characterization of common silicates," (Special Publication 70-101, Pennsylvania State University, Materials Research Laboratory, 1970) 62-70.

31. D. J. Marshall, "Suggested standards for the reporting of cathodoluminescence results," Journal of Sedimentary Petrology, 48 (1978) 651-653.

32. A. N. Mariano, "The use of cathodoluminescence in evaluation of heavy minerals in beach sand," Nuclide Spectra, 10(2) (1977) 2.

33. Anon., "Mineral Separation facilitated by luminescence," (Acton, MA., Nuclide Corporation, publication no. 1178-0971, 1978).

34. J. B. Thomas, "Cathodoluminescence as applied to sandstone petrology: sediment-source rock relationships," Nuclide Spectra, 7(6) (1974) 2.

35. D. F. Sibley, "Extension microfractures in the Tuscarora Orthoquartzite, evidence from luminescence petrography," Nuclide Spectra, 8(1) (1975) 2.

36. R. D. Morton, "Cathodoluminescence applied to uranium exploration," Nuclide Spectra, 11(1) (1978) 2.

37. A. B. Carpenter, and T. W. Oglesby, "Hydrologic significance of manganese, iron, and magnesium in calcite and dolomite cements of the Smackover Formation, eastern Mississippi," Nuclide Spectra, 4(3) (1976) 1.

38. C. E. Barker, and R. B. Halley, "Fluid inclusion and stable isotope evidence for the burial history of the Bone Spring Limestone, southern Guadalupe Mountains, Texas," (Paper presented at the Society of Economic Paleontologists and Mineralogists, First Annual Midyear Meeting, San Jose, CA., August 1984) 10.

39. C. E. Barker, and R. B. Halley, "Fluid inclusion, stable isotope, and vitrinite reflectance evidence for the burial history of the Bone Spring Limestone, southern Guadalupe Mountains, Texas," Special Publication, Gautier, D. L., ed., (Tulsa, OK.: Society of Economic Paleontologists and Mineralogists, 1986) 25 manuscript pages. In press.

40. M. Kastner, "Authigenic feldspars in carbonate rocks," American Mineralogist, 56 (1971) 1403-1442.

41. W. J. Meyers, "Carbonate cement stratigraphy of the Lake Valley Formation (Mississippian) Sacramento Mountains, New Mexico," Journal of Sedimentary Petrology, 44 (1974) 837-861.

42. M. L. Ebers, and O. C. Kopp, "Cathodoluminescent microstratigraphy in gangue dolomite, the Mascot-Jefferson City District, Tennessee," Economic Geology, 74 (1979) 908-918.

43. J. R. Frank, A. B. Carpenter, and T. W. Oglesby, "Cathodoluminescence and composition of calcite cement in the Taum Sauk Limestone (Upper Cambrian), southeast Missouri," Journal of Sedimentary Petrology, 52 (1982) 631-638.

44. R. F. Sippel, and E. D. Glover, "Structures in carbonate rocks made visible by luminescence petrography," Science, 150 (1965) 1283-1287.

45. B. J. Pierson, "The control of cathodoluminescence in dolomite by iron and manganese," Sedimentology, 28 (1981) 601-610.

46. L. Rowan, " A study of fluid inclusions within cathodoluminescent zones in Gangue dolomites, Viburnum Trend Pb-Zn district, southeast Missouri," (Paper presented at the 115th AIME annual meeting, New Orleans, LA., March 4, 1986), 79.

47. A. Matter, S. D. Burley, and J. Mullis, "Quantitative constraints on the timing of reservoir diagenesis: Application of combined cathodoluminescence microscopy and fluid inclusion thermometry," Program and Abstracts, (Sixth Annual Research Conference, Gulf Coast Section, Society of Economic Paleontologists and Mineralogists, 1985) 19-20.

48. R. F. Sippel, "Sandstone petrology, evidence from luminscence petrography," Journal of Sedimentary Petrology, 38 (1968) 530-554.

49. E. S. Sprunt, and A. Nur, "Destruction of porosity through pressure solution," Geophysics, 42 (1977) 726-741.

50. C. E. Barker, Patterns of cementation in hydrothermally altered sediments, Cerro Prieto geothermal field, Baja California, Mexico," Workshop on Cathodoluminescence, Babcock, J., Marshall, D., and Thomas J. B., eds., (Geology Department, University of Tulsa, Tulsa, OK., 1978), 4 p.

51. M. M. Smith, A. Boyde, and S. A. Reid, "Cathodoluminescence as an indicator of growth increments in the dentine in tooth plates of Triassic lungfish," Neues Jahrbuch fuer Geologie und Palaeontologie, 1 (1984) 39-45.

52. S. E. Sommer, "Cathodoluminescence of carbonates; 1. characterization of cathodoluminescence from carbonate solid solutions," Chemical Geology, 9 (1972) 257-273.

53. S. E. Sommer, "Cathodoluminescence of carbonates; 2, Geological applications," Chemical Geology, 9 (1972) 275-284.

54. H.-G. Machel, "Cathodoluminescence in calcite and dolomite and its chemical interpretation," Geoscience Canada, 9 (1986) 139-147.

55. H. Gies, "Activation possibilities and geochemical correlations of photo luminescing carbonates, particularly calcites," Mineralium Deposita, 10 (1975) 216-227.

56. A. N. Mariano, and P. J. Ring, "Europium-activated cathodoluminescence in Minerals," Geochimica et Cosmochimica Acta, 39 (1975) 649-660.

57. L. Rowan, private communication with authors, U.S. Geological Survey, 1 July 1986.

APPENDIX 1

Addresses of Manufacturers

Biddle Instruments, 510 Township Line Rd., Blue Bell, PA 19422. (215) 646-9200.

Buehler Ltd., 41 N. Waukegan Rd., Lake Bluff, IL 60044. (312) 295-6500.

Hillquist, 1545 northwest 49th St., Seattle, WA 98107. (206) 784-1960.

Eastman Kodak Co., Rochester, NY 14650. (800) 225-5352.

E. Lietz, Inc., Rockleigh, NJ 07647. (201) 767-1100.

Nikon, Instrument Division, 623 Stewart Avenue, Garden City, NY 11530 (516)222-0200.

Nuclide Corporation, AGV Divsion, 916 Main St., Acton, MA 01720. (617) 263-2936.

Miller-Stephenson Chemical Co., Inc., George Washington Highway, Danbury, CT 06810. (203) 743-4447.

Technosyn Limited, Coldhams, Cambridge CBI-3EW, England.

Carl Ziess, Inc., One Zeiss Dr., Thornwood, NY 10594. (914) 747-1800.

Bibliography of Cathodoluminescence Petrography

Charles E. Barker

U.S. Geological Survey
Box 25046
Denver, Colorado 80225

This paper contains a bibliography comprised of 424 references that deal with the applications of cathodoluminescence microscopy to carbonate sedimentary rocks, siliciclastic sedimentary rocks, fossils, igneous rocks, metamorphic rocks, ore deposits, mineralogy and chemistry, together with references on methods of cathodoluminescence microscopy, fluorescence petrography, and general references on cathodoluminescence microscopy. The bibliography is intended to augment the other papers on cathodoluminescence microscopy that are contained in this volume. The large number of references contained in this bibliography attests to the importance of the application of cathodoluminescence microscopy in a wide variety of mineralogical studies.

Introduction

This bibliography is to update an eariler compilation of Broecker and Pfefferkorn (1976) who listed 275 papers on cathodoluminescence (CL) that they were aware of in December, 1975. Approximately 10 percent of the references listed by them are from geologic studies. Because of the small number, all references to geology listed by them are included here. Any geologic references that are pre-1976, and were not cited by Broecker and Pfefferkorn, were added. Otherwise, this bibliography includes references published 1976 to 1985, inclusive. If a paper was published in a full-length format, then any related meeting abstract by the author(s), that preceded formal publication is not included. Journal names were spelled out, rather than abbreviated when possible.

Cathodoluminescence (CL) microscopy has become an essential tool in the petrographic description of sedimentary rocks as indicated by the large number of citations from this disipline. CL also has important applications in igneous and metamorphic petrography, ore deposits, and mineralogy. Because of their importance to CL studies, references are included from the related disciplines: scanning electron microscopy (SEM); thermoluminescence; ultraviolet fluorescence petrography; geochemistry of Fe, Mn, Pb, Ni, rare earth elements, etc.; and internal textures of cements.

Some of the citations were found by computer searches of scientific reference data bases (Georef, Petroleum Abstracts, and Chem Abstracts). They appeared on the list because their key word index included the term cathodoluminescence or luminescence and some key word in geology. These references may contain only a mention of cathodoluminescence, but time would not allow checking all the references. Although there are several computer data bases for citations in geology, none are able to compile a complete bibliography of a given subject, simply because the relevant article may not be indexed properly to be identified by the key-word search program. The search can only be as effective as the indexing of the data base and the key-word program. Thus, the value of subject bibliographies is to bring together all of the available references from the data bases, reference lists found in papers on CL, and personal contacts. This is not a perfect system or reference collection. I extend my apologies to any author of a paper that has been omitted and ask that they send the reference to me so that it can be included in future updates of this bibliography.

General References

E. I. Adirowitsch, 1953, Einige Fragen der Lumineszenz der Dristalle. Berlin Akademie, unknown pages.

P. Amieux, La Cathodoluminescence dans les Roches Sedimentaires. Applications Sedimentologiques et Diagenetiques. Ph.D. Thesis, University of Lyon, 1981, 224 pp.

P. Amieux, "Cathodoluminescence: Method of Sedimentological Study in Carbonates", Bulletin des Centres de Recherches Exploration-Production Elf-Aquitaine, 6 (1982) pp. 437-483 [Includes 6 color plates].

W. Broecker, and G. Pfefferkorn, "Bibliography on Cathodoluminescence", Scanning Electron Microsopy, 9 (1976) pp. 725-737.

W. Crooks, "Magnetic Deflection of Molecular Trajectory -- Laws of Magnetic Rotation in High and Low Vacua -- Phosphorgenic Properties of Molecular Discharge", Philosophical Transactions of the Royal Society of London, 170, part 11 (1880) pp. 641-662. [First description of CL phenomena.]

G. F. J. Garlick, "Absorption, Emission, and Storage of Energy in Phosphors", Journal Applied Physics, supplement no. 4 (1955) pp. 85-90.

G. F. J. Garlick, "Cathodo- and Radioluminescence", pp. 685-731 in Luminescence of Inorganic Solids, P. Goldbert, (ed.) Academic Press, New York, 1966.

V. W. Gobel, and W. J. Patzelt, "Moglichkieten der Untersuchung von Festkorpern mit Hilfe der KathodoLuminenzenz", Leitz Mitt. Wiss. Tech. 6, (1976) pp. 263-277.

P. Grant, "The Role of the Scanning Electron Microscope in Cathodoluminescence Petrology", pp. 1-11 in Scanning electron Microscopy in the Study of Sediments, B. W. Whalley, ed., Geo-abstracts, Norwich, 1978.

G. M. Harwood, 1982, "The Geological Applications of Cathodoluminescence". Colloquium Manchester (poster, in press), [as cited by Amieux, 1982].

L. F. Herzog, D. J. Marshall, and R. F. Babione, 1970, The Luminoscope - A new Instrument for Studying the Electron-Stimulated Luminescence of Terrestrial, Extra-Terrestrial and Synthetic Materials under the Microscope, Pennsylvania State University, Materials Research Laboratory, Special Publication no. 70-101, pp. 79-98.

D. B. Holt, M. D. Muir, P. R. Grant, and I. M. Boswarva, Quantitative Scanning Electron Microscopy, Academic Press, New York, 1974, 507 pp.

O. C. Kopp, ed., Geological Applications of Cathodoluminescence, Short Course Notes, University of Tennessee, Knoxville, 1973.

C. O. Kopp, Cathodoluminescence Petrography: A Valuable Tool for Teaching and Research, Journal of Geological Education, 29 (1982) pp. 108-113.

D. H. Krinsley and J. C. Doornkamp, Atlas of Quartz Sand Surface Textures, Cambridge University Press, Cambridge, 1973, 93 pp.

A. F. Kroger, Some Aspects of the Luminescence 1, 2 of Solids, Elsevier, New York, 1948, 310 pp.

W. H. Leverenz, An Introduction to Luminescence of Solids, Wiley, New York, 1950, 569 pp. (Republished by Dover, New York in 1968).

J. V. P. Long, Application of the Electron Microanalyser to Metallurgy and Mineralogy, pp. 279-295 in X-ray Optics and X-ray Microanalysis, H. H. Pattee, ed., 1963 [No publisher given.].

J. V. P. Long, and S. O. Agrell, The Cathodo-luminescence of Minerals in Thin Section, Mineralogical Magazine, 34, (1965) pp. 318-326.

A. S. Marfunin, Spectroscopy, Luminescence and Radiation Centers in Minerals, Springer-Verlag, Berlin, 1980, 352 pp.

D. J. Marshall, 1976, Cathodoluminescence References, Nuclide Corporation, publication no. 1627-0476; 5 pp. Updated version, 1981, publication number 1019-0281.

M. D. Muir, and P. R. Grant, "Cathodoluminescence", Chapter 9 in Quantitative Scanning Electron Microscopy, D. B. Holt, M. D. Muir, P. R. Grant, and I. M. Boswarva, eds., Academic Press, London, 1974.

E. Nickel, The Present Status of Cathode Luminescence as a Tool in Sedimentology, Mineral Science Engineering (Johannesburg), 10, (1979) pp. 73-100.

R. A. Potosky, "Application of Cathodoluminescence in the Field of Petrology", M.S. thesis, University of Tennessee, 1967, 50 pp.

R. A. Potosky, and O. C. Kopp, "Application of Cathodoluminescence to Petrographic Studies", Compass, 47, (1970) pp. 63-69.

P. Pringsheim, and M. Vogel, Luminescence of Liquids and Solids, Wiley Interscience Publishers, New York, 1946, 201 pp.

G. Remond, "Applications of Cathodoluminescence in Mineralogy", Journal of Luminescence, 15 (1977) pp. 121-155.

R. F. Sippel, Geologic Applications of Cathodoluminescence", pp. 2079-2084 in Proceedings, International Conference on Luminescence, Akademiai Kiado, Budapest, 1966.

J. V. Smith and R. C. Stenstrom, 1965, "Electron-excited Luminescence as a Petrologic Tool", Journal of Geology, 73 (1965) pp. 627-635.

G. V. Spivak, G. V. Saparin, M. K. Antoshin, R. A. Bochko, and V. M. Ladygin, "Observations of Rocks by Scanning Electron Microscopy with the Aid of Color Cathode Luminescence", Moscow University, Vestn., Ser. Geol., 28 (1973) pp. 44-47 [in Russian].

P. Weiblen, "Investigation of Cathodoluminescence with the Petrographic Microscope", pp. 245-251 in Developments in Applied Spectroscopy, 4, E. N. Davis, ed., Plenum Press, New York, 1965.

W. Wetzel, "Lumineszenzanalyse und Sedimentpetrographie", Zent Bl. Mineral. Geol. Palaont., Abt. A, no. 8 (1939) pp. 225-247.

W. Wetzel, "Das Lumineszenzmikroskopische Verhalten von Sedimenten", Neus Jahrbuch fuer Geologie und Palaeontologie Abhandlungen, 107 (1959) pp. 261-277.

W. Wetzel, "Lumineszenzmikroskopische Studien an Chilenischen sedimenten", Neus Jahrbuch fuer Geologie und Palaontologie, 6 (1962) pp. 303-308.

W. B. Whalley, Ed., Scanning Electron Microscopy in the Study of Sediments, Norwich Geo-abstracts, 1978, 414 pp.

Sedimentary Rocks

Carbonates

A. E. Adams, and K. Schofield, "Recent Submarine Aragonite, Magnesian Calcite, and Hematite Cements in a Gravel from Islay, Scotland", Journal of Sedimentary Petrology, 53 (1983) pp. 417-421.

S. H. Allen, J. W. Becher, R. L. Folk, C. H. Moore, and J. M. Smitherman, Porosity in Carbonate Rock Sequences, American Association of Petroleum Geologists, Course Notes Series no. 11, part A, 1979, 125 pp.

C. E. Barker, and R. B. Halley, "Fluid Inclusion and Stable Isotope Evidence for the Burial History of the Bone Spring Limestone, Southern Guadalupe Mountains, Texas" [abst.], Abstracts, Society of Economic Palenontologists and Mineralogists, First Annual Midyear Meeting, San Jose, California, 1984, 10 pp.

C. E. Barker, and R. B. Halley, "Fluid Inclusion, Stable Isotope, and Vitrinite Reflectance Evidence for the Burial History of the Bone Spring Limestone, Southern Guadalupe Mountains, Texas", in D. L. Gautier, ed., Society of Economic Paleontologists and Mineralogists, Special Publication.

R. G. C. Bathurst, "Neomorphic Spar Versus Cement in Some Jurassic Grainstones: Significance for Evaluation of Porosity Evolution and Compaction", Journal of the Geological Society, 140 (1983) pp. 229-237.

G. B. Beardall, Jr., and Z. Al-Shieb, "Dolomitization Stages in a Regressive Sequence of Hunton Group, Anadarko Basin, Oklahoma" [abst.]. American Association of Petroleum Geologists, 68 (1984) 452 pp.

R. Borokowski, and J. Mazzullo, "Petrography of the Sandy Dolosparite (unit 7) in (DSDP leg 80) hole 549", pp. 899-903 in Initial Reports of the Deep Sea Drilling Project, 80, U. S. Government Printing Office, Washington, D.C., 1985.

U. Brand, and J. Veizer, "Chemical Diagenesis of a Multicomponent Carbonate System -1: Trace Elements", Journal of Sedimentary Petrology, 50 (1980) pp. 1219-1236.

K. A. Breining, and K. C. Lohmann, "Diagenetic evolution of a Silurian Limestone Reef: Geochemical Documentation of Mixed-water Dolomitization" [abst.], American Association of Petroleum Geologists, 68 (1984) 457 pp.

J. M. Budai, K. C. Lohmann, and R. M. Owen, "Burial Dedolomite in the Mississippian Madison Limestone, Wyoming and Utah Thrust Belt", Journal of Sedimentary Petrology, 54, (1984) pp. 276-288.

R. M. Buyce and G. M. Friedman, "Significance of Authigenic K-feldspar in Cambrian-Ordovician Carbonate Rocks of the Proto-Atlantic Shelf in North America", Journal of Sedimentary Petrology, 45 (1975) pp. 808-821.

A. B. Carpenter, 1975, "Effect of Ferrous Iron on the Stability and Nucleation of Dolomite" [abst.], Geological Society of America, Abstracts with Programs, 7, (1975) 1019 pp.

A. B. Carpenter, and T. W. Oglesby, "Hydrologic Significance of Manganese, Iron, and Magnesium in Calcite and Dolomite Cements of the Smackover Formation, Eastern Mississippi", Nuclide Spectra, 4, no. 3 (1976) 1 pp.

V. B. Carrasco, Cathode Luminescence and Diagenesis of Dolomitized Rudists and its Relation to the Development of Porosity, Revista del Instituto Mexicano del Petroleo, 5 (1973) pp. 5-14 [in Spanish].

L. A. Czerniakowski, K. C. Lohmann, and J. L. Wilson, "Closed System Marine Burial Diagenesis: Isotopic Data from the Austin Chalk and its Components", Sedimentology, 31 (1984) pp. 863-877.

H. G. Churnet, and K. C. Misra, "Zoning in Dolomites as Indicator of a Diagenetic Enviroment Involving Mixing of Fresh and Saline Waters" [abst.]. Abstracts and Programs, The Geological Society of America, 12, no. 4 (1980) pp. 173-174.

P. E. Cowan, and W. J. Meyers, "Diagenesis of Lime Mud, Mississippian-age Bioherms, Sacramento Mountains, New Mexico, American Association of Petroleum Geologists, 66 (1982) pp. 559-560.

C. M. de Reena, and I. K. Kaul, "Diagenesis and Thermoluminescence Characteristics of the Deep Sea Carbonate Oozes from the Indian Ocean", Modern Geology, 7 (1981) pp. 231-241.

J. A. D. Dickson, "Graphical Modelling of Crystal Aggregates and its Relevance to Cement Diagnosis," London Royal Philosphical Society, Transactions, Series A 309 (1983) pp. 465-502.

J. A. D. Dickson, "Diagenesis of Shallow-marine Carbonates," pp. 173-184 in P. J. Brenchley and B. P. J. Williams, Sedimentology: Recent Developments and Applied Aspects, Blackwell, London, (1985).

S. L. Dorobek, "Stratigraphy, Sedimentology, and Diagenetic History of the Siluro-Devonian Helderburg Group, Central Applalachians," Ph.D. thesis, Virgina Polytechnic Institute, 1984, 252 pp.

S. G. Driese, "Sedimentology, Conodont Distribution, and Carbonate Diagenesis of the Upper Morgan Formation (Middle Pennsylvanian), Northern Utah and Colorado," Ph.D. thesis, University of Wisconsin, Madison, 1982 289 pp.

S. G. Driese, "Diagenetic Aspects of Morgan Formation (Pennsylvanian) Shelf Carbonates, Northern Utah and Colorado, Brigham Young University Geology Studies, 30 (1983) pp. 1-18.

B. D. Evamy, "The Precipitational Enviroment and Correlation of some Calcite Cements Deduced from Artifical Staining," Journal of Sedimentary Petrology, 39 (1969) pp. 787-821.

I. J. Fairchild, "Sedimentation and Post-depositional History of the Dalradian Bonahaven Formation of Islay," Ph.D. thesis, University of Nottingham, 1978.

I. J. Fairchild, "Stages in Precambrian Dolomitization, Scotland: Cementing versus Replacement Textures," Sedimentology, 27 (1980) pp. 631-650.

I. J. Fairchild, "Chemical Controls of Cathodoluminescence of Natural Dolomites and Calcites: New Data and Review," Sedimentology, 30 (1983) pp. 579-583.

P. Faupl and A. Beran, "Diagenetic Alterations of Radiolaria and Sponge Spicule-Bearing Rocks of the Strubberg Formation (Jurrassic, Northern Calcareous Alps, Austria)," Neues Jahrbuch Geologie und Palaeontologie, 5 (1983) pp. 129-140.

C. T. Feazel and R. A. Schatzinger, "Prevention of Carbonate Cementation in Petroleum Reservoirs," pp. 97-106 in N. Schneidermann and P. M. Harris, Carbonate Cements, Society of Economic Paleontologists and Mineralogists, Special Publication 36, 1985.

K. F. Ferrigno, "Paleoecology of the Reef of the Holston Limestone," Ph.D. thesis, University of Tennessee, Knoxville, 1973.

K. F. Ferrigno, "The Use of Cathodoluminescence for the Interpretation of Limestone Enviroments and the Identification of Bryozoans," 2 pp. in J. Babcock, D. J. Marshall, and J. B. Thomas, Course Notes, Workshop on Cathodoluminescence, Geology Department, University of Tulsa, Tulsa, Oklahoma, March 8, 1978.

L. A. Fox, "Porosity and Permeability Reduction in the Nugget Sandstone, southwestern Wyoming," M.S. Thesis, University of Missouri, Columbia, 1979, 117 pp.

J. R. Frank, "Vug-filling Calcite of the Taum Sauk Marble (Bonneterre Formation) Upper Cambrian, southeast Missouri: Preliminary report," 2 pp. in J. Babcock, D. J. Marshall, and J. B. Thomas, Course Notes, Workshop on Cathodoluminescence, Geology Department, University of Tulsa, Tulsa, Oklahome, March 8, 1978.

J. R. Frank, "Dedolomitization in the Taum Sauk Limestone," M.A. thesis, University of Missouri, Columbia, 1979, 92 pp.

J. R. Frank, "Dedolomitization in the Taum Sauk Limestone (Upper Cambrian), Southeast Missouri," Journal of Sedimentary Petrology, 51 (1981) pp. 7-18.

J. R. Frank, A. B. Carpenter and T. W. Oglesby, "Cathodoluminescence and Composition of Calcite Cement in the Taum Sauk Limestone (Upper Cambrian), southeast Missouri," Journal of Sedimentary Petrology, 52 (1982) pp. 631-638.

M. H. Frank and K. C. Lohmann, "Cathodoluminescent and Isotopic analysis of Diagenetically Altered Dolomite, Bonneterre Formation, Southeast Missouri," [abst.], Abstracts and Programs, Geological Society of America, 14 (7) (1982), 491 pp.

M. R. Franke, "Natural Porosity, Diagenetic Evolution, and Experimental Porosity Development in Macae Carbonates (Albian-Cenomanian), Compos Basin, Offshore Brazil, Ph.D. thesis, University of Illinois, Urbana, 1981, 141 pp.

T. Freeman, "Morphology and Composition of an Ordovician Vadose Cement," Nature, 233 (1971) pp. 133-134.

G. M. Friedman, "The Problem of Submarine Cement in Classifying Reefrock: an Experience in Frustration," pp. 117-121 in N. Schneidermann, and P. M. Harris, eds., Carbonate Cements, Society of Economic Paleontologists and Mineralogists, Special Publication 36, 1985.

R. A. Garber, "Diagenetic Patterns of Oligocene Reefs in Indonesia," M.S. thesis, Rennsselaer Polytechnical Institute, 1976.

M. George, "Carbonate Equilibrium in the Hosston Formation, Central Mississippi," M.S. thesis, University of Missouri, Columbia, 1977, 83 pp.

H. Gies, "Zur Beziehung zwischen Photolumineszenz und Chemismus Naturlicher Karbonate," Habilitationsschrift, Technische Universitat Clausthal, 1973, 111 pp.

H. Gies, "Activation Possibilites and Geochemical Correlations of Photoluminescing Carbonates, Particularly Calcites," Mineralium Deposita, 10 (1975) pp. 216-227.

H. Gies, "Zur Beziehung Zwischen Photolumineszenz und Chemismus Naturlicher Karbonate," Neues Jahrbuch fur Mineralogie, Abhandlungen, 127 (1976) pp. 1-46.

R. K. Given and K. C. Lohmann, "Derivation of the Original Isotopic Composition of Permian Marine Cements," Journal of Sedimentary Petrology, 55 (1985) pp. 430-439.

E. D. Glover, "Cathodoluminescence, Iron and Manganese Content, and the Early Diagenesis of Carbonates," Ph.D thesis, University of Wisconsin, Madison, 1977, 465 pp.

A. W. Gorody, " The Lower Ordovician Mascot Formation, Upper Knox Group, in North Central Tennessee," Ph.D. thesis, Rice University, Houston, Texas, 1980, 263 pp.

A. W. Gorody, "Dolomitization and Paleohydraulic History of the Lower Ordovician Mascot Formation, Upper Knox Group, in North Central Tennessee [abst.], Geological Society of America, Abstracts with Programs, 12, no. 7, (1980) 435 pp.

M. G. Grammer, "Diagenetic Destruction of Primary Reservoir Porosity in Viola Limestone, South-central Oklahoma, [abst.], American Association of Petroleum Geologists, 69 (1985) pp. 258-259.

J. M. Gregg, "Regional Epigenetic Dolomitization in the Bonneterre Dolomite (Cambrian), Southeastern Missouri," Geology, 13 (1985) pp. 503-506.

A. G. Grover Jr., "Cement Types and Cementation Patterns of Middle Ordovician Ramp-to-basin Carbonates, Virginia," Ph.D. thesis, Virginia Polytechnic Institute, 1981, 229 pp.

G. A. Grover Jr., "Burial Cementation of Carbonates," West Texas Geological Society Bulletin, 23 (1984) 17 pp.

G. A. Grover Jr. and J. F. Read, "Paleoaquifer and Deep Burial Related Cements Defined by Regional Cathodoluminescent Patterns, Middle Ordovician carbonates, Virginia," American Association of Petroleum Geologists, 67 (1983) pp. 1275-1303.

G. A. Grover Jr. and J. F. Read, "Sedimentology and Diagenesis of Middle Ordovician Carbonate Buildups, Virginia," Society of Economic Paleontologists and Mineralogists, Core Workshop no. 4, (1983) pp. 2-25.

D. C. Harris and W. J. Meyers, "Carbonate Cement Stratigraphy of Burlington Limestone (Osagean) of Iowa: Evidence for Eh Gradients in a Regional Mississippian Paleogroundwater System," [abst.], American Association of Petroleum Geologists, 69 (1985) 263 pp.

G. Harwood, "The Application of Cathodoluminescence in Relative Dating of Barite Mineralization in the Lower Magnesian Limestone (Upper Permian), United Kingdon," Economic Geology, 78 (1983) pp. 1022-1027.

P. P. Hearn Jr. and J. F. Sutter, "Authigenic Potassium Feldspar in Cambrian Carbonates: Evidence of Alleghanian Brine Migration," Science, 228 (1985) pp. 1529-1531.

C. M. Hill, "A Diagenetic Study of Carbonate Sediments from the Red Sea Utilizing Cathodoluminescence," M.S. thesis, Rennseller Polytechnic Institute, 1979.

N. P. James and C. F. Klappa, "Petrogenesis of Early Cambrian Reef Limestones, Labrador, Canada," Journal of Sedimentary Petrology, 53 (1983) pp. 1051-1096.

N. P. James and P. W. Choquette, "Diagenesis, Part 9, Limestones - the Meteoric Enviroment, Geoscience Canada, 11 (1984) pp. 161-194.

R. I. Kablanonow, R. C. Surdam, and D. Prezbindowski, "Origin of Dolomites in the Monterey Formation: Pismo and Huasna Basins, California," Society of Economic Paleontologists and Mineralogists, Annual Meeting, Guidebook no. 2 (1984) pp. 38-49.

M. Kastner, "Authigenic Feldspars in Carbonate Rocks," American Mineralogist, 56 (1971) pp. 1403-1442.

J. Kaufman, H. S. Cander, and W. J. Meyers, "Near-Surface and Burial Diagenesis of Mississippian Burlington and Keokuk Formations," [abst.], American Association of Petroleum Geologists, 69 (1985) p. 272.

R. B. Koepnick, "Luminescent Zonation of Carbonate Cements in the Upper Cambrian Straight Canyon and Fera Formations of the Dugway Range, Utah," Nuclide Spectra, 9 no. 2 (1976) 1 pp.

C. Lamiraux, "Contribution a L'etude du Phemomene de Cathodoluminescence Appliquee a des Echantillons de Roche en Lame Mince," Rapport Elf Aquitaine, in press [as cited by Amieux, 1982] (1977) 22 pp.

K. C. Lohmann and W. J. Meyers, "Microdolomite Inclusions in Cloudy Prismatic Calcites: A Proposed Criterion for Former High-Magnesium Calcites," Journal of Sedimentary Petrology, 47 (1977) pp. 1078-1088.

J. H. Lake, "Sedimentology and Paleoecology of Upper Ordovican Mounds of Anticosti Island, Quebec Lake," Canadian Journal of Earth Science, 18 (1981) pp. 1562-1571.

H. G. Machel, "Fazies und Diagenese der Devonischen Riffcarbonate der Borung Romberg (Briloner Riff)," Diplomarbeit Technische Universitat Braunschweig (1979) 231 pp.

H. G. Machel, "Cathodoluminescence of Calcites: Activator Concentrations and Environmental Interpretations," [abst.], 11th International Association of Sedimentologists, International Congress, Hamilton, Ontario, Canada, August 22-27 (1982) Abstracts, 157 pp.

H. G. Machel, "Cathodoluminescence in Carbonate Petrography; some Aspects of Geochemical Interpretation," [abst.], American Association of Petroleum Geologists, 67 (1983) pp. 507-508.

H. G. Machel, "Cathodoluminescence in Calcite and Dolomite and its Chemical Interpretation," Geoscience Canada, 12 (1985) pp. 139-147.

W. L. Manger, W. R. Rice, J. I. Hogue, and S. L. Hokett, "Depositional and Postdepositional History of Stuart City Member, Edwards Limestone (Lower Cretaceous), Washburn Ranch Field, Lasalle County, Texas," [abst.], American Association of Petroleum Geologists, 69 (1985) 283 pp.

D. J. Marshall, "The Status of the Cathodoluminescence Technique in the Study of Carbonates," [abst.], Abstracts, 9th International Congress of Sedimentology, Nice, France (1975) 6 pp.

H. Martin, and H. Zeegers, "Cathode Luminescence and Distribution of Manganese in Upper Tournaisian Limestones and Dolomites, southern Ninant, Belgium," Academy of Science, C.R., series D, 269 (1969) pp. 1922-1924 [In French].

W. J. Meyers, "Carbonate Cement Stratigraphy of the Lake Valley Formation (Mississippian) Sacramento Mountains, New Mexico," Journal of Sedimentary Petrology, 44 (1974) pp. 837-861.

W. J. Meyers, "Chertification in the Mississippian Lake Valley Formation, Sacramento Mountains, New Mexico," Sedimentology, 24 (1977) pp. 75-105.

W. J. Meyers, "Carbonate Cements: Their Regional Distribution and Interpretation in Mississippian Limestones of Southwestern New Mexico," Sedimentology, 25 (1978) pp. 371-400.

W. J. Meyers and K. C. Lohmann, "Isotope Geochemistry of Regionally Extensive Calcite Cement Zones and Marine Components in Mississippian limestones, New Mexico," pp. 223-239 in N. Schneidermann, and P. M. Harris (eds.) Carbonate Cements, Society of Economic Paleontologists and Mineralogists, Special Publication 36, 1985.

J. A. Middleton, "Diagenesis of Reef Materials as seen through Cathodoluminescence," M.S. thesis, Rennsselaer Polytechnical Institute, 1979.

T. Milinkovich, "Dolomitization and Luminescence Stratigraphy of the Irondequoit Formation, Ontario and New York," B.S. thesis, Brock University, 1984, 88 pp.

N. K. Moore and K. R. Walker, "Porosity Occlusion in Ancient Reefal Carbonates, with an Example from the Holston Formation (Middle Ordivician) of East Tennesee," [abst.], Geological Society of America, Abstracts with Programs, 6 (1974) 1049 pp.

D. H. Mruk, "Carbonate Cementation and Dolomitization of the Capitan Limestone, McKittrick Canyon, West Texas," M.S. thesis, University of Colorado, Boulder, 1984.

B. Mukherjee, "Cathodoluminescence Spectra of Indian Calcites, Limestones, Dolomites and Aragonites," Indian Journal of Physics, 22 (1948) pp. 305-310.

S. J. Mugridge, "Cathodoluminescent Zoning in Mississippian Carbonate Cements, Southwestern Manitoba," B.S. thesis, Brandon University, 1981.

M. R. Newton, "Porosity Evolution in Salem Limestone (Mississippian) of South-central Indiana," [abst.], American Association of Petroleum Geologists Bulletin, 69 (1985) 292 pp.

J. C. Neiman and J. F. Read, "Recognition of Unconformity-Sourced Aquifer Cements and later Burial Cements [abst.], American Association of Petroleum Geologists Bulletin, 69 (1985) 293 pp.

A. Northrup, "The Luminescence of Calcite," Rocks and Minerals, 47, 1972, pp. 659-669.

R. D. Ogidan, "Petrography of Caliche," M.S. thesis, Rensselaer Polytechnical Institute, 1976.

T. W. Oglesby, "A Model for the Distribution of Manganese, Iron, and Magnesium in Authigenic Calcite and Dolomite Cements in the Upper Smackover Formation in Eastern Mississippi," M.S. thesis, University of Missouri, 1976, 122 pp.

A. E. Odershaw and T. P. Scoffin, "The Source of Ferroan and Non-ferroan Calcite Cements in the Halkin and Wenlock Limestones," Geological Journal, 5 (1967) pp. 309-320.

P. P. Parekh, P. Muller, P. Dulski, and W. M. Bausch, "Distribution of Trace Elements between Carbonate and Non-carbonate Phases in Limestone," Earth and Planetary Science Letters, 34 (1977) pp. 39-50.

N. R. Pacey, "Luminescent Properties of Calcite in the English Chalk," Modern Geology, (in press).

H. M. Pedley, "The Petrology and Paleoenvironment of the Sortino Group (Miocene) of Southeast Sicily," Evidence for periodic emergence, Journal of the Geological Society, 140 (1983) pp. 335-350.

T. M. Peryt, "Cathodoluminescence in Studies of Carbonate Rocks," Przeglad Geologiczny, 29 (1981) pp. 119-125 [in Polish].

B. J. Pierson, "The Control of Cathodoluminescence in Dolomite by Iron and Maganese," M.S. thesis, University of Kentucky (1977) 85 pp.

B. J. Pierson, "The Control of Cathodoluminescence in Dolomite by Iron and Manganese," Sedimentology, 28 (1981) pp. 601-610.

B. J. Pierson, "Late Cenozoic Geology of the Southeastern Bahama Banks," Ph.D. thesis, University of Miami (1981).

N. E. Pingitore, "The Behavior of Zn^{2+} and Mn^{2+} during Carbonate Diagenesis: Theory and Applications," Journal of Sedimentary Petrology, 48 (1978) pp. 799-814.

H. H. Posey, "Luminescent Cements of the Butterly Formation, Arbuckle Mountains, Oklahoma," 2 pp. in J. Babcock, D. J. Marshall, and J. B. Thomas (eds.) Course Notes, Workshop on Cathodoluminescence, Geology Department, University of Tulsa, Tulsa, Oklahoma, March 8, 1978.

B. M. Radke, and R. L. Mathis, "On the Formation and Occurrence of Saddle Dolomite," Journal of Sedimentary Petrology, 50 (1980) pp. 1149-1168.

N. M. Rees, M. J. Brady, and A. J. Rowell, "Depositional Environment of the Upper Cambrian Johns Wash Limestone (House Range, Utah)," Journal of Sedimentary Petrology, 46 (1976) pp. 38-47.

C. Renfrew and J. S. Peacey, "Aegean Marble: A Petrological Study," The Annual of the British School at Athens, (1968) pp. 45-66.

D. K. Richter and U. Zinkernagel, "Use of Cathodoluminescence in Carbonate Petrography," Geologie Rundschdau, 70 (1981) pp. 1276-1302 [In German].

S. Sathyanarayn and G. Muller, "Origin of Microstructures and Textures in the Carbonate Rocks of the Kaladgi (younger Precambrian) Group, Karnataka, India," [abst.], Congress International de Sedimentologie, 11 (1982) 158 pp.

R. A. Schatzinger, "Phylloid Algal and Sponge-Bryozoan Mound-to-basin Transition: A Late Paleozoic Facies Tract from the Kelly-Snyer Field, West Texas," Society fo Economic Paleontologists and Mineralogists, Core Workshop no. 4 (1983) pp. 244-303.

M. Scherer and H. Seitz, "Rare Earth Element Distribution in Holocene and Pleistocene Corals and their Redistribution during Diagenesis," Chemical Geology, 28 (1980) pp. 279-289.

P. A. Scholle and R. B. Halley, "Burial Diagenesis, Out of Sight, Out of Mind!", pp. 309-334 in N. Schneidermann and P. M. Harris (eds.), Carbonate Cements, Society of Economic Paleontologists and Mineralogists, Special Publication 36, 1985.

J. N. Seitz, "A Diagenetic Study of Miocene Carbonates from Indonesia Utilizing Cathodoluminescence," M.S. thesis, Rensselaer Polytechnical Institute, 1975.

J. A. Shrank, "Formation of Red Sea Lithified Layers as Studied by Cathodoluminescence; Deep Sea Drilling Project Leg 23B," M.S. thesis, Rensselaer Polytechnical Institute (1975).

J. A. Shrank and G. M. Friedman, "Lithified Layers from Sub-bottom Carbonate Sediments of the Red Sea: Deep Sea Drilling Project, Leg 23B," [abst.], 9th International Congress of Sedimentology, Nice, France, Abstracts (1975) pp. 198-206.

R. F. Sippel and E. D. Glover, "Structures in Carbonate Rocks made Visible by Luminescence Petrography," Science, 150 (1965) pp. 1283-1287.

F. D. Smith, R. J. Reeder and W. J. Meyers, "Fluid Inclusions in Burlington Limestone (Middle Mississippian) - Evidence for Multiple Dewatering Events from Illinois Basin," [abst.], American Association of Petroleum Geologists, 68 (1984) pp. 528-529.

S. E. Sommer, "Cathodoluminescence of Carbonates; 2, Geological Applications," Chemical Geology, 9 (1972) pp. 275-284.

R. K. Suchecki and J. F. Hubert, "Stable Isotopic and Elemental Relationships of Ancient Shallow-marine and Slope Carbonates, Cambro-Ordovician Cow Head Group, Newfoundland: Implications for Fluid Flux," Journal of Sedimentary Petrology, 54 (1984) pp. 1062-1080.

L. E. Ten Have, "Relationship of Dolomite/Limestone Ratios to the Structure and Producing Zones of the West Branch Oil Field, Ogemaw County, Michigan," M.S. thesis, Michigan State University (1979) 106 pp.

M. E. Tucker, "Precambrian Dolomites: Petrographic and Isotopic Evidence that They Differ from Phanerozoic Dolomites," Geology, 10 (1982) pp. 7-12.

M. E. Tucker, "Diagenesis, Geochemistry, and Origin of a Precambrian Dolomite: The Beck Spring Dolomite of Eastern California," Journal of Sedimentary Petrology, 53 (1983) pp. 1097-1119.

M. E. Tucker, "Calcitized Aragonite Ooids and Cements from the Late Precambrian Biri Formation of Southern Norway," Sedimentary Geology, 43 (1985) pp. 67-84.

D. S. Ulmer, "Dedolomitization and Calcitization of Gypsum in Mississippian Arroyo Penasico Group, North-central New Mexico," [abst.], American Association of Petroleum Geologists, 68 (1984) 536 pp.

J. Veizer, "Chemical Diagenesis of Carbonates: Theory and Application of Trace Element Technique," in M. A. Arthur and others (eds.), Society of Economic Paleontologists and Mineralogists, Short Course 10, Chapter 3, 1983.

P. D. Wagner and R. K. Matthews, "Porosity Preservation in the Upper Smackover (Jurassic) Carbonate Grainstone, Waler Creek field, Arkansas: Response of Paleophreatic Lenses to Burial Processes," Journal of Sedimentary Petrology, 52 (1982) pp. 3-18.

G. Walkden and J. Davies, "Polyphase Erosion of Subaerial Omission Surfaces in the Late Dinantian of Anglesey, North Wales," Sedimentology, 30 (1983) pp. 861-878.

G. M. Walkden and J. R. Berry, "Syntaxial Overgrowths in Muddy Crinoidal Limestones: Cathodoluminescence Sheds New Light on an Old Problem," Sedimentology, 31 (1984) pp. 251-267.

G. M. Walkden and J. R. Berry, "Natural Calcite in Cathodoluminescence; Crystal Growth during Diagenesis," Nature, 308 (1984) pp. 525-527.

R. A. Walls, E. W. Mountjoy and P. Fritz, "Isotopic Composition and Diagenetic History of Carbonate Cements in Devonian Golden Spike Reef, Alberta, Canada," Geological Society of America Bulletin, 90 (1979) pp. 963-982.

R. A. Walls and G. Burrowes, "The Role of Diagenetic History of Devonian Reefs, Western Canada," pp. 185-220 in N. Schneidermann and P. M Harris (eds.), Carbonate Cements, Society of Economic Paleontologists and Mineralogists, Special Publication 36, 1985.

W. C. Ward and R. B. Halley, "Dolomitization in a Mixing Zone of Near-seawater Composition, Late Pleistocene, Northeastern Yucatan Peninsula," Journal of Sedimentary Petrology, 55 (1985) pp. 407-420.

W. D. Wiggins and P. M. Harris, "Burial Diagenesis in Deep-water Allochtonous Dolomites, Permian Bone Spring Limestone, Southeast New Mexico," Society of Economic Paleontologists and Mineralogists, Core Workshop no. 6 (1985) pp. 140-173.

B. H. Wilkinson, S. U. Janecke and C. E. Brett, "Low-magnesian Calcite Marine Cement in Middle Ordovician Hardgrounds from Kirkfield, Ontario," Journal of Sedimentary Petrology, 52 (1982) pp. 47-57.

B. H. Wilkinson, A. L. Smith and K. C. Lohmann, "Sparry Calcite Cement in Upper Jurassic Limestones of Southeastern Wyoming," pp. 169-184 in N. Schneidermann and P. M. Harris (eds.), Carbonate Cements, Society of Economic Paleontologists and Mineralogists, Special Publication 36, 1985.

L. M. Young, "Dolomites and Early Mississippian Bioherms, Leadville Formation, Molas Lake, Colorado," [abst.], American Association of Petroleum Geologists Bulletin, 69 (1985) 318 pp.

W. Zimmerie and U. Zinkernagel, "Petrography and Diagenesis of a Choice of Cores from the Middle Devonian of Schwarzbachtal-1; Contribution to the Petrographic Studies using Cathodoluminescence," Geologie Landesamt Nordrhein-Westfalen, 63 (1982) pp. 115-139 [In German].

Siliciclastic Rocks

P. A. Allen, M. Mange-Rjetzky and A. Matter, " Dynamic Paleaogeography of the Open Burdigalian Seaway, Swiss Molasse Basin," Eclogae Geol. Helv., 78 (1985) 351-381 pp.

C. E. Barker, "Patterns of Cementation in Hydrothermally Altered Sediments, Cerro Prieto Geothermal Field, Baja California, Mexico," p. 4 in J. Babcock, D. J. Marshall and J. B. Thomas, eds., Course Notes, Workshop in Cathodoluminescence, Geology Department, University of Tulsa, Tulsa, Oklahoma, March 8, 1978.

J. B. Blanche and J. H. M. Whitaker, "Diagenesis of Part of the Brent Sand Formation (Middle Jurassic) of Northern North Sea Basin," Journal of the Geological Society, 135 (1978) pp. 73-82.

J. M. Bodard, V. J. Wall and R. A. F. Cas, "Diagenesis and the Evolution of Gippsland Basin Reservoirs," Australian Petroleum Exploration Association (APEA) Journal, 24 (1984) pp. 314-335.

R. L. Brenner, D. C. Mou and R. Moussavi-Harami, "Diagenetic Studies of Pennsylvanian Sandstones in Kansas and Wyoming using Cathodoluminescence," [abst.], Geological Society of America, Abstracts with Programs, 12, no. 5 (1980) 220 pp.

R. M. Buyce, "Semiquantitative Analysis of Detrital versus Authigenic Potassium-feldspar Contents using Cathodoluminescence," Nuclide Spectra, 7, no. 5 (1974) 1 pp.

L. A. Dengler and E. S. Sprunt, "Cathodoluminescence of some Quartzites from the Bergell Alps," [abst.], EOS, 58 (1977) 1239 pp.

R. J. Dudley, "The use of Cathodoluminescence in the Identification of Soil Minerals," Journal of Soil Science, 27 (1976) pp. 487-494.

H. Fuchtbauer and G. Muller, "Sediment Petrologie: Teil 11, Sedimente und Sedimentgesteine," Stuttgart. E. Schweizerbartsche.

H. Fuchtbauer, R. Leggewie, C. Gockeln, C. Heinemann and P. Schroeder, "Methods of Quartz Investigations applied to Mesozoic and Pleistocene Sandstone and Sands," Neus Jahrbuch Geologie und Palaeontologie, 4 (1982) pp. 193-210.

J. M. Garbarini and A. B. Carpenter, "Albitization of Plagioclase by Oil-field Brines," [abst.], Geological Society of America, Abstracts and Programs, 10, no. 7 (1978) 406 pp.

P. R. Grant and S. H. White, "Cathodoluminescence and Microstructure of Quartz Overgrowths on Quartz," 789-793 pp. in P. O'Hare (ed.), Scanning Electron Microscopy, 1978.

D. W. Houseknecht, "Influence of Grain Size and Temperature on Intergranular Pressure Solution, Quartz Cementation, and Porosity in a Quartzose Sandstone," Journal of Sedimentary Petrology, 54 (1984) 348-361 pp.

S. U. Janecke, B. H. Wilkinson and C. E. Brett, "Fabric and Composition of Marine Phreatic Cements in Middle Ordovician Hardgrounds from Kirkfield, Ontario," Geological Society of America, Abstracts with Programs, 12, no. 7 (1980) 455 pp.

M. Kastner and R. Siever, "Low Temperature Feldspars in Sedimentary Rocks," American Journal of Science, 279 (1979) pp. 435-479.

D. H. Krinsley, "The Present State and Future Prospects of Environmental Discrimination by Scanning Electron Microscopy," pp. 169-179 in W. B. Whalley (ed.), Symposium on Scanning Electron Microscopy in the Study of Sediments, Geoabstracts, Norwich, 1978.

D. H. Krinsley and P. W. Hyde, "Cathodoluminescence Studies of Sediments," Scanning Electron Microscopy, 4 (1971) pp. 409-416.

D. H. Krinsley and N. K. Tovey, "Cathodoluminescence in Quartz Sand Grains," Scanning Electron Microscopy: International Reviews of Advanced Technical Application s of Scanning Electron Microscopy, part 1, (1978) pp. 887-892.

A. N. Mariano, "The Use of Cathodoluminescence in Evaluation of Heavy Minerals in Beach Sand," Nuclide Spectra, 10, no. 2 (1977) 2 pp.

A. Matter, S. D. Burley and J. Mullis, "Quantitative Constraints on the Timing of Reservoir Diagenesis: Application of Combined Cathodoluminescence Microscopy and Fluid Inclusion Thermometry," Program and Abstracts, Sixth Annual Research Conference, Gulf Coast Section, Society of Economic Paleontologists and Mineralogists, (1985) pp. 19-20.

D. C. Mou and R. L. Brenner, "Control of Reservoir Properties of Tensleep Sandstone by Depositional and Diagenetic Facies: Lost Soldier Field, Wyoming," Journal of Sedimentary Petrology, 52, (1982) pp. 367-381.

O. Ogunyomi, R. F. Martin and R. Hesse, "Albite of Secondary Origin in Charny Sandstones, Quebec: A Re-evaluation," Journal of Sedimentary Petrology, 51, (1981) pp. 597-606.

M. R. Owen, "Sedimentary Petrology and Provenance of the Upper Jackfork Sandstone (Morrowan), Quachita Mountains, Arkansas, U.S.A.," Ph.D. thesis, University of Illinois (1984) 167 pp.

M. T. Pagan and J. A. Tweedie, "The Importance of Rock Characterization in North Sea Reservoir Description," Society of Underwater Technology and Oceanography International Symposium, Brighton, England, July 3, 1980, Proceedings Technical Session B, pp. 25-32.

J. K. Pitman and E. S. Sprunt, "Origin and Occurrence of Fracture-filling Cements in the Upper Cretaceous Mesaverde Formation at MWX (Multiwell Experiment), Piceance Creek Basin, Colorado," U. S. Geological Survey, Open File Report, no. 84-757 (1984) pp. 87-101.

E. D. Pittman, "Recent Advances in Sandstone Diagenesis," Review of Earth and Planetary Science, 7 (1979) pp. 39-62.

K. Ramseyer, "A New Cathodoluminescence Microscope and its Application to Sandstone Diagenesis," [abst.], Abstracts, 11th International Association of Sedimentologists, International Sedimentology Congress, Hamilton, Ontario, Canada, August 22-27, 1982, 120 pp.

V. Ranganathan, "The Significance of Abundant K-feldspar in Potassium-rich, Cambrian Shales of the Appalachian Basin," Southeast Geology, 24 (1983) pp. 139-146.

D. K. Richter and U. Zinkernagel, "Petrography of the "Permo-Scythian" of the Jaggl-Plawen Unit (Southern Tyrol, Italy) and a Discussion of the Provenance of the Detritus with Aid from Cathodeluminescence," Geologie Rundschdau, 64 (1975) pp. 783-807 [In German].

L. F. Ruppert, C. B. Cecil, R. W. Stanton and R. P. Christian, "Authigenic Quartz in the Upper Freeport Coal Bed, West-central Pennsylvannia," Journal of Sedimentary Petrology, 55 (1985) pp. 334-339.

I. D. Sanderson, "Recognition and Significance of Inherited Quartz Overgrowths in Quartz Arenites," Journal of Sedimentary Petrology, 54 (1984) pp. 473-486.

P. R. Schluger, "Petrology and Origin of the Red Beds of the Perry Formation, New Brunswick, Canada, and Maine, USA," Journal of Sedimentary Petrology, 46 (1976) pp. 22-37.

P. A. Scholle, "A Color Illustrated Guide to Constituents, Textures, Cements, and Porosities of Sandstones and Associated Rocks," American Association of Petroleum Geologists, Memoir 28 (1979) 211 pp.

D. F. Sibley, "Extension Microfractures in the Tuscarora Orhtoquartzite, Evidence from Luminescence Petrography," Nuclide Spectra, 8, no. 1 (1975) 2 pp.

D. F. Sibley and H. Blatt, "Intergranular Pressure Solution and Cementation of the Tuscarora Orthoquartzite," Journal of Sedimentary Petrology, 46 (1976) pp. 881-896.

R. F. Sippel, "Sandstone Petrology, Evidence from Luminescence Petrography," Journal of Sedimentary Petrology, 38 (1968) pp. 530-554.

G. V. Spivak, R. A. Bochko, G. V. Saparin and M. K. Antoshin, "The Possibility of Determining the Mineral Composition of Clay with Spectral Analysis of Cathodoluminescence and Scanning Electron Microscopy," Moscow University, Vestn., Ser. Geol., 28 (1973) pp. 48-51.

E. S. Sprunt, "Pressure Solution, Cementation, and Cathodoluminescence with Emphasis on Quartz," Ph.D. thesis, Stanford University (1977) 135 pp.

E. S. Sprunt, "Causes of Quartz Cathodoluminescence Colors," Scanning Electron Microscopy, pt. 1 (1981) pp. 525-536.

E. S. Sprunt and A. Nur, "Destruction of Porosity through Pressure Solution," Geophysics, 42 (1977) pp. 726-741.

E. S. Sprunt and A. Nur, "Another Look at Pore Structures; Cathodoluminescence Petrography," SPWLA logging Symposium Transactions, 21 (1980) pp. B1-B16.

P. J. Stalder, "Cementation of Pliocene-Quaternary Fluviatile Clastic Deposits in and Along the Oman Mountains," Geologie en Mijnb, 54 (1975) pp. 148-156.

D. A. V. Stow and J. Miller, "Mineralogy, Petrology, and Diagenesis of Sediments at (DSDP Leg 75) Site 530, Southeast Angola Basin," pp. 859-873 in Initial Reports of the Deep Sea Drilling Project, U.S. Government Printing Office, Washington, 75 (1984) pp. 859-873.

J. M. Taylor, "Pore Space Reduction in Sandstones," American Association of Petroleum Geologists Bulletin, 34 (1950) pp. 701-716.

J. B. Thomas, "Cathodoluminescence as Applied to Sandstone Petrology: Sediment-source Rock Relationships," Nuclide Spectra, 7, no. 6 (1974) 2 pp.

N. K. Tovey and D. H. Krinsley, "A Cathodoluminescent Study of Quartz Sand Grains," Journal of Microscopy, 120 (1980) pp. 279-289.

D. B. Wallem, J. R. Steidtmann and R. C. Surdam, "Depositional Environments and Diagenesis of the Bear River Formation, Western Wyoming," Wyoming Geology Association, Guidebook, Energy Resources of Wyoming (1981) pp. 23-39.

B. Waugh, "Formation of Quartz Overgrowths in the Penrith Sandstone (Lower Permian) of Northwest England as Revealed by Scanning Electron Microscopy," Sedimentology, 14 (1970) pp. 309-320.

J. H. M. Whitaker, "Diagenesis of the Brent Sand Formation: A Scanning Electron Microscope Study," pp. 363-382 in W. B. Whalley, ed., Symposium on Scanning Electron Microscopy in the Study of Sediments, Geoabstracts, Norwich, 1979.

U. Zinkernagel, "Cathodoluminescence of Quartz and its Application to Sandstone Petrology," E. Schweizerhartsche Verlagsbuchhandlung, Contributions to Sedimentology, no. 8 (1978) 69 pp.

U. Zinkernagel, "Framework Alterations in Sandstones in the Course of Diagenesis (Cathodoluminescence Studies)," [abst.], Abstracts, First European meeting, International Association of Sedimentologists, Bochum, Germany (1980) pp. 21-22.

Fossils

B. V. Carrasco, "Cathodoluminescence as an Auxiliary in Studying Diagenesis of Dolomitized Rudistid Shells and Its Relationship to the Development of Porosity," Revista del Instituto Mexicano del Petroleo, 5 (1973) pp. 5-14 [In Spanish].

K. F. Ferrigno, "The Use of Cathodoluminescence in the Study of Limestone and Fossils in Thin Section," in O. C. Kopp (ed.), Geological Applications of Cathodoluminescence, Short Course Notes, University of Tennessee, Knoxville, 1973.

B. M. Low, "Analysis and Interpretation of Fossils in the Onondaga Formation Using Cathodoluminescence as Compared with Petrographic Technique of Examination," B.S. thesis, McMaster University (1974).

M. D. Muir, L. H. Hamilton, P. R. Grant and R. A. Spicer, "A Comparitive Study of Modern and Fossil Microbes, Using X-ray Microanalysis and Cathodoluminescence," in Hall, Echlin, and Kaufmann (eds.), Microprobe Analysis as Applied to Cells and Tissues, Academic Press, London, 1974, pp. 33-58.

D. K. Richter and U. Zinkernagel, "Mn-activated Cathodoluminescence of Echnoid Tests," Proceedings, First European Regional Meeting, International Association of Sedimentologists (1980) pp. 172-176.

M. M. Smith, A. Boyde and S. A. Reid, "Cathodoluminescence as an Indicator of Growth Increments in the Dentine in Tooth Plates of Triassic Lungfish," Neues Jahrabuch fuer Geologie und Palaeontologie, 1 (1984) pp. 39-45.

IGNEOUS - METAMORPHIC ROCKS

V. E. Barnes and S. V. Margolis, "Cathodeluminescence and Microprobe Studies of Libyan Desert Glass and Australites," [abst.], 25th International Geological Congress, Abstracts, Resumes 25, 2, section 15, plantetology (1976) p. 611.

J. L. Berkley, H. G. Brown, K. Keil, N. L. Carter, J.-C. C. Mercier and G. Huss, "The Kenn Ureilite: An Ultramafic Rock with Evidence for Igneous, Metamorphic, and Shock Origin," Geochimica et Cosmochimica Acta, 40 (1976) pp. 1429-1437.

C. J. Derham, J. E. Geake and G. Walker, "Luminescence of Enstatite Achondrite Meteorites," Nature, 203 (1964) pp. 134-136.

D. Dietrich and P. R. Grant, "Cathodoluminescence Petrography of Syntectonic Quartz Fibres," Journal of Structural Geology, 5 (1985) pp. 541-553.

J. Donahue, "Volcanic Ash Correlation by Cathodoluminescence," [abst.], Geological Society of America, Special Paper 121 (1969) 78 pp.

L. v. M. Friberg and R. E. Wymer, "Cathodoluminescence of Metamorphic and Igneous Rocks from a Complex Precambrian Terrain, Wyoming," [abst.], Geological Society of America, Abstracts with Program, 10, no. 7 (1978) 404 pp.

J. E. Geake, A. Dollfus, G. F. J. Garlick, W. Lamb, G. Walker, G. A. Steigmann and C. Titulaer, "Luminescence, Electron Paramagnetic Resonance and Optical Properties of Lunar Material from Apollo 11," Apollo 11 Lunar Science Conference, 3 (1970) pp. 2127-2147.

J. E. Geake, G. Walker, A. A. Mills, and G. F. J. Garlick, "Luminescence of Apollo Lunar Samples," Second Lunar Science Conference, Proceedings, 3 (1971) pp. 2265-2275.

J. E. Geake, G. Walker and A. A. Mills, "Luminescence Excitation by Protons and Electrons Applied to Apollo Lunar Samples," pp. 270-297 in H. C. Urey, and S. K. Runcorn (eds.), The Moon, Proceedings, Symposium of the International Astronomy Union, 47, 1972.

G. R. Green, "Spectroscopic Studies of Transition-metal Luminescence Centres in Silicates," Ph.D. thesis, University of Manchester, 1981.

G. R. Green and G. Walker, "Luminescence Excitation Spectra of Mn^{2+} in Synthetic Forsterite," Phys. Chem. Minerals, 12 (1985) pp. 271-278.

N. N. Greenman and W. B. Milton, "Silicate Luminescence and Remote Compositional Mapping", Proceedings of the 6th Annual Meeting of the Working Group on Extraterrestrial Resources, NASA SP-177 (1968) pp. 55-63.

N. N. Greenman and H. G. Gross, "Luminescence of Apollo 11 and Apollo 12 Lunar Samples," Second Lunar Science Conference, 3 (1971) pp. 2223-2233.

T. R. Greer, V. Vand and J. N. Weber, "Applications of Luminescence Techniques to the Study of the Lunar Surface," in Symposium on the Interpretation of Lunar Probe Data, Huntington Beach, California, (1968) pp. 57-58.

N. Groegler and A. Liener, "Cathodoluminescence and Thermoluminescence Observations of Aubrites," pp. 569-578 in D. J. McDougall (ed.), Thermoluminescence of Geological Materials, Academic Press, London, 1968.

E. D. Hutecheon, I. M. Steele, J. B. Smith and R. N. Clayton, "Ion Microprobe, Electron Microprobe and Cathodoluminescence Data for Allende Inclusions with Emphasis on Plagioclase Chemistry," Geochimica Cosmochimica Acta, Supplement, 10 (1978) pp. 1345-1368.

G. F. Imbusch, "Inorganic Luminescence," pp. 1-92 in M. D. Lumb (ed.), Luminescence Spectroscopy, Academic press, London, 1978.

C. A. Leitch and J. V. Smith, "Striking Cathodoluminescence in the Indarch Enstatite Chondrite," Meteorics, 14 (1979) pp. 469-472.

C. A. Leitch and J. V. Smith, "Two Types of Clinoenstatite in Indarch Enstatite Chondrite," Nature, 283 (1980) pp. 60-61.

R. H. Loeppert and L. R. Hossner, "Reactions of Fe^{2+} and Fe^{3+} with Calcite," Clays and Clay Minerals, 32 (1984) pp. 213-222.

A. N. Mariano, "Cathodoluminescence of Feldspars from Granite Plutons of Southeastern New England," [abst.], Geological Society of America, Abstracts with Programs, 10, no. 2 (1978) 74 pp.

A. N. Mariano, "Enhancement and Classification of Fenitization by Cathodoluminescence," [abst.], Geological Society of Canada-Mineralogical Association of Canada, Program Abstracts, 4 (1979) 65 pp.

S. W. S. McKeever and D. W. Sears, "Meteorites that Glow," Sky and Telescope, 60 (1980) pp. 14-15.

D. B. Nash, "Proton-excited Luminescence of Silicates: Experimental Results and Lunar Implication," Journal of Geophysical Research, 71 (1966) pp. 2517-2534.

D. B. Nash, "Experimental Results on Combined Ultraviolet-proton Excitation of Moon-rock Luminescence," Journal of Geophysical Research, 78 (1973) pp. 3513-3514.

L. E. Page, M. A. Chan and J. D. Tewhey, "Cathodoluminescent Zonation of Anhydrite Veins from the Salton Sea Geothermal Field, California," [abst.], Geological Society of America, Abstracts with Programs, 11, no. 3 (1979) 121 pp.

S. L. Parks, J. W. Erwin, D. C. Noble and E. H. McKee, "Paleomagnetic Stratigraphy, Geochemistry, and Source Areas of Miocene Ash-flow Tuffs and Lavas of the Badger Mountain Area," [abst.], Geological Society of America, Abstracts with Programs, 15, no. 5 (1983) 281 pp.

A. M. Reid and A. J. Cohen, "Some Characteristics of Enstatite from Enstatite Achondrites," Geochimica Cosmochimica Acta, 31 (1967) pp. 661-672.

R. F. Sippel, "Luminescence Petrography of the Apollo 12 Rocks and Comparative Features in Terrestrial Rocks and Meteorites," Second Lunar Science Conference, 1 (1971) pp. 246-263.

R. F. Sippel and A. B. Spencer, "Luminescence Petrography and Properties of Lunar Crystalline Rocks and Breccias," Apollo 11 Science Conference, 3 (1971) pp. 2413-2426.

R. F. Sippel and A. B. Spencer, "Cathodoluminescence Properties of Lunar Rocks," Science, 167 (1970) pp. 677-679.

E. S. Sprunt, "Causes of Quartz Luminescence Colours," Scanning Electron Microscopy, 1 (1981) pp. 525-532.

E. S. Sprunt, L. A. Dengler and D. Sloan, "Effects of Metamorphism on Quartz Cathodoluminescence," Geology, 6 (1978) pp. 305-308.

E. S. Sprunt and A. Nur, "Microcracking and Healing in Granites; New Evidence from Cathodoluminescence," Science, 205 (1979) pp. 495-497.

I. M. Steele and I. D. Hutcheon, "Cathodoluminescence of Carbonaceous Chondrites; Another Petrographic Dimension," [abst.], Meteorics, 17 (1982) pp. 281-282.

I. M. Steele, J. V. Smith and C. Skirius, "Cathodoluminescence Zoning and Minor Elements in Forsterites from the Murchinson (C2) and Allende (C3V) Carbonaceous Chondrites," Nature, 313 (1985) pp. 294-297.

W. F. Thompson and L. M. Friberg, "Cathodoluminescence Colors and Textures of Pre-Cambrian Metasedimentary Rocks from the Black Hills, South Dakota," [abst.], Geological Society of America, Abstracts with Programs, 14, no. 6 (1982) 352 pp.

J. N. Weber, R. T. Greer and V. Vand, "Electron-excited Fluorescence of Serpentines," Planetary and Space Science, 15 (1967) pp. 633-642.

ORE DEPOSITS

A. K. Armstrong, "Petrography and Cathodoluminescence of Carbonate Rocks at Bornite, Alaska," U. S. Geological Survey, Circular 844 (1982) pp. 38-40.

C. E. Barker and J. M. Barker, "A Re-evaluation of the Origin and Diagenesis of Borate Deposits, Death Valley Region, California," pp. 101-135 in J. M. Barker and S. J. Lefond, eds., Borates-Economic Geology and Production, American Institute of Mining, Metallurgical, and Petroleum Engineers, Special Publication, 1985.

L. B. Cobb, "A Study of the Distribution of Trace Elements (Iron and Manganese) and Cathodoluminescent Zonation in Mineralized Fracture Fillings," M.S. thesis, University of Tennessee, 1974, 36 pp.

W. J. Dansart, "A Petrographic Study of the Josephine Breccia in the Metaline District of Northeastern Washington using Cathodoluminescence," M.S. thesis, University of Idaho, 1982.

M. L. Ebers, "Cathodoluminescence Study of Mississippi Valley Type Ore Deposits," [abst.], Geological Society of America, Abstracts and Programs, 5, no. 5 (1973) 395 pp.

M. L. Ebers, "A Study of Gangue Dolomite Mineralization in the Mascot-Jefferson City Zinc District, Tennessee, using Cathodoluminescence," M.S. thesis, University of Tennessee, 1976, 82 pp.

M. L. Ebers, "Cathodoluminescent Zonation Microstratigraphy in Gangue Dolomite in the Mascot-Jefferson City District and its Exploration Application," [abst.], Geological Society of America, Abstracts and Programs, 9, no. 2 (1977) 136 pp.

M. L. Ebers and O. C. Kopp, "Cathodoluminescent Microstratigraphy in Gangue Dolomite, the Mascot-Jefferson City District, Tennessee," Economic Geology, 74 (1979) pp. 908-918.

M. R. Farr, "Regional Isotopic Variation in Bonneterre Formation Dolomite Cements: Implications for Brine Migration Pathways and Sources (abs.)," in Symposium on the Bonneterre Formation (Cambrian), Southeastern Missouri: Stratigraphy, Sedimentology, Diagenesis, Geochemistry, and Economic Geology, J. M. Gregg and R. D. Hagni (eds.), Department of Geology and Geophysics, University of Missouri-Rolla, Rolla, Missouri, p. 8.

J. M. Gregg and R. D. Hagni, "The Use of Cathodoluminescence Microscopy to Reveal Hidden Crystal Faces in Gangue Dolomite Cements, Viburnum Trend, Southeastern Missouri," this volume.

J. M. Gregg and R. D. Hagni, "Irregular Cathodoluminescent Banding in Late Dolomite Cements: Evidence for Complex Faceting and Metalliferous Brines," Geological Society of American Bulletin, in press.

R. D. Hagni and M. Cooper, "The Nature of Phosphorous-bearing Mineral Grains in the Birmingham, Alabama Sedimentary Iron Ores and an Assessment of their Potential Liberation by Beneficiation," 95-117 in R. D. Hagni, ed., Process Mineralogy II, American Institute of Mining, Metallurgical, and Petroleum Engineers, 1982.

R. D. Hagni, "Cathodoluminescence Microscopy Applied to Mineral Exploration and Beneficiation," 41-66 pp. in W. C. Park, D. M. Hausen and R. D. Hagni, eds., Applied Mineralogy, American Institute of Mining, Metallurgical, and Petroleum Engineers, 1985.

R. D. Hagni, "Industrial Applications of Cathodioluminescence Microscopy", this volume.

G. Harwood, "The Application of Cathodoluminescence in Relative Dating of Barite Mineralization in the Lower Magnesian Limestone (Upper Permian), United Kingdom," Economic Geology, 78 (1983) pp. 1022-1027.

L. S. Istas, "Trace Elements in Veins of the Bohemia Mining District, Oregon," Ph.D thesis, University of Washington, 1983, 150 pp.

M. G. Janik, "Cathodoluminescence of Gangue Dolomite near Lead-zinc Occurrences in Stevens County, Northeastern Washington," M.S. thesis, University of Idaho, 1982.

O. C. Kopp, R. M. Larsen and T. L. Ferguson, "A Study of Zoned Dolomite and Calcite Gangue in the Central Tennessee Zinc District using Cathodoluminescence Microscopy and X-ray Fluorescence Analysis," [abst.], Geological Society of America, Abstracts with Programs, 13 (1983) 108 pp.

O. C. Kopp, M. L. Ebers, L. B. Cobb, T. B. Crattie, T. L. Ferguson, R. M. Larsen, R. A. Potosky, and R. T. Steinberg, "Application of Cathodoluminescence Microscopy to the Study of Gangue Carbonates in Mississippi Valley-Type Deposits in Tennessee: The Search for a "Tennessee Trend," this volume.

S. Lindblom, "A Preliminary Study of Cathodoluminescence of Calcite and Sphalerite at Laisvall," ORG 83, Annual Report Ore Research Corporation, Stockholm University, (1983) pp. 61-73. [Abstracted in Fluid Inclusion Research, 1983, 16, p. 154].

M. H. Frank and K. C. Lowmann, "Textural and Chemical Alteration of Dolomite: Interaction of Mineralizing Fluids and Host Rock in a Mississippi Valley-Type Deposit, Bonneterre Formation, Viburnum Trend," this volume.

A. N. Mariano, "The Application of Cathodoluminescence for Carbonatite Exploration and Characterization," 39-57 pp. in J. R. de Andrade RAmos, ed., Preceedings of the First International Symposium on Carbonatites, June, 1976, Pocos de Caldas, Brazil, Published by the Brazil Ministry of Minas Y Energia, 1978.

P. Moller, G. Morteani, J. Hoefs and P. P. Parekh, "The Origin of Ore-bearing Solutions in the Pb-Zn Veins of the Western Hartz, Germany, as Deduced from Rare Earth Element and Isotope Distributions in Calcites," Chemical Geology, 26 (1979) pp. 197-215.

R. D. Morton, "Cathodoluminescence Applied to Uranium Exploration," Nuclide Spectra, 11, no.1 (1978) 2 pp.

L. Rowan, "Cathodoluminescent Zonation in Gangue Hydrothermal Dolomite Cements: Relationship to Mississippi Valley Pb-Zn Mineralization in Southern Missouri and Northern Arkansas," this volume.

M. Shea, "Uranium Migration at some Hydrothermal Veins near Marysvale, Utah; A Natural Analog for Waste Isolation," Materials Research Society Symposia Proceedings, 26 (1984) pp. 227-23?.

R. L. Voss and R. D. Hagni, The Application of Cathodoluminescence Microscopy to the Study of Sparry Dolomite from the Viburnum Trend, Southeast Missouri," Chapter 5 in Mineralogy--Applications to the Minerals Industry (Paul Kerr Memorial Symp. Volume), AIME, New York, NY, pp. 51-68.

MINERALOGY/CHEMISTRY

R. C. G. Adam, T. A. Bielicki and A. R. Lang, "Correlation of Electrostatic Charging Patterns with Internal Structure in Diamonds," Journal of Materials Science, 16, no. 9 (1981) pp. 2369-2380.

I. Balberg and J. I. Pankove, "Cathodoluminescence of Magnetite," Phys. Rev. Getl., 27 (1971) pp. 1371-1374.

G. P. Barsanov and Kh. K. Sarsembaeva, "Luminescence of Iceland Spar," Tr. Mineralog. Muzeya. Akad. Nauk., 13 (1962) pp. 147-152. [In Russian].

G. P. Barsanov, N. E. Sergeeva, N. P. Yushkin, G. V. Spivak, G. V. Saparin and M. K. Antoshin, "Study of the Composition and Inner Structure of Sphalerite by the Use of the Cathodoluminescence Method with Scanning Electron Microscope," Moscow University Geology Bulletin, 29 (1974) pp. 63-66.

R. J. R. S. B. Bhalla, "Electron Beam Damage in Cathodochronic Sodalite," Journal of Applied Physics, 45 (1974) pp. 3703-3709.

R. J. R. S. B. Bhalla, "Intrinsic Cathodoluminescence Emission from Willemite Single Crystals," pp. 71-78 in J. W. Weber and E. White, eds., Space Applications of Solid State Luminescence Phenomena, Pennsylvania State University, Materials Research Laboratory, Special Publication 70-101, 1970.

R. J. R. S. B. Bhalla and E. W. White, "Polarized Cathodoluminescence Emission from Willemite (Zn_2SiO_4(Mn)) Single Crystals," Journal of Applied Physics, 41 (1970) pp. 2268-2269.

R. J. R. S. B. Bhalla and E. W. White, "Intrinsic Cathodoluminescence Emission from Willemite Single Crystals," Journal of Luminescence, 4 (1971) pp. 194-200.

A. I. Blazhevisch, A. V. Lavron and E. I. Panasynk, "Effects of Acceptor-donor Impurities on the Cathodoluminescence of ZnS Single Crystals," Bulletin Acad. Sci, 33 (1969) pp. 906-909.

R. A. Buchanan, K. A. Wickersheim, J. L. Weaver and E. E. Anderson, "Cathodoluminescence Properties of the Rare Earths in Yttrium Oxide," Journal of Applied Physics, 39 (1968) pp. 4342-4347.

R. G. Burns, "Mineralogical Application of Crystal Field Theory," Cambridge University Press, Cambridge, (1970) 224 pp.

R. G. Burns and D. J. Vaughan, "Polarized Electronic Spectra," pp. 39-72 in C. Karr Jr., ed., Infrared and Raman Spectroscopy of Lunar and Terrestrial Materials, Academic Press, London, (1975).

M. Casey and J. Wilks, "Cathodoluminescence in Deformed Diamond," Nature, 239 (1972) pp. 393-394.

E. W. Claffy and R. J. Ginther, "Red-luminescencing Quartz," American Mineralogist, 44 (1959) pp. 987-994.

G. A. Chinner, J. V. Smith and C. R. Knowles, "Transition Metal Contents of AI_2SiO_5 Polymorphs," American Journal of Science, 267-a (1969) pp. 96-113.

A. T. Collins, "Visible Luminescence from Diamond," Industrial Diamond Review, April, (1974) pp. 131-137.

R. Coy-Yll, "Quelques Aspects de la Cathodoluminescence des Mineraus," Chemical Geology, 5 (1970) pp. 243-254.

G. Davies, "Cathodoluminescence of Diamond: A Short Review," Diamond Research 1975, Industrial Diamond Informaition Bureau, (1975) pp. 13-17.

G. Davies, "Cathodoluminescence," pp. 165-181 in J. E. Field, ed., The Properties of Diamond, Academic Press, London, 1979.

J. A. De Ment, "Grading Gems by Cathodoluminescence," Mineralogist, 13 (1945) pp. 112-115.

Hua-Xin Deng, "Cathodoluminescence Study of Apatite," Geochimica, 4 (1980) pp. 368-374.

W. H. Dennen, "Impurities in Quartz," Geological Society of America Bulletin, 75 (1964) pp. 241-246.

G. R. Fonda, "Influence of Activator Environment on the Spectral Emmission of Phosphors," Journal of the Optical Society of America, 47 (1957) pp. 877-880.

R. A. P. Gaal, "Cathodoluminescence of Gem Materials, " Gems and Gemology, 15 (1977) pp. 238-244.

G. F. J. Garlick, "The Kinetics and Efficiency of Cathodoluminescence," British Journal of Applied Physics, 13 (1962) pp. 541-547.

J. G. Geake and G. Walker, "Luminescence of Minerals in the Near Infrared," pp. 73-89 in C. Karr Jr., ed., Infrared and Raman Spectroscopy of Lunar and Terrestrial Materials, Academic Press, London, 1975.

R. Giraud, J. Goni and G. Remond, "Possibilites de la Microanalyse par Sonde Electronique dan la Detection des Elements en Traces, Applications de la Cathodoluminescence a Letude de la Localisation des Elements en Traces Dans les Mineraux," pp. 413-432 in Dosage des Elements a L'etat de Traces dans les Roches et Autres Substances Minerales Naturelles, Coll. Centre Nat. Rech. Sci., 923, 1968 [as cited by Amieux, 1982].

T. B. Gurilenko and G. O. Karapetyan, "Cathodoluminescence of Multicomponent Silicate Glasses Activated with Ce," Zh. Prik Spectrosk., 13 (1970) pp. 85-88. [In Russian].

J. R. Goldsmith and F. Laves, "Potassium Feldspars Structurally Intermediate between Microcline and Sanidine," Geochimica Cosmochimica Acta, 6 (1954) pp. 100-118.

J. C. Goni and G. Remond, "Localization and Distribution of Impurites in Blende by Cathodoluminescence," Mineralogical Magazine, 37 (1969) pp. 153-155.

H. Gorz, R. J. R. S. B. Bhalla and E. W. White, "Detailed Cathodoluminescence Characterization of Common Silicates," Pennsylvania State University, Materials Research Laboratory, Special Publication 70-101, (1970) pp. 62-70.

G. S. Gritsayenko and M. I. Il'In, "Solution Electron Microscopy of Minerals; Main Trends and Possibilities," Akad. Nauk. SSSR, Izv., Ser. Geol., 7 (1975) pp. 21-34. [In Russian].

N. Grogler and A. Liener, "Cathodoluminescence and Thermoluminescence of Aubrites," pp. 569-578 in D. J. McDougall, Thermoluminescence of Geologic Materials, Academic Press, London, 1968.

A. J. Hall, "Post-growth Readjustment of a Cassiterite Twin-boundary Revealed by Cathodoluminescence," Mineralogical Magazine, 42 (1988) pp. 288-290.

M. R. Hall and P. H. Ribbe, "An Electron Microprobe Study of Luminescence Centers in Cassiterite," American Mineralogist, 56 (1971) pp. 31-45.

P. L. Hanley, I. Kiflaw and A. R. Lang, "On Topographicaly Identifiable Sources of Cathodoluminescence in Natural Diamonds," Royal Philosophical Society of London, Transactions, Series A, 284 (1977) pp. 329-368.

D. B. Holi, B. D. Chase and J. B. Steyn, "Scanning Electron Microscopy Studies of Striations in ZnS," Journal of Materials Science, 5 (1970) pp. 546-556.

F. A. Hummel, "Cordierite-indialite: A New Manganese-activated Phosphor," Journal of the Electrochemistry Society, 108 (1961).

F. A. Hummel and J. F. Sarver, "The Cathodoluminescence of Mn^{2+} - and Fe^{3+} - Activated Magnesium Aluminate Spinel," Journal of the Electrochemistry Society, 111 (1964) pp. 252-253.

P. Jeanrot and G. Remond, "Application of Electron Microscopy to Mineralogy; II. Scanning Electron Microscopy and Associated Microanalysis by Dispersive X-ray Spectroscopy," Bulletin de Mineralogie, 101 (1978) pp. 287-304.

C. C. Klick, "Divalent Manganese as a Luminescent Centre," British Journal of Applied Physics, Supplement no. 4, (1955) pp. 74-78.

I. Kiflawi and A. R. Lang, "Polarised Infrared Cathodoluminescence from Platelet Defects in Natural Diamonds," Nature, 267 (1977) pp. 36-37.

Y. Kitano and R. Fujyoshi, "Selective Chemical Leaching of Cadmium, Copper, Manganese and Iron in Marine Sediments," Geoch. J., 14 (1980) pp. 113-122.

K. Kremling, "Trace Metal Fronts in European Shelf Waters," Nature, 303 (1983) pp. 225-227.

E. Lell, N. J. Kreidl and J. R. Hensler, "Radiation Effects in Quartz, Silica, and Glasses," In Progress in Ceramic Science, 4 Pergamon, London, (1966) pp. 1-93.

A. R. Lang, "Internal Structure," pp. 425-469 in J. E. Field (ed.), The Properties of Diamond, Academic Press, London, 1979.

S. Larach, "Cathode-ray-excited Emission Spectroscopy Analysis of Trace Rare Earths," Part 1: Qualitative Studies, Analyt. Chim. acta, 47 (1968) pp. 189-195.

K. R. Laud, E. F. Gibbons, T. Y. Tien and H. L. Stadler, "Cathodoluminescence of Ce^{3+} and Eu^{2+} Activated Alkaline Earth Feldspars," Journal of Electrochemistry Society, 118 (1971) pp. 918-923.

G. Lehmann and H. U. Bambauer, "Quartz Crystals and Their Colors," Agnew. Chem. International Edition, 12 (1973) pp. 283-291.

B. M. Loeffler and R. G. Burns, "Shedding Light on the Colors of Gems and Minerals," American Scientist, 64 (1976) pp. 636-647.

A. S. Marfunin, "Spectroscopy, Luminescence, and Radiation Centers in Minerals," Springer Verlag, Berlin, (1979) 352 pp.

A. N. Mariano, J. Ito and R. J. Ring, "Catholuminescence of Plagioclase Feldspar," [abst.], Geological Society of America, Abstracts and Programs, 5, no. 7 (1973) 726 pp.

A. N. Mariano and P. J. Ring, "Europium-activated Cathodoluminescence in Minerals," Geochimica Cosmochimica Acta, 39 (1975) pp. 649-660.

G. Masse, J. P. Alcardi and J. P. Leyris, "Study of the Yellow Emission of Natural Alpha-HgS," Journal of Luminescence, 17 (1978) pp. 29-48.

J. Mazzaschi, J. Barrau, J. C. Brabant, M. Brousseau and F. Voillot, "Cathodoluminescence Studies of Defects in Natural Diamonds," Rev. Phys. Appl., 15 (1980) pp. 9-14.

R. A. McCauley and F. A. Hummel, "Luminescence as an Indication of Distortion in $A_2^{3+}B_4^{2+}O_7$ Type Phyrochlores," Journal of Luminescence, 6 (1973) pp. 105-115.

W. L. Medlin, "Thermoluminescence Properties of Calcite," Journal of Chem. Phys., 30 (1959) pp. 451-458.

W. L. Medlin, "Thermoluminescence in Dolomite," Journal of Chemical Physics, 32 (1961) pp. 627-677.

W. L. Medlin, "Thermoluminescence in Quartz," Journal of Chemical Physics, 38 (1963) pp. 1132-1143.

W. L. Medlin, "Emission Centres in Thermoluminescent Calcite, Dolomite, Magnesite, Aragonite, and Anhydrite," Journal of the Optical Society of America, 53 (1963) pp. 1276-1285.

M. J. Mendelssohn, H. J. Milledge, E. R. Vance, E. Nave and P. A. Woods, "Internal Radioactive Halos in Diamond," Diamond Research, (1979) pp. 31-36.

H. J. Milledge and others, "Isotopic Variations in Diamond in Relation to Cathodoluminescence," Acta Crystallographica, 40, supplement (1984) 255 pp.

K. M. Mohammed, G. Davies and A. T. Collins, "Uniaxial Stress Measurements on Optical Transitions in Yellow Luminescencing Brown Diamonds," Journal of Physics, Solid State Physics, 15 (1982) pp. 2789-?.

A. M. Moore, "Optical Studies of Diamonds and Their Surfaces; A Review of the Late Professor Tolansky's Work," pp. 245-277 in J. E. Field (ed.), The Properties of Diamond, Academic Press, London, 1979.

K. Nassau, "The Origins of Color in Minerals," American Mineralogist, 63 (1978) pp. 219-229.

K. Nassau and B. E. Precott, "Smoky, Blue, Greenish Yellow, and Other Irradiation-related Colors in Quartz," Mineralogical Magazine, 41 (1977) pp. 301-312.

V. V. Osiko and G. V. Maksimova, "Valence of the Manganese Activator in Crystal Phosphor," Optics and Spectroscopy, 9 (1960) pp. 478-481.

A. M. Protnov and B. S. Gorobets, "Luminescence of Apatite from Different Rock Types," Doklady Akademiya Nauk SSSR, 184 (1969) pp. 110-113. [In Russian].

A. M. Reid, T. E. Bunch and A. J. Cohen, "Luminescence of Orthopyroxenes," Nature, 204 (1964) pp. 1292-1293.

G. Remond and P. Perrot, "Distribution of Impurities in Minerals made Visible through Cathodoluminescence," [abst.], Dan. Geol. Foren. Medd., 19, part 3 (1969) pp. 332-333.

G. Remond, "Exemples d'indification et de Localisation des Elements en Traces dans des Mineraux Luminescents (Cassiterites) a L'aide de L'analseur Ionique," Bull. Soc. Fr. Mineral. Cristallogr., 96 (1973) pp. 183-198.

G. Remond, S. Kimoto and H. Okuzumi, "Use of the SEM in Cathodoluminescence of Natural Samples," Scanning Electron Microscopy, (1970) pp. 33-40.

G. Remond, S. Kimoto and H. Okuzumi, "Applications of Scanning Electron Microscope to the Study of Cathodoluminescence of some Minerals, Limits of Resolution and Sensitivity," pp. 611-617 in G. Shinoda (ed.), Proceedings of the International Conference X-ray Microanalysis, 1972.

G. Remond, C. Le Gressus and H. Okuzumi, "Electron Beam Effects Observed in Cathodoluminescence and Auger Electron Spectroscopy in Natural Materials: Evidence for Ionic Diffusion," Scanning Electron Microscopy, part 1 (1979) pp. 237-242.

P. H. Ribbe and M. R. Hall, "Microprobe Cathodoluminescence and X-ray Emission Studies of Cassiterites," [abst.], Geological Society of America, Special Paper 87 (1966) pp. 135-136.

M. Ross, J. J. Papike and P. W. Weiblen, "Exsolution in Clinoamphiboles," Science, 159 (1968) pp. 1099-1102.

D. D. Saksena and L. M. Pant, "Cathodoluminescence of Crystalline Quartz," J. Chem. Phys., 19 (1951) pp. 363-372.

J. E. Schulman, L. W. Evans, R. J. Ginther and K. J. Murata, "The Sensitized Luminescence of Manganese-activated Calcite," Journal of Applied Physics, 18 (1974) pp. 732-739.

G. V. Spivak, G. V. Saparin, M. K. Anioshin, I. V. Nesterov and G. P. BARSANOV, "Observation of Molybdoscheelite with Cathodeluminescence and Scanning Electron microscopy," Moscow University, Vestn. Ser. Geol., 28 (1973) pp. 40-43.

E. S. Sprunt, "Effects of Impurities on Quartz Cathodoluminescence," [abst.], EOS, 59 (1978) 1216 p.

S. E. Sommer, "Characterization and Application of Cathodoluminescence from Mn Activated Carbonate Minerals," Ph.D. thesis, Pennslyvannia State University, 1969, 124 p.

S. E. Sommer, "Cathodoluminescence of Carbonates; 1. Characterization of Cathodoluminescence from Carbonate Solid Solutions," Chemical Geology, 9 (1972) pp. 257-273.

N. Sumida and A. R. Lang, "Cathodoluminescence Evidence of Dislocation Interactions in Diamond," Philosophical Magazine, a, 43, (1981) pp. 1277-1287.

D. J. Telfer and G. Walker, "Optical Detection of Fe^{3+} in Lunar Plagioclase," Nature, 258 (1975) pp. 694-695.

D. J. Telfer and G. Walker, "Lignad Field Bands of Mn^{2+} and Fe^{3+} Luminescence Centres and their Site Occupancy in Plagioclase Feldspars," Modern Geology, 6 (1978) pp. 199-210.

G. Walker, "Luminescence Centres in Minerals," Chem. Britian, 19 (1983) pp. 824-831.

G. Walker, "Mineralogical Applications of Luminescence Techniques," pp. 103-140 in F. J. Berry and D. J. Vaughan (eds.), Chemical Bonding and Spectroscopy in Mineral Chemistry, Chapman and Hall, London, 1985.

B. C. Weber, "Luminescent Phenomena and Zirconia Research," Interceram, no. 2 (1963) pp. 90-96.

S. West, "Luminescent Iithology," Science News, 114 (1978) pp. 316-319.

S. West, "Luminescent Iithology," Journal of the Fluorescent Mineral Society, 9 (1980) pp. 9-22.

K. A. Wickersheim, R. A. Buchanan and L. E. Sobon, "Cathodoluminescence as an Analytical Techinque in the Determination of Rare Earths in Yttria," Analytical Chemistry, 40 (1968) pp. 807-809.

T. R. Wildeman, "The Distribution of Mn^{2+} in some Carbonates by Electron Spin Resonance," Chemical Geology, 5 (1970) pp. 167-177.

F. E. Williams, "Theory of Activator Systems in Luminescent Solids," British Journal of Applied Physics, supplement no. 4 (1955) pp. 97-102.

G. S. Woods and A. R. Lang, "Cathodoluminescence, Optical Absorption and X-ray Topographic Studies of Synthetic Diamonds," Journal of Crystal Growth, 28 (1975) pp. 215-226.

G. S. Woods, "Electron Microscopy of 'Giant' Platelets on Cube Planes in Diamonds," Philosophical Magazine, 34 (1976) pp. 993-1012.

Methods

D. Allen, "A One-stage Precision Polishing Technique for Geological Specimens," Mineralogical Magazine, 48 (1984) pp. 298-300.

G. C. Amstutz, "The Preparation and Use of Polished Thin Sections," American Mineralogy, 45 (1960) pp. 1114-1116.

J. Babcock and D. J. Marshall, "Introduction to Cathodoluminescence," 5 p. in J. Babcock, D. J. Marshall and J. B. Thomas (eds.), Course Notes, Workshop on Cathodoluminescence, Geology Department, University of Tulsa, Tulsa, Oklahoma, March 8, 1978.

C. E. Barker and T. J. Reynolds, "Preparing Doubly Polished Sections of Temperature Sensitive Rocks," Journal of Sedimentary Petrology, 54 (1984) pp. 635-636.

C. E. Barker and T. Wood, "Notes on Cathodoluminescence Microscopy using the Technosyn Stage, and a Bibliography of Applied Cathodoluminescence," U.S. Geological Survey Open File Report 86-85, (1986) 25 p.

C. E. Barker and T. Wood, "A Review of the Technosyn and Nuclide Cathodoluminescence Stages and their Application to Sedimentasry Geology," this volume.

R. T. Bell, J. R. Smyth and E. K. Hege, "A Device for the Study of Cathodoluminescence," [abst.], Virginia Journal of Science, 17, (1966), p. 260.

E. F. Bond, D. Beresford and G. H. Haggis, "Improved Cathodoluminescence Microscopy," Journal of Microscopy, 100 (1974) pp. 271-282.

M. Brenner, "Computer Enhancement of Low Light Microscopic Images," American Laboratory, 15 (1983) pp. 30-35.

W. Broecker, H. Hoehling, W. A. P. Nicholson, E. R. Krefting, J. Schreiber, Schlacke, and B. Drueen, "Comparison of the Methods of Cathodoluminescence, Electron Probe Microanalysis, and Calcium Staining, Applied to Human Aorta with Isthmus Stenosis," Pathol., Res. Pract., 163 (1978) pp. 310-322.

G. R. Brumby and T. J. Sheperd, "Improved Sample Preparation for Fluid Inclusion Studies," Mineralogical Magazine, 42 (1978) pp. 297-298.

J. G. Delly and W. F. Wills, Jr., "How to Buy a Compound Microscope: An Update," American Laboratory, 17 (1985) pp. 66-115.

J. A. D. Dickson, "Carbonate Identification and Genesis as Revealed by Staining," Journal of Sedimentary Petrology, 36 (1966) pp. 491-505.

W. J. Furbish, "Polished Thin Sections for Cathode-luminescent Study," Nuclide Corporation, Publication no. 1488-1174, (1974).

R. Gaal, "Cathodoluminescence by the Luminoscope," Guilds - Conclave Issue, Boston, Massachusetts, (1976) pp. 13-15.

V. Gobel, "Rapid Examination and Petrography of Bulk Samples by Cathodoluminescence," p. 1 in J. Babcock, D. J. Marshall and J. B. Thomas (eds.), Course Notes, Workshop on Cathodoluminescence, Geology Department, University of Tulsa, Tulsa, Oklahoma, March 8, 1978.

G. S. Gritsaenko, V. E. Sonyushkin, M. Ilin, V. V. Ermilov and O. I. Luneva, "Certain Problems of Electron Petrography (The State-of-art and Methodology)," Lithology and Mineral Resources (USSR), 12 (1977) pp. 443-456.

R. T. Greer and E. W. White, "Microprobe Attachment for Quantitative Studies of Cathodoluminescence," Transactions of the Second National Conference on Electron Microprobe Analysis, paper no. 51, 1967.

T. D. S. Hamilton, I. H. Numro and G. Walker, "Luminescence Instrumentation," pp. 149-238 in M. D. Lumb, ed., Luminescence Spectroscopy, Academic Press, London, 1978.

W. M. Hanusiak, "Low Temperature Cathodoluminescence of Crystalline Silica for Use in the Characterization of Respirable Dusts," Pennsylvania State University, Ph.D. thesis, 1975.

W. M. Hanusiak, "SEM Cathodoluminescence for Characterization of Damaged and Undamaged Alpha Quartz in Respirable Dusts," pp. 125-131 in O. M. Johari (ed.), Scanning Electron Microscopy, 1975.

D. B. Holi and S. Datta, "The Cathodoluminescence Mode as an Analytical Technique: Its Development and Prospects," Scanning Electron Microscopy, 1 (1980) pp. 259-278.

R. G. Hurley, "Innovations in Soft X-ray Spectroscopy and Cathodoluminescence for Quantitative Phase Analysis and Chemical Bonding Characterization of Materials," Ph.D. thesis, Pennsylvania State University, 1973.

V. W. Gobel and W. J. Patzelt, "Cathodoluminescence - Investigations of Solids with the Leitz Contrasting Device," Scientific and Technical Information, Leitz Wetzlar, 6 no. 7 (1976) pp. 263-267.

R. N. Kniseley, F. C. Laabs and V. A. Fassel, "Analysis of Rare Earth Materials by Cathodoluminescence Spectra Excited in an Electron Microprobe," Analytical Chemistry, 41 (1969) pp. 50-53.

R. N. Kniseley and F. C. Laabs, "Applications of Cathodoluminescence in Electron Microprobe Analysis," pp. 371-382 in C. A. Anderson (ed.), Microprobe Analysis, New York, John Wiley, 1973.

O. C. Kopp, "An Inexpensive Means for Increasing the Magnification Range of a Luminoscope," Journal of Sedimentary Petrology, 51 (1981) pp. 667-668.

A. N. Mariano, "Macro-photography of CL," Nuclide Corporation, Publication number 1080-1083, (1983).

D. J. Marshall, "Suggested Standards for the Reporting of Cathodoluminescence Results," Journal of Sedimentary Petrology, 48 (1978) pp. 651-653.

D. J. Marshall, J. Giles, and A. Mariano, "Combined Instrumentation for EDS Elemental Analysis and Cathodoluminescence Studies of Geological Materials," this volume.

A. Mazzucotelli and R. Vannucci, "Determination of Traces of Cerium in Silicates by Cathodoluminescence on a $CaO-CaSO_4$ Matrix after Separation by Ion Exchange Chromatography," Soc. Ital. Mineral. Petrol., Rend., 35 no. 2 (1979) pp. 609-618.

W. Moran, "Radiation Safty Considerations in Cathodoluminescence Observation," 1 p. in J. Babcock, D. J. Marshall, and J. B. Thomas (eds.), Course Notes, Workshop on Cathodoluminescence, Geology Department, University of Tulsa, Tulsa, Oklahoma, March 8, 1978.

S. J. Mugridge and H. R. Young, "Rapid Preparation of Polished Thin Sections for Cathodoluminescence Study of Carbonate Rocks," Canadian Mineralogy, 22 (1984) pp. 513-515.

A. H. Munsell, A Colour Notation, Munsell Colour Co., Inc., Baltimore, 1975.

D. Pye and D. H. Krinsley, "Petrographic Examination of Sedimentary Rocks in the SEM (Scanning Electron Microscope)," Journal of Sedimentary Petrology, 54 (1984) pp. 877-888.

D. E. Ryan and J. P. Szabo, "Cathodoluminescence of Detrital Sands; A Technique for Rapid Determination of the Light Minerals of Detrital Sands," Journal of Sedimentary Petrology, 51 (1981) pp. 669-670.

R. F. Sippel, "Simple Device for Luminescence Petrography," Review of Scientific Instruments, 36 (1965) pp. 1556-1558.

S. C. Sommer, "Selected Area X-ray Luminescence Spectroscopy with the X-ray Milliprobe," Applied Spectrography, 26 (1972) pp. 557-558.

K. J. Spray, "How to Prepare Relief-free Polished Surfaces of Geological or Refractory Specimens," Industrial Diamond Review, 30 (1970) pp. 182-186.

J. B. Steyn, P. Giles and D. B. Holt, "An Efficient Spectrocopic Detection System for Cathodoluminescence Mode Scanning Electron Microscopy (SEM)," Journal of Microscopy, 107 (1976) pp. 107-128.

J. P. Szabo and D. E. Ryan, "Determination of Quartz/Feldspar Ratios in the Fine Sand Fraction of Tills using Cathodoluminescence," [abst.], Geological Society of America, Abstracts with Programs, 12 (1980) 258 pp.

J. E. Taggart Jr., "Polishing Technique for Geologic Samples," American Mineralogist, 62 (1977) pp. 824-827.

J. L. Woodbury and T. A. Vogel, "A Rapid, Economical Method for Polishing Thin Sections for Microprobe and Petrographic Analyses," American Mineralogist, 55 (1970) pp. 2095-2102.

Fluorescence Petrography

J. J. Dravis and D. A. Yurewicz, "Enhanced Carbonate Petrography using Fluorescence Microscopy," Journal of Sedimentary Petrology, 55 (1985) pp. 795-805.

S. H. Linwood and W. A. Weyl, "The Fluorescence of Manganese in Glasses and Crystals," Journal of the Optical Society of America, 32 (1942) pp. 443-445.

D. Maier and W. Wetzel, "Fluoreszenzmikroskopie Geolischer und Palaontologisher Objevte," Zeiss Mitt., 1 (1958) pp. 127-131.

Process Mineralogy Applications to Precious Metal Deposits

PARTICULATE GOLD OCCURRENCES

IN THREE CARLIN CARBONACEOUS ORE TYPES

D. M. Hausen, J. W. Ahlrichs, W. Mueller
Newmont Exploration Limited
44 Briar Ridge Road
Danbury, CT 06810

W. C. Park
10453 South Knoxville Street
Tulsa, OK 74137

Fine particulate gold (1-10 µm) was observed by conventional ore microscopy in flotation and gravity concentrates from three types of Carlin carbonaceous ores, which were designated as carbonate ore, argillaceous ore, and siliceous ore. Finer gold down to 0.1 µm was detected by scanning electron microscopy, and probably extends down into even finer colloidal sizes in significant proportions in these three ore types.

Although flotation recoveries were relatively poor, and only a small portion of the total gold could be accounted for by microscopic observation, it is proposed that significant proportions of the total gold in these select carbonaceous samples may occur as fine particulate gold.

Introduction

The nature of fine gold occurrences in Carlin-type carbonaceous ores has been of considerable interest to both mineralogists and metallurgists since the discovery of Newmont's Carlin gold mine in 1964. Because of the extremely fine-grained characteristics of the gold in carbonaceous ores, microscopically-visible gold has been rarely observed. The textural forms and associations of the gold are poorly understood. Much of the gold has been considered to be finely associated with pyrite and/or the carbonaceous matter, based on test work that indicates most of the gold (>80%) to be amenable to cyanidation after oxidizing chlorination treatment. High levels of gold were reported by J. D. Wells and T. E. Mullins (1) in arsenian pyrites in rock samples from the Carlin mine. A. S. Radtke and B. J. Scheiner (2) had reported earlier that primary gold was contained mostly in the carbonaceous matter.

The purpose of this paper is to present data and observations for gold occurrences identified in specially-prepared concentrates from three different Carlin carbonaceous ores. These concentrates, obtained by flotation, desliming and centrifuging methods, have provided the opportunity to observe particulate forms of gold down to about one micrometer in size by conventional ore microscopy and electron microprobe techniques. Even finer colloidal sizes down to 0.1 micrometers (μm) were detected by high resolution scanning electron microscopy (SEM). Preliminary observations of fine gold occurrences in pyrite and carbonaceous matter were also attempted with various probe and SEM methods of analyses.

Description and Composition of Samples

The three samples of carbonaceous ore types used for these studies based on the major mineral components included the following:

1. Carbonate ore taken from the Carlin No. 1 crusher plant,

2. Siliceous ore collected from near a silicified fault at the Carlin Main Pit,

3. Argillaceous ore from the Maggie Creek West Pit.

Assays and semiquantitative mineralogical data for the above ores are compared in Table I. Petrographic classifications of the rocks based on examination of polished thin sections are also included.

Analytically, head samples of the three carbonaceous ores contain roughly equivalent amounts of organic carbon (0.4-0.5%) as carbonaceous matter, and sulfide sulfur (0.33-0.41%) as pyrite.

The three samples are significantly different, as shown by the comparison of mineralogy and rock types. The carbonate ore sample contains appreciable amounts of dolomite (24%) and calcite (13%), in addition to quartz and illite, and was classified petrographically as dolomitic siltstones and argillites. The siliceous ore contains mostly quartz (77%) as massive microcrystalline chert, which was apparently derived hydrothermally via an adjacent fault. The argillaceous ore contains major amounts of illitic clay (46%) and fine silty quartz as bedded argillaceous shales.

TABLE I

Head Assays and Semiquantitative Mineralogy for Three Carlin Carbonaceous Ore Types

Ore Type	Carbonate Ore	Siliceous Ore	Argillaceous Ore
Assays			
Au (oz/t)	0.174	0.444	0.304
Organic Carbon (%)	0.4	0.4	0.5
Sulfide Sulfur (%)	0.36	0.33	0.41
As (%)	0.04	0.05	0.08
Semiquantitative Mineralogy (Wt %)			
Quartz	42	77	45
Calcite	13	Tr	Tr
Dolomite	24	12	Tr
Illite	16	9	46
Kaolin	1	-	-
Barite	0.5	0.6	-
Jarosite	-	-	7
Pyrite	0.7	0.6	0.8
Ferruginous Oxides	0.3	0.2	0.1
Carbonaceous Matter	0.5	0.5	0.6
Petrographic Rock Type	Dolomitic Siltstones + Argillites	Massive Chert	Argillaceous Shale

Procedures for Separations

Flotation, desliming and centrifuging procedures were used in attempts to obtain separate concentrates of metallic gold, carbonaceous matter and pyrite. Only partial success was achieved because of the following:

1. The fine-grained nature of all the gold, much of the carbonaceous matter, and pyrite prohibited their liberation.

2. Similar flotation characteristics of carbonaceous matter and metallic gold.

3. Slime interference from illitic clays, causing dilution of concentrates by illite agglomerations.

Flotation Tests for Two-Kilogram Charges

Two-kilogram charges for each of the three ore types were ground in a Dorr-Oliver laboratory ball mill. The carbonate and siliceous ore types were ground to approximately 65% minus 200 mesh based on screen sizing of tailings. Due to the soft nature of the argillaceous ore type, and attrition during flotation screen analyses of the flotation tailing indicated a finer grind (80% minus 200 mesh) than 65% minus 200 mesh.

The ground materials were floated in a Galigher Agitair cell. Rougher concentrates were produced and cleaned twice, resulting in a carbon recleaner concentrate, a carbon recleaner tailing, a carbon cleaner tailing, and a rougher flotation tailing. Highest concentrations of carbonaceous matter and pyrite occurred in the carbon recleaner concentrate and carbon cleaner tailing, respectively, but these were heavily diluted by illite. Consequently, these two products were deslimed, producing sands and slimes fractions. This helped somewhat to upgrade the pyrite and carbonaceous matter in the sands fractions, but fully satisfactory concentrates of the two were not achieved.

Flotation and Centrifugal Heavy Liquid Separations

Evaluation of the flotation products from the two-kilogram tests indicated that a concentrate of the carbonaceous matter could be obtained by centrifuging the carbon recleaner concentrate of the carbonate ore type. In order to obtain sufficient amounts of carbon recleaner concentrate for gravity separations by the centrifugal technique, a twenty-kilogram charge was ground and floated. Two charges of ten kilograms each were floated in a one cubic foot Denver cell.

Approximately 30 grams of carbon recleaner concentrate were obtained by flotation and were deslimed. The sands from the deslimed carbon recleaner concentrate were separated into float and sink gravity fractions, using a 2.4 specific gravity liquid.

Separations were conducted in 50-ml centrifuge tubes in an International Centrifuge, Size 1, Model CM, at 2500 rpm. Tubes filled with heavy liquid containing dispersed sample were centrifuged four to five times for 5 to 10 minutes each, until clear separations of sink and float minerals were achieved. To ensure efficient separations, the float fractions were redispersed with a glass rod after each time interval.

At the end of each separation, the tubes were removed and the float fractions decanted into one beaker; the sink fractions from the bottom of the tubes were likewise washed out into a second beaker. Both heavies and lights fractions were washed four or five times with acetone to remove all traces of heavy liquids, and allowed to dry under a ventilated hood.

The two fractions, e.g., floats (<2.4 Sp.G.) and sinks (>2.4 Sp.G.), were weighed and analyzed for gold and organic carbon, as were the flotation products for the sample.

Flotation of Cyanided Pulp

To evaluate "Preg Rob" gold loading on inherent carbonaceous matter, a 24-hour cyanidation test was conducted on the carbonate ore, followed by flotation. Comparisons of assays obtained on flotation products from the cyanided and uncyanided tests are described in this paper under "Gold Loading...".

Flotation Products for the Three Ore Types

The flotation products and sand-slime fractions for two-kilogram charges of the three carbonaceous ore samples were analyzed semiquantitatively by X-ray diffraction, followed by assays for gold, organic carbon, and sulfide sulfur. Small amounts of each separation product were mounted in epoxy for microscopic examination, and select products were evaluated by an electron microprobe.

Assays and Semiquantitative Mineralogy

Assays, distributions and semiquantitative mineralogical data for the flotation and sand-slime fractions from the three carbonaceous samples are given in Tables II, III and IV.

Calculated gold recoveries for the combined concentrates of the carbonate and siliceous ore types are very low (13.5-16.1%), but increase to 44.7% in the argillaceous ore. Generally poor recoveries are also shown for the carbonaceous matter (9.9-26.0% organic C) and pyrite (22.1-35.3% sulfide S).

The sands fractions for the carbon recleaner concentrates contain highest concentrations of gold (5.54-41.0 oz/t) and organic carbon (7.2-13.9%), but even these deslimed fractions are heavily diluted with illite (30-40%). Illite occurred as agglomerations with fine occlusions of various gangue components, as well as microscopic particulates of metallic gold.

There are two forms of pyrite in Carlin ores; one is an euhedral cubic variety, and the other is framboidal. The euhedral variety is usually coarser grained and was the major form recovered in the sands fractions of the carbon cleaner tailings. The framboidal variety, which is usually considered to have the highest gold content, is finer grained, and was largely lost in the flotation tailings due to poor liberation.

Moderate concentrations of pyrite were obtained for the carbon cleaner tailing sand fractions from the carbonate and siliceous ores. The respective sulfide sulfur assays for these fractions were 20.0% and 11.4%, which equate to about 38% and 21% pyrite, mostly as the euhedral variety. Gold assays of 0.82 oz/t and 1.83 oz/t were relatively low, as compared to most of the other fractions, especially the products from the carbon recleaner concentrates and the carbon recleaner tails. This suggested that euhedral pyrite contained very little of the total gold. For the argillaceous ore, very little pyrite was concentrated in any of the products, and no correlation can be shown with regard to pyrite and gold contents.

Inspection of organic carbon versus gold assays shows a nearly linear relationship. The first impulse would be that most of the gold is associated with carbonaceous matter. However, additional tests and studies were conducted that indicated very little, if any, of the gold is indeed associated with carbonaceous matter, as will be discussed later in the paper.

TABLE II

Assays and Semiquantitative Mineralogy
for Flotation Products from Carbonate Ore Sample

Assays and Distributions

	Wt %	Au oz/t	Au Dist(%)	Organic C %	Organic C Dist(%)	Sulfide S %	Sulfide S Dist(%)
Assay Head		0.174		0.4		0.36	
Calculated Head	100.00	0.171	100.0	0.45	100.0	0.33	100.0
Carbon ReCl Conc							
Sand Fraction	0.15	5.54	4.9	10.1	3.3	4.27	2.0
Slime Fraction	0.10	2.35	1.4	4.3	0.9	1.83	0.6
Carbon ReCl Tail	0.59	2.04	7.0	5.3	6.9	4.36	7.8
Carbon Cl Tail							
Sand Fraction	0.15	0.82	0.7	0.9	0.3	20.0	9.1
Slime Fraction	0.59	0.60	2.1	1.5	1.9	1.50	2.7
Flotation Tail	98.42	0.146	83.9	0.4	86.7	0.26	77.8
Combined Concs	1.58	1.74	16.1	3.7	13.3	4.61	22.2

Semiquantitative Mineralogy (Wt %)

Products	Quartz	Cal-cite	Dolo-mite	Illite	Kaolin	Barite*	Pyrite*	FeOx*	Carbon. Matter*
Carbon ReCl Conc									
Sand Fraction	34	3	5	35	2	-	8	-	13
Slime Fraction	40	4	6	41	1	-	3	-	5
Carbon ReCl Tail	40	5	9	30	1	-	8	Tr	7
Carbon Cl Tail									
Sand Fraction	24	7	19	10	1	Tr	38	Tr	1
Slime Fraction	44	8	17	24	2	-	3	Tr	2
Flotation Tail	42	13	26	16	1	0.4	0.5	0.3	0.5

* Calculated from assays.

TABLE III

Assays and Semiquantitative Mineralogy
for Flotation Products from Siliceous Ore Sample

Assays and Distributions

	Wt %	Au oz/t	Au Dist(%)	Organic C %	Organic C Dist(%)	Sulfide S %	Sulfide S Dist(%)
Assay Head		0.444		0.4			0.23
Calculated Head	100.000	0.441	100.0	0.4	100.0	0.29	100.0
Carbon ReCl Conc							
Sand Fraction	0.075	18.9	3.2	13.9	2.4	4.2	1.1
Slime Fraction	0.042	13.1	1.3	8.8	0.8	1.0	0.1
Carbon ReCl Tail	0.401	6.33	5.7	4.7	4.3	4.6	6.3
Carbon Cl Tail							
Sand Fraction	0.342	1.83	1.4	1.5	1.2	11.4	13.4
Slime Fraction	0.341	2.41	1.9	1.6	1.2	1.0	1.2
Flotation Tail	98.799	0.386	86.5	0.4	90.1	0.23	77.9
Combined Concs	1.201	4.96	13.5	3.6	9.90	5.36	22.1

Semiquantitative Mineralogy

Products	Quartz	Cal-cite	Dolo-mite	Illite	Barite*	Pyrite*	FeOx*	Matter*
Carbon ReCl Conc								
Sand Fraction	35	Tr	6	34	-	8	-	17
Slime Fraction	35	Tr	6	46	-	2	-	11
Carbon ReCl Tail	42	Tr	7	36	1	9	Tr	5
Carbon Cl Tail								
Sand Fraction	50	Tr	11	15	1	21	Tr	2
Slime Fraction	55	Tr	9	31	1	2	Tr	2
Flotation Tail	76	Tr	13	9	0.3	0.5	0.2	0.5

* Calculated from assays.

TABLE IV

Assays and Semiquantitative Mineralogy
for Flotation Products from Argillaceous Ore Sample

Assays and Distributions

		Au		Organic C		Sulfide S	
	Wt %	oz/t	Dist(%)	%	Dist(%)	%	Dist(%)
Assay Head		0.304		0.5		0.41	
Calculated Head	100.000	0.303	100.0	0.6	100.0	0.44	100.0
Carbon ReCl Conc							
Sand Fraction	0.075	41.0	10.2	7.2	0.9	3.7	0.6
Slime Fraction	0.049	21.1	3.4	3.0	0.2	2.1	0.2
Carbon ReCl Tail	1.698	2.70	15.1	2.8	7.7	3.5	13.5
Carbon Cl Tail							
Sand Fraction	0.609	0.20	0.4	0.6	0.6	3.2	4.4
Slime Fraction	6.062	0.78	15.6	1.7	16.6	1.2	16.6
Flotation Tail	91.507	0.183	55.3	0.5	74.0	0.31	64.7
Combined Concs	8.493	1.59	44.7	1.9	26.0	1.83	35.3

Semiquantitative Mineralogy (Wt %)

Products	Quartz	Calcite	Dolomite	Illite	Jarosite*	Pyrite*	FeOx*	Carbon Matter*
Carbon ReCl Conc								
Sand Fraction	42	Tr	Tr	40	2	7	-	9
Slime Fraction	40	-	-	50	2	4	-	4
Carbon ReCl Tail	36	-	-	50	3	7	-	4
Carbon Cl Tail								
Sand Fraction	50	-	-	41	2	6	Tr	1
Slime Fraction	35	-	-	55	4	2	Tr	4
Flotation Tail	48	Tr	Tr	47	4	0.6	0.1	0.6

* Calculated from assays.

Microscopically-Visible Gold

Epoxy mounts for all separation products of the three carbonaceous ores were examined in detail up to magnifications of approximately 600X. At these magnifications, the lower size limit for the identification of gold is about 1 µm.

The three products and the sands and slimes fractions, which compose the combined flotation concentrates, contain moderate amounts of illitic agglomerations. Illitic clays serve as a binder for particles of pyrite, carbonaceous matter, non-opaque gangue, and nearly all of the detectable microscopically-visible gold.

Most of the gold detected in these products was in the 1 to 3 µm sizes, although several particles approached 10 µm. Particulate gold was detected as discrete tiny spheres (Fig. 1) and as aggregates of fine gold (Fig. 2) throughout the illitic agglomerations. Individual granules within these clusters may range down to sizes of less than 1 µm, suggesting that even finer sized gold could be present (Fig. 3).

Epoxy mounts were systematically traversed to evaluate the micro-copically-visible gold by a modified gross-counting technique, similar to a method used by Jones and Gavrilovic (3). The numbers and sizes of particles identified in each product were recorded, then calculated into percentages utilizing specific gravity factors of 19 for gold and 2.7 for the other components; microprobe studies indicated that these particles were nearly pure gold (>95% Au). The weighted factors for gold were converted to ounces per ton for comparison with the assayed values. Results for these counts are compared in Table V. It must be emphasized that the values obtained by this technique should be considered as rough approximations, until a statistical evaluation can be formulated.

Particulate microscopically-visible gold (>1 µm) was abundant, especially in the sands and slimes for the carbon recleaner concentrate. However, after calculations, the microscopically-visible gold accounted for only about 0.6 to 3% of the total gold in the overall samples. These studies indicated that particulate gold detected at lower limits of conventional microscopy probably extended into finer sizes, which was subsequently confirmed by scanning electron microscopic studies, discussed in a later section of this paper under "High Resolution SEM....".

Electron Microprobe Studies

Occurrences of submicroscopic gold were further pursued by electron microprobe analysis. These studies were considered preliminary, because probe analyses of the pyrites in the epoxy mounts were limited to particle sizes of 30 µm or coarser due to the instability of finer sizes in the electron beam. Additional probe studies are planned for whole rock sections because the fine pyrite in the rock matrix will be more stable to the electron beam than pyritic particles in epoxy.

Quantitative electron microprobe analyses of cleaner sand fractions were conducted mostly on the coarser euhedral pyrite, which makes up most of these fractions. An ETEC Autoprobe at the Foote Mineral laboratory was used. This instrument is equipped with three computer-driven spectrometers for quantitative analyses. A Bence-Albee program was used for automatic data collection and operation of the spectrometers.

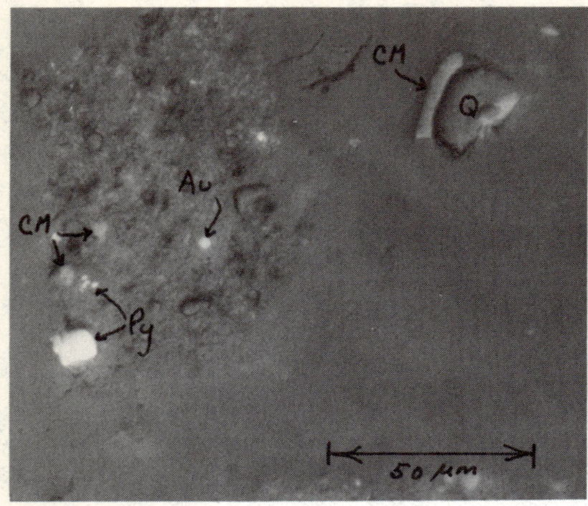

Figure 1.
Fine gold sphere (1μm) in illitic agglomerate with fine carbonaceous matter (CM), pyrite (Py) and quartz (Q). Note the coarser carbon matter on quartz in upper right hand corner.
(X666, incident light)

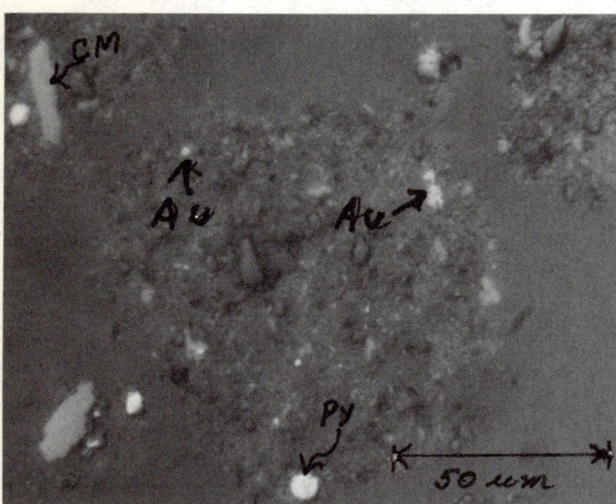

Figure 2.
Single sphere and aggregates of gold in illite agglomerate. Some carbonaceous matter (CM) and a spheroidal pyrite (Py) are also shown.
(666, incident light)

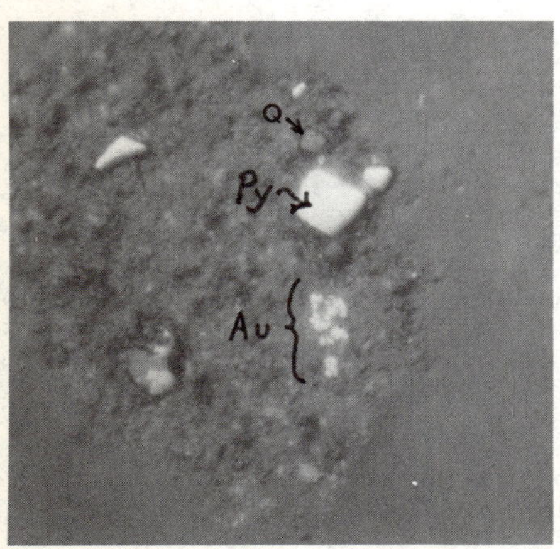

Figure 3.
Numerous fine gold spheres (Au) as clusters in illitic agglomerate with pyrite (Py), quartz (Q) and extremely fine carbon matter.
(X666, incident light)

TABLE V

Microscopically-Visible Gold (>1 μm) in Products

Carbonate Ore

Products	No. of Visible Au Particles	Visible Au (oz/t)	Au Assay (oz/t)	% of Total as Visible Au
Carbon ReCl Conc				
Sand Fraction	60	0.4	5.54	7.
Slime Fraction	33	0.2	2.35	4.
Carbon ReCl Tail	4	0.03	2.04	2.
Carbon Cl Tail				
Sand Fraction	0	0.00	0.82	0.
Slime Fraction	2	0.01	0.60	2.
Flotation Tail	0	0.00	0.146	0.
Calculated Head		0.001	0.171	0.6

Siliceous Ore

Products	No. of Visible Au Particles	Visible Au (oz/t)	Au Assay (oz/t)	% of Total as Visible Au
Carlin ReCl Conc				
Sand Fraction	338	2.	18.9	11.
Slime Fraction	198	0.9	13.1	7.
Carbon ReCl Tail	81	0.1	6.33	3.
Carbon ReCl Tail				
Sand Fraction	13	0.1	1.83	6.
Slime Fraction	31	0.2	2.41	8.
Flotation Tail	0	0.0	0.386	0.
Calculated Head		0.004	0.551	0.9

Argillaceous Ore

Products	No. of Visible Au Particles	Visible Au (oz/t)	Au Assay (oz/t)	% of Total as Visible Au
Carbon ReCl Conc				
Sand Fraction	524	3.6	41.0	9.
Slime Fraction	224	1.2	21.1	6.
Carbon ReCl Tail	38	0.2	2.70	7.
Carbon Cl Tail				
Sand Fraction	0	0.0	0.20	0.
Slime Fraction	21	0.04	0.78	5.
Flotation Tail	0	0.0	0.183	0.
Calculated Head		0.009	0.303	3.

The best sensitivity was achieved by an LiF crystal, using the Lα gold line. The instrument was operated at 20 kv, with a count time of 90 seconds. Background readings were taken for every particle analyzed. The background offset was determined from scans across the gold peak in a 2% Au-98% Ag alloy, prepared for this study. Eight gold-silver alloys were prepared as standards for this study, with gold contents of 0.000%, 0.011%,

0.031%, 0.045%, 0.11%, 0.49%, 0.99% and 2.00%. The lower detection limit was 0.045% (13 oz Au/t). This was not a deterring factor because the small percentages of pyrite in the ores (<1%) would require high concentrations of gold, if significant amounts of gold were associated with the pyrite.

Of 179 pyrite particles probed, only 28 were suspected to be barely within the detection limits of 13 oz Au/t. These results generally confirmed those obtained for the sands fractions of the cleaner tailings, described earlier. Very little gold, if any, was detected in association with the coarse euhedral pyrite in the carbonate and siliceous ores.

Gold Associations with Carbonaceous Matter and Pyrite in Carbonate Ore

Flotation of twenty kilograms of carbonate carbonaceous ore produced the same products as tests for the two-kilogram samples. The main difference was that the larger twenty-kilogram sample made it possible to obtain more carbonaceous matter and pyrite for further separations. The carbonaceous matter and pyrite obtained by flotation were further concentrated by centrifuging with heavy liquids, as described in a previous section.

Various methods, including optical microscopy, the electron microprobe, and the scanning electron microscope, were used to investigate the association of gold with forms of organic matter in primary carbonaceous gold ores. Gold has been shown to concentrate with organic carbon in flotation concentrates, and was identified microscopically as fine free metallic particles in flotation products from the three ore types (4). Particulate gold has also been detected in mutual contact with organic matter in polished thin sections of carbonaceous ore (5).

However, the actual concentration of adsorbed or dispersed submicroscopic forms of gold with the organic carbon phase, itself, is <u>not known</u>. According to Wells and Mullins, microprobe analyses indicated less than 0.06% Au content in carbonaceous matter from Carlin and Cortez (1).

An alternate approach has been to separate and prepare high concentrations of carbon and pyrite by means of flotation and specific gravity methods, followed by the examination and analysis of concentrates. A relatively good gravity separation of pyrite and organic matter was achieved from the carbonate ore sample, utilizing the centrifuge with heavy liquids. Fifty-six percent organic carbon was contained in a float (<2.4 Sp.G.) fraction and about 8% finely occluded pyrite (Table VII), compared with a 62% pyrite concentrate with 4.5% organic carbon in a sink (>2.4 Sp.G.) fraction. The 56% organic carbon is roughly equivalent to 70% carbonaceous particles, based on an 0.8% factor of carbon in organic matter (5). The pyrite percentages in the sink fraction were determined by microscopic point-counting. Principal occluded diluents in both fractions include quartz and minor illite.

Assays and distributions of gold and carbon in the various flotation and gravity products are listed in Table VI.

The high carbon fraction (<2.4 Sp.G.) containing 55.8% organic carbon assays only 1.08 oz Au/t. This indicates that the maximum amount of gold associated with organic carbon could be about 1.5 oz Au/t. Since the total carbonate ore sample contains 0.5% carbonaceous matter, the organic fraction of this ore would contain only about 0.008 oz/t gold, representing only 4.6% (0.008/0.174 x 100) of the total gold.

TABLE VI

Assays and Distributions

	Wt %	Au oz/t	Organic C %	Distribution Au	Distribution Organic C
Assay Head		0.174	0.4		
Calculated Head	100.00	0.174	0.4	100.00	100.00
Carbon Recleaner Conc					
Sand Fraction					
Sinks >2.4 Sp.G.[a]	0.02	2.30	4.5	0.27	0.20
Floats <2.4 Sp.G.[b]	0.01	1.08	55.8	0.06	1.26
Slime Fraction	0.12	4.68	1.9	3.23	0.52
Carbon Recleaner Tail	0.65	2.26	7.7	8.45	11.29
Carbon Cleaner Tail	2.35	1.48	4.0	20.01	21.20
Flotation Tail	96.85	0.122	0.3	67.98	65.53
Combined Concentrates (Calc.)	3.15	1.77	4.8	32.02	34.5

a. 62% pyrite by microscopic point-counting.
b. 70% carbonaceous matter by calculation.

The carbonaceous matter occurs mostly as angular fragments, ranging predominantly from about 10 μm up to 200 um or larger in size (Fig. 4). Most of the carbon is free, with minor amounts as lockings with gangue.

The pyrite is relatively fine, ranging mostly from 2 μm up to about 30 μm. In the sink (>2.4 Sp.G.) fraction, most of the pyrite (62%) is cubic and well crystalline, with minor amounts of the fine-grained spheroidal variety (Fig. 5). The fine pyrite remaining in the organic carbon concentrate appears to be largely of the spheroidal type.

The following assumptions might be made about the gold content in this concentrate of mostly cubic pyrite. By assuming 100% pyrite, with all of the gold in the pyrite, the maximum gold content would be about 3.7 oz/t. Pyrite in the head was 0.7%, which by calculation equals 0.026 oz Au/t, representing approximately 15% of the gold in the overall carbonate ore sample.

Gold Loading on Organic Matter in Carbonaceous Carbonate Ore Sample

This section of the paper describes the ability of the carbonaceous materials to adsorb gold from the cyanide solutions. The term "gold-loading" is used here, but "preg-robbing" is also applicable.

Prior to flotation, cyanidation treatment was conducted on the carbonate ore for 24 hours by means of a standard NEL cyanidation procedure, using 1.15 lbs of NaCN per ton of ore at a pH of about 11.0, adjusted with lime. The sample was not chlorinated prior to cyanidation.

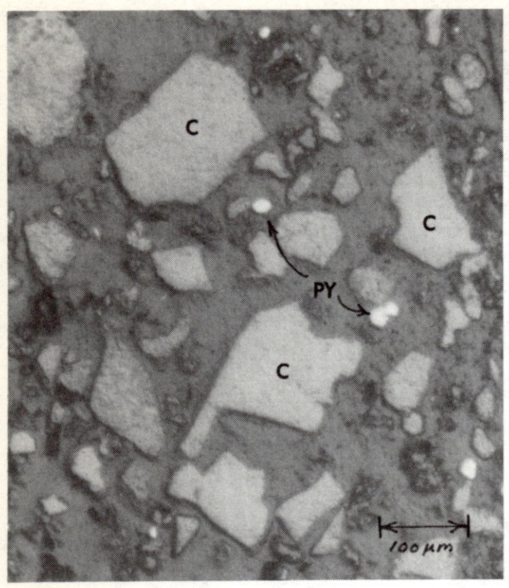

Figure 4.
Organic carbon (C) with minor fine pyrite (Py) in float fraction (<2.4 Sp.G.) of de-slimed carbon recleaner sands from carbonate type ore.
(X165, incident light)

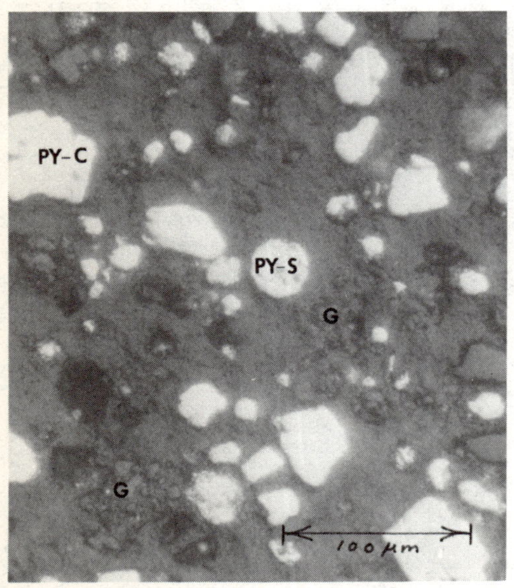

Figure 5.
Pyrite with gangue (G) and minor organic carbon (C) in sink fraction (>2.4 Sp.G.) of deslimed cleaner sands. Pyrite is mostly cubic crystalline variety (Py-C) with minor spheroidal type (Py-S).
(X333, incident light)

Mineralogically, the flotation products consisted mostly of quartz, illite, carbonates, pyrite, and unidentified amorphous phases believed to be mostly carbonaceous matter and poorly crystalline clays. Illitic clays were noted to be relatively high, noticeably in the recleaner concentrate fractions.

Highest amounts of carbonaceous matter (28.26-42.53% Organic C) occurred in the carbon recleaner concentrate slimes and sands (Table VII). Highest clays (approx. 40%) were detected in the recleaner concentrate slimes fraction, which assayed the highest gold (12.8 oz/t).

Flotation products were dried, weighed, and assayed for gold by fire assay. Assays and distributions are listed in Table VII below, and can be compared with data in Table VI for flotation of uncyanided carbonate ore.

TABLE VII

Gold Assays and Distributions

	Wt %	Au oz/t	% Organic C	Au Dist (%)
Assay Head		0.174	0.4	
Calculated Head	100.00	0.155		100.0
Carbon Recleaner Conc				
Sands	0.001	10.8	Insuff.Sample	0.1
Slimes	0.045	12.8	42.53	3.7
Carbon Recleaner Tail	0.180	7.38	28.26	8.6
Carbon Cleaner Tail	1.094	2.91	11.80	20.5
Float Tail	98.68	0.105	0.40	67.1
Combined Concentrates	1.32	3.86	15.	32.9

Significantly higher amounts of gold (10-13 oz Au/t) were concentrated in the cyanided sands and slimes fraction of the carbon recleaner concentrates, as compared with similar products from the uncyanided ore. Since no free gold is expected to occur in the concentrate after cyanidation, this high gold is attributed to the adsorption (or loading) of gold cyanide onto the organic carbon. This is indicated by the correlation of gold with carbon content after cyanidation (Fig. 6). In addition to the ability of the carbonaceous matter to adsorb gold, the following was indicated:

1. If considerable amounts of soluble gold could be adsorbed by carbonaceous matter, very little primary gold was associated with it originally,

2. Even though the sample was not oxidized, a large amount of gold was leached, suggesting significant amounts of gold occurrences other than with pyrite.

The recleaner concentrate was significantly smaller in amount and more slimy (clayey) in composition than that produced from the untreated flotation pulp, suggesting that select coarser sized ore particles may have been rejected during flotation of the cyanided pulp. This was also indicated by the relative abundance of select coarser sized carbon particles in the recleaner and cleaner tail fractions, which were detected microscopically in polished epoxy mounts of these fractions. Free gold was not detected in

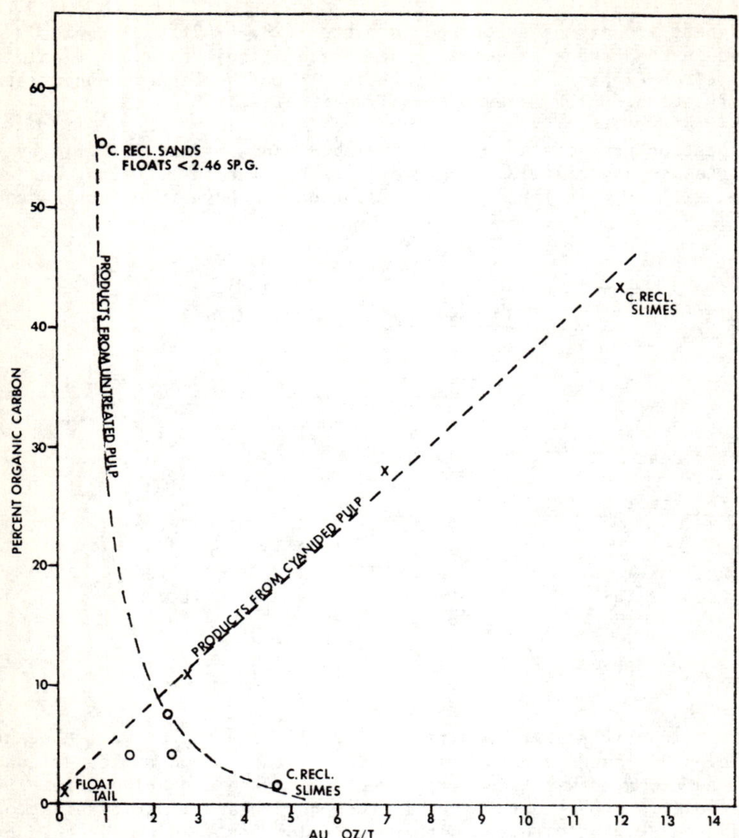

Figure 6. Correlation plots of gold (oz/t) against organic carbon (%) in products from untreated and cyanided pulps of carbonate ore.

the recleaner tail, which contained 7.38 oz Au/t. Insufficient sample was available for microscopic examination of the sands and slimes fractions of the recleaner concentrates. However, free gold was not expected in these products because of the cyanidation treatment.

Smaller amounts of recleaner concentrate were collected than in previous tests, indicating that some rejection of select particles may be attributed to cyanidation. The overall distribution of gold in recleaner concentrates (approx. 4%), is roughly equivalent to that reported for the uncyanided ore. The distribution of gold in the float tails (67.1%) is also equivalent to that reported for the uncyanided ore (approx. 68%).

High Resolution SEM Studies of Fine Gold

Mineralogic studies on the three carbonaceous ore samples have revealed an abundance of fine metallic gold (approx. 1 to 10 µm) in recleaner flotation concentrates. Most of this microscopically-visible gold was near optical detection limits (approx. 1-3 µm) in agglomerations of illitic clay. It was determined semiquantitatively by means of a modified "gross count" procedure that only about 3 to 11 percent of the total gold occurs as metallic grains coarser than about 1 µm.

During the examination of flotation products, the clay-rich slime fractions were separated from the coarser sand fractions in an attempt to concentrate the gold in coarser fractions, more amenable for heavy liquid methods of mineral separation. However, slime fractions from all three ore types assayed high in gold, ranging from about 5 to 25 oz Au/ton; the gold occurred in an essentially submicroscopic form.

Since the particulate gold described from the three carbonaceous ore samples represented mostly the visually resolved gold (>1 µm) in heavy gravity sand fractions, an effort was made also to account for the submicroscopic gold in the high grade slime fractions of each ore type. To this end, portions of the slime concentrate from each ore type were examined by means of a high resolution JOEL Scanning Electron Microscope (SEM), equipped with a Kevex energy dispersive spectrometer (EDS). Samples were mounted on brass sample holders and coated with either carbon or organic polymer. Particulate gold was readily identified in all three slime samples by both secondary and backscattered electron image methods. A number of gold particles were examined, most of which were free, equant in shape and submicrometer in size, ranging from less than 3 µm down to about 0.1-0.2 µm, and averaging about 0.5 µm in diameter. From a practical standpoint, the SEM detection limits for this study were about 0.1 µm.

The following three types of occurrences were noted for the fine particulate gold in order of relative abundance:

1. Free gold particles, completely unattached to other mineral phases (Figs. 7 and 8),

2. Gold covered with thin illitic clay (Figs. 9 and 10),

3. Gold attached to pyrite surfaces (Figs. 11 and 12).

Locked particulate gold in gangue mineral phases was not detectable. All of the gold particles identified appear to be accessible to cyanide treatment. X-ray elemental spectra were run on a number of discrete gold particles, confirming their predominant gold composition.

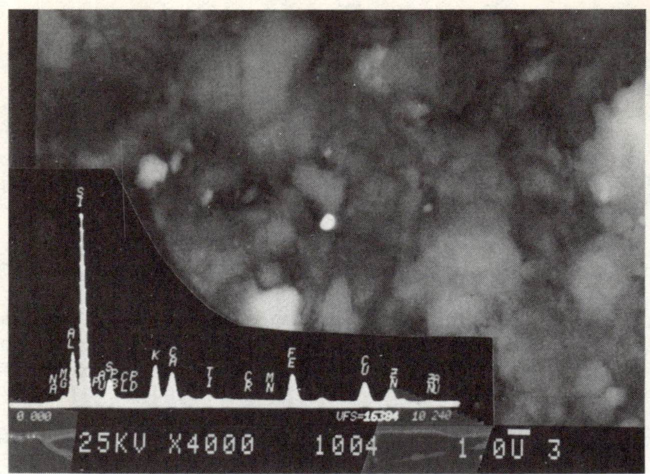

Fig. 7. Backscattered electron image at 4,000X. Free gold particle is white dot measuring about 0.5 μm. X-ray elemental spectra of gangue particles include mainly silicon, aluminum and potassium due to quartz and illite.

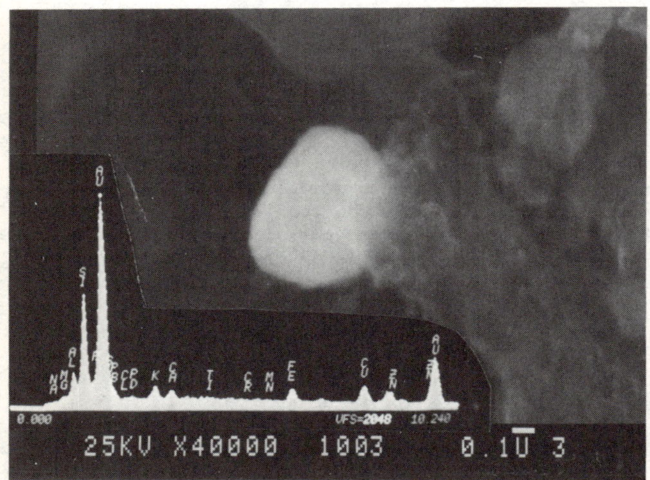

Fig. 8. Backscatter electron image of above at 40,000X, showing gold partially covered by semi-transparent film of illite or carbonaceous matter. X-ray elemental spectra of gold show major gold with minor silicon, aluminum and potassium from illite, and calcium and magnesium from carbonates.

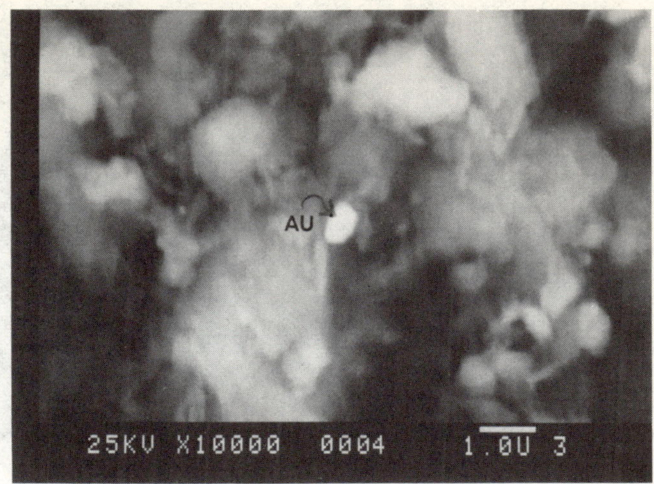

Fig. 9. Scanning electron image of rounded oval-shaped gold grain (0.3 x 0.5 μm) in slimes from siliceous ore. Gold is associated closely with illitic clay on all sides. Magnification at 10,000X permits the examination of textural associations of gold with illitic clay matrix. However, it cannot be determined whether the gold is primarily or secondarily agglomerated with illite.

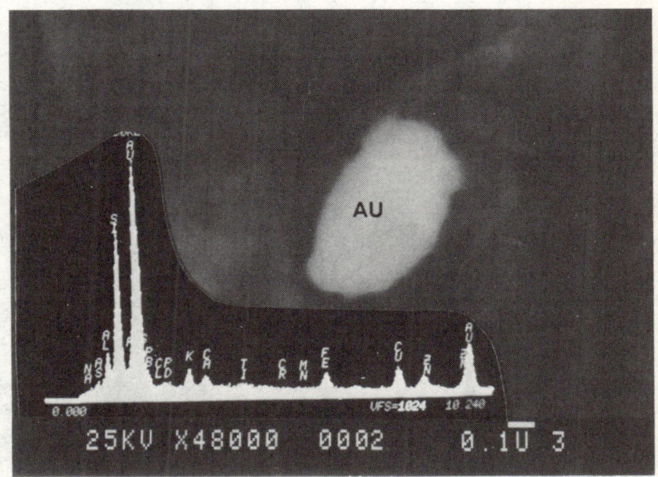

Fig. 10. Scanning electron image at higher magnification (48,000X) of gold grain shown in Figure 9, displaying the intimate textural contacts of the illite with the gold.

X-ray spectra of the gold, showing a high background of Si, Al and K from the illite, along with the major Au peaks.

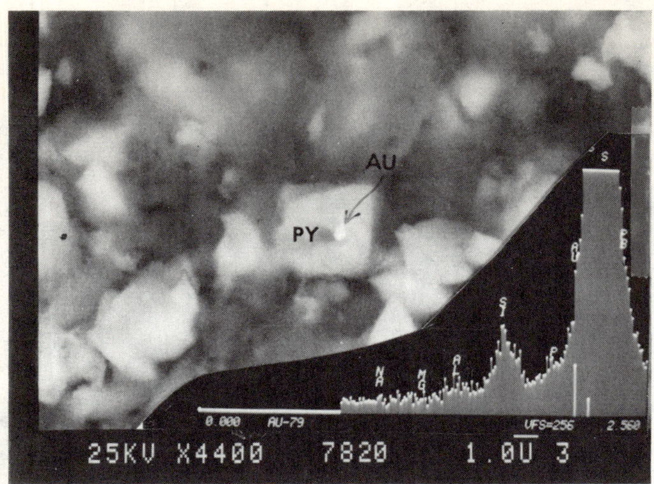

Fig. 11. A backscattered electron image from carbon-coated sample of slimes from argillaceous ore at 4,400X magnification, showing a small (0.3-0.4 μm) grain of gold on pyrite. Most of the angular particle fragments in this field are pyrite. Similar associations of fine particulate gold attached to pyrite surfaces were noted in Carlin carbonate ore. X-ray spectra of the pyrite matris showing major sulfur peak interferes with detection of gold in pyrite matrix. Minor impurities, Si, Al, Mg, are from illitic clays.

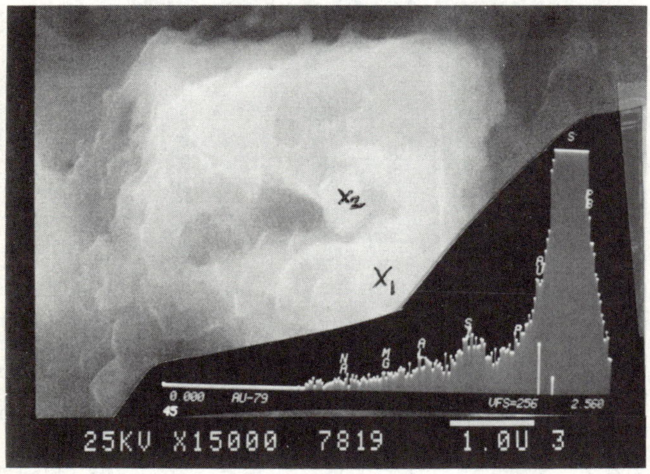

Fig. 12. A higher magnification (15,000X) secondary electron image of the gold on pyrite described in Figure 11, showing the locations of X-ray analysis of the gold particle and of the adjacent pyrite matrix. X-ray spectra of the gold particle (at 200,000X), showing gold peak discernible as a shoulder on major sulfur peak.

X-ray spectra indicate a relatively pure gold composition for most particles, with the exception of one particle that was reported to contain several percent palladium. Most spectra show anomalous silica, alumina and potassium, attributed mostly to illitic clay coatings. It could not be determined whether such illite was formed as primary coatings or secondary agglomerations during test work.

The predominance of submicrometer free gold in the three slime fractions examined indicates that forms of particulate gold extend well into the submicroscopic size ranges. Although it was not possible to estimate the fraction of total gold in this submicrometer form, its very presence as fine grains in relative abundance for detection by SEM methods serves to confirm its metallurgical importance in the three carbonaceous ores.

Free particulate gold was detected by high resolution SEM down to about 0.1 μm. Gold was detectable (and distinguishable) from other phases because it is a strong backscattered electron producer, relative to most other elements in the sample. However, each bright spot required confirmation for gold by means of the energy dispersive spectrometer (EDS).

A total of nine gold particles were noted in the three slime samples examined, requiring approximately 12 hours of instrument time. Although the three samples differed significantly in gold content (approx. 5-25 oz/t), it was not possible to distinguish this concentration difference from SEM scanning, because of the time-consuming distances from particle to particle at these high magnifications. In effect, there are instrumental limitations to the SEM in achieving any sort of semiquantitative estimations of relative amounts of gold from one sample to another. Particulate gold probably extends down into sizes below 0.1 μm, possibly requiring methods of transmission electron microscopy (TEM and STEM) to investigate colloidal gold in finer sizes.

Conclusions

Although carbonaceous Carlin-type disseminated gold ores have been known for more than twenty years, the actual gold occurrences are not well understood. Special separation products of high gold concentrations from three ore types have provided an opportunity to identify particulate forms of gold down to colloidal sizes near 0.1 μm. Since fine gold was detected down to 0.1 μm, it is likely that even finer sizes may occur.

Many gold particles detected by conventional ore microscopy (1-10 μm), and by scanning electron microscopy (<1 μm) are associated with illite agglomerations. Most of these agglomerations were formed during the flotation process, but the possibility for primary gold occurrences with illite should be investigated further. While generally poor gold recoveries were attained, numerous separation products were of high gold grades, coincident to high concentrations of illite. It seems unlikely that gold of such fine sizes could be completely liberated for flotation.

The distribution and concentration of submicroscopic gold within pyrite in these three ores were not completely defined by these studies. Very little of the gold appears to be associated with coarse pyrites (>30 μm), especially in the euhedral variety. Since auriferous varieties of pyrite have been reported by Wells and Mullins (1) for select specimens from Carlin and Cortez, investigations should be continued.

Carbonaceous materials in the carbonate ore were shown to contain very little gold. If, however, the carbonaceous surfaces are not deactivated, cyanided gold is readily adsorbed by the mechanism called "Preg Robbing," so that the gold-loaded carbon becomes refractory to cyanidation. Deactivation of carbonaceous surfaces is accomplished at Carlin by chlorination prior to cyanidation treatment.

References

1. Wells, J. D. and Mullins, T. E., "Gold-Bearing Arsenian Pyrite Determined by Microprobe Analyses, Cortez and Carlin Gold Mines, Nevada," Econ. Geol., 1973, Vol. 68, pp. 187-201.

2. Radtke, A. S. and Scheiner, B. J., "Studies of Hydrothermal Gold Deposition, Nevada: The Role of Carbonaceous Materials in Gold Deposition," Econ. Geol., 1970, Vol. 65, pp. 87-102.

3. Jones, M. P. and Gavrilovic, J., "Automatic Searching Unit for the Quantitative Location of Rare Phases by Electron Probe X-Ray Microanalysis," Trans. Inst. Min. Metall., Vol. 77, 1968, pp. B137-B143.

4. Ahlrichs, J. W. and Mueller, W., Unpublished Company Report.

5. Hausen, D. M., "Process Mineralogy of Select Refractory Carlin-Type Gold Ores," C.I.M.

6. Hausen, D. M. and Park, W., "Observations on the Association of Gold Mineralization with Organic Matter with Carlin-Type Ores," Paper Presented at DREGS Symposium, Denver, Colorado, May 1985.

ORIGIN AND EXTENT OF THE GOLD, SILVER, ANTIMONY AND TUNGSTEN MINERALIZATION IN THE FAIRBANKS MINING DISTRICT, ALASKA

P.A. Metz and B.M. Hamil

School of Mineral Engineering
University of Alaska - Fairbanks
Fairbanks, Alaska 99775

Abstract

The Fairbanks Mining District in central Alaska was discovered in 1902. Approximately 7.5 million troy oz of placer gold, 250,000 troy oz of lode gold, and several thousand short tons of antimony and tungsten have been produced from an area of about 400 square miles. Five main types of mineralization are recognized in the district:

I. Conformable, stratabound volcanogenic zinc, antimony, lead, and copper sulfides in places containing gold and tungsten;

II. Complex precious metal-bearing lead sulfosalt-quartz veins in Cretaceous granitic intrusives;

III. Tungsten skarns adjacent to granitic intrusives;

IV. Gold-bearing polymetallic sulfide-quartz veins crosscutting metavolcanics; and

V. Antimony-bearing gash veins along axial plane shears in metavolcanics.

Fluid inclusions and isotopic data suggest different sources for each type of deposit, with a major metamorphic component for the mineralizing fluid. Detailed description and evaluation of these occurrences are in progress.

<u>Introduction</u>

The Fairbanks Mining District is located just north of Fairbanks, Alaska (see Figure 1). It extends approximately 30 miles northeast-southwest by 10 miles northwest-southeast with its center about 15 miles north of the city center. It is bounded on the north by the Chatanika River and on the south by the Tanana River. It extends from Ester Dome on the southwest to Gilmore Dome and Pedro Dome on the northeast with the Goldstream Valley forming the approximate northeast-southwest axis of the district. The domal areas have been the locale of major gold placer operations as well as small hardrock mines and prospects. The newly opened Grant Mine is located on Ester Dome. The Chatanika, Chena, and Goldstream Valleys and their tributary drainages have been the site of extensive placer operations in the past. A half-dozen small placer mines are working in the district at the present time.

The first placer location was made in 1902 in the north-central part of the district. The greatest production occurred between 1923 and 1962 when the Fairbanks Exploration Company operated 13 dredges in the drainages between Chatanika and Ester. The district has produced over 7,500,000 troy oz of placer gold.
The lode mines developed in the areas of Ester, Pedro and Gilmore Domes as prospectors attempted to trace the source of the placers. Ester Dome is structural feature while Pedro Dome and Gilmore Dome are the topographic expression of Mesozoic granitic intrusives which intrude the surrounding metavolcanic terrane. Lode mining activity has resulted in the production of about 250,000 troy oz of gold and several thousand short tons of antimony and tungsten.

Figure 1. Location map of the Fairbanks mining district, Alaska.

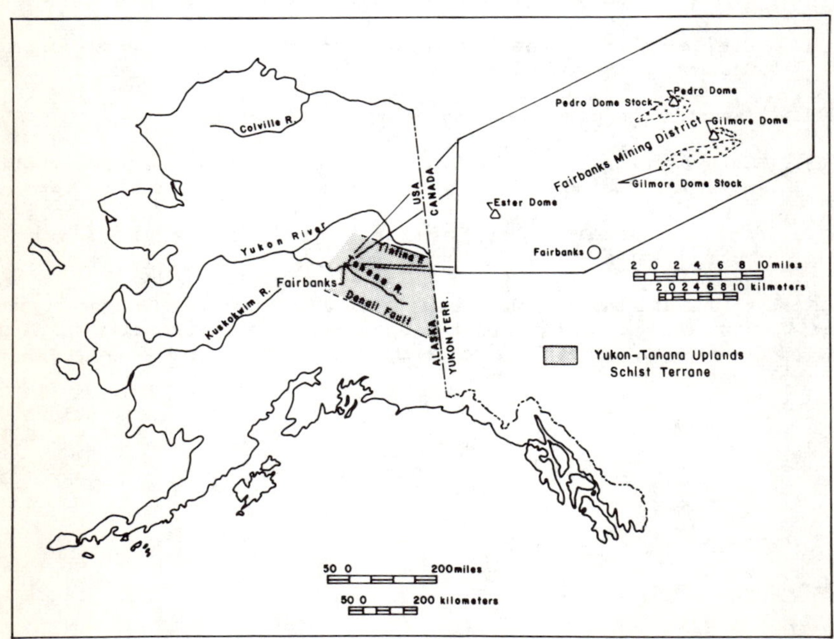

Regional Geology and Petrology Yukon-Tanana Uplands Schist Terrane

The Fairbanks Mining District is located within the northwestern portion of the Yukon-Tanana Uplands Schist Terrane in central Alaska (Figure 2). The terrane is bounded on the south by the Denali Fault and on the north by the Tintina Fault (Foster et al. 1973). The terrane is composed of Precambrian and Paleozoic metasedimentary and metavolcanic rocks. The northern and central part of the terrane consists of an enormous tract of multiple metamorphosed and penetratively deformed quartz-mica schist and gneiss, quartzite, quartz-rich grit, gneissic plutonic rocks, metavolcanic rocks, marble, and calc-schist. The marble locally containes Devonian fossils and portions of the gneissic plutons are dated as Devonian to Mississippian. Mineral assemblages indicate metamorphic grades ranging from lower greenschist facies through amphibolite, eclogite and granulite facies. The Yukon-Tanana Uplands Schist Terrane metamorphic rocks were formerly called the Birch Creek Schist (Mertie, 1937). These metamorphic rocks are unconformably overlain by Paleozoic, Mesozoic and Tertiary sedimentary and volcanic rocks. The metamorphic and volcanic sequences are intruded by Paleozoic, Mesozoic, and Tertiary igneous rocks ranging in composition from peridotite to granite.

Figure 2. Generalized geological map of the Yukon Tanana Uplands Schist Terrane.

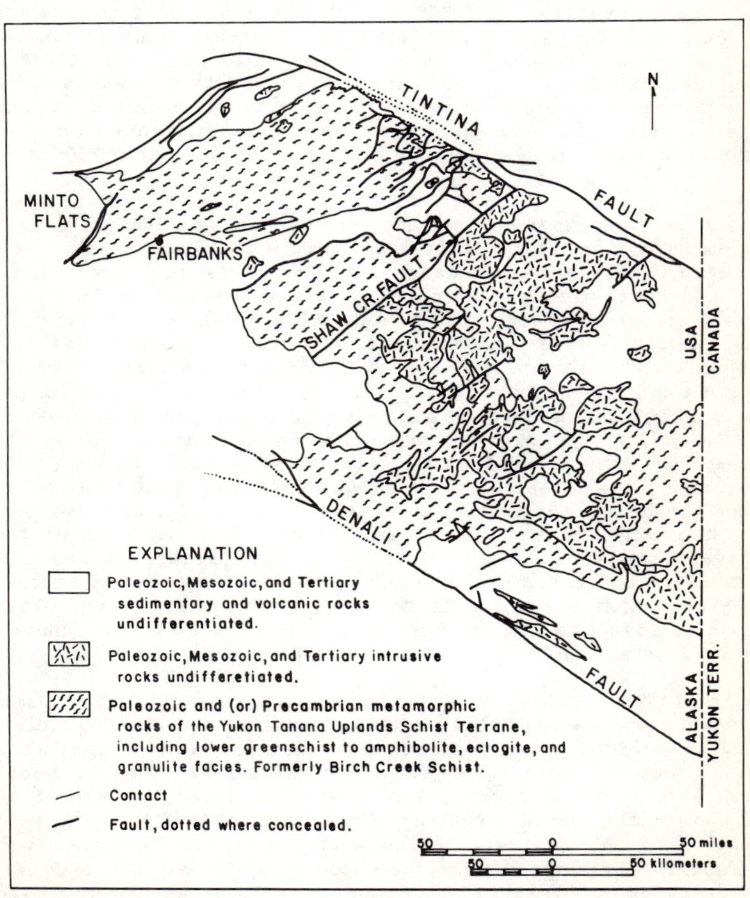

Five groups of significant metalliferous lode mineral deposits occur in the northern and central part of the Yukon-Tanana terrane.

1. Au-Sn-W epigenetic replacement vein deposits throughout the northern and central parts of the terrane.

2. Several Cu and Cu-Mo prophyry deposits occur throughout the terrane.

3. Volcanogenic gold sulfide and gold quartz vein deposits occur throughout the terrane but are concentrated in the Fairbanks mining district in the Cleary Summit and Ester Dome areas.

4. A series of W-Au skarn deposits occur in the Fairbanks district in the Gilmore Dome area and in the eastern part of the terrane.

5. Several Sn greisen deposits occur in the eastern part of the terrane.

The epigenetic replacement deposits do not occur near any known granitic rocks and can only be inferred to have formed during hydrothermal activity associated with intrusion of Late Cretaceous or Early Tertiary granitic plutons, dikes and sills. The gold quartz vein deposits appear to be related to regional metamorphic and deformation events which culminated with intrusions of Cretaceous granitic plutons, dikes and sills. The skarn deposits are related to an extensive suite of discontinuous Cretaceous granitic plutons and related igneous rocks. The volcanogenic gold sulfide, gold quartz vein, and W-Au skarn deposits of the Fairbanks mining district have produced considerable amounts of gold since the early 1900's as noted above.

The first general description of the bedrock and surficial geology of the Fairbanks district was provided by Prindle and Katz (1913). Subsequent investigators have retained many of the original rock units defined by Prindle and Katz. The major contribution by Prindle and Katz was their detailed description of the gold placer deposits which they related to the gold-quartz-sulfide veins, the formation of which, in turn, was attributed to the intrusion of the granitic rocks of Pedro Dome and Gilmore Dome. Smith (1913) gives detailed descriptions of several lode deposits which support the earlier observations of Prindle and Katz (1913). Chapin (1914, 1919), Mertie (1918), and Hill (1933), also described the lode deposits of the district and noted the close spatial relationship between the placer deposits and the lode occurrences. These investigators considered the gold-quartz veins to be the sole lode source and the veins were thought to be genetically related to the granitic intrusive rocks. Chapman and Foster (1969) also related the gold-quartz vein deposits of the district to the Cretaceous-age granitoid intrusions and concluded that the quartz veins were the sole source of the placer gold. Mineralization in the area of Ester Dome was related to a complex of aplite dikes which may be offshoots of an as-yet unrecognized underlying intrusive body.

Metz (1977) and Metz and Robinson (1980) suggest that the antimony-tungsten and associated lode gold mineralization of the Fairbanks district is related to previously unrecognized metavolcanic rocks in the Yukon-Tanana Uplands Schist. Subsequent 1:24,000 scale geologic mapping by Metz (1982), Bundtzen (1982), and Robinson (1982) has delineated the extent of these metavolcanic rocks, named the Cleary Sequence by Metz (1982). A generalized compilation of this mapping is shown in Figure 3. Two non-mineralized pelitic schist sequences comprise the majority of the exposed bedrock geology of the Fairbanks District - the Goldstream Sequence and the Birch Hill

Figure 3. Generalized geological map of the Fairbanks Mining District, Alaska.

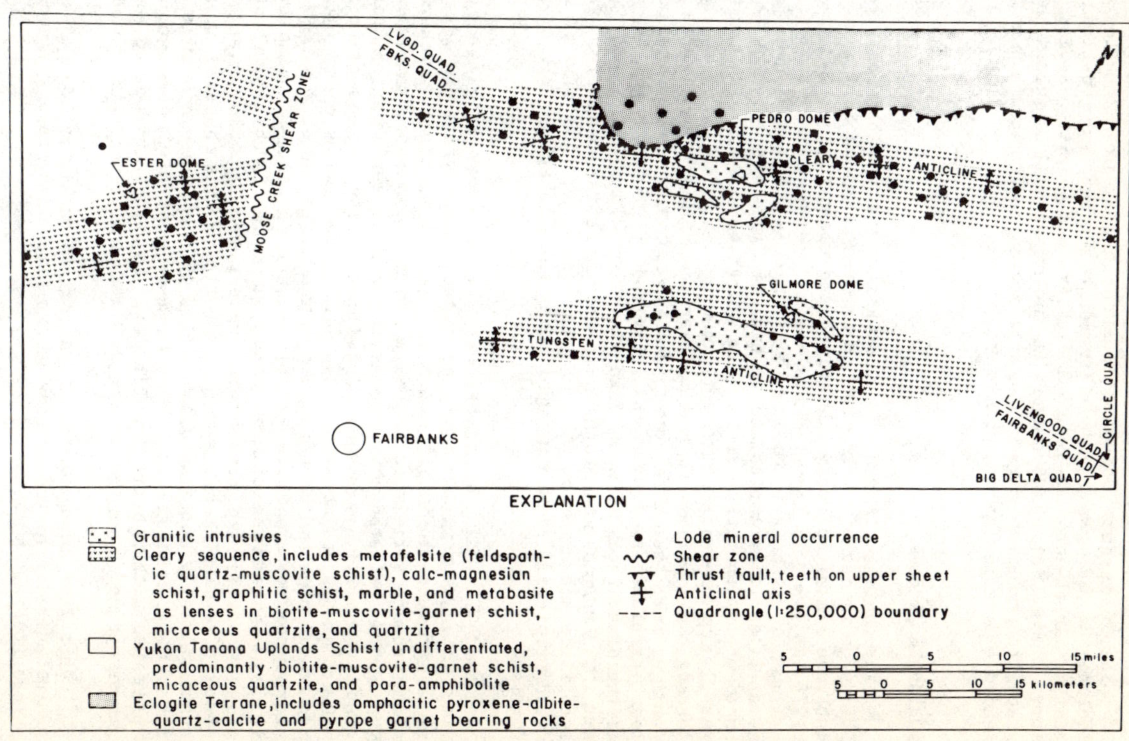

Sequence. A majority of the 236 known lode mineral occurrences in the Fairbanks mining district are found within the Cleary Sequence, however eight of the occurrences are in the eclogitic rocks of the structurally lower Chatanika Terrane also a subunit of the Yukon-Tanana Uplands Schist Terrane. Most of the mineral occurrences in both the Chatanika Terrane and the Cleary Sequence are not spatially associated with exposed intrusive rocks.

The Goldstream Sequence is structurally above the Cleary Sequence and includes banded para-amphibole, tremolite marble, garnet-biotite-muscovite schist, garnet-chlorite-muscovite schist, and graphitic schist. Structurally above the Goldstream Sequence is the Birch Hill Sequence which includes graphitic phyllite, graphitic-quartz schist, calcareous schist, and pyritic-banded quartzite (metachert).

Chatanika Terrane

The Chatanika Terrane is structurally and probably stratigraphically the lowest sequence of rocks exposed in the Fairbanks district. The sequence includes eclogitic rocks, described and classified by Swainbank and Forbes (1975) as group "C" metabasite eclogite of Coleman et at. (1965). These eclogitic rocks are in thrust contact with lower greenschist facies metapelites of probable late Precambrian or early Paleozoic age that occur north of the Chatanika River. The eclogites are also in thrust contact with the Cleary Sequence which outcrops to the south along Cleary Creek. The thrusts strike east-west in the Clear Creek/Cleary Summit area and dip to the south at less than 45°. The eclogitic rocks of the Chatanika Terrane have been correlated by Erdmer and Helmstaedt (1983) with the type "C" eclogites of the Nisutlin Allochthon, Yukon Territory (Faro, Last Peak, Campbell Range, and Mt. Hundere) and Pinchi Lake area, British Columbia. These eclogite localities are evidence of a 1000 km long subduction zone of late Paleozoic or Mesozoic age that extended along the western margin of ancient North America.

Unit EC - The eclogitic rocks can be divided into at least three distinct types including : a) garnet-clinopyroxene rocks; b) garnet-clinopyroxene-amphibole rocks; c) garnet-amphibole rocks. The garnet-clinopyroxene variety occurs as mylonitic, layered, laminated, and massive types, and is characterized by presence of pale-pink pyrope garnet and pale-green omphacitic pyroxene. Untwinned albite is present in the mylonitic variant and is usually associated with garnet and chlorite. Quartz generally occurs as granoblastic aggregates with calcite and clinozoisite. The garnet-clinopyroxene-amphibole variety occurs as finely laminated and massive variants characteized by zoned garnet, light-green, poikioblastic clinopyroxene, and pleochroic, pale-green amphibole that becomes more absorbtive near garnet grains. Calcite is more abundant than quartz. The subunit locally contains disseminated pyrrhotite, pentlandite, chalcopyrite, sphalerite, and galena. The garnet-amphibole rocks occur as massive varieties with incipient foliation. Amphibole varies in composition and garnet occurs as porphyroblasts with euhedral overgrowths on subhedral cores.

Unit IM - Impure Marble is a gray to dark gray, buff weathering coarse grained massive carbonate rock with minor clinopyroxene, garnet, amphibole, phlogopite, and epidote.

Unit CS - Muscovite Calc-Schist is a light brown to red brown finely laminated rock composed of calcite, quartz, muscovite, with or without biotite, garnet, and pyrite.

Unit MQS - Garnet-Feldspar-Muscovite-Quartz-Schist is a medium to dark brown coarse grained rock with garnet porphyroblasts up to a cm in diameter. The unit contains minor biotite and chlorite and at one locality kyanite and staurolite have been reported (Brown, 1962). The feldspar is plagioclase (An_{15}) in a granoblastic fabric.

Unit MQ - Micaceous Quartzite is a dark gray to black fine grained quartzite with fuchsite, graphite, and locally biotite, garnet, and chlorite. The protolith for the unit was probably a graphitic siltstone or chert.

Cleary Sequence

The Cleary Sequence includes at least 300 meters of predominantly bimodal metavolcanic/volcaniclastic rocks with minor intercalated metasedimentary units that outcrop in a belt 10 km wide and 50 km along strike. The "type locality" for the Cleary Sequence is a section along the Steese Highway from Skoogy Gulch over Cleary Summit to Mile Post 25. The basal contact of the Cleary Sequence is a thrust contact, the Cleary Creek Thrust, with the eclogitic rocks of the Chatanika Terrane to the north. This thrust strikes NE-SW in the Cleary Summit area and dips to the south. The upper contact of the Cleary Sequence is also a thrust contact (Goldstream Creek Thrust) with the Goldstream Sequence. These thrusts are generally parallel to sub-parallel.

The lower 200 meters of the Cleary Sequence is composed primarily of calcareous actinolitic greenschist (CAG) and chloritic schist (CAC) with a probable metabasite protolith while the upper 100 meters contains the economically significant lenses of metapyroclastic rock (MFV and MRT) with a metafelsite association.

Unit CAG - Calcareous Actinolitic Greenschist is composed of actinolitic-hornblende and andraditic granet with minor calcite, ablite-oligoclase, clinozoisite, magnetite, pyrrhotite, pentlandite, and chalcopyrite. The protolith of the greenschist was probably a basalt.

Unit CAC - Chlorite Schist is composed of penninite, albite-oligoclase (An_5-An_{15}), epidote, calcite, and quartz. Locally the chlorite is manganiferous clinochlore. The chlorite schist is well foliated with the foliation parallel to compositional layering which ranges from a few millimeters to several centimeters thick. The chlorite schists are intercalated with dark gray to black metachert. Both the manganiferous chloritic schists and metacherts contain disseminations and bands of pyrite, arsenopyrite, sulfosalts (primarily jamesonite and boulangerite), sphalerite, and minor galena, magnetite, ilmenite, rutile, and sphene. The protolith of the chloritic schist was probably a mafic tuff.

Unit MC - Quartz Pebble Metaconglomerate consists of dark gray, cobble to boulder size rounded to subrounded clasts of quartz with minor pebble size clasts of dark gray to black shale. The matrix is silt to very coarse sand size clasts of quartz, feldspar, and carbonate. The unit has a limited areal distribution; never exceeds 10 meters in thickness; and locally is iron stained and contains minor pyrite.

Unit CS - Calcareous Muscovite Schist unit is a dark brown medium grained rock consisting of calcite, pale green muscovite, quartz, albite-oligoclase (An_5-An_{15}) with or without andraditic garnet and dark brown to red-brown biotite. The unit does not contain major sulfides and probably represents a metamorphosed calcareous pelitic sediment. This unit is intercalated with all the following units of the Cleary Sequences but generally increases in abundance in the upper part of the section.

Unit MFV - K-Feldspar Quartz Schist is comprised of pink porphyroblasts of potassium feldspar and quartz blastophenocrysts in a foliated matrix of albite-oligoclase, and clear to yellow muscovite with minor biotite, pistacite, penninite, and zircon. Locally the unit contains disseminated arsenopyrite and pyrite. The unit has a limited areal distribution and occurs primarily as rubblecrop on upper Pedro Creek, Solo Creek, Bear Creek and on Cleary Summit. The protolith was probably a rhyolitic flow or sill.

Unit MRT - Muscovite Quartz Schist is composed predominantly of yellow white to clear muscovite, albite and quartz but locally it contains tourmaline, zircon, and blastopyroclasts of potassium feldspar and lithic fragments. The unit outcrops as discontinuous lenses up to 2 meters thick and 1000 meters along strike and may grade upward to a laminated quartzite (metachert). The unit contains lenses up to 25 cm thick of massive sulfides including jamesonite, boulangerite, tetrahedrite, sphalerite, arsenopyrite, pyrite, and galena. Locally the unit has gold and silver contents in excess of 3 ppm and 600 ppm respectively. The protolith was probably a felsic pyroclastic and or exhalite chert.

Unit GS - Graphitic Schist unit is composed of dark gray to black fine grained graphite-quartz schist with minor sulfides including arsenopyrite and pyrite. The protolith was probably a carbonaceous chert.

Unit IM - Impure Marble unit varies from a green to red, brown, or buff colored medium to coarse-grained carbonate rock with calcite in excess of dolomite and minor quartz, garnet, and chlorite. Where the unit is in contact with the Pedro Dome and Gilmore Dome intrusive rocks the mineral assemblage is altered to calcite, garnet, diopside, idocrase, and hornblende, and locally fluorite, scheelite, molybdenite, and pyrrhotite.

Unit M - Marble unit is a dark-gray, green to buff fine grained siliceous carbonate rock with calcite predominating over dolomite. Locally the unit contains sphalerite, galena, bornite, chalcopyrite, and pyrite conformable with compositional layering. The unit is at least 10 meters thick at most outcrops and exceeds 30 meters at two outcrops.

The extremely light oxygen isotopic values of the marble, -25 to -30 per mil, may be indicative of a volcanic hot spring origin for the carbonate while sulfur isotopic values of 0 to +5 per mil for the sulfides also suggest a volcanic origin for the copper, lead, and zinc mineralization (Metz, 1984b). The marble unit is thus designated as the upper most unit of the volcanogenic Cleary Sequence and is the only major marker horizon for the sequence.

Units MQS and MQ - Muscovite-Quartz-Schist and Micaceous Quartzite are two pelitic units that are intercalated with the volcaniclastic and volcanic sedimentary units of the Cleary Sequence. Where units MQS and MQ comprise over 50 percent of the stratigraphic column the rock units have been designated the Fairbanks Schist. Unit MQS, muscovite-quartz schist, is a light brown to buff medium grained rock composed of light gray to brownish muscovite, and gray quartz with or without garnet and biotite. Unit MQ, Micaceous Quartzite, is a light brown to buff fine grained massively bedded quartzite with gray to light brown muscovite. These units occur throughout the Cleary Sequence but occur in greater abundance in the upper one third of the sequence.

Granitic Rocks

The granitic rocks of the Pedro Dome and Gilmore Dome areas are composed of four major phases. The relative ages of the phases are based on cross-cutting and textural evidence and from oldest to youngest these phases include: fine-grained granodiorite, porphyritic granodiorite, porphyritic quartz monzonite and aplite-pegmatite dikes. the fine-grained granodiorite is the predominant phase of the Pedro Dome stock while the Gilmore Dome stock is composed primarily of porphyritic quartz monzonite.

Pedro Dome Plutons

The fine-grained granodiorite accounts for 90 percent of the Pedro Dome pluton while porphyritic granodiorite and porphyritic quartz monzonite each account for 5 percent of the exposed intrusive complex.

The fine-grained granodiorite is composed of equigranular grains of plagioclase (45%), quartz (22%), biotite (13%), hornblende (11%), and potassium feldspar (7%) with accessory augite, hypersthene, sphene, zircon, apatite, rutile, and opaques. Grain sizes range from 0.5 to 2.0 mm. Locally mafic xenoliths from 1 cm to 30 cm in diameter are present (Blum, 1982).

Porphyritic granodiorite and porphyritic quartz monzonite occur in a small pluton approximately 3 km southeast of the main Pedro Dome intrusive. The porphyritic granodiorite is composed of plagioclase (49%), quartz (26%), potassium feldspar (16%), biotite (8%), and hornblende (1%) with accessory muscovite, sphene, apatite, zircon, rutile, and opaques. Potassium feldspar and quartz megacrysts range up to 20 mm and 10 mm in diameter respectively, while the matrix of feldspar and quartz has an average grain size of about 2.0 mm.

Porphyritic quartz monzonite is composed of plagioclase (36%), quartz (30%), potassium feldspar (28%), and biotite (6%) with accessory muscovite, sphene, apatite, zircon, rutile, and opaques. The grain size distribution of the megacrysts in the prophyritic quartz monzonite is similar to that in the porphyritic granodiorite, however the matrix in the quartz monzonitic rocks has an average grain size of 4.0 mm.

Aplite dikes occur in and adjacent to the porphyritic phases. The aplites are composed of potassium feldspar (38%), quartz (31%), plagioclase (30%), muscovite (1%), and biotite (1%) with accessory garnet, sphene, pyrite, arsenopyrite, apatite, zircon, and rutile (Blum, 1982). The aplite dikes range from very fine-grained rocks to medium to coarse grained pegmatite rocks.

Gilmore Dome Plutons

Porphyritic quartz monzonite accounts for 80 percent of the Gilmore Dome plutons while porphyritic granodiorite accounts for 19 percent of the outcrop/rubble crop area. Aplite-pegmatite occurs as a large 0.5 km^2 area in the prophyritic quartz monzonite and as numerous dikes throughout the porphyritic phases. The compositions of the porphyritic rocks are similar to the compositions in the Pedro Dome plutons. Aplite dikes are much more abundant in the Gilmore Dome plutons and the dikes in these plutons contain up to several percent sulfides, primarily as pyrite and arsenopyrite.

Potassium-argon radiometric dating of the Pedro-Gilmore Dome plutons has yielded an age of 90.7 ± 5.1 m.y. (Britton, 1970). The age estimate is based on hornblende from fine-grained granodiorite. Rubidium-strontium techniques have yielded an age of 91.0 ± 0.7 m.y. and initial $^{87}Sr/^{86}Sr$ ratio of 0.71238 ± 0.00010 (Blum, 1982). From the $^{87}Sr/^{86}Sr$ initial ratios, mineralogical and textural relationships, and whole rock major oxide analyses of Blum (1982), these granitic rocks can be classified as S-Type intrusives (after Chappel and White, 1974).

Mineral Deposits

Detailed investigations of the lode mineral deposits have resulted in definition of five major types of mineralization in the district by Metz and Halls (1981).

I. Volcanogenic stratabound mineralization in which intergrowths of arsenopyrite, jamesonite, boulangerite, galena, sphalerite, pyrite, chalcopyrite, gold, and scheelite occur as conformable laminae and lenses parallel to both foliation and compositional banding in the metavolcanic host rocks. The Cu-W mineralization is associated with metabasic host rocks. The As-Sb-Zn-Pb-Au mineralization is hosted in metafelsic and metabasic rocks.

II. Lead sulfosalt-bearing quartz sulfide veins with argentiferous galena, sphalerite, chalcopyrite, arsenopyrite, trace gold, and stibnite occurring in shear zones within the Cretaceous-age intrusives.

III. Tungsten skarn mineralization in which scheelite-bearing calc-silicate mineralization occurs adjacent to the Gilmore and Pedro Dome granitic stocks.

IV. Gold-bearing polymetallic quartz sulfide veins which occur as fracture fillings in shear zones that cross-cut the metavolcanic host rocks of the schist sequence. The veins contain quartz, boulangerite, jamesonite, robinsonite, semseyite, galena, sphalerite, arsenopyrite, tetrahedrite, chalcopyrite, and gold.

V. Stibnite gash veins and fracture fillings associated with axial plane shears in the metavolcanic host rocks. The mineralization consists of monomineralic, massive subhedral stibnite with minor traces of gold.

This classification system is based initially on mineralogic, petrologic, and field mapping data. It was refined by Metz (1984b) from fluid inclusion, sulfur isotope, and additional field studies. The discussion which follows encompasses the mineral assemblages and intermineral relationships of the five classification types.

Type I Mineralization. The metabasites of both the Chatanika Terrane and the Cleary Sequence contain disseminated mineralization that at least predates the last metamorphic event and is probably syngenetic. The garnet-clinopyroxene-amphibole variety of ecologitic rocks of the Chatanika Terrane contains disseminated pyrrhotite, pentlandite, chalcopyrite, and locally sphalerite and galena. The copper and iron sulfides occur as 0.01 mm to 0.5 mm grains within the matrix of plagioclase, calcite, and quartz while the sphalerite and galena occur as prophyroblasts up to 5mm in diameter. Total sulfides rarely exceed 1 percent of the rock and never exceed 3 percent. The Chatanika Terrane and the eclogitic rocks in particular have relatively low arsenic, antimony, and gold contents compared to the metafelsic rocks of the Cleary Sequence (Metz, 1984a), and no visible gold has been detected in any of the ore sections from the sulfide bearing eclogitic rocks. Placer gold has been produced from several streams that drain only the Chatanika Terrane and the gold/silver ratios for these deposits are significantly different from the gold/silver ratios of the placer deposits formed above the Cleary Sequence (Metz, 1983). Lead/Lead isotopic ratioa (Metz, in preparation) for the Chatanika Terrane are significantly different than those from the Cleary Sequence and its included vein systems. These data suggest a different source of gold for the Chatanika Terane and this source is not currently well delineated.

The calcareous actinolitec greenschist of the Cleary Sequence contains disseminations and bands of pyrrhotite ± chalcopyrite, pentlandite, and scheelite. The sulfide grains range from 0.01 to 0.10 mm while the scheelite ranges from 0.01 to 0.05 mm in diameter. Sulfide contents average approximately 1 percent of the rock while tungsten contents may exceed 50 ppb (Metz, 1984a). The CAG unit of the Cleary Sequence generally has a gold content of less than 100 ppb and no visible gold was noted in any ore section examined under reflected light or by electron microprobe methods.

Unit CAC of the Cleary Sequence as noted above contains both disseminated and banded sulfides of iron, arsenic, antimony, zinc, and lead. The disseminated sulfide grains range from 0.01 to 1.0 mm. Pyrite plus arsenopyrite account for 90 percent of the total sulfides and occur parallel to the compositional banding in the schist. Arsenic concentrations in the CAC unit generally exceed 1000 ppm and may locally exceed 1.0 percent. Native gold occurs as 0.01 to 0.10 mm size inclusions in arsenopyrite. The gold to silver ratio of the inclusions, as determined by electron microprobe techniques, range from 2.5/1 to 4/1. Limited sampling of the CAC unit that contains disseminated sulfides indicates that gold contents may range from 40 ppb to 30,000 ppm.

Unit MRT of the Cleary Sequence contains the major sulfide and precious metal mineralization of the metavolcanic sequence. The lenses of iron, arsenic, antimony, zinc, and lead sulfides locally contain native gold as inclusions in arsenopyrite and silver as argentiferous tetrahedrite. The sulfides range from 0.01 to 5.0 mm in diameter and occur as lenses up to 25 cm thick and several hundred meters in length along strike. The paragenetic sequence is pyrite -arsenopyrite (with or without gold) - sphalerite - galena - sulfosalts (including jamesonite, boulangerite, freibergite). Within the metapyroclastic rocks the sulfides are associated with muscovite and albite and are not segregated into sulfide compositional bands. In the siliceous metachert horizons, the sulfides are segregated into compositional bands up to a cm thick. These bands exhibit the same general paragenetic sequence as is evidence from the textural and grain boundary relations of the non-banded sulfides. The banded sulfides may exhibit a rhythmic pattern that is repeated for four or five cycles over a 25 cm lense.

Gold contents of the MRT unit vary from 40 ppb to 3 ppm over two meter intervals with higher values restricted to the sulfide rich portions of the unit. Silver contents range from 100 ppb to 600 ppm but are restricted to the sulfide bodies within the lenses of the MRT unit (Metz, 1984a).

In the Cleary Summit area the MRT units in aggregate may account for more than 30 meters of the upper 100 meters of the Cleary Sequence. In this area at the crest of the Cleary Anticlinorium, the lenses of the MRT unit are nearly flat lying and can be traced for up to a km along strike. Interburden between lenses of MRT includes units of MQS and MQ and this interburden may vary from a few centimeters to several meters. Thus, in the Cleary Summit area the MRT unit represents a potential low grade bulk mineable gold/silver target.

Type II Mineralization. The fine-grained and porphyritic granodioritic phases of the Pedro Dome stock have been subjected to several brittle shear events. The resulting shear zones contain fracture fillings and replacement type sulfide vein mineralization. The shear zones are the focus of potassic, sericitic, argillic, and propylitic alteration which extends for less than a meter into the unsheared rock.

The sulfide vein paragenesis includes pyrite - arsenopyrite - chalcopyrite - sphalerite - galena and/or argentiferous tetrahedrite. Calcite is the primary gangue mineral with subordinate quartz. Vein widths are usually less than 0.5 meters and the veins are not continouus over a few tens of meters along strike or down dip. The veins are notably deficient in gold but locally may contain up to 20 troy ounces/ton silver.

Locally the shear zones develop into a calcite and quartz stockwork with individual veinlets up to a centimeter wide. Locally this stockworks contain traces of molybderite and scheelite (Sherman, 1983).

Type III Mineralization. Scheelite bearing skarns occur as discontinuous bodies along the margins of the porphyritic quartz monzonite phase of the Gilmore Dome intrusive complex. The exoskarns are hosted in the metasediments and metabasites of the Cleary Sequence. The skarns can be classified into two calc-silicate types. Type A is a marble unit altered to pyroxene (Hd_{70}) + garnet (Alm_{10}) + scheelite which may be overprinted by quartz + hornblende + calcite + scheelite ± idocrase ± flourite ± pyrrhotite ± molybdenite. Type B is marl altered to quartz + amphibole + calcite + scheelite ± chlorite ± clinozoisite ± biotite (Allegro, 1984). Minor veinlets cross cut the skarns and these structures contain quartz + scheelite + amphibole + chlorite + mica and only traces of gold.

Type IV Mineralization. The majority of the lode gold production from the district has come from discordant gold-quartz-sulfide veins that are hosted primarily by Cleary Sequence metavolcanics and metaexhalites. The district historic production of 250,000 troy ounces has come from veins that averaged or exceed 1 troy ounce/ton. The gold-quartz structures occur as shear zone and replacement type veins that range from 10 cm to 5 meters wide with strike length of over 1 km. Most of the structures have been developed over a vertical distance of less than 100 meters.

The vein mineralogy is highly variable and appears to be controlled by the mineralogy of the host rocks. Gold quartz veins hosted in sulfide poor banded quartzite or banded-muscovite quartzite (metaexhalites) contain gold, arsenopyrite and only traces of pyrite, sphalerite, galena, hematite, magnetite, and sulfosalts. The major gangue mineral is quartz and the quartz is coarse grained to vuggy and was introduced in less than five discrete episodes.

The quartz veins in the metarhyolites and chlorite schists contain up to 5 percent sulfides. The paragenetic sequence of these veins can be summarized as follows: early quartz - pyrite - arsenopyrite (± gold) - sphalerite - chalcopyrite - bornite - gold and/or galena, argentiferous tetrahedrite, boulangerite, jamesonite, - late quartz ± ankerite. Antimony only rarely occurs as stibnite which is always associated with late quartz. The quartz is coarse grained to vuggy and was introduced in less than five (usually three) discrete episodes. High grade gold samples (greater than 1 troy ounce/ton) usually contain abundant ankerite.

The grain size distribution of gold in the veins ranges from 1 micron size inclusions in arsenopyrite to 5 mm diameter grains and wires replacing early sulfides and quartz. There is a positive correlation between grade and grain size in the quartz veins.

Type V Mineralization. Monomineralic stibnite veins occur as fracture fillings and replacement structures either above known gold-quartz ore shoots or as isolated structures. The stibnite generally fills or replaces the entire structure and is not associated with major quartz or other gangue mineralogy. Traces of pyrite, arsenopyrite, kermesite, stibconite, and gold are found in or adjacent to the stibnite. The stibnite can occur as fine-grained massive ore or as very coarse grained aggregates with subhedral grains up to 5 cm in diameter.

The stibnite lodes range from several centimeters to two meters in width and are highly irregular with strike lengths of less than 100 meters. Individual structures have produces from 100 to several thousand tons of ore. The most productive and the largest number of stibnite veins are hosted in the quartzites and metarhyolites of the Cleary sequence, however, several occurrences are within the Chatanika Terrane.

Most of the Type IV and V vein type mineralization occurs in structures that strike NE-SW and dip to the south at 45° to 55°. These structures are parallel to the major thrust faults in the district and are probably axial plane shears within imbricate thrust sheets.

Trace Element Data

In order to delineate potential sources of the gold mineralization of the Fairbanks mining district, one or more samples were collected from each of the major mappable rock units from each of the above described terranes. Samples were selected based on the absence of field identifiable major post metamorphic alteration and on the absence of included veins or mineralized structures. Thin sections of hard specimens from the sample sites were examined prior to sample preparation and analysis. Sample size was standardized to 20 pounds and the entire sample was crushed and pulverized in a shatter box to 100 % minus 200 mesh prior to sample splitting. The samples were analyzed by induced neutron activation tehchniques for gold plus 18 elements and the results are reported in Table 1. The detection limit for gold is 5 ppb while detection limits for the other elements range from 0.1 to 200 ppm. Major oxide and rare earth elemental analyses have been completed on the sample suite and are the subject of a separate paper (Metz, in preparation).

The eclogitic rocks of the Chatanika Terrane and their intercalated metasediments have gold contents of less than 5 ppb to 20 ppb. These rocks have low Sb, As, and W contents relative to the Cleary Sequence, however, the eclogitic rocks contain 3 times the Ba, Cr, and Ni contents of any other sequence of rocks in the district.

The metavolcanic and metaexhalites of the Cleary Sequence have gold contents that generally range from above 30 ppb to several thousand ppb. Stratiform sulfide horizons within the sequence contain 3 to 30 ppm gold and up to 600 ppm silver (Metz, 1984a). The average Sb (67 ppm) and As (1440 ppm) contents of the metavolcanics/metaexhalites are two orders of magnitude greater and the average W (22 ppm) cotent is an order greater than that of any of the metasedimentary rocks of the district.

The correlation coefficients of Au-As, Au-Sb, and Au-W are 0.84, 0.31, and -0.12, respectively. These data confirm field observations and ore/silicate rock petrologic examinations that show a strong relationship

Table 1 Trace Element Content of the Major Rock Units of the Fairbnks Mining District

Terrane/ Sample Number/ Description	Element Units	Au PPB	Sb PPM	As PPM	Ba PPM	Cd PPM	Cr PPM	Co PPM	Hf PPM	Fe PCT	La PPM	Mo PPM	Ni PPM	Na PCT	Ta PPM	Th PPM	W PPM	U PPM	Zn PPM
Chatanika																			
3830 Garnet clinopyroxenite		9	0.9	1	<100	12	390	59	5	7.0	48	<2	270	0.66	3	6.1	<2	1.7	1100
3831 Garnet clinopyroxenite		5	7.2	50	420	<5	55	14	6	2.1	38	<2	<50	0.31	4	13.0	4	3.5	<200
3832 Garnet amphibolite		<5	2.2	55	440	<5	64	<10	11	2.3	38	<2	<50	0.15	5	15.0	4	4.2	<200
3833 Muscovite-hornblende schist		19	45.6	46	3100	<5	190	<10	2	1.4	13	13	66	0.41	<1	5.3	<2	4.6	380
3841 Garnet-calc schist		<5	0.4	2	3200	<5	540	75	6	8.4	65	<2	370	0.92	5	8.9	<2	2.8	<200
3842 Garnet amphibolite		<5	1.6	3	1200	<5	530	67	6	10.0	99	<2	380	1.60	5	12.0	<2	2.6	<200
3843 Muscovite-K-Feldspar-schist		<5	3.8	22	530	<5	75	12	8	3.2	38	<2	<50	0.09	<1	17.0	30	2.8	<200
3846 Garnet amphibolite		8	2.7	4	1500	<5	430	39	5	7.5	55	<2	180	2.40	5	8.4	<2	3.0	<200
21342 Marble		<5	1.6	8	1800	<5	300	25	4	5.6	37	<2	75	1.30	3	4.8	4	2.6	<200
21343 Impure marble		<5	1.2	1	5800	<5	370	29	4	5.7	40	<2	100	1.30	3	5.7	<2	1.8	<200
21344 Graphitic metachert		14	14.0	153	1600	31	420	<10	2	1.1	18	24	130	0.26	<1	3.5	18	13.0	670
	$\bar{x}$	8	7.4	32	1790	8	306	32	5	4.9	44	5	156	0.85	3	9.1	7	3.9	341
	s	5	13.3	46	1685	8	182	25	3	3.1	23	7	127	0.73	2	4.5	9	3.2	291
Cleary Sequence Metavolcanics/Metaexhalites																			
197 Banded-muscovite quartzite		110	24.6	54	<100	<5	240	<10	5	0.9	11	<2	<50	<0.05	<1	3.3	88	1.0	<200
198 Banded-muscovite quartzite		46	39.6	369	1300	<5	53	<10	6	1.3	31	8	<50	0.12	<1	12.0	15	1.6	<200
3818 Banded-muscovite quartzite		1260	71.4	2410	150	<6	85	<10	5	7.9	12	<2	<50	<0.40	<1	6.2	30	1.4	<200
3819 Banded-muscovite quartzite		2300	59.2	8920	<100	<7	<50	<10	3	1.5	13	<2	<50	<0.34	<1	5.5	13	<0.5	<200
3824 Banded-muscovite quartzite		81	232.0	839	1900	<9	70	<10	5	2.3	51	<2	<50	<1.10	2	19.0	<8	1.7	830
3835 Metarhyolitic tuff		54	339.0	5310	270	<14	170	<10	3	2.8	27	<3	<50	<1.60	1	7.1	<13	<0.6	<200
3836 Garnet ortho-amphibolite		52	2.8	81	450	<5	690	56	6	10.0	42	<2	240	0.47	4	4.1	<2	1.0	<200
3837 Garnet ortho-amphibolite		110	3.0	86	460	<5	360	51	7	9.4	49	<2	160	0.45	5	4.3	9	1.1	<200
3839 Metarhyolitic tuff w/sulfide		230	25.1	1060	620	<5	71	11	7	3.6	39	2	54	0.12	1	16.0	17	3.0	<200
21151 Feldspathic-chlorite schist		12	3.4	20	180	<5	280	37	<2	6.5	5	<2	130	2.00	<1	0.6	<2	<0.5	<200
21152 Chlorite-talc schist		1570	324.0	6630	920	<14	120	15	5	4.8	48	<3	<50	<1.50	2	25.0	<10	5.2	<200
21155 Ortho-amphibolite		460	1.8	17	260	<5	730	55	7	10.0	62	<2	200	1.10	5	8.1	4	2.1	230
21330 Calc-chlorite schist		29	17.0	403	310	<5	<50	<10	<2	1.1	16	<2	<50	<0.05	<1	5.1	<2	0.8	<200
21331 Feldspathic-muscovite quartzite		11	75.1	234	580	<5	<50	<10	9	2.0	29	<2	<50	0.06	<1	13.0	<2	1.3	<200

Table 1 (continued)

Terrane/ Sample Number/ Description	Element Units	Au PPB	Sb PPM	As PPM	Ba PPM	Cd PPM	Cr PPM	Co PPM	Hf PPM	Fe PCT	La PPM	Mo PPM	Ni PPM	Na PCT	Ta PPM	Th PPM	W PPM	U PPM	Zn PPM
21335 Banded muscovite quartzite		35	2.8	25	870	<5	67	<10	6	0.9	24	<2	<50	0.10	1	19.1	64	2.4	<200
21337 Graphitic schist		19	20.0	28	2100	<5	160	13	5	4.7	64	<2	<50	0.23	2	26.0	14	4.7	<200
21346 Banded muscovite quartzite		33	6.1	17	120	<5	92	25	9	5.4	37	<2	<50	<0.05	2	17.0	48	5.1	<200
21347 Muscovite-garnet schist		32	2.0	4	680	<5	99	18	7	5.8	46	2	<50	0.90	2	18.0	74	4.8	<200
82SH02 Feldspathic-muscovite schist		42	20.8	856	990	<7	<50	17	6	4.4	54	<2	<50	0.08	2	18.0	6	5.8	<200
82CH02 Chloritic schist w/sulfides(1)		>30000	54.8	4970	200	<17	360	<10	<3	0.8	6	<7	<50	<0.06	<1	2.8	11	<1.3	<200
82HY01 Chloritic schist w/sulfides(1)		16400	704.0	>30000	320	<30	180	<10	<3	4.6	20	<8	<50	<3.10	<1	11.0	44	2.8	220
	x̄	341	66.8	1440	650	6	184	20	6	4.5	35	2	78	0.56	2	12.0	22	2.4	235
	s	642	107.3	2593	586	3	205	16	2	3.1	18	1	59	0.61	1	7.7	26	1.8	144
Cleary Sequence Metasediments																			
185 Biotite-muscovite-garnet schist		<5	0.6	8	900	<5	220	18	7	5.2	60	4	<50	1.20	2	19.0	3	3.1	<200
192 Biotite-muscovite-garnet schist		6	3.7	72	390	<5	120	<10	7	5.3	54	<2	<50	0.45	1	19.0	13	3.9	<200
3810 Para-amphibolite		8	0.3	<1	1000	<5	81	16	11	3.8	50	<2	<50	1.20	2	24.0	<2	2.5	<200
3823 Marble		11	26.6	13	<100	<5	<50	<10	<2	<0.5	<5	<2	<50	<0.05	<1	0.7	<2	1.2	<200
3825 Para-amphibolite (?)		<5	0.9	10	<100	<5	680	25	8	10.0	100	<2	190	0.69	6	14.0	4	4.4	<200
3826 Muscovite quartzite		<5	1.0	10	420	<5	<50	<10	5	1.8	17	<2	<50	0.32	<1	11.0	<2	1.8	<200
3827 Marble		6	0.3	2	120	<5	<50	<10	<2	0.7	11	<2	<50	0.14	<1	2.6	<2	1.0	<200
3838 Biotite-muscovite-garnet schist		9	2.8	17	280	<5	190	26	5	7.6	20	9	84	0.71	<1	6.4	<2	1.3	<200
20506 Metaconglomerate		12	4.4	65	850	<5	180	<10	7	2.7	33	<2	<50	0.87	<1	14.0	3	2.0	<200
20510 K-Feldspar-biotite schist		<5	3.0	62	1200	<5	110	18	6	5.1	61	<2	<50	2.50	2	21.0	<2	3.3	<200
21142 Quartzite		7	0.8	4	410	<5	<50	18	5	2.9	16	3	<50	0.18	<1	8.2	17	2.7	<200
21147 Quartzite		9	0.6	4	430	<5	<50	24	6	2.7	17	5	<50	0.25	<1	8.6	<2	2.8	<200
21154 Muscovite schist		11	0.5	2	830	<5	100	20	6	5.7	55	<2	53	0.68	2	20.0	<2	4.7	<200
21320 Muscovite quartzite (1)		<14	373.0	26	1100	<12	250	<10	6	2.8	35	<4	<50	1.30	<1	15.0	<3	2.8	<200
21334 Feldspathic biotite garnet schist		13	4.9	25	780	<5	76	<10	7	4.6	52	<2	<50	0.40	1	20.0	12	2.9	<200
	x̄	8	5.4	22	558	5	143	16	6	4.2	39	3	63	0.69	2	13.5	5	2.5	<200
	s	3	7.9	26	362	0	165	6	2	2.6	27	2	38	0.64	1	7.4	5	1.3	0

Table 1 (continued)

Terrane/ Sample Number/ Description	Element Units	Au PPB	Sb PPM	As PPM	Ba PPM	Cd PPM	Cr PPM	Co PPM	Hf PPM	Fe PCT	La PPM	Mo PPM	Ni PPM	Na PCT	Ta PPM	Th PPM	W PPM	U PPM	Zn PPM
Fairbanks Schist Metasediments																			
3828 Garnet-chlorite-muscovite schist		5	0.9	47	650	<5	<50	<10	7	3.2	33	<2	<50	0.24	<1	14.0	2	2.6	<200
20505 Biotite-muscovite-quartz schist		<5	0.8	2	820	<5	200	<10	7	4.0	58	<2	<50	0.76	2	16.0	<2	3.0	<200
20508 Muscovite-quartz schist		<5	0.6	3	110	<5	110	<10	10	1.7	10	<2	<50	0.78	<1	12.0	<2	2.6	<200
	$\bar{x}$	5	0.8	17	527	5	120	10	8	3.0	34	2	50	0.59	1	14.0	2	2.1	200
Goldstream Sequence																			
3809 Biotite-muscovite schist		26	0.3	10	1000	<5	63	<10	8	2.4	41	<2	<50	1.00	<1	17.0	<2	1.4	<200
3814 Garnet-chlorite-muscovite schist		<5	3.1	217	<100	<5	<50	12	7	1.6	16	3	<50	0.90	<1	7.8	<2	1.2	<200
3815 Garnet-biotite-muscovite schist		8	5.8	160	87	<5	110	<10	9	3.4	61	<2	<50	1.30	2	24.0	3.	5.3	<200
3816 Garnet-amphibolite		<5	0.4	2	<100	<5	63	43	5	10.0	18	<2	<50	2.80	1	1.4	<2	0.6	<200
3817 Calcareous schist		<5	0.2	5	260	<5	<50	<10	3	1.2	16	<2	<50	0.71	<1	6.1	<2	<0.5	<200
3840 Marble		<5	1.3	1	860	<5	<50	<10	<2	1.1	10	<2	<50	0.20	<1	3.4	<2	0.6	<200
3844 Para-amphibolite		<5	0.5	3	400	<5	200	17	8	3.6	34	<2	110	1.30	2	11.0	<2	2.3	<200
21324 Graphitic-quartz schist		21	41.3	80	1200	<5	230	<10	4	3.4	29	<2	57	0.07	<1	7.6	13	2.5	300
	$\bar{x}$	10	6.6	60	500	5	102	15	6	3.3	28	2	58	1.04	1	9.8	4	1.8	212
	s	8	14.2	85	452	0	73	12	3	2.9	17	0	21	0.85	1	7.4	4	1.6	35
Birch Hill Sequence																			
3811 Graphitic-quartz schist		<5	0.8	7	<100	<5	<50	<10	5	1.5	<5	<2	<50	<0.05	<1	5.2	<2	0.9	<200
3820 Calcareous schist		<5	0.7	8	1300	<5	52	<10	3	2.5	28	<2	<50	1.00	<1	11.0	<2	0.9	<200
3821 Graphitic phyllite		<5	14.0	38	2000	<5	76	<10	<2	1.1	20	7	<50	<0.05	<1	2.7	<2	4.6	<200
3822 Graphitic quartzite		<5	4.2	9	1200	<5	97	<10	<2	1.0	6	5	<50	<0.05	<1	1.0	<2	1.7	<200
20503 Carbonaceous quartzite		<5	4.3	7	2700	<5	210	<10	3	0.7	18	<2	<50	0.06	<1	6.2	4	2.6	<200
20504 Pyritic-banded quartzite		8	6.5	47	690	<5	170	15	3	4.4	22	<2	<50	0.07	<1	8.3	12	1.4	<200
	$\bar{x}$	6	5.1	19	1332	5	101	11	3	1.9	16	3	50	0.21	1	5.7	4	2.0	200
	s	1	4.9	18	924	0	77	2	1	1.4	9	2	0	0.39	0	3.6	4	1.4	0

Table 1 (continued)

Terrane/ Sample Number/ Description	Element Units	Au PPB	Sb PPM	As PPM	Ba PPM	Cd PPM	Cr PPM	Co PPM	Hf PPM	Fe PCT	La PPM	Mo PPM	Ni PPM	Na PCT	Ta PPM	Th PPM	W PPM	U PPM	Zn PPM
Aplite Dikes																			
3834 Pyritic Aplite		<5	2.9	17	1200	<5	<50	13	9	5.5	65	<2	<50	2.10	1	19.0	8	5.5	<200
20507 Pyritic Aplite		8	8.9	118	1700	<5	<50	<10	9	4.1	65	<2	<50	1.10	2	20.0	<2	6.6	200
	$\bar{x}$	6	5.9	68	1450	5	50	12	9	4.8	65	2	50	1.60	2	19.5	5	6.1	200
Skarns/Hornfels																			
175 Calc-silicate		250	19.0	13	<100	<7	80	16	<19	8.1	27	32	<50	0.10	1	9.2	2240	4.7	250
186 Hornblende hornfels		7	21.9	62	230	<6	700	44	8	9.2	77	<2	180	0.63	6	9.2	12	2.8	<200
190 Calc-silicate		470	24.2	11	<130	<12	96	13	<64	6.8	14	120	<50	<0.05	<1	2.6	5720	4.5	<200
200 Calc-silicate		37	2.2	47	<100	<6	<50	190	5	15.0	18	4	67	<0.10	<1	5.8	3720	6.1	<200
3812 Calc-silicate		110	1.7	5	390	<5	<50	15	<19	5.2	59	<2	<50	0.72	2	17.0	1600	2.4	<200
3813 Calc-silicate		6	1.1	20	790	<5	<50	<10	4	1.3	21	2	<50	2.20	<1	13.0	2	4.9	<200
3829 Garnet-vesuvianite skarn		220	42.1	11	<100	<9	<50	<10	<53	5.9	30	57	<50	0.07	1	5.9	4590	4.9	<200
3845 Sheared hornfels		20	41.9	162	<100	<5	<50	10	4	4.3	16	2	<50	<0.05	<1	5.5	15	2.6	<200
21149 Calc-silicate		<5	1.3	9	520	<5	73	17	4	4.1	37	3	<50	0.77	<1	12.0	85	3.1	<200
21162 Calc-silicate		100	31.8	54	180	<13	53	59	<21	16.0	42	223	<50	<0.33	3	7.9	19800	4.5	290
21341 Epidote hornfels		10	20.1	78	670	<5	150	13	7	3.1	47	22	<50	0.88	<1	20.0	11	14.0	<200
21349 Altered hornfels		130	65.8	233	520	<20	130	39	<73	13.0	59	100	<50	0.15	1	8.7	8260	3.7	<200
	$\bar{x}$	114	22.8	59	320	8	128	36	23	7.7	37	47	62	0.50	2	9.7	3838	4.8	362
	s	141	20.2	71	250	5	183	51	25	4.8	20	69	37	0.62	2	5.0	5711	3.1	517

1) Note: Not included in mean and standard deviation calculation

between gold and arsenopyrite in all rock units; the presence of major sulfosalts in the metabasic, metafelsic, and metacherts of the Cleary Sequence with or without gold mineralization; and the concentration of W in the metacherts and metabasic rocks and relatively low W contents in the gold enriched metafelsic rocks.

The Au contents of the metasediments of the Cleary Sequence average 8 ppb while the metasediments of the Fairbanks Schist, Goldstream Sequence, and Birch Hill Sequence average 5 ppb, 10 ppb, and 6 ppb, respectively. The average Sb contents of these rock types are 5 ppm, 0.8 ppm, 6.6 ppm, and 5.1 ppm, respectively; the arsenic contents are 22 ppm, 17 ppm, 60 ppm, and 19 ppm, respectively; and the W contents are 5 ppm, 2 ppm, 4 ppm, and 4 ppm, respectively.

Two pyritic aplite dikes were sampled to determine the gold content of these late phases of the intrusive complexes. The average Au, Sb, As, and W contents are 6 ppb, 5.9 ppm, 68 ppm, and 5 ppm, respectively. Although these limited data do not confirm that the late phase granitic rocks are not the sources of these elements in the mineral deposit types described above, the data does support observed field and petrologic evidence that demonstrates little spatial, mineralogical, or petrological association of the Types I, II, IV, and V mineralization with the granitic rocks.

The samples of the tungsten skarns of the Type III mineralization have average Au, Sb, As, W contents of 114 ppb, 22.8 ppm, 59 ppm, and 3838 ppm, respectively. These relatively small bodies of exoskarn are all within Cleary Sequence rocks in contact with or adjacent to the Gilmore Dome porphyritic quartz monzonite.

Stable Isotopic and Fluid Inclusion Data

In order to further constrain the mineral classification system discribed above, fluid inclusion and sulfur isotopic analyses were conducted on 28 of the known 236 mineral occurrences in the Fairbanks district (Metz, 1984b). The results of this investigation are summarized in Figure 4.

Stratabound sulfides from Type I mineralization have a mean ^{34}S of 2.0 per mil excluding two samples of sulphides from a single disseminated occurrence in eclogite which have a mean of 8.8 per mil. Sulphides from both Type II mineralization (endoshears) and Type IV mineralization (exoshears) have a mean of 1.4 and 2.2 per mil, respectively. Stibnite from late antimony veins (Type V mineralization) have a mean of -4.2 per mil. The mean per mil values for the stratabound and vein mineralization are approximately equivalent to the expected values for volcanic sulphide and modern hydrotherms respectively (Nielsen, 1979). The sample means of Types I, II, and IV mineralization are not significantly different as are the variances of 1.19, 1.23 and 1.21 per mil respectively, however the small number of samples particularly from Type II mineralization preclude rigorous hypothesis testing.

The presence of disseminated galena and sphalerite in the eclogitic rocks was not expected nor can descriptions of analogues of that type of mineralization be found in the literature. Lead isotopic analyses of galena from the disseminated mineralization in the Eclogite Terrane and from stratabound and vein mineralization in the Cleary Sequence indicate two distinctive lead isotope populations for the two tectono-stratigraphic units (Metz, in preparation). Although sulphur isotopic data for the Eclogite terrane is very limited, there is a difference of about 7 per mil in the mean values of the Cleary Sequence and Eclogite Terrane sulphides. Although these

Figure 4. Sulfur isotopic and fluid inclusion data from the Fairbanks mining district, Alaska.

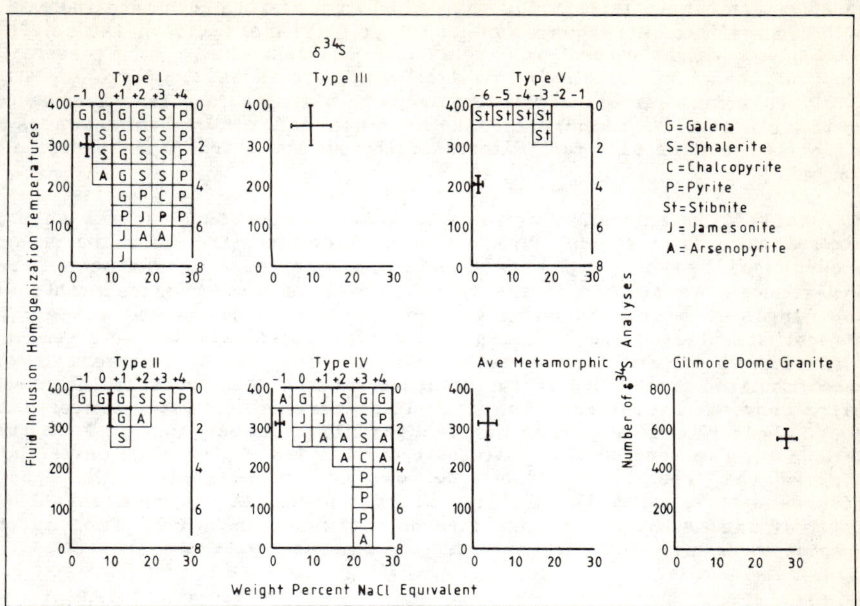

two tectono-stratigraphic units both contain stratabound mineralization, the modes of origin of the mineralization were probably significantly different. In addition, the two rock units and hosted syngenetic mineralization were subsequently subjected to radically different pressure-temperature conditions during one or more regional metamorphic events.

Mineral deposit Types II and IV show the largest difference in sulphur isotope sample means, however, the sample means are not significantly different above the 90 percent confidence interval. Given the order of isotopic exchange rate galena - sphalerite - chalcopyrite - pyrite (Brown et al., 1975) it may be expected that the metamorphosed Type I mineral occurrences would show narrower ranges for galena than pyrite as compared to Types II and IV mineralization. This is not the case as shown in Figure 4. Therefore it can be concluded that metamorphism has had no significant effect on the sulphur isotope values and the degree of equilibrium or desequilibrium in the systems are a function of the original conditions of ore deposition.

The stibnite from Type V mineralization contains significantly lighter sulphur than the other mineral deposit types. The antimony bearing phase in Type V is always restricted to stibnite. The stibnite contains only trace amounts of arsenic and gold. Conversely in Types I, II, and IV mineralization, the antimony bearing phases are complex sulphosalts and stibnite occurs only in trace amounts in a limited number of occurrences.

Figure 4 includes fluid inclusion homogenization and salinity data (Metz, 1984b) for Types I (except the eclogite occurrence) thru V mineral occurrences as well as selected metamorphic and intrusive igneous rocks from the mining district. Fluid inclusion studies from quartz in the various types of mineralization show the following homogenization temperatures and contrasting compositions: Type I mineralization, 280 to 330°C, 5 to 6 weight percent NaCl equivalent, and 4 to 8 mole percent CO_2; Type II and III mineralization, 300 to 400°C, 12 to 14 weight percent NaCl equivalent; Type IV mineralization, 290 to 360°C, 2 to 3 weight percent NaCl equivaltne, greater than 8 mole percent CO_2 in samples with visible gold, and indications of boiling and/or effervences of CO_2; Type V mineralization, 190 to 220°C, less than 2 weight percent NaCl equivalent and less than 4 mole percent CO_2. The granitic rocks have homoginization temperatures in the range of 500 to 600°C, 20 weight percent NaCl equivalent, occassional daughter minerals, and no visible CO_2. Metamorphic rocks have homogenization temperatures and salinities similar to Type I mineralization but lower and highly variable CO_2 content.

Table 2 is a tabulation of calculated temperatures fromm sulphide mineral pairs from several Type I mineral locations (including the eclogite occurrence) and one Type II mineral occurrence. Four of the Type I occurrences have sulphide pairs that indicate temperatures compatible with the ranges of temperatures estimated from the sulphide and metamorphic mineral assemblages and by fluid inclusion homogenizations. The remaining three Type I sulphur isotope pairs indicate temperatures incompatible with those obtained from fluid inclusion measurements. The two sulphide mineral pairs from the Silver Fox Mine (Type II occurrence) indicate temperatures 200°C above the fluid inclusion homogenization temperatures. From these data, it can be concluded that isotopic equilibrium may have been attained in some of the Type I occurrences but was not approached in the Type II occurrence. No suitable sulphur isotope mineral pairs were available for Types IV and V mineralization, thus no estimate can be made from sulphur species concerning equilibrium conditions in these systems.

Table 2. Calculated temperatures from sulphur isotopic compositions of sulphide mineral pairs and fluid inclusion homogenization temperature ranges from the Fairbanks Mining District, Alaska.

Mineral Occurrence	Sulphide Mineral Pair	Calculated Temp. (°C)	Deposit Type	Range of fluid Inclusion Homogenization Temp. (°C)
Chatham Creek Prospect	Py-Ga	700	I	280-360
Christina Adit	Py-Sl	300	I	
Newboy Ext Prospect	Py-Ga	700	I	
	Sl-Ga	630	I	
Ridge Prospect	Sl-Ga	290	I	
Silver Fox Mine	Py-Ga	580	II	300-380
	Sl-Ga	600	II	
Steese Eclogite Prospect	Sl-Ga	330	I	?
Wackwitz Prospect	Py-Ga	400	I	280-360
	Py-Sl	100	I	
	Sl-Ga	100	I	
Willow Creek Prospect	Py-Ga	630	I	
	Sl-Ga	1200	I	

Types I, II, and IV mineral occurrences contain quartz with CO_2 rich fluid inclusions. Types II and IV mineral occurrences often contain variable density CO_2 inclusions that homogenize at the same temperature, thus suggesting effervescence of CO_2 during vein formation. Type V mineralization contains relatively minor CO_2 (3 mole percent maximum) compared to Types I, II, and IV mineralization (13+ mole percent). Type V mineral occurrences do not show any evidence of CO_2 effervescence. The effervescence of CO_2 will result in an increase in pH and an increase in pH should shift the sulphur species to more positive ^{34}S values. A decrease in temperature or an increase in oxygen partial pressure should cause a shift to lighter ^{34}S (Ohmoto, 1972). The disequilibrium conditions in some of the Type I mineral occurrences and in the Type II deposit probably reflects rapid changes in pH, temperature, and partial pressure of volatile phases.

The sample means of Types I and V; and IV and V mineralization differ by approximately 6 per mil. Ohmoto and Rye (1979) and Robinson and Farrard (1982) suggest that equilibrium stibnite values are about 4 per mil lower than H_2S in an ore fluid at the temperature range of 150 to $200^{\circ}C$. Fluid inclusion homogenization temperatures for Type V mineralization indicate ore deposition took place between 180 and $220^{\circ}C$, thus sulphur from either Types I, II, or IV mineralization could account for the ^{34}S values in Type V mineral deposits. In addition, it is possible that Type I mineralization could have been the source of sulphur for Types II and IV mineral deposits.

Genetic Ore Deposit Model

From field observation and the above petrochemical data the following is a synopsis of an ore deposit model for the Fairbanks mining district: (1) Precambrian bimodal submarine volcanism in a rift environment in a back arc basis or narrow epicontinent ocean; (2) Deposition of volcaniclastic and exhalite rocks and sulfide deposits enriched in Au, Sb, As, and W; (3) Compression of the ocean basin accompanied by regional polymetamorphism and deformation; (4) Emplacement of post tectonic Cretaceous anatextic granitic plutons and concurrent skarn formation; (5) Mobilization and mixing of metamorphic fluids due to temperature gradients imposed by the granitic plutons; (6) Deposition of metals in favorable structural settings due to decreases in pressure, temperature, and increases in the fugacity of oxygen as a result of the mixing of metamorphic fluids with oxygenated low salinity meteoric fluids. Fluid inclusion data and vein mineralogy suggest that the metamorphic fluids were enriched in CO_2, H_2S, Sb, and As complexes, and that the vein type mineralization resulted from volatile release or boiling accompanied by an increase in pH at the boiling interface.

The stratabound Sb-W mineralization in the lower Cleary Sequence is similar to the stratabound Hg-Sb-W mineralization in Austria (Maucher, 1976), South Africa (Muff, 1978), and Spain (Arribas, 1984). The stratabound Au-As-Sb mineralization of the upper portion of the Cleary Sequence has strong mineralogic and petrologic similarities to the Archean stratabound gold mineralization in southern Africa (Anhaeusser, 1976), western Australia (Fehlberg and Giles, 1982), Canada (Godwin, 1982), and Zimbabwe (Foster and Wilson, 1982). The stratabound Au mineralization of the Fairbanks district can be analogued with the recent gold discoveries in the Hemlo area (see Patterson, 1984) of the Archean Abitibi Greenstone Belt of northwestern Ontario Canada.

References

1. Allegro, G.L., 1984, Gilmore Dome stratiform tungsten occurrences, Fairganks mining district, Alaska (Abstr.): in Abstract with Programs, 80th Annual Meeting Cordilleran Secture Meeting of the Geological Society of America, Anchorage, Alaska, May 30 - June 1, 1984, p. 266.

2. Anhaeusser, C.R. (1976b) The nature and distribution of Archaean gold mineralization in southern Africa. Mineral Sce. Eng. Vol. 8, 46-84.

3. Arribas, A., 1984, Los Yacimientos de tungsteno de La Zona de Morille: Separata de Boletin Geologico y Minero, Madrid, 1980, 20 p.

4. Blum, J.D., 1982, Petrology, geochemistry, and isotope geochronology of the Gilmore Dome and Pedro Dome plutons, Fairbanks mining district, Alaska: University of Alaska unpub. M.S. thesis, 107 p.

5. Britton, J.M., 1970, Petrology and petrography of the Pedro Dome plutons, Fairbanks, Alaska: University of Alaska unpub. M.S. theses, 52 p.

6. Brown, J.M., 1962, Bedrock geology and ore deposits of the Pedro Dome area, Fairbanks mining district, Alaska: College, Alaska Univ. M.S. thesis, 137 p.

7. Brown, P.R.L., Rafter, T.A., Robinson, B.W., 1975, Sulphur isotopic variations in nature. Part II: Sulphur Isotope ratios in sulphides from Broadlands geothermal field, New Zealand, N.Z. Journ. Sci. V. 18, 35-40.

8. Bundtzen, T.K., 1982, Bedrock geology of the Fairbanks mining district, western sector, Alaska: Alaska Division of Geological and Geophysical Surveys Open-file Report 155, 2 plates, scale 1:24,000.

9. Chapin, Theodore, 1914, Lode mining near Fairbanks: U.S. Geol. Survey Bull. 592-J, p. 321-355.

10. Chapin, Theodore, 1919, Mining in the Fairbanks district: U.S. Geol. Survey Bull. 692-F, p. 321-327.

11. Chapman, R.M., and Foster, R.L., 1969, Lode mines and prospects in the Fairbanks district, Alaska: U.S. Geol. Survey Prof. Paper 625-D, p. D1-D25.

12. Chappel, B.W., and White, A.J.R., 1974, Two contrasting granite types: Pacific Geology v.8, p. 173-174.

13. Coleman, R.G., Lee, D.E., Beatty, L.B., and Brannock, W.W., 1965, Eclogites and eclogites: their differences and similarities: Geol. Soc. of America Bull., 76, p. 483-508.

14. Erdmer, P. and Helmstaedt, H., 1983, Eclogite from Central Yukon: a record of subduction at the western margin of ancient North America: Can. Journ. Earth Sci., v. 20, p. 1389-1408.

15. Fehlberg, B., and Giles, C.W., 1982, Archaean volcanic exhalitive gold mineralization at Spargoville western Australia: in Gold '82 Geology, Geochemistry and Genesis of Gold Deposits, R.P. Foster (editor), A.A. Balkema Rotterdam, p. 285-304.

16. Foster, H.L., Weber, F.R., Forbes, R.B., et al., 1973, Regional geology of Yukon-Tanana Upland, Alaska in Arctic Geology: Am. Assoc. Petroleum Geologists Mem. 19, p. 388-395.

17. Foster, R.P. and Wilson, J.F., 1982, Geologic setting of Archaean gold deposits in Zimbabwe: in Gold '82 Geology Geochemistry and Genesis of Gold Deposits. R.P. Foster (editor), A.A. Balkema, Rotterdam, p. 521-552.

18. Goodwin, A.M., 1982, Archaean greenstone belts and mineralization, Superior Province, Canada: in Gold '82: The Geology Geochemistry and Genesis of Gold Deposits. R.P. Foster (editor), A.A. Balkema, Rotterdam, p. 521-551.

19. Hill, J.M., 1933, Lode deposits of the Fairbanks destrict, Alaska: U.S. Geol. Survey Bull. 849-B, p. 29-163.

20. Maucher, A., 1976, The stratabound cinnabar-stibnite-scheelite deposits: in Wolff, K.G., ed., Handbook of stratabound and stratiform ore deposits, Elsevier Scientific Publishing Company, Amsterdam, p. 477-503.

21. Mertie, J.B., Jr., 1918, Lode mining in the Fairbanks district, Alaska: U.S. Geol. Survey Bull. 662-H, p. 403-424.

22. Mertie, J.B., Jr., 1938, The Yukon-Tanana region, Alaska: U.S. Geol. Survey Bull. 872, 276 p.

23. Metz, P.A., 1977, Comparison of Hg-Sb-W mineralization of Alaska with stratabound cinnabar-stibnite-scheelite deposits of the circum-Pacific and Mediterranean regions, in Short Notes on Alaskan Geology 1977: Alaska Division of Geological and Geophysical Surveys, Geologic Report No. 55, p. 39-41.

24. Metz, P.A., 1982a, Bedrock geology of the Fairbanks mining district, northeast sector, Alaska: Alaska Division of Geological and Geophysical Surveys Open-file Report 154, 1 sheet, scale 1:24,000.

25. Metz, P.A., 1982b, Ore petrology and geochemistry of the Fairbanks mining district, Alaska: abstr. in Proc. Northwest Mining Assoc. 88th Annual Meeting, Spokane, Washington, Dec. 1982.

26. Metz, P.A., 1983, Bedrock stratigraphic, structural and surficial depositional controls of the gold placer deposits of the Fairbanks mining district, Alaska: Abstr. in Proc. 34th Alaska Science Conference, Whitehorse, Yukon Terr., 28 Sept. - 1 Oct. 1983, p. 93.

27. Metz, P.A., 1984a, Statistical analysis of the stream sediment, pan concentrate and rock geochemical data of the Fairbanks mining district, Alaska: University of Alaska, Mineral Industry Research Laboratory, Open-file Report 84-1, 40 p., maps, 9 sheets, scale 1:63,360.

28. Metz, P.A., 1984b, Sulphur isotopic evidence for the genesis of the Au-Ag-Sb-W mineralization of the Fairbanks mining district, Alaska: British Geological Surveys, Isotope Geology, Laboratory Open-file Report, 20 p.

29. Metz, P.A., and Halls, Christopher, 1981, Ore petrology of the Au-Ag-Sb-W-Hg mineralization of the Fairbanks mining district, Alaska: Abstr. in Proc. Mineralization of the precious metals, uranium and rare earths, University College, Cardeff, Wales, 15-18 Dec. 1981.

30. Metz, P.A., and Robinson, M.S., 1980, Investigation of mercury-antimony-tungsten metal provinces of Alaska: University of Alaska, Mineral Industry Research Laboratory, Open-file Report 80-8, p. 153-190.

31. Muff, R., 1978, The antimony deposits of the Murchison range of the notheastern Transvaal, Republic of South Africa: Mon. Ser. Miner. Deposits, No. 16, 90p.

32. Nielson, H., 1979, Sulfur isotopes in Jager, E., and Hunziker, J.C., eds., Lectures in isotope geology: Springer-Verlag, New York, N.Y., p. 283-312.

33. Ohmoto, H., 1972, Systematics of sulfur and carbon isotopes and ore genesis: A review: Econ. Geol. V. 69, p. 826-842.

34. Ohmoto, H. and Rye, R.O., 1979, Isotopes of sulphur and carbon: In: Barnes, H.L. ed., Geochemistry of hydrothermal ore deposits, Wiley Interscience, New York, p. 509-567.

35. Patterson, G.C., 1984, Field trip guidebook to the Hemlo area, Ontario Geol, Surv., Misc. Paper no., 118, 33p.

36. Prindle, L.M., and Katz, F.J., 1913, Geology of the Fairbanks district, in Prindle, L.M., A geologic reconnaissance of the Fairbanks quadrangle, Alaska: U.S. Geol. Survey Bull. 525, p. 59-152.

37. Robinson, B.W., and Farrand, M.G., 1982, Sulfur isotopes and the origin of stibnite mineralization in New England, Australia: Mineralium Deposita V 17, p. 161-174.

38. Robinson, M.S., 1982, Bedrock geology of the Fairbanks mining district, southwest sector, Alaska: Alaska Division of Geological and Geophysical Surveys Open-file Report 146, 1 sheet, scale 1:24,000.

39. Sherman, G.E., 1983, Geology and mineralization of the Silver Fox Mine, Fairbanks District, Alaska: University of Alaska unpub. M.S. thesis, 84 p., maps, 3 sheets, scale 1:4,800.

40. Smith, P.S., 1913, Lode mining near Fairbanks: U.S. Geol. Survey Bull. 542-F, p. 137-202.

41. Swainbank, R.C., and Forbes, R.B., 1975, Petrology of eclogitic rocks from the Fairbanks District, Alaska: Geol. Soc. America Special Paper 151, p. 77-213.

MINERAL ZONING RELATED TO FOSSIL GEOTHERMAL SYSTEMS,

CASTROVIRREYNA Ag-Pb-Zn DISTRICT, CENTRAL PERU

Demetrius C. Pohl[1], Frederico Llerena[2], Victor Quirita[3]

1 Department of Mineral Sciences, American Museum of Natural History, Central Park West at 79th Street, New York, NY 10024

2 Corporacion Minera Castrovirreyna S.A., Jiron Quilca 538 Lima, Peru

3 Castorvirreyna Compania Minera S.A., Gregorio Escobedo 710, Jesus Maria, Lima, Peru

Abstract

The polymetallic silver deposits of the Castrovirreyna district produce 1800 tpd ore grading 240-400 Ag/t, 2% lead and zinc with minor copper and gold. Recent mapping has shown the veins to be part of a fossil geothermal system that was active during the waning phases of a late Miocene andesite volcanic complex. Regionally the volcanic rocks are propylitized. Both surficial tabular silica bodies and silica breccias grade downward through a parallel vein stockwork into the mineralized quartz veins. The veins occur in fault-related structures that strike 075-115° and 130° and vary in length from 200 m to 4 km. Mineralization extends for more than 500 m vertically. The veins and silica bodies are enveloped by a zone of intense argillic alteration with disseminated pyrite whose thickness is dependent on the permeability of the host volcanics. Immediately adjacent to and within the veins, host rocks are variably silicified and sericitized.

The mineralogy of the veins consists essentially of galena, sphalerite, and tetrahedrite and a complex assemblage of silver and lead sulfantimonides and sulfarsenides. Based on identification of the dominant silver-bearing phase it has been possible to establish a zoning pattern that appears consistent throughout the district. From top to bottom, the zones are, native silver, "ruby silver", "black silver", tetrahedrite and polymetallic. This zoning pattern can in turn be related to a depth range referenced to the original paleosurface datum. The mineral zones have characteristic tenors that diminish with depth. However, the tetrahedrite zone appears to show constant but low grades over a great but indeterminate vertical extent. The silver-mineral zoning as well as the more detailed paragenetic relations among the silver sulfantimonides appears to be a response to a declining P/T gradient and progressive oxidation of the geothermal fluid on its rise to the surface.

Introduction

The Castrovirreyna district in central Peru (Figure 1) has produced considerable amounts of silver together with minor amounts of base-metals and gold from pre-Columbian times to the present. Current production of approximately 1800 t.p.d. ore from four major centers grades 240-400 g/t Ag, 1-2% Pb and Zn and 0.5-5% Cu and minor gold. From east to west these are El Palomo (EP), San Genaro (SG), Caudalosa Grande (CG), and Reliquias (RQ). Each center exploits several associated veins. Other areas which have produced in the past but are not presently active, are La Virreyna, Bonanza (BZ), Astohuaraca (AS) and several smaller areas. The district lies almost at the Andean crest in south central Peru approximately 220 km southeast of Lima.

The vein deposits have been described in part by Masias (1924) and more completely by Lewis (1956, 1964) who related the source of mineralization to the underlying late Cretaceous to early Tertiary Andean batholith, and also presented an asymmetrical regional zoning scheme for the mineralization. However mapping by (Salazar 1970, no date) and K-Ar age dating of several rock units by Noble et al. (1972, 1974), indicate that the age of the veins and silver mineralization must be younger than mid-Miocene.

The ore-bearing veins occur in a thick sequence of intermediate to silicic lavas and are spatially related to subvolcanic dyke and plug complexes which dome a relatively flat-lying section of massive andesite lavas, flow-breccias and tuffs.

Regional Geology

The Castrovirreyna district lies within an extensive Eocene to Pliocene andesite volcanic province in Central Peru (Salazar, 1970, no date; Noble et al., 1972, 1974; Noble and McKee, 1975; Petersen et al., 1977; Megard, 1978; Noble et al., 1979; McKee and Noble, 1982). Basement consists of middle to upper Cretaceous marine and volcanic rocks intruded by Late Cretaceous to early Tertiary granodiorites and tonalites. These basement rocks are exposed only on the extreme northeastern and southwestern flanks of the district. The total thickness of the overlying volcanic rocks is at least 3 km.

The Tertiary volcanic pile is part of a regional NW trending synclinorium in which the oldest units have been strongly folded into large NW trending isoclines and dated at 41 m.y. (Noble, McKee and Megard, 1979; Noble et al., 1974). These rocks are unconformably overlain by a sequence of gently folded intermediate to silicic ashflow tuffs which has a K-Ar age of 21.4 m.y. (Noble et al., 1974). Uncomformably overlying this unit is a section of 13.9 m.y. old silicic ashflow and airfall tuffs (Noble et al., 1974). The Caudalosa volcanics (Salazar, 1970) overly this section with a slight angular unconformity and consist of more than 700m of thick andesite lavas and flow breccias with subordinate tuff horizons and subaqueous volcanoclastic sediments. They have been gently folded about NNW trending axes into broad open flexures. Locally, throughout the district, the andesite flows have been intruded and domed by andesite subvolcanic plugs and dike complexes, which may be the temporal equivalents of dome complexes in the Huachacolpa district 30km to the east dated at 8-10 m.y. by McKee et al., (1975).

Small diorite and hybrid dacitic stocks and dikes intrude the Caudalosa Volcanics. The latter are compositionally inhomogeneous and contain clasts of mineralized vein material. The last intrusive event in the district produced isolated vitrophyric andesite plugs and domes. They may be equivalent to unaltered rhyodacite dikes and plug domes having 4 m.y. K-Ar ages in the Huachocolpa district (McKee et al., 1975).

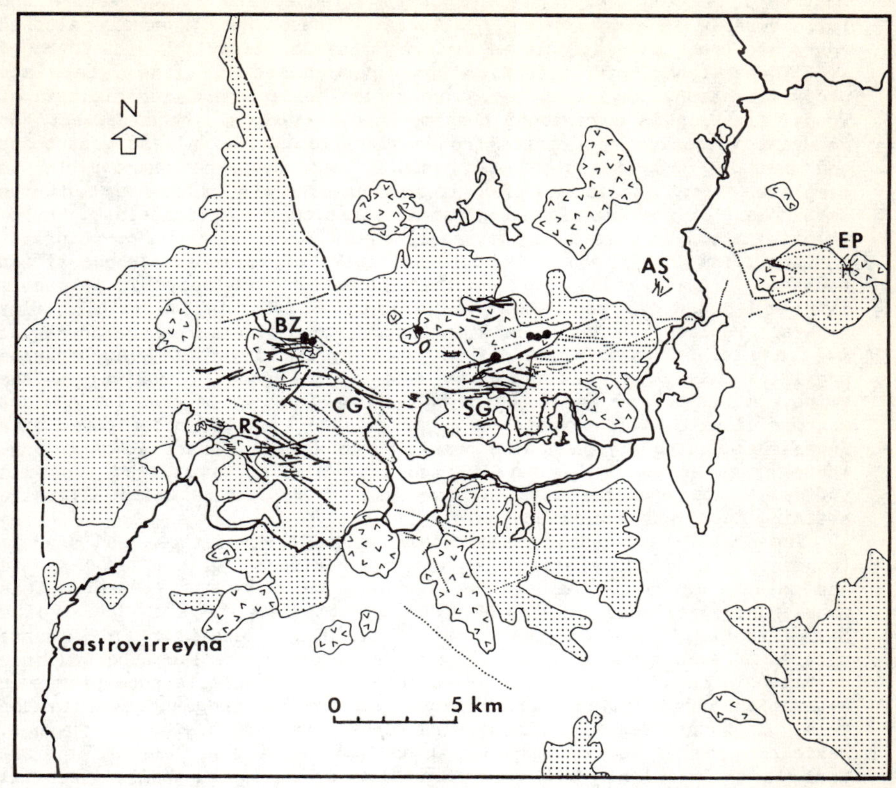

Figure 1. Castrovirreyna district showing distribution of intrusive rocks (angle symbol) and Caudalosa Volcanics (stippled). Dashed lines - faults, solid lines - veins and vein projections, dotted lines - photolinears, filled circles - silica bodies. Mining Centers RQ - Reliquias, BZ - Bonanza, CG - Caudalosa Grande, SG - San Genaro, AS - Astohuaraca, EP - El Palomo.

Alteration

Except for the latest intrusives and andesite flows, all the volcanic rocks in the area are propylitised to greater or lesser degrees. This suggests a hiatus between the deposition of the youngest volcanics and the Caudalosa Volcanics. Tuffaceous members of the section are more strongly altered than massive lava flows and dyke rocks. In addition to regional propylitisation, extensive areas around some of the dome complexes, as at San Genaro, Bonanza, and Reliquias, have been intensely hydrothermally altered to bright yellow-brown clay-limonite rocks. Igneous textures have been largely obliterated by this later alteration and subsequent weathering. Tuffaceous units and ash flow breccias are preferentially altered over wide areas adjacent to faults and vein structures. In the more massive lava flows, alteration is generally confined to 1-5 m selvedges about the veins. Alteration of vein wall rocks follows the same pattern throughout the district. Intensity of alteration appears to be largely a function of porosity and permeability.

Massive flows and intrusives are least affected and flow breccias and tuffaceous units most strongly altered. These hydrothermally altered zones are frequently associated with major mining activity.

Three types of alteration are recognized in the district: propylitization, argillization, and silicification and sericitization. Propylitization is a regional phenomenon affecting all rocks except the youngest volcanics and varies from devitrification of glassy groundmass and partial chloritization of amphibole and pyroxene phenocrysts to complete chloritization of ferromagnesian minerals, bleaching and the development minor calcite and sericitization of plagioclase. The chloritization becomes more intense 10 to 100 meters from major veins.

Argillization is confined to the vicinity of veins and around silica bodies. It may extend for more than 30 meters from veins in permeable lithologies but is generally confined to a zone 1-2 m wide. Clay minerals replace the entire rock, although igneous textures are preserved in fresh distal samples. Disseminated pyrite invariably accompanies argillic alteration, making up to 1-5% of the rock. Proximal to the veins, all textures are destroyed and the rock becomes a compact montmorillonite-silica-pyrite rock. Argillization also occurs over wide areas surrounding silica bodies resulting in areas of subdued topography. These areas are much lighter colored than similar argillized zones in tuffs or flow breccias adjacent to veins which have abundant limonite staining on joints and fractures.

Immediately adjacent to and within veins, the wallrocks and breccia fragments are variably silicified and sericitized. Disseminated pyrite and zoisite as well as minor barite and carbonate may be present. This zone rarely extends more than 20 centimeters beyond the vein. Silicification may take the form of flooding, resulting in a dense mottled to even colored wallrock or it may totally replace the wallrock over an distance of 1-15 cm. Irregular patches of zoisite accompany such alteration and is particularly evident in breccia fragments within the vein. In such cases sericite appears to be lost, but pyrite is retained. Sericite appears to become more abundant in the lower levels of the Candelaria vein at Caudalosa Grande and the Matacaballo vein at Reliquias.

Vein Structures

All veins in the Castrovirreyna district occupy fault-related structures which tend to have two dominant strike directions of 085° to 115° and 130°. These can be related to the dominant compressional direction in this region of the Andes which is directed 040°, based on a major NW trending faulted anticline a Chonta Pass approximately 17 km NE of San Genaro and is normal to the Peru trench axis. The dextral sense of offset by the major 085° trending veins as well as the dextral en echelon termination of many veins supports this interpretation (Figure 1).

A photo linear interpretation of the structures in the district (Figure 1) shows marked increase in the density of these major fractures in a zone approximately 10 km wide and 30 km long, extending from the El Palomo mine in the east to the Reliquias mine in the west. This zone of increased faulting and fracturing may indicate a major conjugate shear system related to the major NW-SE faults and which controlled the locus of emplacement of the Caudalosa Volcanics, dome complexes and quartz veins. The density of photolinears falls off markedly away from this zone. Also noteworthy is the close association of veins and linears with domal intrusive centers within the Caudalosa Volcanics.

The veins generally have steep dips 70-80° to the north or southwest. Small sections of dip reversal within a vein are common. In the San Genaro area, veins to the north of the subvolcanic andesite intrusive complex tend to dip to the south and those to the south tend to dip to the north. In the Reliquias and Caudalosa mines most veins dip steeply to the south. The veins vary in thickness from less than 20 cm to over 5 m and average about 1 m. Strike lengths of individual veins range from 200 m to more than 3 km. Veins tend to branch and step off en echelon at their terminations and toward nearby veins. The Candelaria vein system forms a cymoid loop in its central portion.

The veins in the Castrovirrevna area show a varietv of textures from banded, crustified open space filling, to replacement breccia veins. In several instances silicified breccia fragments are incorporated in banded and crustified veins. Early minerals in the veins may be cross-cut by veins of late minerals such as orpiment or barite, and existing vein material may be brecciated and re-cemented by later minerals. Both the footwall and hangingwall of the veins are generally marked by a thin 1 to 15 cm zone of clay gouge and friable rock meal. In areas of the Poder vein at San Genaro thin seams of mineralized clay gouge cut and brecciate the ore indicating faulting contemporaneous with mineralization. Lewis (1964) noted similar relations in the Trabajo vein at San Genaro and in the Caudalosa vein. The fault structures generally have a very consistent width (2-3 m) and the veins pinch and swell within the structures.

Frequently vein walls show two generations of slickensides, an early steeply plunging set of striae and a sub-horizontal to low-angle set overprinting it. Lewis (1964) notes a few steeply dipping post ore faults which cut veins at high angles and show 2-3 m of left lateral dislacement. They generally strike 045°-050°, but one striking 090° and another at 105° were also observed.

In general, all vein-fault displacements are unknown or small with a maximum recorded horizontal and vertical displacements of about 30 m (Lewis, 1964).

Silica Bodies

Recent mapping by the author and mine staff of the Castrovirreyna Compania Minera S.A. has outlined several massive silica bodies up to 200x200 m in diameter (Figure 1). These take the form of crudely tabular, thickly bedded, approximately horizontal bodies (Figure 2) and pipe-like bodies of massive saccharoidal and chalcedonic silica breccias up to 100 m diameter. They are invariably found at higher elevations than the nearby silver veins. The silica bodies occur in areas of subdued topography surrounded by extensive areas of intensely argillized outcrops.

Approximately one kilometer north of San Genaro a large bedded silica body can be seen to grade laterally into steeply dipping breccia vein outcrops of the Jofre vein. A similar occurrence is exposed on the saddle between the Caudalosa Grande and Bonanza mines. At this location, and in the Carmela area 4 km to the northwest of San Genaro, numerous subparallel veins appear to underlie and grade into the overlying silica cap. Major silica breccias are exposed about 1.5 km to the north northwest of the Caudalosa Grande - Bonanza watershed.

All tabular silica bodies have a crudely bedded appearance with numerous cavities lined with drusy quartz crystals and crusts of colloform silica. Occasionally veinlets and coxcomb aggregates of white barite crystals may be present. The rock consists of massive to vitreous porous, grey to white, saccharoidal quartz with a ghost-like tabular to angular brecciated texture recemented by several generations of silica. The more porous silica rock consists of granular quartz and pale yellow-brown fluffy clay infillings. The clay-filled cavities are quite irregular and do not appear to be pseudomorphic replacements of preexisting feldspar or sulfides, although the latter mechanism cannot be ruled out. Relict sulfides are extremely uncommon and the silica bodies generally have very low silver and base metal contents with occasional high assays up to 120 g/t Ag.

The occurrence of the tabular silica bodies at elevations higher than nearby ore veins, in conjunction with their gradation into sheeted stockwork breccia veins, as well as outcrops of massive silica breccias, suggests that the silica bodies are surficial silica deposits with their associated leached substrates, solution collapse breccias and feeders. The silica bodies occur over a narrow elevation range throughout the district, from 4800-4950 meters above sea-level, suggesting a relatively flat or only slightly undulating paleosurface.

Figure 2. Crudely bedded, tabular silica body surrounded by intensely argillised andesite which forms the low ground. This and other tabular bodies are interpreted as paleosinters or as intensely leached and silicified surficial rocks of a geothermal discharge area.

Zoning

A large number of sulfides, sulfantimonides and sulfarsenides occur in the veins of the Castrovirreyna district. However, total sulfide content in the veins is generally less than 5%, but occasional ore pods may contain as much as 25% total sulfide. In spite of the locally complex mineralogy, the dominant sulfides are pyrite, galena and sphalerite. Tetrahedrite, chalcopyrite and enargite may achieve high local concentrations. Lewis (1964) records a long list of other minerals that include native silver, gold, acanthite, the silver-sulfantimonides pyrargyrite, miagyrite polybasite and aramayoite, lead sulfantimonides such as geocronite, semsey ite and zinkenite, as well as bournonite, stibnite, famatinite, chalcostibite and wurtzite. Gangue minerals are: quartz, barite, kaolinite and montmorillonite, rhodocrosite, siderite, calcite, pyrite and hematite. A plot of Ag/Pb and Ag/Cu in base-metal ore samples by Lewis (1964) shows a good correlation between silver and copper, suggesting that much of the silver in these ores is bound in tetrahedrite rather than galena where no other silver minerals are present.

The general paragenetic sequence of sulfide deposition in the veins appears to be as follows: pyrite I --- sphalerite I --- galena I --- chalcopyrite and tetrahedrite --- enargite --- tetrahedrite II and bournonite --- galena II --- lead and silver sulfantimonides--- sphalerite II--- pyrite II. The complete sequence is never observed and generally an early sequence or only a portion of the sequence will be veined or replaced by a later portion. In addition to extensive and complex mineralogy in the veins, the veins have undergone multiple episodes of mineralization with the consequence that ore minerals show extensive replacement, veining and rimming of early by later minerals.

In his excellent mineragraphic study of the Castrovirreyna district, Lewis (1964) outlined a zoning scheme whereby early-formed sulfides such as galena, tetrahedrite and argentite were progressively replaced by more antimonian phases throughout the depositional history of the veins. He also grouped major ore types into protore, hypogene enriched, and supergene enriched ores, and further subdivided the district into subareas of base metal, lead-antimony and silver-antimony vein deposits.

In detail and with the exposures generated by a further twenty years of mining, this regional zoning pattern for the Castrovirreyna district requires modification. Over the entire district, silver is the main economic mineral and the degree of base metal production is generally a function of which veins and which levels in any vein are producing ore. With the greater vertical exposures now available (200-600 m), a macro-zonation pattern based on examination of seven major vein systems and the silver-bearing phases within them has emerged which appears consistent throughout the district.

In handspecimen it can be readily seen that more antimonian silver minerals replace less antimonian ones or are deposited in sequence of increasing antimony content, that is tedrahedrite or argentite/acanthite is replaced by polybasite and pyrargyrite which is in turn replaced by miargyrite (Figure 3). A similar sequence was also noted in lead minerals (Lewis, 1964), where galena is replaced by geocronite -- semseyite -- zinkenite.

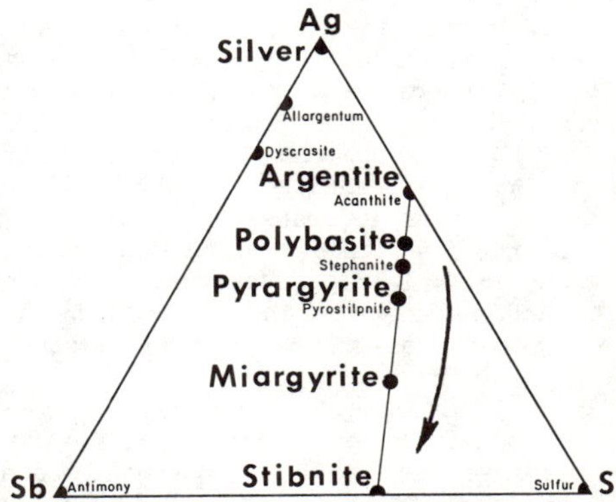

Figure 3. Minerals in the Ag-Sb-S system. Bold type indicates phases recorded in the quartz veins of Castrovirreyna district. Note that polybasite contains essential copper and is a projection from the apex of the Cu-Ag-Sb-S tetrahedron onto the base. Arrow indicates general paragenetic sequence of silver sulfantimonides.

Because the ores are medium to coarse-grained and the silver sulfantimonides are distinctive and relatively easy to identify in hand-specimen, it was found that Lewis' replacement sequence could be crudely outlined over the vertical extent of several veins. The zoning pattern outlined below is based on hand-specimen identification of the dominant silver-bearing phase and the associated gangue minerals. It must be stressed that each of the indicator minerals, except possibly native silver, have a considerable vertical range and that the zones are gradational into one another. Furthermore, as previously noted, late veinlets and minerals cut earlier assemblages and this serves to mask clear cut assignment of any single handspecimen to a particular zone. However, observations integrated over several samples from different locations within the same level or drift are generally consistent. The recognition of paleosinters over some of the veins systems establishes a zero surface datum to which this zoning pattern may be related. Figure 4 shows the schematic arrangement of various zones, their characteristic gangue minerals and their relations. From top to bottom within any vein system the zones are as follows:-

Leached zone. This zone includes relict silica sinters, solution collapse breccias and the extreme upper brecciated sections of some veins. The dominant minerals are granular to chalcedonic quartz, kaolin and occasional barite appearing in veins and cavities toward the base of the zone. The leached zone appears to correspond to the "acid-altered area" of Schoen, White and Hemley (1974) which encompasses a zone from the surface down to or just below the water table at Steam Boat Springs, Nevada. Sporadic native silver and silver chlorides may appear in this zone but in general silver values are very low, with occasional high values to 120 g Ag/t. The silica caps vary from 5-15 m thick; collapse breccias have similar dimensions and the leached zone may total greater than 30 m. This zone is typified by outcrops in the Carmela area, the Bonanza-Caudalosa Grande saddle, and areas to the north and northeast of San Genaro.

Native Silver Zone. This zone lies immediately beneath the leached zone and consists of parallel and/or anastomosing quartz-barite breccia veins. Clay minerals and barite are abundant and native silver is the dominant silver mineral. The type area for this zone is the Astohuaraca mine in which the silver occurs as coarse crystalline wires intergrown with barite. Pyrargyrite, galena, sphalerite and pyrite are minor accessories. A similar occurrence is reported to have been mined at the top of Cerro Reliquias (Masias, 1924; Lewis, 1964). This stockwork zone with native silver is generally less than 30 m thick. The zone corresponds to Lewis' "supergene enriched zone". However, the occurrence of coarse-grained native silver in coarsely crystalline quartz and barite with included platelets of hematite strongly suggests that the native silver is of hypogene origin. However, the presence of sooty acanthite/chalcocite and cerargyrite indicates that supergene processes have been important in the present near surface exposures, and may have enriched the ores.

"Ruby Silver" Zone. This zone is named after the characteristic red color of the dominant silver mineral, pyrargyrite, the "plata roja" or "rosicler" of the local miners. The ore generally consists of massive veins, veinlets, and granular pods and trains of pyrargyrite with subordinate miagyrite and polybasite in vein quartz and barite. Coarse crystals of pyrargyrite line vugs in vein quartz and cement breccia fragments. Other sulfides are galena and low-iron sphalerite and minor pyrite which may be overgrown by pyrargyrite. Gangue minerals are quartz, barite and clays. Minor amounts of small spherical aggregates of siderite occasionally grow over the pyrargyrite, together with minute crystals of late low-iron sphalerite and pyrite. Several generations of pyrargyrite form overgrowths on earlier crystals. In the San Genaro mine

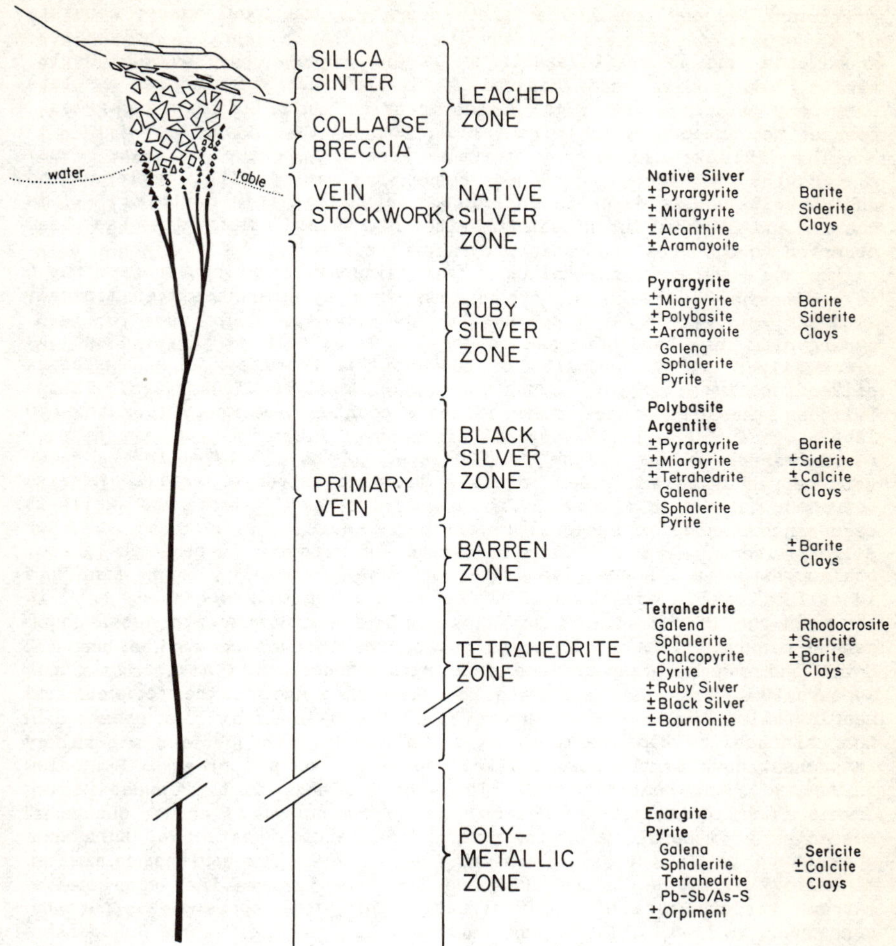

Figure 4. Idealized schematic diagram of silver mineral zoning in the quartz-silver veins of the Castrovirreyna district. Quartz is present in all veins as the dominant gangue mineral. "Clays" refers to both kaolinite and montmorillonite.

the "ruby silver" zone extends for approximately 90 m from below the A level to the 70 level. In the Matacaballo vein in the Reliquias area the vertical extent of this zone is approximately 70 m from the peak of Cerro Reliquias to the 768 level. This zone and the underlying "black silver" zone approximately correspond the "hypogene enriched zone" of Lewis (1964).

"Black Silver" Zone. The "plata negra" of the local miners consists of a complex of dark silver bearing minerals, miagyrite, polybasite, aramayoite and argentite/acanthite. The gangue consists of quartz, barite, hematite and clay minerals. Calcite and siderite may be late accessory minerals in vugs. Accompanying sulfides may be galena, sphalerite, chalcopyrite, tetrahedrite, bournonite and pyrite.

The "black silver" minerals generally occur as late drusy overgrowths on porous corroded aggregates of pyrite, galena and chalcopyrite or as discrete individual crystals within cavities in late quartz and perched upon tetrahedrite crystals. Miagyrite has been observed to overgrow and replace pyrargyrite.

In the Poder vein the "black silver" minerals first appear from the 0 level to the 70 level, but pyrargyrite is the dominant silver mineral, although miargyrite and polybasite are common. In the 70 level pyrargyrite is subordinate and on the 120 level both pyrargyrite and the black silver are subordinate to abundant tetrahedrite. Thus the black silver zone in this vein extends for approximately 60-100 meters. In the Reliquas mine area "black silvers" are encountered commonly from the 670 level to the 560 level, a vertical distance of 110 m.

Tetrahedrite Zone - From the 120 level to the 230 level in the Poder vein tetrahedrite and galena are the dominant silver minerals. Massive tetrahedrite is intergrown with sphalerite and galena and well as occurring as complex euhedral crystals in cavities with galena. Other silver minerals are either subordinate or absent. Green sphalerite, chalcopyrite and bournonite make up the remaining ore minerals identifable in handspecimen. Pyrite is a ubiquitous accessory both in the vein and in the alteration haloes. Barite and quartz are the major gangue minerals in the 120 level but barite is much reduced on the 170 level and massive fine-grained pale pink rhodocrosite makes its first appearance. Clay alteration in the form of gouge on the footwall and hangingwall as well as knots and eyes in the vein is still present. In the Matacaballo vein (Reliquias mine) a similar handspecimen mineralogy is present down to the lowest level (440) and for a further 100 m below in the adjacent Beatita vein. This zone extends to the lowest mining levels at Reliquias and San Genaro with no diminution of grade, change of character or mineralogy of the vein. The vertical extent of this zone appears to be in excess of 200 m. Lewis (1964) describes a similar mineralogy from the Madona area and from the Carmen-Lira veins at the extreme western edge of the district. This zone appears, in part to correspond to Lewis' (1964) protore zone.

Barren zone - In the Matacaballo, Perseguida and Sacasipeudes veins of the Reliquias mine, a zone 20-50 m deep is essentially unmineralized. This zone is centered approximately at the 520 level. The quartz vein does not change in character except that sulfide minerals are reduced to trace amounts. Crustification, banding and brecciation are still visible. The zone appears to be situated between the "black silver" and tetrahedrite zones. A suggested mechanism for the sulfide impoverishment in this zone is the dumping of silica relative to sulfides. Such an effect could be the result of a constriction in the vein and throttling of fluid flow with subsequent rapid pressure drop and dumping of silica. The onset of boiling could also lead to a similar effect.

Polymetallic Zone - A further mineral assemblage, in addition to those mentioned above, occurs in the Caudalosa and Candelaria veins of the Caudalosa Grande mine. The main silver mineral appears to be tetrahedrite but is accompanied by lead and copper sulfosalts of which enargite in the most conspicuous and abundant. Pyrite is a major constituent of the ore as are galena and sphalerite. The enargite and other sulfosalts replace brecciated, strongly silicified wallrocks and pyrite.

In the upper levels of the Candelaria vein (712 level) the main minerals are sphalerite, galena, pyrite, tetrahedrite and lead-antimony sulfosalts, placing it in the Tetrahedrite Zone. However enargite occurs in some areas and becomes a major constituent in the 610 levels and below. In many instances galena-sphalerite-tetrahedrite veins appear to postdate the enargite-pyrite mineralization as do rare late orpiment-realgar veins.

Because of the appearance of abundant enargite, bournonite and chalcopyrite (together with galena and sphalerite) in the assemblage, this zone is referred to as the polymetallic zone. The main gangue minerals are quartz, clay and sporadic calcite.

The precise position of the polymetallic assemblage is indeterminate in that the mineralization is arsenic-rich rather than the more general antimonian assemblage elsewhere and because it is confined to the Caudalosa and Candelaria veins at Caudalosa Grande. The tetrahedrite zone mineralization, although apparently overlying it, also appears to overprint it. From structural considerations, the Caudalosa and Candelaria vein systems appear to be emplaced below the mineralization of the nearby Reliquias area and therefore the polymetallic zone is tentatively placed below the tetrahedrite zone.

Discussion

The morphology and textures of the several tabular silica and breccia bodies, together with their spatial relations to the underlying argentiferous quartz veins, strongly suggests that they are relicts of surficial discharge areas of a fossil geothermal area. The large areal extent of these silica bodies is markedly different from the limited silicification associated with veins, even where the latter intersect permeable tuffaceous units. The general morphology and textures of the tabular silica bodies are permissive of interpretation as a sinter although fine laminar banding and opaline silica frequently seen in modern sinters is absent. It is possible that recrystallization of opaline silica since deposition may have destroyed many fine scale features, leaving only the crudely bedded morphology, but an alternative interpretation is that they are silicified bedded tuffs. Subhorizontal elongate breccia fragments and cavities appear to support the sinter interpretation but may also be due to the solution of compressed pumiceous clasts in a tuff unit.

Schoen, White and Hemley (1974) describe a surficial zone of residual silicification due to complete dissolution and leaching of all other rock components by downward percolating acids associated with geothermal discharge at Steamboat Springs, Nevada. This residual silica cap is underlain by a zone of intense argillic alteration which terminates at or just below the water table. Together the zones are referred to as "acid altered areas". In the Castrovirreyna district, the tabular silica bodies are also surrounded by a wide area of intense argillic alteration. The vertical extent of this alteration is unknown. Thus, although it may be not be possible to decide if the silica bodies are true sinters or surficial silica alteration zones, the bodies appear to have been deposited on or near the paleosurface.

The silica breccia bodies are interpreted to have formed just slightly deeper in the discharge conduits. Both types of silica bodies would be deposited in strongly leached acid environments and therefore silver and other metal values would be very low relative to the underlying veins.

The geothermal system responsible for the generation of the metalliferous quartz veins in the Castrovirreyna district appears to have been related to the waning stages of the volcanic event that produced the Caudalosa Volcanics because the veins cut the subvolcanic dome complexes. Only minor dacitic dikes appear to post-date the veins, followed by the final intrusion of unaltered andesite domes and plugs. The youngest intrusives, because of their general lack of propylitic alteration, must have been intruded after a considerable hiatus in volcanic and geothermal activity. Potassium-argon dates on intermediate composition lava flows at Huachocolpa (25 km to the NE) give ages from 10-4 million years (Noble et al., 1972; McKee et al., 1975), but correlation of these lavas with those in the Castrovirreyna district is uncertain. Radiometric dating of the Caudalosa Volcanics and their associated intrusives is required to constrain the age of these rocks.

The sequence of crystallization and replacement of early silver sulfides by progressively more antimonian silver phases suggests that there is increasing activity of an antimony sulfide component in solution with time during ore deposition. Figure 3 shows the relative compositions of silver sulfantimonides found in the Castrovirreyna area on a ternary Ag - S - Sb plot, as well as the fact that they all fall on the Ag_2S-Sb_2S_3 join. At the same time, the appearance of silver sulfantimonides in the veins coincides with the abundant appearance of barite as a gangue mineral, implying that oxidized sulfur species were becoming important in the ore solutions. Data in Garrels (1960) and Brookins (1972) show that reduced aqueous sulfur species are oxidized at lower fO_2 than the polysulfantimonide ion (Sb_4S^{2-}). Although the data are for 25 C and 1 atmosphere, this qualitative relation is expected to be maintained within the epithermal temperature range. A scenario in which oxygenated meteoritic water is entrained into the rising geothermal fluid containing both H_2S and Sb_4S^{2-} species and progressively oxidizes them to SO_4^{2-} and $HSbO_2(aq)$ could provide the mechanism for the observed paragenetic sequence. H_2S would oxidize prior to Sb_4S^{2-}, and thus the activity of antimony sulfide would increase with time. The occurrence of wurtzite in the Caudalosa mine (Lewis, 1960) and the San Genaro mine (Palache et al. 1944), may possibly indicate low H_2S activity in the ore fluid (Scott, 1974). As oxidation goes to completion, any remaining silver in solution can be reduced to the native element by the following reaction (Boyle, 1968).

$$Ag^{+} + Fe^{2+} = Ag + Fe^{3+} \tag{1}$$

The presence of hematite, blackening barite intergrown with wires of native silver, supports this interpretation, as well as requiring that the native silver zone occur at the uppermost and most oxidized levels of the system.

From fluid inclusion studies, Sawkins and Rye (1974) infer declining temperatures from around 300°C during early sphalerite deposition to less than 200°C for late stage sphalerite and barite. They also note very variable salinities from 4-18 wt% NaCl equivalent and fluctuating D values, which they attribute to mixing of magmatic and meteoric components. Thus the zoning pattern in the silver minerals observed in the Castrovirreyna district reflects a response to both a declining P/T gradient within the veins as well as the progressive oxidation of the geothermal fluid due to mixing with meteoric waters during its migration up the geothermal conduits to the surface.

Conclusions

1) Veins and ore mineralization postdate intrusion of andesitic subvolcanic complexes that domed the Caudalosa Volcanics and pre-date several small inclusion-rich dacitic dykes and plugs. However, there is a close spatial relation of the veins with the subvolcanic dome complexes. The intrusives and Caudalosa Volcanics are younger than 13.9 million years (Noble et al., 1974) but correlation with younger intermediate lavas having K-Ar ages of 10-4 million years (Noble et al., 1972; McKee et al., 1975) at Huachocolpa (about 30 km to the northeast) is uncertain. Radiometric dating of the Caudalosa Volcanics and their associated intrusives is desirable.
2) Tabular and brecciated silica bodies occur at higher elevations than nearby veins and can be shown to grade downward into them and they are interpreted as siliceous sinters, relict silica caps and collapse breccias. Recognition of several discrete areas of paleosinters and solution collapse breccias suggests that the ore veins of the Castrovirreyna district were produced at various centers within a major geothermal field that stretched over 30 km.
3) It has been possible to establish a broad vertical mineral zoning based on handspecimen recognition of the dominant silver-bearing ore and gangue minerals within the veins. This zoning sequence appears to be consistent at this scale throughout the Castrovirreyna district and may be applicable to other silver-bearing vein systems. The mineralogical zoning may in turn be spatially related to certain portions of the geothermal system and to the original paleosurface.
4. The silver-mineral zoning and the more detailed paragenetic relations among the silver sulfantimonides appears to be a response to a declining P/T gradient and progressive oxidation of the geothermal fluid on its rise to the surface.
5) The zoning of silver minerals may be an important tool in estimating the economic potential of as yet unexploited veins within the Castrovirreyna district in that it may allow assessment of the grade potential and vertical extent of unexplored veins. In vein deposits in which there is high variance in grade distribution and consequently high exploration costs, this may be particularly significant.

Acknowledgements: This work was supported by the Economic Mineralogy Fund of the American Museum of Natural History, New York. Without the active cooperation and logistical field support of Castorvirreyna Compania Minera S.A. and Corporacion Minera Castrovirreyna, S.A., this project would not have been possible. I would like to thank the geological mine staffs of both companies for their helpful attitudes and frank discussions; in this respect Ingenieros A. Vargas, S. Sanchez and J. Canales of C.C.M.S.A. and J. Gomez, and A. Paricahua of C.M.C.S.A. deserve special thanks. Professor U. Petersen is thanked for his constructive criticism and review.

References

(1) R.W., Boyle, "The geochemistry of silver and its deposits," Geological Survey of Canada Bulletin, 160 (1968) 264.

(2) D.A., Brookins, "Stability of stibnite, metastibnite and some probable dissolved antimony species at 298.15 K and 1 atmosphere," Economic Geology, 67 (1972) 369-372.

(3) R.M., Garrels, Mineral Equilibria (New York, NY: Harper & Brothers, 1960)

(4) R.W., Lewis, Jr., "Geology and Mineralogy of the Castrovirreyna district, Huancavelica," Sociedad Geologica del Peru, Boletin, 30 (1956) 217-2244.

(5) R.W., Lewis, Jr., "The geology, mineralogy, and paragenesis of the Castrovirreyna lead-zinc-silver deposits, Peru, (Ph.D. thesis, Stanford University, 1964).

(6) A., Masias, Boletin de la Sociedad Geologica del Peru (1924) Tomo 3.

(7) E.H., McKee, Noble, D.C., and Petersen, U., et al., "Chronology of late Tertiary volcanism and mineralization, Huachacolpa District, Central Peru," Economic Geology, 70(2)(1975), 388-39.

(8) F., Megard, "Etude geologique des Andes du Perou central," Office de Recherche Scientifique et Technique Outre-Mer, Memoire,86 (1978) 310.

(9) D.C., Noble, and McKee, E.H., "Cenozoic stratigraphic and tectonic framework of the Andes of Peru," Geological Society of America, Abstracts with Programs, 7 (1975) 1214.

(10) D.C., Noble, McKee, E.H., Farrar, E., and Petersen, U., "Episodic volcanism and tectonism in the Andes of Peru," Earth and Planetary Science Letters, 21 (1974) 213-220.

(11) D.C., Noble, McKee, E.H. and Megard, F., "Early Tertiary 'Incaic' tectonism, uplift and volcanic activity, Andes of central Peru," Geological Society of America Bulletin, 90(10)(1979), 1903-1907.

(12) D.C., Noble, Petersen, U., McKee, E.H., Arenas, F.M. and Benavides, Q.A., "Cenozoic volcano-tectonic evolution of the Julcani-Huachocolpa-Castrovirreyna area, central Peru," Geological Society of America, Abstracts with Program, 4(7)(1972), 613.

(13) C., Palache, Berman, H., and Frondel, C., Dana's System of Mineralogy, Seventh Edition, vol. 1 (New York, NY: John Wiley and Sons, Inc., 1944)

(14) U., Petersen, Noble, D.C., Arenas, M.J. and Goodell, P.C., "Geology of the Julcani Mining District, Peru," Economic Geology, 72(6) (1977) 931-949.

(15) D.H., Salazar, "Mapa geologica de Departamento de Huancavelica, escala 1:250,000," Servicio Nacional de Geologica y Minera (Peru), Lima (1970).

(16) D.H., Salazar, "Mapa geologica del Castrovirreyna-Departamento de Huancavelica-Peru !preliminary1, escala 1:100,000," Servicio Nacional de Geologica y Minera !Peru1, Lima (no date).

(17) F.J., Sawkins, and Rye, R.O., "Fluid inclusion and stable isotope studies indicating mixing of magmatic and metoric waters, Caudalosa silver deposit, central Andes, Peru," Fourth IAGOD Symposium, Varna, II (1974).

(18) R., Schoen, White, D.E., and Hemley, J.J., "Argillization by descending acid at Steamboat Springs, Nevada," Clays and Clay Minerals, 22 (1974) 1-22.

(19) S.D., Scott, "Stoichiometry of sulfides," in Ribbe, P.H. editor, Sulfide Mineralogy, Short Course Notes, vol. 1 (Mineralogical Society of America, 1974)

Process Mineralogy Applications to Mineral Deposits:

Exploration, Predictive Metallurgy

PREDICTIVE METALLURGY AND BENCH FLOTATION TESTS OF MASSIVE SULFIDE ORES FROM TALLY POND, NEWFOUNDLAND

J.L. Jambor[1], D.J.T. Carson[2], A. Stemerowicz[1] and W. Petruk[1]

[1]CANMET, Department of Energy, Mines and Resources,
555 Booth Street, Ottawa, Ontario, Canada K1A 0G1

[2]Noranda Exploration Company Limited, 55 Yonge Street,
Toronto, Ontario, Canada M5E 1J4

Abstract

The Tally Pond deposit near Buchans in central Newfoundland contains about 450,000 tonnes of pyritic massive sulfides grading about 3% Cu and 3% Zn distributed in three zones designated the North, South, and Southeast. Chalcopyrite and sphalerite account for the Cu and Zn; the chief source of the modest silver values in the deposit is hessite (Ag_2Te), and low grades of Pb are present as galena. Supergene alteration in the form of digenite and covellite is weak and spatially restricted in the North and Southeast zones, but the South Zone has been affected to the extent that unacceptably low metal recoveries are anticipated. Liberation of the chalcopyrite and sphalerite in the North Zone and Southeast Zone was predicted to be at economically satisfactory levels; bench flotation tests of North Zone ore produced a copper concentrate containing 26.0% Cu and a zinc concentrate containing 50.9% Zn.

Introduction

The Tally Pond massive sulfide deposit is in central Newfoundland (Fig. 1), about 35 km southeast of Buchans (Fig. 2). At the time this study was begun (1984) the Tally Pond property was held jointly by Noranda Exploration Company Limited and Abitibi-Price Incorporated, with Noranda as the operator. Recently most of the Abitibi-Price mineral holdings in the Buchans area, including Tally Pond, were purchased or optioned by BP-Selco.

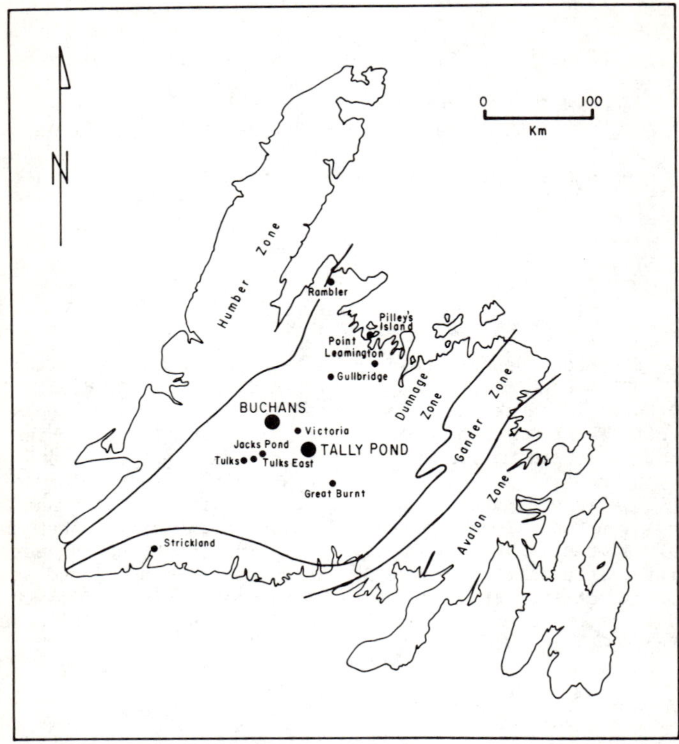

Figure 1 - Location of the Tally Pond and other post-ophiolite volcanogenic massive sulfide deposits with respect to the tectono-stratigraphic zones of Newfoundland.

The Tally Pond area is underlain principally by coarse felsic volcanic breccia that is part of the Ordovician Victoria Lake Group (1). Pyritic massive sulfides have been intersected in several diamond-drill holes in three areas designated the North Zone, the Southeast Zone, and the South Zone (Fig. 3). Possibly these zones represent sulfide deposition peripheral to a central dome structure. Geological correlations (Fig. 4) are based mainly on the assumed continuity of massive sulfides intersected in drillholes; the continuity, however, has not been demonstrated unequivocally and thus even the interpretation of the basic geological framework is tenuous. Although additional drilling is necessary to establish firmly the geometry of the sulfide lenses, current assumptions are that three or four discontinuous massive lenses are present and in aggregate contain about 450,000 tonnes grading approximately 3% zinc and 3% copper, with low values in lead, and negligible gold. Silver values in

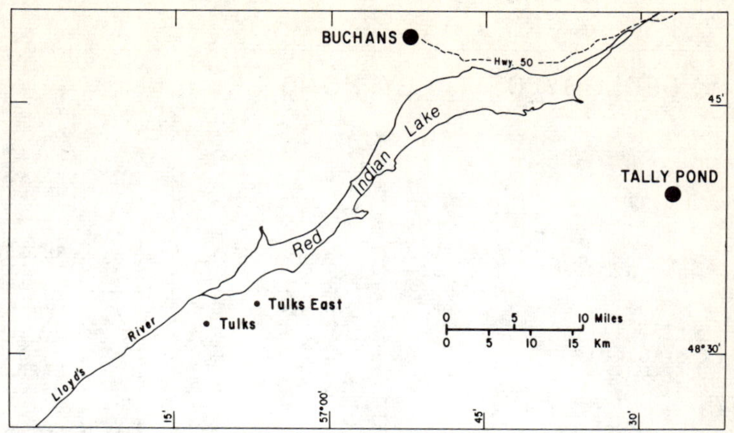

Figure 2 - Detailed location map of the Tally Pond deposit.

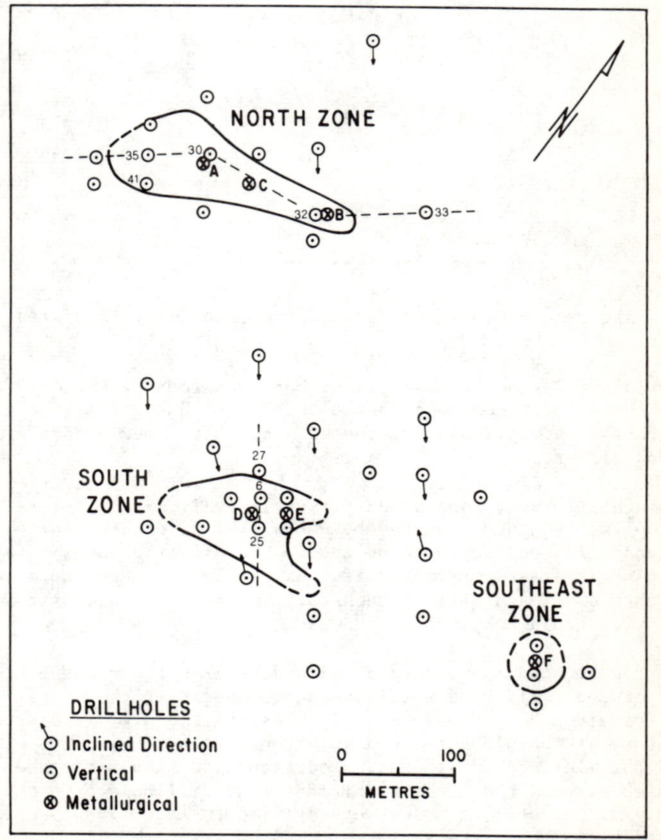

Figure 3 - Designations and approximate outlines of the sulfide zones, and locations of the drillholes sampled for this study. Dashed lines in the North and South zones refer to cross-sections shown in other Figures.

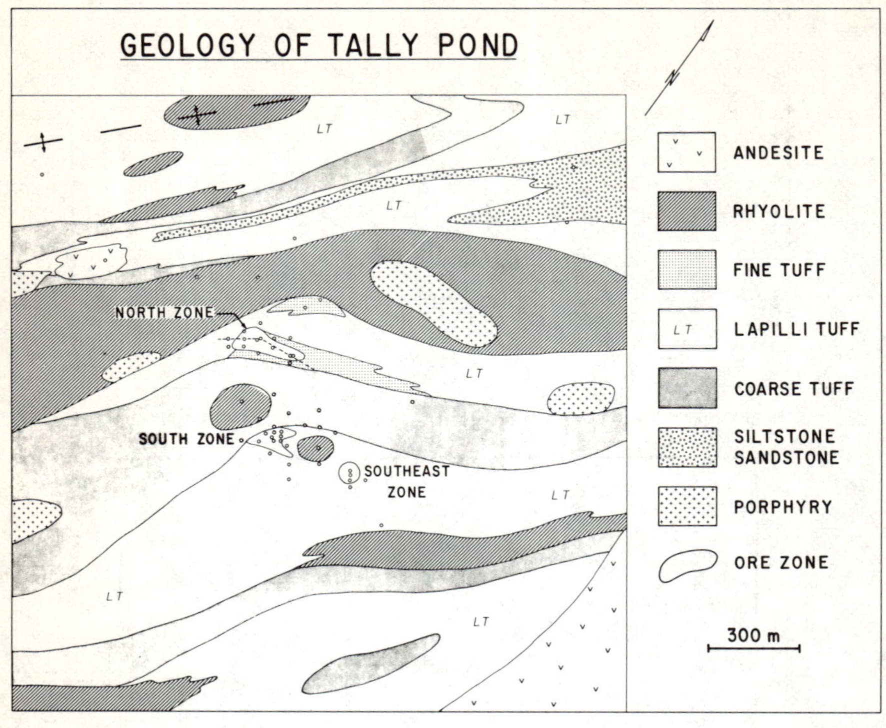

Figure 4 - Generalized geology of the Tally Pond area.

2-m assay intervals are generally low, but occasionally exceed 34 g (1 oz)/ton. About three quarters of the total ore tonnage is in the North Zone, which has an above-average copper grade but lower than average grades of zinc and silver.

Although the Tally Pond deposit is relatively small, further exploration is being done and the potential for finding additional massive sulfide lenses is considered to be good. The existing tonnage is, however, of particular interest because of its potential as a short-term source of feed for the now closed mill at Buchans. Mining of the deposit would be by open pit, with haulage by truck to Buchans.

The present study was undertaken to determine the mineralogical character of the Tally Pond sulfides and to assess whether metallurgical difficulties might be encountered in processing the ore. The drill cores were examined at the property, and an extensive suite of cores was collected for microscopic studies. Additional samples were taken from a trench which exposed the massive sulfides near drillhole 6 in the South Zone (Fig. 3). As well, 6 holes were drilled in late 1984 to obtain cores for metallurgical bench tests.

General Mineralogy

Pyrite

The Tally Pond massive sulfides consist largely of pyrite, and the base-metal values reflect the presence of sphalerite,, chalcopyrite, and galena. The pyrite varies from fine-grained framboidal (Fig. 5) to coarse euhedral grains more than 1 mm along an edge. Most of the pyrite occurs as crystals and polycrystalline aggregates, but colloform textures, growth zones in crystals, and framboidal masses are common. Lamination of the massive sulfides occurs mainly in North Zone samples, and in these much of the layering arises from the alternation of framboidal layers versus those rich in pyrite crystals and base metal sulfides. Whereas the North Zone contains abundant framboidal pyrite, the opposite extreme is represented by the Southeast Zone wherein only individual framboids or small clusters were occasionally seen but no framboidal layers were observed.

Other textural features of the pyrite are mentioned in the description of the sulfides of the North Zone. Many of the coarse pyrite grains in the deposit have undergone in-situ brecciation as is indicated by the matched grain boundaries of separated fragments. Also notable is that appreciable amounts of pyrite in the deposit consist of grains deposited as sedimentary debris (Fig. 6). Some features of the pyrite, such as undeformed framboids (single as well as groups), delicate colloform textures, and intricate growth zones indicate a lack of metamorphic recrystallization. In contrast, some samples contain coarse-grained, euhedral crystals with sieve-texture inclusions, and other samples contain polycrystalline pyrite aggregates with triple-joint grain boundaries; these features are indicative of strong metamorphic recrystallization. The distribution of the contrasting textures might be useful in assessing or delimiting the role of local dynamic metamorphism, but this aspect was not pursued.

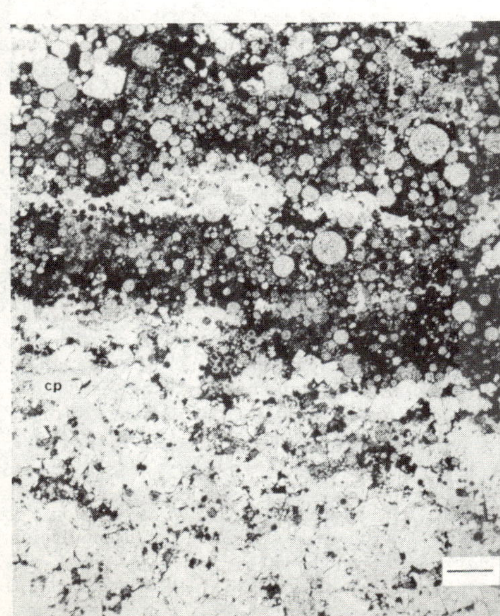

Figure 5 - Sulfide layering arising from contrasting textures and mineral assemblages: top layer consists mainly of framboidal pyrite with the finer grained framboidal matrix blackened from acid etching; bottom layer consists of pyrite crystals in a chalcopyrite (cp) matrix. Reflected light, oil immersion. Drillhole 35 at 25.4 m (North Zone). Bar scale represents 0.05 mm.

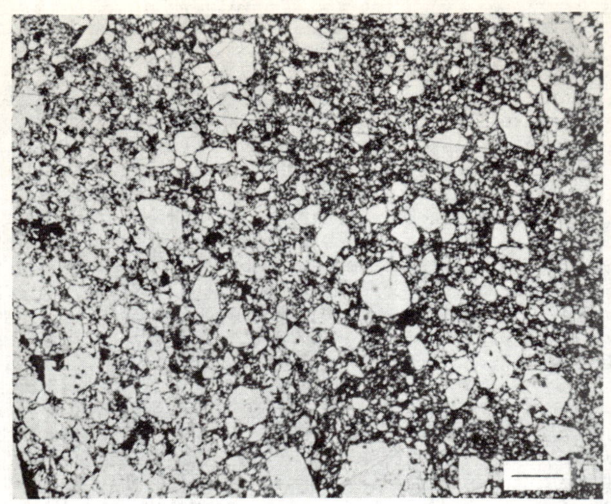

Figure 6 - Angular detrital pyrite fragments in a fine-grained pyrite-rich matrix. Reflected light, oil immersion. Drillhole 32 at 13.6 m (North Zone). Bar scale represents 0.05 mm.

Sphalerite, Chalcopyrite, Galena

Sphalerite and chalcopyrite occur chiefly in the interstices among pyrite grains, and chalcopyrite in particular is commonly a veinlet mineral in fractured pyrite. Both minerals also occur as megascopic patches and streaks containing only minor amounts of other minerals. Some of the core pieces from the South and Southeast zones consist largely of massive sphalerite accompanied by only a few per cent pyrite; this type of occurrence was not observed in samples from the North Zone.

Microprobe analyses of sphalerite in two samples gave iron contents ranging from 2.2 to 4.6% wt % (Table 1). Most of the sphalerite throughout the deposit is free of emulsion-type bleb inclusions of chalcopyrite (commonly referred to as the "chalcopyrite disease"). Intergrowths of this type are widespread but are volumetrically small because of their absence in coarse-grained sphalerite and their erratic presence in fine grains. The strongest and most persistent occurrences of the chalcopyrite disease are in the Southeast Zone.

Galena is present in the majority of Tally Pond samples, usually as only a low fraction of a percentage of the sulfides, and usually as fine-grained intergrowths locked in pyrite. In the few assay intervals with lead contents exceeding 2%, the galena is predominantly coarser grained and a high proportion is in clean grains amenable to separation.

Other Sulfides

Pyrrhotite occurs in trace amounts as bleb inclusions, mostly about 5 micrometers in diameter, in coarse-grained pyrite. The largest bleb is 60 x 30 micrometers. Arsenopyrite is rare and tetrahedrite was not found. Marcasite was not observed; X-ray powder diffraction patterns of spheroidal porous pyrite showed these to be homogeneous rather than fine-grained mixtures of pyrite and marcasite.

Table 1. Microprobe Compositions of Tally Pond Sphalerite

		Southeast Zone drillhole 46, 16.6 m			South Zone drillhole 25, 19.9 m		North Zone drillhole 35, 30.8 m		
		1*	2	3	1	2	1	2	3
wt %	Zn	62.4	62.1	62.2	64.5	64.4	64.5	64.4	64.2
	Fe	4.6	4.4	4.6	2.2	2.1	2.3	2.4	2.3
	Cd	0.3	0.3	0.3	0.1	0.1	0.2	0.3	0.2
	S	32.9	33.2	33.0	33.3	33.1	33.0	33.0	33.1
		100.2	100.0	100.1	100.1	99.7	100.0	100.1	99.8

*Different areas in the same polished section.

Covellite and digenite are locally abundant supergene alteration products whose distributions are discussed in the descriptions of each sulfide zone. Minute amounts of enargite are associated with the secondary copper sulfides in a few samples. The enargite occurs mainly as thin rims around pores in altered chalcopyrite, and as thin seams along the middle of covellite veinlets.

Other Minerals

Meagerly distributed blebs and laths of tellurides, up to 40 micrometers in longest dimension, occur in galena in several specimens and less commonly in sphalerite and chalcopyrite. The tellurides are altaite (PbTe), hessite (Ag_2Te), an unidentified bismuth telluride, and two other phases whose homogeneity is questionable: a lead-bearing bismuth telluride and a lead-bearing silver telluride. Many of the telluride grains are complex intergrowths, and no attempt has been made to further specify the mineral species. The local presence of silver tellurides in the polished sections shows an extremely good correlation with the assay intervals highest in silver. As no other silver minerals have been found in the deposit, hessite presumably is the principal silver ore mineral.

Seams and particles of graphite were observed in some polished sections, especially those with framboid-rich layers from the North Zone. The seams parallel the layering and are up to 30 μm wide and 500 μm long.

The principal non-sulfide minerals are chlorite, muscovite, quartz, siderite and dolomite. Needles and aggregates of rutile are common wherever silicate minerals are present. The North Zone massive sulfides have an extremely low non-sulfide content consisting mainly of interstitial siderite and dolomite.

North Zone

The possible configuration of the drill-intersected massive sulfides in the North Zone is shown in Fig. 7. Many of the samples from this Zone consist almost wholly of tightly packed sulfides. This feature which is a manifestation of very low non-sulfide gangue contents, is so characteristic of North Zone sulfides that their megascopic polished surfaces alone are usually sufficient to distinguish these polished sections from those of the southern sulfide zones.

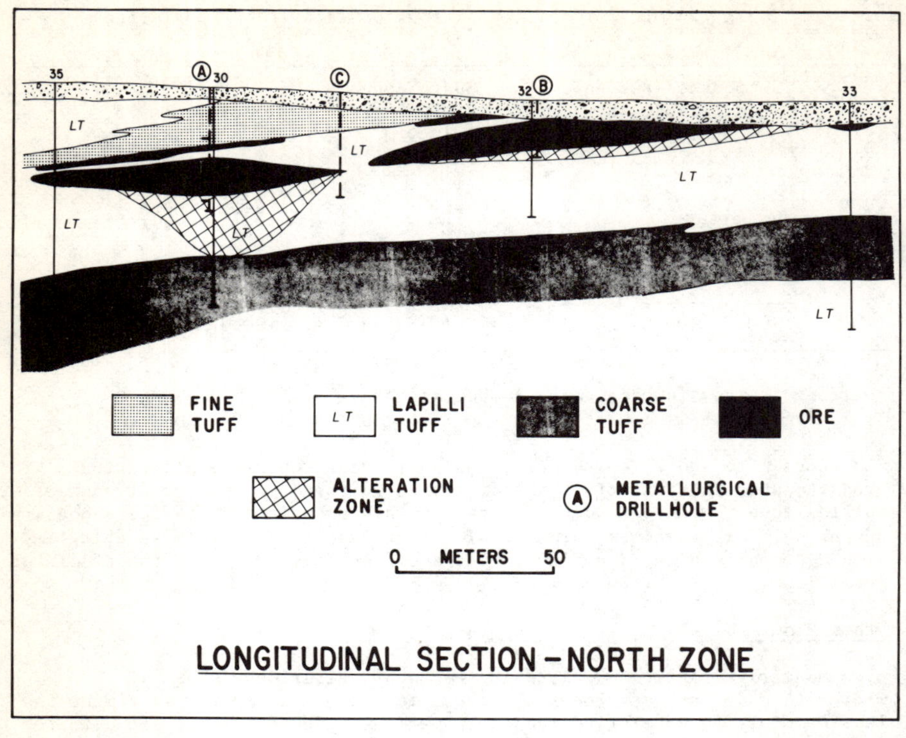

Figure 7 - Longitudinal cross-section showing the possible configuration of the massive sulfides in the Tally Pond North Zone.

Ore Microscopy

Samples from drillholes 41 and those shown in Fig. 7 were studied microscopically. Pyrite varies in habit from coarse euhedral crystals to fine-grained polyframboidal masses. Ovoids of porous pyrite resembling "bird's-eye" pyrite occur occasionally but are not sufficiently abundant to be of metallurgical concern. X-ray powder diffraction patterns of the ovoids did not reveal the presence of marcasite, and thus excess acidity derived from the breakdown of fine-grained pyrite-marcasite intergrowths should not be a problem.

Chalcopyrite is abundant in nearly all North Zone samples; sphalerite is abundant in some samples and merely of incidental presence in many others. Galena is abundant in relatively few samples, and only in these are the grain sizes and textures appropriate to the attainment of adequate lead recovery; in most samples the galena is not only sparse and fine-grained, but the grains are either locked as non-recoverable bleb inclusions in pyrite, or are in complex intergrowths that should appear as pyrite-galena middlings. It was anticipated that better than 50% Pb recovery would be obtained only from the few assay intervals that average more than 0.5% Pb.

Chalcopyrite in the North Zone occasionally occurs in large, relatively homogeneous patches. By far the commonest forms, however, are veinlets which healed fractured pyrite, and grains which filled the interstices among pyrite grains (Fig. 8). In a few cases chalcopyrite replaced pyrite to form fine-grained, sieve-like textures not readily amenable to separation; nevertheless, the highest proportion of chalcopyrite-pyrite middlings will be derived from thin veinlets of chalcopyrite in pyrite. Sphalerite and chalcopyrite are closely associated and usually have simple mutual grain boundaries. It was thought that liberation percentages of sphalerite would probably be similar to those of chalcopyrite.

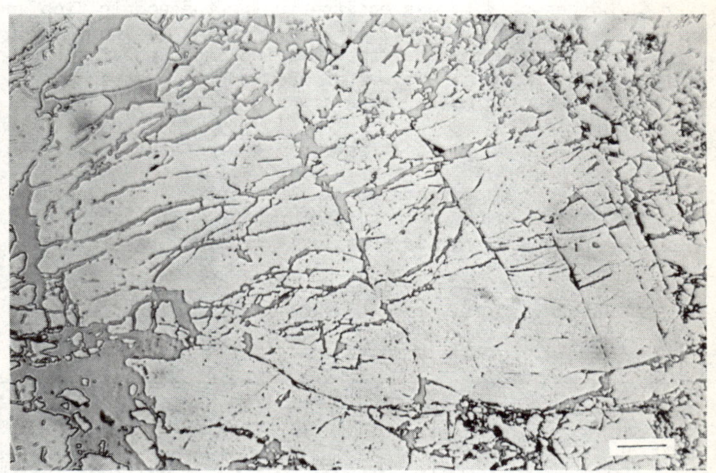

Figure 8 - Fine-grained chalcopyrite as fracture fillings and replacements of coarse-grained pyrite. Reflected light. North Zone drillhole 30 at 17.85 m. Bar scale represents 0.035 mm.

Small amounts of graphite in the North Zone are associated principally with framboidal pyrite. Although graphite is known to disrupt normal flotation, of greater concern was the recognition of thin rims of covellite and digenite on chalcopyrite. This type of supergene alteration occurs at the tops of both main lenses in drillholes 35 and 41, at the top contact of the thin sulfide zone in drillhole 30 (Fig. 7), and at the initial intersection in drillhole 30. The alteration is weak and sporadic, is confined mainly to a thin zone within about 0.6 m of the sulfide contacts, and seems to decrease from west to east. Even within these thin alteration zones, however, overall covellite-digenite content averages much less than 0.1 vol. %, and thus the detrimental effects on sphalerite-chalcopyrite flotation selectivity should be suppressable by cyanide addition (2).

Image Analysis Study and Bench Flotation Tests

Introduction. Ore previously processed from deposits near the Buchans mill required a grind of 100% minus 25 micrometers (minus 500 mesh) to achieve complete liberation, and thus the possible need for a fine grind to liberate the Tally Pond sulfides was not a serious concern. For the initial tests for the North Zone, eleven selective flotation experiments were done on material from the thick massive sulfide zone intersected by metallurgical drillhole A (Fig. 7). Half of the crushed minus 10 mesh material was divided into 19 portions, each weighing about 2 kg. One of the portions was used as a head sample and assayed Cu 5.27%, Zn 3.35%, Pb 0.29%, Fe 38.00%, S 42.05%, Ag 35 ppm, Au 0.4 ppm. A 100-gram portion of the head sample also was used for image analysis study.

Initial Flotation Tests. Table 2 summarizes the conditions used and the best results obtained from the initial 11 flotation tests. The best combination of copper concentrate grade and recovery was achieved using the high lime + aeration method (Test A-10). However, as is indicated by the high zinc content in the copper concentrate, Cu/Zn selectivity was poor. Much better Cu/Zn selectivity was obtained using the combination soda ash + NaCN + aeration (Test A-5, Table 2). The difference in copper recovery in the above two tests was attributed to the grind employed (Table 2), and it was postulated that a copper recovery of 90% should be attainable using the soda ash + NaCN + aeration method at a finer grind.

The main problem in concentrating the ore was the inability to produce a zinc concentrate higher in grade than about 42% zinc (about 65% sphalerite). The zinc concentrate contained about 10% chalcopyrite and 2% galena with the balance made up largely of pyrite. It was thought that the pyrite might be present as pyrite-sphalerite middling particles.

Table 2. Summary of Best Results from Initial Flotation Tests of Tally Pond North Zone Ore

Test no.	Copper selective flotation method	Grind K_{80} e_m	Remarks	Conc grade %		Recovery		Sep. Eff. %
				Cu	Zn	Cu	Zn	Cu/Zn
A-5	Soda ash + NaCN + aeration	45	Best Cu/Zn separation efficiency	20.30	1.86	81.0	13.1	67.9
A-10	High lime + aeration	33	Best Cu conc grade-recovery combination	20.58	5.11	90.5	36.9	53.6
A-4	Soda ash + NaCN + aeration	45	Best Zn conc grade-recovery combination	2.75	41.31	2.7	75.5	-

K_{80} for the grind is the 80% passing size.

Image Analysis Study. An image analysis study was done for the bench test feed crushed to minus 10 mesh (1651 micrometers) to predict the minimum and optimum grinds for liberating the minerals. As well, the concentrate and tailings from bench test A-10 (Table 2) were studied to define the behavior of the minerals.

The bench test feed was screened into 1651-833, 833-417, 417-208, 208-105, 105-53, 53-25, and minus 25 micrometer fractions, and polished sections were prepared from each screened fraction. The polished sections were analyzed to determine liberations for chalcopyrite, sphalerite and galena by measuring the proportions of each mineral (in area %) that are present as free and unliberated particles. Particles containing more than 90 area % of a particular mineral are considered to be free.

The percentages of free chalcopyrite and sphalerite in each screened fraction of the bench feed are plotted in Fig. 9. The results show similar liberations for chalcopyrite and sphalerite. Relatively small proportions of the minerals are free in the coarse screen fractions, with 35% being free in the 208-105 fraction, and 70 to 80% in the 53-25 micrometer fraction. From these data the interpretation is that about 50% of the mineral would be liberated at a grind of 100% minus 208 micrometers (minimum grind), but high liberation would not be obtained until the ore was ground into particles that are smaller than 100% minus 53 micrometers (optimum grind). The liberation data for chalcopyrite in the minus 25 micrometer fraction is lower than in the 25-53 micrometer fraction, thus suggesting that chalcopyrite liberation at a grind of 100% minus 25 micrometers would not be much higher than that obtainable at a grind of 100% minus 53 micrometers.

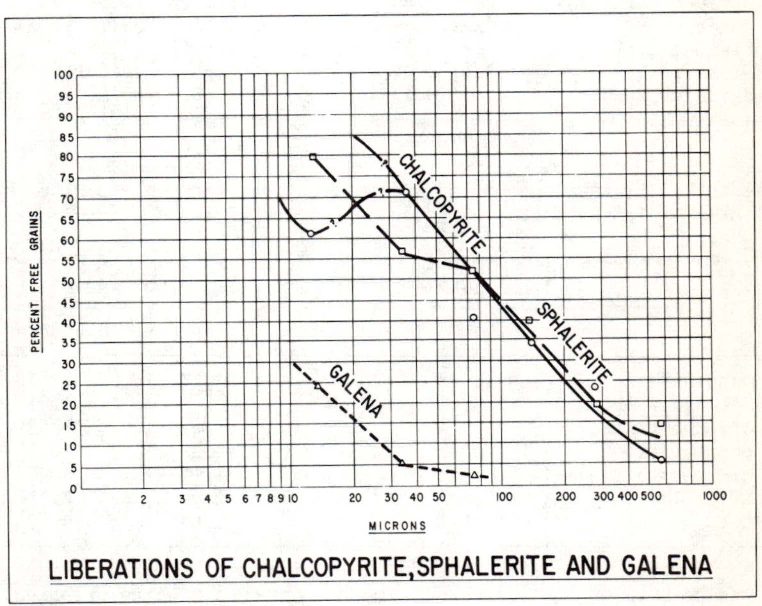

Figure 9 - Liberations of chalcopyrite, sphalerite and galena in screened fractions of ore from the North Zone.

For the mill products from flotation test A-10, the quantities of pyrite, chalcopyrite, sphalerite and galena were determined with the image analyzer and were found to be in good agreement with mineral proportions calculated from the assay data (Table 3). Liberation characteristics were determined for the chalcopyrite and sphalerite by measuring the area percent of the respective mineral that is present in each particle. The size distribution showed that the feed had been ground to 67% minus 25 micrometers (99% minus 53 micrometers), and Table 4 shows the distribution of free and unliberated chalcopyrite and sphalerite in the various mill products. The final tails sample contains 0.9 wt % chalcopyrite and 0.2 wt % sphalerite, which represent 3.85% of the chalcopyrite and 2.77% of the sphalerite in the feed. All of the chalcopyrite and sphalerite in the tails occur as unliberated grains in particles that contain less than 70% of either chalcopyrite or sphalerite; the chalcopyrite inclusions are smaller than 13 micrometers in diameter, and those of sphalerite are less than 4.6 micrometers.

Table 3. Assay Data and Metal Distributions for Metallurgical Test Samples A-10, Tally Pond North Zone

Mill Product	XRF Assays			Metal Distributions		
	Cu	Pb	Zn	Cu	Pb	Zn
Cu concentrate	20.58	0.36	5.11	90.5	30.14	36.88
Cu cleaner tailing 2	3.50	0.44	4.26	0.76	1.83	1.51
Cu cleaner tailing 1	2.02	0.37	2.13	1.88	6.54	3.24
Zn concentrate	1.94	1.73	29.67	2.16	36.72	54.25
Zn cleaner tailing	1.71	0.44	1.65	0.85	4.17	1.35
Final tail	0.32	0.09	0.14	3.85	20.60	2.77
Feed (calculated)	5.21	0.27	3.18	100.00	100.00	100.00

Table 4. Distributions of Free and Unliberated Chalcopyrite and Sphalerite in Products from Test A-10

Mill Product	Chalcopyrite		Sphalerite	
	% Free	% Unliberated	% Free	% Unliberated
Cu concentrate	69.5	21.0	23.27	13.61
Cu cleaner tailing 2	0.07	0.76	0	1.51
Cu cleaner tailing 1	0.67	1.21	1.81	1.43
Zn concentrate	1.02	1.14	44.32	9.93
Zn cleaner tailing	0.05	0.80	1.11	0.24
Final tail	0.00	3.85	0.00	2.77
Feed	71.24	28.76	70.51	29.49

The results of the image analysis study showed that an excellent recovery of free (98%) and of unliberated (72%) chalcopyrite was obtained in the copper concentrate, with large chalcopyrite grains being preferentially recovered. Only a trace amount of fine-grained free chalcopyrite was lost to the tailings. The major problem, as was detected in the flotation tests, is that the chalcopyrite concentrate contains an unusually high amount of sphalerite; as indicated in Table 3, the copper concentrate contains 5.1% Zn (7.9 wt % sphalerite) which is equivalent to 36.9% of the sphalerite in the ore. This high proportion of sphalerite suggests that the mineral was preferentially pulled into the copper circuit, and the problem is in flotation selectivity rather than middlings.

The zinc circuit recovered 98% of the free sphalerite and 77% of the unliberated sphalerite from the zinc circuit feed. The zinc concentrate, however, contains about 46 wt % pyrite and only 46 wt % sphalerite. The image analysis study showed that about 2/3 of the pyrite is free and could be discarded without much loss of sphalerite. To obtain upgrading above about 40 wt % zinc, the concentrate would have to be reground to liberate sphalerite inclusions that range from 6.5 to 37 micrometers in diameter.

Final Flotation Tests. Test A-5 (Table 2) was modified by using a finer primary grind (80% minus 33 micrometers) and regrinding the copper rougher concentrate to about 80% minus 20 micrometers prior to cleaning. This resulted in a slightly better copper concentrate grading 21.06% Cu and 1.71% Zn, and a copper recovery of 83.7%. In subsequent tests involving NaCN and different samples, however, some unusual and highly erratic results were obtained, and thus the use of cyanide was eventually abandoned.

An alternative Cu/Zn selective flotation technique is to employ sulfur dioxide at an acid pH for sphalerite depression. Although poor selectivity and evolution of hydrogen sulfide gas from the pulp were encountered in the first trial tests, gas evolution ceased and a dramatic improvement in Cu/Zn selectivity occurred when the ground pulp was intensely aerated prior to the addition of sulfur dioxide. The best results using the sulfur dioxide flotation technique at a pH of 4.5-5.0 are given in Table 5.

Table 5. Results Obtained Using SO_2 Selective Flotation Technique

Mill Product	wt %	Assays (%)		Distribution (%)		Sep. Eff. % Cu/Zn
		Cu	Zn	Cu	Zn	
Cu cleaner conc 3	19.00	26.00	1.10	90.01	6.53	83.48
Cu cleaner tail 3	2.69	4.00	2.49	1.96	2.09	
Zn concentrate	4.35	2.38	50.89	1.89	69.20	
Zn cleaner tail*	6.91	0.99	7.25	1.25	15.68	
Final tail	67.05	0.40	0.31	4.89	6.50	
Feed (calculated)	100.00	5.49	3.20	100.00	100.00	
Cu cleaner conc 2**	21.69	23.27	1.27	91.97	8.62	83.35

*Combined (1+2+3), **First 2 products combined

The primary grind employed to obtain the Table 5 results was 80% minus 33 micrometers, and both the copper and zinc rougher concentrates were reground prior to cleaning. The first and second stage copper cleaner tailings were added to the pulp just before zinc rougher flotation.

Although the copper concentrate (Table 5) is probably close to the optimum achievable for North Zone ore, higher grade zinc concentrates are probably attainable by additional cleaning, or by "reverse flotation", i.e., by floating pyrite and chalcopyrite away from the sphalerite. The zinc concentrate produced in the individual tests amounts to less than 100 g, too little to permit additional flotation steps.

Southeast Zone

The Tally Pond Southeast Zone possibly has the configuration shown in Fig. 10. The massive sulfides consist of the same mineral assemblage that occurs in the North Zone, but the Southeast Zone has a lower average copper grade, a higher zinc grade, and more galena. Features detrimental to good ore-mineral recoveries (such as sieve textures, thin fracture fillings, and fine-grained intergrowths with pyrite, Fig. 11) occur only locally and are volumetrically small. As a whole, the textures of the ore minerals are simple and the grain sizes are sufficiently coarse that liberation should be as high, and probably higher, than that obtained for the ore minerals in the North Zone. As is the case in the North Zone, small amounts of graphite occur among the sulfides of the Southeast Zone. Weak supergene alteration consisting of hairline veinlets and rims of digenite with traces of associated covellite were noted to occur over a 1-meter interval in drillholes 46 and 51 (Fig. 10). This alteration is not at the massive sulfide contacts, but is instead centered in the middle of the sulfide zone in drillhole 46, and in a similar position in the lower sulfide zone in drillhole 51.

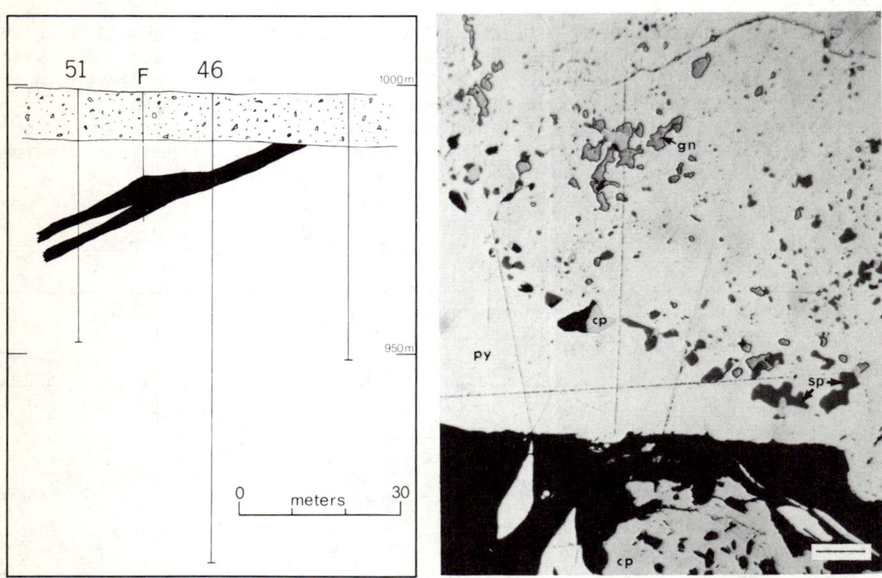

Figure 10 (left) - Probable configuration of the massive sulfides in the Southeast Zone.

Figure 11 (right) - Fine-grained inclusions and minute blebs of galena (gn), sphalerite (sp) and minor chalcopyrite (cp) locked in coarse pyrite (py). Black area at bottom is gangue containing coarser chalcopyrite. Drillhole 32 at 16.1 m, North Zone. Reflected light. Bar scale represents 0.035 mm.

Bench Flotation Tests

Bench tests of the Southeast Zone sulfides were based on the drill-core sample obtained from metallurgical drillhole F (Fig. 10). The core from the upper part of this hole consisted of massive sulfides from 15.3 to 18.1 m, and semi-massive sulfides from 18.1 to 23.5 m. These were designated as test samples 5 and 6, respectively, except that test sample 5 also included 0.5 m of hangingwall waste. Sample 5 assayed 1.52% Cu, 1.25% Pb and 8.55% Zn and had a calculated sulfide content of about 81%. Sample 6 assayed 2.69% Cu, 0.64% Pb and 3.83% Zn and had a calculated sulfide content of about 63%.

Both samples contained graphite which was effectively depressed by addition of Depramin (sodium carboxymethylcellulose) used in conjunction with the sulfur dioxide selective flotation technique that was employed for the North Zone ore. In test sample 5, however, a good copper concentrate could not be obtained because excessive amounts of pyrite floated with the chalcopyrite. In one case the copper rougher concentrate, after three stages of cleaning with lime added to depress the pyrite, was upgraded to 16.4% Cu but at the expense of a drop in copper recovery from 81% in the rougher concentrate to 55% in the cleaner concentrate. Additional work to produce a good copper concentrate could not be done because of the lack of material; nevertheless, the same flotation technique was used for sample 6 and resulted in a copper rougher concentrate assaying 8.54% Cu and 2.35% Zn. After only two stages of cleaning, the concentrate was upgraded to 22.20% Cu and 1.46% Zn with copper and zinc recoveries of 79.4% and 4.2%, respectively, to give a Cu/Zn separation efficiency of 75.2%.

In summary, test sample 5 from the upper part of the Southeast Zone gave a poor copper concentrate, whereas test sample 6 from the lower part of the zone gave a satisfactory product. In contrast to these mixed results, sphalerite from both test samples floated readily; after two stages of cleaning the zinc concentrates graded in excess of 50% Zn with more than 70% of the zinc recovered in the concentrate.

No attempt was made to concentrate the galena. Generally it was distributed randomly among all the test products, though the best concentration was in the copper cleaner tailings. In a continuous plant operation it is expected that some of the galena will be concentrated with the sphalerite to give a lead content of several percent in the zinc concentrate.

Microscopic Examination of Test Products

Microscopic study of a polished section of the graphite concentrate from test sample 5 showed that most of the concentrate consists of free grains of pyrite with minor free grains of sphalerite and chalcopyrite, and scattered free grains of anhedral graphite or hydrocarbon. To confirm the identification of the carbon species, a portion of the concentrate was leached in hot aqua regia to remove the sulfides, and then leached in hydrofluoric acid to remove quartz. The residue was examined by X-ray powder diffraction patterns and was found to consist of poorly crystallized graphite.

The test sample 5 copper concentrate mentioned above as assaying 16.4% Cu was examined in polished section to see why poor copper concentration was obtained. The concentrate also contains 1.69% Pb and 6.17% Zn, and the polished section shows that abundant free pyrite and sphalerite are the main contaminants (Fig. 12). Middling particles of various sulfide combinations are common, and small amounts of free galena and a few particles of covellite and rare graphite are present. Although the possible detrimental effects of covellite cannot be discounted, neither the presence of the small amounts of this mineral nor the incomplete liberation seems to account adequately for the failure to achieve good concentrate grades from this sample. It is thought that satisfactory concentration is attainable and that the present problems relate simply to the need to optimize the grind and flotation conditions.

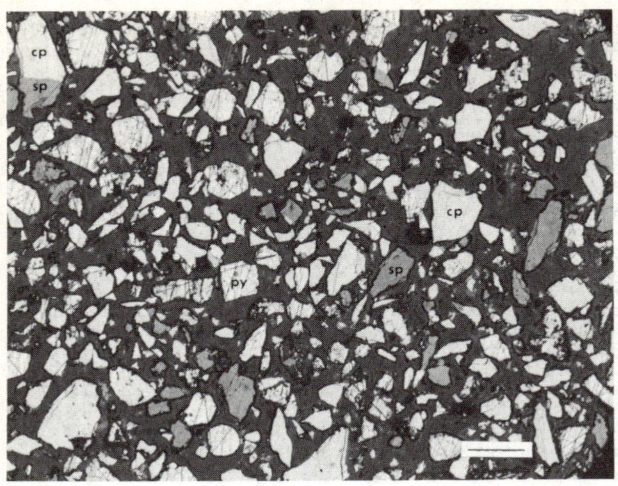

Figure 12 - Copper concentrate from test sample 5 (upper part of Southeast Zone) showing dilution by abundant free pyrite (py) and grey sphalerite (sp), and some middling particles (e.g., sp-cp). Scratchy pyrite surfaces were obtained by etching with nitric acid. Reflected light. Bar scale represents 0.05 mm.

South Zone

Cores from six drillholes in the South Zone were sampled for microscopic studies. The main sulfide intersections and possible configuration of the massive sulfide zone is shown in Fig. 13. In drillhole 6, the near-surface sulfide intersection shows oxidation and 1 to 1.5 m of strong supergene alteration evident from covellite-digenite rims and veinlets which have attacked both chalcopyrite and sphalerite. Similar alteration occurs sporadically through most of drillhole 25, and in one polished section from near the bottom of the massive sulfides about 50% of the chalcopyrite grains are rimmed by covellite; the associated sphalerite contains covellite veinlets up to 25 micrometers wide and 1 mm long, commonly with medial seams of enargite.

Supergene alteration in drillholes 26 and 28 (Fig. 13) is much weaker and much more restricted in spatial distribution. Chalcopyrite, sphalerite and galena in the massive zone occur predominantly as large homogeneous masses and as coarse interstitial grains; liberation of all three minerals should be excellent.

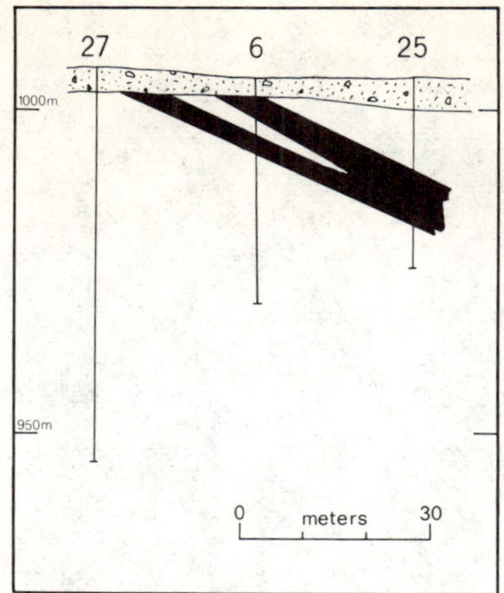

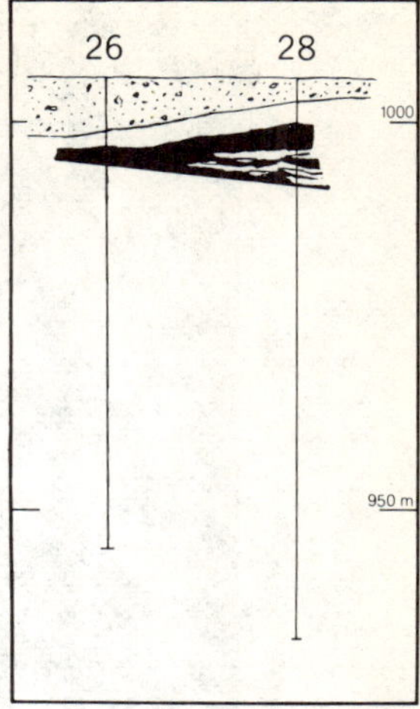

Figure 13 - Possible configuration of the sulfides in the South Zone.

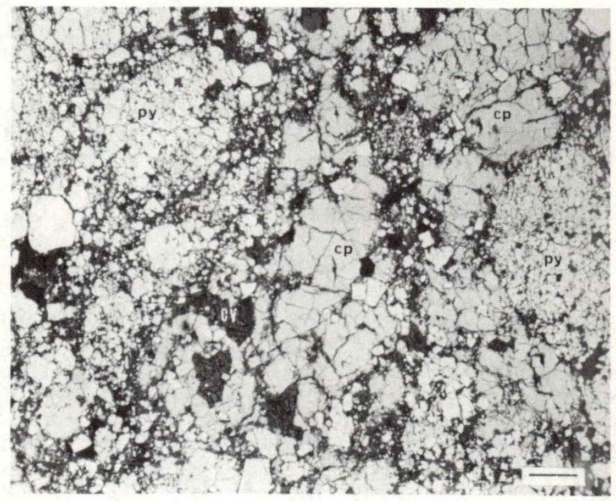

Figure 14 - Residual chalcopyrite (cp) strongly altered by dark rims and hairline veinlets of covellite which also occurs in patches (cv) and is the main dark matrix throughout the areas of pyrite (py). South Zone. Reflected light, oil immersion. Bar scale represents 0.05 mm.

Examination of samples from a trench which exposed massive sulfide bedrock near drillhole 6 has confirmed that covellite development is locally intense. All of the trench samples show extreme alteration of both chalcopyrite and sphalerite (Fig. 14,15), commonly to the extent that covellite rather than chalcopyrite is the main source of the copper values.

Figure 15 - Chalcopyrite (white, cp) and sphalerite (grey, sp) extensively replaced by covellite (cv). South Zone trench. Reflected light, oil immersion. Bar scale represents 0.05 mm.

Current assumptions are that the South Zone contains only 50,000-60,000 tonnes of ore. The microscopic studies have indicated that a significant part of this ore has been affected by supergene alteration of sufficient intensity that conventional processing will be rendered ineffective. Although reconnaissance bench flotation experiments are planned to test the cores from the metallurgical drillholes, favorable results would not be economically significant because of the small size of the South Zone and the need for selective mining of its less altered parts.

Acknowledgements

The cooperation and assistance of personnel from Noranda Exploration Company, Abitibi-Price Incorporated, and the Mineral Development Division of the Newfoundland Department of Mines and Energy made this study possible. Technical assistance at CANMET was provided by D.R. Owens, J.M. Beaulne and P. Carriere.

References

1. H.S. Swinden and B.F. Kean, "Volcanogenic sulfide mineralization in the Newfoundland Central Mobile Belt," Mineral Deposits of Newfoundland - a 1984 Perspective, Newfoundland Dept. Mines and Energy, Report 84-3 (1984), pp. 55-80.

2. D.C. McLean, "Upgrading of copper concentrate by chalcocite derimming of pyrite with cyanide," Process Mineralogy III, Petruk, W., ed., TMS-AIME (1984), pp. 3-13.

THE COMPOSITION AND ORIGIN OF THE BEEMERVILLE CARBONATITE, SUSSEX COUNTY, NEW JERSEY

K. D. Seborowski and A. H. Vassiliou

Woodward-Clyde Consultants
201 Willowbrook Blvd., Wayne, New Jersey 07470
Department of Geological Sciences
Rutgers University, Newark, New Jersey 07102

Abstract

Carbonatite is part of the Beemerville alkalic rock complex of late Ordovician age and is one of only few occurrences in eastern North America. The Beemerville carbonatite exhibits several modes of occurrence as well as mineralogical, geochemical, petrological and textural variation. It outcrops as dikes within the Martinsburg Formation, as xenoliths within volcanic diatremes and as diatremes. Based on petrographic analysis, both dike-type and idatreme-type occurrences are classified as "silicocarbonatite". Based on electron microprobe and neutron activation analyses, it is further classified as an "apatite-magnetite variety" due to its relatively low rare-earth content. Field and analytical data suggest that the carbonatite resulted from the differentiation of an alkalic parental magma which probably gave rise to the carbonatite fluid phase and an alkaline residuum.

Introduction

The Beemerville carbonatite is part of the Beemerville alkalic rock complex which is located approximately 1.7 miles north of Beemerville, Sussex County, New Jersey (Fig. 1). The complex consists primarily of a large lobate body of nepheline syenthe with associated extrusive equivalent rocks such as phonolite and tinquaite and relatively rarer alkalic rocks such as bostonite and micromelteigite. This alkalic rock complex represents a unique such occurrence in the eastern United States because of the presence of the associated carbonatite.

The rocks of the complex intrude into the Marlinsburg Formation of upper Ordovician age. This contact relationship plus a radiometric age determination of 435 $\pm$ 20 million years (1) suggest that the complex is of late Ordovician age.

The carbonatite outcrops as dikes within the Martinsburg Formation, as xenoliths within volcanic diatremes, and as the matrix material encasing xenoliths of hornfelsic Martinsburg Formation, Kittatinny Group, Jacksonburg Formation and Precambrian gneiss in association with the volcanic diatremes.

Previous geologic investigations in the Beemerville area have centered on petrologic aspects of the complex in terms of rock association and origin (2,3,4,5). In addition to field and petrographic data, this study used wet chemical, x-ray, electron microprobe and neutron activation analysis data in order to provide additional information on the composition of the carbonatite, especially its rare earth element content, as well as on its origin.

Associated Country Rocks

The Beemerville alkalic rock complex is within a region characterized by a series of northeast-southwest trending valleys and ridges which are part of the Appalachian Valley and Ridge physiographic province. The study area (Fig. 1) is represented by a sequence of highly folded and faulted miogeosynclinal Paleozoic sedimentary rocks unconformably overlying Precambrian gneiss and marble.

A Precambrian suite forms the basement underlying the study area. The suite consists of a variety of high-grade quartzo-feldspathic orthogneiss, paragneiss, amphibolites and marbles which were locally migmatized and intruded by sodic granites, hornblende granites and alaskites (7,8,9). The intense metamorphic and structural deformation of these rocks is believed to have taken place during the Grenville orogeny (10).

In the vicinity of the study area (Fig. 1), miogeosynclinal strata of Cambrian age (Hardyston Quartzite) and Ordovician age (Kittatinny Limestone, Jacksonburg Formation and Martinsburg Formation) record the separation and convergence of the North American continent with Africa and Europe; Silurian strata (Shawangunk Conglomerate) record post-Taconic deposition. The Hardyston Quartzite (5-200 feet) marks the base of the Cambrian and unconformably rests on Precambrian basement (11). Conformably overlying the Hardyston is the Kittatinny Limestone (2700-3000 feet) which consists of dolomite and calcitic dolomite (11,12). Unconformably overlying the Kittatinny is the Jacksonburg Formation (135-150 feet) which consists of calcirudite, calcarenite and calcilutite (7,11,13,14). The contact between the Jacksonburg and the overlying formation, the Martinsburg (6000-9000 feet), is conformable and gradational from a shaley limestone (Jacksonburg) to a non-calcareous formation (Martinsburg) consisting of slate and shale (6,13).

The carbonatite and the associated alkalic rocks are intrusive into the Martinsburg which is the principal paleozoic formation exposed in the study area and vicinity (Fig. 1). The Shawangunk Conglomerate (1800 feet) overlies the Martinsburg unconformably (11) and forms a steep escarpment in the study area and vicinity (Fig. 1).

Associated Alkalic Rocks

The intrusive rocks forming the carbonatite-alkalic rock complex at Beemerville have been studied intermittently for a period of more than one hundred years (2,3,5,15). Figure 2 represents a detailed outcrop map of the associated alkalic rocks in the study area.

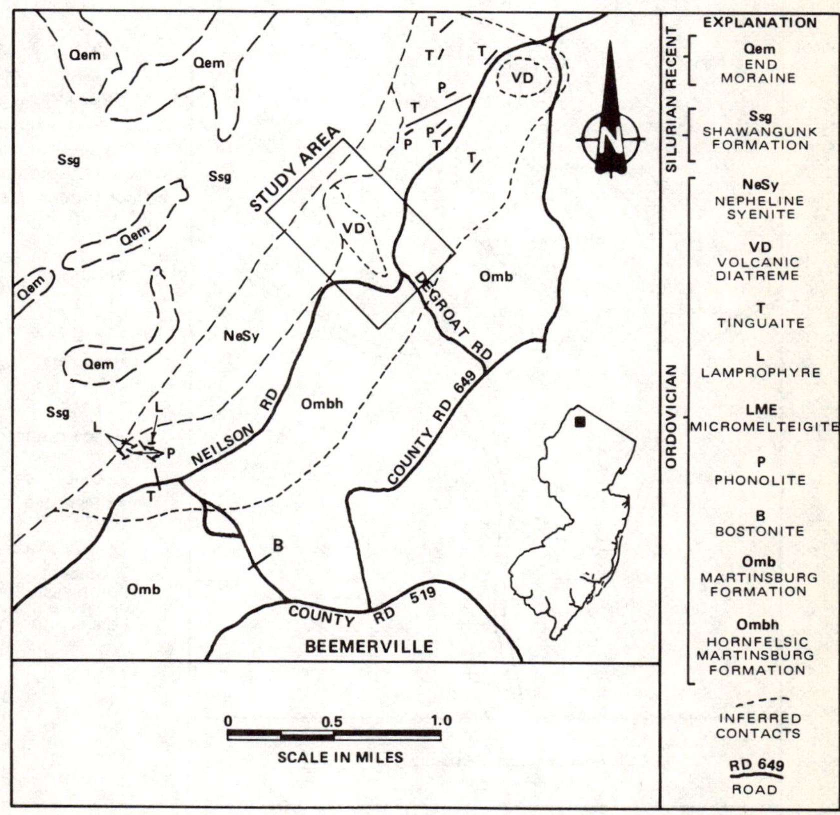

Figure 1 - Geologic map of the Beemerville alkalic complex Sussex county, New Jersey (adapted from 5 and 6)

Nepheline Syenite

This intrusive is the major constituent of the complex (Fig. 2). It is a large lobate body, approximately two miles long and 1500 feet wide, which strikes northeast-southwest or parallel to the structural trend of features characteristic of the Valley and Ridge physiographic province.

Nepheline and orthoclase, primarily as phenocrysts, and associated aegerine-augite, biotite and magnetite make up the major components of this medium to coarse grained intrusive (5).

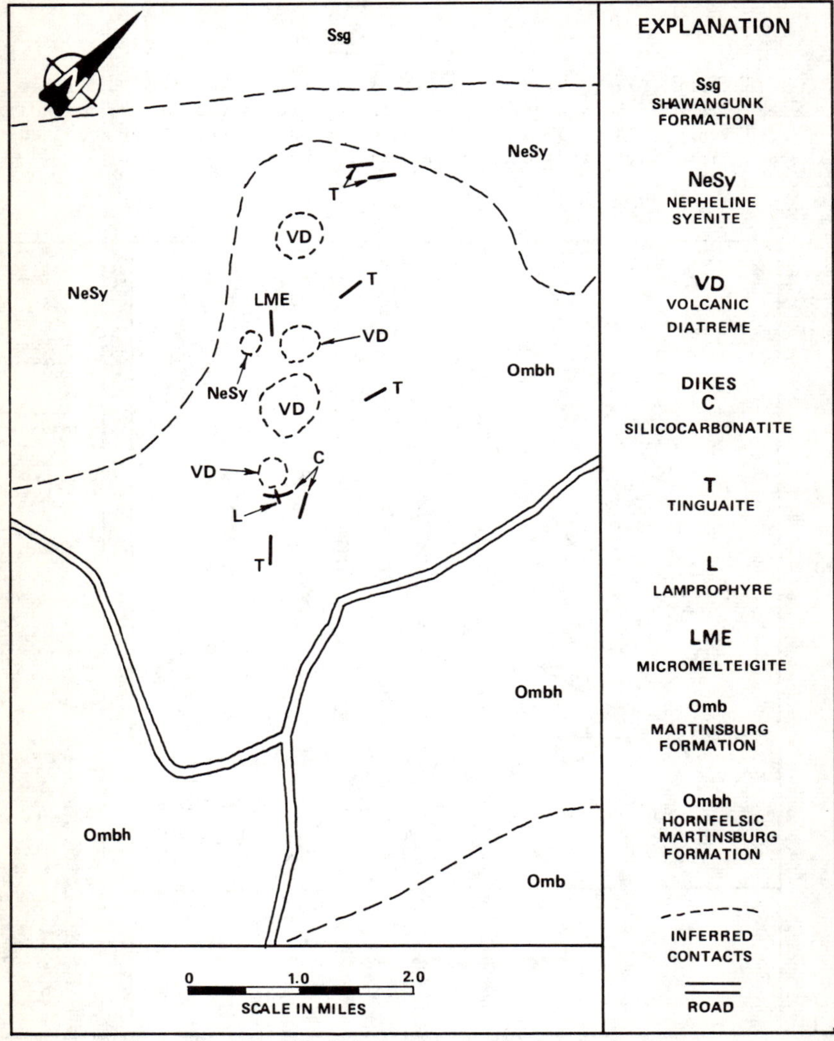

Figure 2 - Detailed outcrop map of the study area as outlined in Figure 1.

There has been considerable debate concerning the mode of occurrence of this pluton which has been described as a lenticular sill (16,17,18), a flat laccolith (16), a dike (2,3,15,17) or a stock (5). Furthermore, the lack of good contact exposures between the nepheline syenite and most of the country rocks prohibits definitive age determinations on the basis of emplacement relationships. However, all igneous rocks known to be in place in the area are found intruded into the Martinsburg Formation. Based on this, the radiometric age determination of 435 + 20 million years for Beemerville biotite (1) and a paleomagnetism study (19) which indicates that secondary hematite in the nepheline syenite is largely the product of pre-Shawangunk weathering, it appears that the maximum age of the intrusive is late Ordovician.

Associated with and intrusive within the nepheline syenite and the Martinsburg Formation are numerous dikes and possibly sills of varying petrologic type:

Phonolite

This fine-grained extrusive equivalent of nepheline syenite occurs as dikes which disect the southwestern lobe of the nepheline syenite intrusive and also intrude the Martinsburg Formation and diatremes (Fig. 1) (5,20).

Tinguaite

This fine-grained variety of phonolite containing acicular pyroxene crystals, forms dikes which disect the nepheline syenite as well as the Martinsburg Formation (Figures 1 and 2) (3,16,20,21).

Bostonite

This fine-grained alkalic syenite containing phenocrysts of feldspar, occurs as dikes within the Martinsburg Formation (Fig. 1) (6,20).

Micromelteigite

The presence of lamprophyric dikes intruding the Martinsburg Formation has been reported earlier (5). Micromelteigite is an iron-magnesium rich lamprophyre that contains nepheline; it intrudes the Martinsburg Formation between the two lobes of nepheline syenite (Fig. 2). Other dikes, located to the south of the southern lobe, are believed to be of similar composition although intensely altered.

Diatremes

These breccia-filled volcanic pipes, formed by gaseous explosions, occur within the Martinsburg Formation (Figures 1 and 2). They were originally described (3) as having a matrix of ouachitite, a lamprophyre rich in iron-magnesium phenocrysts contained in a groundmass of glass or analcime. Carbonatite occurs as xenoliths or as matrix within the diatremes. Xenoliths representative of the underlying geology consisting of Martinsburg Formation, Kittatinny Group, Jacksonburg Formation and Precambrian gneiss are also found in the diatremes.

The Beemerville Carbonatite

The carbonatite within the Beemerville alkalic rock complex is the only such occurrence in the eastern United States. Its occurrence and a brief description of its mineralogy were first reported towards the end of the last century (2) but the carbonatite dike was described as a band of dark-gray, fine-grained limestone.

Reports on the composition and origin of the Beemerville carbonatite are relatively scarce. The first chemically analyzed carbonatite sample representing a carbonatite xenolith from a large diatreme was reported in 1952 (4). More recently, two studies (5, 22) provided analytical data and postulated on the origin of carbonatite xenoliths in the diatremes as well as on carbonatite dikes associated with the alkalic rock complex.

Figure 3 shows the spatial distribution of carbonatite outcrops investigated in this study. Based upon field, petrographic and geochemical examination of these outcrops, two distinct modes of occurrence are revealed. Outcrops I-A and I-B occur as carbonatite dikes intruded into the Martinsburg Formation and are referred to a "dike-type" carbonatites; outcrops II-A, II-B and II-C occur as carbonatite matrix in the diatremes and are referred to as "diatreme-type" carbonatites.

Field Relationships

Outcrop I-A. This outcrop exposes a carbonatite dike 20 feet by 5 feet intruded into the Martinsburg Formation which is locally metamorphosed to a hornfels (Fig. 3). The aerial extent of contact metamorphism attributed to this intrusion cannot be determined due to the widespread metamorphism incurred during the intrusion of the nepheline syenite west of the area and, to a lesser degree, the formation of the volcanic diatremes in the immediate area.

The intrusive is dark gray in color and possesses a fine-grained texture. The ferromagnesian fraction, biotite and magnetite, is readily discernible with the aid of a hand lens, with grains encased in a matrix of calcite; traces of pyrite are also visible. The dike also contains minor amounts of angular Martinsburg Formation xenoliths.

Outcrop I-B. This dike which measures 26 by 6.5 feet intruded into the Martinsburg Formation approximately 150 feet southwest of outcrop I-A (Fig. 3). Although the color is slightly darker and the texture coarse-grained, the mineralogy of this dike is identical to that of outcrop I-A described above. In addition, autoliths of pure calcite 1 to 1.5 inches in diameter as well as angular xenoliths of Martinsburg Formation are also visible in the dike. Furthermore, the dike itself is cut by a smaller lamprophyric dike. Local segregation and alignment of the ferromagnesian fraction and calcitic fraction in the dike produced a banded appearance in this dike.

Outcrops II-A, II-B, II-C. A distinct megascopic change occurs at these diatreme-type carbonatite outcrops (Fig. 3). Although contact exposures are lacking due to the glacial cover, the presence of volcanic breccia in these areas indicates their association with volcanic diatremes.

Carbonatite II-A exhibits a dense, very fine-grained texture and is gray in color. The decreased grain size is attributed to a rapid crystallization of the diatreme material. There is a marked increase in the amount of xenolithic Martinsburg Formation fragments and lesser amounts of xenolithic dolomite and gneiss representative of the underlying geology. No segregation of the ferromagnesian or calcitic fractions is discernible. Occasional grains of biotite 1 to 1.5 inches in diameter are also encountered.

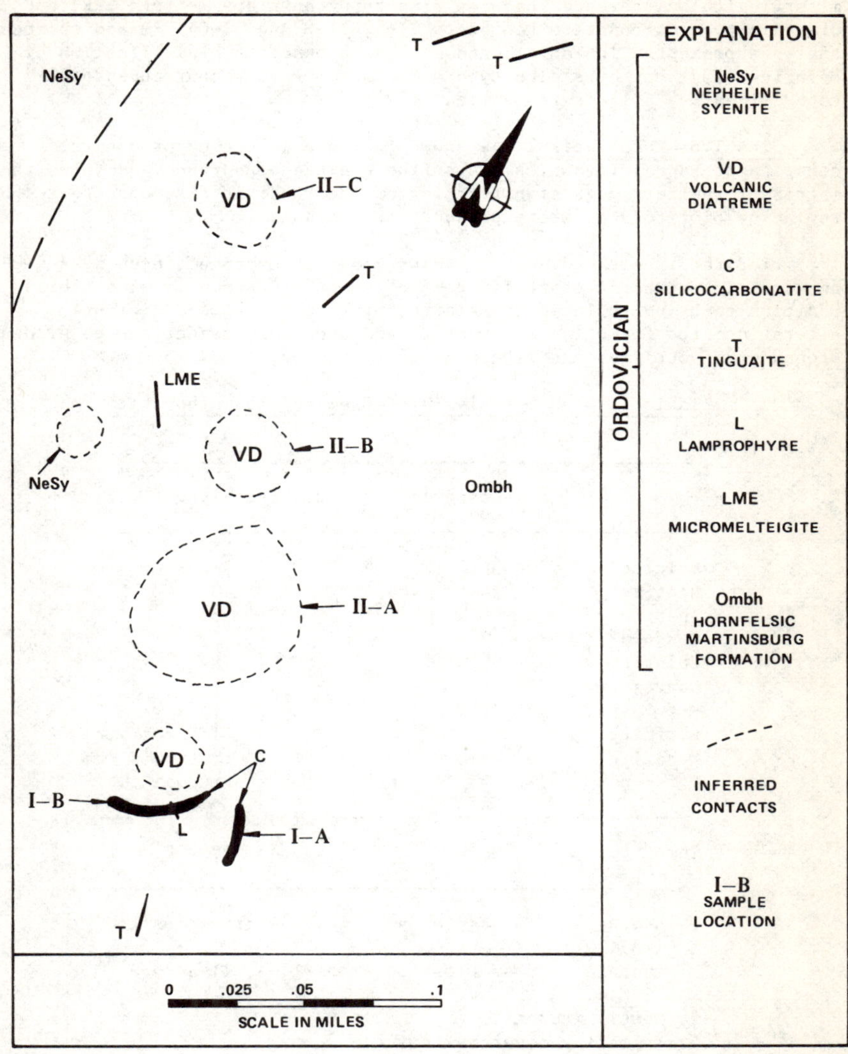

Figure 3 - Location of outcrops of carbonatite sampled for this study. (See also Figures 1 and 2.) I-A and I-B: dike-type; II-A, II-B, II-C: diatreme-type

Outcrops II-B and II-C are practically identical to outcrop II-A in terms of overal texture and composition. However, in outcrop II-B, the segregation of the ferro-magnesian fraction from the calcitic fraction is faintly discernible.

Petrographic Analysis

Petrographic analysis of carbonatite samples reveals quantitative mineralogical differences between dike-type (outcrops I-A and I-B) and diatreme-type carbonatite (outcrops II-A, II-B and II-C). Based on modal analyses presented in Table I and on the carbonatite classification of Heinrich (23), both the dike-type and diatreme-type occurrences are categorized as "silicocarbonatite."

Dike-Type Carbonatite. As shown in Table I, the major mineral constituent is calcite which has a fine-grained appearance and forms the matrix of the ferromagnesian fraction of the rock; it also appears to be replacing biotite and, where present, orthoclase.

Biotite, the second most abundant mineral component, occurs as anhedral to subhedral crystals, some of which are strongly pleochroic. In addition to being replaced by calcite, biotite exhibits incipient alteration; the formation of chlorite and secondary magnetite are evident around grain outlines and within grain fractures.

Table I. Modal Analysis of Beemerville Carbonatite

	Dike-type [1]	Diatreme-Type [2]
Calcite	56	33
Biotite	24	40
Apatite	5	8
Opaques[3]	8	7
Chlorite	5	8
Quartz	tr	1
K-feldspar	tr	1
Sericite	tr	1
Monazite	tr	tr
Pyroxene[4]	tr	tr
TOTAL	98	99

1. Average of 13 samples analyzed (outcrops I-A and I-B).
2. Average of 9 samples (outcrops II-A, II-B, and II-C).
3. Mostly magnetite.
4. Primarily aegerine-augite.

Magnetite, an abundant accessory, occurs primarily as euhedral to anhedral grains of primary origin. Colorless apatite is also a common accessory. Minerals in trace amounts include primary pyrite, aegerine-augite, monazite, quartz and orthoclase. As demonstrated experimentally (24), it is not possible to form orthoclase and calcite from the same magma. Both orthoclase and quartz in the dike are probably xenocrysts that appear to be the result of the partial assimilation of country rock by the magma.

The rare occurrence of a carbonatite autolith was found in outcrop I-B. The autolith is composed of nearly pure coarse-grained calcite with trace amounts of biotite, magnetite and apatite.

Diatreme-Type Carbonatite. The mineralogy of these samples is qualitatively similar to that representative of dike-type carbonatite samples; however, the two types differ substantially in terms of quantitative mineralogy (Table I).

The major mineral constituent is biotite which is pleochroic and ranges from very fine-grained anhedra to large euhedra. It shows pronounced alteration to chlorite and to secondary magnetite along grain rims and within fractures in the grains. Calcite, the second most abundant constituent, is very fine-grained and is seen replacing biotite as well as occur along fractures of both apatite and magnetite grains. Accessory magnetite occurs mainly as euhedral to subhedral crystals of primary origin. Accessory apatite and trace amounts of primary pyrite and monazite are also present.

The relatively large amounts of biotite and quartz, as well as the presence of orthoclase, sericite, and aegerine-augite may be explained in terms of the partial assimilation of country rock by the carbonatite magma. The decreased grain size of the primary minerals is the result of rapid crystallization of the molten material.

Bulk Chemical Analysis

Table II presents geochemical data for the Beemerville carbonatite as well as for carbonatites in general. Trace-element data are shown in Table III.

The compositional differences between the Beemerville carbonatite and carbonatites in general (Table II) are believed to have resulted from the partial assimilation of country rock by the carbonatite magma in the Beemerville area. For instance, the relatively higher amounts of SiO_2, Al_2O_3. Fe_2O_3, TiO_2, Na_2O, and K_2O in the Beemerville carbonatite can be attributed to assimilation of the underlying Precambrian gneiss and Ordovician Martinsburg shale. It is also important to note that since the percent content of CaO, CO_2 and particularly MgO is less for the Beemerville carbonatite than for Gold's (25) average carbonatite, the assimilation of the underlying limestone and dolomite is not believed to have been significant.

Other Analyses

Since carbonatites are usually associated with relatively high rare-earth element concentrations (23), various types of analyses were conducted in order to determine both the qualitative and quantitative presence of these elements in the Beemerville carbonatite. Samples representing both dike-type and diatreme-type occurrences were analyzed by

Table II. Chemical Composition of Beemerville Carbonatite

	A	B	C	D	E
SiO_2	22.73	14.87	15.70	24.51	12.10
TiO_2	2.23	2.44	2.95	1.78	0.80
Al_2O_3	6.30	6.28	4.25	7.46	3.55
Fe_2O_3	13.02	14.73	13.49	12.33	6.90
MnO	0.56	0.25	-	0.43	0.61
MgO	5.12	3.53	5.25	3.58	5.64
CaO	24.79	29.78	30.10	25.77	35.12
Na_2O	2.19	2.40	2.57	1.53	0.42
K_2O	2.32	2.12	2.90	2.62	1.49
H_2O+	1.87	1.10	1.00	-	1.39
H_2O^-	0.47	-	0.71	-	-
P_2O_5	2.75	2.38	-	-	2.06
CO_2	15.08	21.28	18.10	19.08	28.73
BaO	0.25	-	-	-	0.26
SrO	0.48	-	-	-	0.40
F	-	-	-	-	0.88
S	-	-	-	-	0.62
TOTAL	100.16	101.16	97.02	99.09	100.97

A. Average of 8 analyses (this report).
B. From Milton (4).
C. From McKague and Levendosky (22).
D. Average of 3 analyses from Maxey (5).
E. Average for carbonatites in general. From Gold (25).

Table III. Trace-element content of Beemerville Carbonatite

	Dike-type[1] (ppm)	Diatreme-type[2] (ppm)
Ba	3340	1200
Sr	6020	2734
Th	122	40
U	198	294
Zn	62	47

1. Average of 5 analyses from different outcrops.
2. Average of 3 analyses from different outcrops.

x-ray fluorescence as well as by electron microprobe techniques. However, no rare-earth elements were detected by x-ray fluorescence, suggesting that, at best, these elements were present in amounts below the detectable capability of the technique. On the other hand, the electron microprobe analysis suggested the presence of lanthanum and cerium in one of the dike-type carbonatite samples tested.

Neutron Activation Analysis. A sample of dike-type carbonatite was analyzed by this method (26). All rare-earth elements detected in the sample are shown in Table IV.

It has been reported (23) that rare-earth elements in carbonatites showing a marked enrichment of the light over the heavy rare-earth fraction, occur in rhombohedral carbonates such as calcite as well as in apatite and monazite. This applies to the Beemerville carbonatite since the light rare-earth fraction is approximately 18 times more abundant than the heavy fraction and since calcite, apatite and monazite are present in the rock. It has also been reported (27) that the carbonatite rare-earth mineral assemblages show a tendency to have an increase of the cerium to yttrium ratio with a decrease in age. This ratio in the Beemerville occurrence is approximately 12 to 1, suggesting late stage crystallization of minerals containing rare-earth concentrations.

Based on a provisional carbonatite classification (28), the Beemerville occurrence may be referred to as an "apatite-magnetite" variety rather than a "rare-earth" variety of carbonatite since its total rare-earth content is only about 0.5 percent.

Discussion

The age relationship of the various rock types of the Beemerville carbonatite-alkalic rock complex is ascertained by certain field relationships observed here and in other studies (5,22). These relationships indicate a sequence of events for the area, from oldest to youngest, as potassic syenite, carbonatite, lamprophyre, volcanic diatreme, nepheline syenite, bostonite, phonolite and tinguaite based upon

Table IV. Neutron Activation Analysis of Beemerville Carbonatite

Rare Earth Element		Content (ppm)
Yttrium	(Y)	166.00
Lanthanum	(La)	1120.00
Cerium	(Ce)	1960.00
Praseodymium	(Pr)	210.00
Neodymium	(Nd)	879.00
Samarium	(Sm)	131.00
Europium	(Eu)	30.90
Gadolinium	(Gd)	166.00
Terbium	(Tb)	12.70
Dysprosium	(Dy)	49.60
Holmium	(Ho)	6.41
Erbium	(Er)	29.60
Thulium	(Tm)	2.43
Ytterbium	(Yb)	24.20
Lutetium	(Lu)	2.21

the following: xenocrysts of partially assimilated orthoclase interpreted as originating from potassic syenite occur within silicocarbonatite inclusions; one of the carbonatite dikes is cut by a small dike of lamprophyre; potassic syenite, carbonatite and lamprohyre micromelteigite occur as xenoliths within the volcanic diatremes; nepheline syenite and bostonite intrude the volcanic diatremes; and phonolite and tinguaite intrude the nepheline syenite.

Based on the above and the radiometric age determination of late Ordovician (1), it is believed that the Beemerville occurrence resulted from the differentiation of a deep-seated alkalic parental magma which gave rise to magmas of carbonatitic and syenitic composition during the Taconic orogeny. The genetically related intrusives of the Cortland Complex (near Peekskill, New York) are also believed to have originated from a subcontinental alkalic magma source that resulted from compressional stresses during the Taconic orogeny (29, 30).

During the final stages of the Taconic Orogeny, continental convergence and associated crustal subduction occurred somewhere to the east of the present coast of North America. Pressure changes at the crust-mantle interface due to the stress relief resulting from crustal upwarp could have allowed partial melting of mantle material and the formation of an alkalic magma (33,34). Through processes of assimilation and magma fracturing, this magma worked its way into the upper crust where it came to occupy a magma chamber at some depth to the east of the present complex.

It is here that, through the processes of differentiation and fractional crystallization with gravity settling and magma degassing, the formation of a highly mobile carbonate fraction with an alkaline residuum occurred. As the processes continued, the volatile content increased and the carbonate fraction through assimilation and magma fracturing moved upward and intruded carbonatite dikes into the country rock. Pressure accumulation in the country rock resulted from magmatic intrusion, upward migration of volatiles from the magma chamber and the conversion of groundwater to steam. When this pressure exceeded the lithostatic pressure exerted by the country rock this material was explosively ejected and volcanic diatremes were formed. These conduits to the surface allowed the rapid release of the carbonatite magma. The presence of angular xenolithic fragments including carbonatite, Paleozoic shale limestone and dolomite, and Precambrian gneiss within a fine-grained carbonatite matrix indicate: 1) the rapid upward movement of this material from a depth of several miles; and 2) assimilation of the country rocks by the carbonatite magma was minor.

The associated fracturing of the country rock by the magmatic intrusion created paths and possibly other conduits. These passages provided a readily available access for the upward migration of the more viscose alkaline fraction resulting in the intrusion of the nepheline syenite and associated alkalic rock-types present in the Beemerville area.

Conclusions

Field, petrographic and chemical analytic data presented in this study suggest the following on the general nature of the Beemerville carbonatite.

1. Dike-type and diatreme-type are two distinct modes of occurrence for the Beemerville carbonatite. The former occurs mostly as dikes intruding into the Martinsburg Formation and the latter as the fine-grained matrix material in association with ouachitite in volcanic diatremes.

2. Both types of Beemerville carbonatite may be classified as the "apatite-magnetite" variety of "silicocarbonatite".

3. The Beemerville carbonatite and the associated alkalic rock complex resulted from the differentiation of a deep-seated alkalic parental magma which gave rise to magmas of carbonatitic and syenitic composition during the compressional tectonics of the Taconic orogeny.

References

1. R. E. Zartman et al., "K-Ar and Rb-Sr Ages of Some Alkalic Intrusive Rocks from Central and Eastern United States," Am. Jour. Science, 265 (1967) 848-870.

2. B. K. Emerson, "On a Great Dike of Foyaite or Elaeolite-Syenite Cutting the Hudson River Shales in North-western New Jersey," Am. Jour. Science, 23 (1882) 302-308.

3. J. F. Kemp, "On Certain Porphyritic Bosses in Northern New Jersey", Am. Jour. Science, 38 (1889) 130-134.

4. C. Milton, "Dikes of Special Petrologic Interest in Sussex County, New Jersey" (Pennsylvania Geologists Annual Field Conference, 18th, Trip A, 1952), 28.

5. L. R. Maxey, "Petrology and Geochemistry of the Beemerville Carbonatite-Alkalic Rock Complex, New Jersey," Geol. Soc. Am. Bull., 87 (1976) 1551-1559.

6. W.J. Spink, "Stratigraphy and Structure of the Paleozoic Rocks of Northwestern New Jersey" (Ph.D. Thesis, Rutgers University, 1967), 311.

7. A. A. Drake, Jr., "Precambrian and Lower Paleozoic Geology of the Delaware Valley, New Jersey-Pennsylvania," Geology of Selected Areas in New Jersey and Eastern Pennsylvania and Guidebook of Excursions, ed. S. Subitzky (New Brunswick, N.J.: Rutgers University Press, 1969), 51-131.

8. A. A. Drake, Jr., "The Reading Prong of New Jersey and Eastern Pennsylvania: An Appraisal of Rock Relations and Chemistry of a Major Proterozoic Terrane in the Appalachians," Geol. Soc. Am. Special Paper, 194 (1984) 75-109.

9. D. R. Baker and A. F. Buddington, "Geology and Magnetite Deposits of the Franklin quadrangle and part of the Hamburg quadrangle, New Jersey," U.S. Geol. Survey Prof. Paper, 638 (1970) 73.

10. C. Scotese et al., "Paleozoic Base Maps," Jour. Geol., 87 (1979) 217-278.

11. J. L. Lewis and H. B. Kummel, The Geology of New Jersey (N. J. Dept. of Conserv. and Econ. Div., Div. of Geol. and Topog., Bull 50, 1940) 203.

12. H. B. Kummel and S. Weller, "Paleozoic Limestones of the Kittatinny Valley, New Jersey," Geol. Soc. Am. Bull., 12 (1901) 147-161.

13. A. A. Drake, Jr., "Carbonate Rocks of Cambrian and Ordovician age, Northampton and Bucks Counties, Eastern Pennsylvania and Warren and Hunterdon Counties, Western New Jersey," U.S. Geol. Survey Bull., 1194-L (1965) 7.

14. R. S. Dietz, "Geosynclines, Mountains, and Continent Building," Sci. American, 1972, 124-132.

15. G. H. Cook, Geology of New Jersey (New Jersey Geol. Survey, 1868) 144-145.

16. M. Aurousseau and H. S. Washington "The Nephelite Syenite and Nephelite Porphyry of Beemerville, New Jersey," Jour. Geol., 30 (1922) 571-586.

17. A. S. Wilkerson, "Nepheline Syenite from Beemerville, Sussex County, New Jersey," Am. Mineralogist, 31 (1946) 284-287.

18. E. S. Davidson, "The Geologic Relationship and Petrography of a Nepheline Syenite near Beemerville, Sussex County, New Jersey" (M.S. Thesis, Rutgers University, 1948), 140.

19. M. S. Proko and R. B. Hargraves, "Paleomagnetism of the Beemerville (New Jersey) Alkaline Complex," Geology, 1 (1973) 185-186.

20. A. S. Wilkerson, "Tinguaite and Bostonite in Northwestern New Jersey," Am. Mineralogist, 37 (1952) 120-125.

21. J. E. Wolff, "Leucite-tinguaite from Beemerville, New Jersey," Harvard Univ. Mus. Comp. Zoology Bull., 38 (1902) 273-277.

22. H. L. McKague and W. T. Levendosky, "Preliminary Examination of Ouachitite Diatremes near Beemerville, N.J.," Abs. Transactions, Am. Geophysical Union, 52 (1971) 374.

23. E. Wm. Heinrich, The Geology of Carbonatites (Chicago, Il.: Rand McNally and Company, 1966) 555.

24. P. J. Wyllie, "Experimental Studies of Carbonatite Problems; The Origin and Differentiation of Carbonatite Magmas," Carbonatites, ed. O. F. Tuttle and J. Gittins (New York, N. Y.: Interscience Pubs., 1966), 311-352.

25. D. P. Gold, "Average Chemical Composition of Carbonatites", Econ. Geol., 58 (1963) 988-996.

26. F. M. Graber et al., "Neutron Activation Analysis Determination of All 14 Stable Rare-Earth Elements with Group Separation and Ge (Li) Spectrometry," Jour. Radioanalytical Chem., 4 (1970) 229-239.

27. A. P. Khomyakov, "Distribution of Rare Earths in Carbonatite-Hematite Veins of Western Tannu-Ola," Geochem. Intern., 1 (1964) 40-43.

28. W. T. Pecora, "Carbonatite - A Review", Geol. Soc. Am. Bull., 167 (1956) 1537-1555.

29. N. M. Ratcliffe, "Cortlandt-Beemerville Magmatic Belt: A probable late Taconian Alkalic Cross Trend in the Central Appalachians," Geology, 9 (1981) 329-335.

30. M. A. Domenick and A. R. Basu, "Sm-Nd Age of the Cortlandt Complex: Implications for Petrogenesis, Crustal Contamination and Tectonics," Geol. Soc. Am. Abstracts With Programs, 13 (1981) 440.

31. N. M. Ratcliffe, et al., "Emplacement History and Tectonic Significance of the Cortlandt Complex and Related Plutons, and Dike Swarms in the Taconide Zone of Southeastern New York based on K-Ar and Rb-Sr investigations," Am. Jour. Science, 282 (1982) 358-390.

32. M. A. Domenick and A. R. Basu, "Age and Origin of the Cortlandt Complex, New York: Implications from Sn-Nd Data," Contributions to Mineralogy and Petrology, 79 (1982) 290-294.

33. H. S. Yoder, Generation of Basaltic Magma (Washington, D. C.: National Academy of Sciences, 1976) 265.

34. Yu L. Kapustin, "On the Origin of Carbonatites," Intern. Geol. Rev., 19 (1976) 997-1008.

Process Mineralogy Applications to Industrial Minerals

EXPLORATION FOR BURIED TALC BODIES

THROUGH ANALYSES OF THE RESIDUAL SOILS

A M. Blount* and B. H. McHugh**

* Department of Geological Sciences, Rutgers University, Newark, NJ 07102
(and Newark Museum, P.O. Box 540, Newark, NJ 07101)
and
** Geology Department, Eastern Washington University, Cheney, WA 99004
(Present address: Dresser Atlas, P.O. Box 508, Williston, ND 58802)

Abstract

Elevated values of talc have been noted in residual soils (residuum) associated with buried talc deposits. Such concentrations are documented for deposits in talc districts in Pennsylvania, Alabama, Montana and Washington. Differing procedures of sampling and analysis are mandated by variations in geology and weathering between the districts. The factors which must be considered in sampling are the depth of the residuum and the size fraction in which the talc is concentrated. When analysing the samples, the problem of varying mass absorption coefficients and the problem of preferred orientation must be considered. The former is inherent to analyses of all mixtures, whereas the latter is specifically severe with layer silicates because of their shape. A number of procedures which can alleviate these problems and which can accomplish the analyses within reasonable time limits are described. Detailed analytical procedures for Montana soils are given in an appendix.

Introduction

A large amount of research has been done in recent years on techniques to discover concealed ore deposits by geochemical methods. Much effort has been directed toward determining if soil, rock, water or even gases may best reflect associated ore and toward determining which method is best for specific types of ore. Where soils have proved useful, investigation has been done to determine which size fraction or horizon contains concentrations of the desired element or of pathfinder elements.

In 1980, Blount and Vassiliou (1) pointed out, as a result of a study of the Winterboro, Alabama, talc deposits, that talc resists weathering and is concentrated in the soil. Further, it was proposed that this phenomenon might be useful in exploration for buried talc bodies. Although this talc is not visible in the soil, it may be easily detected by X-ray diffraction. As with geochemical exploration, however, the question remained as to what portion of the residuum would yield the most useful information. Investigation of talc districts in Pennsylvania, Alabama, Montana and Washington has shown that there is considerable variation in the residuum associated with these deposits. The purpose of this paper is to demonstrate that talc anomalies are present over buried talc bodies, that each district has unique problems and that techniques of sampling and analysis must take into account the individual problems of the district in question. However, because of the broad interest in Montana, the procedures of analysis for soils from Montana are described in the Appendix.

Soil-Talc Anomalies

The soils associated with known talc bodies in all four districts studied show anomalous concentrations of talc. In this section, one or more examples will be shown from each district. It will be noted that the data are presented differently for each district and although this may appear uncoordinated this has been imposed by certain characteristics of the soil, rock and weathering environment inherent to each district.

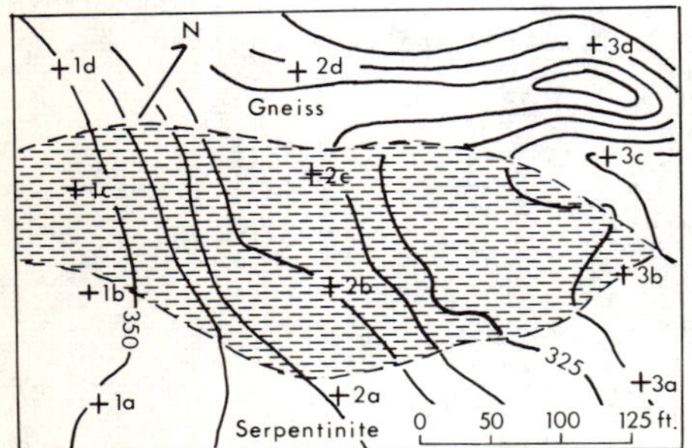

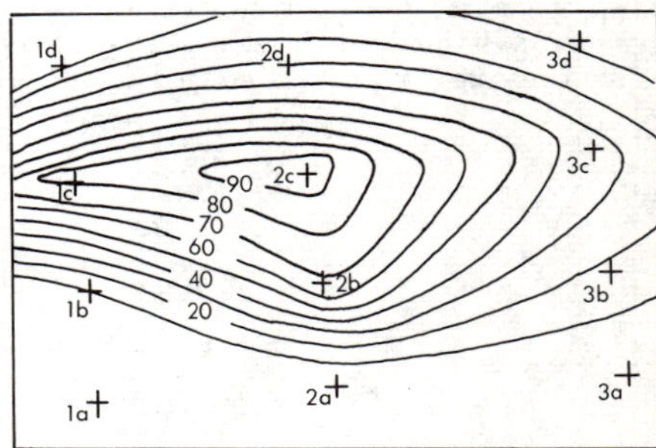

Figure 1. Maps showing geology and percent talc in the soil at a talc deposit near Lancaster, Pennsylvania. The shaded area indicates the extent of the buried talc body as determined by core drilling and trenching. See Table 4 for details of soil sampling and analysis.

The talc in Pennsylvania occurs in the southeastern corner of the state near Lancaster and is associated with serpentinite. The deposit studied occurs as a lens between serpentinite and gneissic country-rock (Fig. 1). The residuum is thin here varying from a foot or less over the serpentinite to several feet over the gneiss. A strong talc anomaly occurs over and slightly downslope from the talc body at this site (2). The values indicated are percent talc in the bulk soil where the samples were obtained from 12 inches depth. Geochemical analysis of the ground water has indicated that the talc in the saturated portion of the soil is not leaching.

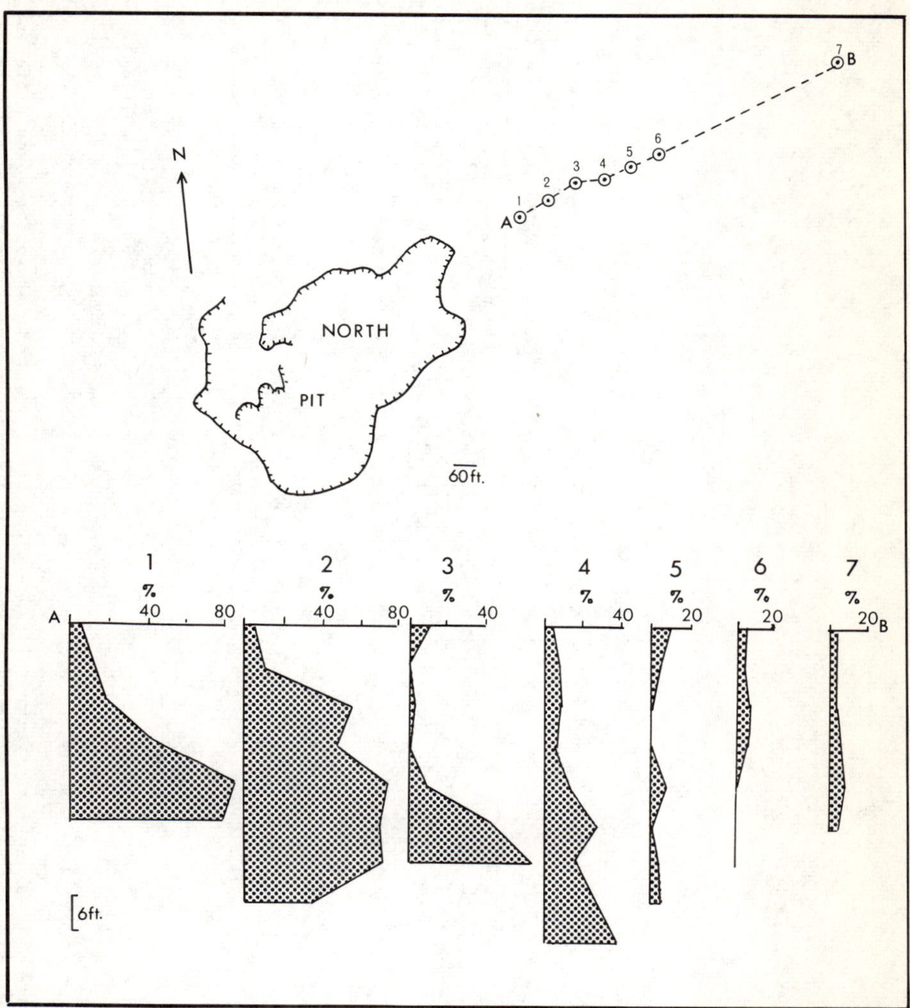

Figure 2. Top: map showing location of auger drill samples in relation to the north pit at Winterboro, Alabama (S 1/2 sec. 3, T20S, R4E). Bottom: the percent talc at various depths is shown for each drill hole.

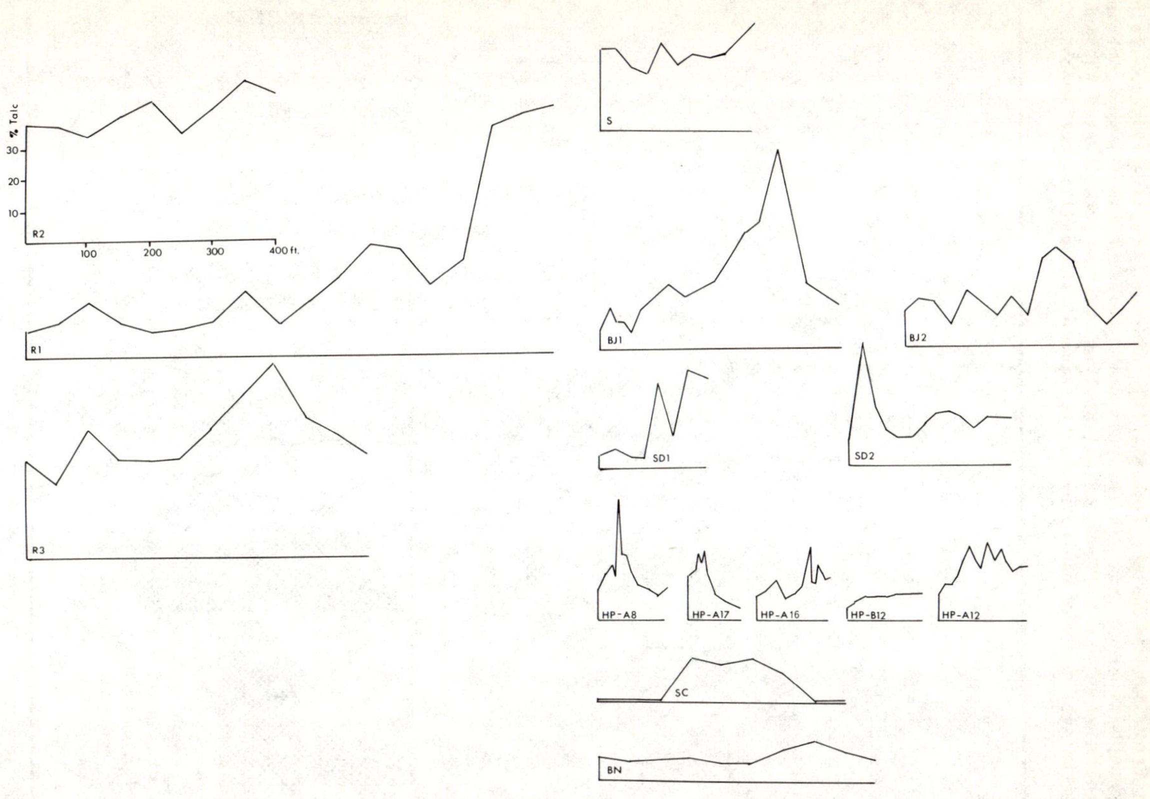

Figure 3. Percent talc versus distance along the traverse is shown for seven different sites. Scale is the same for all and is shown on the top left traverse. See Table 1 for abbreviations.

The talc deposits occurring in Alabama are considerably different in geology and in weathering environment from those in Pennsylvania. Here the deposits are associated with relatively unmetamorphosed dolomite. The north body was originally enclosed in dolomite but had been weathered out and is resting in residual clay. The bedrock is deep, in places seventy or more feet. Results of soil analysis are shown along one traverse away from the ore body. Whereas, the average value of percent talc in the <4 μm fraction of the soil decreases away from the body, it is also noted that there is considerable variation in the values within each profile.

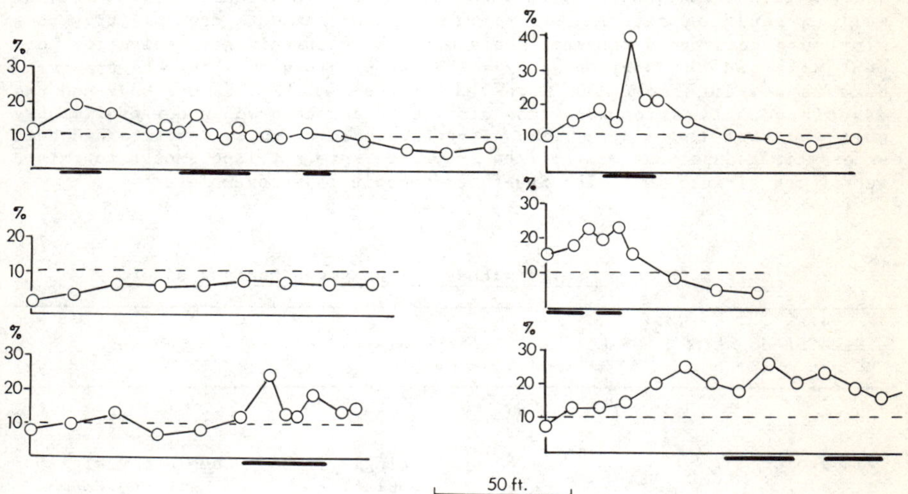

Figure 4. Percent talc is shown for six traverses at Holden Property, Montana. Black bars at the bottom of each traverse show the position of talc lenses. See Table 4 for sampling and analysis procedures. Dashed line is the regional threshold value for talc.

The talc in the soil at Winterboro has not come from the migration of talc from the talc body but represents the weathering product of a zone of hydrothermally altered rock which originally surrounded the ore body. The presence of such a zone has been demonstrated by analysis of the acid-insoluble residue of dolomite core samples obtained from the south body which is enclosed by dolomite bedrock (1). A halo-shaped alteration pattern is not present in the soil and would not be expected. Two factors would tend to disrupt such a pattern. First, the amount of insoluble material in the dolomite varies from place to place such that if 100 ft. of dolomite weathers beneath the altered zone, it might yield five feet of clay in one place and one foot in another. Second, the bedrock surface of carbonate rock, upon which such residuum rests, is generally highly irregular. Carbonate surfaces are notorious for having pinnacles and pits. In fact, drill data from the north pit demonstrates such irregularity at this site. Because of these factors a "mashed down" replica in the soil of the original alteration zone would be highly unlikely. Thus, the irregular pattern of talc content in each profile, observed in Fig. 2, is an inherent property of the talc-soil anomalies surrounding Winterboro-type deposits.

Seven talc bodies or occurrences have been studied in Montana (Table 1). At each of these the talc is associated with dolomitic marble. Soil samples were collected along one to six traverses at each site with the sampling interval selected according to the size of the known body. Depth of the residual soil was generally about twelve inches; however, in a few places it reached to five feet.

Although space does not permit discussion of each deposit, we will show two as examples. Before doing so, however, one observation is in order, that is that there seems to be a direct relationship between the size of the deposit and the amount of talc observed to be present in the soil. Figure 3 and Table 1 list these deposits from largest to smallest, as well as could be determined. Figure 3 shows the data from all traverses plotted according to the same scale with the vertical axis indicating percent talc and the horizontal axis indicating distance along the traverse. Here as at Winterboro the high-talc soils reflect the ore body and the associated alteration zone. One difficulty encountered in our preliminary work was that the size of the alteration halo of these deposits had been underestimated so that most of the traverses were not long enough to extend out of the altered zone. The Sauerbier deposit is an example.

Table 1. Percent talc in the soil at seven Montana sites.

Name of deposit*	Symbol	Average % talc	Local threshold	Max. % talc	Comments
Regal	R	34	50	80	
Sauerbier	S	28	single population	36	Completely within altered zone
Banning-Jones	BJ	19	37	65	
Smith-Dillon ext.	SD	16	30	40	
Holden	HP	13	19	39	
Johnny Gulch	BN	8	**	13	Traverse upslope from deposit
Spring Creek	SC	3	**	14	

* See reference 3 for location of each deposit except Holden which is in Sec. 24, T7S, R7W.

** Insufficient data for calculation.

As with geochemical analysis one needs to determine what values should be considered anomalous. In general, in deposits surrounded by a zone of feeble mineralization, one would expect a plateau of intermediate values indicating the feeble mineralization surrounded by an area of low background values. The central part of an anomaly of high values is closely related to the ore. These sets of values are divided by the regional threshold and the local threshold. To determine these threshold values statistical methods (4,5) have been applied to the data from each deposit although with some reservation because of the small amount of data in some sets. The local threshold values are listed in Table 1. The regional threshold is more uncertain since not many of our samples were obtained from soils over nonmineralized rock. However, from results obtained from soils and acid-insoluble residues of carbonate rocks, it appears that any soil containing 10% or less talc should not be considered unusual. Values somewhat over 10% would belong to a plateau of feeble mineralization or to an anomaly above a small deposit. A 10% regional threshold value is probably conservatively high.

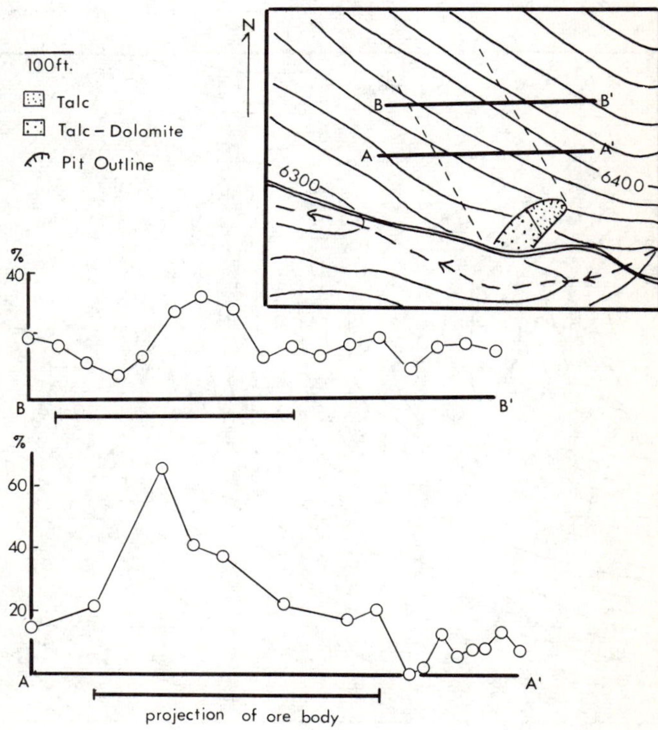

Figure 5. Two profiles from the Banning-Jones mine, Montana, are shown. Dashed lines on the map and bars under the traverses indicate probable projection of talc body. See Table 4 for sampling and analysis procedures.

Six soil traverses were made of the Holden Property (Fig. 4). Each was parallel to a trench that had previously been been dug so that the location of talc layers or seams was known. The talc layers of several inches to several feet were interspersed with layers of hornblende gneiss or amphibolite. To determine how well the soil mineralization reflected that of the underlying rock, soils were sampled at closely spaced intervals, in places at 10 ft. intervals. Figure 4 demonstrates that there is a reasonable correspondence between elevated soil-talc and the underlying small talc layers.

The Banning-Jones site consists of a larger talc body (Fig 5). Two soil traverses were made upslope from the cut. The traverse nearer the body shows a distinct peak in the area of the projected ore body. The second traverse shows a less distinct peak which may indicate that the body is pinching out in that direction. At this site there appears to be some movement of the soil and of the anomaly downslope due to migration of the soils on this slope. Because there is little vegetation at some of the Montana sites, on steeper slopes there can be considerable downslope movement, in some cases as much as 100 ft.

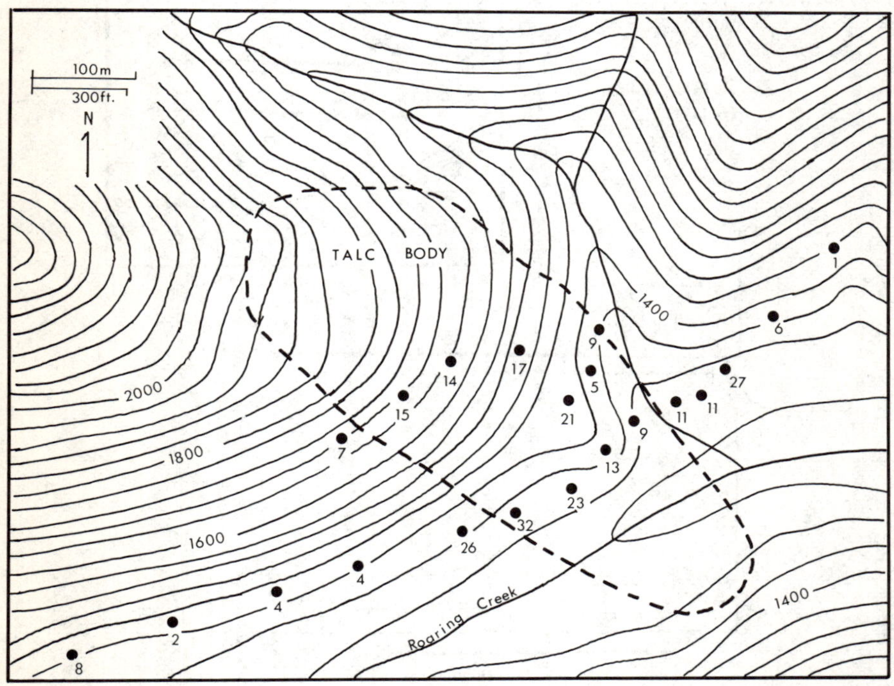

Figure 6. Soil sampling locations are shown for several traverses across a low-grade talc body, Entiat, Washington (SE 1/4 sec. 5 and NE 1/4 sec. 8, T25N, R20E). Values indicated are intensity of the 9.4° talc peak. See Table 4 for sampling and analysis procedures.

Finally, three talc occurrences were studied in Washington State (Table 2). Two were associated with serpentinite and one with dolomite. The latter shows the highest grade material. All these deposits are in rugged country with slopes about 30° and had shallow soils. A typical example is the deposit along Roaring Creek near Entiat, Washington, shown in Fig. 6. Here the talc body consists of talc and tremolite with serpentine, and the country rock is a mixture of schist, amphibolite and gneiss. The values shown are absolute intensity of the 9.4° peak where the amount of sample in the beam remains constant. Values increase over the talc body.

In this section we have attempted to demonstrate that, in the four talc districts studied, significantly elevated values of talc are found in the vicinity of talc bodies. Various methods of sample collection and analysis have been applied to these areas because of variations in geology and weathering conditions inherent to each district. In addition to the various problems presented by the environmental conditions, there are certain problems specific to layer silicates and their analysis by X-ray diffraction procedures which must be addressed to obtain successful results. In the next section techniques of soil sampling will be discussed followed by a discussion of factors and problems affecting the analysis of the material by X-ray diffraction.

Table 2. Average intensity of the 9.4° talc peak for the background and anomaly at three Washington sites.

	Average background	Average anomaly	Maximum intensity	Associated rock
Entiat SE1/4 sec. 5 and NE1/4 sec. 8, T25N, R20E	4	19	32	serpentinite
Londonderry S1/2, SE 1/4 sec. 16, T35N, R11E	7	30	42	serpentinite
Metaline Falls SE1/4, NW1/4 sec. 23, T39N, R44E	3	51	80	dolomite

Factors Affecting Sampling

The residual soils vary considerably from district to district in total depth, particle size distribution and concentrations of talc with depth. Each of these affects how the samples are collected and processed. The residuum in the Alabama talc district is very thick. For this reason a mechanical soil-auger was used to obtain samples up to 72 ft. deep (6). Samples were collected representing each six foot interval, which corresponded to the drill string length. Although there is some contamination of deeper samples from above, this did not cause serious problems. The residuum in the Pennsylvania, Montana and Washington districts was thin. The maximum depth was approximately five feet. Hand-powered soil samplers could be used in these districts. The tube-type sampler proved to be the most useful device (7), although the auger was also used in Pennsylvania (8).

The talc present in the soil is not universally concentrated at one depth or in one size fraction; thus, for each district studied experimental work was performed to determine the optimum size fraction and depth. First, it was necessary to determine if samples from one depth would yield different results from those obtained from another depth. We have previously shown data from the Winterboro deposits, Alabama (see Fig. 2). In some profiles there is considerable variation in percent talc, and, in addition, no one specific depth is necessarily characteristic of the profile as a whole. Since analyzing the complete profile is excessively time consuming, the best solution for the Winterboro-type deposits was to analyze samples from one depth (for example, 30 ft.) but being aware that some variation was to be expected. In the case of the Montana deposits, the soil is thin. Rarely were depths as great as five feet encountered. Table 3 shows nine sampling locations out of 36 at the Banning-Jones site that were greater than one foot in depth. The values at various depths show little variation, and, in all cases studied, the distribution pattern for the soil-type anomaly appears the same whether the entire profile or just the top twelve inches was analyzed. For this reason, it was concluded that the time necessary to sample the residuum to bedrock was not justified and that only the twelve inches are needed. At the Pennsylvania deposit described, the samples were collected to bedrock, but those samples from twelve inches adequately showed the anomaly. Likewise for Washington state deposits, samples were taken directly below root level.

Table 3. Variation in percent talc in the soil versus depth in inches at Banning-Jones site.

	Traverse #1					Traverse #2			
Hole No.	9	15	16	17	18	1	3	4	5
0-12"	59	7	10	12	15	15	16	13	21
12-24"	83	tr	7	8	10	12	16	nd	15
24-36"		10	9	5			13	9	
36-48"			8					14	
48-60"								10	

tr = trace and nd = not detected.

Second, it was necessary to determine what part of the sample should be prepared for analysis. Traditionally sediments, sedimentary rocks and soils have been analyzed by separating out the less-than-2 μm particles and essentially ignoring the bulk (9). This was done for the pragmatic reasons that layer silicate minerals are concentrated in this fraction and that mounts of this size fraction produce highly reproducible results. Larger particle sizes show greater deviation in intensity of X-ray diffraction peaks. For example, Klug and Alexander (10) showed that the 15-50 micron particles of quartz have a mean percent deviation of intensity of 18.2 while the less-than-5 μm particles have a mean percent deviation of 1.2.

An examination of the less-than-2 μm fraction of Montana soils indicated that talc is not appreciably abundant in that fraction (Fig. 7). Smectite was the major constituent. Further examination of several size fractions indicated that talc was most abundant in the bulk soil samples. To obtain reproducible results by X-ray diffraction the particle size was reduced by wet grinding in a micronizing mill. Grinding samples containing layer silicates must be done in a liquid because of the tendency of the layer-silicate structures of some minerals associated with talc to break down during frictional heating of dry grinding. Bulk samples were used in the analysis of the Pennsylvania soils but for a different reason which will be discussed later.

The samples from Alabama deposits were analyzed in the less-than-4 μm fraction. In this case the bulk samples generally contain 50 to 80 percent quartz. Layer silicates, especially talc and chlorite, are indicators of alteration. Use of the bulk samples dilutes this useful part of the sample. At the same time, the less-than-2 μm fraction contains true clays, those which have formed in the weathering environment, also diluting the sample. The <4 μm fraction was a compromise and has proved useful because it also contains unaltered minerals from the parent rock. This permits useful information to be obtained about the geology of the buried surfaces near the talc bodies.

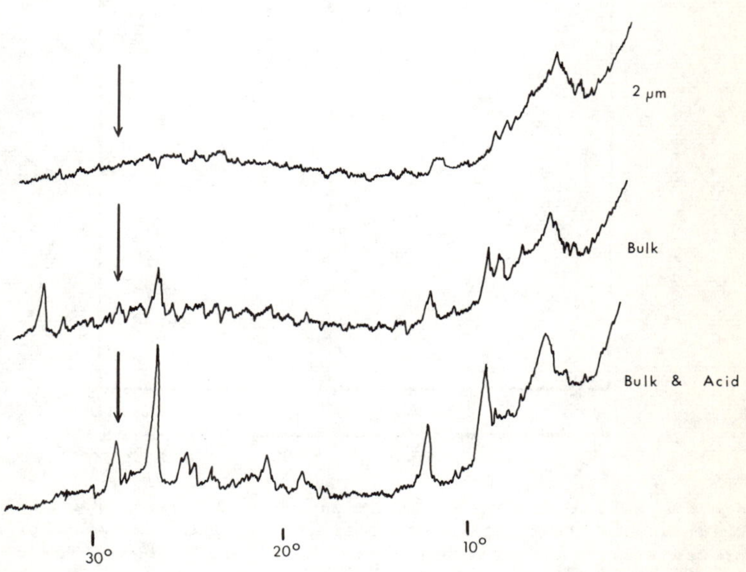

Figure 7. Diffraction patterns of a typical Montana soil prepared (top) by separating the 2 μm fraction, (middle) by grinding bulk sample and (bottom) by grinding and acid treating overnight in 1N HCl. Arrow indicates the talc peak used in quantitative analysis.

Factors Affecting Analysis

To determine the best procedure for analysis of soil samples by X-ray diffraction requires consideration of two factors: (A) variation of mass absorption coefficients among samples that contain varying mixtures of components and (B) tendency of layer silicates to orient themselves into preferred orientations during sample preparation. The first factor is one which is inherent to all analyses of mixtures. Because each mineral has a specific and generally unique mass absorption factor, mixtures containing various proportions of the same minerals will most likely have different absorption coefficients for the mixture. If the mass absorption coefficient is less for one mixture than for another, the X-ray beam will penetrate a greater mass of the lower absorption-coefficient mixture and the peak intensity for an unknown mineral will be larger even if occurring in the same proportion in both mixtures. The second factor is caused by the plate-like shape of the layer silicate minerals. When samples are prepared for analysis, it is difficult to avoid preferred orientation. This preferred orientation causes certain atomic planes to be preferentially oriented to diffract the X-rays, and reflections from these planes become greatly enhanced over others. It is not uncommon for certain peaks to become as much as four to ten times more intense than for samples with random orientation. Unfortunately, in layer silicates, the most intense reflections which are used in quantitative analysis are just those that are affected most by preferred orientation.

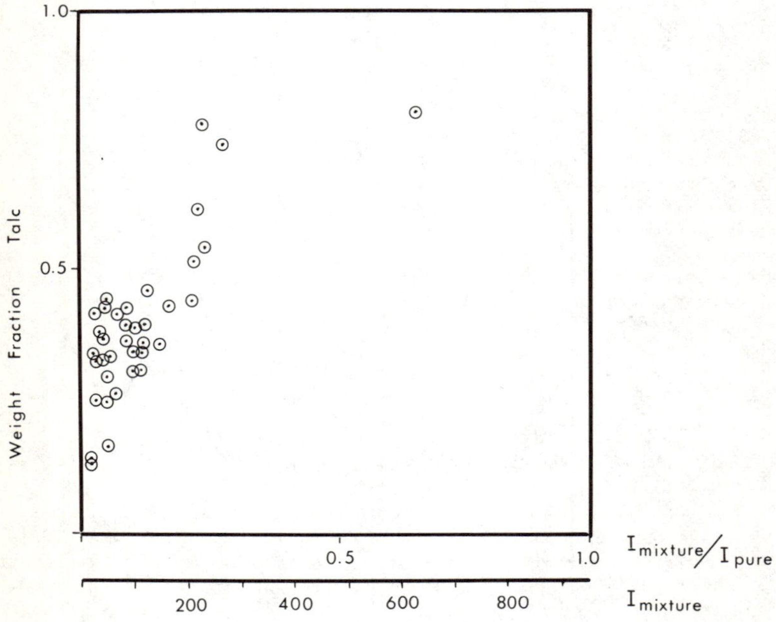

Figure 8. Weight fraction of talc in the soil versus intensity of 28.6° peak for samples from the Regal deposit in Montana. The method of Rex (see text) was used to determine percent, and this accounts for scatter at values less than 0.50.

With these factors in mind let us examine the procedures which were used in each of the four districts studied. The data shown for Entiat, Washington is peak intensity (11). Since the amount of talc is less than 30% not too much error is introduced by simply considering the intensity to be linearly related to amount of talc in the soil. In this range the curve is fairly linear. On the other hand, intensity versus weight percent talc is shown in Figure 8 for the Regal deposit in Montana. Here there is marked lack of linearity. The sample containing 80% talc shows a peak intensity of double what might be expected. The very high value is due not only to the variation in absorption but also to the effect of preferred orientation. At values of 50% or less of talc, equidimensional minerals in the sample tend to keep the layer silicates from going into preferred orientation, but as the percentage of layer silicates increases the probability of preferred orientation increases.

To avoid this problem the Alabama and Montana soils have been analyzed according to the method of Rex (12) with the mounts specifically prepared to enhance preferred orientation. This method involves using ratios so that the mass absorption coefficient for the mixture cancels out. The formula used for the Montana samples is as follows:

$$\text{Weight percent talc} = \frac{I_{talc}K_{talc}}{I_{k+chl}K_{k+chl} + I_{qtz}K_{qtz} + I_{talc}K_{talc}} \times 100$$

where $K_{talc} = 1.00$
$K_{k+chl} = 1.67$
$K_{qtz} = 1.00$

and the intensities are measured at 28.6° 2θ for talc, 26.6° 2θ for quartz (also includes mica) and 24-25° 2θ for kaolinite and chlorite. The K values are determined experimentally by mixing each mineral 50:50 with quartz. In addition, a correction factor was applied to the 28.6° peak because amphiboles and talc overlap at this position. This can be done because the peaks of these minerals do not overlap at low 2θ angles. Since the ratio of intensity for the (28.6° talc peak/9.4° talc peak) equals 0.60 and the intensity of the (28.6° amphibole peak/10.4° amphibole peak) equals 0.60, the portion of the 28.6° intensity belonging to talc is:

$$\frac{I_{9.4°}}{I_{9.4°} + I_{10.4°}}$$

In the procedure of Rex we may ignore amorphous material, organic material and any minerals we so choose (12). Dolomite, calcite and feldspars have been ignored in the Montana soils.

This procedure was considered for the Pennsylvania samples; however, there was one group of minerals which occurred in significant quantity which could not be handled in this way. These were corrensite and the less perfect corrensite-like layer silicates. This mineral group consists of interlayers of chlorite and vermiculite mixed in a proportion nearly 50:50. (The name "corrensite" implies exactly 50:50 and exactly alternating layers.) The intensity of a given reflection, such as *002*, can vary by as much as eight times depending upon the perfection of alternation of the layers (13). A value of K for corrensite could not be determined because of this intensity variation; consequently, the internal standard procedure proved more applicable. In this procedure a specific portion of standard is added to the sample and a ratio of intensity of a certain peak in the standard to a certain peak of talc is obtained. This is compared with a standard curve which had previously been obtained by experimental procedures (see standard texts for details). Dolomite with a peak at 30.9° was used as an internal standard for the Pennsylvania samples because its mass absorption coefficient was near to those of the minerals in the sample and because the equidimensional shape tended to keep the sample in random orientation which was desired for this procedure.

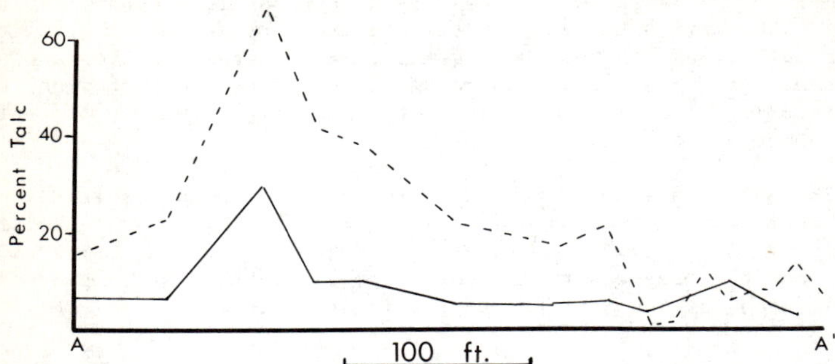

Figure 9. Traverse of Banning-Jones showing percent talc determined by Rex (dashed line) and by internal standard method (solid line).

Finally, both the method of Rex and the method using the internal standard were applied to several traverses in Montana. Since certain minerals are ignored using the former procedure, we wanted to determine if equivalent results would be obtained. Figure 9 shows that the internal standard method yields lower values but the the curves are similar. The lower values for the internal standard method were expected because carbonates and feldspars occur in considerable quantities in these samples, and they were ignored or removed in the method of Rex. Similarity of curves has been found in all cases checked by both methods. Since it is the curve shape and the threshold values determined for each method that are important in interpreting the data, either method can be used. Since the method of Rex is faster in terms of sample preparation it was used when possible in this work.

In the X-ray diffraction analysis, particularly when quantitative results are sought, one seeks to maximize either the tendency of the X-ray diffraction mounts to show preferred orientation or to show random orientation because peak intensities will be intermediate between these extremes if partly oriented. It is not possible to correct for partial orientation effects. With the method of Rex we attempted to attain maximum preferred orientation both of unknowns and standards. Random orientation was used with the internal standard method. This is necessary because it has not been possible to find good plate-like minerals or chemicals to be used as internal standards, and equidimensional standards cause partial orientation unless completely random orientation is specifically prepared.

Mounts giving maximum preferred orientation were obtained by mixing the sample in water, removing a certain size fraction with an eyedropper onto a glass slide and permitting it to dry. The size fraction was separated by gravity settling in water using Stoke's Law to calculate the time and depth at which the sample should be withdrawn from the suspension. Some problems may occur in this method because further gravity separation of particles takes place during the drying of the sample on the glass slide (14). To avoid this, the sample can be separated according to particle size

Table 4. Sampling and analysis procedures found to be best in each of the four districts studied.

District	Sampling		Analysis	
	Depth and Technique	Size fraction	Method	Mount
Pennsylvania	12-18 inches, tube-type sampler	bulk	dolomite internal standard	randomly oriented mount
Alabama	24-30 ft., mechanical auger	<4 microns	method of Rex	oriented mount
Montana	surface to 12", tube-type soil sampler	bulk	method of Rex	oriented mount, carbonate removed
Washington	below root level, tube-type soil sampler	<2 microns	method of Rex (absolute intensity*)	oriented mount

* Method used in the original study.

and then permitted to dry in a beaker. After drying it is mixed with several drops of water to form a paste and spread across a slide with a spatula. The layer silicates become oriented during this process. Both preparation methods were used on some samples in this study with no apparent difference in results. Since the eyedropper method is faster, it was used in subsequent work.

Random mounts were prepared using the Cab-O-Sil method (15). In both cases where this procedure was used the sample was mixed with 50% standard and then mixed with Cab-O-Sil. Although this amount is somewhat large, it was done to insure random orientation. The minerals, dolomite and olivine, used as internal standards are equidiminsional and tend to inhibit the plate-like minerals from attaining parallel orientation.

An additional factor in sample mount preparation is the problem presented by the carbonates. Figure 7 shows the diffraction pattern of the bulk samples with and without carbonate minerals. Carbonates are generally removed by investigators when analyzing layer silicate mixtures because the resulting patterns are greatly improved (16). Much of the problem is due to the carbonates acting as cements and preventing the layer silicates from settling into preferred orientations on the slide mounts. Because significant quantities of carbonates were present in the Montana soils, it was necessary to treat the ground-bulk samples in 1N HCl to dissolve the calcite, dolomite or magnesite. Samples from the other districts did not show appreciable carbonates and therefore were not treated with acid.

Conclusions

Although elevated talc values are found above and surrounding talc bodies, there was considerable variation in procedures used in obtaining this information. As indicated these variations in techniques are imposed by differences in weathering conditions and geology of the districts. These variable factors fall into two groups: those involving sample collection and those involving sample analysis. While we have attempted to indicate in the previous sections the reasons why different techniques are needed, it may be unclear what procedures were actually used in each district. Table 4 shows what we have found to be the most useful methods in our investigation of each district. The data shown in Figures 1 through 6 were obtained with these techniques.

Finally, very little has been said about sampling intervals along traverses or grids. The reason is that this study involved known bodies and occurrences of talc. The traverse intervals were selected so that at least one sampling location would overlie the mineralized body. For example, if the mineralized body was 30 ft. wide, a sample interval of 25 ft. was used. In area of unknown bodies in carbonate terranes, there are two possible approaches. First, in the case where all field work is done before the laboratory analysis, the sampling interval should be selected as in this study. After deciding the size of target which is acceptable, the interval is chosen so that it is not wide enough to skip across the target without getting at least one sample. Second, if analyses can be done concurrently with sampling, a large grid interval can be selected. The samples when analyzed will indicate certain areas are above the regional threshold value for talc. Within those areas a smaller sampling interval can be used to determine the target concentration with values above the local threshold value.

APPENDIX

Analytical Procedures for Montana Soils

The analysis of talc content in soils is a geochemical exploration technique despite the fact that it is different from the typical trace element investigations generally done. As such one must recognize that sampling and analytical methods must sacrifice precision for speed and that, as a consequence, an isolated result has little meaning in and of itself. With these limitations in mind, the techniques which have proven most useful for Montana soils will be described. The true accuracy of the results is not known and is probably highly variable. The standard deviation of replicate runs is 15% of the amount measured at 10 to 20% talc and decreases as the measured amount increases. At 80 to 90% talc the standard deviation is 2%. The procedures can be broken down into the following steps: (1) obtaining samples, (2) sieving and splitting samples, (3) grinding and removing carbonates (organics) and (4) running and analyzing diffraction patterns.

The soils encountered in a typical investigation include those from the weathering of marble, gneiss, amphibolite and schist. The following minerals are found:

Derived from rock	Formed in the soil
quartz	smectite
feldspar	vermiculite
plagioclase	kaolinite
K-feldspar	calcite
mica	
biotite	
phlogopite	
muscovite	
amphibole	
talc	
chlorite	
carbonates	
dolomite	
calcite	
magnesite	
serpentine - rare	
diopside - rare	

The inherited minerals are more abundant in the coarser fractions of the soil than in the finer fractions; whereas, the smectite, vermiculite and kaolinite are more abundant in the finer fractions. Calcite is often deposited in the soil over weathering dolomitic marble ($MgCa(CO_3)_2$). The magnesium ions are flushed from the soil during weathering but the calcium carbonate is deposited. Calcite often coats other rock fragments in the soil. In addition to the minerals present, some soils contain organic material which interferes with the mineralogical analyses of soil.

Sampling

We sampled soils with a tube-type soil sampler of the type used by soil scientists. The sampler is 12 inches long and 13/16 inches in diameter. The sampler has various tips which screw into the end. We found the tip for moist soils (S2) was the best for all Montana locations. A three foot extender rod was placed between the handle and the sampler, and we used a foot jack which fits on the extender rod. This permitted the sampler to be used easily in a standing position for most soils. The sampler was filled once at each location and represented surface to twelve inches (or less).

In addition to the soil sampler, a 12-inch long loaf pan was used so that the soil could easily be dislodged from the sampler without loss. The sample then was transferred to a cloth sample bag. Since the soil was often damp, the fabric container permitted drying.

Sieving and splitting

The soil samples in fabric sample bags were dry by the time they reached the laboratory. These samples consisted of rock fragments as well as soil and, consequently, had to be sieved before splitting. When the sample had formed one or more clumps in the sample bag during drying, the material was disaggregated by gently pounding the unopened bag with a pestle. Then the sample was sieved with a no. 6 and no. 20 sieve so that the resulting particles were all less than 0.84 mm in diameter. An aliquid part of 2 cm^3 or less was obtained using a sample splitter.

The sieving was observed to affect the final results. For example, a sample with a small amount of talc which was present due to downslope movement of talc showed a value of 4% talc when sieved to less than 0.84 mm; whereas, when it was sieved to only less than 4 mm yielded a value of 9% talc. In the latter case, a single fragment of talc could affect the results, and, thus, we felt that it was better to sieve the samples to the smaller 0.84 mm particle size.

Grinding and removing carbonates

The 2 cm^3 aliquid part was ground in 7 ml of water for five minutes. When grinding soil samples two factors must be considered. First, that layer silicates can be destroyed by friction during dry grinding as was indicated in the body of this paper. It is important not to destroy any of the minerals whose peaks are to be used in the relative intensity calculations. For this reason we used wet grinding in a McCrone Micronizing Mill which employs 48 cylindrical grinding elements. Since the grinding elements produce line and plane contact shears rather than point contact blows of the ball mill, it proved to be particularly suited for layer silicates.

Second, the time necessary to yield uniform results needs to be considered. If grinding time is insufficient, the layer silicates show enhanced preferred orientation with resulting increased intensity of basal reflections. To determine the correct grinding time the intensity of *002* versus grinding time for muscovite was determined. Muscovite was chosen because it is more intractible to grind than chlorite or talc. The results are shown in Figure 10. Although there is a continuous decrease in intensity with grinding time, after approximately 4 minutes increased grinding time had only minor effect on intensity. With soils there is a distribution of particle sizes before grinding. For example, there may be mica in the less-than-2 μm fraction and other mica so coarse that it must be ground. The less-than-2 μm fraction would be equivalent to an extremely long grind and would have a basal intensity lower than could be achieved with the coarser mica during any reasonable grinding time. For this reason a compromise in terms of speed and accuracy was made. In our work we used a five minute grinding time on all Montana soil samples. Some error was introduced because of this relatively short grinding time.

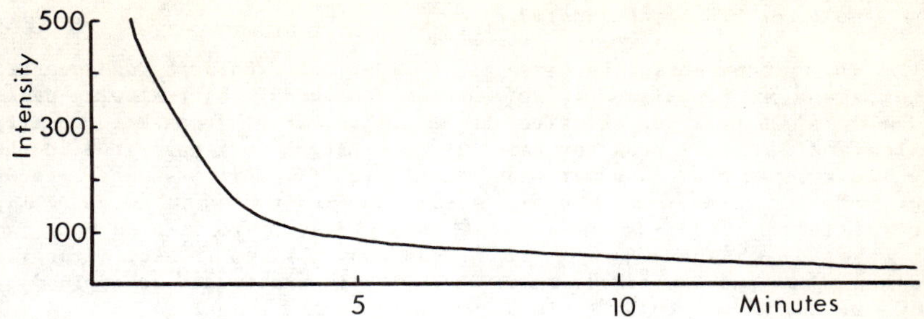

Fig. 10. Intensity of the *002* muscovite peak in samples ground for different time intervals.

The ground sample mixed with water was transferred to a 300 ml beaker. 250 ml of 1N HCl were added. The sample was permitted to react until all the carbonate was removed. This was easily determined because the solution was cloudy as long as CO_2 was being generated because the powdered sample was being carried upward with the bubbles. When the liquid became clear, the acid was decanted or siphoned off and 250 ml of water were added. The sample was permitted to settle until the water was clear and then decanted off. This procedure was done a second time. This second water wash was decanted off and the sample remained at the bottom of the beaker with several ml of water. This was vigorously mixed and a few drops transferred with an eyedropper to a clean glass slide. The sample, when dry, was ready for X-ray diffraction analysis.

During the acid treatment, if the reaction was very slow it was often because magnesite was the carbonate in the soil. Magnesite reacts much less vigorously than calcite or dolomite. We found it useful to speed up this step by keeping the acid solution at 50 to 60° C.

In addition to magnesite another problem was the presence of organic material in some soils. It was particularly prevalent in samples obtained from timbered slopes. Organics could be removed, after splitting, by treating the sample with 12.5% sodium hypochlorite, which is sold under the brand name *Clorox* (note that not all bleach has this strength). When the soil became light in color the bleach was removed and the sample cleaned with water as was done after the acid treatment. We do not recommend removing organics unless the X-ray diffraction patterns are observed to be poor because of high background.

Analysis

The basis of obtaining quantitative information from X-ray diffraction patterns is that the intensity of a peak for a given mineral varies in relation to the amount of that mineral in the sample. However, even if the most intense peak of mineral *X* is considered, it may be more or less intense than the most intense peak of mineral *Y* when each of the minerals is measured at 100% of the sample. In some way, one must obtain a factor to relate these intensities. We chose to use the method of Rex (12) which was developed to handle large volumes of samples obtained during deep-sea drilling. This procedure has the advantage that not all components in the sample need to be considered.

In Montana soils, we chose to consider talc, chlorite and quartz. Kaolinite and mica peaks overlap and were included. The following peaks were used: 25° 2θ peak for chlorite and kaolinite, 26.6° 2θ peak for quartz and mica and 28.6° 2θ peak for talc. In our initial work described in the body of this paper the scans were run from 2° to 35°. This was done because of our unfamiliarity with Montana soils; we did not want to miss any important information. In subsequent work, however, we scanned only two small areas of the pattern: 8° to 11° 2θ and 24° to 29° 2θ. Figure 11 shows the peaks and the measurements made. The intensity was determined by multiplying the height by half width (width of peak a heigth/2).

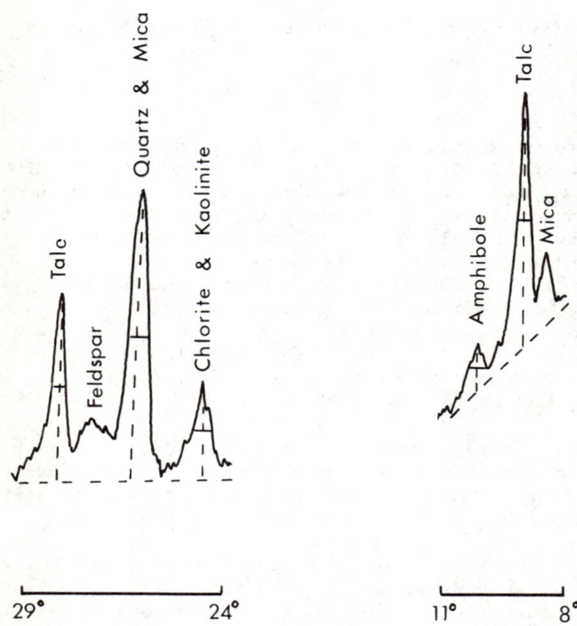

Fig. 11. Typical X-ray diffraction pattern showing measurements made. Solid lines across the peaks are the half widths. Scan speed 2° 2θ/min., chart speed 1 inch/min., 40KV, 25 MA, Cu radiation with Ni filter. 1° Divergence slit and 0.2° Receiving slit.

Percent talc was determined by the following formula:

$$\text{Weight percent talc} = \frac{I_{talc}K_{talc}}{I_{k+chl}K_{k+chl} + I_{qtz}K_{qtz} + I_{talc}K_{talc}} \times 100$$

In the formula *k* indicates kaolinite, *chl* chlorite and *qtz* quartz. It can be seen that in addition to the measured intensities, one needs to determine the K values. The K values must be determined for each set of instrumental settings. The procedure to determine K is simple. The K for quartz is defined to be 1.0. Mixtures were made up containing equal weights of quartz and the mineral in question. For example, suppose we wanted K for chlorite. Chlorite from Montana was ground, dried and mixed 50:50 by weight with quartz. We reground the mixture after weighting to insure that the sample was mixed as uniform as possible.

$$K_{chl} = I_{qtz}/I_{chl}$$

Generally we made up six sample mixtures of each mineral. The values we obtained were:

$$K_{talc} = 1.00$$
$$K_{k+chl} = 1.67$$
$$K_{qtz} = 1.00$$

It cannot be emphasized too strongly that K values are dependent upon experimental conditions and must be determined for each instrument.

There is an overlap of amphibole with talc peaks at 28.6° 2θ, and a correction was made for this. Since the ratio of the 28.6° 2θ peak of amphibole to the 10.4° 2θ peak is 0.6 and the ratio of the 28.6° 2θ peak of talc to the 9.4° 2θ peak is 0.6, the portion of 28.6° 2θ peak which can be attributed to talc is:

$$\frac{I_{9.4^\circ}}{I_{9.4^\circ} + I_{10.4^\circ}}$$

The reason that we use the 9.4° peak to make a correction rather than directly for quantitative evaluation is that the problem of preferred orientation makes it advisable to use peaks in a small 2 theta range, i.e. 28.6° peak. In actual soils, some samples contain more layer silcates than others and will show more preferred orientation than others. A peak at 9° will increase twelve times in intensity as the mineral is changed from random to preferred orientation. At 29° the change will be 4.0 times. The talc intensity was being compared with others at 26.6° and 25°. The latter shows an increase in intensity of 4.6. Thus, if all the peaks are in a small 2 theta range the change in intensity will affect all of them in more nearly the same amount.

A number of simplifications were made to increase the speed of the anlysis. These should be borne in mind when interpreting the results. First, feldspar and all carbonates were eliminated from consideration. Second, although micas were considered because they overlap at the 26.6° peak position, the K value used for this peak was lower than the true value for mica. Experimental results indicate that K for muscovite at 26.6° is 2.5. A correction for the overlap of mica and quartz could have been made by measuring the 20.9° peak for quartz which is one-fourth the intensity of the 26.6° peak for quartz. One could have used that value to determine how much of the 26.6° peak belonged to quartz and how much to mica. We did not chose to do this because the scan time would have increased 50% and because we could not distinguish reasonably quickly which mica was present, biotite, phlogopite or muscovite each having different K values. Observations indicate that samples high in mica are low in talc and, thus, the percent talc in the samples at the low end of the scale may be increased slightly by this simplifications.

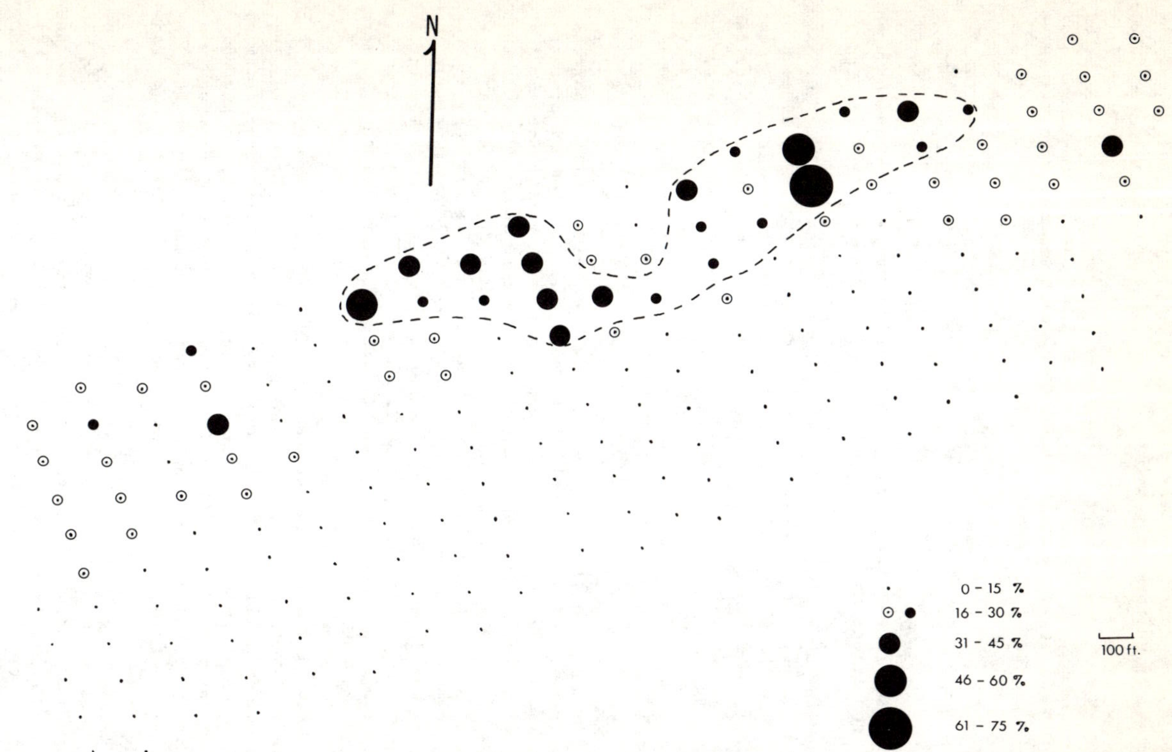

Fig 12. This diagram shows a portion of the area studied in sec. 36, T7S, R7W Madison County, Montana. The solid circles indicate those values above the local threshold of 25% talc which was determined by statistical analysis. The sampling was stopped at the boundary between the marble and the country rock on the north side. The dashed line indicates the anomalous area.

Results

To determine anomalous values, a number of procedures could be used. The simpliest method was to plot the data on a map and contour or otherwise separate the high values from the low values. A cluster of high values has low probability of occurring by chance (Fig. 12). Alternatively, if a large enough number of data points was available, statistical procedures could be used such as those described by Tennant and White (4) and Lepeltier (5). The only modification was that the data plotted on a normal distribution rather than a log-normal distribution found in trace element studies.

Time

Since some readers will be interested in the amount of time necessary to obtain and analyze soil samples, a few words about our experience will be given. It is with some reservations, however, because in many cases we were not using the most efficient methods.

The samples in the field were obtained at a rate of about 15 to 20 per day. This work was done by one geology student with an assistant and included locating the sample positions by brunton and tape. The samples were 100 ft. apart. When samples were obtained at 25 ft. intervals the work went more rapidly.

The preparation of samples, from bag to acid, was done a about 5 samples per hour. In our work grinding was the limiting factor. The time necessary to decant and add water after acid treatment was minor. We did not use a centrifuge in our work since sample gravity settling was compatible with our work schedule. The X-ray diffraction analysis was performed at a rate of eight samples per hour. This was done on an instrument which required each sample to be placed in the beam and the scan range to be set manually.

Most of these procedures could probably be speeded up through use of more efficiently designed laboratories, through use of two micronizers, through use a centrifuge and through use of an X-ray diffraction instrument with automatic sample feed.

References

1. A. M. Blount and Andreas H. Vassiliou, "The Mineralogy and Origin of the Talc Deposits near Winterboro, Alabama," Economic Geology, 75 (1980) 107-116.

2. A. M. Blount and T. Spohn, "Mineralogy of Residual Soils Overlying a Talc Deposit," Mineralium Deposita, 17 (1982) 17-21.

3. Richard H. Olson, "The Geology of Montana Talc Deposits," Eleventh Industrial Minerals Forum (Montana Bureau of Mines, Butte, Montana, 1976) 99-144.

4. C. B. Tennant and M. L. White, "Study of Distribution of Some Geochemical Data," Economic Geology, 54 (1956) 1281-1290.

5. C. Lepeltier, "A Simplified Statistical Treatment of Geochemical Data by Graphical Representation," Economic Geology, 64 (1969) 538-555.

6. Steffan Reed Helbig, "Investigation of a Mineralization Halo Associated with the Talc Deposits Near Winterboro, Alabama, as Expressed in the Residual Soils" (M.S. thesis, Rutgers University, 1983) 34-39.

7. Robert Joseph Piniazkiewicz, "Geology of and Exploration Techniques for Pre-Beltian Talc Deposits of the Malesich Ranch, Ruby Range, Madison County, Montana" (M.S. thesis, University of Arizona, 1984), 75-78.

8. Thomas Spohn, "Mineralogy of the Residual Soils which Overlie a Talc deposit in the State Line Serpentine, Pennsylvania" (M.S. thesis, Rutgers University, 1981) 10-14.

9. K. M. Towe, "Quantitative Clay Petrology: the Trees but not the Forest?," Clays and Clay Minerals, 22 (1974) 375-378.

10. H. P. Klug and L. E. Alexander, X-ray Diffraction Procedures for Polycrystalline and Amorphous Materials (New York, NY: John Wiley and Inc. 1962), 716.

11. Brian McHugh, "X-ray Exploration for Talc Deposits" (M. S. thesis, Eastern Washington University, 1985) 62.

12. R. W. Rex, "X-ray Mineralogy," Initial Reports of Deep Sea Drilling Project, vol. 4 (U.S. Government Printing Office, 1970), 748-753.

13. M. N. A. Peterson, "Expandable Chloritic Clay Minerals from Upper Mississippian Carbonate Rocks of the Cumberland Plateau of Tennessee," American Mineralogist, 46 (1961) 1245-1269.

14. Ronald J. Gibbs, "Error Due to Segregation in Quantitative Clay Mineral X-ray Diffraction Mounting Techniques," American Mineralogist, 50 (1965) 741-751.

15. Alice M. Blount and Andreas H. Vassiliou, "A New Method of Reducing Preferred Orientation in Diffractometer Samples," American Mineralogist, 64 (1979) 922-924.

16. M. L. Jackson, Soil Chemical Analysis--Advanced Course (Madison, Wisconsin: Published by author, 1970), 895.

PRELIMINARY INVESTIGATION OF GRAPHITE RESOURCES IN MICHIGAN

J.Y. Hwang[1], D.H. Carlson[1], A.M. Johnson[2] and J. Van Alstine[3]

[1]Institute of Materials Processing
[2]Mineral Biotechnology Group, Biosource Institute
Michigan Technological University
Houghton, Michigan

[3]Geological Survey Division
Michigan Department of Natural Resources
Marquette, Michigan

Abstract

A preliminary geological, mineralogical, and beneficiation investigation of Michigan graphite resources has been undertaken. More than 3 billion tons of graphite bearing rocks are estimated to occur in a belt near U.S. Highway 41 in Baraga and Marquette Counties. Carbon contents of three field samples collected from different locations range from 17 to 30%. Flotation and heavy liquid separations of these samples were performed at minus 400 mesh. The results vary with the samples. This variation is discussed on the basis of mineralogical examinations.

Introduction

Graphite is an industrial mineral having wide application. For example, it is used for refractories, steelmaking, foundry facings, lubricants, brake lining, batteries, crucibles and pencils. The United States has long relied on foreign countries for its graphite supplies. Madagascar, Sri Lanka and Mexico are the major suppliers.

Economic deposits of graphite have been categorized into five main geological types by Cameron (1) and Graffin (2): (i) Flake graphite disseminated in metamorphosed, silica-rich sedimentary rocks; (ii) Flake graphite disseminated in marble; (iii) Deposits formed by metamorphism of coal or carbon-rich sediments (amorphous); (iv) Veins filling fractures, fissures, and cavities in country rock; and (v) Contact metasomatic or hydrothermal deposits in metamorphosed, calcareous sedimentary rocks.

Graphite is present in the western part of Michigan's Upper Peninsula near L'anse. Small tonnages were mined from the early 1880's until about 1910. At that time the impure graphite was used with no processing except grinding and primarily as a paint for U.S. battleships. A small amount was used for foundry mold facings.

The potential for using this resource to supply domestic needs is not known. Basic information from the geological, mineralogical and beneficiation investigations is necessary for the evaluation. A preliminary study of these investigations is reported here.

Geology

General Description

Outcrops of graphite-rich slate of the amorphous variety (type iii above) have been found over a strike length of 30 miles along the northern margin of the Marquette Trough in Baraga and Marquette Counties (Figure 1). The graphite-rich rock is present in the lower slate member of the Michigamme-Formation, a thick and extensive sedimentary/volcanic formation which consists of five prominent members (3, 4). The straigraphy of the Michigamme Formation is shown in Table I.

Although the geology of the areas has not been completely mapped, the field occurrence of the graphite-bearing strata suggests that the lower slate formation is continuous along the northern margin of the Marquette Trough. The lower slate member is 15 m to 30 m thick.

The area has been regionally metamorphosed which may have played a role in the formation of the graphite. The metamorphism is described by James (5).

Sampling Sites

Three samples, designated as A, B and C, were collected from different locations, as indicated in Figure 1. Sample A was taken from the pit near an old iron ore mine, the Taylor Mine, in Baraga County. The slate here is highly sheared and contains minor lenses and pods of white vein quartz. The cleavage is oriented in an east-west direction and dips about 75 degrees south. The rock is apparently graphitic throughout. A steeply dipping diabase dike is just north of the pit.

Sample B was from a pit (old graphite mine area) about one mile south from sample site A. Here the graphitic slate also strikes east-west and dips south. The rocks at these two sites are similar. A steeply dipping dike which is kaolinized is exposed in the pit at sample site B.

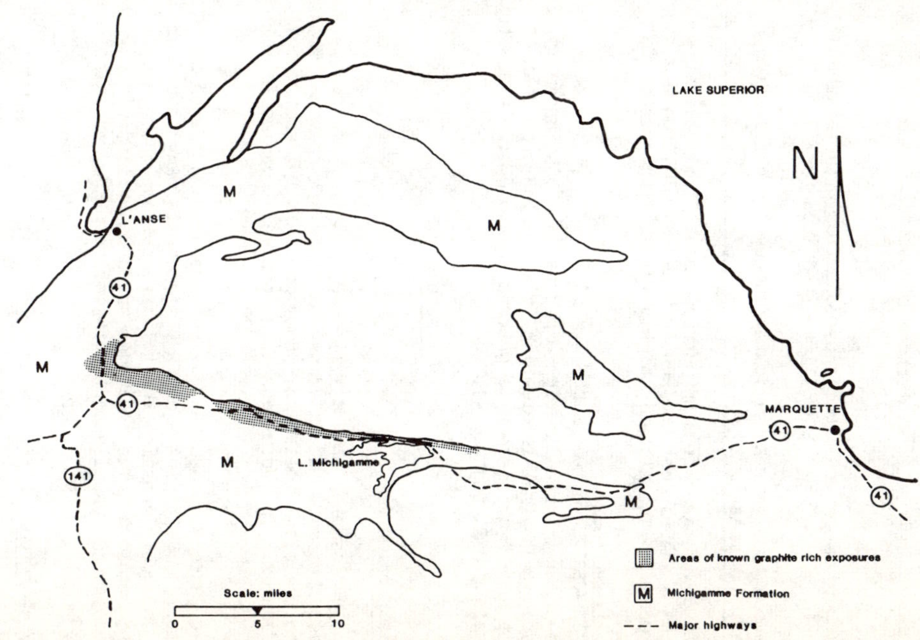

Figure 1. Geologic map of graphite-rich slate and sample collection locations.

Table I. Stratigraphy of the Michigamme Formation

Member	Description
Upper Slate	metamorphosed pelites and graywackes
Bijiki Iron-Formation Member	silicate (cherty) iron formation
Lower Slate	metamorphosed pelites, graywackes with interbeds of pyritic carbonaceous slates (40-90 feet thick)
Clarksburg Volcanics	mafic pyroclastics with interbedded argillites and iron-formation
Greenwood Iron Formation Member	silicate (cherty) iron formation

Sample C was taken from an outcrop about 30 miles east southeast of sample sites A and B, near the Bessie Mine, an old iron ore mine, in west Marquette County. The slate is highly veined with quartz.

Resource Estimation

Areas of known graphite-rich exposures are also shown in Fig. 1. From the available geological information it was calculated that at least 3 billion tons of graphite bearing rock is present in a one-mile wide strip of the 30-mile belt along U.S. Highway 41.

Mineralogy

General Description

All of the samples show a well developed slaty texture, though sample C is much less sheared. Quartz veins and bands are common. Most of them are parallel to the slate foliation. Some occur in folded, lensoid, or other irregular shapes. The thickness of quartz veins varies from 1 cm to less than 1 mm.

Greasy lustered graphite is very common in samples A and B, but it is rare in sample C. Samples A and B also contain many fractures and cavities. Cleavages obliquely crossing the foliation planes and quartz veins can be observed in sample B. The cleavages are often filled with pyrite. Thin limonitic coatings, usually on the foliation plane, are also common in sample B.

Laboratory Analyses

All of the samples were examined by X-ray powder diffraction, optical microscopy, and electron microprobe analyses. The mineral constituents among the three samples are somewhat similar in the major minerals but are quite different in minor minerals. The primary reason for these variations in the minor minerals is probably due to the effect of the intrusive diabase dikes for samples A and B. Some variation may be also attributable to original sedimentary conditions, metamorphism, and/or hydrothermal alteration.

Quartz, clays, mica, and graphite are the major minerals in all three samples. The quartz content is very high in sample C, primarily due to extensive silicification. The clays are mostly illite with lesser amounts of montmorillonite. Kaolinite is present in minor amounts in samples A and C, but it is rich in sample B. This is probably due to the kaolinization of the intrusive dike. Mica is a major component in samples B and C. Lesser amounts were observed in sample A. A qualitative estimation of the relative contents of these minerals based on X-ray diffraction and petrographic studies is presented in Table II. The graphite contents were determined from the carbon analyses.

Table II. Approximate modal analysis of the major mineral contents of three graphite samples.

Mineral	Relative Content (%)		
	Sample A	Sample B	Sample C
Quartz	35	30	55
Clays	20	14	10
Mica	10	13	13
Graphite	27	29	17
Others	8	14	5

Feldspar, chlorite, biotite, and amphiboles are common minor minerals in samples A and B. These minerals probably owe their origin to the intrusive dikes. Pyrite, marcasite, hematite, ilmenite, limonite, rutile, calcite, dolomite, siderite, and apatite are some other minor and trace minerals. These minerals are more common in samples A and B. Sample B especially contains large amounts of hematite. Calcite is relatively enriched in sample A.

Graphite generally occurs as the matrix of the slate, which hosts numerous fine laminations of quartz and silicates. Figure 2 shows a typical appearance of this texture. The gray phases are quartz and silicates (primarily clays and sericite). Graphite is poorly polished and concaved on the section and is dark gray to black on this photograph. A closer look shows that the graphite matrices are aggregates of numerous flaky microcrystalline graphite grains. The sizes of the flaky grains are usually 1 to 5 micrometers. The shapes of the aggregates are irregularly elongated along the slate foliations. The sizes of the aggregates are about 5 to 100 micrometers, but mostly below 30 micrometers.

Quartz occurs in several different forms. It is present in veinlets parallel to and cutting across the foliation planes. Discrete quartz grains are also common. In some samples, silicification is so extensive that most clays are replaced. This silicification also yields very fine quartz, frequently a few micrometers for individual grains, scattered throughout the slate (Figure 3). Some specimens from sample site C even show that clay and graphite occur as remnant inclusions in quartz. Microcrystalline quartz is also common in these inclusions.

Clays and mica generally form thin laminae, interfingered with the graphite. The thicknesses of the laminae vary from a few to more than 100 micrometers. Opaque minerals include pyrite, marcasite, hematite, rutile, and ilmenite. Residual ilmenite is rare and it is frequently replaced by rutile, hematite, and/or pyrite. Hematite and pyrite are most frequently found in the fissures and cavities of the samples. The grain sizes of opaques usually range from 5 to 50 micrometers, though larger grains are also observed. Veins of hematite, replacing the quartz, are particularly evident in sample B.

Figure 2. The photograph shows the texture of graphite-rich slate. Each small division equals 25 micrometers. The upper half contains a quartz band (gray) with scattered opaque minerals (white). The lower part consists of graphite (black) with laminated quartz (gray, higher relief) and clays (dark gray, lower relief).

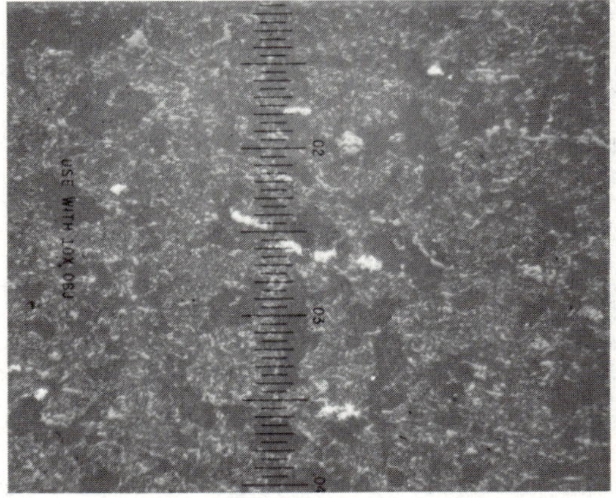

Figure 3. The photograph shows silicification in the graphite-rich slate. Each small division is 12 micrometers. Most clays have been replaced by microcrystalline quarts (gray), as shown by the mottled grains in graphite matrix (black). The white phases include hematite, rutile, and pyrite.

Whole rock chemical analyses of the samples are listed in Table III. These results agree well with the mineralogical study. The strong silicification of sample C is responsible for its high silica content. The introduction of ferromagnesium minerals from the intrusives probably accounts for the higher iron and magnesium contents in samples A and B. The extraordinary high iron content of sample B is explained by the hematite veins and cavity filling.

Table III. Chemical analyses of field samples.

Portion of Head Sample	Component Analyzed	Sample A	Sample B	Sample C
Ignition Loss	Carbon	27.05	29.02	17.00
	Sulfur	0.24	0.28	0.15
	Other	0.78	1.72	0.59
Ash	SiO_2	55.40	50.93	69.93
	Al_2O_3	8.16	7.02	6.64
	CaO	0.145	0.106	0.06
	Na_2O	0.124	0.105	0.35
	MnO	0.02	0.009	0.007
	Fe	1.43	4.72	0.30
	MgO	0.623	0.490	0.16
	Other	6.03	5.61	4.81

Beneficiation

Previous Work

No published work is available on the beneficiation of Michigan graphite. However, a set of laboratory notes dated 1942 was discovered in Michigan Technological University files. This early work showed that samples with head grades of 30% carbon produced flotation concentrates that contained 31-32% carbon. The comments in the file report stated:

> "We did not uncover any reagent or operation which would free the graphite from the gangue and produce a satisfactory concentrate. Also that any amount of grinding would not change the results of the concentrate analysis (grade)".

Current Work

Grinding, heavy liquid separation, and flotation were carried out in the study to gather basic data on the susceptibility of the carbon to beneficiation. This work was done for each of the three samples. These tests were run simultaneously with the mineralogical examination.

Liberation Study. Heavy liquid separations were first conducted in an attempt to determine the liberation size of graphite. Stage-crushed minus 10 mesh charges were dry ground 11 minutes in a laboratory rod mill, and screened into three size fractions; +200, -200 + 400, and -400 mesh. Heavy liquid separation at specific gravity 2.45, utilizing tetrabromoethane adjusted by acetone, were performed to yield sink and float products in a centrifuge. The 2.45 specific gravity was selected since the specific gravities of quartz and graphite are given as 2.65 and 2.3, respectively. The separation results of sample B are shown in Table IV. From these data, it can be determined that the liberation size of graphite is below 400 mesh.

Table IV. Heavy liquid separation of sample B in various size fractions at a specific gravity of 2.45.

Mesh Size	Wt % of Sample	Flotation Fraction Grade, %C	Flotation Fraction Recovery	Sink Fraction Grade, %C	Sink Fraction Recovery
+200	47.5	43.9	18.3	27.1	81.7
-200, +400	17.6	46.2	28.1	24.6	71.9
-400	34.9	60.2	51.8	19.1	48.2
	100.0				

Grinding Test. A series of grinding tests was carried out at various solids contents (70 to 45%) and periods (15 to 60 minutes) to determine the optimum conditions for producing the - 400 mesh products. It was found that wet grinding 400 gram charges at 45% solids for 1 hour in a laboratory rod mill would produce the desired fineness. The size distributions of the ground products, determined by Leeds and Northrup Microtrac, are presented in Figure 4. Sample C had the lowest grindability, although it still resulted in a grinding product with 96.2% passing 31 micrometers and 54.0% passing 11 micrometers. When sample B was ground at the same condition the product was 97.4% passing 31 micrometers and 68.4% passing 11 micrometers.

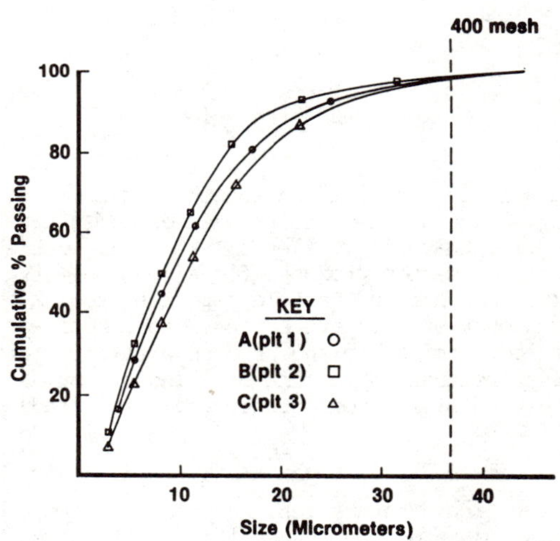

Figure 4. Size distributions of the three samples after one hour grinding.

Separations. Flotation tests were conducted in a Denver laboratory cell. The flowchart of flotation is given in Figure 5. The standard flotation conditions are shown in Table V. Flotation using kerosene as a collector, included a rougher stage and three cleaning stages in each test. No reagent was added in the cleaning stages. The graphs of grade versus recovery of carbon for each sample are shown in Figures 6 through 8. The results indicate that the responses differed for the three samples. Sample B yielded the highest grade (47.5% C) with the highest recovery (55.1%), whereas sample C yielded the lowest grade (38.0% C) with the lowest recovery (16.2%). The flotation response appeared to be unaffected by pH changes through the range of 3 to 9. Poor flotation of sample C at pH 3 was the only exception.

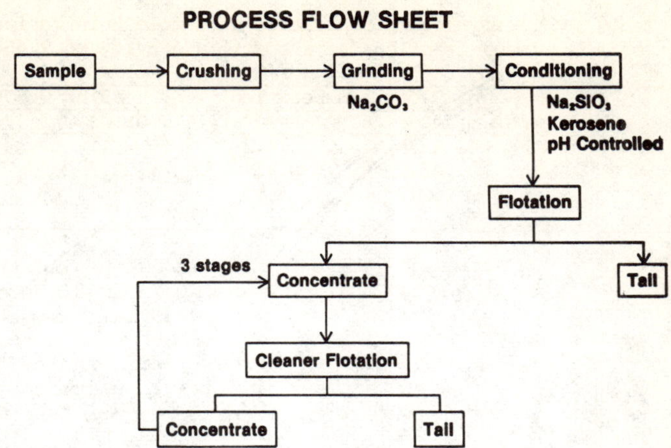

Figure 5. Flowchart of the flotation process.

Table V. Standard conditions for bench scale flotation experiments.

Condition	Remarks
Reagents:	Sodium Carbonate, 2 lb/ton at grinding Sodium Silicate, 1 lb/ton Kerosene, 0.50 lb/ton MIBC, 0.25 lb/ton
pH:	3, 6, and 9; adjusted by NaOH and H_2SO_4
Conditioning Time:	3 minutes
Flotation Time:	3-6 minutes
Temperature:	Room Temperature
Percent Solids:	20%
Impeller Rotation:	1250 rpm
Water:	Deionized Water

The coefficient of separation (CS) was also calculated based on the equation: $CS(\%) = R^1 + R^2 - 100$, where R^1 and R^2 are the recoveries of component 1 (graphite) and 2 (other minerals) in the concentrate and tailing, respectively. The coefficients of separation for sample A, B and C at pH 9 are 5.55, 8.80 and 1.94% , respectively.

Heavy liquid sink-float separations were used to determine the separability limit (maximum grade at maximum recovery). Experiments were performed at three specific gravities (2.45, 2.55 and 2.65) for each ground flotation feed. The results are presented in Figures 6 through 8. Sample B yielded a concentrate containing 62.4% carbon with a 52.9% recovery. The separability of sample C was the poorest as the flotation tests had shown, where the concentrate contained only 54.2% carbon at a very low 6.1% recovery. The coefficients of separation of the heavy liquid tests for samples A, B and C are 32.76, 40.99, and 5.16%, respectively.

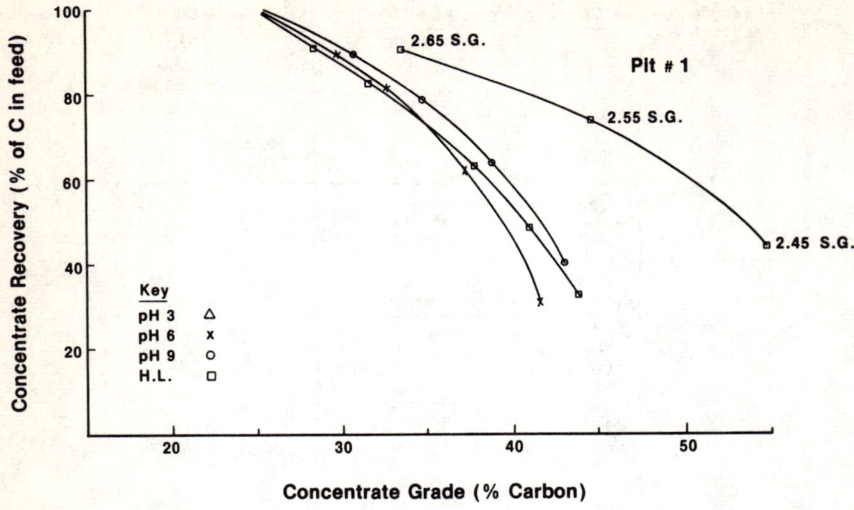

Figure 6. Flotation and heavy liquid separation results of sample A.

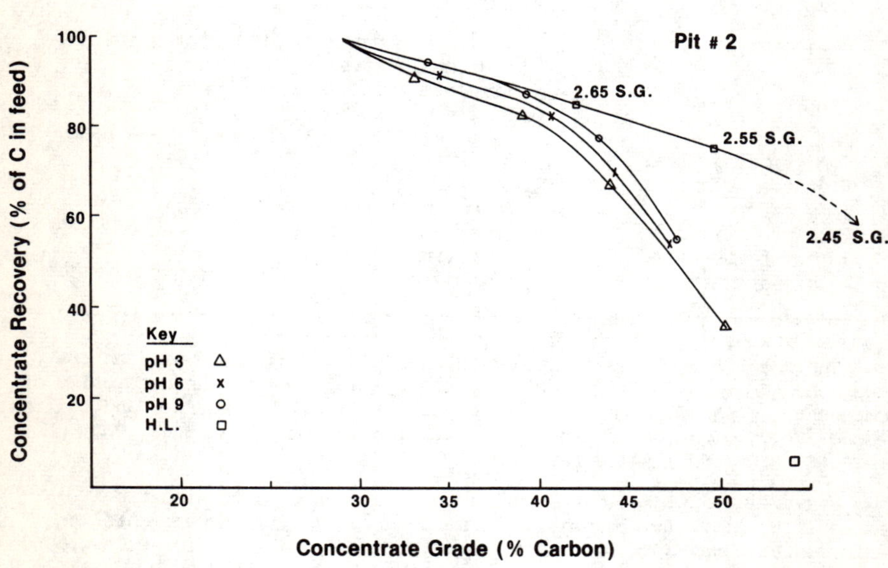

Figure 7. Flotation and heavy liquid separation results of sample B.

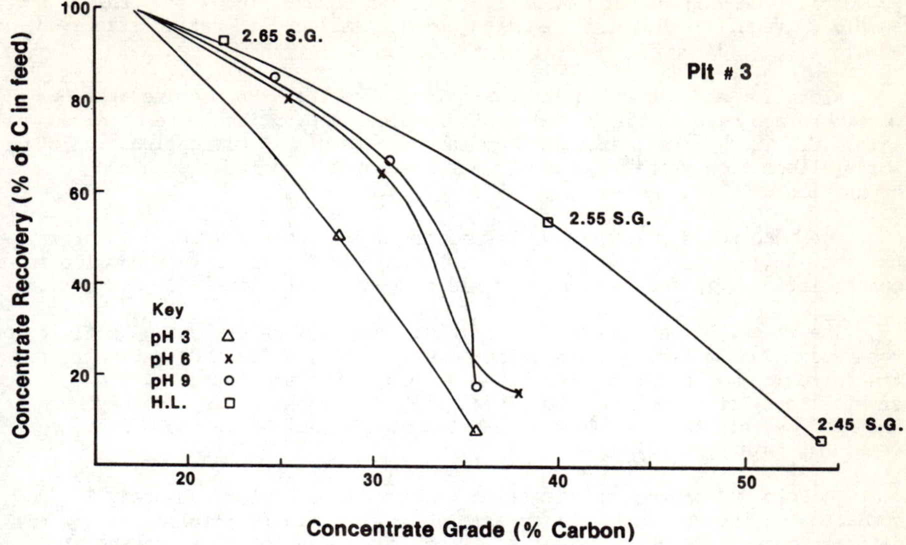

Figure 8. Flotation and heavy liquid separation results of sample C.

Discussion

The graphite resources in the Upper Peninsula of Michigan are interpreted to fall into the third category as described by Cameron (1) and Graffin (2); i.e. deposits formed by the metamorphism of coal or amorphous, carbon-rich sediments (amorphous)." The geological and mineralogical studies both indicated that silicification of these samples is in the order C > A > B. Whole rock chemical analyses also support this order of silicification.

Mineralogical examination shows that silicification has occurred on two scales. More than one half of the quartz is present in coarse veins, with the remainder as fine grained replacements and pore infillings.

Silicification impacts the beneficiation process in at least two ways. First, it causes difficulties in grinding since the soft clays and mica are replaced by hard quartz. The original pores, foliation planes, and fissures are filled with quartz which cements everything together. Greater energy is thus required for grinding. From the particle size distributions of samples ground under the same conditions (see Figure 4), it is obvious that grindability has an inverse relation to silicification. For example, sample C has been subjected to highest degree of silicification and, therefore, yields the lowest fraction of finer particles.

Secondly, silicification results in decreased liberation. The mineralogical study shows that graphite occurs as aggregates containing numerous flaky microcrystalline grains. Usually, the sizes of the aggregates are below 30 micrometers and the sizes of the individual grains are below 5 micrometers. Consequently, it is obvious that the liberation size of graphite has to be far below 30 micrometers, and probably near or below 10 micrometers. Since grinding sample C yields fewer finer particles, less liberated graphite can be expected. The introduction of microcrystalline quartz into the pores of the sample C graphite aggregates requires an even smaller liberation size even more.

Flotation and heavy liquid separations show that the best separations were obtained with sample B whereas the poorest separations were obtained with sample C. This is best explained by the effects of silicification. In other words, liberation probably accounts for most of the variations in the separation results.

The occurrence of two size classes of quartz (coarse veins vs microcrystalline) suggests that two stages of grinding would offer some economics in beneficiation. The coarse quartz could be removed at about 100 mesh size.

The effect of heating of the graphite from intrusive igneous activity on the beneficiation test results is not clear. It is suspected that heat from the intrusion may have reduced the volatile content and increased the crystallinity of graphite. The floatability of graphite has probably been enhanced by this latter effect in samples A and B, although there is no real proof for this assumption.

To help understand the flotation results, a mineralogical study was conducted on the concentrate from sample B. Preliminary results using x-ray diffraction peak height intensities on the feed and flotation concentrate suggest that quartz is preferentially removed relative to clays, mica, and hematite. Possible explanations include the following: (i) these minerals may not be sufficiently liberated from graphite, (ii) the minerals may be coagulated with graphite, (iii) the minerals may not be depressed by the modifiers, (iv) graphite may become smeared on the gangue minerals during the grinding, and (v) there may be mechanical entrapment of some minerals in the froth; e.g., clays and mica. The possibility of the smearing of graphite has been suggested by Zaman (6). Mechanical entrapment of clays and mica in the froth is a common problem in coal flotation. Sodium silicate was utilized as dispersant and depressant in the present study. The flotation results indicate that this reagent was effective in depressing quartz. However, it is recommended that the dosage of this reagent and the introduction of other reagents be considered in future beneficiation tests. Mineralogical examinations indicate that hematite especially is liberated to a greater extent than the flotation results indicate. Therefore, it is apparent that a more detailed study of the flotation chemistry is needed.

Summary and Conclusions

Preliminary geological, mineralogical, and beneficiation investigations were undertaken to evaluate the graphite resources in the Upper Peninsula of Michigan. Graphite was most abundant in the lower slate member of Michigamme Formation, a middle Precambrian sequence of sedimentary and volcanic rocks which has been regionally metamorphosed. This occurrence is interpreted as an amorphous type of deposit. The carbon contents range from 17 to 30%. An initial conservative estimate indicates that the resource amounts to more that 3 billion tons.

Mineralogical and beneficiation test data were collected for samples from three different locations. Beneficiation tests were performed on 400-mesh material. The best result by froth flotation was a 47.5% carbon grade with 55.1% of the carbon recovered in the concentrate. Flotation reagents included sodium carbonate, sodium silicate, kerosene, and methyl isobutyl carbinol (MIBC). Heavy liquid separation of this sample at a specific gravity of 2.45

yielded a 62.4% carbon grade with 52.9% of the carbon recovered in the concentrate.

Possible reasons for the poor separations observed include the problems in liberation, flotation chemistry, froth entrapment and smearing of the graphite or gangue minerals. Insufficient liberation was confirmed by the mineralogical study. Modifications to the flotation chemistry may also be required to improve beneficiation.

Samples showing intense silicification were found to be difficult to beneficiate. The silicification makes grinding and liberation more difficult. This is an important consideration for future research efforts on Michigan graphite.

REFERENCES

(1) E.N. Cameron, "Graphite," in Industrial Minerals and Rocks, 3rd ed., J.S. Gillson et al ed., (New York, NY: AIME, 1960), 455-469.

(2) G.D. Graffin, "Graphite," in Industrial Minerals and Rocks, 5th ed., S.J. Lefond et al ed., Vol. 2 (New York, NY: SME of AIME, 1983), 757-773.

(3) J.S. Klasner and W.F. Cannon, Bedrock Geologic Map of the Southern Part of the Michigamme and Three Lakes Quadrangles, Marquette and Baraga Counties, Michigan, Map I-1078, (U.S. Geological Survey, Reston, VA 22092, 1978).

(4) W.F. Cannon and J.S. Klasner, Bedrock Geologic Map of the Southern Part of the Diorite and Champion Quandranges, Marquette County, Michigan, Map I-1058, (U.S. Geological Survey, Reston, VA 22092, 1977).

(5) H.L. James, "Zones of Regional Metamorphism in the Precambrian of Northern Michigan," Geol. Soc. America Bull., 66 (1955), 1455-1488.

(6) S. Zaman, "Graphite", in SME Mineral Processing Handbook, N.L. Weiss, ed., (New York, NY: SME of AIME, 1985), 28-2.

BARITE ORE POTENTIAL OF TAILINGS PONDS IN

WASHINGTON COUNTY SOUTHEASTERN MISSOURI

Heyward M. Wharton

Missouri Department of Natural Resources
Division of Geology and Land Survey
Box 250, Rolla, Missouri 65401

Abstract

Four large tailings ponds were drilled and evaluated to determine their potential as sources of barite. Over 70 barite tailings ponds are present in the district; many are large and most of them abandoned. There may be as much as 2 million short tons of barite contained in the estimated 40 million tons of tailings in the area. Core samples were analyzed and tonnage-grade estimates prepared for each pond. The most favorable one contained about 1 million tons grading 7 percent barite, or about 82,000 tons of the mineral. Screen tests suggest about half the barite is in minus 400-mesh sizes. Work by the U. S. Bureau of Mines on a Missouri waste-pond sample yielded a marketable barite drilling-mud product after treatment by flotation. The Missouri Survey's study has attracted some attention, but no commercial recovery projects have been attempted to date.

Introduction

The Missouri Geological Survey investigated the ore potential of barite tailings ponds in the Washington County area. Four large ponds were drilled on close centers and tonnage-grade estimates were prepared. Samples were analyzed at the U. S. Bureau of Mines Rolla Research Center. A detailed report of the investigation was issued in 1972 (1). The study indicated there is a substantial tonnage of barite available for recovery in the tailings ponds.

Location and Mining History

The Washington County or Southeastern Missouri barite district is about 50 miles southwest of St. Louis and centers around the town of Potosi, the county seat (figure 1). The barite-producing area lies within the Southeast Missouri Lead district, one of the great metal-mining centers. Shallow mines around Potosi were the principal lead producers in the state prior to 1850. Barite, traditionally called "tiff" by the miners, nearly always accompanies galena in the Washington County lead ores, and was discarded as useless in those days. Tiff mining was initiated around 1850, at about the time that lead mining in the county began its decline (2). The main use for barite at that time was in paints.

In the early days, lead and barite mining usually consisted of sinking shallow pits and shafts in the residual surface clays. Fragments of galena and barite were picked out of the clay and cleaned by hand. Hand digging for barite continued as an important industry in Washington County until 1942. Large strip mines and washer plants have been in use since then. Much of the production today comes from remining old hand diggings and some of the earlier strip-mine areas.

The Washington County district has been a world leader in barite production during much of the 20th century, according to D. A. Brobst, longtime commodity specialist for the U. S. Geological Survey (3). Missouri's annual production ranked either first or second in the United States among the other important producers - Arkansas, Georgia, Nevada, and Tennessee - from 1885 until 1983. The state's cumulative total output of barite from the start of record keeping in 1872 through 1980 was 12,894,688 short tons, the most of any state. However, Nevada took over the lead in 1981 after its mines and plants had produced a record 2,482,000 tons of barite during the year (4). By contrast, Missouri's annual output averaged only about 167,000 tons per year during the 1970's, a considerable decline from the 285,000 tons per year average posted during the 1960's.

Geology and Ore Deposits

The barite ores are of the residual type. Fragments of the mineral occur in red and brown surface clays that have accumulated from weathering of the Potosi and Eminence Dolomites, the uppermost formations in the Upper Cambrian Series (figure 2). Bedrock exposures of these rocks in roadcuts and strip pits often have fractures and other openings filled with barite. The porous horizons that are mineralized appear to result primarily from solution activity. The residual ores average about 10 feet thick, ranging from a few feet up to 25 or 30 feet.

The lead deposits being mined today in the Southeast Missouri district are all in the Cambrian Bonneterre Formation, the lowest carbonate unit in the section (figure 2). The nearest mine sites, at Indian Creek and Viburnum, are shown in the southwestern part of figure 1. Barite is absent in

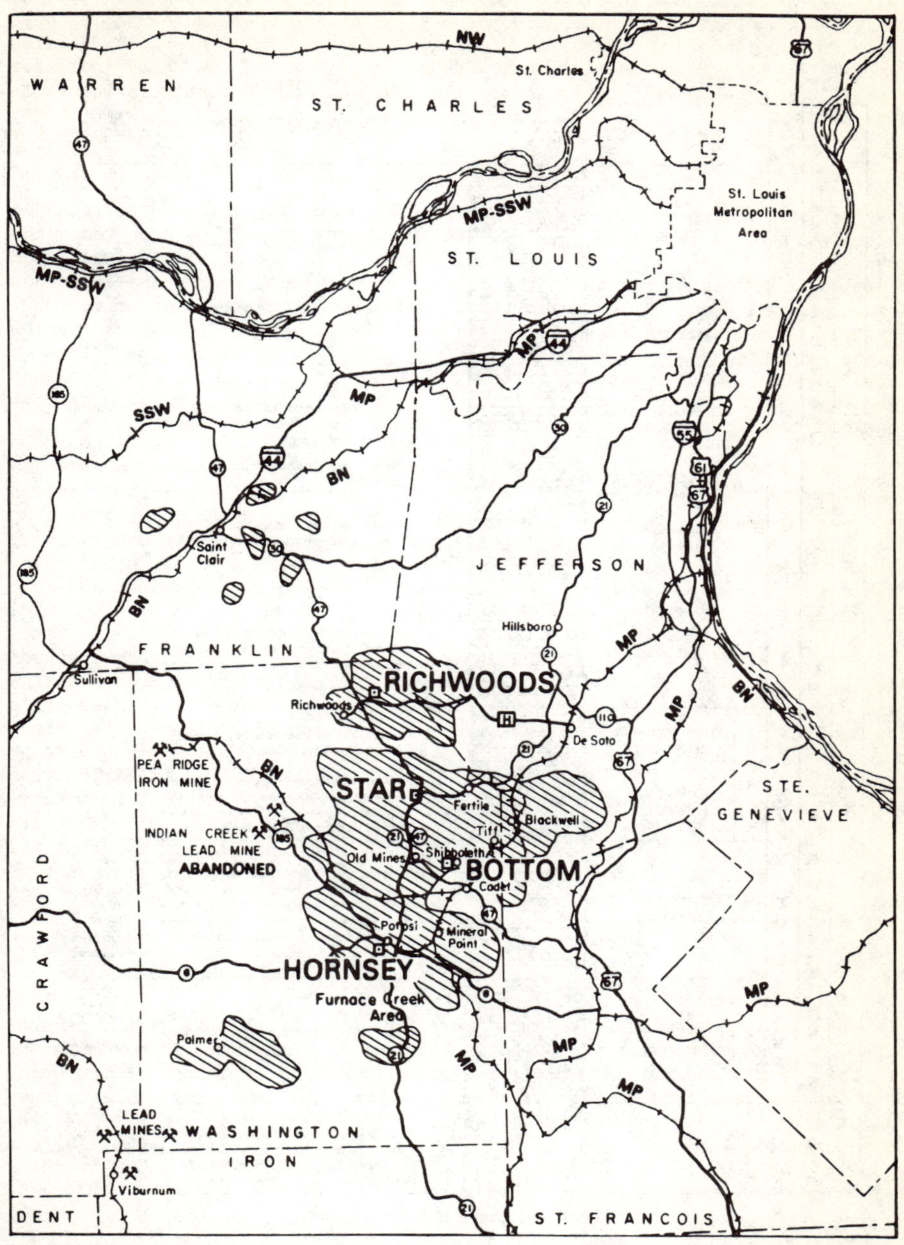

Figure 1. Washington County barite district in southeastern Missouri with locations of the four tailings ponds test-drilled by the Missouri Geological Survey.

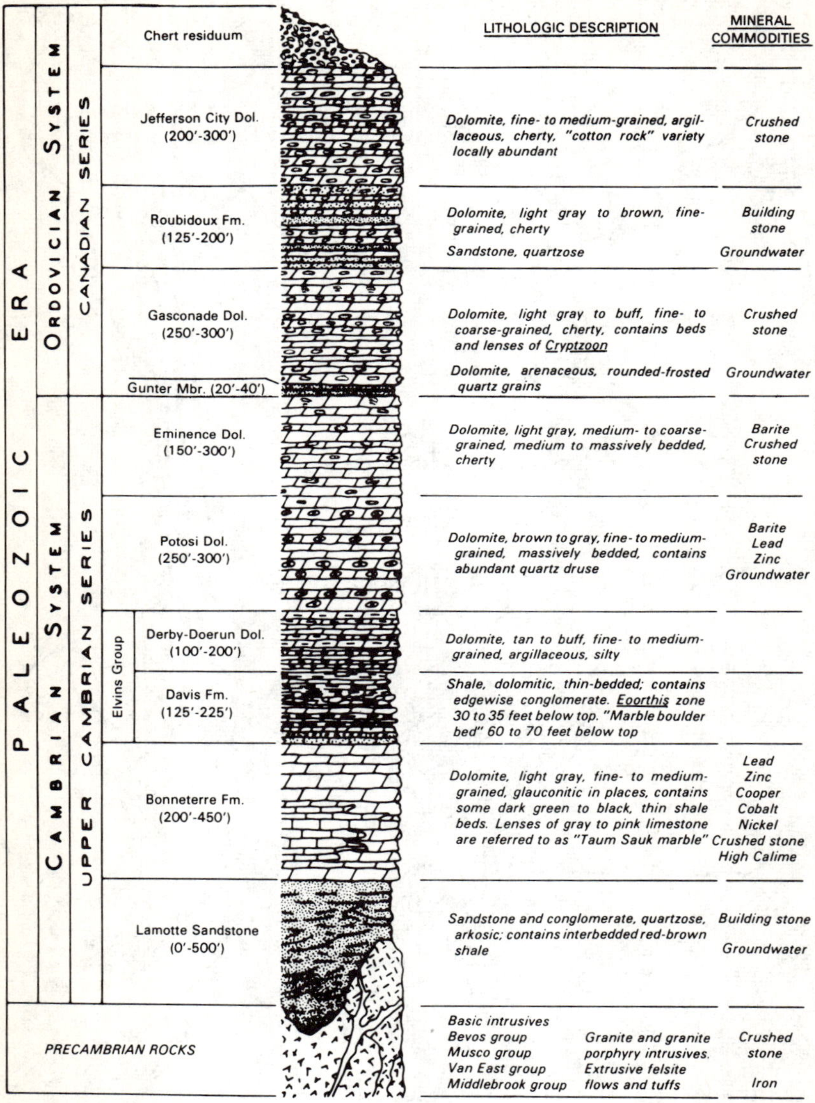

Era	System	Series	Unit	LITHOLOGIC DESCRIPTION	MINERAL COMMODITIES
PALEOZOIC ERA	ORDOVICIAN SYSTEM	CANADIAN SERIES	Chert residuum		
			Jefferson City Dol. (200'-300')	*Dolomite, fine- to medium-grained, argillaceous, cherty, "cotton rock" variety locally abundant*	*Crushed stone*
			Roubidoux Fm. (125'-200')	*Dolomite, light gray to brown, fine-grained, cherty* *Sandstone, quartzose*	*Building stone* *Groundwater*
			Gasconade Dol. (250'-300')	*Dolomite, light gray to buff, fine- to coarse-grained, cherty, contains beds and lenses of <u>Cryptzoon</u>*	*Crushed stone*
			Gunter Mbr. (20'-40')	*Dolomite, arenaceous, rounded-frosted quartz grains*	*Groundwater*
	CAMBRIAN SYSTEM	UPPER CAMBRIAN SERIES	Eminence Dol. (150'-300')	*Dolomite, light gray, medium- to coarse-grained, medium to massively bedded, cherty*	*Barite* *Crushed stone*
			Potosi Dol. (250'-300')	*Dolomite, brown to gray, fine- to medium-grained, massively bedded, contains abundant quartz druse*	*Barite* *Lead* *Zinc* *Groundwater*
			Elvins Group: Derby-Doerun Dol. (100'-200')	*Dolomite, tan to buff, fine- to medium-grained, argillaceous, silty*	
			Elvins Group: Davis Fm. (125'-225')	*Shale, dolomitic, thin-bedded; contains edgewise conglomerate. <u>Eoorthis</u> zone 30 to 35 feet below top. "Marble boulder bed" 60 to 70 feet below top*	
			Bonneterre Fm. (200'-450')	*Dolomite, light gray, fine- to medium-grained, glauconitic in places, contains some dark green to black, thin shale beds. Lenses of gray to pink limestone are referred to as "Taum Sauk marble"*	*Lead* *Zinc* *Cooper* *Cobalt* *Nickel* *Crushed stone* *High Calime*
			Lamotte Sandstone (0'-500')	*Sandstone and conglomerate, quartzose, arkosic; contains interbedded red-brown shale*	*Building stone* *Groundwater*
PRECAMBRIAN ROCKS				*Basic intrusives* *Bevos group* — *Granite and granite* *Musco group* — *porphyry intrusives.* *Van East group* — *Extrusive felsite* *Middlebrook group* — *flows and tuffs*	*Crushed stone* *Iron*

Figure 2. Generalized stratigraphic column of southeastern Missouri.

the Bonneterre lead deposits, but it is the major constituent in the barite-galena deposits hosted by the Potosi and Eminence formations. Total mine production of barite and lead to date from southeastern Missouri are in the same order of magnitude: approximately 13 million short tons of barite versus 17.6 million tons of lead.

Both the lead and the bedrock barite-lead occurrences are stratiform, and are generally considered to be Mississippi Valley-type deposits. Detailed studies in the barite district by Wagner (5) suggest that fault and fracture systems have acted as channelways for the mineralizing solutions, which have then migrated laterally into preferred, porous and permeable horizons where mineral deposition took place. Residual ores were formed when the mineralized horizons were subjected to surface weathering and removal of the soluble carbonates in the host-rock dolomites. There has been speculation about whether the mineralization in the Bonneterre and Potosi-Eminence are related or completely separate. The evidence to date is inconclusive, one way or the other (6, 7).

Mine - Plant Operations and Tailings Ponds

The barite-washer plants are equipped with rotary breakers, log washers, trommel screens and jigs. Typical large plants can treat about 120 cubic yards of ore per hour and require up to 5000 gallons of water per minute. Rated capacities are from 25,000 to 30,000 tons of barite per year. Three twin plants about double that size were built in 1979-1980, but their operations were either suspended or sharply curtailed in 1982. The vigorous washing and tumbling in the processing equipment are such that barite recovery is sacrificed for throughput.

The surficial clays containing the barite are excavated by front-end loaders or power shovels, and loaded into large off-the-road dump trucks for transport to the washer plants. The barite ore is dumped and washed into large rotary breakers using high-pressure water. Material coarser than 3 inches is rejected and the undersize flows into log washers. The log washer overflow, which is mostly finer than 10 mesh, is piped by gravity or pumped to the tailings pond. The underflow is fed into trommel screens where material coarser than about 1 inch is removed and conveyed into a storage bin for road and dike construction material. Undersize material, usually minus 1 inch, is washed through the screens and fed into the jigs where the barite is recovered. Waste gravel from the jigs is also retained for construction uses.

The log washer overflow product makes up the bulk of the material in the tailings ponds. There are about 75 ponds in the district with over 40 classified as large since each is believed to contain half a million tons of tailings or more.

Drilling Project

Four older tailings ponds with stabilized surfaces were selected for drilling (figure 1). A skid-mounted core drill and 3-inch diameter Shelby tubes were used to collect samples. A 30-inch extension was connected above the sampling tube to help prevent contamination by caved and squeezed-in material.

The Bottom pond near Cadet was the largest and most economically attractive of the four tailings ponds that were drilled, so it has been singled out for description in the balance of this report. It has a 52-acre surface area and 77 holes were drilled to test it (figure 3). The holes averaged about 15 feet deep; maximum depth was 30 feet. In the other ponds, maximum depths ranged between 50 and 65 feet.

Grade Zones in the Bottom Pond

The specific gravity of barite is about 4.5, exceptionally high for a nonmetallic mineral. As expected, barite grades determined from the drill samples fall off systematically with distance away from the washer discharge points. Location of the mill and the 4 grades zones in the Bottom pond are shown in figure 3.

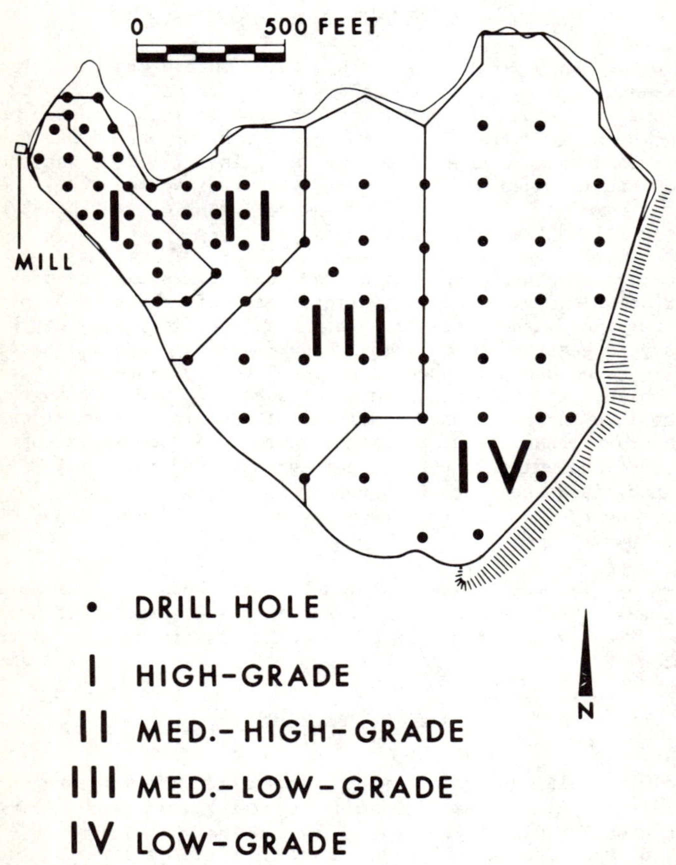

Figure 3. Drill pattern and grade zones on the Bottom pond.

A total of 1,090,000 tons of tailings was calculated for the entire content of the Bottom pond. An estimate of 86,000 tons grading 18 percent barite was made for Area I, 133,000 tons grading 11 percent for Area II, 353,000 tons grading 8 percent for Area III, and 518,000 tons grading about 5 percent for Area IV. Average grade for the pond as a whole was 7.2 percent barite, equivalent to nearly 82,000 tons of the mineral. Estimates of barite contents by grade zones are 16,000 in Area I, 15,000 tons in Area II, 27,000 tons in Area III, and 24,000 tons in Area IV. The tonnage-grade estimates for the pond are summarized in figure 4. The 572,000 tons of tailings in Areas I, II, and III combined, have an average grade of 10.2 percent $BaSO_4$ and contain about 58,000 tons of barite. This material is generally higher grade than average mine-run ore in the district, estimated at around 5 percent barite.

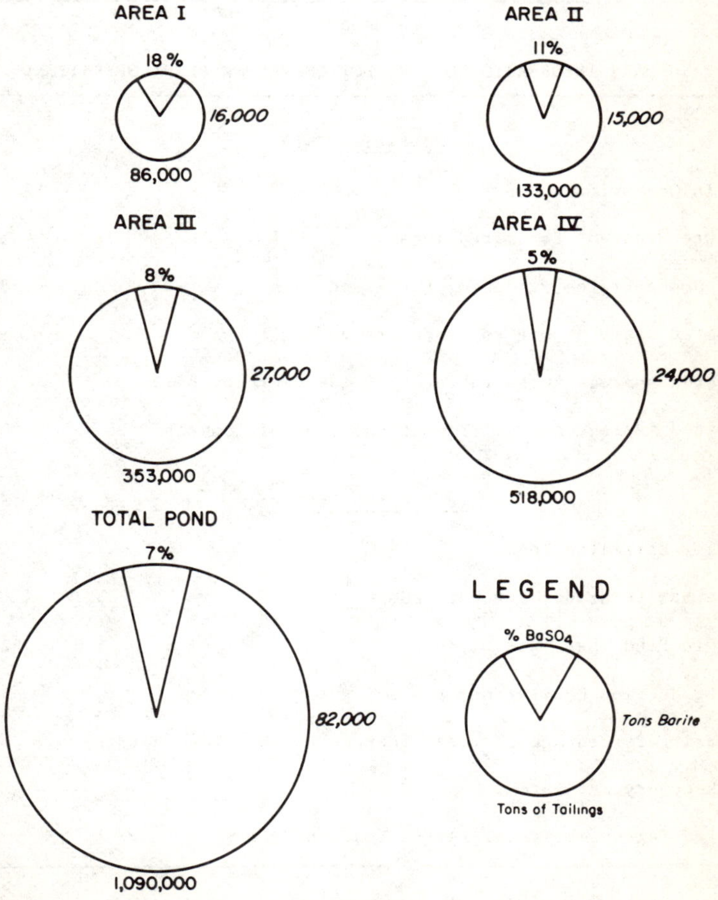

Figure 4. Tonnage-grade estimates for the Bottom pond.

Summary Information on the Bottom Plant and Pond

Over a 15-year period, the Bottom washer plant treated about 4 1/2 million tons of ore and produced roughly 300,000 tons of barite concentrates (table I). The ratio of ore to concentrates was 15 to 1, and recoverable ore grade about 6.7 percent barite for the lifetime of the operations. Plant output averaged around 20,000 tons of concentrates annually. The writer is indebted to Mr. Clarence C. Houk, manager in Missouri, and to the parent NL Baroid organization in Houston, Texas, for granting permission to the Survey to drill this pond, and for supplying the basic operating data given in the table.

Several important parameters for the Bottom pond are listed in the second entry of table I. The tonnage-grade estimates given in the previous section are repeated. The 82,000 tons of barite believed to be present in the pond are equivalent to about 4 years output of the plant. Metallurgical recovery was estimated at 70 percent. It follows that about 20 percent of the barite originally present in the ore is now in the tailings pond; the remaining 10 percent is assumed to have been lost with the gravels and coarser waste rock, much of which is used in the dike construction.

Table I

SUMMARY INFORMATION ON THE BOTTOM WASHER PLANT OPERATIONS

Operating Data	
Years in Operation	15:1945-1959
Crude Ore Treated, Estimated Tons	4,500,000
Barite Concentrates Produced, Estimated Tons	300,000
Calculated Ratio of Ore to Concentrates	15:1
Calculated Average Recoverable Ore Grade, Percent $BaSO_4$	6.7%
Calculated Average Annual Production, Tons of Concentrates	20,000
Tailings Pond Data	
Tailings, Estimated Tons	1,090,000
Average Barite Grade, Percent $BaSO_4$	7.2%
Barite in Pond, Estimated Tons	82,000
Plant Production Equivalent, Years	4
Pond Barite Percentage of Total (Concentrates + Pond Barite)	21%
*Plant Recovery Estimate	70%
Estimated Percent of Total Barite Lost in Pond	20%

*Assumes that about 1/3 of the barite losses are in the gravels and coarser waste rock; the remaining 2/3 in the tailings pond.

Size and Grade Information on the Composite Samples

Thirty or more drill samples from each grade zone in the Bottom pond were composited to make up representative samples of the tailings for metallurgical testing. Screen test results for each of the area samples are given in table II. The composite sample assays agree closely with the barite grades calculated for each area in the tonnage estimates, as indicated in table III. Weight percentages (from table II) and barite assays of the plus and minus 400-mesh fractions of the composite samples are shown in the upper entry in table III. The fall off in grain sizes and barite contents with distance from the washer discharge are well illustrated. The uniformly low-grade assays of the minus 400-mesh fractions are also noteworthy. The latter fraction was calculated to comprise about 84 percent of the tailings in the pond as a whole.

Table II

SCREEN ANALYSES, COMPOSITE SAMPLES OF THE BOTTOM POND

Mesh	Area I High-Grade	Area II Medium-Grade	Area III Med-Low-Grade	Area IV Low-Grade
+10	1.42%	0.65%	0.10%	0.02%
-10 +14	.89	.61	.13	.03
-14 +20	1.34	1.00	.18	.04
-20 +28	2.05	1.61	.28	.06
-28 +35	3.16	2.61	.59	.08
-35 +48	5.53	4.40	1.38	.18
-48 +65	7.56	5.65	2.49	.48
-65 +100	5.71	4.69	2.94	.90
-100 +150	5.15	3.92	2.84	1.27
-150 +200	2.65	2.17	2.01	1.01
-200 +270	2.42	1.77	1.97	1.31
-270 +325	1.80	1.47	1.77	1.37
-325 +400	.62	.53	.61	.36
-400	59.70	68.93	82.71	92.89
	100.00%	100.00%	100.00%	100.00%

Table III

SIZE-GRADE INFORMATION FOR THE COMPOSITE TAILINGS SAMPLES AND FOR THE HEAVY MEDIA BARITE SEPARATES

Barite Assays of Composite Samples and Plus & Minus 400 mesh-Fractions

	Area I		Area II		Area III		Area IV		Pond Total Calculated	
Tonnage Estimate Assays:	18.2% $BaSO_4$		11.3% $BaSO_4$		7.7% $BaSO_4$		4.6% $BaSO_4$		7.2% $BaSO_4$	
Composite Sample Assays:	20.1% $BaSO_4$		11.2% $BaSO_4$		6.7% $BaSO_4$		4.9% $BaSO_4$		7.5% $BaSO_4$	
Mesh	(Wt.%)	Assays	(Wt.%)	Assays	(Wt.%)	Assays	(Wt.%)	Assays	(Wt.%)	Assays
+400	(40)	42% $BaSO_4$	(31)	28% $BaSO_4$	(17)	21% $BaSO_4$	(7)	20% $BaSO_4$	(16)	27.5% $BaSO_4$
-400	(60)	5.3%	(69)	3.7%	(83)	3.7%	(93)	3.7%	(84)	3.7%
	(100)		(100)		(100)		(100)		(100)	

Weight Percentages of Total Barite by Sizes in the HMS Samples

Mesh	Area I	Area II	Area III	Area IV	Pond Total Calculated
+65	25.3%	14.7%	1.1%	0.3%	4.3%
-65 +400	59.0	62.6	53.4	28.9	43.3%
-400	15.7	22.7	45.5	70.8	52.4%
	100.0%	100.0%	100.0%	100.0%	100.0%
*Pounds of Barite Per Cubic Yard of Tailings:	491	260	160	90	149

*Based on the density determinations and barite assays developed in the tonnage-grade estimates.

A heavy media was used to recover barite fractions from each composite sample. Screen tests were then made on these fractions to determine the weight percentages of the barite in each area, in the plus 65 mesh, minus 65 plus 400 mesh, and minus 400-mesh size fractions (table III). The 65-mesh break was chosen because it is near the lower size limit for gravity concentration and the upper size limit for froth flotation. Once again there is a marked decline in grain size, and in effect of grade, with distance from the plant discharge points. There is practically no plus 65-mesh barite in Areas III and IV, which include about 80 percent of the total tons of tailings. In Areas I and II combined, less than 20 percent of the barite is plus 65 mesh. For the pond as a whole, only about 4 percent of the barite is plus 65 mesh, whereas 43 percent is minus 65 plus 400 mesh and about 52 percent is minus 400 mesh.

In the final entry, the pounds of barite per cubic yard of tailings are given for each grade zone and for the pond as a whole. These values are based on the assays and density determinations developed in the tonnage-grade estimates as explained in the footnote. The approximate cutoff grade in the district, 100 lbs of barite per cubic yard, was exceeded in all but Area IV.

Concluding Remarks

The tailings pond study had some important implications for the Missouri barite industry. The uniformly large barite contents of the 4 randomly selected ponds that were tested confirmed that the ponds in the area have high potential as future sources of barite. It follows, then, that the washer-jig plant recoveries are quite low, and probably range between 65 and 70 percent of the feed ore grades, as surmised by Sackett in 1958 (2). A third consideration is the barite particle sizes in the plant discharges and ponds. In the Bottom pond, over 90 percent of the barite was minus 65 mesh in size, and about half was minus 400 mesh (table III). Those grain sizes are too small for recovering by jigs, so it will be necessary to use froth flotation or some of the more selective gravity techniques to treat plant discharges or the ponded tailings.

The roughly 75 barite waste ponds in the Washington County area are estimated to contain as much as 40 million tons of tailings. Assuming an average grade of 5 percent, the ponds would contain at least 2 million tons of barite, equivalent to over 10 years' supply at normal production rates for the district. The economic benefits that would accrue from extending the productive life of the district by this means was in fact the main justification for funding the investigation.

The barite contents in Areas I and II of the Bottom pond are more than double mine-run ore grades in the district (table III). It is surprising, then, that no attempts have been made to exploit any of the high-grade areas in the Missouri ponds; all the more so, since it has been done successfully in the Cartersville area in Georgia. Metallurgical tests were made on splits of the Survey's core samples by NL Baroid and other property owners, but nothing further. In mid-1982, an out-of-state company leased 2 ponds and was preparing to set up a portable flotation plant, but the project was abruptly cancelled because of the serious economic recession.

Another response to the demonstrated poor barite recoveries of the conventional plants took place in the 1980-1982 period by two newly arrived companies. The IMCO Services new Apex washer at Mineral Point, Missouri was equipped with a flotation plant set up to recover barite from plant discharges. Unfortunately, operations at the Apex facility were suspended indefinitely in 1982 before the new unit could be adequately tested. De Soto

Mining Company, operating new plants in the Richwoods, Missouri area, also considered installing flotation circuits, but plans were scrapped when the barite market was decimated by the 1982 recession. Research had been done by the U. S. Bureau of Mines on recovering ultrafine barite from waste ponds at its Tuscaloosa Research Center in fiscal year 1979. Two reports were issued in 1982 (8,9). In the first report, flotation and selective flocculation were used on a barite sample from Nevada which included extremely fine sizes. A high recovery was achieved and a marketable drilling-mud product was prepared. The second report included flotation tests made on a Missouri tailings pond sample among others. Again, results were reasonably good.

The Survey's pond study has served as a reminder that a large, untapped source of barite remains in the Washington County district. Missouri barite operators seem to think the tailings ponds will eventually be reworked, but probably not in the near future, because depressed conditions in the industry are expected to continue for some time. More concern is expressed about the environmental aspects of exploiting the tailings than about satisfactory excavation and recovery techniques.

REFERENCES

1. Heyward M. Wharton, Barite Ore Potential of Four Tailings Ponds, Washington County Barite District, Missouri (Missouri Geological Survey & Water Resources Rept. of Inv. 53, 1972), 91.

2. E. L. H. Sackett, "Barite in Washington County, Missouri - History and Development" (Paper presented at AIME Meeting, St. Louis, Missouri, 24 October 1958), 9.

3. Donald A. Brobst, Barite: World Production, Reserves, and Future Prospects (U. S. Geological Survey Bull. 1321, 1970), 46 p.

4. S. G. Ampian, Barite (Minerals Yearbook vol. I, U. S. Bureau of Mines, 1982), 113-124.

5. R. Joseph Wagner, "Stratigraphic and Structural Controls and Genesis of Barite Deposits in Washington County, Missouri" (Ph.D. thesis, University of Michigan, 1973), 265.

6. J. Ruiz, W. C. Kelly and C. J. Kaiser, "Strontium Isotopic Evidence for the Origin of Barites and Sulfides from the Mississippi Valley-Type Ore Deposits in Southeast Missouri - A Discussion," Economic Geology, 80 (3) (1985), 773-775.

7. C. J. Kaiser, W. C. Kelly, R. J. Wagner, and W. C. Shanks III, "Geologic and Geochemical Controls of Mineralization in the Southeast Missouri Barite District," Economic Geology, 81 (1986), in press.

8. W. E. Lamont and G. V. Sullivan, Recovery of Ultrafine Barite from Mill Wastes (U. S. Bureau of Mines Rept. of Inv. 8668, 1982), 12.

9. W. E. Lamont and G. V. Sullivan, Recovery of High-Grade Barite from Waste Pond Materials (U. S. Bureau of Mines Rept. of Inv. 8673, 1982), 13.

Process Mineralogy Applications to Coal

MINERALOGY AND DEEP-CLEANING OF CANADIAN HIGH-SULPHUR COALS

G. I. Mathieu and P. R. Mainwaring

Because of environmental and energy problems, a revolution is taking place in the physical beneficiation of coal. This requires advanced developments in multidisciplinary science and engineering technologies, in order to solve the variety of problems that face the coal preparation engineer in producing an environmentally acceptable fuel at an economically competitive price.

The processing of fine coal, for instance, has been a long-standing problem in the industry which was partly attributed to insufficient knowledge of the mineralogy of coal and associated minerals, as well as inadequate application of beneficiation methods. A study, was conducted to characterize two Canadian coals using optical and scanning electron microscope techniques, and subsequently using the mineralogical findings to develop the best procedure for ultimate cleaning of coal. The samples under investigation contain (1) 26% ash and 2% sulphur, and (2) 20% ash and 5.2% sulphur.

Using a progressive grinding and processing approach, it was possible to remove more than 95% of the ash and sulphur-bearing minerals from both coals with satisfactory thermal yield and acceptable costs. The end products from the coals analyzed 0.8 to 1.6% sulphur and ash, whereas the treatment methods utilized were successive flotation and magnetic separation when liberation of the minerals progressed as predicted by the mineralogical studies.

INTRODUCTION

The purpose of this study was to investigate a practical method for cleaning Canadian high-sulphur coals by physical means, and to find the best technology for achieving the desired coal size and purity. Coal is abundant in eastern Canada with total reserves of nearly 5 billion tons, but this coal has a high sulphur content at 2-10% weight. The highest concentration of sulphur is found in the Minto coalfield (New Brunswick), which is exploited on a limited scale at 0.5 million tons/year for local power generation. The Cape Breton Development Corporation (DEVCO) is the major coal producer of the Atlantic region with two mines in operation and another under development, all located in the Sydney coalfield of northeastern Nova Scotia (Fig. 1).

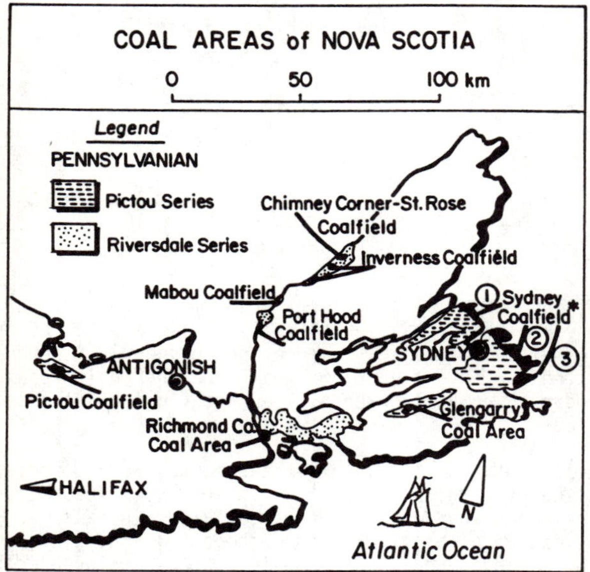

Fig. 1 **Sydney coalfield**

①-Prince, ②-Lingan, ③-Donkin

The research work was initiated on DEVCO coals because of the company's important position in the eastern Canada market and its current expansion program, as is shown by the company's current and predicted production figures for the next decade (Table I). Estimates of the sulphur content of the deposits are also given.

Currently, coal from the Lingan Mine is beneficiated to metallurgical grade by conventional means including heavy media separation, cycloning and flotation. A middling product of thermal grade is also produced. The Prince washery uses only screens and Baum jigs, and all of its production is of thermal quality. The composite thermal coal analyzes 2.5-3.0% total sulphur; pyrite accounts for approximately 65% of the sulphur.

Since part of the future production is intended to be used for the conversion of Nova Scotia's thermal generating stations from oil to coal (possibly in the form of coal-water slurry), lower sulphur products will be required in order to meet the proposed National Guidelines on sulphur

Table I - DEVCO's Coal Production and Expansion

Mines	Reserves (million tonnes)	Total sulphur analysis (% St)	Production (Mt/year) 1985	1987	1995	Life expectancy (years)
Lingan	100	1.8-2.0	1.4	3.0*	3.0*	65
Prince	70	4.5-5.0	1.0	1.0	1.0	70
Donkin	450	3.0-3.5	-	-	1.5	300
Total	620	3.2	2.4	4.0	5.5	-

*Including the Phalen seam - production scheduled to start in 1987.

dioxide (SO_2) emissions, i.e., 0.5-1.6 lb/10^6 BTU (depending on the coal source), which corresponds to 0.4-1.4% sulphur content for 14 000 BTU/lb coal. To satisfy such requirements, essentially all of the pyrite sulphur will have to be removed from the DEVCO coals. Furthermore, if the finished coal product is to be used as a burning slurry, a high reduction of the other mineral matter will also be necessary and a rigid size control will be required.

To characterize the coals, several vertical channels were cut across various faces of the DEVCO's operating mines. Each 0.15 m vertical section was analyzed for ash content and sulphur forms and examined for pyrite-marcasite grain sizes. Furthermore, extensive petrographic and mineralogical studies were conducted on the coal macerals and associated minerals at the Atlantic Coal Institute (1) and CANMET. On the basis of the mineralogical findings, a flowsheet was designed and tested.

General Characteristics of Coals

The sulphur forms in coals are pyrite, organic, and sulphate, all responsible for SO_2 emission when burning coal in the atmosphere. Pyrite sulphur is a constituent of sulphide minerals (pyrite and marcasite, in particular), whereas sulphate is one of their oxidation products. All of these phases are physically removable from the coal provided they are sufficiently liberated by grinding. Conversely, the organic sulphur originates from plant material and is bound to the molecular structure of the macerals. Its removal requires breaking the chemical bonds. Some representative results, which are illustrated in Fig. 2, indicate relatively constant organic sulphur contents of 0.6% for the Lingan seam and of 1.5% for the Prince, but with a random distribution of pyritic sulphur. The sulphate content is low in the Prince coal (0.2%) and almost negligible in the Lingan (0.05%)

Size measurements of the pyrite-marcasite grains in each 0.15 M section were made and dimensions were averaged for the top, middle and bottom zones of the deposits. The channel samples were also submitted to low-temperature ashing (LTA) and semi-quantitative determinations were made for the main residual constituents (Table II).

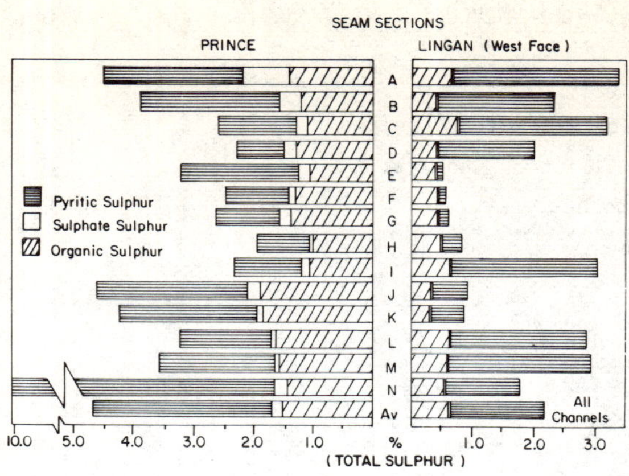

Fig. 2 - Distribution of pyritic, sulphate, and organic sulphur across vertical channels of Prince and Lingan seams

Table II. Pyrite Size and Content, and LTA Weights and Analyses in Mine Channels

Channel sections	Top (A, B, C, D)		Middle (E, F, G, H)		Bottom (I, J, K, L, M)	
Data	Lingan	Prince	Lingan	Prince	Lingan	Prince
Pyrite:						
Mean size (μm)	62	44	13	8	30	31
Assay (% Fe_2S)	4.1	4.1	0.4	2.4	3.0	6.0
LTA:						
Weight	9.1%	8.9%	3.0%	3.6%	9.7%	17.9%
% Fe_2O_3	27.1	51.9	41.7	43.8	27.9	62.6
% Al_2O_3	14.2	24.1	13.7	22.4	13.5	16.0
% SiO_2	50.7	21.5	40.1	25.9	43.0	19.7

Table II shows that:

(1) The dimension of the pyrite grains are directly proportional to its concentration, i.e., the pyrite size is coarser in coal that has a higher FeS_2 per cent;

(2) The pyritic- and ash-minerals both tend to be at low concentrations in the same zones (compare the pyrite contents with the corresponding LTA weights);

(3) The high iron oxide content in the LTA residues indicates the presence of iron-bearing minerals other than pyrite which suggests magnetic separation as a possible method for removing the ash-minerals from the coals.

Devco's Coals: Petrography and Mineralogy

Petrographically, the coals from the Sydney field are of high volatile "A" bituminous rank with a high content of vitrinite and with minor amounts of exinite, semi-fusinite and inertinite (Table III).

Table III. Organic Components of DEVCO Coals

Category/Group		Species		% volume		Texture and morphology
				Lingan	Prince	
Coal macerals	Reactive	Vitrinite		68.0	66.7	1. Massive
		Exinite	Sporinite	6.9	5.8	2. Banded
			Resinite	1.4	5.8	3. Cellular
		Semi-fusinite		1.5	1.2	4. Complex
	Inert	Semi-fusinite		2.9	2.4	[No. 3 and 4, in
		Inertinite	Fusinite	1.6	2.0	particular, have
			Macrinite	0.4	0.4	inclusions of pyrite, kaolinite,
			Micrinite	2.7	1.1	calcite and siderite
			Others	0.6	0.2	ite (~10 µm)].
Organic sulphur				0.6	1.5	Solid solution in coal macerals.

Optical, scanning electron and X-ray studies identified 26 inorganic minerals from which the nine most abundant are listed in various occurrences in Table 4.

Table IV - Minerals Occurring in DEVCO Coals

Category/Group	Species	% volume		Texture and morphology
		Lingan	Prince	
Silicates, carbonates and oxides	Illite Kaolinite Quartz Calcite	11.1 (host rock)	8.8	- Free grains - Bands parallel to coal - Intergrowths with coal
	Clays Quartz Calcite Siderite	1.2 (associated with coal)	1.3	- Filling in cellular structure - Veinlets and fine inclusions
Sulphides	Pyrite-Marcasite	0.4	1.1	- Massive: bands and patches, cell filling (95% + 20 µm)
		0.3	0.7	- Crystal groups: Clusters, framboids Lenses (50% + 20 µm)
		0.2	0.6	- Globules and cauliflowers (60% + 20 µm)
		0.15	0.2	- Crystal grains: Euhedral, subhedral, pinpoint (100% - 20 µm)
Iron sulphate		0.05	0.1	- Alteration of pyrite

In summary, the shape of the minerals vary from large irregular patches to tiny crystals with intermediate forms such as masses, fragments, bands, veinlets, globules, lenses, needles, etc. These are defined hereunder along with their major mineralization. As the size decreases, the massive occurrences are described as patches, masses or fragments; the illite-kaolinite, quartz and calcite in the host rock (as well as most of the coal) are largely found in such irregular shapes. About one third of the pyrite-marcasite also occurs in this category. Minor amounts of the above species are found as bands and veinlets, i.e., as crack and fissure fillings in coal macerals. Finally, pyrite and, to a much lesser degree, some other inorganic minerals occur as (1) globules of round to semi-circular shapes, (2) clusters of crystals generally running as lenses parallel to the coal bedding, (3) thin bladed or needle-like grains, and (4) finely disseminated pinpoint crystals. Occasionally, the above forms are intergrown, agglomerated and coalesced.

A number of microphotographs (Fig. 3-20) have been selected to

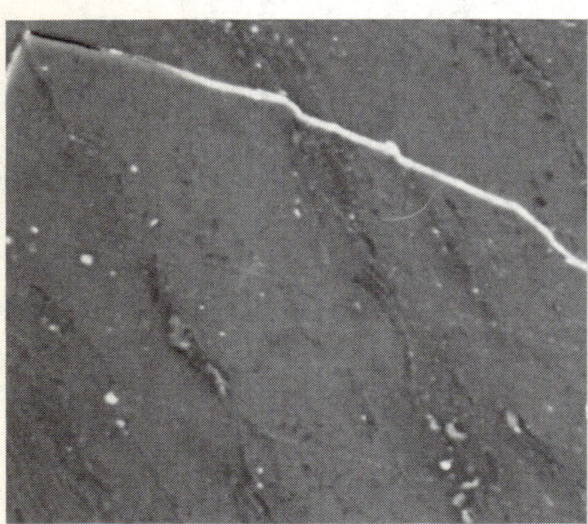

Fig. 3 - Massive coal with scattered inclusions of clay (grey) and pyrite (white).

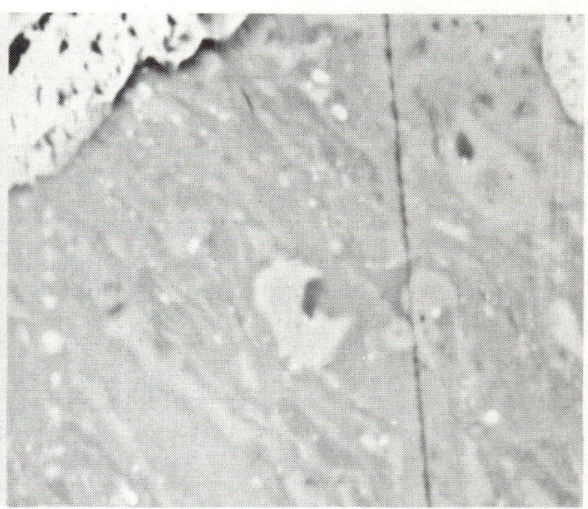

Fig. 4 - Interbanded vitrinite and fusinite with patches of pyrite in top corner.

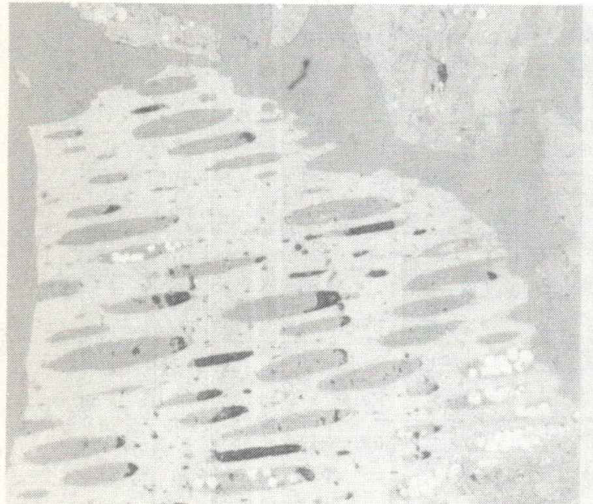

Fig. 5 - Coal cellular structure cavities filled with clay and pyrite grains.

Fig. 6 - Coal exhibiting intergrown macerals with pyrite (white) and clay (light grey) areas.

illustrate the petrographical and mineralogical features. The first set of pictures (3 to 6) shows the main structures of coal, namely massive, banded, cellular and complex, with some mineral inclusions and intergrowths. For instance, Fig. 3 shows massive coal containing scattered inclusions of clay and pyrite, whereas the banded and cellular coals have pyrite patches (Fig. 4) and infillings of crystalline pyrite and amorphous clay (Fig. 5). The complex coal structure (Fig. 6) illustrates the presence of various macerals (dark shades) along with pyrite globules and fragments.

The next series of photographs (Fig. 7-10) are backscattered electron images (BEI) showing, at various magnifications, the major oxide, carbonate and silicate minerals (viz., quartz, calcite and siderite) and the clay constituents, illite and kaolite in particular. Pyrite also is occasionally present as small crystals and irregular fragments.

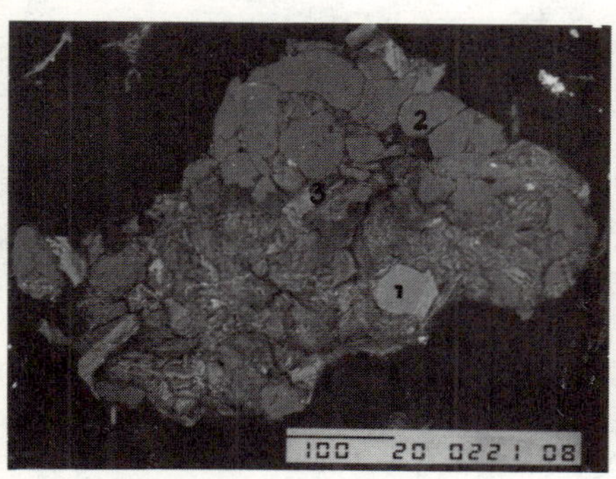

Fig. 7 - Area showing illite (1,2,5), quartz with Fe-silicate (3), pyrite (4), rock chips (6) and a kaolinite-grain (7) in massive coal. (BEI x35)

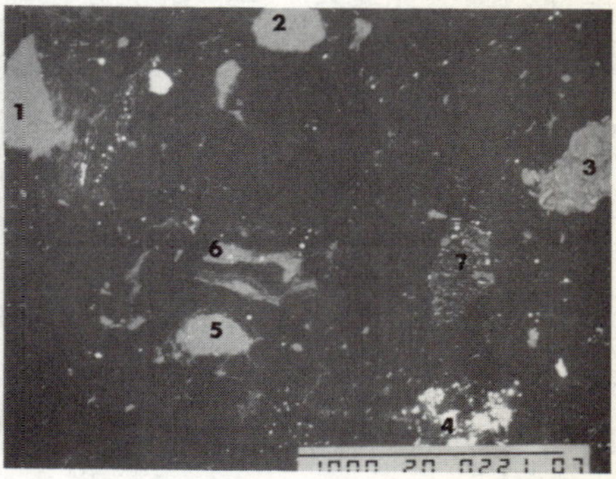

Fig. 8 - Enlarged view of top right grain of Fig. 6 which exhibits iron carbonate, likely siderite (1), associated with quartz (2) and clay minerals (3). (BEI x200)

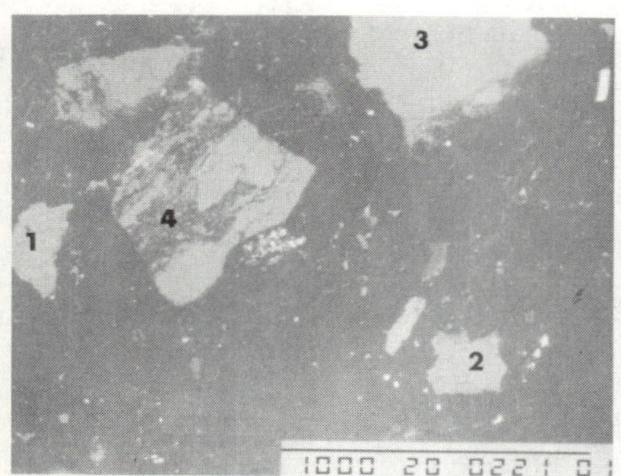

Fig. 9 - Closer view of calcite-quartz quartz rock (4) and illite grains (1,2, surrounded by pinpoint pyrite (white) in a coal matrix. (BEI x50)

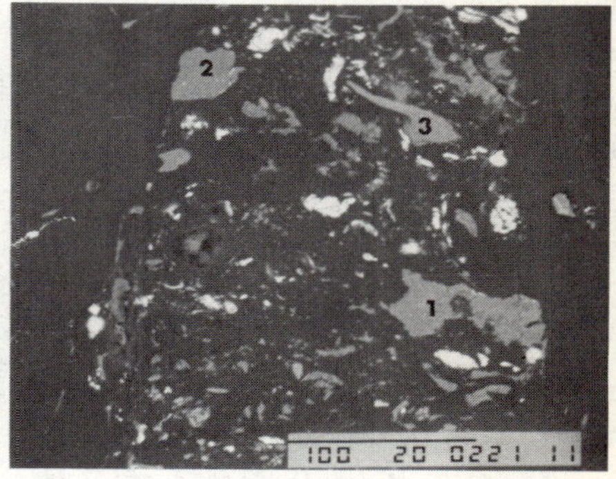

Fig. 10 - Enlargement of center-right zone of Fig. 6 showing 3) kaolinite (1,2,3), and a few pyrite fragments in a coal complex structure. (BEI x350)

A set of four pictures (Fig. 11-14) illustrates the residual inorganic minerals from low-temperature ashes which were produced by oxygen-plasma burning below 150°C. A scanning electron microscope (SEM) equipped with an energy dispersive X-ray (EDX) detector was used for observation, as recommended by Finkeman and Flustoker (1981) in their review of a variety of mineralogical techniques for mineral characterization in coal.

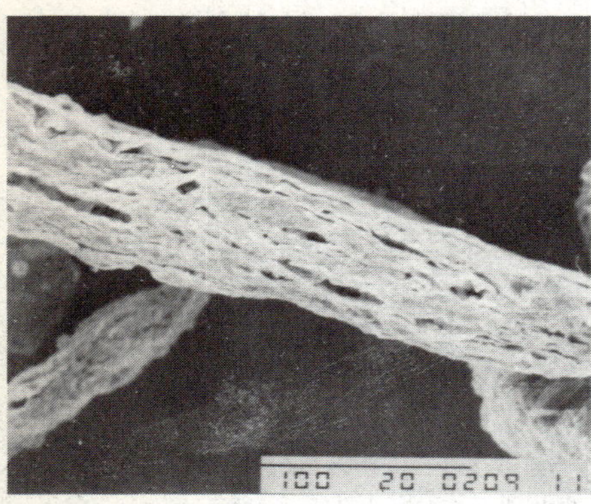

Fig. 11 - Photograph of a coal LTA residue which shows mineral matter lenses (white) and voids left after coal combustion (black). (LTA-SEI x350)

Fig. 12 - Highly magnified view of sample exhibiting an illite field, which contains pyrite crystals (white). (LTA-BEI x5000)

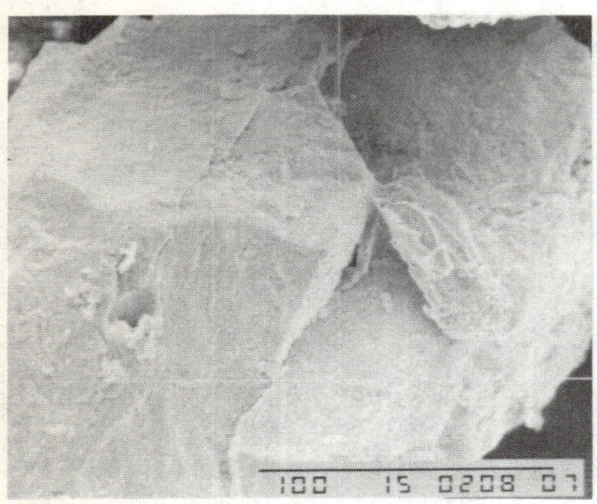

Fig. 13 - Massive iron-bearing mineral (siderite) shown after LTA treatment. (LTA-SEI x500)

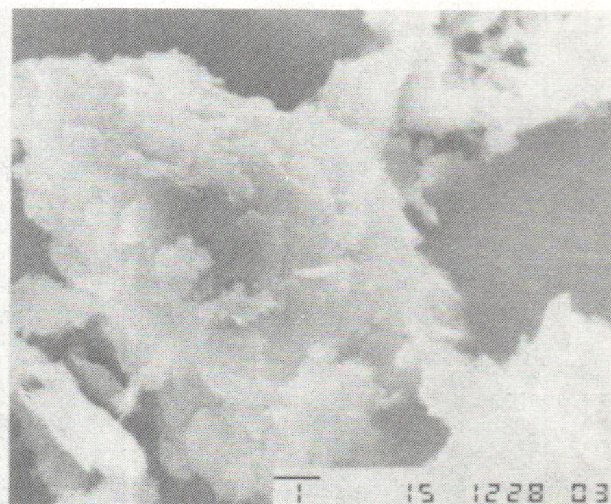

Fig. 14 - Illite/quartz intergrowth in flocculated clays from low-temperature treatment. (LTA-SEI x7500)

For the detailed identification of the pyrite morphology, oil immersion and reflected light microscopy (OI-RLM) was applied to large fields, whereas the backscattered electron image technique was used at high magnification (see Fig. 15-17 and 18-20, respectively). In these micrographs, the main forms of pyrite are shown as masses, framboids, globules, lenses and crystals, either individual or coalesced.

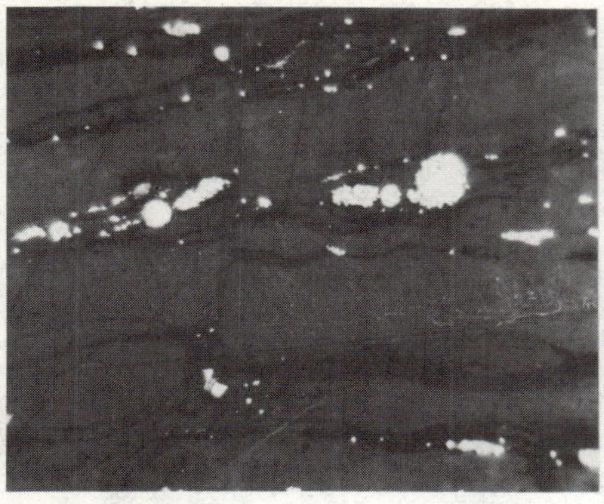

Fig. 15 - Area view showing well-formed FeS_2 framboids, globules and lenses, all embedded in vitrinite groundmass (Lingan: coal OI-RLM).

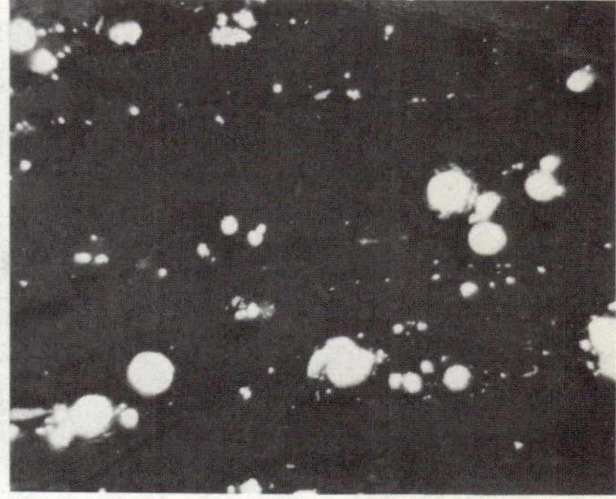

Fig. 16 - Similarly mineralized zone of Prince coal with its most common forms of pyrite, i.e., masses, clusters, globules and pinpoint crystals. (OI-RLM)

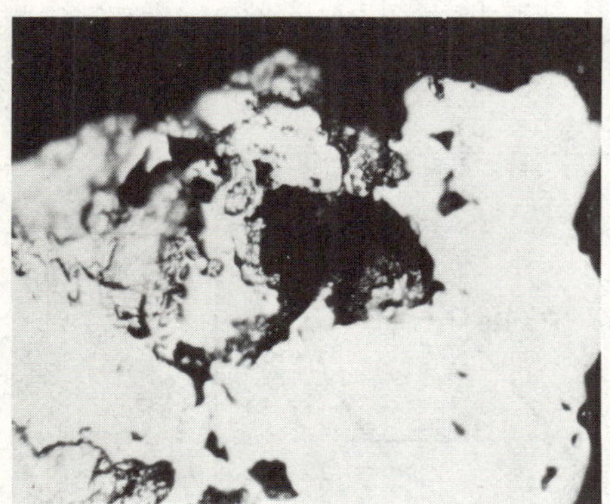

Fig. 17 - Irregular patch of pyrite (white) in Prince coal, probably remnant of a cleat or secondary filling. (OI-RLM).

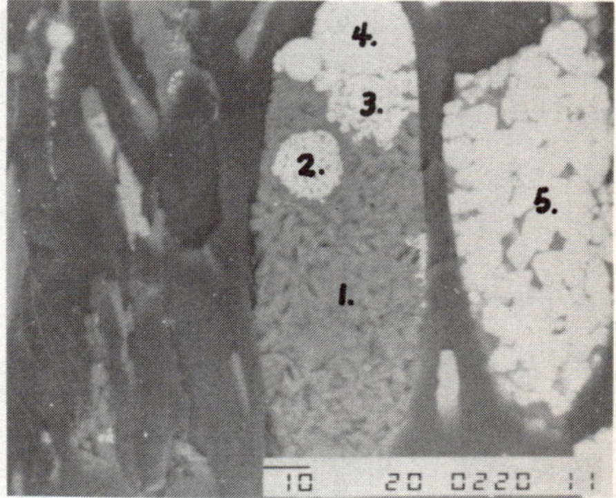

Fig. 18 - Coal pellet with cavities filled with clay (1) and pyrite in forms of agglomerated and coalesced framboids (2,3,4) and crystals (5). (BEI x750)

J. A. R. Stirling of the Atlantic Coal Institute has conducted thorough size-distribution studies of the coal macerals and pyrite in the DEVCO coals, and CANMET mineralogists have made similar measurements and computations for the other ash-bearing minerals. The pyrite grains were found to be considerably finer than the coal macerals. The ash minerals vary extensively in size, with the host rock attaining the dimensions of the coal and part of the clays being as fine as the pyrite pinpoint crystals. Representative size-distribution curves (Fig. 21) illustrate these tendencies. It is noteworthy that the d_{50} line intersects the pyrite, ash-minerals and coal curves at about 90, 250 and 350 μm, respectively. Furthermore, some 60% of the coal macerals and 40% of the ash-minerals grains are plus 295 μm, thus allowing for a primary cleaning at relatively coarse size. Conversely, pyrite liberation would require extremely fine grinding with 20% um fraction in the sample examined.

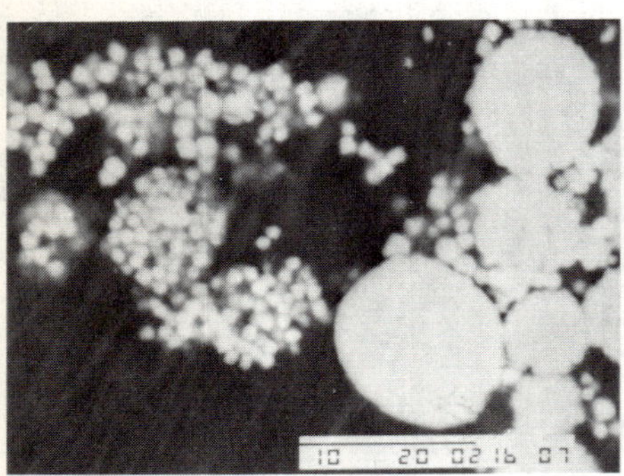

Fig. 19 - Enlargement illustrating framboidal and crystallite colonies with some globules and cauliflowers. (BEI x1500, EDX)

Fig. 20 - Higher magnification of clusters of framboids and crystals, some being intergrown. (BEI x 3500 EDX)

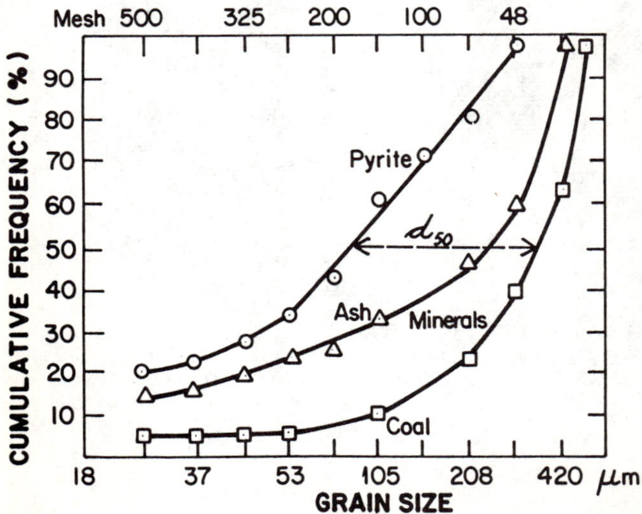

Fig. 21 - Pyrite, ash minerals and coal size distributions.

Further mineralogical work was done to establish a correlation between the size and the morphology of the pyrite grains, as well as to predict their liberation as a function of fineness of grinding. S. G. Whiteway has reported that 30-50% of the total pyrite contained in the Prince and Lingan coals occurs as crystals and grains finer than 20 um in size; the coarser forms comprise mainly patches, cell replacements and fillings, whereas the framboids, clusters, globules and cauliflowers are almost equally split between the plus and minus 20 μm size (Table V).

Table V. Relation between Pyrite Morphology and Size

Group	Morphology	Pyrite size			
		Lingan		Prince	
		+20 μm	-20 μm	+20 μm	-20 μm
Solitary grains	Pinpoint crystals	0.0	3.0	0.0	4.3
	Individual crystals {euhedral	0.0	3.7	0.0	0.7
	Individual crystals {subhedral	0.0	7.9	0.0	3.0
Crystal groups	Needles	0.0	0.0	0.0	1.3
	Lenses	1.2	3.1	3.7	1.3
	Framboids	4.9	4.9	1.7	0.3
	Clusters	7.4	9.2	9.7	5.0
	Globules	4.8	13.0	8.0	11.7
	Cauliflowers	0.6	0.0	4.3	2.3
Massive	Veinlets	0.6	0.0	0.7	0.0
	Cell {filling	6.7	2.4	**	**
	Cell {replacement	12.8	0.6	2.3	0.0
	Irregular patches	11.0	2.4	38.7	1.0
% Total (calculated)		50.0	50.0	69.1	30.9

*Partly on crystalline form.

**None observed in the Prince sample examined.

A method was developed at CANMET by William Petruk for predicting mineral liberation as a function of fineness of grinding. Polished sections of the rock (or coal) samples are microscopically examined and mineral grain sizes measured using a Quantimet image analyzer. A predicted liberation curve is then constructed using an empirical formula based on the actual dimensions of the grains and a factor dependent on the friability of the minerals. In the case of pyrite, for instance, it is considered that a grinding to 100% minus a given size would liberate all the grains coarser than this dimension plus half of the finer grains. When applied to the Prince coal by Louis Cabri and G. I. Mathieu, the method gave the pyrite liberation curve shown in Figure 22. From these data, it is also possible to calculate the per cent of total sulphur removable by physical means by assuming that, in addition to the unliberated pyrite, all the organic sulphur is physically locked in the coal. The results are also plotted in Figure 21. Both curves indicate the necessity of very fine comminution for a maximum refection of the sulphur from the coal values.

Deep-Cleaning of Lingan and Prince Coals

Raw coal samples representative of the current DEVCO mining operations were selected for beneficiation. Chemical analyses for ash content and sulphur types yielded the results given in Table VI.

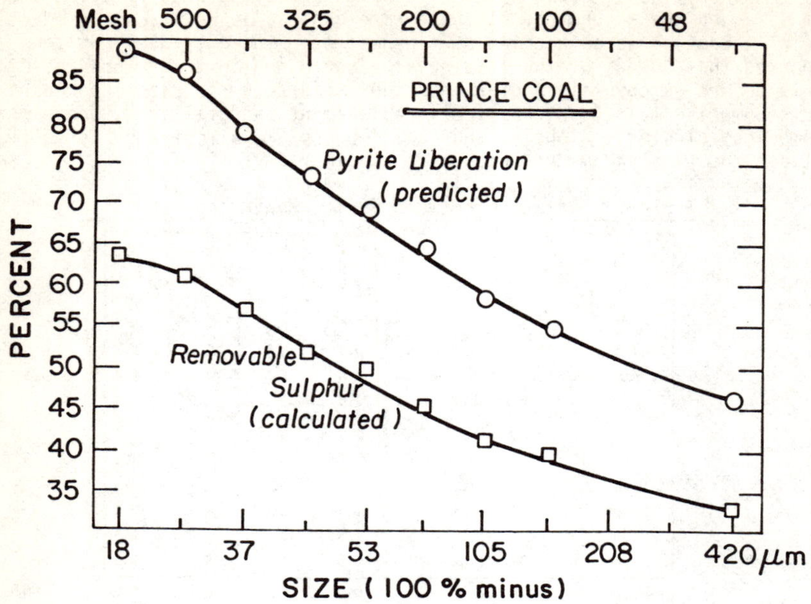

Fig. 22 - Liberated pyrite and removable sulphur as per mineralogy.

Table VI. Analysis of Head Samples

Seam	% Ash	% Sulphur			
		Pyritic	Organic	Sulphate	Total
Lingan	26.0	1.25	0.65	0.05	1.95
Prince	20.5	3.5	1.5	0.2	5.2

The coal cleaning procedure was tailored to the mineralogical findings, i.e., with the host rock to be removed at coarse or intermediate sizes and the fine pyrite and clay minerals at a greater degree of fineness. Initial sample preparation down to minus 3 mm was made by conventional closed-circuit crushing (Figure 23).

The crushed samples were progressively ground and submitted to three levels of cleaning, which are represented graphically in Figure 24 and labelled A, B and C. The first beneficiation stage (A) involved coal rougher-cleaner flotation applied to a pulp of increasing fineness. The flotation concentratyes were then reground for further cleaning by froth flotation (Level B) and magnetic separation (Level C). The three successive beneficiation processes were aimed respectively at (i) rejecting the bulk of the gangue at either coarse or intermediate size, (ii) removing additional ash and pyritic minerals liberated by further grinding, and (iii) attracting the remaining paramagnetic impurities associated with the coal by high-gradient magnetic separation. Each cleaning operation was tested at various degrees of fineness to compare the results as a function of size and

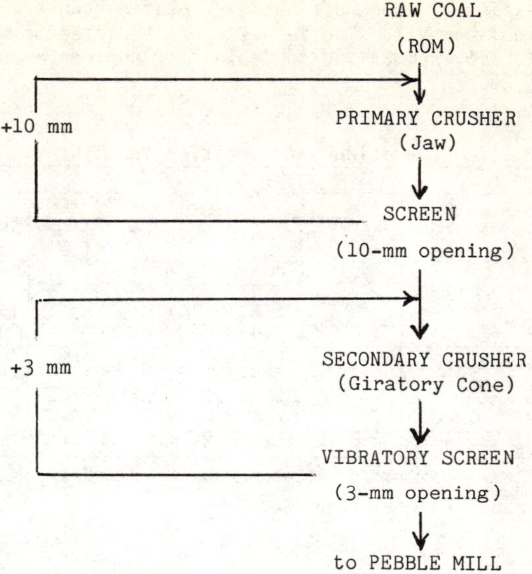

Fig. 23 - Coal crushing procedure

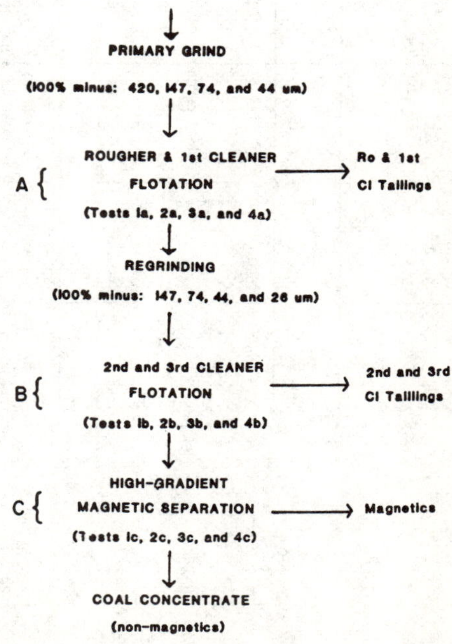

Fig. 24 - Progressive grinding and processing of coal.

level of processing. The conditions of flotation and magnetic separation are reported in Table VII. The results are tabulated in detail in Appendices I to III, summarized in Table VIII and represented graphically in Figures 24 and 25.

Table VII. Conditions of Flotation and Magnetic Separation

Beneficiation Level : Description	Operations	Time (min)	Reagents* kg/t		pH (natural)	
			MICB	Kero-sene**	Lingan	Prince
A : Direct grinding & flotation (DG&F)	Rougher	Cond: 2	0.5-0.8	0.7-0.8	7.3-7.5	6.6-6.9
		Flot: 5-10	0.1-0.3	0.1-0.2	7.4-7.9	6.9-7.3
	1st cleaner	Cond: 1	0.3-0.4	0.3-0.5	7.4-8.0	6.9-7.2
		Flot: 3-7	0.1-0.3	0.1-0.2	7.6-8.3	7.0-7.4
B : Progressive grinding & flotation (PG&F)	2nd cleaner	Cond: 1	0.4-0.9	0.4-0.6	7.5-7.8	6.8-7.1
		Flot: 4-9	0.1-0.2	0.1-0.2	7.8-8.3	6.9-7.6
	3rd cleaner**	Cond: 1	0.4-0.5	0.3-0.4	-	7.0-7.4
		Flot: 3-8	0.1	0.1	-	7.1-7.7
C : High gradient magnetic separation (HGMS): 2.1 Tesla, 20 mm/s						

*The longer flotation times and higher reagent consumptions correspond to finer grinding.
**Used only for Prince coal if necessary.

Table VIII. Summary of Results of Flotation and Magnetic Separation

Grinding size	Analyses & yield (%)	Lingan			Prince		
		DG&F	PG&F	(+HGMS)	DG&F	PG&F	(+HGMS)
-147 μm	Ash	3.2	2.2	(1.3)	5.7	4.4	(2.9)
	S_t	1.5	1.4	(1.1)	3.3	3.1	(2.4)
	yield	92.9	88.4	(81.0)	92.6	84.5	(79.9)
-74 μm	Ash	2.8	2.0	(0.9)	5.2	3.9	(2.3)
	S_t	1.5	1.3	(0.9)	3.1	2.7	(1.9)
	yield	92.0	90.9	(83.4)	91.0	84.5	(79.8)
-44 μm	Ash	2.5	1.9	(0.8)	4.1	3.5	(2.0)
	S_t	1.4	1.2	(0.9)	2.9	2.5	(1.7)
	yield	91.4	89.9	(82.0)	87.9	80.2	(75.3)

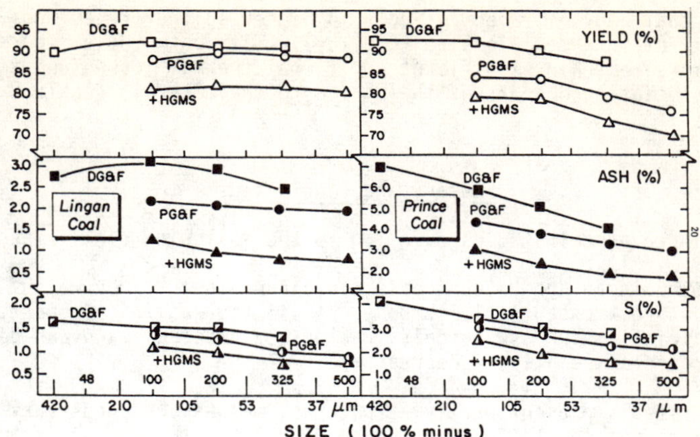

Fig. 25 - Effect of feed size and processing of thermal yield, ash and sulphur content of Lingan and Price coal products.

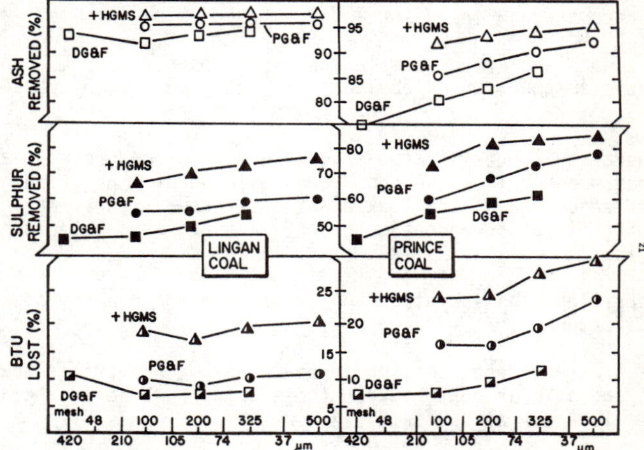

Fig. 26 - Correlation between combustible lost and ash sulphur removed by flotation and magnetic separation.

The results of flotation and magnetic separation were analyzed from two main points of view, namely:

1. the grade of the coal products and their thermal yield as a function of the cleaning level and fineness of grinding (Figure 25), and

2. the rejection of the pyritic and ash impurities versus the corresponding BTU losses as the grinding and cleaning progress (Figure 26).

The ash romoval from the Lingan coal, as compared to the Prince, is higher and less dependent of the grinding fineness, as expected by a greater liberation of its inorganic minerals. Conversely, the pyrite is coarser in the Prince coal and this also reflects in higher sulphur rejections particularly at intermediate size (Fig. 24). With both coals, increases in sulphur removal entrain significant additional thermal losses due to the intimate association of part of the pyrite to the macerals (Table 8 and Fig. 25).

Summary and Conclusions

The mineralogical studies have led to the following conclusions:

1. Both Lingan and Prince coals contained about 20% by weight of host rock essentially composed of illite, kaolinite, quartz and calcite, which are largely free of coal macerals and can be liberated at relatively coarse size.

2. Pyrite is present partly as patches, masses and bands varying in size from 100 to 200 um, and as globules, clusters and framboids averaging about 50 um.

3. Pyrite and the clay minerals occur frequently in vitrinite and fusinite as cell fillings and replacements, as well as fine intergrowths and inclusions of less than 20 um in size.

4. The Lingan and Prince samples analyzed respectively 0.8 and 1.5% organic sulphur, which obviously is not removable by physical means.

This mineralogy suggests a two-stage cleaning procedure with a relatively coarse removal of the host rock followed by fine grinding for liberation and rejection of the locked and intergrown impurities. A study on coal grinding-flotation-magnetic separation along this pattern demonstrated that:

(i) increasing fineness of grinding improved the quality of the coal products,

(ii) the progressive processing approach gave concentrates lower in ash and sulphur content than those obtained by direct grinding and flotation at the same size,

(iii) a magnetic treatment is necessary for maximum elimination of both pyritic and ash impurities, and

(iv) the grade improvements due to supplementary cleaner flotation stages and high-gradient magnetic separation are achieved at the expense of additional thermal losses attaining 9.4 to 11.9% for the Lingan coal and 11.2 to 12.7% for the Prince.

It was found that in general the ash and pyritic minerals behaved as predicted by the mineralogy. Finally, the progressive grinding and processing approach is economically attractive since the fine comminution is applied on only 65% of the feed for significant savings in energy.

References

1. J. J. Laffin, "Coal 2000: Nova Scotia Perspectives", Minister of Mines and Energy Presentation (1982).

2. Atlantic Coal Institute (Staff), "Assessment of Nova Scotia Coal Preparation Problems", Report UP-A-207 (1982).

3. D. Birk and J. Pilgrim, "Development of the Beneficiation Potential for Eastern Canadian Coals", Phase 2 - Analysis of R.O.M. Coals: Chemical and Petrophysical Analysis (Vol. 1), and Mineral and Maceral Analysis (Vol. 2), Atlantic Coal Institute; Sydney, Nova Scotia (1985).

4. J. A. R. Stirling, "Influence of the Mineralogy of Coal, and in Particular of Sulphur Compounds on the Possibility of Sulphur Removal and Ash Reduction", Research and Productivity Council Report 48; New Brunswick (1984).

5. S. G. Whiteway and R. S. Moore, "Beneficiation Techniques Applicable to Nova Scotia Coals", Atlantic Research Laboratory Technical Report 39, Halifax, Nova Scotia (1983).

6. R. B. Finkelman, et al., "Identification and Significance of Accessory Minerals from a Bituminous Coal", Fuel 57 (1978) pp. 763-768.

7. William Petruk, "The Application of Quantitative Mineralogical Analysis of Ore to Ore Dressing", CIM Bulletin 69, pp. 146-153.

8. Louis J. Cabri, "Mineralogical Examination of Coal Samples from DEVCO Prince Mine", CANMET, Energy, Mines and Resources Canada, Mineral Sciences Laboratories, Division Report ERL/MSL 77-200 (IR) (1977).

9. G. I. Mathieu, "Desulphurization and De-ashing or Eastern Coal by High-gradient Magnetic Separation", CANMET, Energy, Mines and Resources Canada, Division Report ERP/MSL 79-80 (OP) (1979).

APPENDIX I

GRADE AND THERMAL YIELD OF COAL PRODUCTS FROM FLOTATION AND MAGNETIC SEPARATION OF LINGAN SAMPLE (26.0% Ash, 1.95% S_t and 10 950 BTU/LB)

Grind (100% minus)	Regrind (100% minus)	Cleaning procedure	Coal analysis % Ash	% S_t	BTU	Thermal yield (%)
420 µm	Nil	Ro & 1st Cl flotation	2.75	1.61	14 622	90.0
147 µm	Nil	Ro & 1st Cl flotation	3.22	1.50	14 574	92.9
420 µm	147	Ro, 1st, 2nd Cl flot:	2.16	1.37	14 778	88.4
		plus HGMS	1.29	1.10	14 975	81.0
74 µm	Nil	Ro & 1st Cl flotation	2.84	1.46	14 642	92.0
147 µm	74 µm	Ro, 1st, 2nd Cl flot:	2.03	1.32	14 811	90.9
		plus HGMS	0.90	0.95	15 065	83.4
44 µm	Nil	Ro & 1st Cl flotation	2.51	1.37	14 718	91.4
74 µm	44 µm	Ro, 1st, 2nd Cl flot:	1.95	1.24	14 848	89.9
		plus HGMS	0.84	0.92	15 086	82.0
44 µm	37 µm	Ro, 1st, 2nd Cl flot:	1.92	1.22	14 847	89.6
		plus HGMS	0.82	0.91	15 090	80.8
	26 µm	Ro, 1st, 2nd Cl flot:	1.90	1.18	14 868	88.8
		plus HGMS	0.80	0.83	15 115	80.1

APPENDIX II

GRADE AND THERMAL YIELD OF COAL PRODUCTS FROM FLOTATION AND MAGNETIC SEPARATION OF PRINCE SAMPLE (20.5% Ash, 5.2% S_t and 11 050 BTU/LB)

Grind (100% minus)	Regrind (100% minus)	Cleaning procedure	Coal analysis % Ash	% S_t	BTU	Thermal yield (%)
420 µm	Nil	Ro & 1st Cl flotation	6.7	4.0	13 443	92.7
147 µm	Nil	Ro & 1st Cl flotation	5.7	3.3	13 827	92.6
420 µm	147 µm	Ro, 1st, 2nd, 3rd flot:	4.4	3.1	14 020	84.5
		plus HGMS	2.9	2.4	14 370	79.9
74 µm	Nil	Ro & 1st Cl flotation	5.2	3.1	13 947	91.0
147 µm	74 µm	Ro, 1st, 2nd, 3rd flot:	3.9	2.7	14 255	84.5
		plus HGMS	2.3	1.9	14 655	79.8
44 µm	Nil	Ro & 1st Cl flotation	4.1	2.9	14 056	87.9
74 µm	44 µm	Ro, 1st, 2nd, 3rd flot:	3.5	2.5	14 363	80.2
		plus HGMS	2.0	1.7	14 770	75.3
44 µm	37 µm	Ro, 1st, 2nd, 3rd flot:	3.2	2.3	14 456	79.8
		plus HGMS	1.9	1.7	14 780	74.5
	26 µm	Ro, 1st, 2nd, 3rd flot:	3.2	2.3	14 454	76.0
		plus HGMS	1.9	1.7	14 785	70.8

APPENDIX III

MASS, ASH AND SULPHUR REJECTED VERSUS BTU LOSSES IN COAL CLEANING BY FLOTATION AND MAGNETIC SEPARATION

Coal seam	Size 100% minus	Impurities rejected (%)									BTU losses %		
		Mass			Ash			Sulphur					
		DG&F	PG&F	(+HGMS)	DG&F	PG&F	(+HGMS)	DG&F	PG&F	(+HGMS)	DG&F	PG&R	(+HGMS)
Lingan	420 µm	32.6	-	-	92.9	-	-	46.4	-	-	10.0	-	-
	147 µm	30.2	34.5	(40.8)	91.4	94.6	(97.1)	46.3	53.0	(66.6)	7.1	11.6	(19.0)
	74 µm	31.2	32.8	(39.4)	92.5	94.8	(97.9)	48.9	54.5	(70.5)	8.0	9.1	(16.6)
	44 µm	32.0	33.7	(40.5)	93.4	95.0	(98.1)	53.2	57.4	(71.9)	8.6	10.1	(18.0)
	37 µm	-	33.9	(41.4)	-	95.1	(98.2)	-	58.4	(71.7)	-	10.7	(19.2)
	26 µm	-	34.6	(42.0)	-	95.2	(98.2)	-	60.4	(75.3)	-	11.2	(19.9)
Prince	420 µm	23.8	-	-	74.1	-	-	41.4	-	-	7.3	-	-
	147 µm	26.0	33.4	(38.6)	79.4	85.7	(91.6)	53.0	60.3	(71.7)	7.4	15.5	(20.1)
	74 µm	27.9	34.5	(39.9)	81.7	87.5	(93.3)	57.0	66.0	(78.0)	9.0	15.5	(20.2)
	44 µm	30.9	38.3	(43.7)	86.2	89.5	(94.5)	61.5	70.3	(81.6)	12.1	19.8	(24.7)
	37 µm	-	39.0	(44.3)	-	90.5	(94.8)	-	73.0	(81.8)	-	20.2	(25.5)
	26 µm	-	41.9	(47.1)	-	91.4	(95.1)	-	74.3	(82.7)	-	24.0	(29.2)

EFFECTS OF SULFUR CONTENTS OF COALS ON THEIR DESULFURIZATION WITH LIME OR LIMESTONE DURING COMBUSTION

M. H. Erten

Department of Mining Engineering
University of Missouri at Rolla
Rolla, Missouri 65401

In this investigation the effects of sulfur contents of coals on their desulfurization during combustion with the use of lime or limestone were studied. The coals tested contained between 2.10 and 6.24 percent sulfurs and the combustion temperatures used varied between 800 C and 1400 C. The best results were obtained with a stoichiometric ratio of four lime (CaO) or limestone ($CaCO_3$) to one of sulfur in the coal. The tests showed that lime is a more effective desulfurization agent at all the temperatures tested. The tests also indicated that it is difficult to make a correlation between the sulfur contents of coals and the percentages of sulfurs emitted or absorbed during free burning or controlled combustion in the presence of absorbing agents.

Introduction

The aim of this research was to compare the SO_2 emissions from low, medium, and high sulfur coals during combustion at different temperatures in the presence of various percentages of lime or limestone used as SO_2 absorbing agent. The three coal samples used in the experiments were taken from a drill hole belonging to an unidentified coal seam, from the Crowburg seam in the Vernon County, and from the Bevier seam in the Randolph County, all in the State of Missouri. The average proximate analyses of these unwashed coal samples are given in Table I (Ref. 1). As can be seen from this table, the samples contained 2.10, 4.83, and 6.24 percent total sulfurs. During the rest of this paper these samples will be identified as Samples #1, #2 and #3.

Table I. Proximate Analyses of the Coals Tested

Sample No.	Name of Seam	%Moisture (Total)	%Ash (Dry)	%S (Dry)	%Vol. Mat. (Dry)	Btu/Lb. (Dry)
1	Unidentified	1.69	7.76	2.10	34.02	13,735
2	Crowburg	7.75	9.76	4.83	41.76	12,533
3	Bevier	6.65	14.95	6.24	37.58	11,557

The forms of sulfur in these three samples are given in Table II (Ref. 1). The sulfate and pyritic sulfurs of each sample were determined by following the ANSI/ASTM D-2492 standard procedure. The organic sulfur percentages were obtained by subtracting sulfate and pyritic sulfur percentages from the total sulfur percentage in each sample.

Table II. Forms of Sulfur in Coals Tested

Sample No.	% Total S	% Sulfate S	% Pyritic S	% Organic S
1	2.100	0.010	1.200	0.890
2	4.830	0.171	2.400	2.219
3	6.240	0.321	3.680	2.239

If these coals were washed in a conventional coal washing plant, probably most of the sulfate and pyritic sulfurs could be cleaned, but the organic sulfurs would remain almost totally in the clean coal.

The limestone used in the tests was chemical grade pure $CaCO_3$. The lime was obtained from this carbonate by calcining it at 950°C. The average particle sizes of both powders were below 5.0 microns, determined with a Fisher sub-sieve sizer.

An acid-base titration Leco sulfur analyzer apparatus was used to run the combustion tests to determine the sulfur emission during each test. As shown in Figure 1, the Leco sulfur analyzer contains a high temperature furnace with a combustion tube and a titrator consisting of the titration vessel and the buret. During each test, about 200 milliliters of 1.15% H_2O_2 solution is added into the titration vessel for reacting with the combustion products coming from the combustion tube. Any SO_2 present in these gases will react with the H_2O_2 in the titration tube to form H_2SO_4 which can be titrated with the 0.05 N NaOH solution from the graduated buret. The end point (green) is indicated with the use of 10 drops of methyl purple solution added to the reaction tube at the beginning of the test. The Leco sulfur analyzer is so designed that if 0.500 gram of coal sample is burned in the combustion tube, the buret is calibrated to read the percentage of sulfur in the coal directly. For total sulfur determination, the furnace temperature is maintained at 1350°C. If due to the addition of lime or limestone into the coal sample, or because of lower temperature in the combustion tube, some sulfur is retained in the coal residue, the buret reading will be lower than that for total sulfur. This indicates the effect of absorbants or temperature on the retention of sulfur during combustion.

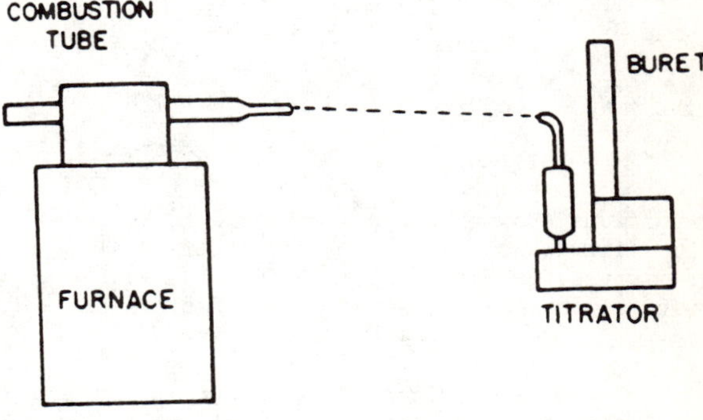

Figure 1. Leco Sulfur Analyzer

Experimental Procedures

There were three parameters affecting the results of these tests. These were:

1. The total sulfur content of coal sample.
2. The percentages of lime or limestone added to the coal.
3. The temperature of combustion.

The coal samples tested contained between 2.10 and 6.24 percent total sulfurs and could be classified as medium to high sulfur coals.

The percentages of lime (CaO) or limestone ($CaCO_3$) added to the half-gram coal charge depended on the total sulfur content of each coal and to the choice of stoichiometric ratio of sorbent to sulfur in each coal sample. The following brief explanation shows the method of calculating the stoichiometric ratios and percentages of sorbents in the combustion test mixtures.

When coal is burned at sufficiently high temperature, all the sulfur is converted into SO_2 gas and the reaction between this gas and any CaO or $CaCO_3$ present in its path is assumed to take place according to the following equations.

$$CaCO_3 + 1/2\ O_2 + SO_2 = CaSO_4 + CO_2;\ CaCO_3/S = 100/32 = 3.125 \quad (1)$$

100 g 32 g (S)

$$CaO + 1/2\ O_2 + SO_2 = CaSO_4;\ CaO/S = 56/32 = 1.750 \quad (2)$$

56g 32g (S)

These equations show that for a stoichiometric ratio of one, we have to add 3.125 milligrams of $CaCO_3$ or 1.750 milligrams of CaO to the coal-sorbent mixture for each milligram of total sulfur in the coal. For stoichiometric ratios other than one, these factors are multiplied by the ratio desired.

For a given stiochiometric ratio, the actual percentages of $CaCO_3$ or CaO in the coal-sorbent test mixtures are calculated as in the example given below. Let us take Sample #2 which contains 4.83% sulfur to explain the following procedure. Each gram of this coal contains 48.3 milligrams (1000 x 0.0483) of sulfur. If all this sulfur is converted into SO_2 and then combined with $CaCO_3$ or CaO to form stable sulfate ($CaSO_4$), from equations (1) and (2), for a stoichoimetric ratio of one, the weight of $CaCO_3$ or CaO required per 0.500g of coal would be:

$CaCO_3$; 48.3 x 0.50 x 3.125 = 75.46 mg.

CaO ; 48.3 x 0.50 x 1.750 = 42.26 mg.

When 75.47 mg of $CaCO_3$ is added to 500 mg of coal, the test mixture will contain:

[75.47 (500 + 75.47)] x 100 = 13.11 % $CaCO_3$.

In the case of CaO, this percentage will be:

[42.26 (500 + 42.26)] x 100 = 7.79 % CaO.

Table III summarizes the percentages of $CaCO_3$ and CaO in the combustion test mixtures of sample #2 for stoichiometric ratios between one and five. Even though they are not shown in this paper, the same type of calculations had to be made for the other two samples before preparing the test mixtures.

Table III. Percentages of $CaCO_3$ and Ca O in the Test Mixtures of Sample #2 For Stoichiometric Ratios Between One and Five.

Stoichiometric Ratios	mg of $CaCO_3$ Per 0.50g Coal	%$CaCO_3$ in Mixture	mg of CaO Per 0.50g Coal	%CaO in Mixture
1.0	75.47	13.11	42.26	7.79
2.0	150.94	23.19	84.52	14.46
3.0	226.41	31.17	126.78	20.23
4.0	301.88	37.65	169.04	25.27
5.0	377.35	43.01	211.30	29.71

Another series of calculations involved the determination of pounds of SO_2 emission per million Btu of the coal burned. EPA Emissions Standards have different maximum limits of pounds of SO_2 emissions, depending on the percentage of sulfur in the coal.

When one pound of sulfur burns according to $S + O_2 = SO_2$ reaction, two pounds of SO_2 are produced. Thus, 100 pounds of a coal containing 1% sulfur will theoretically emit 2 pounds of SO_2. On the other hand a coal like Sample #2, containing 4.83% S, will create an SO_2 emission of 4.83 x 2 = 9.66 pounds of SO_2/100 pounds of coal if allowed to burn freely. However, EPA evaluates the SO_2 emission according to the calorific value of coal and has set up different limits for low and high sulfur coals. EPA's limits are based on the pounds of SO_2 emissions for each million Btu of heat given off in burning a coal. Therefore, the calorific value of coal in Btu/Lb must be determined in addition to its sulfur content before determining the maximum uncontrolled emission limit for a coal. The following example will show the procedure in calculating the SO_2 emission potential of a given coal.

The Sample #2 has a calorific value of 12,533 Btu/Lb on dry basis, determined with a Parr Oxygen Bomb Calorimeter. According to this heat value, to generate 1,000,000 Btu, one needs to use 1,000,000 v 12,533 = 79.80 pounds of coal which will contain 79.80 x 0.0483 = 3.854 pounds of total sulfur. When this amount of sulfur burns freely it produces 3.854 x 2 = 7.708 pounds of SO_2. Therefore, the uncontrolled SO_2 emission capability of this coal is given as 7.708 Lb. SO_2/1,000,000 Btu. The uncontrolled SO_2 emission capabilities of three samples used in these experiments are given in Table IV.

Table IV. The Uncontrolled SO_2 Emission Capabilities of Samples Tested

Sample No.	Btu/Lb. (Dry)	$Lb/10^6Btu$ Coal	%S in Coal (Dry)	$Lb.SO_2/10^6$ Btu
1	13,735	72.81	2.10	3.058
2	12,533	79.80	4.83	7.708
3	11,557	86.53	6.24	10.799

According to EPA New Source Performance Standards (NSPS), when coal-fired boilers burn coal, 60% of the SO_2 produced must be removed from the flue gases if the coal has a capability of producing up to 2 Lb $SO_2/10^6$ Btu. Between 2 and 6 Lb $SO_2/10^6$ Btu, the removal rates vary between 70 and 90 percent. However, after 6 Lb $SO_2/10^6$ Btu, 90% of available SO_2 must be removed from the flue gases (Ref. 2). According to this criteria, our Sample #1 is subject to 80% removal rate which means that only 0.60 Lb $SO_2/10^6$ Btu can be freely emitted into the atmosphere. The other two samples are subject to 90% removal rate which sets the maximum emission limits to 1.20 Lb. $SO_2/10^6$ Btu.

During the combustion tests conducted for this research, depending on the furnace temperature, part of the SO_2 was absorbed by $CaCO_3$ or CaO added to the coal and, therefore, the Lb. $SO_2/10^6$ Btu emissions from different samples were below those given in Table IV. Any SO_2 escaping with the combustion gases was caught in the titration vessel of the Leco analyzer and was titrated with NaOH solution. The buret reading recorded the percentage of sulfur escaping from the coal tested.

The combustion temperatures used were between 800°C and 1400°C. The reaction for sulfate formation in the combustion zone is given by the following equation.

$$CaCo_3 + SO_2 + 1/2\ O_2 \quad = \quad CaSO_4 + CO_2.$$

The general belief is that as the combustion temperature exceeds 1000°C, reverse reaction will dominate, causing most of the SO_2 to escape with the rest of the gases. At temperatures below 1000°C, the retention of SO_2 as $CaSO_4$ is favored. At temperatures lower than 900°C, $CaCO_3$ is not completely calcined and therefore all the sorbent is not utilized.

Test Results

The laboratory scale combustion tests were conducted with the use of $CaCO_3$ or CaO with Samples #1, #2, and #3 with the use of stoichiometric ratios of sorbent to sulfur in the coal between one and five. However, at lower stoichiometric ratios, the SO_2 retention during combustion was very poor and for this reason these results are not reported in this paper. The excess $CaCO_3$ or CaO in the combustion zone created a more suitable environment for better retention of SO_2 as $CaSO_4$. It is claimed that (Ref. 3) the $CaSO_4$ formed over the particles of $CaCO_3$ or CaO during their reaction with SO_2 plugs the pores of unreacted sorbents and stops further reaction. As a result, the amount of sorbent needed is much greater than the theoretical amount calculated for a stoichiometric ratio of one.

Since the best results were obtained with a stoichiometric ratio of at least four, only these results are reported in the following tables.

Table V. Effects of CaO and $CaCO_3$ Addition to Sample #1 Containing 2.10% Sulfur (Stoichiometric Ratio = 4)

Combustion Temp. °C	% $CaCO_3$ or CaO Addition by Weight	% of total S Emitted	% of Total S Absorbed	Lb. SO_2 Emission per 10^6 Btu
	None	85.71	12.29	2.621
800	12.83% CaO	20.95	79.05	0.641
	20.90% $CaCO_3$	22.86	77.14	0.699
	None	85.71	12.29	2.621
900	12.83% CaO	20.95	79.05	0.641
	20.90% $CaCO_3$	26.67	73.33	0.816
	None	86.67	13.33	2.650
1000	12.83% CaO	21.90	78.10	0.670
	20.90% $CaCO_3$	28.57	71.43	0.874
	None	87.62	12.38	2.679
1100	12.83% CaO	28.57	71.43	0.874
	20.90% $CaCO_3$	30.48	69.52	0.932
	None	94.29	5.71	2.883
1200	12.83% CaO	32.38	67.62	0.990
	20.90% $CaCO_3$	49.52	50.48	1.514
	None	95.24	4.76	2.912
1300	12.83% CaO	58.10	41.90	1.777
	20.90% $CaCO_3$	68.60	31.40	2.098
	None	100.00	0.00	3.058
1400	12.83% CaO	80.00	20.00	2.446
	20.90% $CaCO_3$	87.62	12.38	2.680

Table VI. Effects of CaO and $CaCO_3$ Addition to Sample #2 Containing 4.83% Sulfur (Stoichiometric Ratio = 4)

Combustion Temp. °C	% $CaCO_3$ or CaO Addition by Weight	% of Total S Emitted	% of Total S Absorbed	Lb. SO_2 Emission per 10^6 Btu
	None	90.61	9.39	6.984
800	25.27% CaO	7.55	92.45	0.582
	37.65% $CaCO_3$	24.08	75.92	1.856
	None	94.69	4.31	7.299
900	25.27% CaO	8.57	91.42	0.661
	37.65% $CaCO_3$	23.67	76.33	1.824
	None	95.10	4.90	7.330
1000	25.27% CaO	9.39	90.61	0.724
	37.65% $CaCO_3$	20.82	79.18	1.605
	None	95.71	4.29	7.377
1100	25.27% CaO	10.00	90.00	0.771
	37.65% $CaCO_3$	22.45	77.55	1.730
	None	98.16	1.84	7.566
1200	25.27% CaO	13.67	86.33	1.054
	37.65% $CaCO_3$	23.67	76.33	1.824
	None	98.98	1.02	7.629
1300	25.27% CaO	25.31	74.69	1.951
	37.65% $CaCO_3$	31.84	68.16	2.454
	None	100.00	0.00	7.708
1400	25.27% CaO	45.71	54.29	3.523
	37.65% $CaCO_3$	52.65	47.35	4.058

Table VII. Effects of CaO and $CaCO_3$ Addition to Sample #3 Containing 6.24% Sulfur (Stoichiometric Ratio = 4)

Combustion Temp. °C	% $CaCO_3$ or CaO Addition by Weight	% of Total S Emitted	% of Total S Absorbed	Lb. SO_2 Emission per 10^6 Btu
	None	90.25	9.75	9.746
800	30.40% CaO	10.90	89.10	1.177
	43.82% $CaCO_3$	21.47	78.53	2.319
	None	90.83	9.17	9.728
900	30.40% CaO	11.70	88.30	1.263
	43.82% $CaCO_3$	19.87	80.13	2.146
	None	90.87	9.13	9.813
1000	30.40% CaO	12.82	87.18	1.384
	43.82% $CaCO_3$	20.83	79.17	2.249
	None	93.27	6.73	10.072
1100	30.40% CaO	16.19	83.81	1.748
	43.82% $CaCO_3$	26.28	73.72	2.838
	None	94.55	5.45	10.210
1200	30.40% CaO	21.47	78.53	2.319
	43.82% $CaCO_3$	25.64	74.36	2.769
	None	97.76	2.24	10.557
1300	30.40% CaO	32.37	67.63	3.496
	43.82% $CaCO_3$	34.29	65.71	3.703
	None	100.00	0.00	10.799
1400	30.40% CaO	39.26	60.74	4.204
	43.82% $CaCO_3$	42.95	57.05	4.638

Comparisons of Test Results

The combustion tests with coals containing 2.10 to 6.24 percent sulfurs showed that CaO has a greater sulfur absorption ability than $CaCO_3$ when both were used with a stoichiometric ratio of four. The samples containing 2.10% and 4.83% sulfurs behaved similarly with the use of CaO as SO_2 absorbant between 800°C and 1200°C, but as the sulfur content of coal increased to 6.24%, the similarity was lost and the sulfur emissions even at lower temperatures became considerably higher. For the sample containing 2.10% sulfur, the EPA Emission limit of 0.60 Lb. $SO_2/10^6$ Btu was somewhat surpassed even at the low temperatures of combustion that were utilized. However, for the sample containing 4.83% sulfur, the EPA Emissions limit of 1.20 Lb. $SO_2/10^6$ Btu was easy to maintain with CaO up to a combustion temperature of 1200°C. On the other hand, as the sulfur content of coal increased to 6.24%, the EPA Emissions limit of 1.2 Lb. $SO_2/10^6$ Btu was exceeded at all the temperatures tested except that at 800 °C.

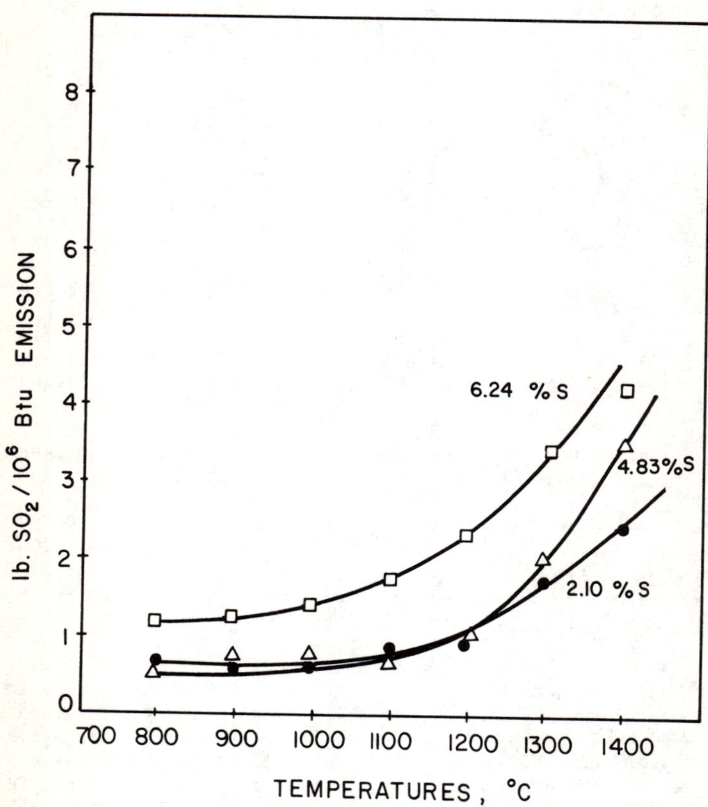

Figure 2. Comparisons of Sulfur Emissions From Three Samples at Various Temperatures With the Use of CaO with a Stoichiometric Ratio of Four.

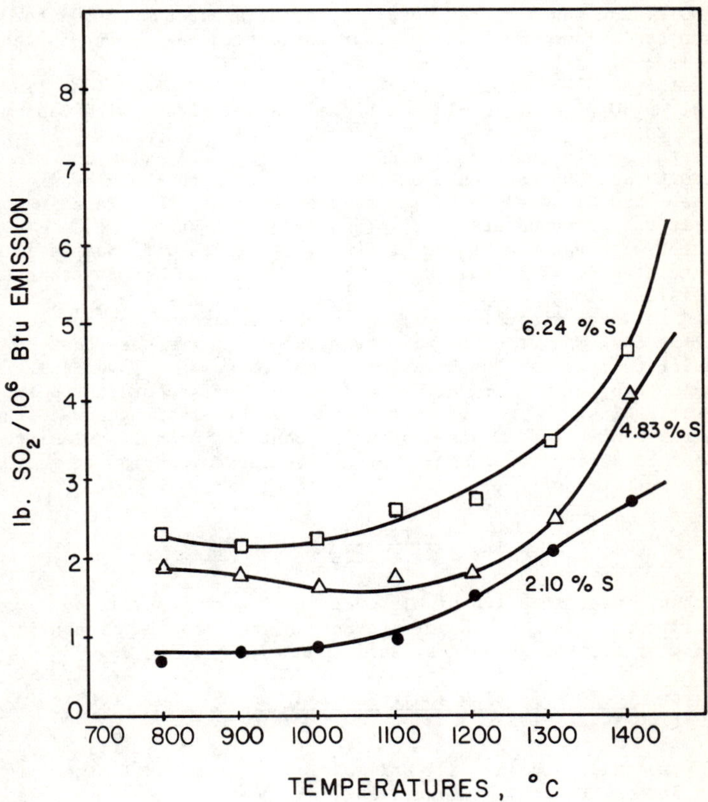

Figure 3. Comparisons of Sulfur Emissions From Three Samples at Various Temperatures With the Use of $CaCO_3$ With a Stoichiometric Ratio of Four.

The use of $CaCO_3$ as sulfur absorbing agent during combustion of coals containig between 2.10% and 6.24% sulfurs indicated that even at low temperatures, the EPA Emissions limit of 0.60 Lb.$SO_2/10^6$ Btu can be easily exceeded with all the samples tested. With coals containing 4.83% and 6.25% sulfurs, even the EPA Emissions limit of 1.2 Lb.$SO_2/10^6$ Btu was surpassed at all the temperatures used during the tests.

Conclusions

From this work the following conclusions were drawn.

1. For the three samples tested at temperatures between 800°C and 1400°C, CaO was a better desulfurization agent than $CaCO_3$ for the same stoichiometric ratio of four sorbent to one sulfur in the coals.
2. For the sample containing 2.10% sulfur, the EPA Emissions limit of 0.60 Lb.$SO_2/10^6$ Btu could not be maintained either with CaO or with $CaCO_3$.
3. For the sample containing 4.83% sulfur, the EPA Emissions limit of 1.20 Lb.$SO_2/10^6$ Btu could only be observed with the use of CaO under 1200°C. When $CaCO_3$ was used as sorbent, the EPA Emissions limit was exceeded at all the temperatures tested.
4. Except with the use of CaO at 800°C, all tests with Sample #3, containing 6.24% sulfur, showed higher emissions of SO_2 than EPA Emissions limit.
5. When the same stoichiometric ratio of one sorbent to four sulfur in the coal was used for all the samples, it was seen that it is easier to meet the EPA Emissions requirements for lower sulfur coals than for higher sulfur coals. However, according to the test results, it is difficult to make a correlation between the sulfur contents of coals and the percentages of sulfurs emitted or absorbed during free burning or controlled combustion of medium to high sulfur coals.

References

1. M. H. Erten, "Beneficiation of Missouri Coals With Emphasis on Desulfurization," Research Report to be Published by the Missouri Department of Natural Resources, Rolla, MO 65401, 1985, 28.

2. T. F. Edgar, Coal Processing and Pollution Control, Gulf Publishing Co., Houston, Texas, 1983, p. 339.

3. L. Grainger and J. Gibson, Coal Utilization, Graham and Trotman, 14 Clifford Street, London, England, 1981, pp. 80-92.

LIBERATION AS A FUNCTION OF SIZE REDUCTION IN THE COAL-ASH-PYRITE SYSTEM

J.W. Perez
Freeport Minerals Company
P.O. Box 26
Belle Chasse, LA 70037

F.F. Aplan
Mineral Processing Section
The Pennsylvania State University
University Park, PA 16802

Abstract

The effect of crushing on liberation behavior in the coal-ash-pyrite system has been evaluated and a classification system developed to describe the different types of feed assemblages. Liberation behavior of the various species is well described in terms of fractional yields within a given size range. A simple negative exponential model has been developed which fits the data well and can be used to simulate yields of coal and pyrite out of raw coal or middlings. Subsequent papers will give further confirmation and applications of the model.

Introduction

Minerals usually occur in ore deposits as heterogeneous aggregates of the various constituent mineral species. Effective mineral beneficiation schemes depend on each mineral species acting, insofar as possible, as an independent entity. It is thus necessary to detach or liberate the desired mineral entity from the undesired species comprising the heterogeneous aggregate. This liberation is achieved by comminuting the rock or aggregate to a size sufficiently small to effect liberation of the dissimilar species. Agricola in his book De Re Metallica (1), published in the 16th century, makes direct reference to this problem: "Although the miners, in the shafts or tunnels, have sorted over the mineral which they mine, still the ore which has been broken down and carried out must be broken into pieces by a hammer or minutely crushed, so that the more valuable and better parts can be distinguished from the inferior and worthless portions".

The need to produce suitable forms of mineral species that meet market conditions gives rise to the two principal objectives of mineral processing: the liberation of mineral species followed by the concentration of the desired species. Thus, liberation is of paramount importance to any concentration process. Economically, it is desirable to achieve maximum liberation at the coarsest possible size, and the degree or extent of liberation will often determine which separation process is to be utilized. In some instances, the purity of the final concentrate must be sacrificed because of the inability of the concentrating device to treat feeds with a high degree of fineness. The comminution step is thus an important variable affecting not only the nature of the concentration process but also the recovery and grade of the final concentrate produced.

The prediction of the degree of liberation and the characteristics of the resulting particle assemblage (which constitutes the feed to the concentration step) is an important consideration for the optimization of any mineral processing operation. Mathematical models capable of predicting liberation as a function of size reduction will fulfill this need. Although several such models are available, none of these, unfortunately, have gained wide industrial application. While they offer hope for the future, these models, at present, generally suffer from problems of mathematical complexity and/or their inadequate state of development and testing. The purpose of this work is, therefore, to describe the size reduction-liberation characteristics of coal from its associated refuse material (ash and pyrite) and to present a simple, empirically based, mathematical model to describe such characteristics.

Background

Historically, the size reduction necessary for mineral liberation has been estimated from grain size measurements of polished sections (2,3,4) and/or done totally pragmatically by grinding the ore to various sizes and evaluating the response of the ground product to the concentrating step (4). Liberation modeling by empirical means has been widely practiced in the form of release analysis curves, separation curves and grade recovery curves (5, 6). These procedures usually consist of developing a series of grade-recovery or grade-yield curves by concentrating the material following grinding to various degrees of fineness.

Gaudin (7) was the first to quantify liberation phenomena in the form of a physico-mathematical model for an ideal binary system. Weigel and Li (8) improved on Gaudin's model to the extent that their model provided information on the nature of the resulting locked particles and removed the limitation imposed by Gaudin on grain arrangements, that is, they assumed a randomized distribution of the mineral grains. The main limitations of both models are the use of a cubic geometry to describe both the individual mineral grains and the aggregates and the use of a cubic fracture lattice which generates cubic particles of uniform size. Another approach to liberation modeling utilizes mass-size balances to describe simultaneously liberation and size reduction in a combined manner. In a binary ore of components A and B, one can describe the rate of appearance and disappearance of species A, B and AB and, simultaneously, the corresponding rates for the different size ranges of each of the species. This approach has been reported by Andrews and Mika (9) and more recently by Peterson and Herbst (10); the latter specifically addressing the problem of liberation in the coal-ash-pyrite system. This approach, not withstanding its attractiveness, is highly complex and requires large computational facilities. Still another approach, is that taken by King (11) Finch and Petruk (12), Klimpel (13) and Meloy and Gotoh (14). Basically this approach utilizes a probabilistic strategy to predict the degree of liberation. It is assumed that the method of breakage does not affect liberation, but that only the degree of size reduction affects liberation.

Criteria for Liberation

The criteria used to define liberation will vary depending on the modeling approach utilized, either for mathematical convenience or because of the demands placed on it by the particular field of application. As a general rule, empirical approaches to liberation modeling use criteria based on recovered total units of mineral value or total yield of a given product. Models such as those of Gaudin (7), Wiegel-Li (8), King (11) and others (12, 14) have generally used a definition based on a volume or number basis; that is, the volume of a free mineral species (V_A or V_B), divided by the volume of the free plus the volume of locked mineral species in question (V_{AB}). This criterion has several problems associated with it: (1) in most cases the actual free (e.g. V_A) and locked (V_{AB}) volumes are difficult to determine, often involving a tedious microscopic counting procedure, (2) in the case where the weights of A and B are used, the specific gravity of A and B must be known and that of the AB mixture estimated, (3) the percentage liberation does not relate to the process variables used in coal processing. For this reason, a new criterion was developed. This criterion is not only easily associated with coal processing variables, but it is equivalent to the previous one if some simplifying assumptions are made.

The criterion proposed here uses a weight basis (thus avoiding microscopic procedures), incorporating the concept that a perfect mineral separation is rarely achievable and relating it to the well-known yield terminology. For a binary system containing species A and B, the fractional yield (Y) of any constituent is defined as follows:

$$Y_A(X) = \frac{W_A(X)}{W_A(X)+W_B(X)+W_{AB}(X)} \quad (1)$$

$$= \frac{W_{A(X)}}{W_{A+AB+B}(X)} \quad (2)$$

W_A is the weight of species A at size X as characterized by a specific gravity range, W_B is the weight of species B and W_{AB} is the weight of the locked fraction. A similar criterion was developed by Muller et al.(15) in their work on tin ore beneficiation.

A special case is worthy of note, that is when the relative abundance of A is much larger than B. This is the case when the liberation of A is massive in comparison to that of B and $W_A >> W_B$:

$$Y_A(X) = \frac{W_A(X)}{W_A(X)+W_{AB}(X)} \tag{3}$$

This case is of practical importance and is typically encountered when coal-ash-pyrite middlings are crushed and large quantities of clean coal are liberated while only minor amounts of ash and pyrite are freed.

Further, since $W_A = \bar{\rho}_A V_A$ and $W_{AB} = \bar{\rho}_{AB} V_{AB}$, where $\bar{\rho}$ is the specific gravity of the mineral entity, then:

$$Y_A(X) = \frac{1}{1 + (\bar{\rho}_{AB}/\bar{\rho}_A) V_{AB}(X)/V_A(X)} \tag{4}$$

In this case, an equivalence between both the volume-number liberation and the fractional yield criteria is easily found and is the liberation (L) for species A and B at any size, X, as given by the expressions:

$$L_A(X) = \frac{1}{1+ \frac{1}{Y_A(X)} -1 \quad (\bar{\rho}_A/\bar{\rho}_{AB})} \tag{5}$$

$$L_B(X) = \frac{1}{1+W_{AB}/W_B(\bar{\rho}_{AB}/\bar{\rho}_B)} \tag{6}$$

A relationship of far more practical consequence is that between fractional yield and total or cumulative yield. The total yield of a given product (phase) is the sum of the products of the individual fractional yields, $Y_A(X)_i$, times the fraction, by weight, $f(X)_i$, contained in each individual size interval (i.e., the fractional yield times the weight density):

$$Y_{A_{TOTAL}} = \sum_{i=1}^{n} Y_A(X)_i f(X)_i \tag{7}$$

Classification of Liberation Systems for Crushed Coal

Before one can consider the liberation of various combinations of coal, pyrite and ash, it becomes necessary to make a systematic classification of the experimental system under study. Such a classification is presented in Table I and shows the equivalence between several of the systems studied. Basically, there are three general groupings. The first, Group I, is one in which a rich matrix has inclusions of minor amounts of one or more lean species. In this group belongs feed samples whose particles are in the 1.4-2.0 and 2.0-2.96 specific gravity ranges. These gravity fractions represent largely carbopyrite particles in the former and essentially free ash, coal-ash or ash-pyrite particles in the latter. Group II corresponds to a material composed of a very rich matrix which contains inclusions of very minor amounts of one or more other mineral species. This group differs from the previous one in the sense that the material may be considered to be essentially liberated from the point of view of the matrix-forming mineral. To this group is assigned the 1.4 float specific gravity feed material and 2.96 sink specific gravity feed. The last group, Group III, corresponds to mixtures of the other two groups. This group encompasses raw coals which may be considered a mixture of species whose specific gravities range from about 1.20 to 5.0. Raw coal then consists of particles of 'free' coal, 'free' pyrite, 'free' ash, and locked particles representing combinations of the preceding species. Besides raw coal this group would also include such products as the 2.0 sink material composed largely of various free and locked ash and pyrite particles with very much smaller amounts of coal.

Table I. Classification of feed samples to be liberated.

Grouping	Specific Gravity of Feed Sample	Composition		
		Coal	Ash	Pyrite
Group I				
Rich matrix with inclusions of lean species.	1.4-2.0	rich	lean	lean
	2.0-2.96	very lean	rich	lean
Group II				
Very rich matrix with inclusions of very lean species.	1.4 float	very rich	very lean	very lean
	2.96 sink	very lean	very lean	very rich
Group III				
Highly variable matrix with inclusions of highly variable species	raw coal	variable	variable	lean to very lean
	2.0 sink	very lean	rich	variable

The specific gravity size ranges selected here to represent 'free' coal (1.4 float), 'free' pyrite (2.96 sink), carbopyrite particles (1.4 x 2.0), etc. are arbitrary but were found to be reasonable for the coals tested here. Other coals and other use specifications may require a somewhat different definition as would the desires of the experimentalist. From Table I, it should be noted that in each group it is possible to study the liberation of either the rich or the lean phase. As will be demonstrated later, the response of Group II is similar to that of Group I, and, for all practical purposes, the very rich phase may be considered to be liberated. For this reason, attention in this paper will be focused principally on the presentation of experimental data representing Group I. A subsequent report will consider the problem of Group II and the prediction of yields out of raw coal (Group III) (16).

One final word about this type of classification system is in order. A similar classification may be applied to metalliferous ores and industrial minerals. For example, a disseminated porphyry copper ore may be classified into Groups I and II depending on the relative abundance of the mineral assemblage under study.

Experimental Methods and Equipment

Sample Preparation

Several kinds of feed samples were used during the course of this work, representing five specific gravity fraction (1.4 float, 1.4-2.0, 2.0 sink, 2.0-2.96 and 2.96 sink) and three different coals. The coals used were: Lower Kittanning (Brookdale), Pittsburgh Seam (Robena) and Upper Freeport (Helvetia). Two different preparation procedures were utilized to prepare the feeds, Procedures A and B (Figures 1 and 2). Both of these procedures were designed to generate feed samples of narrow specific gravity and narrow particle size ranges.

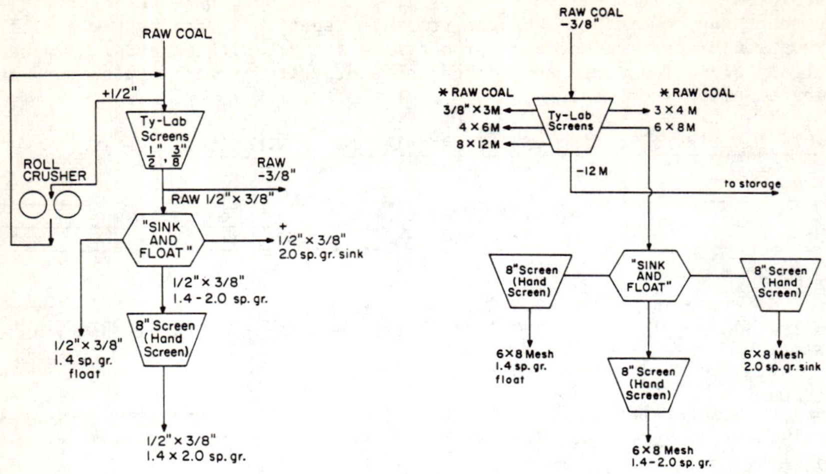

Figure 1. Sample Preparation scheme for 1/2" x 3/8" feeds (Procedure A).

Figure 2. Sample preparation scheme for -3/8" feeds (Procedure B).

Analytical Procedures

The analyses of the raw coals used in this study are shown in Table II. Though not reported here, the washability of each coal was determined using standard float/sink techniques, and these data were used to guide the selection of the specific gravity and particle size ranges to be evaluated (17). To the same end, the pyrite grain size and intercepted chord length distributions were estimated through the use of the automated reflectance microscopy (ARM) (18) on one inch polished sections of each coal.

Table II. Analyses of raw coal samples (dry basis)*.

Analysis	Lower Kittanning	Upper Freeport	Pittsburgh Seam
% volatile matter	17.34	25.35	36.25
% fixed carbon	69.56	54.03	47.61
% ash	13.10	20.62	16.14
BTU/lb, ash free	15,567	15,114	15,116
% sulfur, total	2.95	2.95	2.53
% sulfur, pyrite	2.22	2.59	1.36
% sulfur, organic	0.41	0.35	1.15
% sulfur, sulfate	0.32	0.01	0.02
ASTM rank	LV bit	HVA bit	HVA bit

* Analyses by Warner Labs, Cresson, PA.

Comminution Systems

Two different comminution devices, a roll crusher and a hammer mill, were used in the work reported here. The roll crusher was an 8"ϕx5" laboratory double roll crusher, smooth type, spring mounted (Sturtevant Mill Co., Boston, Mass.), and the hammer mill was an 8"ϕx6" swing hammer mill (O.B. Wise Co., Knoxville, Tenn.) with three swing hammers (stirrup type). Great care was taken to assure that essentially all of the material that was crushed was recovered. Further experimental details may be found in a thesis by one of us (17).

Experimental Results

The liberation response of feed samples in the 1.4-2.0 specific gravity range, as a function of size reduction, was studied extensively. Special emphasis was placed on the fractional yield of the coal fraction (defined here as the 1.4 float) and of the pyritic fraction (approximated by the 2.96 sink) as influenced by coal seam, feed size and crushing device.

Figure 3 illustrates the typical behavior exhibited upon crushing of a 1.4-2.0 specific gravity feed crushed with a smooth roll crusher. Using the fractional yield concept (the yield of a given product in a given size range), it is seen that the coal liberated out of this specific gravity and size range reaches an asymptotic value and appears to follow an exponential form. By way of contrast notice that the pyrite fractional yield does not asymptotically approach some fractional yield value as particle size becomes smaller. This suggests that for coal the liberation attained here is close to the practical maximum, while the pyrite liberation is not even close to its ultimate value.

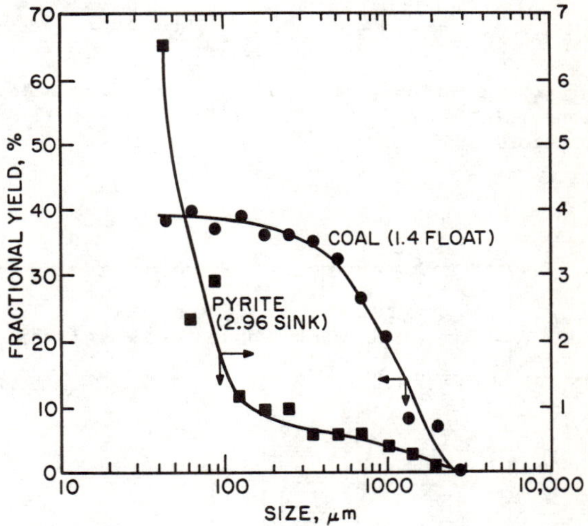

Figure 3. Fractional yield of coal and pyrite from a Lower Kittanning carbopyritic feed (6 x 8 Mesh, 1.4 x 2.0 sp. gr.), crushed in a smooth roll crusher.

Figures 4 and 5 illustrate that this behavior is not unique to the Lower Kittaning coal sample since 'coal' (1.4 float) and 'pyrite' (2.96 sink) liberated out of identical specific gravity fractions of other coals follow a similar pattern. Moreover, this type of response is repeated when the size reduction is carried out in other milling devices (see Figure 7 and discussion below). Feed size does not influence this behavior significantly, as illustrated in Figure 6.

Other specific gravity fraction feeds show a similar behavior, as will be demonstrated subsequently; that is, the predominant phase behaves in a similar fashion to the coal fraction liberated from a 1.4-2.0 specific gravity feed. Likewise the lean phase behaves in a similar fashion to the response shown by the pyritic fraction in Figures 3 and 5.

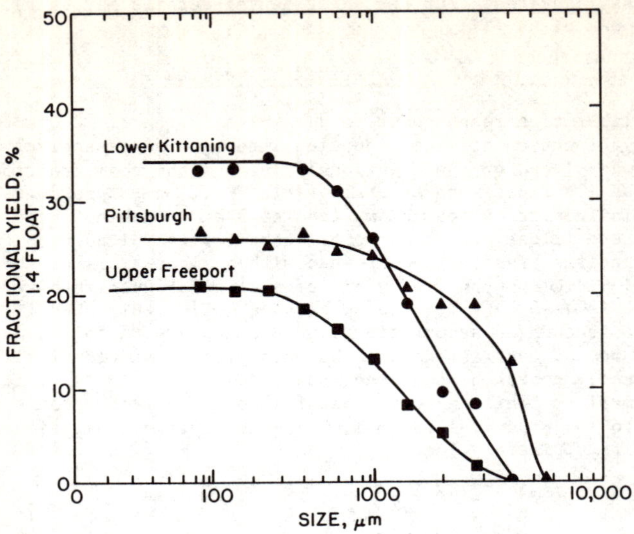

Figure 4. Fractional yield of coal from some Allegheny Series carbopyritic feeds (1/2'x3/8", 1.4x2.0 sp. gr.), crushed in a smooth roll crusher.

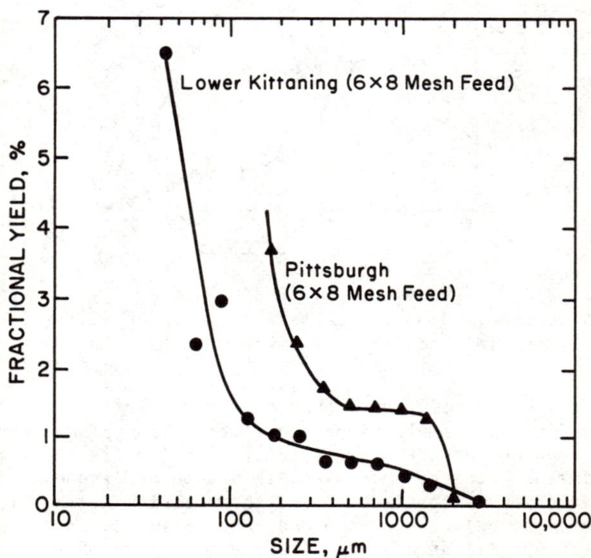

Figure 5. Fractional yield of pyrite from Pittsburgh Seam and Lower Kittanning carbopyrite feeds (6x8 mesh, 1.4x2.0 sp. gr.), crushed in a smooth roll crusher.

At least for the systems studied here, these results indicate that the method of breakage does not greatly affect liberation, with the degree of size reduction being the major determinant, as illustrated in Figures 6 and 7. Kiss and Schonert (19) have reached a similar conclusion for another system. This conclusion and the observed repetitive liberation behavior indicates the possibility of postulating a simple model to describe the liberation phenomena (17).

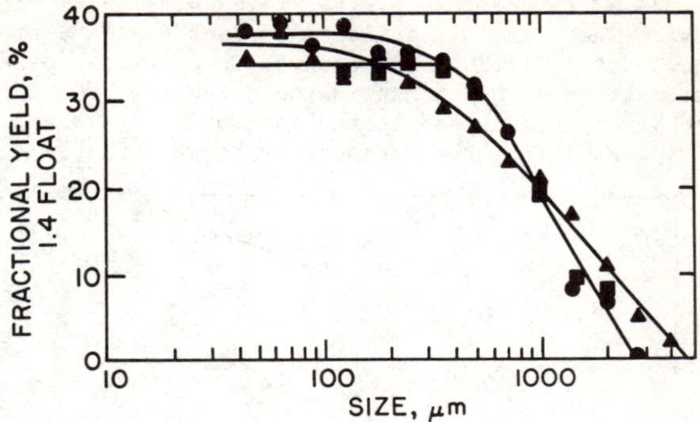

Figure 6. Fractional yield of coal as a function of feed size. Carbopyrite feeds (1.4x2.0 sp. gr.), crushed in a smooth roll crusher.
● 6x8 Mesh, ▲ 3x4 Mesh, ■ 1/2x3/8 Inch.

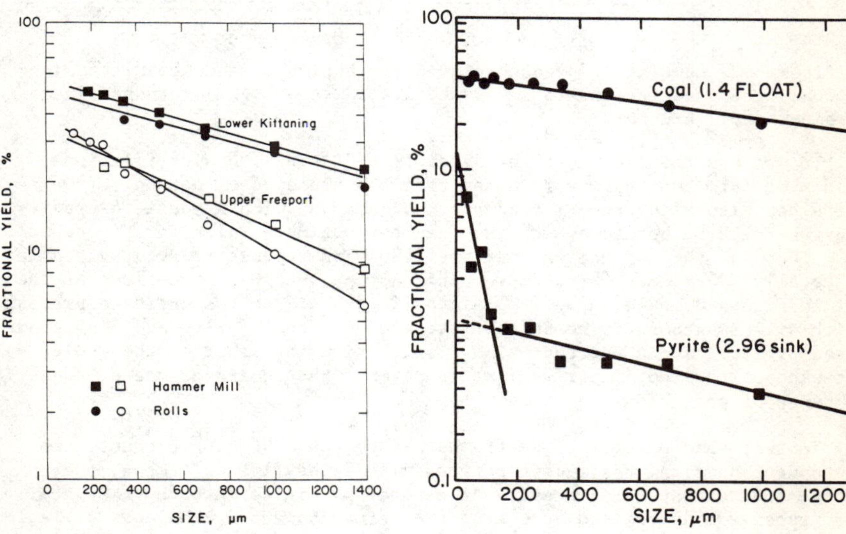

Figure 7. Fractional coal yield using various milling devices. Carbopyrite feed (1/2"x3/8", 1.4x2.0 sp. gr.).

Figure 8. Fractional yield of coal and pyrite from a Lower Kittanning carbopyritic feed (6x8 mesh, 1.4x2.0 sp. gr.). Linearized form.

A Model for Liberation

The fact that the fractional yield versus log size plots for 1.4 float and 2.96 sink specific gravity fractions coming from carbopyrite feeds showed what appears to be an exponential type of response (see Figures 3-6), suggests that straight lines could be obtained if the plotting procedure was reversed. If confirmed, this would provide a simple predictive equation. A plot of linear size versus log fractional yield for 'coal' (Figures 7 and 8) and for 'pyrite' (Figure 8) shows this to be exactly the case. The 'coal' (1.4 float) product yields a single straight line, while the 'pyrite' (2.96 sink) product is seen to separate into two distinct linear regions. The change in the curve at ∿100 μm for the Lower Kittaning is to be expected based on the data of Figure 3 and confirmed by the Schuhmann sizing plot of the crushed pyrite (Figure 9).

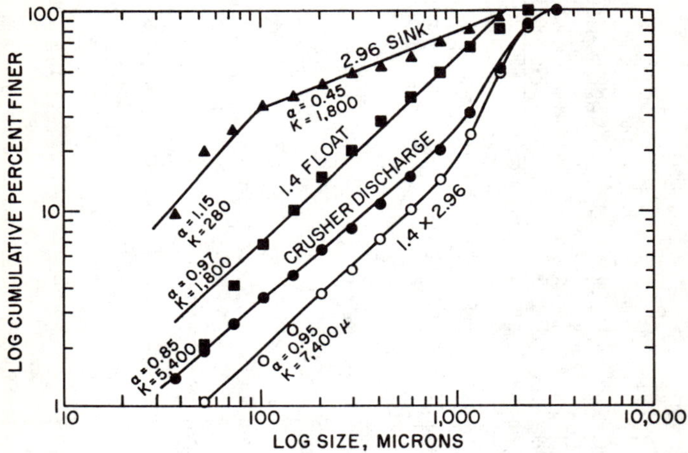

Figure 9. Cumulative product size distribution of a Lower Kittanning Coal carbopyritic feed (6 x 8 Mesh, 1.4 x 2.0 sp. gr.), crushed in a smooth roll srucher. ○ Feed.

It was hypothesized that the separation of the 2.96 sink fractional yield plot into two regions resulted from the presence of not only free pyrite but also high grade locked pyritic material with a specific gravity in excess of 2.96 but somewhat less than the ultimate value for pyrite at 5.0. That is, the assumption that the 2.96 sink represents free pyrite is not totally adequate. Furthermore, this abrupt change in the slope of the plot in Figure 8 could only be obtained if each of the two entities present (which will subsequently be demonstrated to be 'free' pyrite and high grade locked pyrite) have a narrower size distribution than that of the whole, so that each of the two linear regions represents the domain of one of these two types of particles.

To test this premise as to the nature of the 2.96 sink product, two independent physical tests were performed. In the first of these, the average apparent specific gravity of several particles in each of several size ranges was determined by use of the Berman balance. The data of Table III shows that a drastic change in the particle specific gravity occurs at roughly 100 μm for this coal pyrite, about the same size where change was noted in the fractional yield plot.

Table III. Specific gravities of particles from the 2.96 product from crushed 1.4-2.0 sp. gr. sample of Lower Kittanning coal.

Tyler Mesh Size	Average Size, μm	'Apparent' Specific Gravity
8 x 10	1981	3.16
10 x 14	1397	3.17
14 x 20	991	3.29
20 x 28	701	3.33
28 x 35	495	3.16
35 x 48	351	3.38
48 x 65	246	3.50
65 x 100	175	4.27
100 x 150	125	4.59
150 x 200	88	4.73
200 x 270*	63	4.71
270 x 400	44	-

* limit of detection

In the second set of confirmatory tests, samples of the 'free' and the high grade locked pyrite were briquetted, polished and examined by reflected light microscopy. Representative particles from each of these groups are shown in Figures 10 and 11. It can be readily observed that there is a substantial difference between the coarser 48 x 65 mesh (295 x 208 μm) sized particles and those in the finer size range, 270 x 400 mesh (44 x 37 μm).

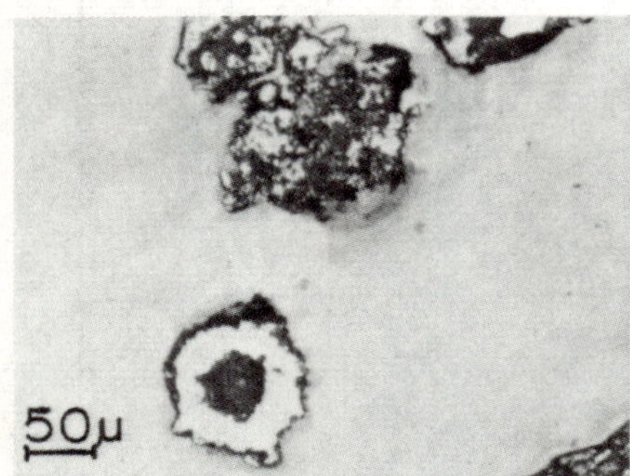

Figure 10. Typical morphological composition of a 48x65 Mesh, 2.96 sink particles from a crushed carbopyrite feed, Lower Kittanning Seam (1/2"x3/8", 1.4x2.0 sp. gr.).

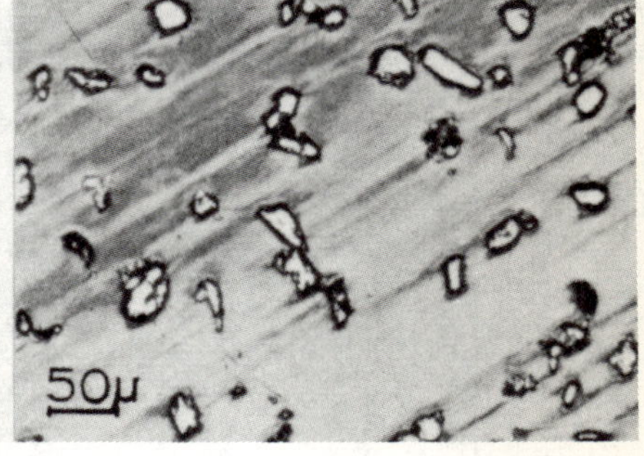

Figure 11. Typical morphological composition of 270x400 Mesh, 2.96 sink particles from a crushed carbopyrite feed. Lower Kittanning Seam (1/2"x3/8", 1.4x2.0 sp. gr.).

The coarse size fraction is composed largely of sponge-like aggregates (high grade locked) with an average specific gravity of ∿3.2 and some free pyrite, while the particles in the finer fraction are coherent and have an average specific gravity of ∿4.7 which is close to the textbook value of 5.05. A similar pattern has been found for the pyrite contained in the three coals studied here as well as that from Lower Freeport seam, with the transition occurring in the 100 - 300 μm range. It may be anticipated that the observed break in the curve will be a function of the pyrite genesis in the particular sample under study and will also be influenced by the nature of the pyrite in any adjacent strata which may become incorporated into the raw coal during mining. Microscopic evaluation of the finer particles using both polished section microscopy (see Figure 11) and transmitted light microscopy convincingly demonstrates the cohesive nature of the fine pyrite particles.

The linearization procedure was also found to apply to the products of size reduction resulting from the use of other milling systems (Figure 7) and of other specific gravity feeds (Figure 12). The crushing of a 2.0-2.96 specific gravity feed illustrated in Figure 12 can generally be regarded as a rich shaly matrix with small inclusions of pyrite and/or coal. For convenience the evaluation of liberation to be discussed here will be limited to the production of clean coal and of pyrite. The response for both of these products is seen to be similar to that observed for carbopyritic feeds. Nevertheless, two points of difference are observed: (1) the asymptote value for the 2.96 sink fraction is lower (this is not unexpected since this feed contains only about 4% by weight of pyrite in constrast to ∿6% pyrite in the carbopyritic feed) and (2) the plot for the clean coal fraction has a much steeper slope than that noted for coal liberated out of the coal-rich carbopyrite (1.4 x 2.0) fraction.

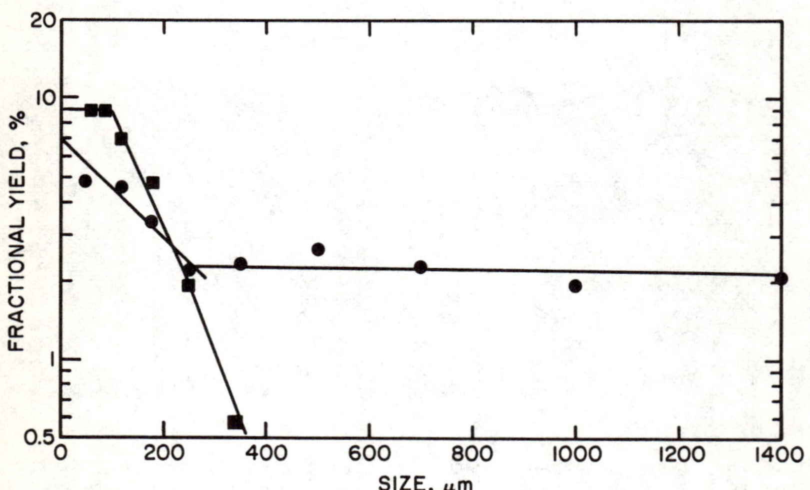

Figure 12. Fractional yield of coal and pyrite from Lower Kittanning concentrates (6 x 8 mesh, 2.0 x 2.96 sp. gr.), crushed in a smooth roll crusher.
■ 'coal', 1.4 float ● 'pyrite', 2.96 sink.

This linearization implies that the liberation behavior of the different species in the coal-ash-pyrite system can be described with the aid of a simple negative exponential model, with an equation of the form:

$$\text{Fractional Yield of A} = ae^{-X/u} \qquad (8)$$

where a is a constant, X is the particle size and u an empirical constant related to the average grain size of species A. In the case where the curve breaks into two linear regions, as with the 2.96 sink material, the equation can be modified in the following manner:

$$\text{Fractional Yield of B} = be^{-X/u} + b'e^{-X/u'} \qquad (9)$$

where b and b' are constants, X is the size and u and u' are empirical constants. This is the form that should be used for the liberation of pyrite out of a carbopyrite (1.4 x 2.0 sp.gr.) feed or for any species that still contains particles showing significant amount of locking. The evaluation of these constants is rather simple, since the constants a, b, and b' can be determined from the corresponding y intercepts of the lines, and 1/u and 1/u' are the corresponding slopes. The prediction of total or cummulative yield from both selected specific gravity feeds and raw coals (which, for all practical purposes, are mixtures of various specific gravity feeds) will be addressed in a future publication (16). Furthermore, this model has also been successfully applied to describe the liberation of metallic mineral species from an ore and this is also slated for future publication.

Conclusions

1. A simple classification system has been developed to describe the nature of various feeds to be liberated in the coal-ash-pyrite system. The system correlates morphology, composition and specific gravity in such a way that feeds with a similar response to liberation upon crushing are classified within the same group.

   Group I Rich matrix with inclusions of lean species.
   Group II Very rich matrix with inclusions of very lean species.
   Group III Highly variable matrix with inclusions of highly variable species.

2. An analysis of the data generated in this work showed that fractional yield (the yield of a given constituent out of a given size range divided by the total amount of material in that size range) is an especially attractive way of expressing liberation in the coal-ash-pyrite system. In addition to describing the liberation of a given species out of a locked multiple species assemblage, fractional yield can also be used to determine the total yield of a given species by a summation process.

3. The liberation behavior of different species in the coal-ash-pyrite system can be described by a simple, empirical, mathematical model of the form:

$$\text{Fractional Yield of A} = ae^{-X/u}$$

where a and u are empirical constants easily determined from a linear size-log fractional yield plot of the data, X is the size and A is the species in question.

For the situation where liberation of a given species is largely but not totally complete (often typified by pyrite liberated out of coal), the size - fractional yield plot is seen to break into two linear segments each described by an equation of the above form.

4. The liberation behavior of different species out of various coal fractions, on crushing, can be predicted by use of the negative exponential model, the concept of fractional yield and the categorization of feed samples described here.

5. Future work will demonstrate that this technique can be used to predict liberation of a given species out of raw coal and ore samples.

Acknowledgements

The authors wish to acknowledge the Pennsylvania Science and Engineering Foundation and The Appalachian Regional Commission for their financial support. The help of Professor Alan Davis and Dr. A. Moza of The Pennsylvania State University, with the reflected light and automated microscopy techniques, is also gratefully acknowledged.

List of Symbols

A Phase A in a mineral aggregate, usually the abundant phase

B Phase B in a mineral aggregate, usually the lean phase

AB Locked particle containing phases A and B

L Liberation of a given species

V Volume of a given species

W Weight of a given species

X Particle or grain size

f(X) Fraction, by weight, of size X

Y Fractional yield of a given species

i Index indicating number of size interval

$\bar{\rho}$ Specific gravity of a given species or aggregate

a,b Empirical constants determined from the y intercept of the linear size log fractional yield plot

u,u' Empirical constants determined from the reciprocal of the slope of the same plot

References

1. Georgius Agricola, De Re Metallica, Translated from the first Latin edition of 1556 by Herbert C. and Lou Henry Hoover (New York, NY: Dover Publications, 1950).

2. R.E. Head, A.L. Crawford, F.E. Thackwell and G. Burgener, "Detailed Statistical Microscopic Analysis of Ore and Mill Products of the Utah Copper Co.," (USBM RI 3288, 1935).

3. J.T. McCartney, H.J. O'Donnell and S. Ergun, "Pyrite Size Distribution and Coal-Pyrite Particle Association in Steam Coals", (USBM RI 7231, 1969).

4. F.F. Aplan, "Mineral Processing - Evaluation to Indicate Processing Approach", SME Mining Engineering Handbook, ed. A.B. Cummings and I.A. Given (New York, NY: AIME, 1973) Vol. II, Section 27, 15-28, 86-89.

5. C.C. Dell, "Release Analysis, A New Tool for Ore Dressing", Recent Developments in Mineral Processing (London: Inst. Min. and Met., 1953) 75-84.

6. A.A. Jowett, "A Mathematic Form of Mineral Separation Curves", Inst. of Min. and Met., 78 (1969) C185-190.

7. A.M. Gaudin, Principles of Mineral Dressing, (New York, NY: McGraw-Hill, 1939), 70-84.

8. R.L. Weigel and K. Li, "A Random Model for Mineral Liberation by Size Reduction", Trans. AIME, 238 (1967) 179-189.

9. J.R.G. Andrews and T.S. Mika, "Comminution of Heterogeneous Material: Development of a Model for Liberation Phenomena". XI International Mineral Processing Congress, Cagliari, Vol. I (1975), 59-88.

10. R.D. Peterson, R.D. and J.A. Herbst, "Estimation of Kinetic Parameters of a Grinding-Liberation Model", Int. Jnl. of Min. Process., 14 (1985), 111-126.

11. R.P. King, "A Model for the Quantitative Estimation of Mineral Liberation by Grinding", Int. Jnl. of Min. Process., 6 (1979), 207-220.

12. J.A. Finch and W. Petruk, "Testing a Solution to the King Liberation Model", Inst. Min. and Met, 87 (1978) C272-277.

13. R.R. Klimpel, "Applications of a Model for the Analysis of Liberation from a Binary System", Powder Tech., 39 (1984), 117-128.

14. T.P. Meloy and K. Gotoh, "Liberation in a Homogeneous Two-Phase Ore", Int. Jnl. of Min. Process., 14 (1985), 45-55.

15. L.D. Muller, K.J. Henley and R.E.K. Benjamin, "Applied Mineralogy in Tin Ore Beneficiation", Second Technical Conference on Tin, Bangkok, ed. W. Fox, Vol II (1969), 559-598.

16. J.W. Perez and F.F. Aplan, "Yield Prediction by a Simple Liberation Model", manuscript in preparation.

17. J.W. Perez, "Liberation as a Function of Size Reduction in the Coal-Ash-Pyrite System", (Ph.D. Thesis, The Pennsylvania State University, 1981).

18. A. Davis and F.J. Vastola, "Developments in Automated Reflectance Microscopy of Coal", Journal of Microscopy, 109 (Part 1) (1977), 3-12.

19. L. Kiss and K. Schonert, "Liberation of Two-Component Material by Single particle compression and Impact Crushing", Aufberectungs. Technick., 5, (1980), 223-230.

Process Mineralogy Applications to Liberation

PREDICTING AND MEASURING MINERAL LIBERATIONS IN ORES AND MILL PRODUCTS, AND EFFECT OF MINERAL TEXTURES AND GRINDING METHODS ON MINERAL LIBERATIONS

W. Petruk

CANMET, Energy, Mines and Resources, 555 Booth Street, Ottawa, Canada

Abstract

Mineral liberations that would be obtained when an ore is ground can be predicted by analysing the minerals in an unbroken ore and applying a liberation model. The mineral liberations that are obtained in a broken ore can be measured by analysing mill products with an image analyser. A series of tests was performed to assess the predictions and measuring techniques. In most instances the correlation between the predicted and measured values was close, but for galena and cassiterite in base metal ores, and for chalcopyrite lamellae in sphalerite, the measured values were significantly higher than predicted. It is interpreted, from the anamalous values, that mineral liberation is controlled by the breakage characteristics of the host mineral. If the host mineral breaks into particles whose size distribution is the same as the screen analysis for the broken ore, the observed liberation will correspond to the predicted liberation. If, on the other hand, the host mineral breaks into particles that are smaller than the size distribution for the ore, the observed liberation will be higher than the predicted liberation.

Introduction

Mineral liberation studies in connection with mineral processing involve analysing the minerals in an unbroken ore to predict the percentage of mineral that would be free when the ore is ground, and measuring the percentage of mineral that is free in ground products. If the basis for predicting the mineral liberations is valid, and if accurate measurements are made from samples that are representative of the ore and mill products, there should be a close correlation between the predicted and observed mineral liberations. In the Process Mineralogy Section at CANMET we routinely analyse polished sections of unbroken ores with an image analyser to predict mineral liberations that would be obtained when the ore is ground, and measure mineral liberations in the ground ore. Close correlations are usually obtained, but the observed mineral liberations for galena, chalcopyrite and cassiterite in base metal ores are often significantly higher than predicted. A study was conducted to find the reason for this observation by evaluating the methods of predicting and measuring mineral liberations, by correlating the mineralogy, ore textures and mineral grindability to mineral liberations, and by considering grinding methods.

Liberation Models

Many liberation models have been developed for predicting mineral liberations that would be obtained when an ore is ground, and others are being developed (1,2,3). The models are generally applied by determining the size distributions of minerals in an unbroken ore and using the data to predict mineral liberations; some models incorporate mineral textures (4). The size distributions can be determined by measuring either the intercept length or the surface area of the grains with an image analyser. When the intercept length is measured it is considered to be equal to the grain diameter. When the surface area of grains is measured the grain diameter can be either measured separately or calculated from the surface area. Grain diameters that can be determined are the maximum diameter, minimum diameter, equivalent circle diameter and the equivalent square diameter. Workable models for determining size distributions by image analysis were developed by King (5) for using intercept lengths, and by Klimpel (6) and Petruk (7) for measuring surface areas. King's and Klimpel's liberation models predict the percentage of mineral that would be free in each screened fraction, whereas Petruk's liberation model predicts the minimum grind, optimum grind and the total liberation that would be obtained at a particular grind. King's liberation model was simplified by Finch and Petruk (8) so that it requires only one parameter. The simplified King liberation model and Petruk liberation model were used in this study.

Simplified King Liberation Model

The model assumes random breakage of an ore. It was developed for the measurement of the mean intercept length of a mineral in an unbroken ore. The model predicts the fractional degree of mineral liberation in a screened fraction by the relationship:

$$Lm(D) = 1 - \frac{D}{2\ \mu m + D}\left[1 - \exp-\left(\frac{2\ \mu m + D}{\mu m}\right)\right]$$

where $Lm(D)$ = fractional degree of liberation of mineral m, for mesh size D,

D = geometric mean of mesh size D

μm = mean intercept length of mineral in unbroken ore

Petruk Liberation Model

The model assumes that a mineral will be liberated when an ore is broken into particles that are the same size as or smaller than the size of the mineral grains in an unbroken ore. the model was developed for predicting the minimum and optimum grinds in connection with mineral dressing, and for predicting the apparent total liberation that would be obtained at a particular grind. It involves measuring the surface areas of grains in an unbroken ore, using the equivalent square diameter as the grain diameter, and plotting the cumulative size distribution of a mineral on a log-linear graph on the basis of area % of particles within each size range. The minimum grind is defined as the grind at which the first reasonable mineral recovery (generally about 50 to 70%) can be obtained in an acceptable grade concentrate by normal concentration techniques. Measurements have shown that about 50 to 60% of the mineral is liberated at this grind, but the liberation model assumes 50%. The optimum grind is the point at which increased grinding will not significantly improve liberation or metal recovery. The model defines the minimum grind as a grind which gives a size distribution of particles (80% passing) that is equivalent to the grain size distribution for the mineral in question. the optimum grind is the point where the cumulative size distribution curve for the mineral, plotted on a log-linear scale, flattens. The approximate total liberation when the grind is finer than the size distribution of the mineral is (100 - (value at 95% point on size distribution curve for mineral) + 1/2 (value at 95% point on size distribution curve for mineral). The approximate total liberation when the grind is coarser than the size distribution of the mineral is 1/2 (point on size distribution curve of grind which corresponds to the size of the 95% point on size distribution curve of mineral).

Methods of Analysing Mineral Grains in Unbroken Ore

Intercept lengths

The size distribution of the intercept lengths and the mean intercept length can be determined with an image analyser by scanning along lines that are spaced at such intervals that the same grain is not intersected twice.

Surface Areas

When the minerals are equidimensional and occur as separate grains, their surface area, maximum diameter, minimum diameter, the equivalent circle diameter and the equivalent square diameter can be measured with most image analysers. The equivalent square diameter, which is the square root of the surface area, is used at the CANMET laboratories because it gives the best correlation to screen analyses. The size distribution of a mineral in an unbroken ore is obtained by measuring the surface area of each grain, determining the appropriate grain diameter, and calculating the percentage of surface area that is covered by grains within each size range. All grains within the fields of view must be analysed. When analysing grains with an image analyser, touching grains would be measured as one large grain instead of two small ones, and a wrong size distribution would thus be obtained. It is possible, however, to separate the touching grains before analysis with most image analysers, by using erosion-dilation and boolean operation techniques. If the grains in the unbroken ore are interconnected with veinlets or elongated grains, the image of the veinlets and elongated grains must be separated from the image of irregular near equidimensional grains and saved in a separate image memory. This has been done with a Kontron image analyser and is shown as an example in Fig. 1. The surface area and equivalent square diameter would be determined for each feature in the image of the near equidimensional grains, and the surface area and minimum diameter would be determined for each feature in the image of veinlets and elongated grains. The data from the two measurements would be combined to produce a size distribution curve.

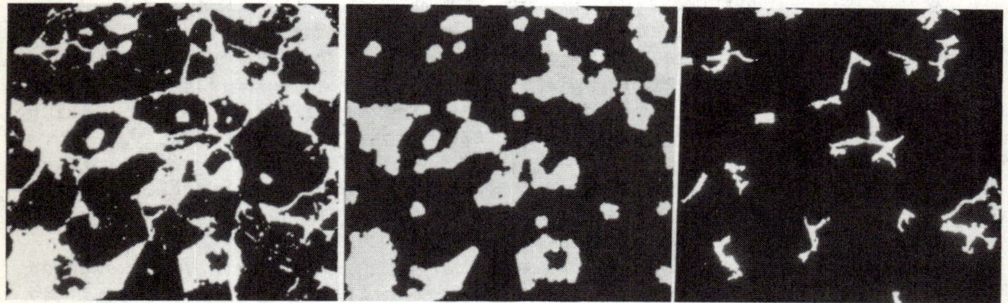

Fig. 1. SEM-BSE photograph on screen of image analyser: (a) large chalcopyrite grains interconnected with narrow veinlets, (b) image of large chalcopyrite grains, (c) image of interconnecting veinlets

Measurements of Mineral Liberations in Ground Ore

Each screened fraction should be analysed when determining mineral liberations in a ground ore. Two types of liberation analysis can be performed; one is the percent of the mineral that is completely free, and the other percent of mineral that is present in each particle. The second type of analysis will classify the particles as containing 0.1-10, 10-20, 20-30, 30-40, 40-50, 50-60, 60-70, 70-80, 80-90, 90-99.9 and 100% of the mineral. Particles in the 0.1 to 99.9% class contain unliberated grains of the mineral, and particles in the 100% class are free grains.

Mineral liberations determined by analysing polished sections are apparent mineral liberations, because the true mineral liberations cannot be determined due to the "slicing effect" which occurs during preparation of polished sections. This effect should produce a higher apparent mineral liberation than the true liberation. Petruk (9) conducted a series of tests by mounting unliberated grains in polished sections and analysing them. He measured up to 15 area % free. However, routine analyses at the CANMET laboratories, on randomly distributed grains in polished sections, generally produce liberation values of 1 to 10% for samples that are known to contain small amounts of free mineral. Furthermore, for most suites of samples analysed at CANMET laboratories, the best materials balances are obtained by using the raw apparent mineral liberation data. It is also

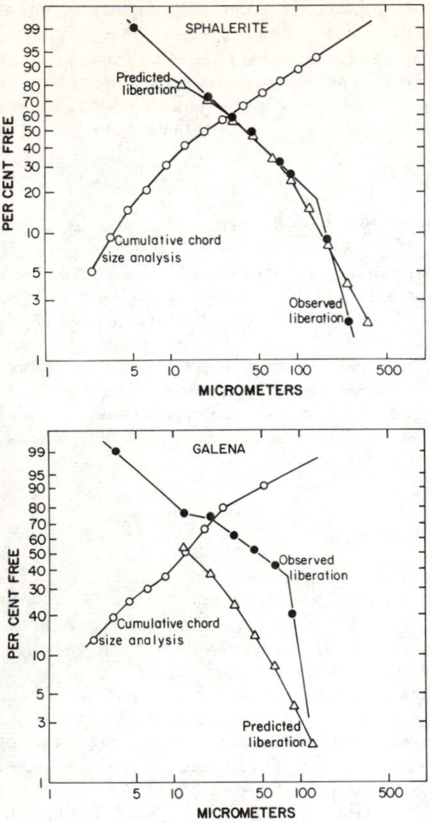

Fig. 2. Relationships between predicted and observed liberations for sphalerite, chalcopyrite and galena in base metal ores

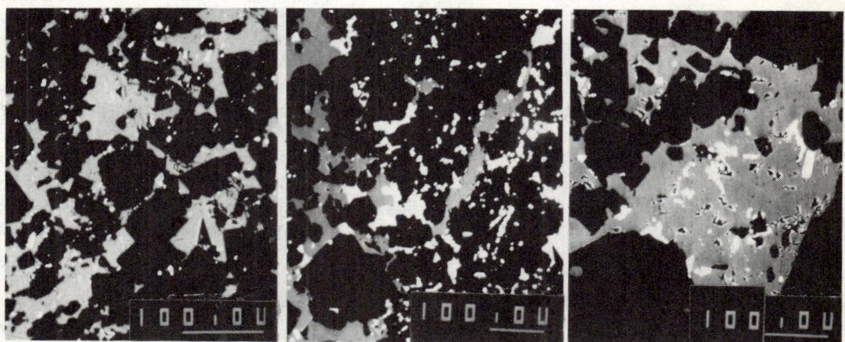

Fig. 3. Textures of minerals in base metal ores (a) sphalerite (grey) in pyrite (black), (b) chalcopyrite (grey) in pyrite (black) and sphalerite (white), (c) galena (white) in sphalerite (grey), pyrite is black

noted that, in Fig. 2, where the predicted liberation is less than 5%, there is a close correlation between the predicted and observed liberation. These observations indicate that the apparent mineral liberation in specially designed experiments are considerably higher than the true mineral liberations, but in randomly distributed grains of a mixture in polished section the apparent liberation seems to be close to the true mineral liberation. The reason for this behaviour is not known. The apparent mineral liberation is now used without corrections at the CANMET laboratories.

Relationship Between Predicted and Observed Mineral Liberations

The relationship between the observed and predicted liberation was tested using both Petruk liberation model and the simplified version of King liberation model. The tests were conducted by measuring the size distributions of minerals in unbroken ores, and the apparent mineral liberations in a series of screened fractions of the ground ores. It was assumed that the observed (apparent) liberation in screened fractions is close to the true liberation. In general, close correlations between predicted and observed mineral liberations were obtained with both models, but there are several exceptions for base metal ores.

Liberation of Sphalerite, Galena and Chalcopyrite

In tests on sphalerite, galena and chalcopyrite liberation from base metal ores, good correlations between predicted and observed liberations were obtained for the sphalerite, higher observed than predicted liberations were obtained for the galena, and in some cases higher observed than predicted liberations were obtained for chalcopyrite (Fig. 2). In order to assess these results the behaviour of minerals during grinding must be understood. It is interpreted that, when the host mineral breaks into particles that are either the same size as or smaller than the size of the enclosed mineral, the enclosed mineral is released. The released mineral could then break into smaller particles and release daughter inclusions. If the breakage characteristics of the host mineral are the same as the breakage characteristics of the ore (i.e. host mineral = major gangue mineral) the observed liberation of the enclosed mineral should be close to the predicted liberation, because the existing liberation models consider size distribution of the enclosed mineral. If, on the other hand, the host mineral does not have the same breakage characteristics as the breakage characteristics of the ore, the mineral liberation will be different than predicted by the existing liberation models. It follows from this interpretation that the observed liberation for sphalerite was close to the predicted liberation because of the mode of occurrence of sphalerite; the sphalerite occurs as masses and as inclusions in pyrite (Fig. 3a). The host mineral, pyrite, has the same breakage characteristics as the breakage characteristics of the ore, hence it released the sphalerite when it broke into sizes predicted by the liberation model. The galena, on the other hand, occurs as inclusions in sphalerite and pyrite with sphalerite being the main host mineral (Fig. 3c). it has been observed that during grinding sphalerite breaks into smaller particles than pyrite (10,11). Hence more galena would be liberated than predicted by the model because the host mineral, spalerite would be in smaller particles than indicated by the screen analysis. The chalcopyrite is more complex because some occurs along fractures in pyrite and some as iclusions in pyrite, pyrrhotite and sphalerite (Fig. 3b). Whenever most of the chalcopyrite occurs along fractures in pyrite and/or as inclusions in sphalerite and pyrrhotite higher observed than predicted liberations are obtained, but not when it occurs as inclusions in pyrite. This suggests that the network of chalcopyrite-filled fractures in pyrite is significant for predicting chalcopyrite liberation. These observations suggest that, in addition to grain size distribution of the wanted mineral in an unbroken ore, the composition and texture of the host mineral has a bearing on mineral liberation.

Liberation of Chalcopyrite Lamellae From Sphalerite

A test was conducted to determine the liberation of minute chalcopyrite lamellae from sphalerite. Analyses showed that 1.4% of the sphalerite in a zinc-lead-copper ore contained chalcopyrite lamellae (Fig. 4), the lamellae accounted for an average of 1.65% of each chalcopyrite-bearing sphalerite grain. These lamellae ranged from 0.5 to 5 μm in width, with the mean width being 0.9 μm. The sphalerite would, therefore, have to be broken into grains smaller than 5 μm to liberate the chalcopyrite. It was found by analysing the concentrates and tailings from the concentrator that 22% of the sphalerite was broken into particles smaller than 5 μm. Hence over 75% of the sphalerite in the concentrates plus tailings occurs as particles that are too large to liberate the chalcopyrite lamellae. It was determined, however, that only 0.15% of the sphalerite in the concentrates plus tailings still contained the chalcopyrite lamellae. The lamellae were up to 2 μm wide and accounted for an average of 2.2 area % of the chalcopyrite-bearing sphalerite grains. Hence nearly 90% of the chalcopyrite lamellae in the ore had been liberated, even though only 22% of the sphalerite was broken into small enough particles to liberate the chalcopyrite lamellae. It is noted that the chalcopyrite lamellae occur along cleavage planes; it is interpreted, therefore, that during grinding the sphalerite broke preferentially along these planes and liberated the chalcopyrite lamellae.

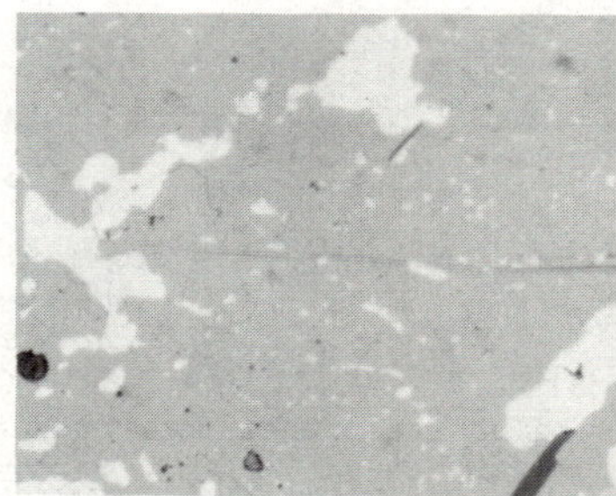

Fig. 4. Chalcopyrite (white) grains and lamellae in sphalerite (grey)

Cassiterite in Base Metal Ores

Preliminary analyses indicate that several fold more free cassiterite is present in the tailings from base metal ores than liberation models permit. The cassiterite in these ores occurs as minute inclusions in sphalerite. It has been noted above that during grinding sphalerite breaks into smaller particles than pyrite, but even when the size distribution of the sphalerite in the zinc concentrate is considered a higher observed than predicted cassiterite liberation is obtained. This suggests that like the chalcopyrite lamellae, the cassiteite occurs along zones of weakness in sphalerite, either along grain boundaries or cleavage planes.

Effect of Grinding Method

The liberation that was obtained for sphalerite when a base metal ore was ground in commercial rod and ball mills was tested by Petruk (12). He observed that the total liberation produced within each mill was equal to the total predicted liberation, but the grain size distribution of the free sphalerite from each mill was significantly different. It was observed that, in the rod mill, large sphalerite grains were liberated and were broken into very small grains (Fig. 5), apparently by the intense action of the rods. By contrast, in the ball mill large sphalerite grains were liberated, but their size was reduced only a small amount (Fig. 5). Consequently, the percent of sphalerite that is free in screened fractions from a rod mill is different than in sceened fractions from a ball mill. This indicates that the method of grinding would have no effect on the total liberation of a mineral whose host has the same grinding characteristics as the grinding characteristics of the ore, but could have an effect on the liberation of daughter minerals from the enclosed grains.

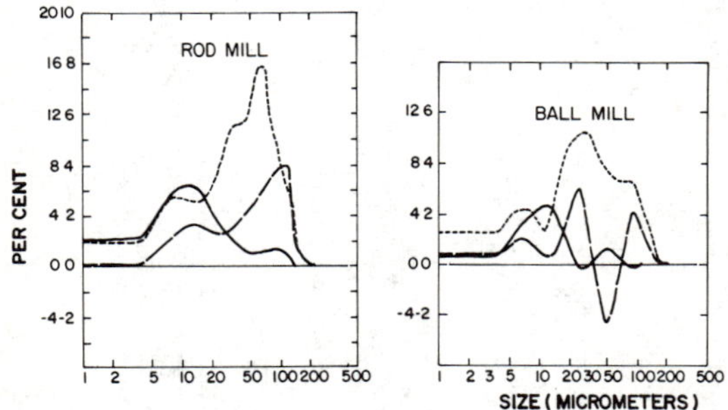

Fig. 5. Quantity and size of free sphalerite produced in a rod mill and a ball mill. Dotted curve = size distribution of unliberated sphalerite (refer to numbers on left); dashed curve = quantity and size of sphalerite released during grinding (numbers on right); solid curve = quantity and size free sphalerite produced during grinding (numbers on right)

Breakage of Host Mineral

It has been shown above the mineral liberation is dependent on the size of the wanted mineral and, in some cases, on the breakage characteristics of the host mineral. During grinding the host mineral will likely break along zones of weakness such as fractures, cleavage planes, grain boundaries, etc. until the grains have been separated. It is suggested that if the host mineral is a hard mineral, such as pyrite or quartz, it will tend to break along the zones of weakness and release any minerals that occur along them. The density of the zones of weakness in the hard mineral will determine the size to which the ore will break readily during grinding and release the enclosed mineral. Further breakage will occur across grain boundaries and will require higher amounts of energy. It follows that each ore should have a "natural grind" which would be controlled by the size distribution of the zones of weakness in the host mineral. The size distributions of these zones can be measured with an image analyser by utilizing a grain reconstruction technique (Fig. 6).

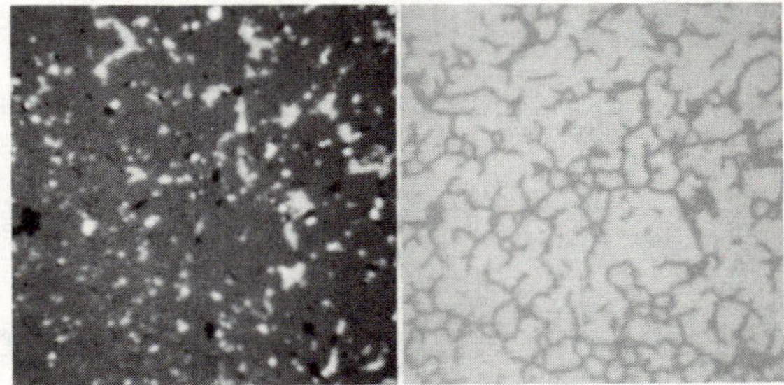

Fig. 6. (a) SEM-BSE diagram of pyrite with minute inclusions of sphalerite, galena and chalcopyrite along grain boundaries, and minute pits along grain boundaries and fractures, (b) grain reconstruction; grain boundaries and fractures reconstructed with Kontron image analyser.

Conclusions

(1) The apparent mineral liberations, determined by measuring mineral liberations from polished sections of randomly distributed grains, are close to the true mineral liberations. For practical purposes, the apparent mineral liberation can be used without corrections.

(2) The existing liberation models do not consider enough variables to accurately predict the mineral liberations that would be obtained when an ore is ground. They do, however, consider the most significant variable, size distribution of the mineral, and do provide a guide for estimating the grind required to obtain mineral liberations. This is adequate for most practical purposes. In general, the models will predict the proper grind for the minerals that occur primarily as inclusions in minerals which have the same grinding characteristics as the grinding characteristics of the ore.

(3) Mineral liberation also depends upon the breakage characteristics of the host mineral. These characteristics are affected by (a) grain size distribution, (b) abundance of cleavage planes, and (c) incipient fractures.

(4) To accurately predict mineral liberations it is necessary to analyse the significant mineral parameters and to develop appropriate liberation models. The mineral parameters are:
(a) Percentage of mineral in each host mineral
(b) Size distribution of mineral in each host
(c) Size distributions of zones of weakness in host mineral (grain boundaries, fractures, cleavage planes, etc.)
The parameters can be measured in the unbroken ore pieces with an image analyser.

References

1. C.L. Lin, J.D. Miller, J.A. Herbst, J.E. Sepulveda, and K.A. Prisbrey, "Prediction of Volumetric Abundance From Two-dimensional Mineral Images," Applied Mineralogy, Ed. W.C. Park, D.M. Hausen and R.D. Hagni (Warrendale, PA: The Metallurgical Society, 1984), 157-170.

2. J.E. Sepulveda, J.D. Miller, and C.L. Lin, "Generation of Iregularly Shaped Multiphase Particles for Liberation Analysis," Proceedings XV International Mineral Processing Congress, (Cedex, France: Bureau de Recherches Geologiques et Minieres, 1985), 120-132.

3. G. Barberry, "Nouvelle methode pour caracteriser la liberation minerale par analyses d'imagess au moyen de mesures dans l'espace a une et deux dimensions. Theorie et exemple d'applications," Proceedings XV International Mineral Processing Congress, (Cedex, France: Bureau de Recherches Geologiques et Minieres, 1985), 20-30.

4. G. Barberry, "Particle Composition Measurement and Prediction in Mineral Processing," Proceedings 18th Annual Meeting of Canadian Mineral Processors, CIM, (1986) in press.

5. R.P. King, "A Model for the Quantitative Estimation of Mineral Liberation by Grinding," Int. Jour. Mineral Processing, 6, (1979), 207-220.

6. R.R. Klimpel, "Some Practical Approaches to Analysing Liberation from a Bindary System," Process Mineralogy III, ed. W. Petruk (New York, N.Y.: The Society of Mining Engineers of A.I.M.E., 1984), 65-81.

7. W. Petruk, "The Application of Quantitative Mineralogical Analysis of Ores to Ore Dressing," CIM Bull. 69(767) (1976), 146-153.

8. J.A. Finch and W. Petruk, "Testing a Solution to King Liberation Model," Int. Jour. Mineral Processing, 12, (1984), 305-311.

9. W. Petruk, "Correlation Between Grain Sizes in Polished Section with Sieving Data and Investigation of Mineral Liberation Measurements from Polished Sections," Inst. Mining and Metallurgy, Trans. C, 87, (1978), C272-278.

10. J.R. Craig, R.H. Yoon, R.M. Harlick, W.Z. Choi, and T.C. Pong, "Mineralogical Variations During Comminution of Complex Sulfide Ores," Process Mineralogy III, ed. W. Petruk (New York, N.Y.: The Society of Mining Engineers of A.I.M.E., 1984), 51-63.

11. W. Petruk, R.G. Pinard, and J. Finch, "Relationship Between Observed Mineral Liberations in Screened Fractions and in Composite Samples," Mineral and Metallurgical Processing Jour., A.I.M.E. (1986), in press.

12. W. Petruk, "Image Analysis Studies to Determine Liberation of Sphalerite, Chalcopyrite and Galena During Grinding of Fine Grained Pyritic Zn-Pb-Cu-Ag Ores," Proceedings XV International Mineral Processing Congress, (Cedex, France: Bureau de Recherches Geologiques et Minieres, 1985), 77-87.

EFFECT OF GRAIN SIZE DISTRIBUTION AND GRAIN TYPE ON LINEAR AND AREAL GRADE DISTRIBUTIONS

C.L. Lin and J.D. Miller

Department of Metallurgy and Metallurgical Engineering
University of Utah
Salt Lake City, Utah 84112

Abstract

Linear and areal grade distributions from the results of image analysis have been used to estimate the corresponding volumetric grade distribution to more accurately establish liberation features of multiphase particulate systems. The characteristics of the linear and areal grade distributions are influenced by the external structure and internal texture of multiphase mineral particles. A computer program (PARGEN) has been developed to grow, irregularly shaped, multiphase particles which can be probed in order to establish the relationship between volumetric grade distribution and linear or areal grade distributions. Previous studies based on a fixed grain size, indicate that the volumetric grade and dispersion density of the particles have a significant effect on linear and areal grade distributions. In this paper, the effect of grain size and type is evaluated using a modified PARGEN program. Four grain size distributions, uniform, exponential, normal and Weibull, are evaluated. Results suggest that grain size distribution has only a modest effect on linear and areal grade distributions. Also, the significance of grain type (external or internal) is of relatively little importance.

Introduction

Randomly oriented, irregularly shaped, multiphase particles have been generated and analyzed using the PARGEN program in order to predict the volumetric grade distribution from linear and areal grade distributions for liberation analysis (1,2). These studies have established that volumetric grade and dispersion density have a significant effect on linear and areal grade distributions, whereas particle shape has only a slight effect (3). In addition, these simulations have allowed for the determination of a general transformation matrix and the solution of the transformation equation to predict the volumetric grade distribution from one and/or two dimensional information (4,5).

$$f(g_i) = \int_0^1 h(g_i | g, Nn) p(g) dg \tag{1}$$

where

$f(g_i)$ = linear or areal grade density function

$h(g_i | g, Nn)$ = general transformation function, a conditional probability function

$p(g)$ = volumetric grade density function.

Using this approach the prediction of volumetric grade distributions has been accomplished and verified both by computer simulation and by experimental depth profile measurements.

Although this methodology involving the transformation equation has been rather successful, it can be criticized for not explicitly taking into consideration the grain size distribution of the dispersed phase. In past work, the dispersed phase has been characterized by the volumetric grade and the dispersion density (number of equisized grains per particle). Nevertheless, with this approach, the simulation of a complete distribution of grade classes results in the variation of grain sizes, although not explicitly specified, from one grade class to another. Of course the maximum grain size will occur for a dispersion density of one grain per particle and will be limited by the particle grade.

In this regard, research efforts have been directed to attempt to evaluate the significance of grain size distribution on the formulation of the transformation matrix and the solution to the general transformation equation for the prediction of volumetric grade distribution from one and/or two dimensional information. Further, the effect of the exposed grain type, generated from random nuclei at the surface of the particles, is compared with the previous results obtained for the particels having an internal grain type generated from random nuclei inside the particle.

Grain Size Distributions

All the results obtained previously for establishing the general transformation matrix were based on the multiple probe or section analysis for several particles of a fixed grain size (determined by the volumetric grade and dispersion density). In an actual situation, for a population of particles, the grain size distribution may or may not influence the characteristics of the linear/areal grade distributions. In this regard, four kinds of grain size distribution; uniform, exponential, normal and Weibull, have been evaluated. The PARGEN program was modified so that the grains of dispersed phase generated inside each particle were not of fixed size but rather were constrained by a specified distribution function. As a result, in this analysis a large number of particles had to be grown in order that the total number of grains in the system was statistically significant. The necessity of a random number generator to simulate different distributions arises in solving this problem. The technique for grain size generation variates which obey the desired probability density function is described.

Random Number Generator

A uniformly distributed number generated artificially in a computer is called a pseudo-random number or simply a random number. Most computers, except microcomputers, have a built-in pseudo-random number generator. The Fortran program for the generation of such pseudo-random numbers is described in detail in the literature (6,7). The generation of these random numbers results in a uniform distribution between zero and one. The other distributions (exponential, normal and Weibull) were obtained by the inverse transformation method (7). This method utilizes the fact that the cumulative distribution function of these distributions can be inverted by using values selected from the uniform distribution, U(0,1), obtained from the pseudo-random number generator. The functional forms', probability function f(x), used for computing the random number are as follows.

For the exponential distribution, the probability density function is specified,

$$f(x) = ae^{-ax} \tag{2}$$

where a=1 for this study.

For the normal distribution, the probability density function is specified,

$$f(x) = \frac{1}{v\sqrt{2\pi}} \exp\left[\frac{(x-u)^2}{2v^2}\right] \tag{3}$$

where u=1.0 and v=0.2 for this study.

For the Weibull distribution, the probability density function is specified,

$$f(x) = \frac{vx^{v-1}}{u^v} \exp\left[-(\frac{x}{u})^v\right] \tag{4}$$

where u=0.8 and v=0.5 for this study.

Simulation Scheme

A series of simulations with the use of the modified PARGEN program were carried out to investigate the effect of grain size distribution on the transformation function. Simulations were performed for three levels of volumetric grade (g=10%, 30% and 50%) and two levels of dispersion density

(Nn=3 and 10) for linear and areal grade distributions. For each simulation 50 particles were generated and analyzed to develop a statistically significant picture of the overall behavior of the grain size distribution. The generated grain size distribution for each simulation was recorded and compared with the expected distribution. Further investigation of the accuracy of the generation of the grain size distributions was confirmed by examining the sectioned images of the generated particles with computer graphics.

Grain Types

Grain type is the other variable which may influence the characteristics of the transformation function. Two grain types can be specified for, and generated by, the PARGEN program. Internal grains are generated from random nuclei at the surface of the particle. The effect of grain type was studied for particles with volumetric grades of 10%, 30% and 50% and a dispersion density of 1.

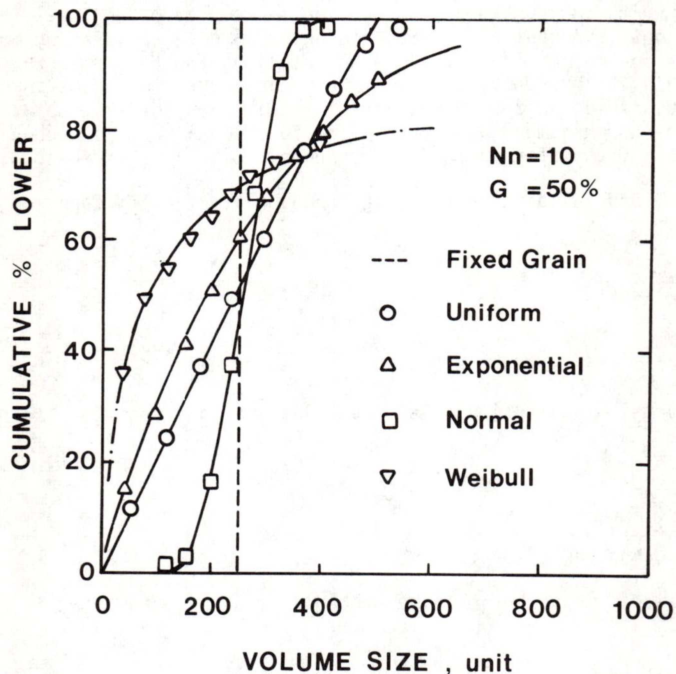

Figure 1. Grain size distributions of the dispersed phase for 50 particles generated by PARGEN each with a volumetric grade of 50% and a dispersion density, Nn=10. Solid lines are the expected distributions and the symbols represent the generated distributions.

Results and Discussion

Effect of Grain Size Distribution

The simulated grain size distributions were tested by using the corresponding random number generator to generate a large number sequence and plotting the resulting grain size distribution. The number sequence which establishes the grain size distribution depends on the dispersion density and the total number of particles being generated. Figure 1 illustrates one of these grain size distributions obtained from 50 generated particles at 50% volumetric grade and a dispersion density of 10. The simulated and expected grain size distributions appear equivalent. Although the simulated grain size distribution compared favorably against the expected distribution, the random process for the generation of the nuclei of the dispersed phase grains may change the outcome of the grain size distribution. In this regard the plane sections through the generated particles with different grain size distributions were examined. Figure 2 shows an example of the sectioned images of these grain size distributions. As can be seen from Figure 2, grain coalescence is quite common and even if the grains were set as convex, the resulting coalesced grain may be concave.

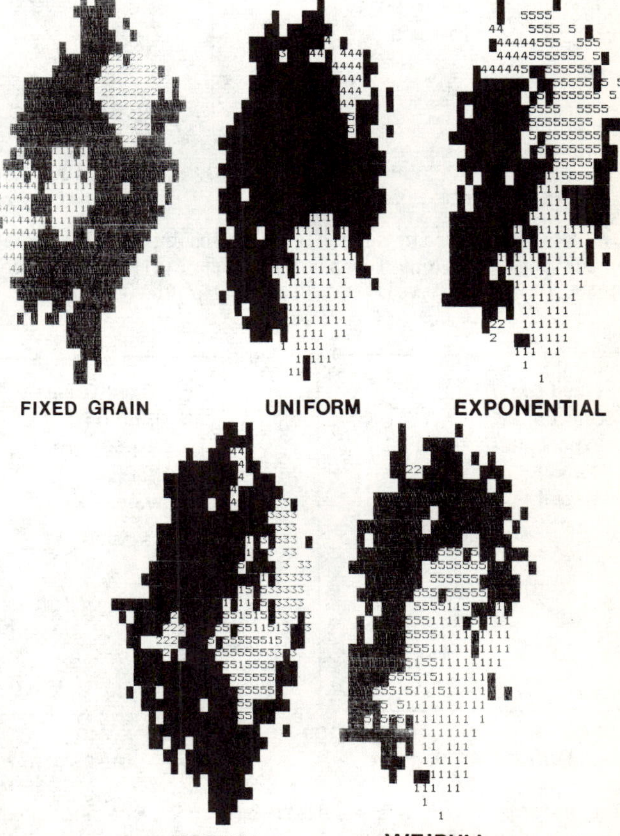

Figure 2. Selected sections of irregularly shaped PARGEN particles with different grain size distributions.

Figure 3 presents the results of different grain size distributions for particles with a volumetric grade of 50% and a dispersion density of 3. The variations observed are quite small. At a higher dispersion density, Nn=10, the comparison of grain size distributions for the linear and areal grade distributions are given in Figure 4 for particles with a dispersed phase volumetric grade of 50%. Results suggest that the grain size which follows the Weibull distribution has a modest effect on both linear and areal grade distribution. Greater apparent liberation of both continuous and dispersed phase is observed for the Weibull grain size distribution, however, the effect is insignificant compared to the effect of volumetric grade and dispersion density. At lower volumetric grades 10% and 30% the agreement is even better.

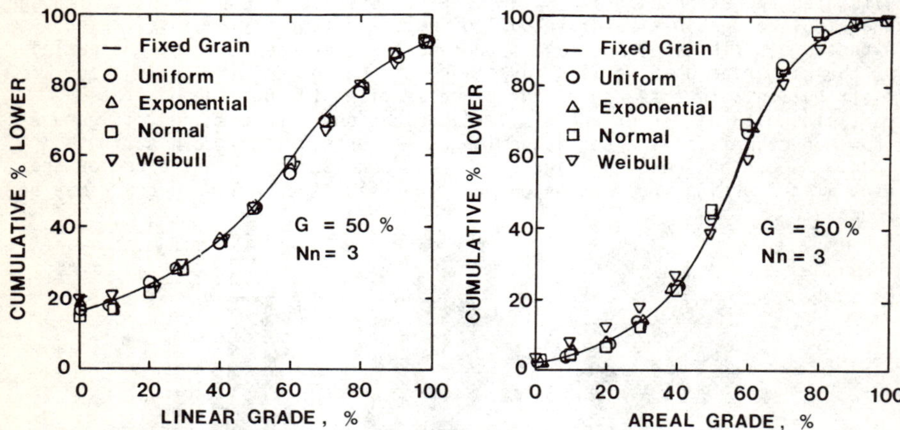

Figure 3. The effect of grain size distribution on the linear and areal grade distributions for simulated particles with a volumetric grade of 50% and a dispersion density of 3.

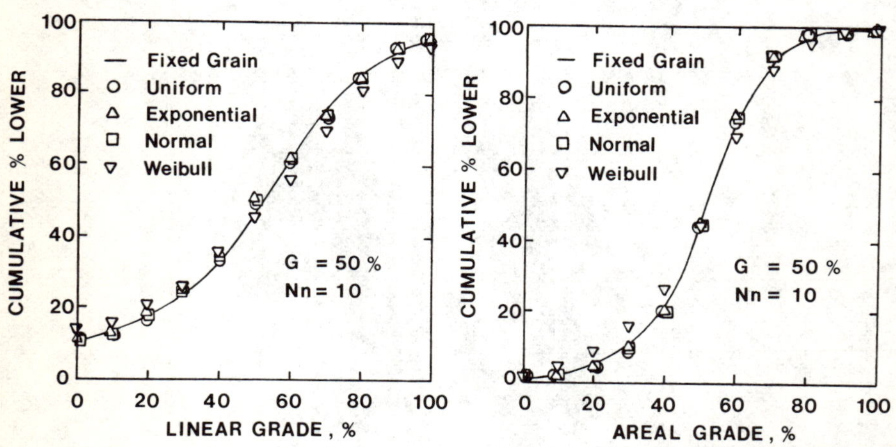

Figure 4. The effect of grain size distribution on the linear and areal grade distributions for simulated particles with a volumetric grade of 50% and a dispersion density of 10.

Effect of Grain Type

Theoretically, the extent of liberation will be proportional to the ratio of grain interfacial area to the particle surface area for a convex particle containing a single grain as discussed by Steiner (8). For a spherical particle, it is evident that the apparent liberation of the dispersed phase for a fixed composition decreases in the order shown in Figure 5. The results for linear grade analysis have been confirmed from Monte Carlo Simulation by Horton (9) at Imperial College. Recently, the probability of apparent liberation for an isotropic uniform random (IUR) probe through the sphere/plane particle (Type a of Figure 5) has been solved by Moore (10). It is clear that the plane interface for a spherical particle is the extreme case for apparent liberation, because the plane interface has the smallest interfacial area.

The effect of grain type was examined using simulated particles generated by the PARGEN program. The exposed grain type of particle (random nuclei of dispersed phase are located at the surface of particle) has been analyzed and the results compared with the internal grain case in Table I. As expected, the exposed grain particle shows a slightly higher degree of apparent liberation. All the above results are based on particles containing a single grain. Of course, for multiple grain particles, an even smaller apparent liberation is expected. It is noted that internal grain particles actually have grain exposure at the surface. The effect might be even greater if the internal grain was truly an internal grain.

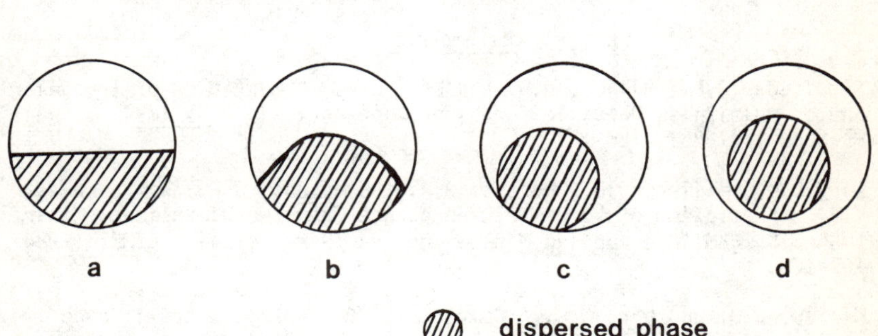

Figure 5. Schematic representation of the variation in grain type for spherical particles.

Table I. Comparison of Apparent Liberation of Dispersed Phase as Percent for Linear Grade Analysis at Nn=1

Grain Type	Volumetric Grade, %		
	10	30	50
Internal	0.20	1.90	7.36
Exposed	0.30	2.29	8.04

Conclusions

The characteristics of the linear and areal grade distributions are influenced by the external structure and internal texture of the multiphase mineral particles. Previous studies based on a fixed grain size, indicate that the transformation functions for linear and areal grade distributions mainly depend on the volumetric grade and dispersion density of the particles. Four grain size distributions, uniform, exponential, normal and Weibull, have been evaluated. It was found that both linear and areal grade distributions are rather insensitive to the grain size distributions. The conclusion of this exercise is that no matter what the grain size distribution, the analysis can be made by assuming a fixed grain size and a specified dispersion density.

Results suggest that the apparent liberation for linear grade analysis is influenced by grain type. The exposed grain particles in which the nuclei of the dispersed phase are located at the surface of the particle have a slightly higher degree of apparent liberation. The difference in apparent liberation between grain types decreases as the number of grains in the particle increases.

Acknowledgment

The authors would like to acknowledge the support of the USBM (under grant No. G1125149) through the Utah Generic Center in Comminution.

References

1. Sepulveda, J.E., Miller, J.D. and Lin, C.L., "Generation of Irregularly Shaped Multiphase Particle for Liberation Analysis," XV IMPC, Cannes, France, 1 (1985), 120-132.

2. Lin, C.L., Miller, J.D., Herbst, J.A., Sepulveda, J.E. and Prisbrey, K.A., "Prediction of Volumetric Abundance from Two-Dimensional Mineral Images," ICAM '84, Applied Mineralogy, ed. Park, et al., AIME, (1984), 157-170.

3. Miller, J.D. and Lin, C.L., "Comparison of Linear and Areal Grade Distributions as Estimates of the Volumetric Grade Distribution in Liberation Analysis," (paper presented at 16th Annual Fine Particle Society Meeting, Miami, Florida, April 1985) and to be published in the Conference Proceedings, Hemisphere (1986).

4. Miller, J.D. and Lin, C.L., "Treatment of Polished Section Data for Detailed Liberation Analysis," paper presented at the Engineering Foundation Conference on Recent Developments in Comminution, Hawaii, Dec. 1985 and submitted for publication , Int. J. Mn. Proc. 1986.

5. Lin, C.L., Miller, J.D. and Herbst, J.A., "Solutions to the Transformation Equation for Volumetric Grade Distribution from Linear and/or Areal Grade Distributions," paper presented at International Particle Tech. Conference, Nurnberg, West Germany, April 1986.

6. Carnahan, B., Luther, H.A. and Wilkes, J.O., Applied Numerical Methods, John Wiley and Sons, Inc., New York, 1969.

7. Lewis, T.G., Distribution Sampling for Computer Simulation, Lexington Books, Lexington, Massachusetts, 1975.

8. Steiner, H.J., "Liberation Kinetics in Grinding Operations," XI IMPC, Cagliari, Rome (1975), 35-58.

9. Horton, R., "Determination of the Volumetric Composition of Idealized Composite Mineral Particles by Random Intercepts," Ph.D. Thesis, University of London, 1978.

10. Moore, S.W., "Stereology of Random Lines Through Idealised Composite Particles of Variable Texture," Ph.D. Thesis, University of London, 1985.

PREDICTION OF PARTICLE COMPOSITION DISTRIBUTION AFTER FRAGMENTATION OF HETEROGENOUS MATERIALS: INFLUENCE OF TEXTURE

G. Barbery and D. Leroux

GRAIIM, Department of Mining and Metallurgy
Laval University, Québec G1K 7P4, Canada

Abstract

The problems encountered in developing prediction models for composition distribution of particles obtained by the breakage of multiphase materials are presented. The needs to characterize texture by methods that can be calibrated using image analyzers, to characterize particle production by similar methods, to relate particle production and texture, to use integral geometry methods and stochastic geometrical processes to solve the complex probability equations in the three dimensional space, are stressed. A presentation is made of a model which incorporates the required elements. Special cases of the model, combining a description of texture based on Poisson polyhedra or Boolean models with primary Poisson polyhedra grains, a description of particle production calibrated on screen fractions extracted from ground ores, independence of texture and breakage, are presented. Applications of the model to data presented by other researchers give evidence of its value and its calibration potential from laboratory physical separations. A modification of the model of multitextural materials is presented. The modification removes limitations due to independency of texture and breakage.

Introduction

Particle composition distribution prediction is required in attempting to describe the properties of particles populations, when these particles are obtained from the breakage of multiphase materials. It is at times presented by various authors as "liberation" prediction, but the information required is better described by the complete four words, just like particle size distribution prediction would be considered a nonsense if researchers were only attempting to predict the relative proportion in, say, the fine or the coarse size range. It is a very difficult subject, and it is not surprising that over the years, various researchers have tried to relate multiphase material texture and breakage pattern to the resulting particle composition distribution. In the present paper, an attempt is made at assessing the various models that have been presented, and at developing and testing a model which fulfills what are believed to be essential characteristics of a model for predictive purposes, and which can be understood with reference to Figure 1.

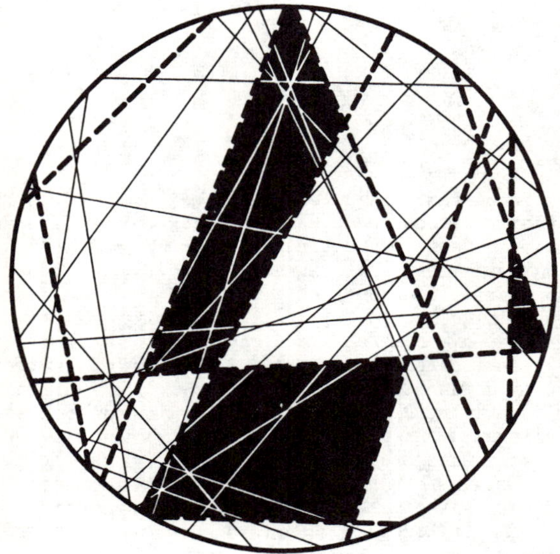

Figure 1 - Example of a plane section through a biphasic material (white and black phases) having a texture (in broad dashed lines, corresponding to a random Poisson polyhedral process, broken (fine lines) by another random Poisson polyhedral process. The two random processes are independent.

The model incorporates the following features, which are felt to be essential in any model of this type:

- an ore texture characterization method which can be calibrated by image analysis techniques, and which involves parameter estimation techniques based on the principles of stochastic geometry

- a breakage analysis characterization method which can be calibrated using conventional sizing techniques (sieving) and particle shape description which can be quantified using image analyzers

- a correct description of the limitations of the model for the interrelationship between ore texture and breakage pattern

- a correct assessment of the statistical geometrical properties involved, especially taking into account integral geometry for convex particles, in order to predict particle distribution in the three dimensional space.

The model has already been presented in its main features by Barbery and Leroux (1,2,3). A complete presentation of historical developments is included in (2), with a critical evaluation of existing models. The present paper insists on the potential modification of the general model to cope with multitextural materials.

Multiphase Materials Texture Characterization

The statements to be made in the paper concern biphasic textures. Some of the concepts can be extended to multiphase materials, as have been pointed out by Serra (4) and Davy (5).

The essential problem in trying to quantify texture, is to retain a texture characterization method that should be simple enough to be measured by instrumental methods, yet, at the same time, be able to describe the fantastic complexity of natural materials, that are resulting from complex deterministic and random events. Such an approach has been put forwards by Matheron (6) and Serra (4), and is conceived around the principles of instruments such as image analyzers.

Basically the method extends the concept of autocorrelation functions to complex three dimensional features. It will be assumed that the material has isotropic stationary properties. Let the material consist of phases 0 and 1, at a respective volumetric concentration π_o and π_1 (their sum being equal to 1). The covariance function is defined by the conventional equation:

$$C(r) = E\{[f(x) - \pi_1].[f(x + r) - \pi_1]\} \tag{1}$$

In this expression f(x) is the indicating function of the material, being equal to 1 when point x is in phase 1, the mathematical expectation being taken for the complete material volume. Under this condition $\pi_1 = E[f(x)]$. Equation 1 is taken for all pairs of points at a distance r and all directions. Serra (4, pp. 275-283) explains how C(r) can be measured on plane sections of the material, using image analyzers. C(r) has the following properties:

$$C(0) = \pi_o\pi_1; \quad C(\infty) = 0; \quad C'(0) = -0.25\ S_{01}$$

where S_{01} is the specific surface area of contact between phases 0 and 1 per unit volume of material.

Various examples of C(r) have been given (4). A large number of natural or synthetic materials have been described by these functions, such as rocks, metals, wood, ceramics, etc.. A very useful concept which has also been introduced by Serra is the modelling of texture, in order to limit the number of model parameters to be estimated. Barbery (1) has used a description of texture based on Poisson polyhedra, such as given on Figure 1. In fact these random textures are not often encountered. For ores, it would seem more appropriate to model texture based on what has been called Boolean schemes (4). The concept can be illustrated in the following manner:

- points are placed according to a three dimensional Poisson process in space; let θ be the density of the process (number of points per unit volume)

- each point of the Poisson process is the seed for the growth of a crystal. If two crystals meet, there will be no disturbance in their respective growth, which will stop independantly for each component. Figure 2 is a plane section through a Boolean model in which the primary grains consist of Poisson polyhedra having the same parameter λ_3. Obviously any growth pattern can be retained. It is simpler, for reasons to be seen later, to retain convex primary grains. It must be noted that, in spite of the fact that primary grains are convex on Figure 2, the resulting texture is far from convex!

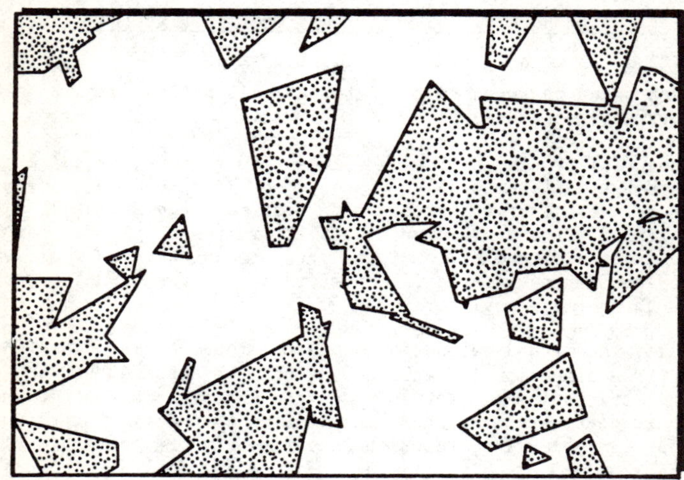

Figure 2 - Example of a Boolean process with Poisson polyhedra as primary grains. Plane section through a simulated process.

Serra (4) has derived most interesting properties of Boolean models. If phase 0 consists of primary grains, then

$$\pi_1 = \exp\{-\ \theta E[\text{Mes } X']\} \tag{2}$$

where Mes X' is the Lebesgue measure of the volume of primary grains.

$$C(r) = \pi_1{}^2\{\exp[\theta K(r)] - 1\} \tag{3}$$

where K(r) is the covariance of the primary grains, defined by:

$$K(r) = \int k(x).k(x + r)dx$$

taken for all x and all directions on the primary grains.

In the case of monodisperse spheres of diameter D:

$$K(r) = \frac{\pi D^3}{6}\{1 - 1.5(r/D) + 0.5(r/D)^3\} \tag{4}$$

In the case of Poisson polyhedra of parameter λ_3:

$$K(r) = \frac{6}{\pi^4 \lambda_3{}^3} \exp(-\lambda_3 r) \qquad (5)$$

For primary convex grains, the cumulative distribution of intercept lengths in phase 1 is:

$$F(\ell) = 1 - \{\exp \theta K'(0)\ell\} \qquad (6)$$

where $K'(0)$ is the value of the first derivative of $K(r)$ at the origin. It is equal to $-0.25\ E[S(X')]$, where $S(X')$ is the surface area of primary grains. The exponential relationship given by Equation 6 has been encountered experimentally (7,8) in the assessment of mineral grain intercept distribution in ores. It forms the basis of a model presented by King (7).

Intercept length distribution in phase 0 has been presented (4). The expression is rather complex and does not provide a closed form solution.

At the present stage of our development, and taking into account the applications that researchers around Serra have made of this model, for pores in sandstone, coke texture, iron ore sinter characteristics, it is felt that this model, which can be calibrated by image analysis techniques, presents major interests.

Particle Production Characterization

The particle production characterization methods should be amenable to calibration using real particles obtained from comminution devices, and not only pure geometrical shapes selected only for their simplicity. The shape retained should have, however, some limitation that make solvable the complex general equations which describe the interaction of texture and fragmentation. It is felt that the hypothesis of convexity is not too limitative for practical applications. This simplification enables one to use the remarkable properties of convex bodies, in integral geometry, and in particular a theorem that has been introduced by Crofton, and redeveloped by Gilbert (9). Since this theorem, and its modifications due to Davy (5), forms the basis of the present prediction method, it will be explained in some detail here.

The basic concept lies in the following presentation:

Assume that the particles consist of compact non interactive convex solids placed at random in space, under isotropic conditions, and call $P(\ell)$ the probability that two points separated by a distance ℓ are present in the same particle. The particles are assigned the function of phase "0" as indicated above. $P(\ell) + \pi_0{}^2$ at the origin, and the results expressed above can be used, especially for $C(0)$ and $C'(0)$. The relationship between intercept length distribution and $P(\ell)$ is:

$$f(\ell) = \frac{4\ E(V)}{E(S)\pi_0} \frac{d^2P(\ell)}{d\ell^2} \qquad (7)$$

in which $f(\ell)$ is the probability distribution function of intercept length, $E(V)$ and $E(S)$ are respectively the mathematical expectation of particle volume and surface.

Let us consider the following problem. Take a point x inside a particle i, and call y a point seen inside the particle from x. Obviously, due to convexity the set consisting of all points y is identical to the set of x, and the average volume seen is the mean by measure of V_i, $M(V)$, where

$$M(V) = \frac{E(V^2)}{E(V)}$$

For convex and disjoints particles, M(V) can be assessed in the following manner: the volume element $\ell^2 d\Lambda d\ell$ belong to the same particle as x with a conditional probability equal to $\frac{1}{\pi_o} P(\ell,\Lambda)$, where Λ is an arbitrary direction.

$$M(V) = \frac{1}{\pi_o}\iint \ell^2 P(\ell,\Lambda)d\ell d\Lambda = \frac{4\pi}{\pi_o}\int_0^\infty \ell^2 P(\ell)d\ell \text{ under isotropy} \tag{8}$$

Taking into account Equation 7, and integrating it by part to obtain the relationship between the moments of $f(\ell)$ and $P(\ell)$:

$$E(L) = \int_0^\infty \ell f(\ell)d\ell = \frac{4E(V)}{E(S)}$$

$$E(L^4) = \int_0^\infty \ell^4 f(\ell)d\ell = 12\ \frac{E(L)}{\pi_o}\ \ell^2 P(\ell)dl$$

and thus:

$$M(V) = \frac{\pi}{3}\ \frac{E(L^4)}{E(L)} \tag{9}$$

For particles that fill in space completely, Gilbert (9) and Davy (5) make use of Equation 8 with $\pi_o = 1$.

$P(\ell)$ functions can be measured on image analyzers, for particles mounted in a matrix and respecting the hypothesis of isotropy. At the limit, simple distributions can be taken, such as the one derived (6) for Poisson polyhedra. Monosized spheres (1) can also be taken in a first approach. The $P(\ell)$ function is then K(r) given in Equation 4.

In the present paper, the particle size distribution corresponding to real particles obtained by grinding an iron ore, as given by King (10,11), is used. The function given is the intercept length probability density function for particles having a "convenient size" D, where size is defined as sieve size.

$$f(\ell) = \frac{1}{1.2D}\ (2 - \frac{\ell}{1.2D})\ \exp\ (\frac{-\ell}{1.2D})$$

for $\ell < 1.2D$, and $f(\ell) = 0$ for $\ell > 1.2D$

The specific surface area per unit volume of these particles is 4/E(L), which gives e/0.3D. The particle volumetric shape factor, obtained by writing $V = kD^3$ is 0.5133, as derived from Equation 9.

The function $P(\ell)$, which will be essential in deriving particle composition distribution is obtained by integration of Equation 7 and taking the proper limit values during integration:

$$P(\ell) = 1 - \frac{\ell}{1.2D}\ \exp\ (\frac{-\ell}{1.2D}) \tag{10}$$

Integration of Texture and Breakage

Davy (5) uses, as it has been introduced by all the other authors before (2), the concept of independence of texture and particle production.

It is then only necessary to consider one function $P(\ell)$, independent of the phase boundaries across length ℓ. A derivation can be made (5), which uses analogies with known results of integral geometry (9). Using the concept of the present paper, an outline presentation can be made for monosized particles, as already introduced (1).

Consider a breakage mechanism which only produces particles having the same volume V. Each particle will consist of phase 0 at a volume V_o and 1 at V_1 ($V = V_o + V_1$), let m be the volumetric ratio of phase 0. The function of interest in that case becomes the grade probability distribution function g(m). The following mathematical expectations will be calculated:

$$E(V_oV) = E(V^2)\int_0^1 m\ g(m)\ dm = E(m)\ E(V^2) = \pi_0 E(V^2) \tag{11}$$

$$E(V_1V) = E(V^2)\int_0^1 (1 - m)\ g(m)\ dm = E(1 - m)\ E(V^2) = \pi_1 E(V^2) \tag{12}$$

$$E(V_oV_1) = E(V^2)\int_0^1 m(1 - m)\ g(m)\ dm = (m_1 - m_2)E(V^2) \tag{13}$$

where $m_1 = \pi_o$ and m_2 are the first two moments from the origin of g(m).

Davy (1984) gives the following general equations (also valid when particle size distribution is polydisperse):

$$\frac{E(V_iV)}{E(V)} = \pi_i\ 4\pi\int_0^\infty \ell^2 P(\ell) d\ell \tag{14}$$

taking Equation 8 into account, it is easy to see the equivalence between Equations 11, 12 and 14. The consequence of this relationship is trivial, since it states that the mean grade of size fraction does not change with particle size, an obvious consequence of the independence of texture and breakage.

Equation 13 contains much more information, since it gives access to the variance of particle composition.

Equation 13 can be rewritten:

$$E(V_oV_1) = E(m)E(V^2)-E(V_1^2)$$

division by E(V) provides:

$$\frac{E(V_oV_1)}{E(V)} = \pi_1\frac{E(V^2)}{E(V)} - \pi_1\frac{E(V_1^2)}{E(V_1)} \tag{15}$$

In Equation 15, the second term on the right hand side must be calculated. For this, use will be made of Equation 8, and of conditional probability. The probability that two points separated by a distance ℓ belong to the same phase, "1", has been presented above. It corresponds to the global covariance:

$$K(\ell) = C(\ell) + \pi_1^2$$

The probability that two points belong to the same particle after breakage is $P(\ell)$. The resulting probability that two points initially in phase 1 will be in the same particles when the distance between them is ℓ is thus:

$$L(\ell) = P(\ell)\ [C(\ell) + \pi_1^2]$$

And use can be made of Equation 8:

$$\frac{E(V_1{}^2)}{E(V_1)} = \frac{1}{\pi_1} 4\pi \int_0^\infty \ell^2 L(\ell) d\ell$$

Equation 15 can then be developed to give, after solution:

$$\frac{E(V_0 V_1)}{E(V)} = 4\pi \int_0^\infty \ell^2 P(\ell) [\pi_1 \pi_0 - C(\ell)] d\ell \qquad (16)$$

This last equation, which is Equation 16 in Davy (5), provides thus a mean to evaluate the second moment of the distribution of g(m), when the functions $P(\ell)$, $C(\ell)$ are known, in addition to π_1.

An incomplete Beta function can be fitted to g(m), on the interval [0,1] (1) and the various "washability" curves corresponding to size fractions can be produced. Figure 3 gives an example of such a separability curve, using a texture description based on random Poisson polyhedra, as given on Figure 1, and monosized spherical particles.

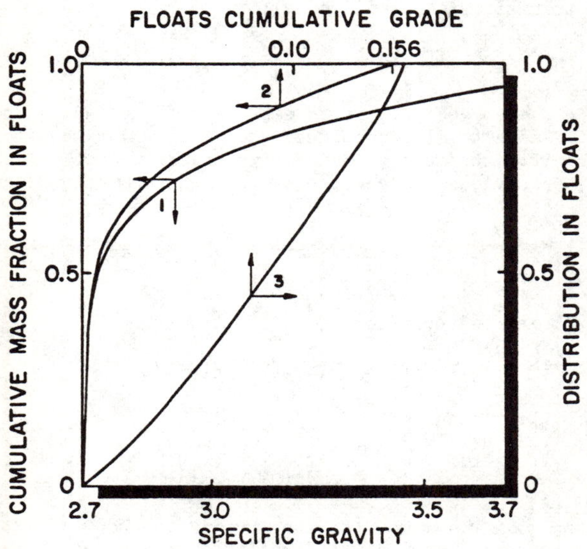

Figure 3 - Example of separability curves obtained by the method developed. Texture: random polyhedral, 0.1 by volume for phase 0 (specific gravity = 4.5), 0.9 phase 1 (s.g. = 2.7). Particle generation: spheres at a diameter = D_{so} of the Poisson polyhedra. Curve 1: mass fraction in "floats" against s.g., Curve 2: mass fraction in floats against floats grade in 0, Curve 3: distribution of phase 0 in floats against floats grade in 0.

Examples of Calibrations of the Method

At the present time, there is little experimental data which can be fully used, since few authors provide functions such as $P(\ell)$ (or $f(\ell)$), or $C(\ell)$! In order to illustrate the validity of a model combining a Boolean scheme with primary Poisson polyhedra, to particles that follow $f(\ell)$ presented by King (10,11), a program called BOOKING has been written, which predicts particle composition distribution for monodisperse particulate systems. The extension to multidisperse material simply combines the results of BOOKING to those of a screen size distribution.

Since BOOKING can provide simulated separability curves, the existence of such data for practical examples, as determined by physical separations in the laboratory (using either gravity separation or magnetic separation), enables one to use such curves for calibrating the model. Programs have been written to estimate BOOKING parameters based on heavy liquid or magnetic separation tests.

Example of the Data Provided by Wiegel (12)

In his analysis of liberation in magnetic iron ores, Wiegel (12) tests his liberation model against experimental results consisting of Davis tube separation on screened fractions of magnetic taconites ground at various sizes. First of all, an examination of Wiegel's data shows that the hypothesis of independence between texture and breakage is acceptable, since magnetite content does not vary markedly with particle size. Taking into account the characteristics of the separation device, it has been assumed that particles will not be retained in the Davis tube when their magnetite concent is lower than 5% by volume. Knowing π_1 (non magnetic minerals) and that limit, incomplete Beta functions can be tested; for the model to be valid, a single parameter must apply for texture characterization. In the BOOKING case, it consists of λ_3, the parameter of the Poisson polyhedra, or its D_{80}, as it has been put forwards by Barbery (1). Figure 4 presents the results of the liberation predictions for Wiegel's model, and for BOOKING; the latter model giving an overall better estimate of liberation, especially at coarse sizes (right hand side of the graph).

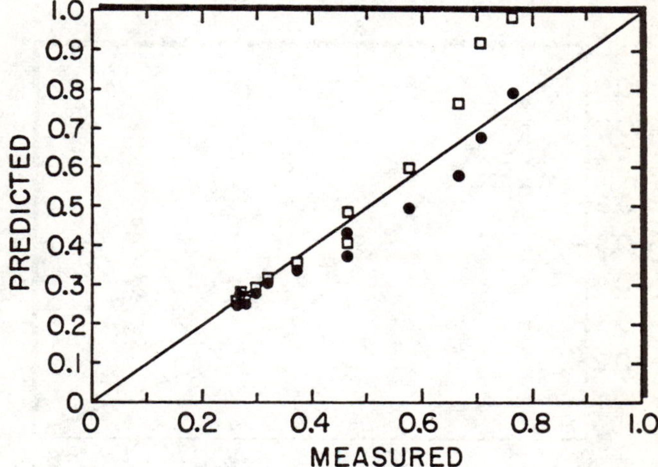

Figure 4 - Comparison of Wiegel (1976) experimental results with his model prediction (□) and those obtained by BOOKING (●). Figures indicate volumetric fraction of a particle size range going to the concentrate in a magnetic separator. Points on the right hand side represent coarse particles.

Example of the Pyrite Fraction in Witswatersrand Gold Deposits, Given by King (7).

King (7), in the illustration of his linear model of liberation gives some experimental data on the liberation of pyrite in South African gold ores. Surprisingly, the pyrite content of the starting material is not given. It is assumed here to be low, as is stated in many descriptions of the mineralogy of these ores, and for the purpose of this exercise, it will be taken as 5% by volume. King (7) gives an intercept length distribution of pyrite in situ, which is very close to an exponential function. We have retained a Boolean process with Poisson polyhedra primary grains; when the volume content in the grain phase is low, it can be assumed that the interaction between primary grains is small, so that intercept distribution across phase 0 corresponds to the one that could be obtained on isolated primary grains. In that case, the solution is known:

$$F(\ell) = 1 - \exp(-\pi\lambda_3\ell)$$

giving a λ_3 parameter of 0.0026 (equivalent value: D_{80} = 246 μm), for the data provided by King.

The BOOKING program was used on King's measured liberation data, reported on Figure 6. In the experimental determination, a heavy liquid separation at a density of 3.3 was used. If, as is common in South African gold ores, pyrite (density 5.0) is associated with low density silicates (density 2.7), then the material analyzed by King consists of particles having a volumetric pyrite content higher than 0.26. It is interesting to note that King, in the demonstration of his model, had to prove that particles separated were "essentially liberated", whereas the present model does not make this assumption.

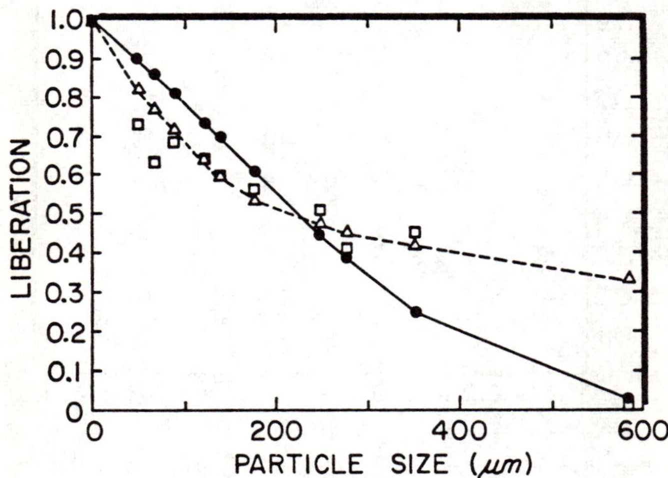

Figure 5 - Comparison of King (1979) experimental results (□), with King's model (Δ), and BOOKING model (●). Note agreement of experimental value and Booking prediction at 600 μm.

The model was calibrated using the information reflecting the evolution of pyrite distribution in particles having a volumetric grade higher than 0.26, by estimating parameter λ_3 of the primary Poisson polyhedra. The model gave an estimate of 0.0025 (D_{80} = 258 μm), and the predicted distributions are given on Figure 6, together with King's model prediction. Again, BOOKING gives a much better estimate of liberation at coarse sizes than King's model, which, as the author had pointed out, fails at sizes coarser than texture elements, mainly due to its one dimensional aspect which systematically overestimates liberation. The good agreement between measured and estimated λ_3, even under the rough assumptions made, is of interest.

Modification of the Model

As it was stated clearly, the BOOKING model only applies to pure transgranular breakage, and, as such, would have limited validity. An elementary check of transgranular breakage is revealed by a size-assay analysis: it would show itself as random variations in the assay of the various size fractions around the average corresponding to the head grade. The example of Wiegel's data was used since it had that property.

This property is rarely encountered, and any method claiming to be general for the prediction of particle composition distribution should approach this limitation. During the course of work in progress on the use of coal washability data, it became clear that coal does not have the transgranular breakage property, as is indicated by data available in published papers (13).

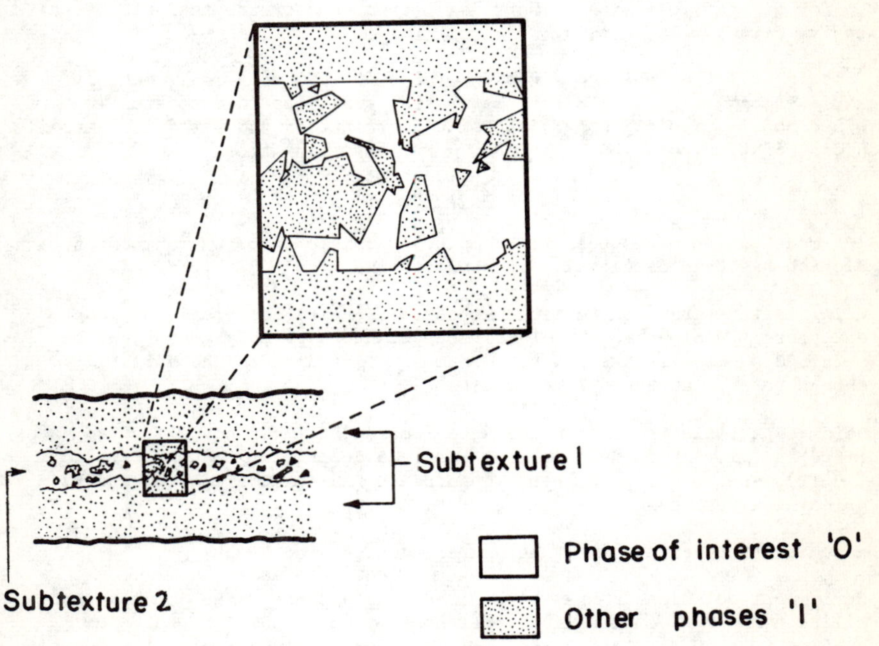

Figure 6 - Example of a bitextural material

A modification of the model which could cope with coal and similar materials has been developed, and can be presented with reference to Figure 6. The overall material is assumed to consist of two different subtextures, one not containing the mineral of interest (coal in the present case), the other corresponding to associations between the mineral of interest and the other minerals. As is hinted from Figure 6, a Boolean model with Poisson polyhedra as primary grains can be retained for this second subtexture. Breakage of each subtexture is assumed to be random isotropic, and, for the case of the second subtexture, would correspond to pure transgranular breakage. The limitation which is removed in the original model is the similar breakage property of each subtexture. In other words, it is possible to allow for preferential breakage of shale compared to the coal seams or coal seamlets.

Model calibration and parameter estimation is more complex than in BOOKING, since, in addition to the textural identification problems for the Boolean phase, the volumetric proportion of the second phase present in the first subtexture needs to be determined for each particle size.

Parameter estimation can be obtained using washability data. Since the first subtexture does not contain coal, addition or removal of pure shale to any size fraction does not affect the relationship between floats cumulative coal recovery/floats cumulative coal grade. Parameters of the BOOKING model for the Boolean subtexture can thus be estimated, by fitting an incomplete Beta function to this relationship. Second phase volumetric proportion in the Boolean subtexture is then obtained, and, through a mass balance, volumetric proportion of the first subtexture. For the model to be valid, it must provide a unique texture assessment for the Boolean subtexture, for all particle sizes. Work is in progress for the analysis of the extensive data available (13).

It is felt that the modification of the BOOKING model will make it flexible enough to cope with real materials, and to be incorporated in combined models predicting particle size distribution and particle composition distribution.

Conclusions

A model which makes possible complete particle distribution composition prediction has been developed. It consists in:

- using texture characterization method that can be calibrated on image analyzers. Boolean models of Poisson polyhedra primary grains can be retained as models having a small number of parameters and versatile enough to cope with complex textures

- using a particle production characterization that can be calibrated using image analyzers. By default, the particle descriptor measured by King (10,11), can be selected, corresponding to real particles produced by grinding a real ore

- assuming that texture and breakage are not related

- deriving a solution in the three dimensional space of the complex probability equations, on the lines presented by Davy (5). The solution enables to predict particle composition distribution at any given particle size

- the model has been tested against experimental data presented by other researchers (7,12), and can be calibrated with data obtained through conventional laboratory physical separations

- the main model limitation lies in the assumption of pure transgranular breakage

- a modification is presented, which consists in splitting the material into two subtextures having different breakage property, one not containing the mineral of interest, and breaking by pure transgranular fragmentation. Parameter estimation for this modified model is in progress for coal, and uses conventional washability data.

References

1. G. Barbery, "Random sets and integral geometry in comminution and liberation of minerals" (Paper presented at AIME Annual Meeting, New-York, February 1985, accepted for publication in Mineral and Metallurgical Processing, 1986).

2. G. Barbery and D. Leroux, "Prediction of particle composition distribution after fragmentation of heterogeneous materials" (Paper presented at Engineering Foundation Meeting, Hawaii, December 1985, submitted for publication in International Journal of Mineral Processing).

3. G. Barbery, "Particle composition distribution: measurement and prediction in mineral processing", (paper presented at Annual Meeting of Canadian Mineral Processors, Ottawa, January 1986, to be published in the Proceedings, 1986).

4. Jean Serra, Image analysis and mathematical morphology (London: Academic Press, 1982).

5. P.J. Davy, P.J., "Probability models for liberation", J. Applied Probability, 21 (1984) 260-269.

6. Georges Matheron, Random sets and integral geometry (New-York, N.Y.: J. Wiley and Sons, 1975).

7. R.P. King, "A model for the quantitative estimation of mineral liberation by grinding", Int. J. Mineral Processing, 6 (1979) 207-220.

8. J.A. Finch and W. Petruk, Testing a solution to the King liberation model", Int. J. Mineral Processing, 12 (1984) 305-311.

9. E.N. Gilbert, "Random subdivisions of space into crystals", Ann. Math. Statist., 33 (1962) 958-972.

10. R.P. King, "Determination of the size distribution of irregularly shaped particles from measurements on sections or projected areas", Powder Technology, 32 (1982) 99-112.

11. R.P. King, "Measurement of particle size distribution by image analyzer", Powder Technology, 39 (1984) 279-289.

12. R.L. Wiegel, "Integrated size reduction mineral liberation model", Trans. AIME, 260 (1976) 147-152.

13. J.T. Wizzard, R.P. Killmeyer and B.S. Gottfield, "The Department of Energy's Coal Washer Performance Computer Program" in Proc. Inst. Conf. on Use of Computers in the Coal Industry, Y.J. Wang and R.L. Sanford Eds (New York, N.Y.: AIME, 1983), 216-221.

Process Mineralogy Applications to Mineral Processing

PLATINUM-IRON ALLOYS IN THE DULUTH GABBRO COMPLEX:

MINERALOGY AND BENEFICIATION

T. Sabelin, I. Iwasaki and K. J. Reid

Mineral Resources Research Center
University of Minnesota
56 East River Road
Minneapolis, MN 55455

Abstract

Platinum group minerals (PGM) were found in a drill core from the mineralized zone in the basal portion of the Duluth Gabbro Complex. They occur within oxide-rich troctolite and gabbroic rocks with disseminated sulfide mineralization. The PGM are mostly Pt-Fe alloys and occur as discrete grains or in association with base metal sulfides. Magnetic separation upgraded the PGM in the nonmagnetic tails.

Introduction

The Duluth Gabbro Complex is a large body of dominantly mafic igneous rocks in northeastern Minnesota with significant Cu-Ni mineralization along a 50 mile length of the basal contact (1,2,3,4). Mafic and ultramafic rocks are typically the sites for platinum group mineral (PGM) occurrences, particularly where associated with Cu-Ni mineralization. Although the Duluth Gabbro Complex has been extensively studied in terms of its Cu-Ni resources, relatively little attention has been given to its platinum group element (PGE) contents and mineralogy. In a previous study of massive sulfide rock, sperrylite ($PtAs_2$) was found in nickel arsenide and recovery of arsenide minerals by flotation has been investigated (5,6). Recently, an announcement was made of the discovery of significant Pt-Pd mineralization, with peak assays of more than 9 g/t (0.27 oz per st), in a drill core containing disseminated sulfides 10 miles southeast of Ely, Minnesota (7,8). This paper describes the mineralogy and mode of occurrence of PGM from within and adjacent to the mineralized zone and evaluates their potential for beneficiation.

Sample Selection and Preparation

The samples examined in this study were collected from a drill hole in a minor mafic intrusion, known as the South Kawishiwi intrusion, at the western margin of the Duluth Gabbro Complex (Figure 1). The section of drill core studied occurs at 2400-2440 ft depth which is near the base of the intrusion.

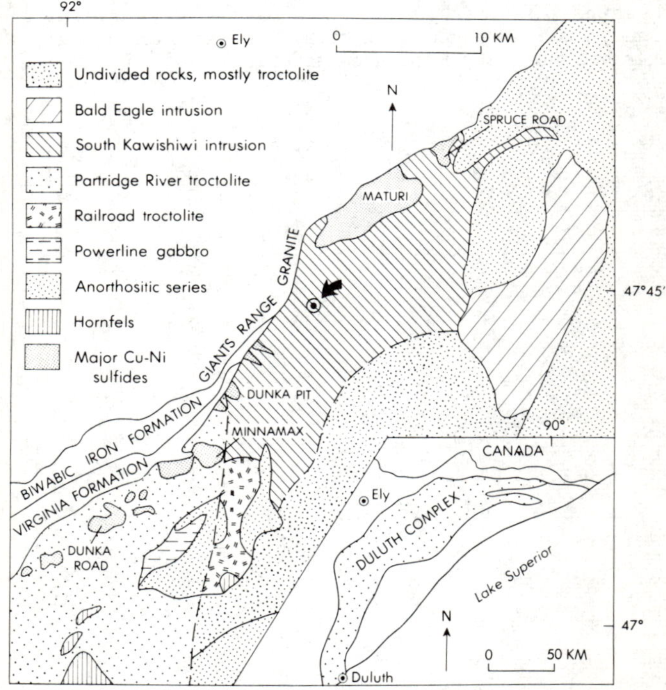

FIGURE 1. GENERALIZED GEOLOGIC MAP OF THE DULUTH GABBRO COMPLEX NEAR ELY, MINNESOTA SHOWING DRILL HOLE LOCATION. INDEX MAP SHOWS THE LOCATION OF THE COMPLEX IN NORTHEASTERN MINNESOTA (AFTER FOOSE AND WEIBLEN (9)).

Chemical analyses were made on one ft intervals of 1/4 BX-size core crushed to minus 100 mesh from the depth interval 2400 to 2430 ft, and selected number of samples were used in beneficiation tests. Mineralogical studies were made on selected horizons both within this interval and below it using standard petrographic polished sections and polished thin sections made from the remaining core.

Chemical analysis consisted of determination of Cr and Ag by atomic absorption spectrometry, and determination of Pt, Pd and Au by a combined fire assay-DC plasma technique. Pt, Pd, Au and Ag analyses were performed by a commercial laboratory which reported detection limits of 15 ppb Pt, 2 ppb Pd, 1 ppb Au and 0.1 ppm Ag per 20 g sample. Limited availability of sample material necessitated analyzing samples of less than 5 g weight resulting in detection limits of 66 ppb Pt, 9 ppb Pd, 4 ppb Au and 0.1 ppm Ag. Duplicate analyses could not be made but comparison of assay values of crude samples with those of magnetic separation tests showed that they were consistent with each other. The concentration methods used in beneficiation were Davis magnetic tube tests and sink-and-float tests using tetrabromoethane with a specific gravity of 2.96. Further testing, such as the effect of mesh-of-grind and recovery of Cu and Ni sulfides closely associated with PGM using bulk sulfide flotation, was not carried out due to the scarcity of samples. Mineralogical analysis was made by optical microscopy and scanning electron microscopy (SEM). PGM identifications were confirmed by qualitative x-ray analysis using energy dispersive analysis (EDS) on the SEM. Chemical compositions of some of the oxide minerals associated with PGM occurrences were obtained by an electron probe microanalyzer (EPMA) and a wavelength dispersive x-ray system (WDS).

Mineralogy

Oxide-rich troctolite, consisting essentially of oxide minerals, plagioclase and olivine, is the predominant rock type in the drill core interval examined. A thick layer or zone of oxide-rich troctolite occurs at 2408 to 2418 ft depth. Several thinner layers occur below this zone interstratified with coarse-grained gabbroic rocks. The troctolites contain as much as 30 to 65 percent oxides with higher concentrations observed locally. The oxides occur as subhedral rounded grains, 0.1 to 1.0 mm dia., and are typically enclosed in plagioclase. The major oxide mineral is an Fe-rich chromian spinel mineral dominated by the magnetite component Fe_3O_4 and, for simplicity, will be referred to in this report as magnetite. The magnetite has high TiO_2 contents (6-12%) and variable MgO (1-3%) and Al_2O_3 (2-6%) contents. It is associated with minor amounts of ilmenite and an Al-rich chromian spinel mineral which appear to have exsolved from the magnetite. The Al-rich chromian spinel mineral is dominated by the hercynite component $FeAl_2O_4$ and will be referred to as hercynite although it has significant MgO (9-15%) contents. Minor amounts of biotite are also present in the troctolites, occurring chiefly as rims on oxide grains. There is very little alteration of the troctolites; if present, it is chiefly as serpentinization of olivine.

The gabbroic rocks are generally coarser-grained than the troctolites and contain very little oxides. They are composed of plagioclase, olivine and clinopyroxene. In some gabbroic layers the secondary minerals amphibole, calcite, chlorite and serpentine are observed. Gabbroic rocks also host the richest sulfide horizons (up to 1.5% Cu, 0.3% Ni).

The Cu-Ni mineralization in these rocks is of the disseminated type and sulfide minerals are present only in trace amounts. Chalcopyrite is the most abundant sulfide and occurs in several textural varieties: as small- to medium-size grains locked with oxides and silicates; as fine-grained inclusions in plagioclase; and in myrmekitic intergrowth with ilmenite and hercynite. Pyrrhotite, pentlandite, bornite and Cu sulfides are present as fine-grained inclusions in plagioclase and/or in association with chalcopyrite. PGM are closely associated with the base metal sulfides and a number are found also as inclusions in plagioclase.

PGM Chemistry and Mineralogy

Chemical analyses show unusually high concentrations of Pt and Pd in the main oxide-rich zone at 2408 to 2418 ft depth (Figure 2). Average values for

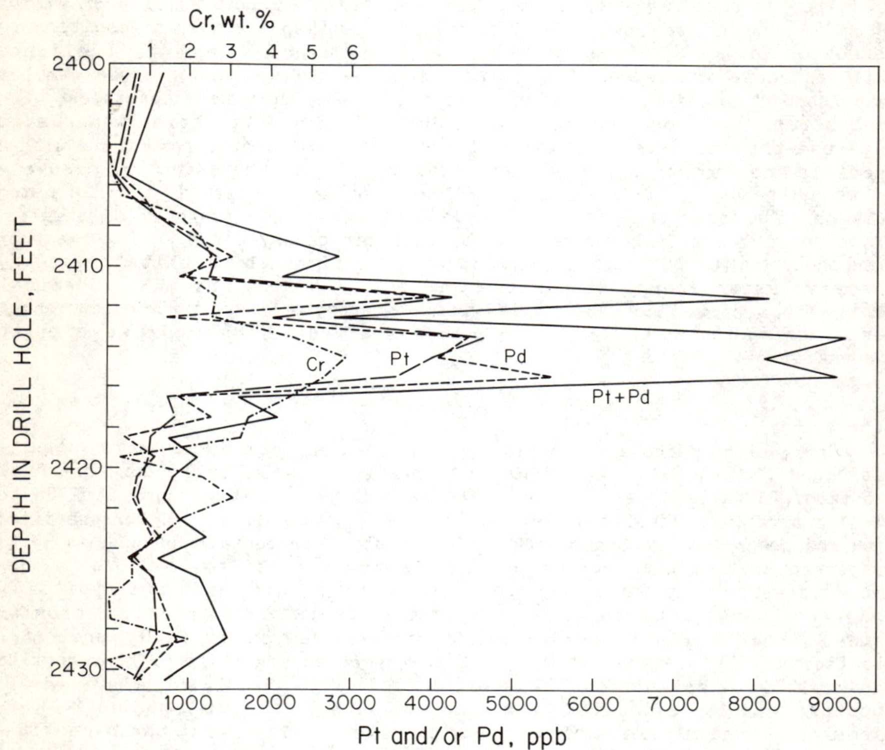

FIGURE 2. ELEMENTAL CONCENTRATION PROFILE FOR INTERVAL OF DRILL CORE CONTAINING THE Pt-Pd ZONE

Pt and Pd contents over the 10 ft interval are 2520 ppb and 2580 ppb, respectively, or 5.1 g/t (0.15 oz/st) Pt plus Pd. Pt plus Pd contents from individual one ft intervals are as high as 9100 ppb or 9.1 g/t (0.27 oz/st) and a 5 ft interval in the richest part of the zone averages 7.4 g/t (0.22 oz/st) Pt plus Pd. The Pt-to-Pd ratio is nearly 1:1 over the entire zone and the Pt and Pd correlate closely with Cr which ranges from 1.9 to 5.3 percent. The variation in the whole rock Cr content corresponds to the variation in Cr content of the spinels, which are the major Cr-bearing minerals.

Within the Pt-Pd mineralized zone the most common PGM found are Pt-Fe alloy and Pd alloys. Semiquantitative EDS analysis shows that Pt-Fe alloy grains have variable composition with Pt contents varying from approximately 73 to 82% by weight and Fe from 14 to 20% by weight. Pt-Fe alloy also contains minor amounts of Pd, Cu and Ni. The Pt-Fe alloy is distinguishable from other PGE minerals and the base metal sulfides in reflected light by its high reflectivity and white color. Most grains are in the size range of a few micrometers. A few larger grains were found, measuring 7, 10, and 35 μm diameter (Figures 3a and 3b). The Pd alloy grains are on the order of a few micrometers in diameter and were identified by qualitative EDS analysis. The data suggest that Pb, Cu and Ag occur with the Pd.

Pt-Fe alloy was also found in an oxide layer below the Pt-Pd mineralized zone at 2440 ft depth. It occurs in a composite grain with three compositionally distinct Pd minerals. The Pt-Fe alloy is similar in composition to the other Pt-Fe alloy grains found and constitutes 85% of the grain (Figure 3c). The grain occurs with another composite grain containing complex fine-scale intergrowths of several PGM. Irarsite ((Ir,Ru)AsS), Pt sulfarsenides, Pd arsenides and Pt-Fe alloy have been identified by EDS in the second grain. Its texture suggests the presence of additional phases but they were too fine-grained (< 1 μm size) to be analyzed by SEM.

Beneficiation Tests

Native platinum and Pt-Fe alloys constitute major PGE minerals in Bushveld pegmatites and placer deposits and are commonly processed by gravity separation. Since gravity separation methods become less effective below 200 mesh, magnetic separation methods may be considered. Cabri and Feather (10) classified Pt-Fe alloys into four groups: native platinum (> 80 at.% Pt), ferroanplatinum (20-50 at.% Fe), isoferroplatinum (Pt_3Fe), and tetraferroplatinum (PtFe). Tetraferroplatinum is very ferromagnetic, whereas the others have variable magnetic susceptibilities depending on their composition. Isoferroplatinum is nonmagnetic when the lattice is ordered and is ferromagnetic when disordered. Cold working during comminution, therefore, may induce ferromagnetism. The more dilute alloys are reported to be ferromagnetic to 2 at.% Fe (11). In fact, magnetic separation of copper-nickel flotation tailings at Norilsk was reported to improve the recoveries of Pt and Pt-Fe alloys (12).

Davis magnetic tube tests were performed on selected drill core pulp samples containing different levels of Cr contents to determine whether simple magnetic separation may be used to preconcentrate the PGM. Table I presents the test results. Pt recovery ranged from 86 to 98 percent and Pd recovery from 88 to 99 percent in nonmagnetic tailings with weight recoveries of 23 to 84 percent. In view of the magnetic properties of Pt-Fe alloys, and their abundance in the oxide-rich Duluth Complex samples, it is puzzling that virtually all the Pt is found in the nonmagnetic tailings. The recoveries of Au and Ag for the samples in Table I were variable and averaged 63 to 99 percent for Au and 58 to 94 percent for Ag.

Mineralogical examination of drill core also showed that a number of PGM grains occur as inclusions in plagioclase and in alteration minerals. Thus a sink-and-float test was made on one of the nonmagnetic tailings products to determine whether Pt and Pd may be further upgraded. The results are presented in Table II. Microscopy examination of the products showed the float to be composed primarily of plagioclase with lesser amounts of biotite and chloritic alteration minerals. The cleaner float was also composed primarily of plagioclase with fine-grained hercynite and some biotite. The cleaner sink contained mostly hercynite and some composite oxide-biotite and oxide-plagioclase grains. The assay values showed that Pt is distributed evenly among the three products whereas Pd is concentrated somewhat in the cleaner float and cleaner sink products. Therefore, the analytical results

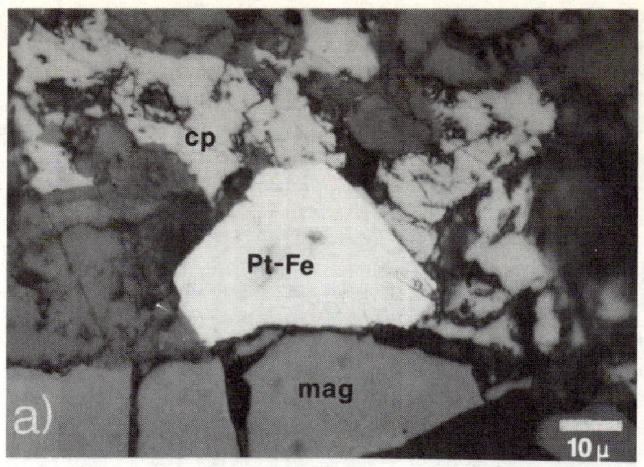

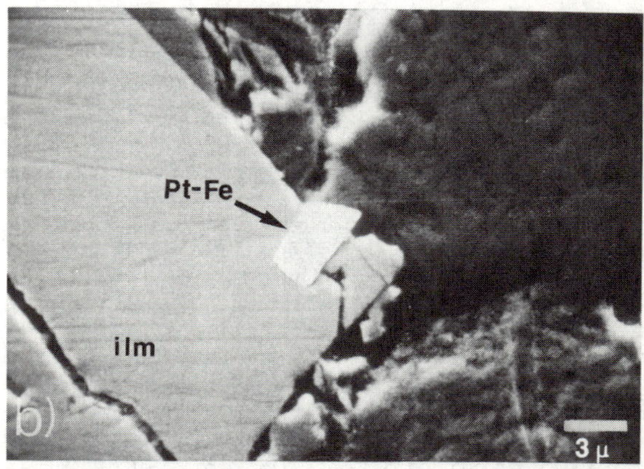

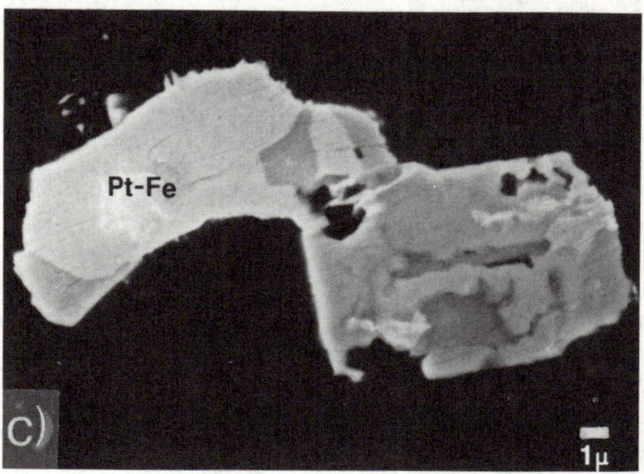

FIGURE 3. Pt-Fe ALLOY GRAINS (Pt-Fe) a) ASSOCIATED WITH MAGNETITE (mag), CHALCOPYRITE (cp) AND SILICATE ALTERATION MINERALS, REFLECTED LIGHT; b) IN ILMENITE (ilm), SEM; AND c) IN COMPOSITE GRAIN WITH Pd MINERALS, SEM

Table I. Davis Magnetic Tube Separation Results for Samples at Variable Depth in Drill Core

Sample Depth (ft)	%Cr Head	Product	% wt	Assay ppb Pt	Assay ppb Pd	Assay ppm Au	Assay ppm Ag	Recovery, % Pt	Recovery, % Pd	Recovery, % Au	Recovery, % Ag
2410-2411	1.99	Mag	41.2	180	70	20	0.2				
		NonMag	58.8	1340	1240	30	0.5	91	96	68	78
		Composite	100.0	860	760	30	0.4				
2411-2412	2.66	Mag	70.3	340	160	10	<0.1				
		NonMag	29.7	6190	9600	120	0.4	88	96	84	(63)
		Composite	100.0	2080	2960	40	(0.2)				
2413-2414	4.5	Mag	76.5	240	120	<4	0.1				
		NonMag	23.5	6400	9300	30	0.3	89	96	(70)	58
		Composite	100.0	1690	2280	(10)	0.1				
2415-2416	5.37	Mag	58.8	170	130	20	0.3				
		NonMag	41.2	5130	12520	630	2.2	95	99	96	83
		Composite	100.0	2210	5230	270	1.1				
2417-2418	3.47	Mag	57.3	30	20	5	0.3				
		NonMag	42.7	2680	3540	670	2.6	98	99	99	87
		Composite	100.0	1160	1520	290	1.3				
2420-2421	2.38	Mag	36.3	140	90	80	0.7				
		NonMag	63.7	620	670	80	1.7	89	93	64	81
		Composite	100.0	450	460	80	1.3				
2421-2422	3.10	Mag	48.2	90	70	4	0.3				
		NonMag	51.8	530	460	50	1.6	86	88	93	86
		Composite	100.0	320	270	30	1.0				

-continued-

Table I. Davis Magnetic Tube Separation Results for Samples at Variable Depth in Drill Core (continued)

Sample Depth (ft)	% Cr Head	Product	% wt	Assay ppb Pt	Assay ppb Pd	Assay ppb Au	Assay ppm Ag	Recovery, % Pt	Recovery, % Pd	Recovery, % Au	Recovery, % Ag
2423-2424	1.28	Mag	23.9	40	30	5	0.7				
		NonMag	76.1	560	590	40	1.5	98	98	96	94
		Composite	100.0	440	460	30	1.2				
2425-2426	0.66	Mag	15.8	220	390	130	0.9				
		NonMag	84.2	510	570	40	1.1	92	89	63	87
		Composite	100.0	460	540	50	1.1				

showed no further separation of the PGE's and suggest that plagioclase and alteration minerals account for only some of the sites in which PGM occur. It becomes of interest to study the beneficiation behaviors of the Pt-Fe alloy grains, particularly their magnetic and flotation properties and the effect of compositional changes on those properties, for recovery from nonmagnetic tailings.

Table II. Sink-and-Float Test Results for 2413-2414 ft Depth Nonmagnetic Sample at S.G. 2.96

	% Wt	Pt, ppb	Pd, ppb
Float	26.2	5600	5040
Cl Float	12.8	6400	10200
Cl Sink	61.0	5800	9760

Summary

PGM were found in a drill core within oxide-rich troctolite rock with disseminated sulfide mineralization. One troctolite zone is over 10 ft thick and contains significant Pt-Pd mineralization with peak assay values of over 9 g/t (0.27 oz/st) Pt plus Pd. The Pt-to-Pd ratio over the entire 10 ft interval is nearly 1:1, and Pt and Pd contents show a close correlation with Cr content. Pt and Pd occur mostly as alloys and are closely associated with Cu, Ni and Fe sulfides. Magnetic separation concentrated Pt and Pd into nonmagnetic tails with recoveries exceeding 88 percent. Au and Ag recoveries are variable with Au ranging from 63-99 percent and Ag from 58-94 percent.

References

1. B. Bonnichsen, "Sulfide Minerals in the Duluth Complex," Geology of Minnesota: A Centennial Volume, ed. P.K. Sims and G.B. Morey (St. Paul, MN: Minnesota Geological Survey, 1972), 388-393.

2. M. L. Boucher, "Copper-nickel Mineralization in a Drill Core from the Duluth Complex of Northern Minnesota," Rep. Invest. US Bur. Mines 8084 (1975) 55.

3. S. N. Watowich, "A Preliminary Geological View of the MINNAMAX Copper-Nickel Deposit in the Duluth Gabbro," Proc. 51st Annual Meeting Minnesota Section AIME, 39th Annual Mining Symposium, 1978, 19.1-19.11.

4. P. W. Weiblen and G. B. Morey, "A Summary of the Stratigraphy, Petrology, and Structure of the Duluth Complex," Am. J. Sci., 280-A (1980) 88-133.

5. I. Iwasaki, P. W. Weiblen, K. J. Reid, P. J. Ryan, H. Nakazawa, and A. S. Malicsi, "Platnum Group Minerals and Their Recoveries from Copper-Nickel Bearing Duluth Gabbro," (Paper submitted to Trans. SME-AIME for publication, 1986).

6. P. J. Ryan and P. W. Weiblen, "Pt and Ni Arsenide Minerals in the Duluth Complex," 30th Annual Institute on Lake Superior Geology, Wausau, Wisconsin, April 24-28, 1984, 58-60.

7. "Chromium and Platinum Found in Northeastern Minnesota," Mining Engineering, 37 (9) (1985), 1115.

8. T. Sabelin, "Platinum Group Element Minerals in the Duluth Complex," 31st Annual Institute on Lake Superior Geology, Kenora, Ontario, May 6-11, 1985, 83-84.

9. M. Foose and P. W. Weiblen, "The Physical and Petrologic Setting and Textural and Compositional Characteristics of Sulfides from the South Kawishiwi Intrusion, Duluth Complex, MN, USA," (Paper submitted to Minnesota Geological Survey, St. Paul, MN, for publication, 1986).

10. L. J. Cabri and C. E. Feather, "Platinum-iron Alloys: A Nomenclature Based on a Study of Natural and Synthetic Alloys," Can. Mineral., 13 (1975) 117-126.

11. J. Crangle, "Some Magnetic Properties of Platinum-rich Pt-Fe Alloys," J. Phys. Rad., 20 (1959) 435-437.

12. "Improvements in Cu-Ni Ore concentration at Norilsk," World Mining, 35 (3) (1982) 77.

FLOTATION KINETICS AND ORE MICROSCOPY OF VIBURNUM LEAD ORES

Michael G. Schroer[1], Scott L. Volner[2] and John L. Watson[3]

[1] Allied Mineral Products, Columbus, Ohio.

[2] Dee Gold Mining Co., Elko, Nevada.

[3] University of Missouri-Rolla, Rolla, Missouri.

Abstract

The flotation response of several lead/zinc ores from the Viburnum Trend of S.E. Missouri was investigated and related to ore mineralogy. A standardized kinetic flotation test was devised and demonstrated to clearly characterize flotation response and flotation problem areas. Mineralogical evidence was then identified to explain the flotation difficulties of several of the ores and these difficulties were found to arise from a variety of sources. A possible cause of low lead recoveries of some ores was shown to be the presence of galena particle coatings in the form of cerrusite or goethite. In other cases complex locking at fine particle sizes constituted a cause of lead losses. The presence of hydrocarbons, possibly as pyrobitumen, was demonstrated for one ore and in this case multiple mineral activation could be the cause of the poor lead selectivity found in this ore. In general, however, the cause of low Pb/Zn selectivity ratios for the ores investigated was attributed to the presence of copper ions in solution and their tendency to activate sphalerite. Copper ions in solution were shown to be the result of the leaching of oxidized copper minerals, with the leaching being more effective in the presence of cyanide. In addition it was demonstrated that no other significant metal solubility occurred in the wet grinding of these Viburnum ores.

Introduction

The lead/zinc ores of the Viburnum Trend in S.E. Missouri have been mined and beneficiated for approximately 20 years. The geology of the Trend is well documented (1) and the major minerals are well defined as galena, sphalerite, chalcopyrite and pyrite with dolomite as the major gangue component. At the present time several mining companies are producing lead, zinc and copper concentrates (2,3) from the ores of the Viburnum Trend by the processes of comminution, classification and flotation. For many years the major economic mineral of the trend has been galena and the flotation of galena has been the subject of considerable research. However one research aspect which has not been investigated in detail, deals with an ore mineralization which does not appear amenable to galena flotation. In these ores the galena appears chemically identical to the free floating form of galena, but its presence, together with the other minerals in the ore, can cause considerable flotation problems with low lead recoveries and concentrate grades.

There are a variety of possible, but unproven, causes of these flotation problems and briefly stated they include:

a) surface oxidation - galena can oxidize to cerrusite and anglesite and oxidation products have been reported for Viburnum Trend galena (3,4). The oxidation is often promoted by the presence of other sulfides especially marcasite. It has been shown (5) that the susceptibility of sulfides to oxidation increases in the following manner:

$$Cu_2S < PbS < CuFeS_2 < FeS_2$$

This suggests that pyrite and chalcopyrite will oxidize prior to galena in mixed sulfide ores.

b) activation of other minerals - low flotation concentrate grades often result, not from the lack of galena recovery, but from the flotation of other minerals activated by some component of the ore. Examples of such activation may be the result of soluble copper mineralization (6) or the presence of hydrocarbons, both natural and man made. Copper ions will activate other minerals even in the presence of cyanide and hydrocarbons can cause blanket flotation of all minerals present in the ore.

c) crystal habit - galena has a variety of crystal habits with the predominant forms being cubic and octahedral (7) and little is known regarding their individual response to flotation.

At the St. Joe Mining Corporation operations in Viburnum, MO there have been numerous occasions when plant lead recoveries and grades have fallen due to some undetected change in the ore. To investigate this overall problem two research projects were initiated in the Department of Metallurgical Engineering at the University of Missouri - Rolla. The first project set out to define the problem by devising a standardized flotation test which would yield kinetic data. Analysis of this data from a free floating and a problem ore would then permit the ore differences to be characterized and the underlying causes of the flotation difficulties to be related to the ore mineralogy. The second project was designed to investigate the relationship between flotation response and mineralogy for several ores and to identify the origin and character of the minerals contributing to the flotation problems.

Kinetic Theory

The technique of assessing the flotation process by kinetic analysis is well established (8) and it represents a valuable tool in the operation and understanding of flotation. Briefly stated, kinetic analysis requires that flotation be considered analogous to a first order reaction where the rate of flotation (dCi/dt) of a given mineral species (i) is proportional to the concentration (Ci) of that mineral in the pulp. This may be stated as-

$$dCi/dt = -Ki\ Ci \tag{1}$$

which when integrated yields-

$$Ci(t)/Ci(0) = e^{-Ki\ t} \tag{2}$$

The constant Ki is defined as the specific flotation rate constant for the mineral species i and it has the units of inverse minutes. Many flotation systems have been analysed as first order rate processes and it has been found, in general, that the best representation of flotation is achieved when the mineral species is considered to consist of two components (9). One component is taken to be fast floating and one slow floating with three parameters defining the system. The rate constants of the components are defined as Ki_f and Ki_s with the fast floating fraction of the mineral species being designated F_f. Thus the process is then represented by-

$$Ci(t)/Ci(0) = F_f\ e^{-Ki_f\ t} + (1-F_f)\ e^{-Ki_s\ t} \tag{3}$$

The values of the three kinetic parameters may be determined from a simple laboratory batch flotation test in which the concentrate is recovered in time increments. A subsequent plot of Ci(t)/Ci(0) against time then provides the kinetic parameters as illustrated in Fig. 1.

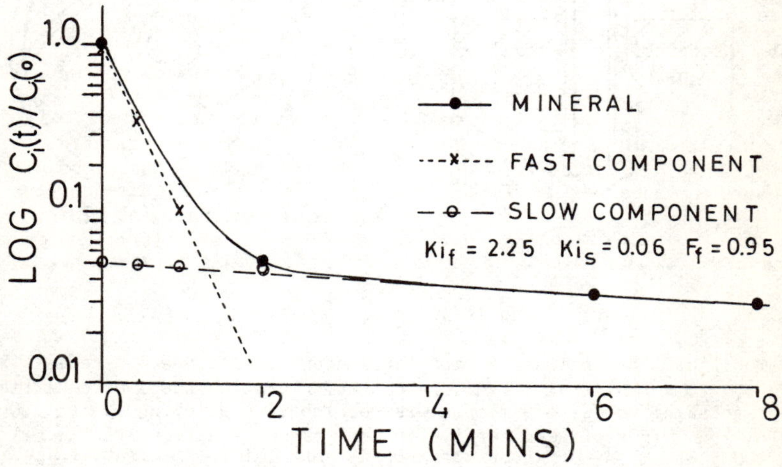

Figure 1 - Semi-log Plot of Unfloated Mineral against Time.

Once the kinetic parameters are established for each mineral species present in the ore, a model may be constructed and used to predict flotation performance based upon these laboratory derived parameters. However there are important differences between laboratory tests and industrial flotation banks and hence a model (10) for a bank of cells uses a modified form of equation 3. This modified equation assumes perfect mixing in a single cell of a bank of n cells and an average cell residence time of T minutes and it may be written as-

$$Ci(t)/Ci(0) = F_f/(1 + Ki_f\ T)^n + (1-F_f)/(1 + Ki_s\ T)^n \tag{4}$$

Equation 4 may then be utilized to predict the behavior of a bank of cells in terms of laboratory kinetic data but this data must be gathered in a carefully designed and operated test. In this research program, the flotation cell design and operation was developed to permit the kinetic characterization of the Viburnum ores and hence identify the problem areas prior to mineralogical examination.

Experimental

The research was divided into two sections with the first test (A) examining 1) flotation kinetics and 2) mineralogy. Test B investigated the general relationships between flotation and ore properties.

Test A 1) Flotation

A Fagergren laboratory flotation machine was modified as illustrated in Fig. 2 to permit standardized kinetic data to be determined for Viburnum Ores.

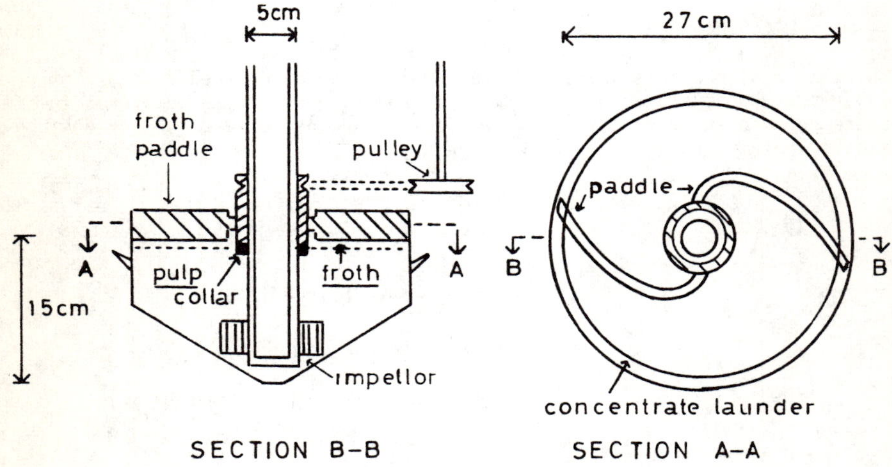

Figure 2 - Modified Laboratory Flotation Cell.

The modifications included a circular concentrate launder, a radial froth scraper and a froth/pulp level controller. Flotation tests were carried out after grinding to 80% -75 micron with ore charges weighing 1.5 to 2.0kg and containing 70-90g of lead metal. Typical reagent additions to the pulp were 200-1000g of amyl xanthate (collector) / tonne of lead in the ore (ie.10-50 g/tonne of ore for a 5% lead ore), 500g of zinc sulfate (depressant) / tonne of ore and 10 g of sodium cyanide (depressant) / tonne of ore. It was determined that the cell operated most consistently when the pulp surface was maintained at 16 mm below the cell lip with a methyl isobutyl carbinol (frother) addition of 35g / tonne of ore. A standard kinetic flotation test consisted of concentrate samples being collected after 0.5, 1, 2, 4, and 8 minutes of operation, with the flotation taking place at the natural pH of the ore (ie. 7.9-8.4).

The chemical analyses of the ores and flotation products were undertaken by the laboratories of the St Joe Mining Corporation at Viburnum using wet techniques for the elements of lead, zinc, copper and iron. The resulting data were then analyzed by a microcomputer program to calculate the kinetic parameters.

Test A 2) Mineralogy

To determine the mineralogy of the ores associated with these kinetic parameters, flotation feed and product samples were mounted in plastic and polished down to one micron fineness. The samples were then examined using a Nikon Opiphot polarizing, binocular microscope to permit mineralogical identification and photography.

Test B 1) Flotation

Six ores from the Viburnum and Fletcher mines of St. Joe were subjected to a series of tests designed to characterize their floatability. Feed samples of 500g were ground in a 20 X 20 cm batch laboratory mill at 50% solids by weight for 12 minutes. Chemical additions to the mill included 1000g of sodium isopropyl xanthate per tonne of lead in the ore (50g/tonne of 5% Pb ore), 500g of zinc sulfate per tonne of ore and 0-1000g of sodium cyanide per tonne of ore. The flotation was carried out in a standard Denver 1.4 liter cell at the natural pH of the ore using 40g of Aerofroth 72 per tonne of ore and a concentrate was removed from the cell over a period of 7 minutes.

Test B 2) Mineralogy

Feed and product samples of the six ores were mounted in plastic, polished and examined under the microscope as detailed above. This time , however, point counts were undertaken, where appropriate, to quantify the mineralogy.

Test B 3) Ion Solubility

To determine the solubility of the major metal sulfides, present in the ores, 250 g samples were ground at 50% solids by weight in a 20 X 20 cm mill for 12 minutes. 500 g of sodium isopropyl xanthate per tonne of lead in the ore and 0-1000g of sodium cyanide per tonne of ore were added to the mill water, which after grinding was analyzed by AA for lead, copper, zinc and iron.

Results and Discussion

Test A 1) Kinetics

The following data in Table I characterize a free floating galena ore and hence establish the kinetic parameters for a Viburnum ore which is amenable to flotation.

Table I Kinetic Data for a Free Floating Ore.

Feed Ore (5.0% Pb, 0.3% Zn, 0.2% Cu, 1.7% Fe)

	Pb	Zn	Cu	Fe	Gangue
K_s	0.013	0.001	0.009	0.001	0.001
K_f	2.50	1.35	1.91	1.53	1.24
F_f	0.94	0.08	0.64	0.07	0.01

These results illustrate that 94% of the lead mineralization is fast floating and that the fast rate constant is relatively high compared with the rates of the other minerals present. In addition the zinc, iron and gangue minerals are predominantly slow floating while the copper mineral in this test has bulk lead /copper concentrate. Using a simple flotation model based on equation 4 the response of a rougher bank of 6 cells, fed with this ore, was simulated and the results are presented in Table II.

Table II Lead Rougher Bank Simulation of a Free Floating Ore.

Cell No.	Conc Wt.	Predicted Grades Pb	Zn	Cu	Fe
1	5.3	69.9	0.3	1.8	1.6
2	1.3	58.7	0.4	1.9	2.0
3	0.4	39.3	0.5	1.6	2.3
4	0.3	18.3	0.4	1.1	2.1
5	0.2	7.6	0.3	0.8	1.8
6	0.1	4.4	0.3	0.6	1.7
Total Con	7.6	62.1	0.3	1.7	1.7
Tail	92.4	0.3	0.3	0.1	1.7

These predicted values are what would be expected of a rougher bank flotation of a free floating Viburnum ore, with a lead recovery of 94.6%. In comparison Tables III and IV present equivalent data for an ore identified as being difficult to float.

Table III Kinetic Data for a Difficult Ore.

Feed Ore (2.9% Pb, 0.1% Zn, 1.5% Cu, 5.2% Fe)

	Pb	Zn	Cu	Fe	Gangue
Ks	0.002	0.004	0.002	0.000	0.000
Kf	2.00	1.71	1.65	1.55	1.57
Ff	0.38	0.34	0.29	0.10	0.01

The major difference between the two ores, in terms of the feed grade, is that the problem ore has a higher sulfide mineral content, especially pyrite. The kinetic data illustrate that the ore has a high fraction of slow floating galena with a rate of 0.002 inverse minutes. The problem appears to be the galena not floating rather than the excessive flotation of any of the other minerals, although the sphalerite response is much higher than for the free floating lead ore. This can be seen by comparing the kinetic parameters for each mineral species in Tables I and III. Overall the value of kinetic analysis in identifying problem areas is clearly indicated by this simple example.

Table IV Lead Rougher Bank Simulation of a Difficult Ore.

Cell No.	Conc Wt.	Predicted Grades Pb	Zn	Cu	Fe
1	2.6	30.8	0.9	11.6	13.6
2	0.9	26.1	0.8	11.3	13.9
3	0.2	19.9	0.8	10.0	12.7
4	0.2	13.1	0.6	7.6	9.6
5	0.1	8.8	0.5	5.4	6.6
6	0.1	7.1	0.4	4.4	5.1
Total Con	4.1	27.5	0.8	11.0	13.1
Tail	95.9	1.8	0.1	1.1	4.9

Tables III and IV clearly define the problems of low concentrate grade (27.5% Pb) and recovery (39.0% Pb) that would be encountered if this ore was treated in the plant, but they do not indicate the cause of the large fraction of slow floating galena or the reason for the low slow rate. To assist in the determination of these latter properties a mineralogical microscope was used to examine the mineralogy of the ores and flotation products.

The use of kinetic data for flotation simulation of ores of the Viburnum Trend is fully detailed elsewhere (11) together with full experimental details of the research reported here.

Test A 2) Mineralogy

In a normal or free floating lead ore, the concentrate should consist essentially of galena as is shown in the photomicrograph in Fig. 3. The illustrated presence of chalcopyrite in the concentrate is expected as this test simulated a lead-copper float. The tail is displayed in Fig. 4 where dolomite, sphalerite and pyrite grains can be identified. In the case of the problem ore, Fig. 5 shows the presence of galena in the tail together with a grain of covellite, while Fig. 6 illustrates the presence of a goethite coating on a galena particle in the tail. Both Figs. 5 and 6 suggest possible reasons for the poor flotation performance. The presence of an iron coating on the surface of a galena particle is sufficient to prevent that particle adsorbing the xanthate collector and hence the particle is rejected to the tail. The presence of covellite, which is an oxidation product of chalcopyrite, in the tail indicates oxidizing conditions in the ore and hence galena may be oxidized to either cerrusite or anglesite. The presence of cerrusite is shown in Fig. 7 and again a cerrusite coating is capable of preventing the flotation of a galena particle. Thus the high slow floating galena fraction of the difficult ore is therefore probably due to galena coatings of cerrusite or goethite which render galena particles hydrophylic. The enhanced sphalerite flotation response can be explained by the copper mineral oxidation providing copper ions in solution which can activate the sphalerite.

The Test A data illustrate the manner in which flotation kinetics and mineralogy complement each other in the effective analysis of the flotation process. The kinetic data highlight the problem areas of a flotation and the mineralogy determines the underlying causes of the problems.

Test B 1) Flotation

Of the six ores tested, one (a) is an ore that can be considered free floating and the other five are problem ores identified by the mill personnel. Table V gives the flotation results under the standard conditions outlined above with no cyanide addition.

Table V Flotation Results for Six Viburnum Ores.

	LEAD %			ZINC %			COPPER %			IRON %			SEL
Ore	Head	Tail	Rec	Head	Tail	Rec	Head	Tail	Rec	Head	Tail	Rec	Pb/Zn
(a)	5.2	.14	97	.56	.54	12	.01	.01	58	1.46	1.49	6	16.6
(b)	8.1	.75	92	.31	.19	47	.37	.05	88	1.48	1.55	8	3.6
(c)	10.5	.67	94	.09	.08	29	.08	.02	81	1.49	1.58	9	6.4
(d)	5.0	.52	90	.35	.13	67	.37	.03	93	1.48	1.37	16	2.2
(e)	3.0	.93	71	.22	.03	87	.68	.36	50	1.38	1.38	6	0.6
(f)	9.7	.58	95	.83	.53	52	2.77	.38	90	4.93	5.26	19	4.4

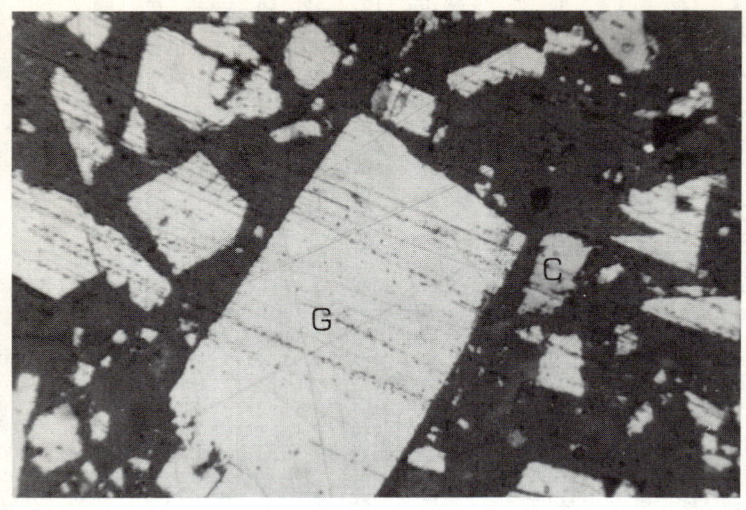

Figure 3 - Photomicrograph of Galena/Chalcopyrite Concentrate (X200).

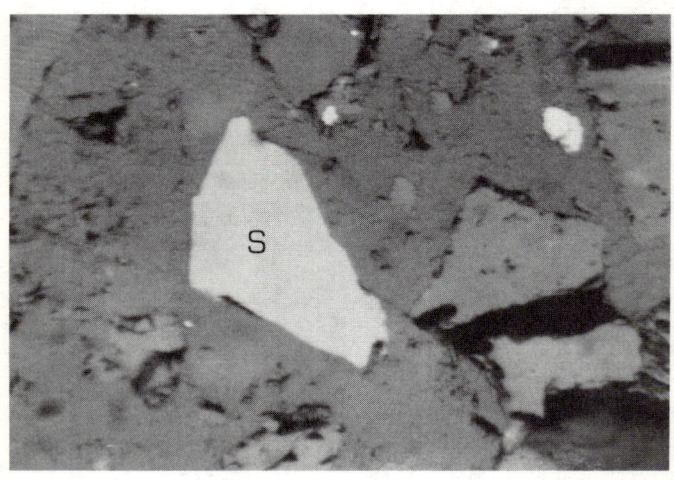

Figure 4 - Photomicrograph of Flotation Tailing Showing Sphalerite (X200).

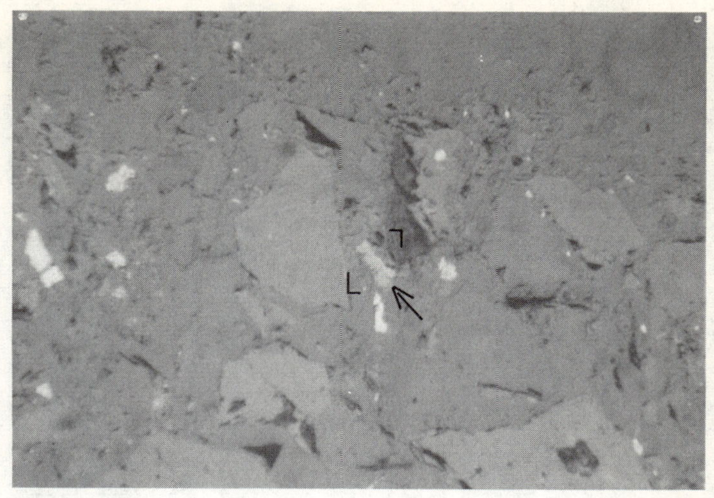

Figure 5 - Photomicrograph of Problem Ore Tailing Showing Covellite (X200).

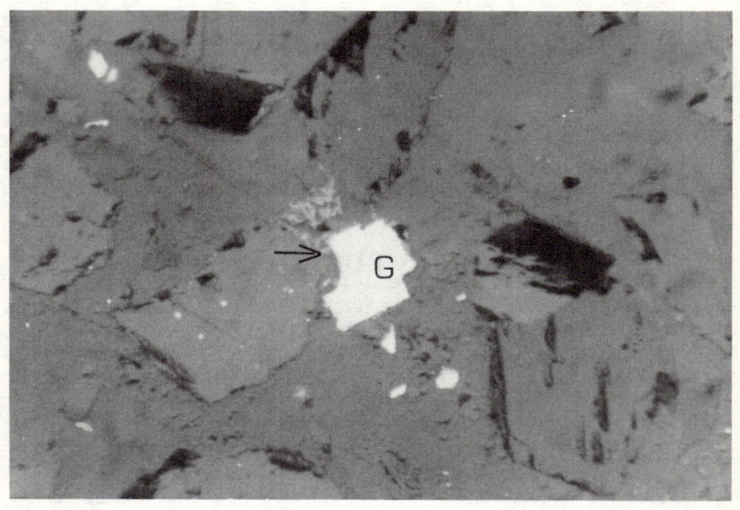

Figure 6 - Photomicrograph of Goethite Coating on a Galena Grain in Problem Ore Tailing (X200).

Figure 7 - Photomicrograph of Galena Grain Showing Cerrusite Coating (X200).

Ore (a) has the expected low lead tail and high lead recovery with low zinc and iron recoveries. The copper in this ore is present in very low amounts but it still is mainly recovered to the lead concentrate. The other ores all display high lead tails and lower lead recoveries which characterize problem ores. In addition they also show increased zinc recoveries and this is demonstrated by the Pb/Zn selectivity factor given in Table V. It is apparent that in ores (b) - (f) some galena depressing mechanism is at work together with activation of the sphalerite. Ore (e) is shown to have the lowest lead recovery and highest zinc recovery, both of which are reflected in the lowest Pb/Zn selectivity value of 0.6 which suggests sphalerite is being floated preferentially to galena.

When the flotation tests were repeated with increasing quantities of sodium cyanide, the Pb/Zn selectivities did increase as the cyanide helped to depress the sphalerite and pyrite. In general the cyanide did not effect the galena flotation and above additions of 500g/tonne of ore, the cyanide tended to hurt the Pb/Zn selectivity. While these results do characterize the ores in terms of flotation, they do not indicate the cause of the problems, but merely the effects.

Test B 2) Mineralogy

The mineralogical examination of the six ores revealed that many different ore properties were responsible for the flotation results given above. The free floating ore (a) may be taken as a standard and the other ores compared with it to illustrate the differences in mineralogy. A gravity concentrate of ore (a) is shown in Fig. 8 and it can be seen that the galena and sphalerite minerals are free and relatively coarse. This results in the free floating characteristics of this ore. Ore (b) is illustrated in Fig. 9 and here the presence of hydrocarbons (possibly pyrobitumen) is immediately obvious together with some particle locking. The hydrocarbons have the capability of activating minerals indescriminantly and this can explain the

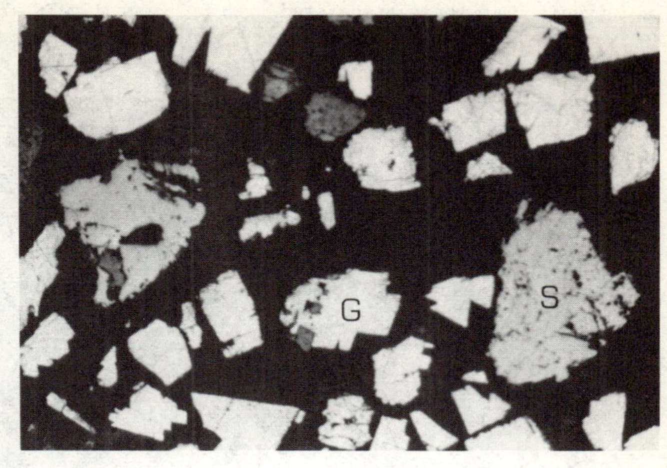

Figure 8 - Photomicrograph of Ore (a) Showing a Gravity Concentrate of Free Galena and Sphalerite Grains (X200).

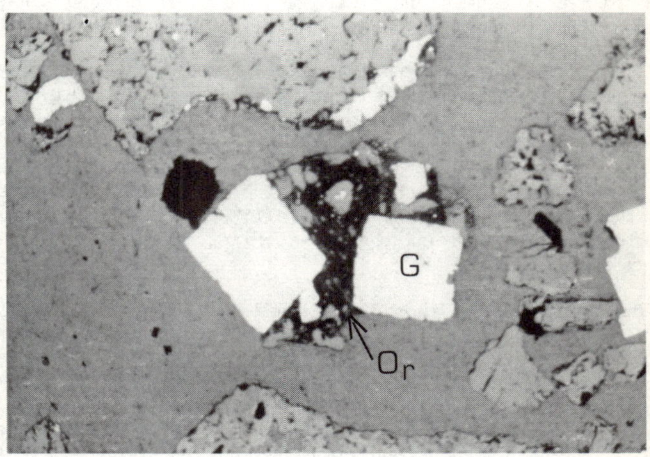

Figure 9 - Photomicrograph of a Galena Grain from Ore (b) Showing an Associated Organic Rich Phase (X200).

increased recovery of sphalerite, chalcopyrite and pyrite in this ore. In addition the composite particles can cause a loss of galena to the tail resulting in lower lead recoveries. Figure 10 shows the presence of pyrite flecks in an organic rich phase associated with octahedral galena in ore (b).

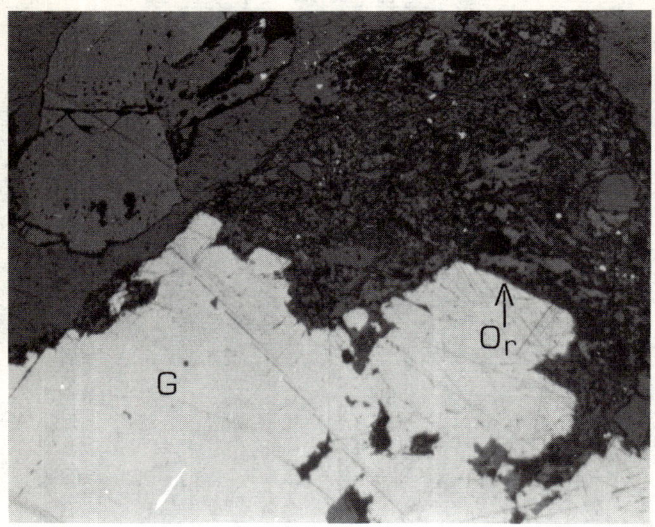

Figure 10 - Photomicrograph of an Octahedral Galena Grain Associated with an Organic Rich Phase in Ore (b) (X200).

Ore (c) displays considerable oxidation with some galena grains having oxidation rims of 40 microns, as illustrated in Fig. 11. This feature can therefore explain the low lead recovery but no evidence of hydrocarbons or copper mineral oxidation was determined in this ore to explain the poor selectivity. However for ore (d) the increased zinc recovery is explainable in terms of the oxidation of the copper mineralization illustrated in Fig. 12. The oxidation of chalcopyrite to covellite will result in some copper ions entering the solution and thus sphalerite can adsorb these ions and then behave as a copper mineral.

It was noted earlier that ore (e) had the poorest flotation results, but no evidence of galena oxidation, hydrocarbon presence or copper mineral oxidation was found for this ore. Figure 13 displays the flotation tail for ore (e) and it is apparent that galena is present in the tail in coarse, unoxidized particles. The only exceptional property of this ore is that the marcasite to pyrite ratio is higher than the other ores, and past evidence suggests (3) that under such conditions galena oxidation is likely. In general oxidized minerals will not float with a xanthate collector and it is evident from Table V that the lead and copper recoveries are the lowest in ore (e) of any of the ores. In addition the zinc recovery in ore (e) is the highest of any of the ores and this suggests oxidation and subsequent activation of sphalerite.

The final ore (f) demonstrates, in Fig. 14, considerable locking but displays no evidence of oxidation or hydrocarbon presence. Therefore the flotation problems in this case are suspected to be a function of liberation

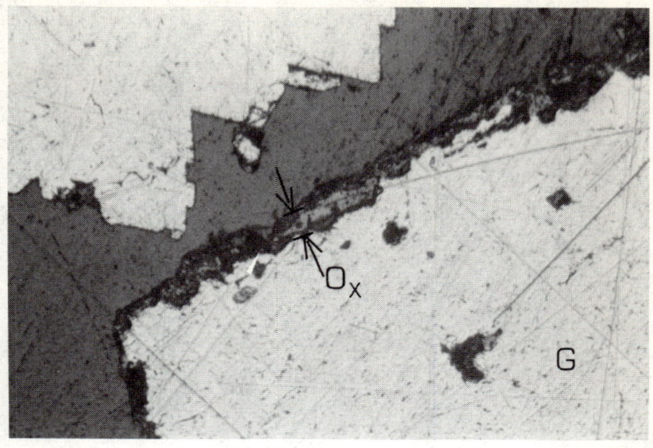

Figure 11 - Photomicrograph of a Galena Grain from Ore (c) Showing an Oxidation Rim (X200).

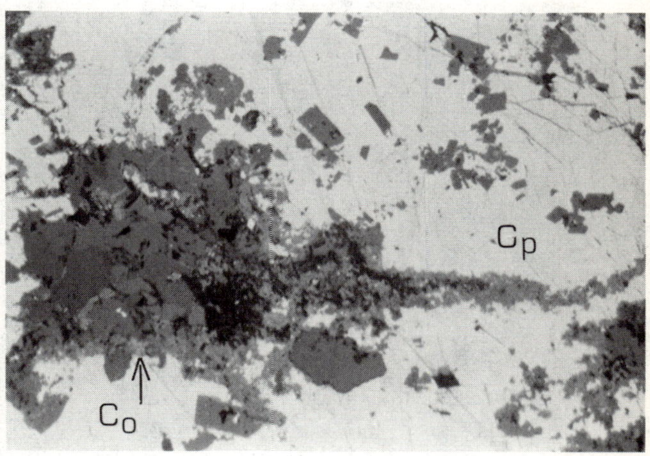

Figure 12 - Photomicrograph of Covellite Replacement of Chalcopyrite in Ore (d) (X200).

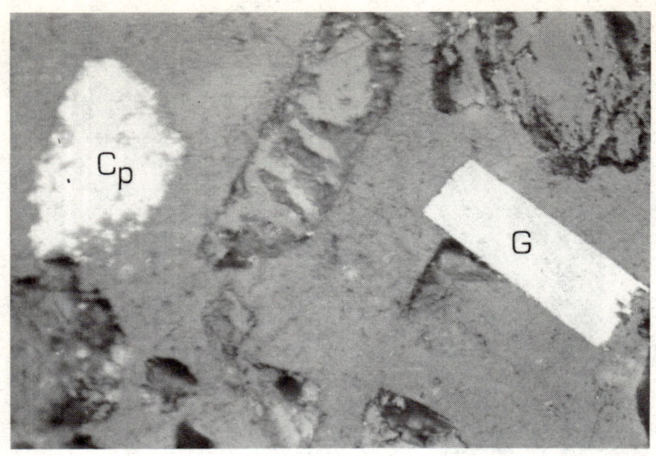

Figure 13 - Photomicrograph of Oxidized Chalcopyrite and Free Galena in the Tailing of Ore (e) (X200).

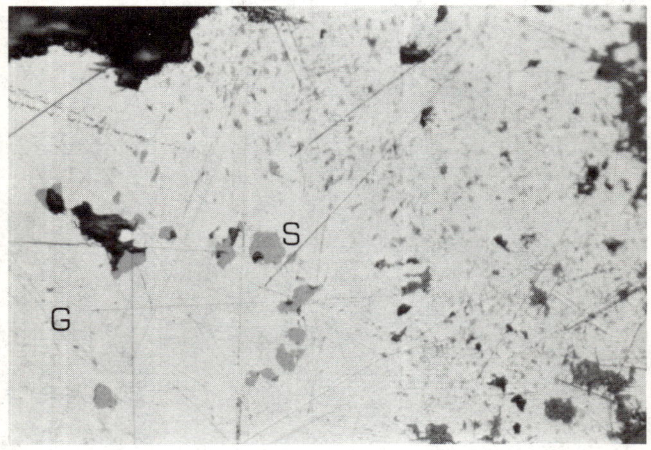

Figure 14 - Photomicrograph of Galena Grain Illustrating Sphalerite Inclusions of Varying Sizes in Ore (f) (X500).

and it was determined that 35% of the galena particles contained sphalerite or chalcopyrite inclusions of a size less than 30 microns.

Test B 3) Ion Solubility

The analyses of the mill water from the grinding tests are listed in Table VI for the six ores.

Table VI Metal Ion Concentration of Mill Water Samples.

Ore	NaCN g/tonne	Metal Ion Conc (mg/1) Pb	Zn	Cu	Fe
(a)	0	.2	.05	.2	.1
	500	.2	.05	1.5	.2
	1000	.3	.05	3.1	.4
(b)	0	.2	.05	.2	.1
	500	.4	.05	1.7	.2
	1000	.2	.05	4.8	.5
(c)	0	.2	.05	.2	.1
	500	.3	.05	1.3	.3
	1000	.4	.05	8.6	.1
(d)	0	.2	.05	.2	.1
	500	.2	.09	2.5	.3
	1000	.2	.25	14.0	.5
(e)	0	.2	.05	.2	.1
	500	.6	.05	4.8	.1
	1000	.2	.08	12.0	.1
(f)	0	.2	.05	.2	.1
	500	.4	.05	.2	.1
	1000	.3	.12	.6	.1

It can be seen that the copper concentration is the only concentration which varies significantly between ores and with cyanide concentration. The results suggest that the presence of copper ions for activation in flotation is more likely in ores (d) and (e). This supports the flotation results and the mineralogical explanations given above and hence confirms that these ores have poor Pb/Zn selectivity, due to the activation of sphalerite by copper ions resulting from the soluble copper mineralization. Of the remaining ores, only (c) displays copper ions significantly above the levels of ore (a) and hence likely to cause activation.

Further details of Test B are reported elsewhere (12) together with a description of other tests relating to the flotation problems and mineralogy of Viburnum ores.

Conclusions

The use of a standardized kinetic flotation test has been demonstrated to clearly identify the flotation character of some Viburnum lead/zinc ores and to indicate the problem areas if flotation difficulties exist. The kinetic parameters of fast and slow mineral specific flotation rate constant and fast floating mineral fraction were explained in terms of ore mineralogy. For the kinetic flotation tests, the flotation problems were identified as galena oxidation and copper activation leading to a low fast floating galena fraction and a high sphalerite fast floating fraction, respectively.

For the six Viburnum ores, in which general flotation response was related to ore mineralogy, it was shown that the flotation difficulties resulted from several different causes. Galena depression, and hence low lead recovery, was linked to coatings masking the galena particle surface and to the composite nature and intimate locking of some galena particles. The presence of hydrocarbons (pyrobitumen) were a possible cause of multiple mineral activation and hence poor lead selectivity in one ore. In general, however, poor lead selectivity resulted from the activation of sphalerite by copper ions in solution. These copper ions were derived from the oxidation of copper mineralization in some ores and such copper ion production was found to increase in the presence of cyanide.

Acknowledgements

The authors wish to acknowledge the St. Joe Mining Corporation for their permission to undertake and publish this work and for their assistance with samples and sample analysis. In particular Mr. J. J. Jones and Mr. Q. Schmidt are thanked for their cooperation during the project. The funding for the overall project was provided by the St. Joe Mining Corporation and the Missouri State Mines Institute and both parties deserve special thanks for their support.

References

(1) H. M. Wharton, "Total Mine Production and Value from the Viburum Trend Mines", Missouri Department of Natural Resources Report, 1984.

(2) D. A. Heitman, "The Buick Concentrator", Paper Presented at the AIME Conference, Arizona, 1979.

(3) F. H. Sharp, "Lead-Zinc-Copper Separation and Current Practice at the Magmont Mine", Flotation , M. C. Fuerstenau (ed), AIME-SME, 1976, pp.1215-1231.

(4) K. L. Sutherland and I. W. Wark, Principles of Flotation , Aus.IMM, 1955.

(5) R. W. Smith, "Oxidation of Galena in South East Missouri Ores", St Joe Lead Mining Co., In-House Paper, 1984.

(6) N. P. Finkelstein, and S. A. Allison,"The Flotation of Zinc Sulfide : A Review", Flotation , M.C.Fuerstenau (ed), AIME-SME, 1976, pp.414-457.

(7) C. W. Clendenin, and M. M. Mouat, "Geology of the Ozark Lead Company Mine", Economic Geology , 72, 1977, pp.398.

(8) R. Schuhmann,"Flotation Kinetics 1 - Methods for Steady State Study of Flotation Problems", J.Phy. Chem. , 46, 1942, pp.891-902.

(9) W. R. Bull,"The Rates of Flotation of Mineral Particles in Sulphide Ores", Proc. Aust. I.M.M. , 220, 1966, pp.69-78.

(10) A. J. Lynch, Mineral and Coal Flotation Circuits , Elsevier Sci Pub, 1981, pp.57-96.

(11) S. L. Volner,"Flotation Kinetics of a Viburnum Missouri Lead Ore", M.S. Thesis, University of Missouri - Rolla, 1984, pp.149.

(12) M. G. Schroer,"A Study of Problem Lead/Zinc Ores In S.E.Missouri in Terms of Mineralogy and Galena Oxidation", M.S. Thesis, University of Missouri - Rolla, 1986, pp.95.

Process Mineralogy Applications to Agglomeration

EVOLUTION OF PELLETS PROCESSED FROM MAGNETIC ORE CONCENTRATES

IN THE GRATE-KILN-COOLER SYSTEM

Tsu-Ming Han

The Cleveland-Cliffs Iron Company, Research Laboratory
Ishpeming, Michigan

This paper covers a general description of the scheme for pelletizing magnetic ore concentrates by the grate-kiln-cooler system. The physical and chemical environments in the units of the system and within the pellets during processing are interpreted and discussed. The products discharged from each of the units are characterized by means of macroscopic and microscopic examinations, wet chemical and Satmagan analyses, and physical property determinations.

The evolution of the green pellets as they travel through the drying, preheating, indurating, and cooling stages involves three successive processes: pseudomorphic oxidation, recrystallization, and retrograde oxidation. These processes not only lead to changes in the color, size, weight, porosity, composition, magnetic property, and compressive strength of the pellets but they also create two distinctive internal structures: core cracking and concentric textural-mineralogical zoning. These structures have direct bearings on both the physical and reduction properties of the pellets, and are not normally seen in pellets made from nonmagnetic concentrates. The major factors affecting the magnitude of these changes are concentrate size distribution, green pellet size, preheat conditions, indurating temperature, cooling rate, and retention time.

Based on the data presented herein, the cooling rate is almost as important as the indurating temperature in pelletizing magnetic ore concentrates. Consequently, the cooler of the grate-kiln-cooler system not only functions as a unit for cooling the pellets but also as a device for oxidizing and indurating the pellets.

Introduction

Iron ore pellets are now one of the principal blast furnace feed materials for iron making throughout the world. Most of these pellets are produced from the magnetic concentrates derived from finely ground low-grade iron ores by magnetic concentration, with or without a final upgrading step by flotation.

In the early 1950's, these ore concentrates were largely pelletized through either the straight grate or the shaft furnace system. Since the early 1960's, the grate-kiln-cooler system has become a popular system although Dravo grate and circular grate systems have also been adopted by some iron ore companies. The data relative to the design, production, capacity, and operating condition of these systems, as well as the quality of the final pellet product, are widely published and reported (3,4,5,6,7,8). However, little information is available in relation to the following:

1. The physical and chemical environments in the systems and within the pellets.
2. The macrostructural, microtextural and mineralogical changes in the pellets after each stage of processing, and their effects on the quality of the final pellet product.

This paper reports the physical and chemical conditions in each unit of the grate-kiln-cooler system, the thermal and chemical environment within the pellets, and the characteristics of the pellets discharged from each unit of the system. It also discusses the processes of pellet evolution and their effects on product properties and the major factors affecting the physical and chemical qualities of the final pellet product.

The data presented herein are based mainly on studies of pellets produced at the Empire Mine in northern Michigan. Through 1986, the pellet plant, which now consists of four grate-kiln-cooler systems, had produced over 92 million long tons of pellets since start-up in 1964 as a single pelletizing system.

Scheme for Green Pellet Preparation and Induration

The green pellet feed generally referred to as "balling feed" consists of magnetic ore concentrate with a desirable amount of bentonite as a binder. The amount of bentonite added is a function of bentonite quality, distribution within the concentrate matrix, and also the size distribution and the moisture content of the ore concentrate. There are four stages of green pellet preparation: mixing, balling, sizing, and cleaning (Figure 1-A).

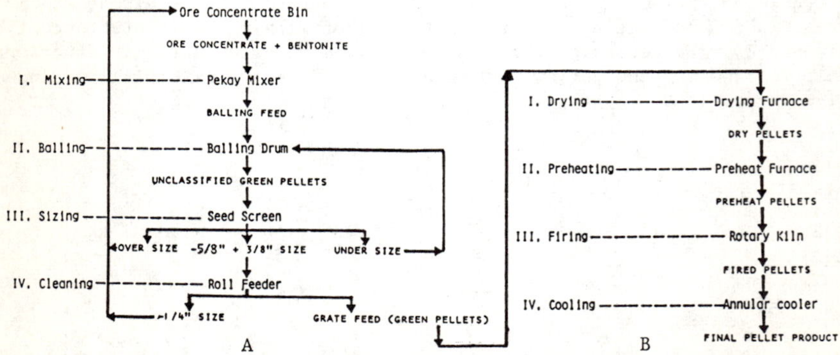

Figure 1 - A simplified scheme for green pellet preparation and induration.

The quality of the green pellets plays a key role in determining the production rate, the plant operating condition, and the final product quality. Green pellets of good quality should not be too sticky, brittle, or plastic. Ideally, the green pellets should be closely sized, smooth-surfaced spheres and should have a high survivability rate during transfer and in the drying stage. Iron makers generally request a high amount of final product in the -1/2"+3/8" size range. The finished green pellets are then subjected to four successive stages of induration by the grate-kiln-cooler system. These stages are drying, preheating, firing, and cooling as shown in Figure 1-B.

General Description of the Grate-Kiln-Cooler System

A. Grate-Kiln-Cooler System

There are two flows passing through the system during the plant operation: the pellet flow and the airflow. Their directions are shown in Figure 2.

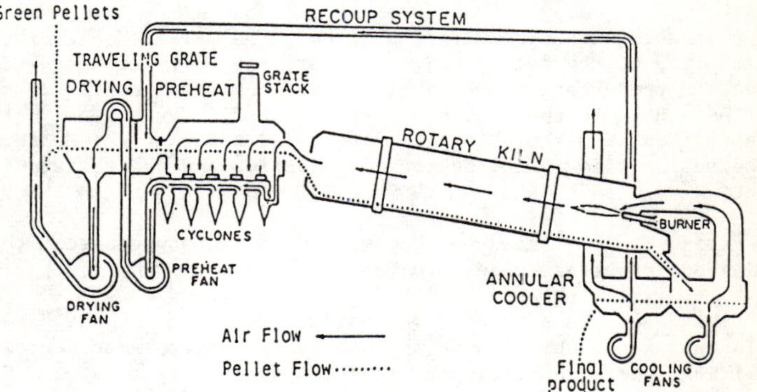

Figure 2 - The grate-kiln-cooler system.

1. Pellet Flow. The green pellets are fed onto the traveling grate at a speed varying from 140 inches to 200 inches per minute. The pellet bed on the grate generally ranges from 4 to 6 inches in depth. The retention time is generally about 4 minutes for drying and about an equal amount of time for preheating. The preheat pellets are directly discharged into the rotary kiln which rotates at 1 to 1.5 revolutions per minute. The retention time for the pellets in the kiln may vary from 25 to 30 minutes. The fired pellets are then discharged onto the circular grate of the cooler with a bed thickness of about 30 inches. The grate travels about 45 to 60 inches per minute. There are two cooling stages with a retention time of about 15 minutes for each stage.

2. Airflow. The airflows in the system are derived from: (a) "primary air" which is introduced from the kiln burner located at the pellet discharge end of the kiln, (b) "secondary air" which originates from the cooler pellet bed during the first stage of cooling, and (c) "recoup air" which also originates from the cooler pellet bed, but during the second cooling stage. The secondary air temperature is approximately 2250°F, while the recoup air is about 1100°F to 1200°F.

The primary air combines with the secondary air and flows countercurrent to the pellet flow up the kiln and down through the preheat furnace pellet bed. The peak temperature of the air occurs at the "burning zone" of the kiln and generally ranges from 2300°F to 2400°F. Because of the heat exchange between the air and pellets, this same air exits the kiln at a temperature of 1800°F to 2000°F. However, the temperature in the preheat furnace is normally about 2050°F at the top and about 1500°F at the bottom of the pellet bed. This increase in temperature is apparently attributable to the oxidation of magnetite in the preheat furnace.

The airflow utilized for the first stage of drying is by downdraft and is derived from the preheat furnace through the cyclone-and-fan system. At this stage, the pellet bed is about 600°F to 700°F at the top and 250-300°F at the bottom. The recoup air serves as the principal heat source for the second stage of drying. At this stage, the temperature at the top of the pellet bed is about 1000°F to 1100°F and about 500°F to 600°F at the bottom.

B. Physical Environments in the System

1. Drying and Preheat Stages. The green pellets on the grate remain relatively stationary during the drying and preheat stages, although some trickling and interpellet attrition occur as the grate moves.

Due to the effect of downdraft airflow, some pellet decrepitation may occur at the top of the pellet bed. However, the most striking and significant effect is the creation of a thermal gradient from the top to the bottom and from the center to the side of the pellet bed. Consequently, the pellets at the top differ from those at the bottom, and those at the center differ from those on the side in the degree of oxidation and the resistance to wear and impact.

2. Firing Stage. Based on observation and interpretation, the kiln may be divided into four successive zones:

a. Feed Zone. This zone is generally referred to as the "approaching zone." In this zone, the preheat pellets are immediately subjected to wear, impact, mixing, and size sorting as the kiln constantly rotates. Some fines and pellet chips may be generated.

b. Settling or Depositional Zone. This zone is marked by the presence of circular ridges of ore fines and pellet chips. The size and distribution of these ridges may vary with the indurating temperature, the flame pattern, the type of fuel employed, and the nature of the ore concentrate. The fines may be picked up by pellet chips and defective pellets as they travel through this zone.

c. Burning or Firing Zone. This zone represents the section of the kiln with the highest temperature. The pellets are subjected to this temperature for about 8 to 10 minutes as they travel toward the discharge zone. Because of the effect of pellet size sorting created by the dynamic rotary action of the kiln, the larger pellets tend to roll in the front of the pellet flow, have more of a chance to be in direct contact with the kiln lining, and are probably subjected to a higher temperature than the smaller pellets. Therefore, most of the larger pellets will likely exhibit a higher degree of induration than the smaller ones. This may be one of the major factors accounting for the low porosity of large kiln pellets.

d. Discharge Zone. This zone may be designated as the "cooling zone" in which the temperature may drop to 2250°F from the firing temperature of 2350°F. At this temperature, the centripetal oxidation of pellets may be rejuvenated. It may take 4 to 6 minutes for the pellets to travel through this zone before discharging onto the cooler.

3. Cooling Stages. In addition to the cooling zone in the kiln, there are two cooling zones in the cooler - the first cooling zone and the second cooling zone. Cooler air is introduced from the bottom of the cooler by updraft. As a result, the rate of cooling for the pellets at the bottom of the pellet bed is much faster than that at the top.

Figure 3 graphically illustrates the approximate temperature and retention time in each of the physical environments of the grate-kiln-cooler system.

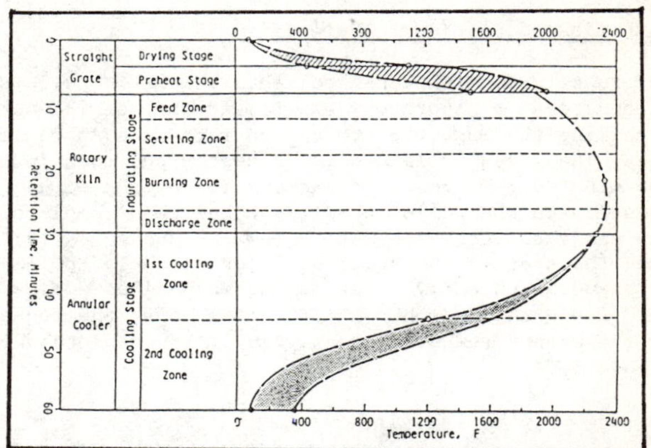

Figure 3 - Physical environments in each unit of the grate-kiln-cooler system.

C. Chemical Environments in the System

The chemical environments are oxidizing; however, they vary in magnitude between the units of the system. This is evidenced by the percent oxidation or the magnetite content of the green pellets as they travel through the four successive stages of the indurating process (Figure 4).

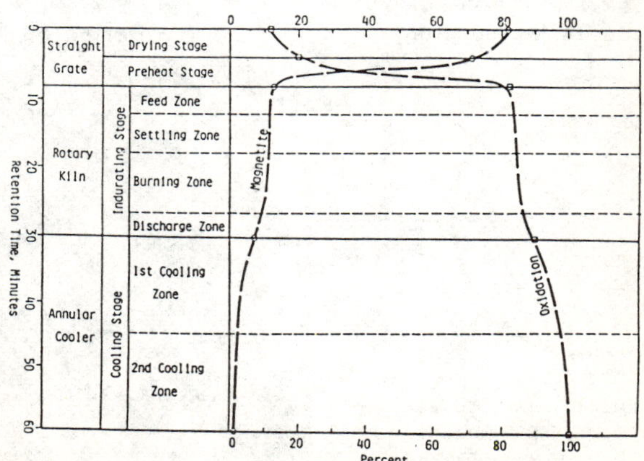

Figure 4 - The degree of oxidation and magnetite content of products discharged from each unit of the grate-kiln-cooler system.

Based on the data presented in Figure 4, the chemical environment in the drying and preheat furnaces is considered as strongly oxidizing. In the kiln, it varies with the indurating temperature in the four zones classified. At a firing temperature of 2350°F, it is probably moderately to strongly oxidizing in the feed zone, moderately oxidizing in the depositional zone, slightly oxidizing to neutral in the firing zone, and moderately oxidizing in the discharge zone. The chemical environment in the cooler is, of course, strongly oxidizing in which the size, porosity and the magnetite content of the pellets discharged from the kiln are noticeably reduced, and the compressive strength of these pellets is greatly increased.

D. Thermal Chemical Condition in Pellets

In order to determine the chemical environment in pellets under the pelletizing condition, a laboratory experiment was carried out. The test was done by balling the magnetic ore concentrate with approximately 3% specular hematite. The green pellets were preheated and then indurated at about 2400°F for 30 minutes. Microscopic examination of the fired pellets reveals that the magnetite in the pellet shell zone has been completely oxidized to hematite, whereas the specular hematite remains unchanged (Figure 5-A). In the core zone, the specular hematite has been slightly to completely reduced to magnetite pseudomorphically, whereas the magnetite remains as magnetite (Figure 5-B). Along the periphery of the core zone, the reduced specular hematite and the magnetite have been partially to completely oxidized to hematite (Figure 5-C).

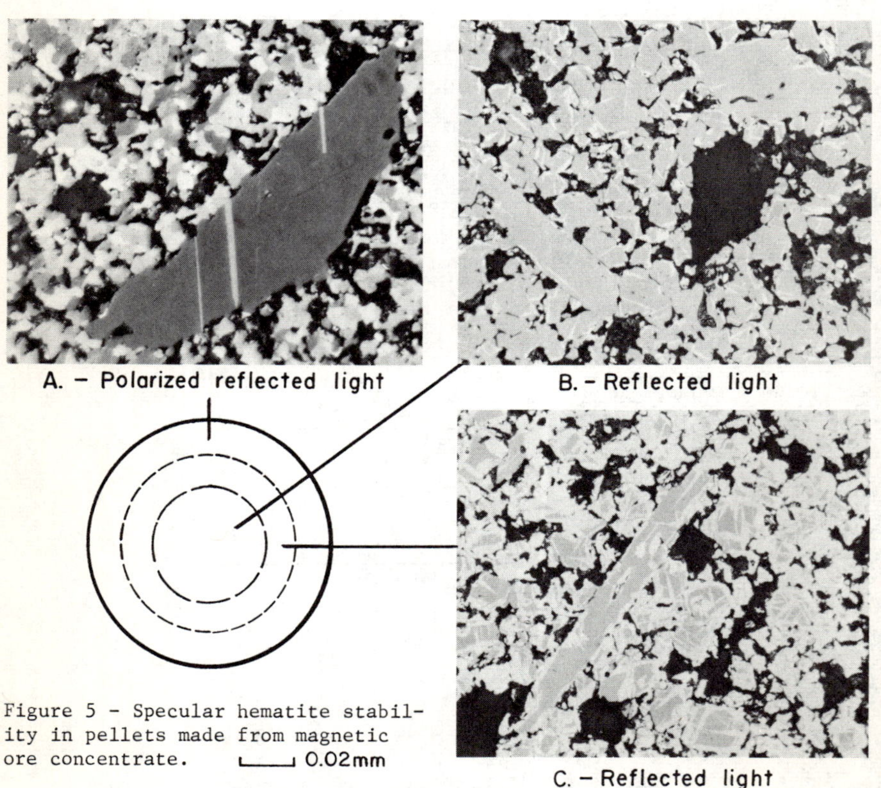

Figure 5 - Specular hematite stability in pellets made from magnetic ore concentrate.

Based on the above observation, it is believed that the core portion of the pellets is probably slightly reducing, while the shell of the pellets is strongly oxidizing.

Characterization of Pellet Products

A. Macroscopic Features

Several macroscopically visible changes occur as the green pellets are successively subjected to drying, preheating, indurating, and cooling. The major changes are described below:

1. Pellet Color. The initial color of the green pellets is charcoal gray. It turns to dark reddish gray after drying and to earthy red after preheating. These preheat pellets subsequently change to dull steel gray after indurating and cooling. Internally, the charcoal gray cores are typically present in the drying and preheat furnace discharges, and the dark gray cores are typically present in the pellets discharged from the kiln. The cooler pellets are mostly free of these cores.

2. Pellet Size. Visually, there is not much change in size between the green pellets and the preheat pellets. The size reduction becomes rather noticeable after indurating and cooling. Laboratory data show that the most important and significant size decrease takes place in the cooler, which explains why the cooler pellets are much lower in porosity and stronger in compressive strength than the kiln pellets.

3. Internal Structures. Concentric textural and mineralogical zoning and core cracking are two of the most striking and characteristic features normally observed in the pellets processed from magnetic concentrates, regardless of how they were commercially produced. Figure 6 shows the development of these two features in the products discharged from each unit of the grate-kiln-cooler system.

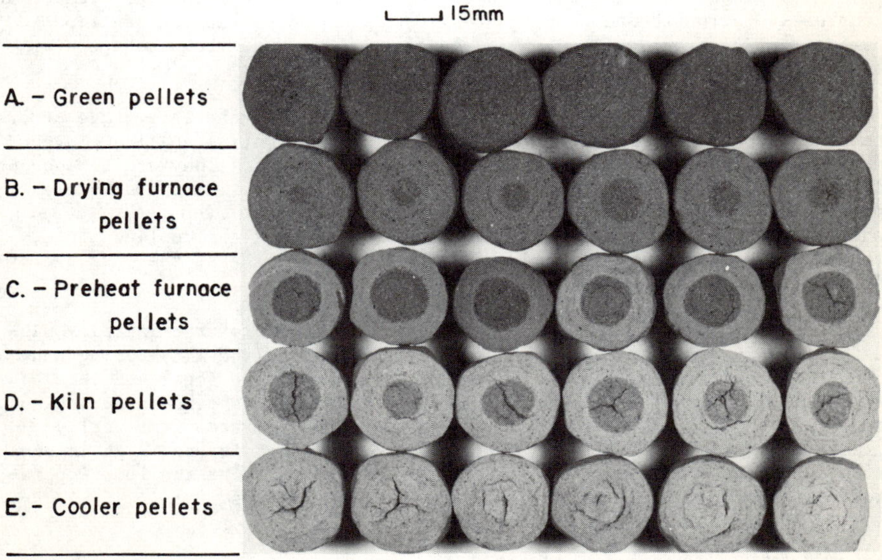

15mm

A. - Green pellets

B. - Drying furnace pellets

C. - Preheat furnace pellets

D. - Kiln pellets

E. - Cooler pellets

Figure 6 - Internal structures of pellet products.

a. Concentric Zoning. The zonal structure in the pellets is created by the oxidation of magnetite which occurs during each of the processing stages. The pellets discharged from the drying furnace are generally composed of three zones that may be visually recognized as a reddish dark gray shell, a relatively dense reddish dark gray transitional zone, and a nearly spherical charcoal gray core (Figure 6-B). This zonal structure represents the effects of centripetal oxidation of the green pellets at about 1100°F for about two minutes.

The preheat furnace pellets are typically composed of three distinctive and sharply divided zones: an earthy red shell, a relatively dense, grayish red inner shell, and a charcoal gray core (Figure 6-C). This zonal structure represents a product resulting from the centripetal oxidation of the drying furnace pellets which takes place in the preheat furnace at about 2050°F for about four minutes. It shows no genetic relationship to, but is superimposed on, that of the drying furnace pellets. The kiln discharge pellets are composed of essentially the same three zones as those observed in the preheat pellets, except the inner shell zones are thicker and the cores are smaller and more extensively cracked (Figure 6-D). The cooler pellets are characterized by the presence of two to three concentric zones. The two-zonal pellets are composed of a dull gray hematite shell and a dense cracked steel gray core (Figure 6-E). The three-zonal pellets are composed of a dull gray hematite shell, a dense, cracked and steel gray inner shell, and a dark gray, commonly cracked unoxidized core.

b. Core Cracking. Cracks are, as a rule, distributed in the core portion of the pellets. They are commonly dendritic or crescent-shaped. The former extend outward and terminate at the boundary between the inner and the outer hematite shell, whereas the latter are usually located along the boundary between these two shells. The outer shell of the pellets is free from cracks, whereas the inner shell and core are almost universally cracked. These cracks are the principal structural weaknesses of the pellets, which not only determine the pellet strength but also govern the nature of pellet breakdown upon transfer and low temperature reduction. This also explains why the pellets made from the magnetic concentrates are relatively weaker in compressive strength and tend to break down into chips rather than fines during reduction and shipment as compared to those made from the specular hematite concentrates.

The cracks are believed to largely develop during the late stage of preheating, since the drying furnace pellets and the hematite shell achieved during the preheating are practically free from cracks. They are subsequently modified and intensified by the indurating process. These cracks are probably a result of differential volumetric changes between the oxidized shell and the unoxidized core upon induration. During the cooling stage, these cracks tend to be widened and shortened by healing at the ends and by pulling apart in the middle.

Both the textural-mineralogical zoning and the core cracking play significantly important roles in determining the physical quality during handling. Other factors being constant, pellets with three zones and those extensively internally cracked are more susceptible to breakdown during shipment than those with two zones and those least to moderately internally cracked. They also affect the reduction properties at low and high temperature conditions, such as survivability and reducibility at the low temperatures and softening and melting at the high temperatures.

B. Microscopic Features

1. Green Pellets. The green pellets are made of a magnetic concentrate processed from a fine-grained magnetic sedimentary iron formation by magnetic separation supplemented by amine flotation (Figure 6-A). This iron formation consists of two principal ore types with two distinctive mineral assemblages:

I. Magnetite+quartz+carbonates
II. Magnetite+quartz+carbonates+iron silicates

The magnetite in Mineral Assemblage I is mostly euhedral, generally ranging from 15 to 30 microns in size, and occurs as disseminated grains, coalesced octahedra, and thin laminae. It is believed to be derived from siderite through oxidation and enrichment (1,2). The magnetite in Assemblage II is mostly subhedral and euhedral, generally ranging from 10 to 20 microns in size, and occurs as highly concentrated massive and laminated layers. It is believed to be derived from hematite through pseudomorphic replacement supplemented by overgrowth (3). The size and iron distribution of the ore concentrate produced from these ores are shown in Table I.

Table I. Size and Iron Distribution of an Ore Concentrate Produced from the Two Major Ore Types

Micron	% Wt	% Fe	% Fe Distribution
+37	1.76	24.8	0.65
-37+25	4.75	45.1	3.19
-25+10	61.86	69.0	63.46
-10+5	20.16	71.0	21.28
-5	11.47	67.0	11.42
Total	100.00	67.26	100.00

Figure 7 shows the size range and the morphological character of the ore particles in a green pellet. Note that some of the larger grains are actually octahedral crystals. Compositionally, partially oxidized magnetite and pellet fragments and fines (ground induration returns) are present as minor parts of the green pellets.

Figure 7 - Shows the morphology and size range of the magnetite in a green pellet. 0.01mm

2. <u>Drying Furnace Pellets</u>. Most of these pellets are featured with a core-and-shell structure (Figure 6-B). The shell is composed of slightly oxidized magnetite particles (Figure 8-A). The core represents the nonoxidized portion of the pellets (Figure 8-B). There is almost always a thin transitional zone between the shell and the core, in which the magnetite particles appear to have a higher degree of oxidation than those distributed in the shell zone (Figure 8-C).

The oxidation initiates either along the peripheries or along the cleavage planes of the magnetite particles. All of the mineral particles in the pellets still retain their original surface texture and morphology.

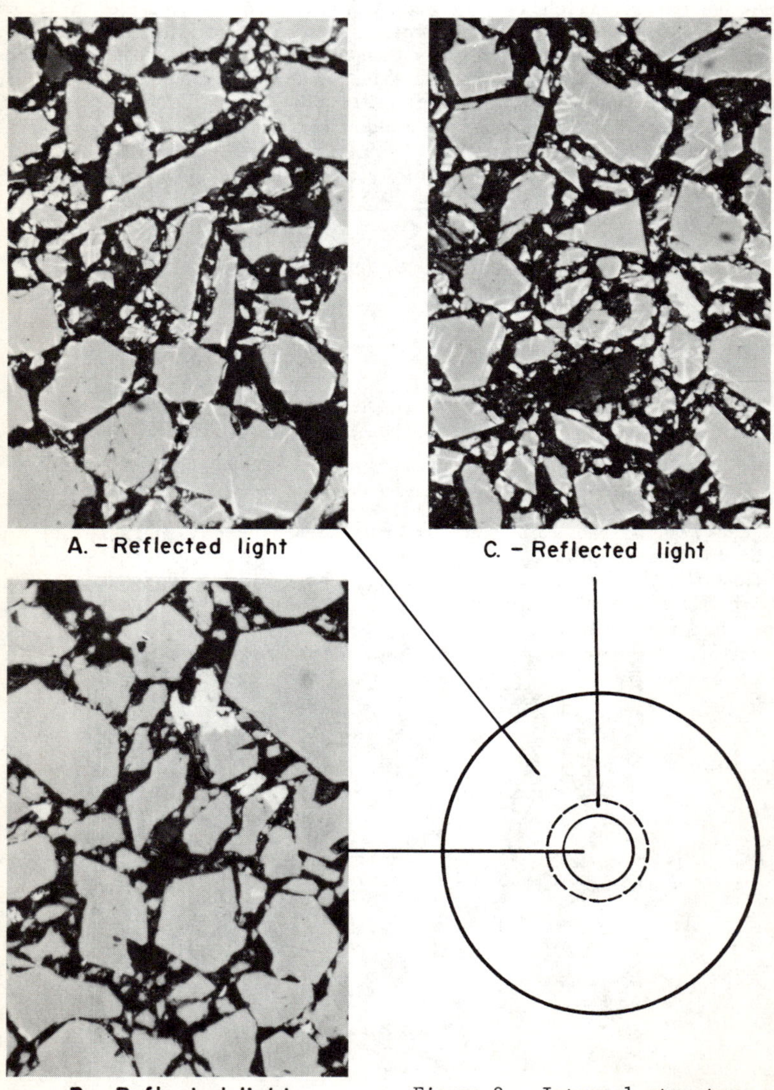

Figure 8 - Internal structure of drying furnace pellets.

3. Preheat Furnace Pellets. These pellets may be divided into three concentric zones: the outer shell, the inner shell, and the core. The shell zones are principally composed of hematite pseudomorphs and some quartz particles. Under polarized light, the hematite pseudomorphs are actually individual hematite aggregates containing hematite grains of different size, shape and orientation. The hematite grains in the pseudomorphs from the outer shell zone (Figure 9-A) are ill-formed and smaller than those in the pseudomorphs from the inner shell zone (Figure 9-B). The core zone is composed of magnetite with some quartz particles, both of which still retain their original texture and morphology and show no visible difference from those in the core zone of the drying furnace pellets (Figure 9-C). There is no microscopically visible ore bridge development between the pseudomorphs in the shell zones and the magnetite in the core zone.

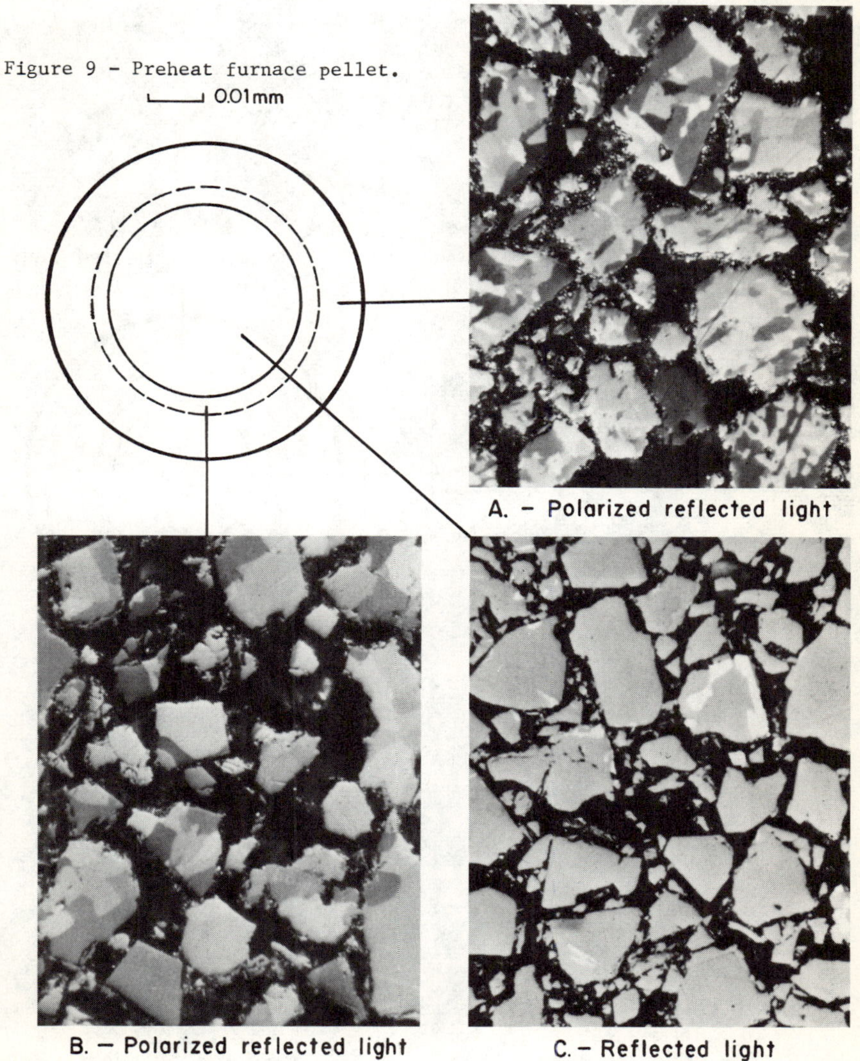

Figure 9 - Preheat furnace pellet.

A. - Polarized reflected light

B. — Polarized reflected light

C. - Reflected light

4. Kiln Pellets. These pellets show essentially the same internal structure and mineralogical composition as the preheat furnace pellets, except for some modifications. The hematite grains in and out of the pseudomorphs of the preheat furnace pellets have been recrystallized into subangular to subrounded granules. The pseudomorphs themselves tend to lose their original identities and are linked together by tiny hematite granules (Figure 10-A and B). In the core zone, the magnetite fines have grown into tiny granules and function as bridges between their coexisting large magnetite particles. These large particles have also been affected by marginal diffusion and tend to lose their original morphology (Figure 10-C). The development of hematite porphyroblasts along the outer rim of the magnetite core is evident (Figure 10-D). These porphyroblasts are probably a retrograde oxidation product formed in the discharge zone shortly after the pellets' departure from the firing zone.

Figure 10 - Kiln pellets.

0.01 mm

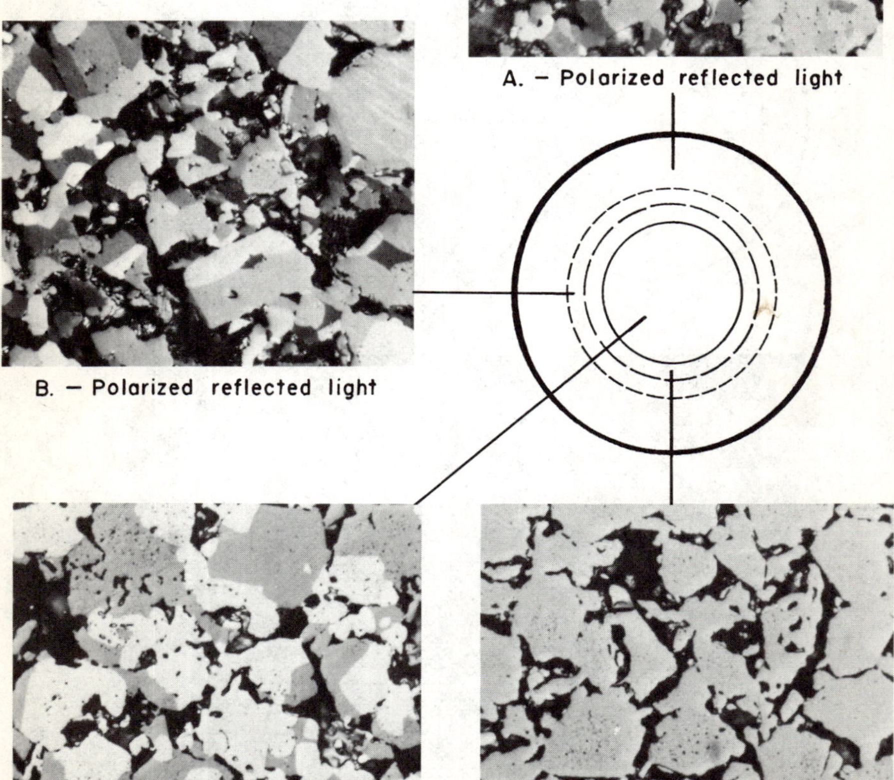

5. Cooler Pellets. Most of the pellets here have a hematite aggregate shell and a porphyroblastic hematite core. A magnetite core may be present in some of the pellets. In the latter case, the porphyroblastic hematite becomes the inner shell of the pellets. Other factors being constant, the microtexture of these ore minerals varies with the indurating temperature employed.

Based on the degree of induration, pellets from the same sample may be classified as fairly well-indurated and well-indurated.

a. Fairly Well-indurated Pellets. These pellets are essentially composed of hematite pseudomorphs with substantial minute hematite granules functioning as bridges. However, the pseudomorphs in the shell zone are aggregates made up of numerous, closely packed, tiny hematite granules (Figure 11-A). Those in the transitional zone contain much fewer hematite granules than those in the shell zone (Figure 11-B). Those in the core zone not only contain some to appreciable amounts of slag inclusions but also form large, porous, mutually contacted irregular islands as observed under polarized light (Figure 11-C). Each of these islands is composed of several hematite pseudomorphs having the same extinction position. Furthermore, a single hematite pseudomorph may belong to more than one of these islands. The magnetite, if present, generally retains its original particle outlines and is distributed in the core of the pellets.

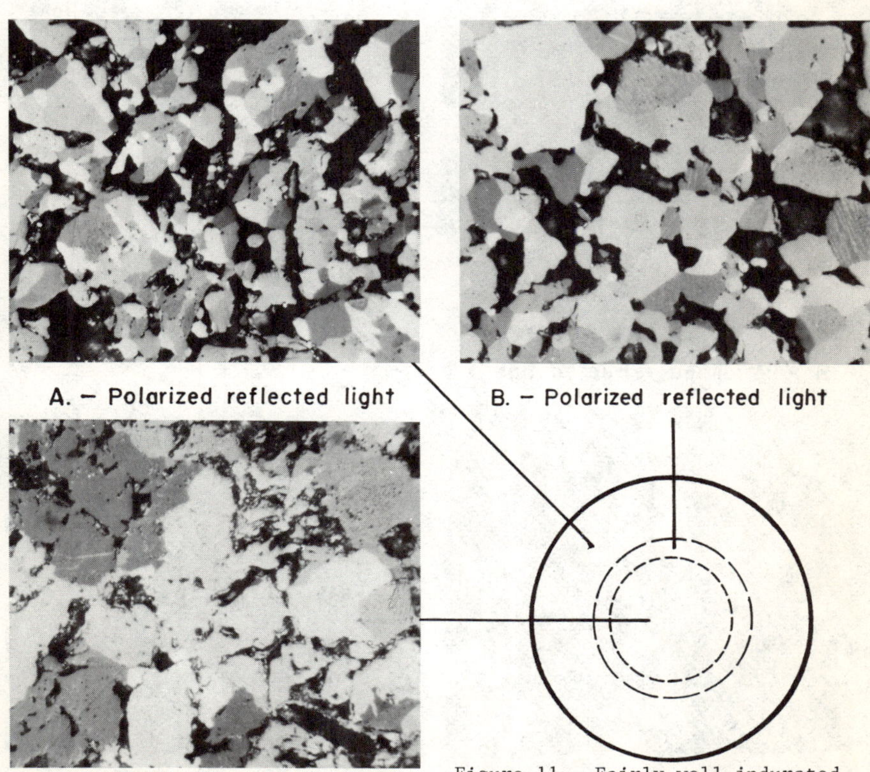

Figure 11 - Fairly well-indurated cooler pellet.

b. Well- and Overindurated Pellets. In these pellets, most of the hematite pseudomorphs in the shell zone have lost their original identities and display an equigranular texture. Twinings are commonly observed in these individual grains (Figure 12-A). The porphyroblastic hematite grains are large, slag-bearing, fairly porous, and commonly twinned, and are in a mutual contact relationship (Figure 12-B and C). Magnetite, if present, may exhibit two modes of occurrence, one of which is present as nearly equigranular grains with interstitial slag and is distributed in the core zone (Figure 12-D). The other occurs as thin coatings on the pellet surface or as slag-bearing octahedra sparsely scattered throughout the outer part of the pellet shell. The latter mode of occurrence is not commonly observed in the commercially produced pellets.

Quartz grains with fused rims are commonly present in the shell zone but are not often seen in the core zone of the pellets.

Figure 12 - Well-indurated cooler pellet. 0.01mm

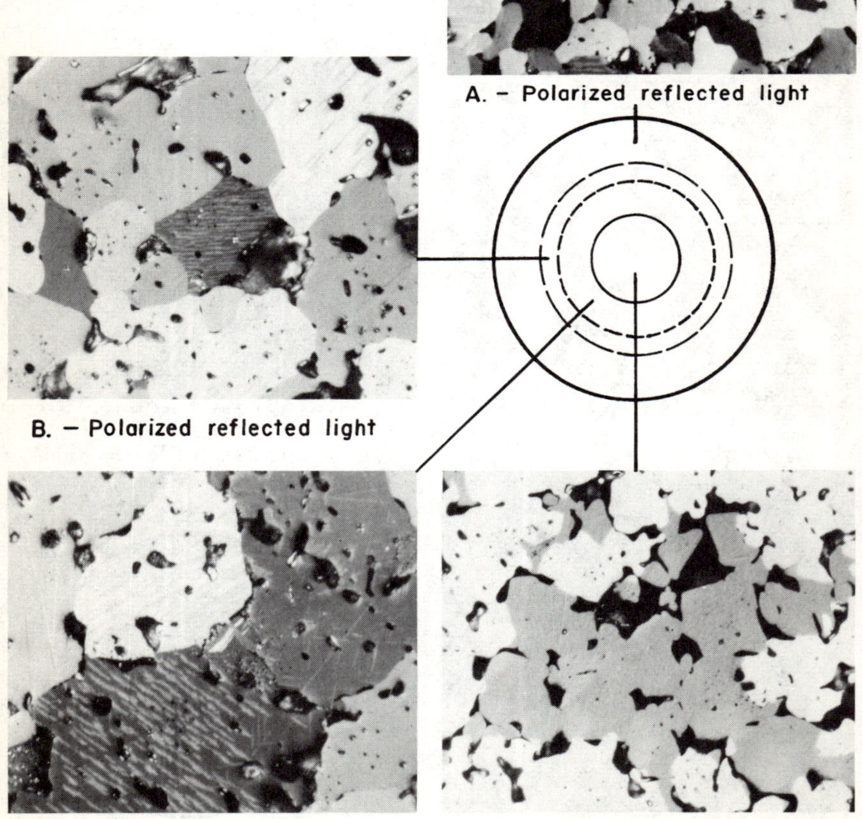

Evolution of Pellets in the Grate-Kiln-Cooler System

A. Process of Pellet Evolution

On the basis of the macroscopic and microscopic features described, the evolution of pellets produced from magnetic concentrates by the grate-kiln-cooler system may involve the following processes:

1. Progressive Oxidation Followed by Recrystallization. The oxidation initiates from the surface to the center of the magnetite particles and from the shell to the core of the pellets as soon as the temperatures reaches approximately 662°F. It progresses with an increase in temperature prior to the indurating stage. This process takes place principally in the preheat furnace. The product achieved in each of the pellets is a shell composed of hematite pseudomorphs of different sizes, and a core composed of essentially magnetite particles that have the same texture and composition as those in the green pellets. The pseudomorphs from this type of oxidation are in aggregate forms composed of a great many ill-formed, minute hematite grains, which exhibit essentially the same microtexture as those produced from the oxidation of magnetite to martite in nature.

The recrystallization of these fine hematite grains in the pseudomorphs becomes evident as soon as the temperature increases to about 2100°F. These tend to grow larger, to increase in roundness, and to segregate from each other as the indurating temperature increases up to 2450°F. Eventually, the pseudomorphs lose their original identities, and the hematite grains become nearly equigranular. In the normal pellet product, the pseudomorphs still retain their original particle outlines and are strongly held together by a substantial number of small hematite granules.

2. Direct Recrystallization Followed by Retrograde Oxidation. During the progressive oxidation stage, the magnetite particles in the core zone show little change, both compositionally and morphologically. However, during the indurating stage, the large magnetite particles tend to lose their original identities by marginal diffusion and the small ones tend to grow into magnetite granules through direct recrystallization. Equigranular magnetite interstitially filled with slag may be seen in the core portion of some pellets fired at 2450°F or higher.

The oxidation of the recrystallized magnetite probably initiates after the pellets have left the burning zone on their route to the cooler. It may therefore be designated as "retrograde oxidation." The hematite produced by this process is slag-bearing and occurs either as mutual contact, large anhedral grains or as individual euhedral porphyroblasts embedded in their slag-bearing magnetite host. This process is responsible for the appreciable reduction in pellet porosity and the considerable increase in pellet compressive strength.

B. Products from the Evolution Processes

In plant practice, the green pellets may generate two types of drying furnace products: (a) pellets containing essentially slightly oxidized magnetite grains throughout and (b) pellets containing a shell essentially made up of slightly oxidized magnetite grains and a core composed mainly of magnetite. The former product is not commonly seen in the samples investigated. The drying furnace discharge may also generate two types of preheat pellets: (a) pellets containing essentially hematite pseudomorphs throughout and (b) pellets with a shell essentially composed of hematite pseudomorphs and a core composed mainly of magnetite. The former type is practically nil in the commercially produced preheat pellet samples studied.

The preheat pellets with the hematite pseudomorphs will directly undergo grain growth and recrystallization to hematite which is present either as mutually contacted, nearly equigranular grains or as mosaic aggregates bridged by minute hematite granules. Pellets with either of these microtextures are designated as "unizonal" pellets. The preheat pellets with a hematite pseudomorph shell and a magnetite core will be recrystallized and either completely or partially oxidized to hematite after the indurating and cooling stages. The completely oxidized pellets contain hematite with two entirely different microtextures. These pellets are designated as "bizonal" pellets. The incompletely oxidized pellets are characterized by the presence of magnetite remnants in the core portion of the bizonal pellets. These pellets are designated as "trizonal" pellets.

Based on the pellet internal structure and mineralogical composition, one may interpret the approximate location of each of the pellets on the preheat grate and on the circular grate of the cooler. For instance, pellets with a thin aggregate hematite shell are more likely from the bottom bed on the preheat grate and those with a thick shell are more likely from the top. Pellets with a bizonal structure are more likely from the top bed of the cooler grate, and those with a trizonal structure are from the bottom. This is illustrated by Figure 13.

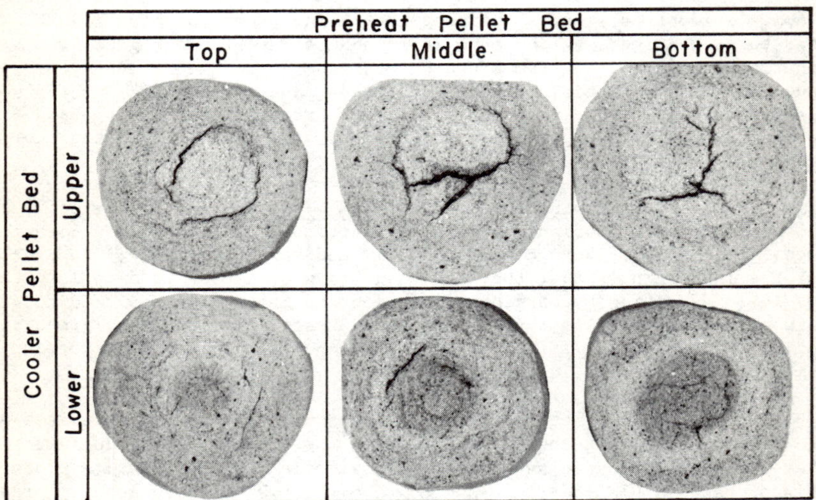

Figure 13 - Pellet internal structure and mineralogical composition versus preheat and cooler bed level. 30mm

C. Effect of Evolution Process on Product Properties

In addition to the changes in color and size of the pellets, the other major changes may be generally described below:

1. Pellet Weight. The weight of the dried green pellets progressively increases as the magnetite in the pellets continuously oxidized to hematite. Consequently, the percent weight gain is determined by the quantity and the degree of oxidation of the magnetite in the pellets. Laboratory tests showed that the average dried green pellets gain about 2.20% after about 99.00% of the magnetite in the pellets has changed to hematite.

2. Magnetic Property. This property is determined by a Binson magnetic susceptibility meter and is presented as the magnetic susceptibility meter reading per gram of a pellet. It is directly proportional to the magnetite content of the pellets (Figure 14-A). Consequently, it decreases as the pellets travel from the grate through the rotary kiln, annular cooler to the dump point of the cooler, and has a direct bearing on the pellet quality. Figure 14-B shows its effect on the pellet compressive strength.

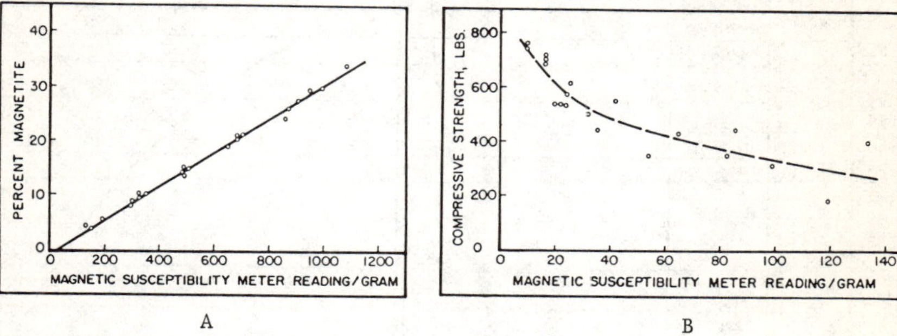

Figure 14 - The relationship of magnetic susceptibility (MSMR/gram of pellet) to magnetite content in pellets (A) and its effect on pellet compressive strength (B).

3. Pellet Porosity and Compressive Strength. These properties are primarily determined by the degree of grain growth and recrystallization and the magnitude of slag development, which are, in turn, governed by the indurating temperature (Figure 15-A and B). At the normal plant operating temperature, about 2300°F to 2350°F, the porosity is inversely and the compressive strength is directly proportional to the indurating temperature. Higher temperatures not only lower the pellet porosity but also promote slag and magnetite formation. The latter reduces the pellet compressive strength and is detrimental to the reduction properties of the pellets.

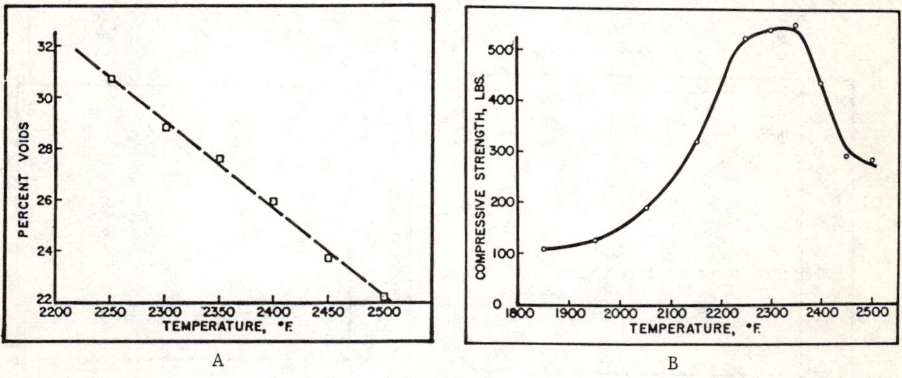

Figure 15 - Effect of temperature on porosity (A) and compressive strength (B) of pellets.

In plant practice, the porosity of the pellets discharged from the preheat furnace is insignificantly different from those discharged from the drying furnace. However, it decreases appreciably as the preheat pellets travel through the grate-kiln-cooler system (Figure 16-A). Strengthwise, there is a noticeable increase in compressive strength between the pellets discharged from the preheat furnace and those discharged from the drying furnace. It progressively increases considerable as the pellets pass the indurating and cooling stages (Figure 16-B).

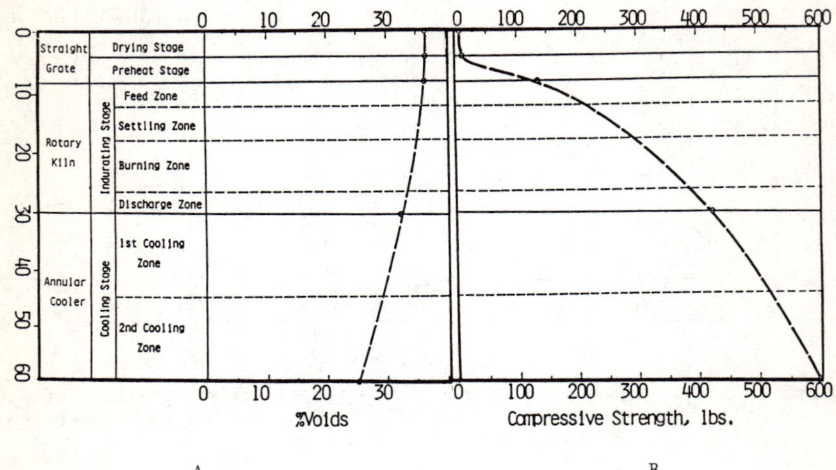

A B

Figure 16 - Magnitudes of decreasing porosity (A) and increasing compressive strength (B) as the pellets travel through the grate-kiln-cooler system.

4. <u>Chemical Composition</u>. The magnetite in the green pellets progressively oxidizes to hematite as the pellets travel through the grate-kiln-cooler system. This oxidation not only leads to an increase in pellet weight and a decrease of pellet magnetic property but also leads to a lowering of the content of chemical constituents, such as iron (Figure 17-A) and silica (Figure 17-B).

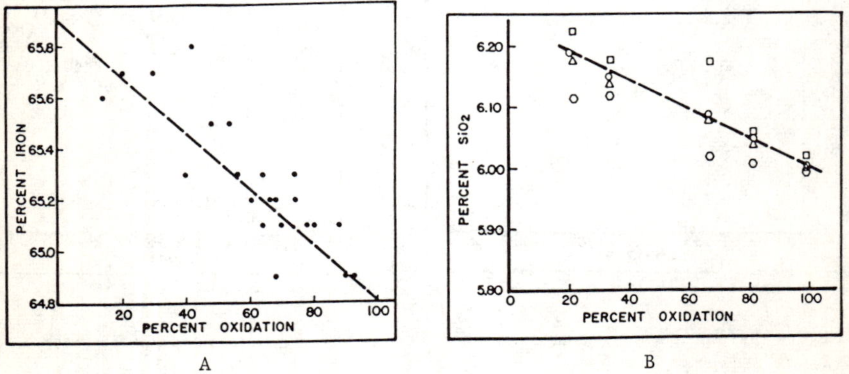

A B

Figure 17 - Effect of percent oxidation on iron (A) and silica content.

The magnitude of these decreases is, of course, largely determined by the magnetite content and the degree of oxidation and can also be affected by the presence of minor amounts of carbonates and iron silicates.

D. Classification of Final Pellet Product

As a general rule, the pellets produced from the grate-kiln-cooler system are smooth-surfaced, more uniformly fired, and less dusty compared to those produced by other processes. However, there are still noticeable differences in the degree of induration and the percentage of oxidation between pellets from the very same sample. In general, the pellets in a sample may be classified into the following groups:

Table II. Pellet Types in Final Pellet Product

CLASSIFICATION OF PELLETS			
OXIDATION	STRUCTURE	INDURATION	QUANTITY
Completely oxidized	Uni-zonal	Well-indurated	Rare
		Fairly well-indurated	
	Bi-zonal	Well-indurated	Major
		Fairly well-indurated	
Incompletely oxidized	Tri-zonal	Well-indurated	Minor
		Fairly well-indurated	

Major Factors Affecting Macrostructure, Microtexture and Mineralogical Composition of Pellets

In addition to the retention time factor, the major factors affecting the macrostructure, microtexture, and mineralogical composition of the pellets processed from the magnetic ore concentrates by the grate-kiln-cooler system are: (a) the size distribution of the ore concentrate, (b) the green pellet size, (c) the indurating temperature, and (d) the cooling rate. Since the macrostructure is essentially a product of progressive and retrograde oxidation, and the porosity and compressive strength are the results of the microtextural and mineralogical transformation, they may therefore be referred to as the major factors affecting the percentage of oxidation, porosity, and compressive strength of the pellets.

A. Size Distribution of Ore Concentrates

Table III shows the size distribution of three different magnetic ore concentrates which may be designated as Concentrate A, B, and C respectively. These concentrates differ noticeably in balling characteristics and produced green pellets with different qualities. It was reported that green pellets of desirable quality were difficult to produce with Concentrate A. Concentrate B produced green pellets that provided good plant operating conditions, whereas the green pellets produced from Concentrate C tended to generate a dusty kiln condition during operation.

Table III. Size Distribution of Three Ore Concentrates

Size Microns	Concentrate A % Wt	Concentrate A % Cum. Wt	Concentrate B % Wt	Concentrate B % Cum. Wt	Concentrate C % Wt	Concentrate C % Cum. Wt
25	1.40	1.40	2.28	2.28	5.28	5.28
25+10	59.35	60.75	61.07	64.35	66.59	71.87
-10	39.25	100.00	35.65	100.00	28.13	100.00
Total	100.00		100.00		100.00	

Based on the configuration of the DTA curves (Figure 18), these concentrates may be respectively indexed as "1234", "1324", and "3421". Under the same operating condition, the pellets produced from "1234" are expected to be different from those produced from "1324" and "3421" with respect to percent oxidation, porosity, and compressive strength.

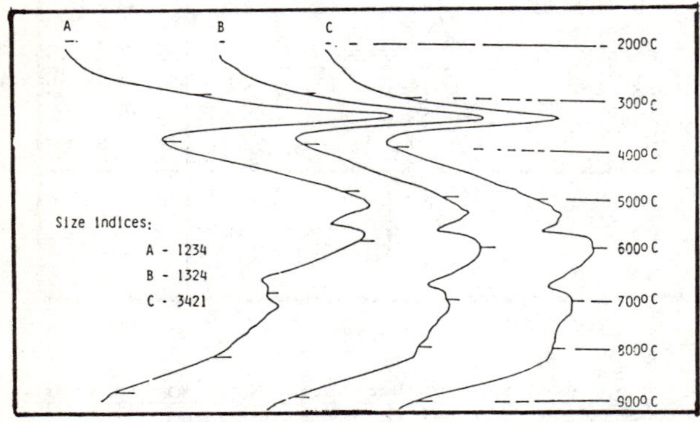

Figure 18 - Configuration of DTA curves, an indicator of particle size in balling feed (ore concentrate).

Two plant green pellet samples respectively made from "1234" and "3421" were processed by preheating at 2000°F with retention times of 3, 5, and 7 minutes, firing at 2450°F for 30 minutes and cooling to room temperature. Figure 19 shows the differences in the the degree of oxidation, the magnitude of cracking, and the distinctiveness of the concentric zoning in pellets processed from two concentrates with a different particle size distribution.

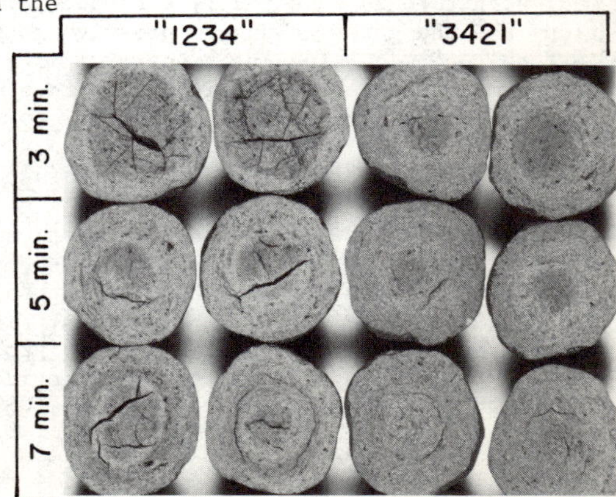

Figure 19 - Shows the internal macrostructures of pellets processed from two concentrates with a different particle size consist.

15mm

Table IV shows the percent oxidation and the compressive strength of these pellets.

Table IV. Effect of Concentrate Size Distribution and Retention Time on Percent Oxidation and Compressive Strength

Retention Time, Min.	Percent Oxidation "1234"	Percent Oxidation "3421"	Compressive Strength, Lbs "1234"	Compressive Strength, Lbs "3421"
3	70.28	95.46	386	1172
5	94.88	97.97	865	1173
7	98.80	99.53	755	942

The data above reveal that the size distribution of the ore concentrate not only affects the balling condition, but also determines the physical property and chemical composition of the final pellet product.

B. Green Pellet Size

The size of the green pellets is one of the key factors affecting the physical and chemical qualities of the final product under the established pelletizing conditions. As a general rule, the trizonal pellets are mostly large pellets which are more susceptible to breakdown during transfer and under reduction conditions.

Due to the firing and cooling mechanism of the grate-kiln-cooler system, the effect of pellet size on the preheat and cooler product differs from that on the kiln product. Figure 20 represents a set of test data obtained from a series of products discharged from the grate, kiln, and cooler.

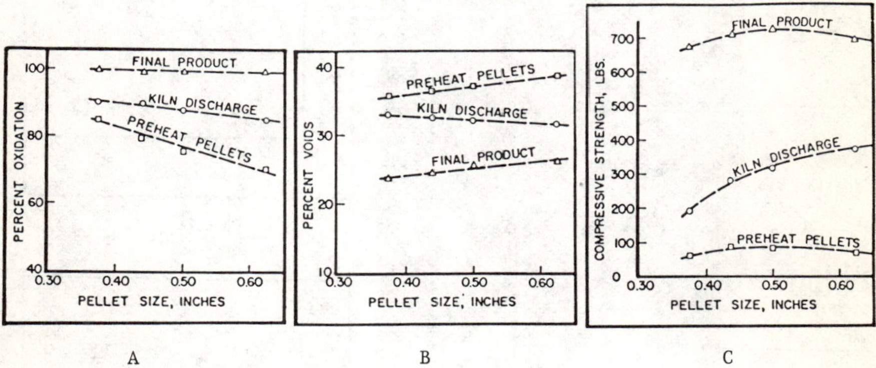

Figure 20 - Effect of pellet size on percent oxidation (A), porosity (B) and compressive strength (C) of pellets discharged from the grate, kiln, and cooler.

The figure shows:

1. The degree of oxidation in each of the products increases with a decrease in pellet size (Figure 20-A).
2. The porosity of the preheat and the cooler pellets increases, whereas that of the kiln pellets decreases as the pellet size increases (Figure 20-B). This can probably be attributed to the preferential induration of the large pellets resulting from the size sorting effect in the kiln.
3. The compressive strengths of the larger preheat and cooler pellets are

often lower than those of the relatively smaller ones, whereas those of the larger kiln pellets are often noticeably stronger than of the relatively smaller ones (Figure 20-C). Retention time and the location of the pellets on the grate and in the cooler are believed to be the chief factors accounting for the weaker strength of the larger pellets discharged from the grate and from the cooler.

C. Preheat Condition

Due to the effect of downdraft heating and the short retention time of the pellets on the preheat grate, the pellets from the upper part of the pellet bed are higher in degree of oxidation, lower in porosity, and stronger in compressive strength than those from the lower part. This statement is supported by the macroscopic features and the test data of a set of pellet samples collected from three levels of the pellet bed on the grate. Figure 21-A shows that the magnitude of core cracking decreases and the size of the unoxidized core increases as the bottom of the pellet bed is approached. Figure 21-B shows that the percent oxidation of the pellets decreases, not only toward the grate but also with the increase in pellet size. Figure 21-C shows that the compressive strength of the pellets decreases not only toward the grate but also shows that 1/2" pellets are noticeably weaker than 7/16" pellets and some of the 3/8" pellets as well.

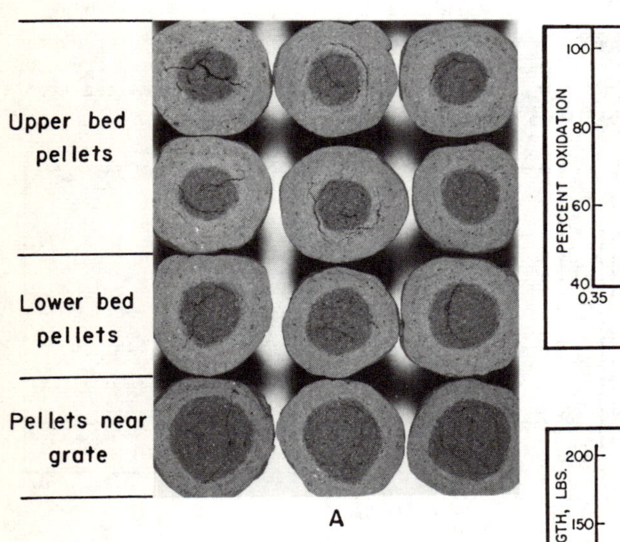

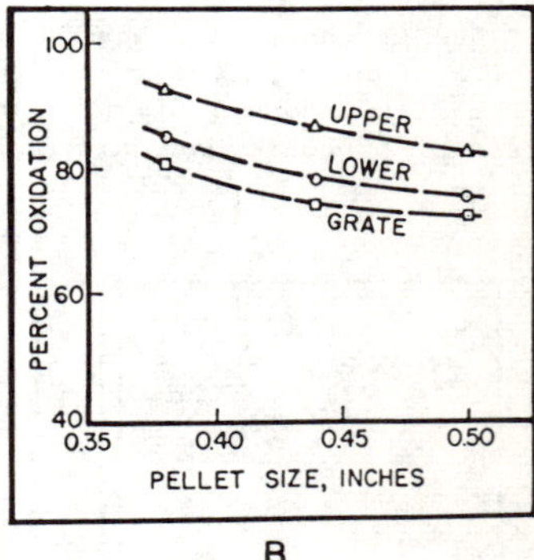

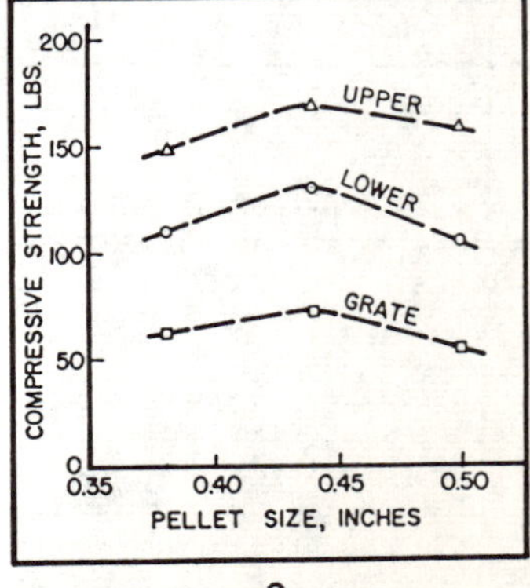

Figure 21 - Macroscopic features (A), percent oxidation (B), and compressive strength (C) of preheat pellets from different bed levels. 15mm

It is believed that most of the pellets from the lower part of the pellet bed may never be fully indurated at the designated temperature after their discharge from the rotary kiln. As a result, the same pellet sample may contain pellets with different degrees of induration and oxidation which directly lead to the differences in the physical properties and chemical composition of the final pellet product.

D. Indurating Temperature

Laboratory tests reveal that the pellets produced in the 2150°F-2250°F range are unizonal - composed of mosaic hematite aggregates with an insufficient number of ore bridges. Those fired at 2350°F are mostly bizonal - composed of a shell made up of strongly bridged, mosaic hematite aggregates, and a core made up of porous, interlocked, slag-bearing hematite porphyroblasts. Those fired at 2450°F are mostly trizonal and are composed of an equigranular hematite shell, a porphyroblastic hematite inner shell, and a slag-magnetite core. These differences in macrostructure, microtexture, and mineralogical composition have direct bearings on the physical and reduction quality of the final pellet product.

The plant test results showed that, in the 2350°F-2450°F indurating temperature range, an increase in temperature will result in a decrease in percent oxidation and porosity, but will result in an increase in the compressive strength of the kiln pellets. However, the increase in temperature will not only lower the percent oxidation and the porosity, but also the compressive strength of the cooler pellets.

Table V. Effect of Indurating Temperature on Physical and Chemical Properties of Final Pellet Product

Indurating Temperature $^{\circ}$F	Kiln Pellets			Cooler Pellets		
	% Oxidation	% Void	Crush Lbs	% Oxidation	% Void	Crush Lbs
2350	87.67	32.35	316	99.09	26.65	721
2400	84.27	29.15	350	97.25	24.31	712
2450	80.09	27.52	356	87.71	20.18	617

Figure 22 shows the internal structure and the size of the unoxidized cores in these pellets.

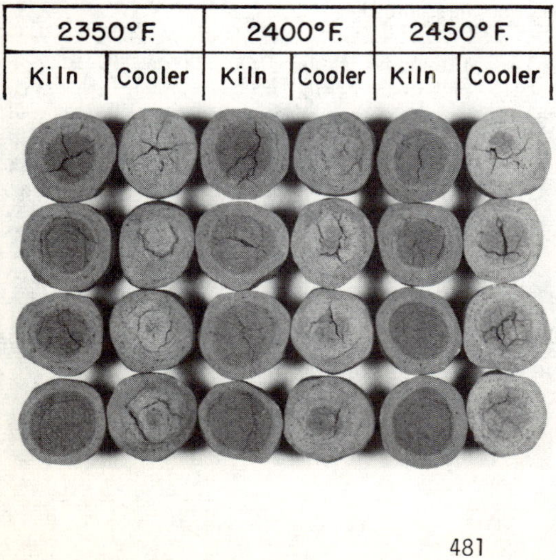

2350°F.		2400°F.		2450° F.	
Kiln	Cooler	Kiln	Cooler	Kiln	Cooler

Figure 22 - Effect of Indurating temperature on the internal structure and mineralogical composition of kiln and cooler pellets. 10mm

E. Cooling Rate

The cooling rate is almost equally important as the indurating temperature in determining the final quality of the final pellet product. This can be demonstrated by the following two examples:

1. Kiln Pellets vs. Cooler Pellets. The kiln pellets represent an indurated but quickly cooled product, whereas the cooler pellets represent an indurated and slowly cooled product. Other factors being constant, the former differs considerably from the latter in percent oxidation, porosity, and compressive strength (Table V).
2. Pellet Location in Cooler. Due to the effect of the updraft atmospheric air, the pellet cooling rate at the lower part of the cooler is faster than that at the upper. This explains why the trizonal pellets are frequently observed in the samples collected from the lower part of the cooler bed, and the pellets from the upper part of the bed are practically all bizonal. Qualitywise, the pellets from the upper part of the cooler bed are higher in degree of oxidation, lower in porosity, and stronger in compressive strength than those from the lower part. The data below was obtained from the pellets collected from the top, middle, and bottom of the cooler bed at the pellet dump point when the firing temperature was 2400°F.

Table VI. Effect of Cooling Rate on % Oxidation and Physical Properties of Final Pellet Product

Cooler Bed	% Oxidation	% Voids	Crush, lbs
Top	98.59	23.44	707
Middle	98.13	23.65	723
Bottom	97.04	25.25	590

Macroscopically, the unoxidized core is practically absent in the pellets from the top. It may be frequently seen in those from the middle and becomes a common feature in those from the bottom of the pellet bed. The study also shows that the internal cracks in the well-indurated and highly oxidized pellets are wider and more conspicuous than in the partially oxidized and fairly well indurated pellets.

Acknowledgment

The author is indebted to his colleagues, Messrs. R. Harma, P. D. Nora, K. G. Oja, and R. E. Rex for reading and commenting on the manuscript. The author also expresses his thanks to Mrs. Marion DeLarye for the typing and to Mr. Randy Leppala for preparing the graphs. This study is part of the low-grade iron ore beneficiation program being carried out by The Cleveland-Cliffs Iron Company; gratitude is therefore due to Management for permission to publish this paper.

References

1. Tsu-Ming Han, "Diagenetic-Metamorphic Replacement Features in the Negaunee Formation of the Marquette Iron Range, Lake Superior District," Soc. Mining Geol. Japan, Spec. Issue 3 (1971), 430-438.

2. Tsu-Ming Han, "Microstructures of Magnetite as Guides to Its Origin in Some Precambrian Iron-Formations," Fortschritte der Mineralogie, 56 (1978), 105-142.

3. W. A. Knepper, ed., Agglomeration (New York, NY: Interscience Publisher Inc., 669-714. D. M. Urich and Tsu-Ming Han.

4. R. A. Koski and P. D. Nora, "Pelletizing of Various Ore Types in Grate-Kiln System," (Paper presented at the SME-AIME Annual Meeting, Salt Lake City, Utah, 1969).

5. M. R. Ranade and P. D. Nora, "Selection of Agglomeration Process Variables for Increasing the LTD and Reducibility of Magnetite-based Acid Pellets," (Paper presented at the SME-AIME Annual Meeting, New York, NY, 1985).

6. K. V. Sastry, ed., Agglomeration 77, vol. 1 (New York, NY: American Institute of Mining, Metallurgical and Petroleum Engineers, Inc.), 1-22, A. English and R. D. Frans. 46-73, R. A. Koski.

7. L. White, ed., E&MJ Second Operating Handbook of Mineral Processing (New York, NY: E&MJ Mining Informational Services, McGraw-Hill, Inc., 1980), 2, 3, 8, 12, and 13, R. W. Hoppe and R. A. Thomas.

8. J. Yang and W-K Lu, "Comparative Studies of Commercial Iron Ore Pellets," Proceedings of the 44th Ironmaking Conference, Detroit, Michigan, 1985.

EXTRUSION AND INDURATION OF IRON ORE USING POLYMERS AND CALCIUM CARBONATE AS PLASTICIZER/BOND

A.K. Kuriakose, D.H.H. Quon, K.E. Bell and J.A. Soles,

Mineral Processing Laboratory, Mineral Sciences Laboratories,
CANMET, Energy, Mines and Resources Canada, Ottawa, Canada

Abstract

A method is described for the agglomeration of iron ore concentrates by extrusion using organic polymers and calcium carbonate as plasticizer and binder. Although these extrudates exhibit satisfactory green and dry crushing strength, they have poor strength at low and middle induration temperatures (400-800°C). This is due to the decomposition of the polymers at about 400°C. However, at high induration temperatures (1250-1300°C), the strength of extruded iron ore agglomerates made with polymers and calcium carbonate surpasses that of samples made with the addition of bentonite only.

Introduction

The blast furnace process is the major iron production technology in the world. This process requires iron ore in a coarse form, to form a porous bed to allow passage of the reducing gases, and essentially one which is free of fines, so that it is not blown out by the blast. With the relatively low-grade iron ores common to North America, it is first necessary to fine-grind and beneficiate the ores to upgrade quality, then to agglomerate the fines by pelletization and induration before shipping to the steel producers. The pelletizing is usually carried out in a disk or drum pelletizer by mixing the ore powder with 0.3 to 1.0 wt. % bentonite and the required amount of water for balling. The role of bentonite in the agglomeration is to give the pellets sufficient strength during sintering.

The presence of bentonite in the iron ore pellets has a detrimental effect on steel production, as it introduces silica, alumina and alkali into the system. Alkali and acid gangue contamination are the most common problems. Because of this and also because of the claimed improvement in the reducibility of pellets made by incorporation of certain organic binders, iron ore companies are searching for new binders for the production of iron ore pellets.

It has been claimed (1-6) that a cellulose based polymer (Peridur XC-3*) at quantities ranging from 0.1 to 0.4 wt. % could be used as a substitute for bentonite in the production of balled iron oxide pellets. Pilot plant tests conducted in Cia Vale Do Rio Doce (CVRD), Vitoria, Brazil, using Peridur binder have reportedly produced fired pellets of acceptable physical characteristics as specified by the ISO standard (7). A study was hence undertaken at CANMET in order to understand the influence of various organic binders in the pelletizing process and to optimize the conditions for pellet production. An extrusion process was used for the pellet fabrication because this method had the potential of producing uniformly sized, high density pellets with properties much desired by the blast furnace operators.

Varying amounts of lime were also incorporated into the mixtures, added as lime, calcium hydroxide or calcium carbonate, in order to increase the high temperature strength of the pellets. The results of these experiments are the subject of this paper.

General Characteristics of Binders

Bentonite is considered to be one of the best low- and high-temperature binders for pelletizing iron ore. It provides sufficient green and fired strength to the pellets. In the extrusion process, the bentonite also serves as a plasticizer to aid in the extrusion. There are very few inorganic binders having physical properties similar to bentonite; however a variety of organic binders have been used as plasticizers and to provide green strength for clay free materials so that bodies can be molded into desirable shapes before firing. These binders have been used extensively in ceramics processing, including dry pressing, extrusion, tape casting and injection molding. Descriptions of the various binders have been documented in the literature (8-10).

---

*Reg. TM of Industrial Colloids, Enka, Holland

A binder for iron ore extrusion must be readily soluble or dispersable in a suitable liquid (either organic or water). A uniform dispersion of the liquid phase is important for dispersing the binder throughout the particulates and to provide fluidity for slips or plasticity for extrusion. Dry green strength in the body is developed by the evaporation of liquid. The binder is retained in the body and provides organic bond between particles.

The binders can strongly affect the rheology of the liquid phase (11). They increase the viscosity and change the flow characteristics from Newtonian to pseudoplastic in most cases. Also, with some binders gels can develop under various conditions. The rheology of the solution directly affects the behaviour of suspensions and pastes formed by adding particulates to the solution. It is readily apparent that the applicability of a given binder for a specific process is very much dependent on the rheological characteristics of the binder solution.

Experimental Procedure

Raw Materials

Mt. Wright iron ore concentrates, which were composed of essentially specular hematite with traces of magnetite, feldspar and quartz, were used in this study.

The organic binders selected were: carboxymethyl cellulose (CMC), methyl cellulose (MC), hydroxyethyl cellulose (HEC), and ammonium alginate. Inorganic binders used for the experimental work were calcium oxide, calcium hydroxide and calcium carbonate. The amount of these additives used was normally less than 3 wt. %. Table 1 gives a description of these binders.

Table I. Source and Description of Binders

Binder	Chemical Identity	Source
Aqualon	carboxyl methyl cellulose	Hercules
Alginate (Na)		Kelco
Calcium hydroxide		Fisher Chemical
Calcium carbonate		Fisher Chemical
Calcium oxide		Fisher Chemical
Carboxyl methyl cellulose (Na)		Hercules
CMC-7H	carboxy methyl cellulose	Hercules
CMC 66TL	carboxy methyl cellulose	Hercules
Methocel	methyl cellulose	Dow Chemical
Methocel F4M	hydroxypropyl methyl cellulose	Dow Chemical
Natrosol	hydroxyethyl cellulose	Hercules
Superloid	ammonium alginate	Kelco

A knowledge of the plasticity of the iron ore with the binder system is essential for making the mixture of proper consistency for the extrusion process. Therefore the plasticity of the iron ore with binder or binder system was first determined by using a Brabender plastograph (12). This equipment measures the plasticity of the material in terms of consistency, which is a measure of the viscosity of the flowing plastic dispersion. Factors affecting plasticity that were considered included amounts of binder, binder combination, and the water content to produce optimum consistency for extrusion. For successful extrusion, the minimum consistency value should be about 450 m.g of torque on the plastograph.

The mixtures for extrusion were prepared by mixing appropriate amounts of the constituents with the required amount of water (as determined from the plastograph data) in a blender until a uniform dispersion of the components was obtained. An auger type extruder (Fig. 1) was used to produce pellet batches of about 4 kg. A vacuum of about 63 cm of Hg was maintained in the extrusion chamber. The die was designed to simultaneously extrude three columns, each having a diameter of about 2.5 cm. A piano-wire device was used to cut cylindrical pellets of desired length as soon as the extrudate emerged from the die.

Fig. 1. Extruder used for pellet making

The strength of the green extrudates was determined by a drop test (deformation behaviour when dropped from a fixed height of 45 cm) and the dry crushing strength was measured after drying them at 120°C for 24 h. Some of the oven-dried extrudates were then heat-treated in a resistance furnace at various fixed temperatures ranging from 500 to 1300°C for periods of 10 to 20 minutes. Usually, the furnace was preheated to the required temperature and a quantity of the dry pellets (contained in an alumina boat) was introduced into the hot zone; the time of heating of the pellets from room temperature to the test temperature was normally less than 10 minutes. On completion of the induration cycle, the pellets were cooled and tested for crushing strength.

The reaction of the binders and iron ore at various temperatures was determined using the differential thermal analysis technique (DTA).

The microstructure of the indurated extrudates was examined using an optical microscope.

Results and Discussion

Plasticity of the Organic and Inorganic Binders

The relationship between the amount of MC added to the ore and the resultant consistency values is illustrated in Fig. 2. The consistency of the mixture increases continuously with increasing MC content, reaching as high as 1600 m.g at an MC content of 1.5 wt. %. An MC addition level of only 0.25% produces a torque of 750 m.g, which results in excellent extrusion behaviour.

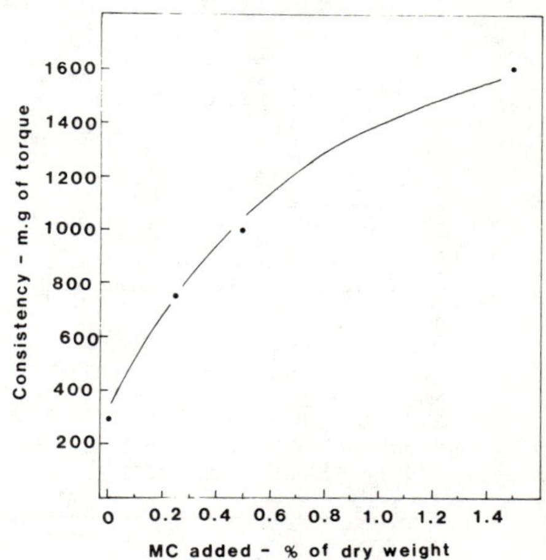

Fig. 2. The effect of MC concentration on the consistency of iron ore

One of the concerns of using organic binders for iron ore pelletizing is that the organic material decomposes or burns off below about 400°C. This results in weakening of the pellet so that it will lack sufficient strength to withstand the pressure and abrasion during the sintering of the pellets in the moving bed and rotary furnaces that are normally used in industry. In order to develop adequate pellet strength during the induration and also during subsequent shipping to the steel manufacturers, an additional inorganic binder that promotes sintering at higher temperatures is required. A possible candidate for this purpose other than bentonite is lime (CaO), because it can react with the ingredients in the iron ore to produce calcium ferrites and slags that can act as sintering aids through a liquid phase mechanism (13).

Because of the above considerations, the combined effect of organic binders and lime (added as such or as calcium hydroxide or as calcium carbonate) on the extrudability of such mixtures was investigated by measuring their consistency values in the plastograph.

The consistency versus water added to iron ore containing 0.25 wt. % MC and 1 to 2 wt. % CaO, $Ca(OH)_2$ and $CaCO_3$ is shown graphically in Figs. 3 and 4. A common characteristics of these curves is a sharp peak at a water content level of 10-12 wt. %. In general, a higher consistency is obtained from a mixture containing $CaCO_3$ with an organic binder than from one containing CaO or $Ca(OH)_2$ with the same organic binder (Fig. 4).

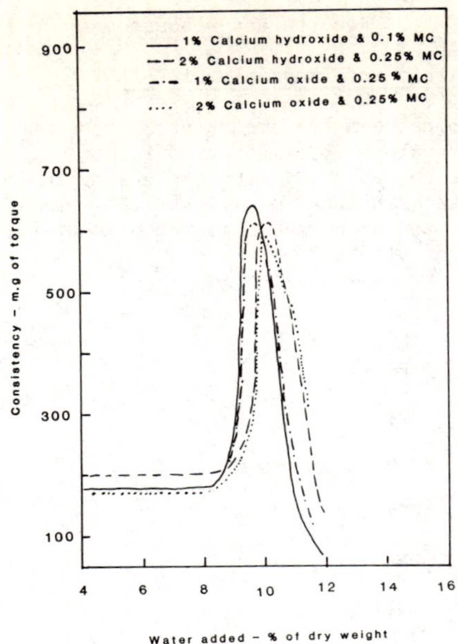

Fig. 3. Effect of water content on the iron ore consistency with calcium oxide, calcium hydroxide and MC addition

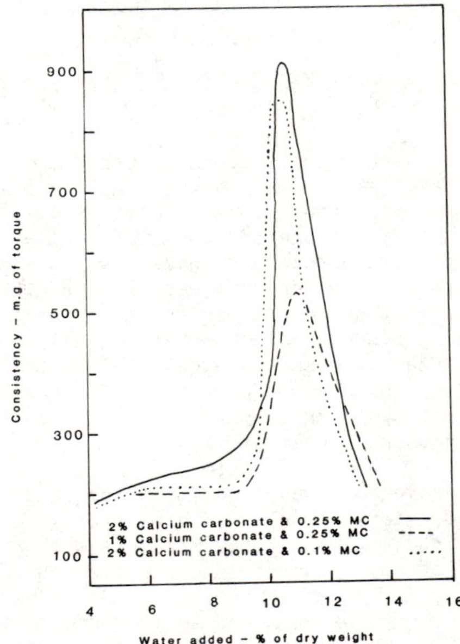

Fig. 4. Effect of water content on the iron ore consistency with calcium carbonate and MC addition

The results obtained using HEC or ammonium aliginate with $CaCO_3$ are shown in Fig. 5. It is observed that better plasticity is developed in mixtures containing low (0.25 wt. %) HEC. However, similar consistencies can be obtained at higher alginate (0.5 wt. %) content. The difference in the plasticity development appears to relate to the viscosity difference of these two binders.

The effect of different grades of CMC on the consistency of the iron ore containing $CaCO_3$ is shown in Fig. 6. The lower viscosity grades of CMC develop a lower consistency than the CMC-7H of higher viscosity.

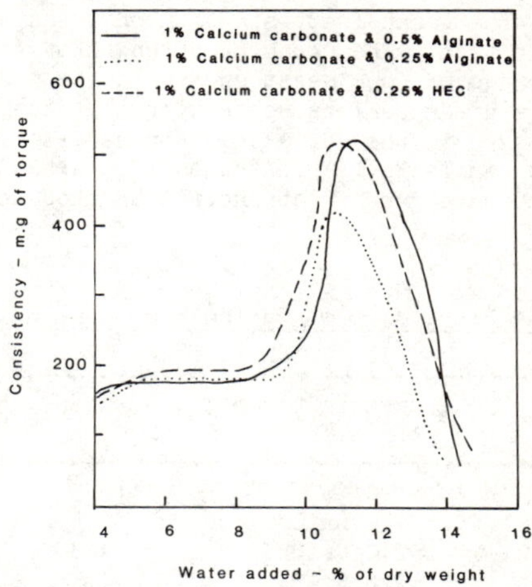

Fig. 5. Effect of calcium carbonate, HEC and alginate on the consistency of iron ore

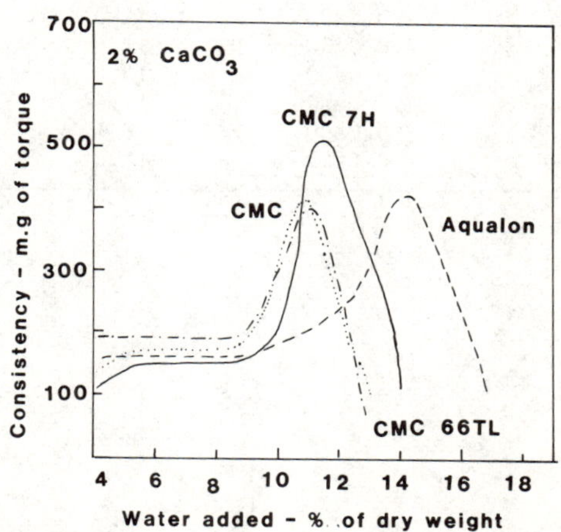

Fig. 6. Effect of additions of calcium carbonate and various grades of CMC on the consistency of iron ore

Of the different calcium compounds examined as possible replacements for bentonite, calcium carbonate was selected for further studies as it improves the consistency of the iron ore containing the smallest quantities of organic binders. As previously reported a minimum consistency value of 450 m.g of torque is required for iron ore extrusion, and the combination of $CaCO_3$ with any of several organic binders is suitable for this purpose.

Extrusion of the Pellets

Iron ore extrusion feed (cf. Table II) containing MC, CMC, HEC or alginate and 1 to 2 wt. % $CaCO_3$, based on the results of the plastograph experiments, were extruded and cylindrical pellets of about 2.5 x 2.5 cm in size were obtained. The green (wet) strength of these extrudates was estimated by the drop test. The crushing strength of the dry samples was determined as an average of four measurements in the radial direction. It was observed that the green pellets containing alginate, MC and CMC have sufficient strength even in the wet state, for handling. The dry strength of the products shown in Table III indicates that extrudates containing alginate, HEC and CMC have greater strength than those containing MC.

Table II. The Plasticity of Various Iron Ore Extrusion Mixes

Organic binder, %	Inorganic binder, %	Water %	Consistency %
0.25 MC	2.0 CaO	10.5	580
0.25 MC	1.0 $Ca(OH)_2$	9.4	640
0.25 MC	2.0 $Ca(OH)_2$	9.8	600
0.25 Methocel F4M	2.0 $CaCO_3$	10.5	720
0.25 MC	1.0 $CaCO_3$	10.8	850
0.25 MC	2.0 $CaCO_3$	11.9	910
0.50 Alginate	1.0 $CaCO_3$	11.5	510
0.25 HEC	1.0 $CaCO_3$	11.0	510
0.25 CMC	1.0 $CaCO_3$	11.0	395
0.25 CMC 66TL	2.0 $CaCO_3$	11.0	400
0.25 Aqualon	2.0 $CaCO_3$	14.5	420
0.25 CMC-7H	2.0 $CaCO_3$	11.5	505
0.25 Am. alginate	1.0 $CaCO_3$	8.1	410
0.25 Maniflo	2.0 $CaCO_3$	11.0	375

Table III Strength Data on Dry Pellets

Extrusion No.	Binders and Water	Dry Strength, N
1	1% $CaCO_3$, 0.5% alginate & 11.0% water	825
2	1% $CaCO_3$, 0.5% alginate & 10.0% water	897
3	1% $CaCO_3$, 0.25% HEC & 10.5% water	618
4	2% $CaCO_3$, 0.25% MC & 10.0% water	390
5	" " " (no vacuum)	278
6	1.0% $CaCO_3$, 0.5% CMC & 12.0% water	1328
7	2.0% $CaCO_3$, 0.25 CMC-7H & 12.0% water	1281

Some Reactions within the Extruded Pellets

Prior to the induration of the extrudates, DTA was used to examine the reactions of the various binders and iron oxide in the pellets. The resulting curves, shown in Fig. 7, indicate that the organic binder is oxidized at a temperature below 400°C, as revealed by the exothermic peak at about 350°C. The endothermic peaks at 576, 680 and 800°C represent the transformation of quartz, Curie temperature of the iron oxide and decomposition of the $CaCO_3$ respectively.

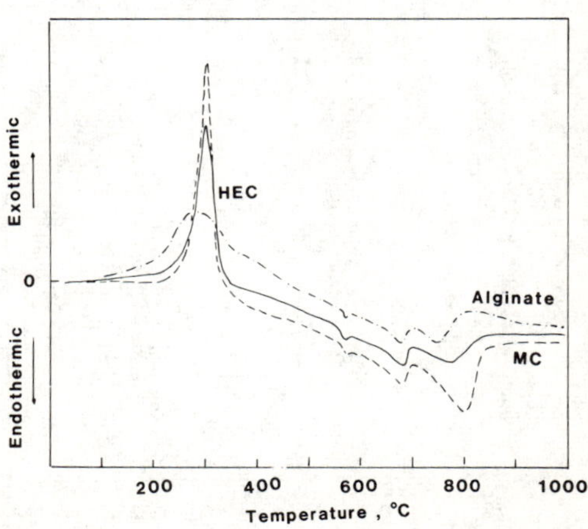

Fig. 7. DTA curves for iron ore containing different binders

Induration and Strength Development of the Pellets

Samples of the pellets made from various compositions were heat treated at temperature levels of 500, 800, 1050, 1150, 1250 and 1300°C for 10 min and 20 min and their crushing strength was determined.

The effect of various binder/binder systems on the strength of pellets at various temperatures ranging from 120°C to 1250°C is illustrated in Fig. 8. The data show that in the case of bentonite or bentonite plus organic binder (as a reference standard), the strength increases regularly with the increase in temperature whereas in the case of organic binder plus calcium carbonate the strength first decreases as the temperature is raised to about 500°C and then continues to increase with increase in temperature above 800°C. The reduction in strength from 120° to 500°C is due to the loss of the organic binder. It appears, however, that at about 800°C a bridging of iron oxide grains through the formation of calcium ferrite has already commenced. Of the four organic binders used, the weakest low temperature strength is observed in mixtures containing MC and the highest for alginate and CMC. At an induration temperature of 1050°C, the strength is still less than, or nearly equal to that of the dry pellet, in all the cases except for CMC. Nevertheless, this is a tremendous increase from the low strength at 500°C, which occurred through the loss of the organic binder. At higher temperatures, further significant strengthening occurs through the formation of sufficient quantities of liquid phases, presumably by the reaction between iron oxide, CaO and the free silica in the system.

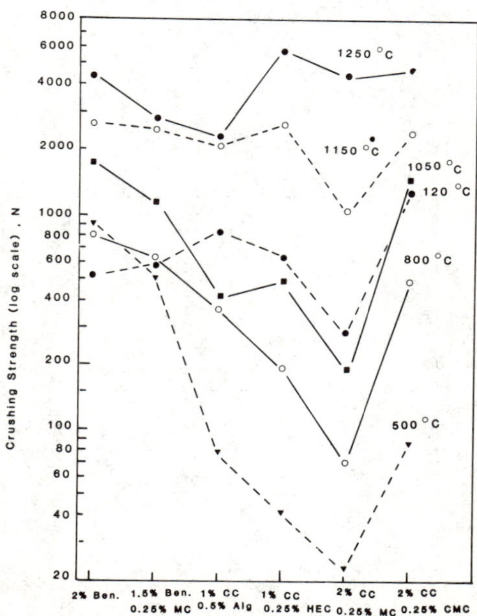

Fig. 8. Strength of pellets with various binders as a function of the induration temperature for 20 min

The pellet strength development with temperature and time, as shown in Figs. 9 and 10, indicates that both MC and HEC develop the highest strength of about 6000N among the various specimens at 1300°C. Also, for extrudates containing alginate, the pellet strength is the lowest when sintered at 1300°C. The reason for this is not clear. A crushing strength of 2000N for the fired pellets is sufficient for transportation and also for use in the blast furnace operation and this can be achieved by sintering the iron ore pellets containing 1 to 2% calcium carbonate and as low as 0.25% of the organic binders MC, CMC or HEC at about 1150°C for 20 min or at about 1250°C for 10 min.

Typical microstructures of the indurated pellets are represented in the photomicrographs in Figs. 11 and 12. On sintering, the large specular hematite grains are changed from angular to a rounded shape. This is due to the reaction of the grains with CaO and free silica, resulting in the formation of a liquid phase that bonds the grains together. Extensive microstructure modification of the matrix phase also can be seen in the photomicrographs. In the matrix, the fine-grained iron oxide is also bonded by the liquid phase, which results in interstitial material between the large platelets of the specular hematite.

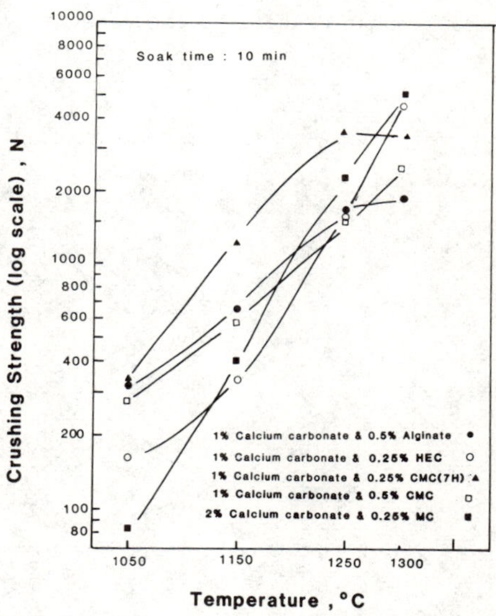

Fig. 9. Strength development of pellets indurated at various temperatures for 10 min

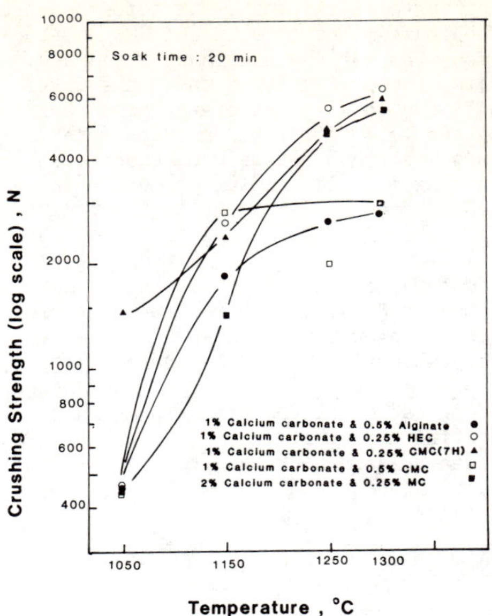

Fig. 10. Strength development of pellets indurated at various temperatures for 20 min

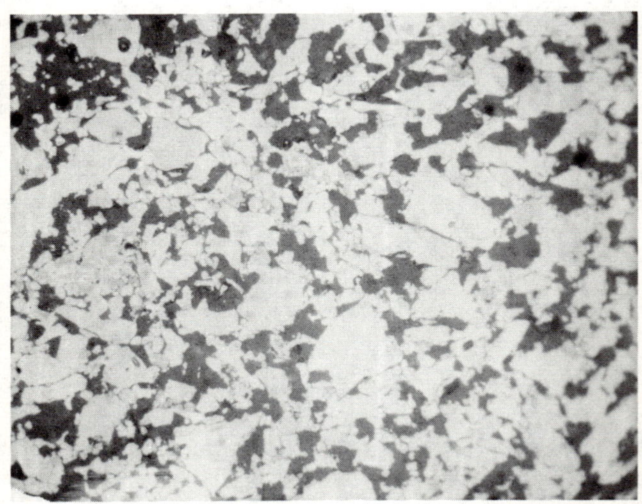

Fig. 11. Microstructure of pellet containing MC and calcium carbonate indurated at 1300°C for 20 min

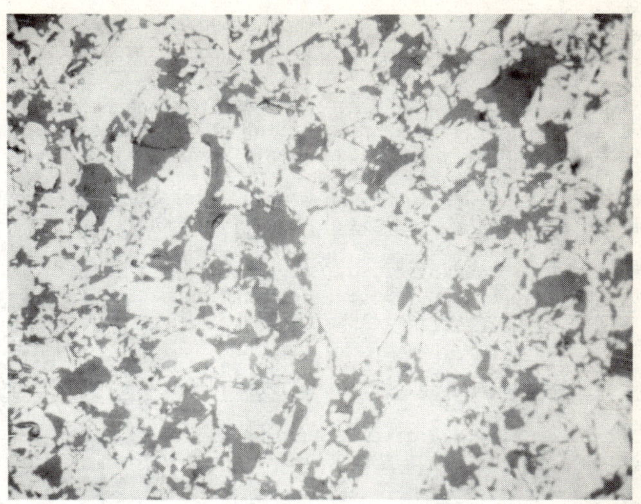

Fig. 12. Microstructure of pellet containing alginate and calcium carbonate indurated at 1300°C for 20 min

Conclusion

Specular hematite from Mt. Wright, when mixed with a polymer solution, behaves as a plastic dispersion mass in much the same way as clay and water. This means that the system would have a liquid phase that "lubricates" the iron oxide particles and has sufficient viscosity to prevent the individual particles from sticking to each other.

Strong pellets can be successfully extruded by using a combination of polymer and $CaCO_3$. Although polymers provide good dry strength for the pellets, the decomposition of the polymer when heated, drastically reduces the pellet strength at temperatures in the order of 500°C. However, a high temperature sintering begins to set in at about 800°C and consequently the strength recovers and then increases rapidly with temperature from thereon up to the temperature range used in this study. When properly indurated, pellets containing polymer and $CaCO_3$ tend to develop strengths even higher than those containing bentonite and bentonite-polymer mixtures and therefore it appears that the manufacture of iron ore pellets by the extrusion process investigated in this study by using calcium carbonate-polymer mixture as extrusion/bonding agents is worth further evaluation. Future research should be directed to improvements in the low- and mid-temperature strength of the pellets so that they can be manufactured in high yields in a moving bed or rotary furnace.

References

1. de Souza, R.P., Mendonca, C.F. and Kater, T. "Production of acid iron ore pellets for direct reduction using an organic binder"; Mining Eng 36:1437-1440; 1984.

2. Steeghs, H.R.G. and Kater, T. "Recent experience with organic binder for iron ore pelletizing"; Skillings Mining Rev 12:6-11; 1984.

3. Kater, T. and Steeghs, H.R.G. "Organic binders for iron ore pelletization"; 57th Annual Meeting of the Minnesota Section, AIME 45th Annual Mining Symposium; Jan. 1984.

4. Sastry, K.V.S. and Fueerstenau, D.W. "Agglomeration 77"; Proc II Intern Symp on Agglomeration, Vol. 1, 381-401; 1977.

5. Roorda, H.J., Burghardt, O., Kortmann, H.A., Jipping, M.H. and Kater, T. "Organic binders for iron-ore agglomeration"; XI Intern Miner Process Congr. Gagliari 139-158; 1975.

6. de Souza, R.P. "The production of pellets in CVRD using hydrated lime as binder is growing up fast"; Ironmaking Proc 35:182-195; 1976.

7. "Iron Ore Pellets - determination of crushing strength" - Third draft proposed ISO/DP 4700; ISO/TC 102/3 N456; Jan. 12, 1979.

8. Trischel, C.C. and Emrich, E.W. "Study of several groups of organic binders under low pressure extrusion"; Am Ceram Soc Bull 29:129-132; 1946.

9. Eugelleitner, W.H. "Pelletizing as applied to the ceramic field"; Am Ceram Soc Bull 58:206-107; 1975.

10. Hopfner, P. "Water soluble cellulose ethers as binding and plasticizing agents"; Interceram 19:2:149-151; 1970.

11. Onoda Jr., G.Y. "The rheology of organic binder solutions in ceramic processing before firing"; (Edited by Onoda Jr., G.Y and Hench, L.L.), 235-250, John Wiley & Sons; 1978.

12. Quon, D.H.H. and Bell, K.E. "Extrusion of Mt. Wright iron ore with crganic polymers and bentonite as plasticizer"; Division Report MRP/MSL 85-72 (IR), CANMET, Energy, Mines and Resources Canada; 1985.

13. Osborn, E.F. and Muan, A. "System of $CaO-Fe_2O_3-SiO_2$"; in Phase Diagrams for Ceramists, Am Ceram Soc; 1964.

CHARACTERIZATION AND EVALUATION OF MIDWEST CLAYS FOR IRON ORE PELLET BONDING

Larry A. Haas, Jeffrey A. Aldinger, Rolland L. Blake, and Matthew N. Plis

Twin Cities Research Center
Bureau of Mines
U.S. Department of the Interior
Minneapolis, MN

The Bureau of Mines evaluated the potential of using Minnesota glacial lake clay as a binder for iron ore pellets. The best clay evaluation procedures for prediction the binder effectiveness were the cation exchange capacity (CEC) using the methylene blue method and the plate water absorption test (PWAT). The CEC method was the fastest method, and several samples were evaluated in 1 hour.

The best Minnesota clay binder was found in the Red River Valley which appeared to contain the most smectite. However, compared to a typical western bentonite (WB-17) at the 0.5-pct level, about 2 pct Minnesota clay was required to obtain comparable unfired (green) pellet strengths. The addition of soda ash only slightly improved the binding properties of Minnesota clay, but considerable improvement was obtained by the addition of a small quantity of water absorbing organic compounds. With a mixture of 0.5 pct Minnesota clay (IV-15C) or paint rock (ITS-46) and 0.1 pct pellet properties were about the same as with 0.5 pct of typical western bentonite (WB-17). The reduction rate at the 40 pct level [9dR/dt)$_{40}$] and the percentage of +6.3 mm particles from the reduction disintegration index (RDI) test were also about the same. Silica contamination with the paint rock binder mixture was considerably less than with an equivalent quantity of western bentonite.

Introduction

Historically, the majority of domestic iron ore has been supplied from the Lake Superior region of northern Minnesota and Michigan. This area currently accounts for 96 pct of total U.S. production, with those States responsible for 71 and 25 pct, respectively (1). Sixty million tons of iron ore pellets were shipped from Minnesota in 1979 while, as a result of the severity of the recent economic decline, only 24 million tons were shipped in 1982 (2). A pelletizing plant has closed as recently as May 1985, partly owing to a rise in imports (3). It is imperative that processing costs be reduced if the domestic iron ore industry is to remain competitive with foreign suppliers.

Approximately one-eighth of the pelletizing cost has been attributed to the cost of the binder. The cost of binder-grade bentonite on the Mesabi Iron Range has tripled since 1974 and the quality has not been consistent (4). Roughly two-thirds of the total binder cost to the plants is due to transportation. Obviously, a sizeable savings could be realized if a source of suitable native clay binder could be located near the pelletizing plants.

The Bureau of Mines, as part of its mission to investigate alternative technology options for processing domestic iron ores, conducted research to determine if clay deposits located near the Mesabi Iron Range could be completely or partly substituted for bentonites obtained from the Western States. Some research (5-6) has been done using local Minnesota clays as iron ore binders, but to date none have been found to replace western bentonites.

Bentonites from the Western States are currently used as a binder during the pelletizing process to agglomerate Minnesota iron ore concentrates. Only limited quantities of high quality bentonite are economically mineable. Bentonite was formed by the deposition of extensive ash falls related to volcanism that occurred during the Cretaceous period. The largest deposits occur in Wyoming and Montana. The western bentonites are predominantly sodium montmorillonite clays with small quantities of associated minerals such as quartz, feldspar, biotite, and locally gypsum.

A major concern of the iron ore pelletizing industry has been to develop meaningful methods for evaluation the overall quality of binder materials (7). Standardized quality acceptance tests and binder specifications for organic and inorganic materials are currently nonexistent. Tests are needed that are relatively fast, simple, and reproducible. Historically, the relative importance of some of the suggested binder specifications has not been well defined, and additional research is required to identify the key parameters.

In general, all taconite companies specify a western (or Na) bentonite with high swelling characteristics because it is known to have better binding properties than southern (or Ca) bentonite. Other specifications that have been suggested are moisture contents, particle sizes (colloid content) and grit contents. In 1983, ASTM (8) published the plate water absorption test (PWAT), which is presently the most widely utilized specification for bentonite acceptance.

The goals of this research were (A) to determine under what conditions local midwestern clays could be used as a binder for iron ore concentrates and (B) to determine the relative usefulness of some clay characterization tests for evaluation different clay binders.

Raw Materials, Sources, and Sampling

Midwest Clay Samples

Four types of natural clay exist in Minnesota: (A) marine kaolinitic in the Ordovician Decorah shale, (B) kaolinitic clay from residual weathering during the Cretaceous period of Precambrian igneous and metamorphic rocks, (C) Pleistocene clay in glacial deposits including clay interstratified in tills and reworked tills and glacial lake clay, which was deposited in six major lakes formed in front of melting continental glaciers when the ice blocked the natural northward drainage of the water, and (D) Holocene clay at the bottom of current lakes. The marine clays and the Cretaceous weathering clays are kaolinitic (9). The marine clays are interstatified with shale, and the Cretaceous clays are covered by about 30 m of glacial deposits. Clays in glacial till are intimately mixed with silt, sand, and gravel. The glacial lakes are gone, but their deposits of clay, silt, sand, and gravel are widespread (Figure 1). Glacial lake clays are exposed at the depth to over 30 m. A few samples examined previously had shown some smectite content and binder potential (5-6). The Holocene clays at the bottom of current lakes are scattered, thin, and covered by water or soil. Therefore, glacial lake clays were selected for the major testing studies as a pellet binder.

Sampling of glacial lake clays in this study involved three methods: (A) Auger samples were obtained by contracting with commercial drillers. (B) Smaller split-tube samples (2.5- and 7.5-cm-diam) were obtained from the Minnesota Department of Transportation, the Minnesota Geological Survey, the North Dakota Geological Survey, and the U.S. Army Corps of Engineers. (C) Shovel samples were obtained by Bureau personnel from nonglacial-lake clay outcrops and pits. The source locations of the samples are shown in Figure 1. The area code column contains Roman numerals for augered samples and county abbreviations for split-tube and shovel samples.

Auger Sampling. Auger sampling areas of glacial lake clays were first chosen within about 100 km of the operating taconite plants (areas I-V in Figure 1). As the study progressed and the results of clay characterization and pellet strength tests indicated a rather marginal response, other areas were selected (areas VI-IX in Figure 1).

Specific field augering sites were chosen in each area by a study of maps, aerial photographs, and field notes and experience of the geologists.

The 15-cm-diam augering operation consisted of removing 1.5-m section samples sequentially to depths up to 22 m. Each 1.5-m section (about 30 kg) was kept separate and placed first in a plastic bag and then a cloth bag; a 0.1-kg sample was also placed in a separate plastic bag for an "as received" moisture determination. All of the 1.5-m samples were spread out on laboratory benches and several samples from each hole were selected for the binder characterization and pelletizing tests. These grayish or brownish selected samples consisted of three or four 1.5-m section samples from each auger hole that were essentially free of grit, pebbles, peat, and soil; and that were located nearest the top, middle (one or two samples), and bottom of the hole.

The material in each 1.5-m selected section was subdivided into about 5-cm-diam pieces and placed randomly in a pile. About one-fifth of the pile was dried at 75° C overnight and the remainder was placed in a sealed bag. The dried pieces were ground to -6 mm in a gyratory cone crusher and

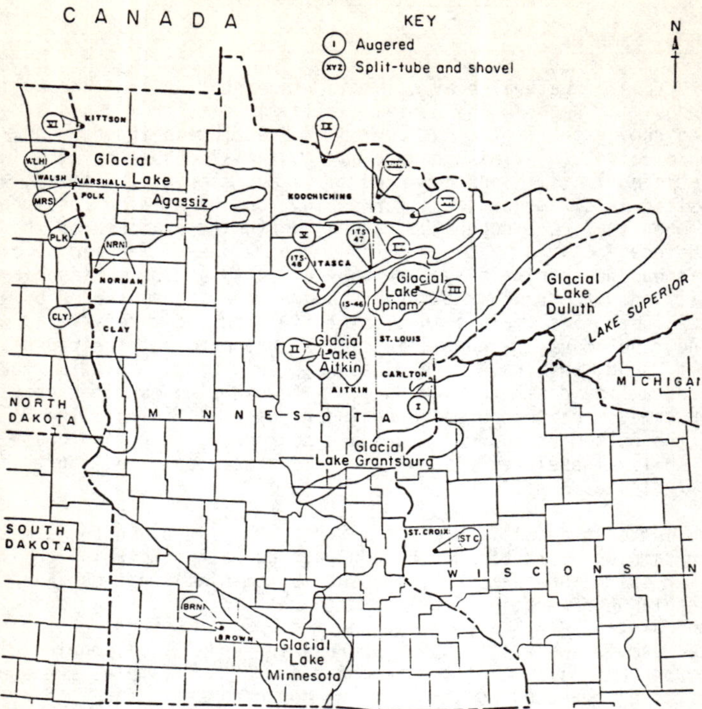

Fig. 1 - Map of former major glacial lakes of Minnesota and locations of augered, split-tube, and shovel sampling clay areas.

then pulverized in a 20-cm-diam[1]. The chemical analyses of some selected augered samples are shown in Table 1.

Split-Tube and Shovel Sampling. Several split-tube and shovel samples (about 1 kg each) were obtained during the course of this study. The location of these selected sampling sites were based on information from the U.S. Department of Agriculture Soil Conservation Service, Minnesota Department of Transportation, and other sources. These sub-soil samples were dried and pulverized by the same methods as the augered samples. The chemical analyses of these samples were about the same general range as those shown in Table 1.

Commercial Samples

Various materials were obtained from commercial sources for the clay characterization and the pellet binder studies. These powdered materials were dried by the same methods as the other samples. Chemical analyses of the commercial samples are given in Table II.

The commercial organic materials were used as received. The carboxyl methyl cellulose (CMC) samples, designations of "P" and "H", had sodium

[1]Reference to specific brand names is made for identification only and does not imply endorsement by the Bureau of Mines.

TABLE I. - Chemical Analyses of Augered Clay Samples[1]

Sample		Si/Al	Partial analysis, pct										LOI at temp., pct		
Code	Footage	ratio	Si	Al	Ca	Mg	Fe	K	Na	C	P	S	105° C	510° C	1000° C
I-7A....	15-20	2.15	21.3	9.9	4.0	2.8	4.6	3.1	0.7	1.4	0.10	0.01	1.80	6.40	11.50
I-7B....	40-45	3.29	23.7	7.2	8.4	3.2	1.8	1.3	0.8	5.0	0.13	0.30	0.26	1.70	16.70
I-7D....	72-75	2.69	23.4	8.7	3.8	2.7	4.2	2.5	0.9	1.9	0.11	0.08	1.20	4.60	9.70
II-32A..	5-10	3.23	25.5	7.9	6.8	2.7	2.7	1.8	1.1	1.7	0.12	0.03	2.00	4.20	9.20
II-32C..	45-50	2.67	24.0	9.0	3.6	2.1	3.0	1.6	0.7	3.4	0.09	0.16	1.20	4.30	14.60
III-9A..	8-10	3.88	28.3	7.3	4.8	2.2	2.2	1.4	1.7	2.3	0.13	0.20	0.79	1.90	7.70
III-9B..	25-30	2.91	25.6	8.8	3.0	2.1	3.4	1.8	1.5	1.2	0.09	0.04	1.30	3.50	7.50
IV-15A..	5-10	2.43	22.1	9.1	7.3	2.7	2.8	2.2	1.2	2.4	0.12	0.01	NE	5.70	14.30
IV-15B..	20-25	2.28	21.4	9.4	7.0	2.7	3.0	2.1	1.0	2.4	0.10	0.02	NE	6.60	14.90
IV-15C..	45-50	2.43	21.9	9.0	6.9	2.5	3.0	2.1	1.2	2.7	0.10	0.02	NE	5.70	13.90
V-23A...	10-15	3.06	24.2	7.9	7.9	3.0	1.8	1.3	0.8	3.9	0.13	0.04	0.93	3.20	14.40
V-23C...	40-45	2.85	23.1	8.1	8.0	3.0	1.9	1.2	0.8	5.0	0.13	0.09	0.58	3.00	15.10
VI-36A..	15-20	2.57	25.2	9.8	2.5	1.9	4.0	1.9	0.5	0.9	0.09	0.22	4.00	7.90	12.90
VI-36B..	30-35	2.35	24.0	10.2	2.6	1.9	4.2	1.9	0.4	1.8	0.08	0.11	3.90	8.60	13.80
VI-36C..	70-75	2.73	25.1	9.2	3.5	2.0	3.6	2.0	0.5	1.5	0.10	0.13	4.20	9.10	15.90
VI-37A..	25-30	2.54	24.4	9.6	2.9	1.9	3.8	2.0	0.5	1.5	0.10	0.11	3.40	7.80	13.70
VI-37C..	70-75	2.82	25.4	9.0	3.8	2.1	3.5	1.9	0.5	1.9	0.10	0.17	3.00	6.90	14.00
VII-38A.	5-10	3.03	21.2	7.0	5.4	2.4	3.2	2.0	1.2	2.4	0.10	0.02	2.80	7.80	13.90
VII-38C.	45-50	2.94	20.3	6.9	7.1	2.8	3.1	1.8	0.9	1.3	0.10	0.02	2.70	7.90	16.50
VIII-39A	15-20	3.33	21.3	6.4	7.7	2.8	2.6	1.8	1.3	3.1	0.11	0.08	3.20	7.10	16.80
VIII-39C	45-50	3.10	20.8	6.7	7.1	2.7	2.9	1.8	1.1	3.1	0.11	0.12	2.40	6.30	15.40
IX-40A..	15-20	3.12	18.1	5.8	13.0	3.2	1.9	1.5	0.8	4.1	0.15	0.09	1.50	3.90	18.00
IX-40B..	45-50	3.38	20.3	6.0	9.3	3.0	1.9	1.5	0.8	3.6	0.14	0.13	1.70	4.30	16.10
IX-41A..	15-20	4.09	22.1	5.4	8.5	3.2	1.7	1.4	1.0	4.0	0.14	0.06	1.30	3.00	16.90
IX-41B..	45-50	3.41	20.1	5.9	9.2	3.0	1.9	1.5	0.9	3.8	0.13	0.12	1.50	4.30	15.80
BRN-43..	NE	3.88	32.2	8.3	0.3	0.6	2.7	1.6	0.3	4.6	NE	0.09	NE	NE	7.30
ITS-46..	NE	1.26	3.4	2.7	0.2	1.6	48.7	<0.1	0.2	0.6	0.15	0.01	2.50	10.70	12.10
WLH-52..	NE	3.38	24.0	7.1	3.6	1.9	3.5	1.6	0.5	3.0	0.10	0.19	2.10	6.10	11.30
STC-53..	NE	3.83	30.2	7.9	1.3	1.3	7.9	1.0	0.4	0.2	0.05	<0.01	NE	3.20	5.60

LOI Loss on ignition.
NE Not evaluated.
[1]The manganese percentage was <0.1 in all the samples.

TABLE II. - Chemical Analyses of Commercial Inorganic Samples

Sample		Si/Al	Partial analyses, pct										LOI at temp., pct		
Name	Code	ratio	Si	Al	Ca	Mg	Fe	K	Na	C	P	S	105° C	510° C	1000° C
Chlorite..........	CH	0.78	9.4	12.0	<0.1	4.3	17.6	<0.1	<0.1	0.58	<0.01	<0.01	<0.1	1.4	8.5
Meta bentonite.....	M-B	1.57	20.9	13.3	5.4	2.5	1.4	4.8	0.3	1.80	0.08	0.01	1.4	4.8	12.4
Silica gel.........	SG	>9.99	40.5	<0.2	<0.1	<0.1	<0.1	<0.1	<0.1	0.56	<0.01	0.05	5.9	9.3	11.3
Zeolite............	Z	1.05	17.2	16.4	<0.1	<0.1	0.2	0.1	11.6	<0.10	<0.01	0.02	4.4	15.4	17.5
Kaolinite..........	KA-1	0.98	21.3	21.7	<0.1	<0.1	0.4	0.1	0.1	0.27	<0.01	0.04	0.2	3.0	4.4
Vermiculite........	VE	2.33	9.8	4.2	10.2	9.0	0.4	<0.1	<0.1	8.60	0.07	0.02	1.9	14.2	34.3
Fireclay...........	KA-2	2.09	27.4	13.1	0.1	0.8	1.2	1.7	0.1	1.60	0.03	0.14	1.1	7.9	9.7
Attapulgite........	AT	3.74	25.4	6.8	2.1	3.2	2.4	0.4	<0.1	1.30	0.16	0.12	6.2	14.1	16.1
Illite.............	I-S	2.51	25.9	10.3	0.4	1.6	4.5	3.8	0.4	2.00	0.13	0.25	1.5	7.1	8.9
Al_2O_3 - colloidal..	Al-C	<0.01	0.1	35.4	1.4	<0.1	<0.1	<0.1	0.1	4.50	<0.01	0.05	5.5	27.4	29.6
Beidellite.........	BE	2.55	25.8	10.1	0.7	1.5	4.3	1.7	0.8	0.44	<0.01	0.02	5.0	10.0	10.9
Southern bentonite.	SB	3.53	28.6	8.1	2.0	1.7	2.3	0.6	0.2	1.40	0.03	0.22	6.8	11.5	14.5
Bentonite/mica.....	B-M	2.54	22.9	9.0	0.3	1.3	3.1	0.1	1.9	1.50	<0.01	0.13	8.1	9.6	14.2
Western bentonite..	WB-3	4.85	32.5	6.7	1.0	1.6	1.0	0.2	1.5	0.13	0.05	0.11	6.2	7.9	11.0
Do..............	WB-4	4.68	33.2	7.1	1.0	1.0	2.2	0.4	1.5	0.45	<0.01	0.04	2.6	4.3	7.9
Do..............	WB-5	4.29	32.2	7.5	1.4	1.2	2.4	4.0	1.3	0.25	<0.01	0.06	2.8	5.3	9.8
Do..............	WB-6	2.72	28.8	10.6	1.4	1.4	3.2	0.4	1.4	0.83	<0.01	0.15	1.4	NE	7.1
Do..............	WB-7	3.62	29.7	8.2	1.3	1.1	2.5	0.4	2.2	1.10	<0.01	0.05	4.6	6.3	9.8
Do..............	WB-8	2.41	25.3	10.5	2.7	1.7	2.9	0.6	1.7	1.80	0.03	0.14	4.4	7.4	14.4
Do..............	WB-9	3.01	25.6	8.5	2.2	0.9	4.4	0.3	2.0	1.70	0.17	0.08	6.7	7.4	13.1
Do..............	WB-10	3.38	29.4	8.7	1.2	1.0	3.1	0.4	1.9	0.38	NE	0.15	5.0	6.6	10.5
Do..............	WB-11	3.87	30.6	7.9	1.0	1.2	3.5	0.3	0.9	0.29	<0.01	0.03	4.3	6.6	9.9
Do..............	WB-12	3.13	29.4	9.4	1.1	1.3	2.6	0.5	2.1	0.29	NE	0.15	4.1	5.6	9.5
Do..............	WB-13	2.72	28.0	10.3	0.9	1.4	2.8	0.4	1.7	0.27	NE	0.18	5.5	7.0	11.5
Do..............	WB-14	3.01	28.3	9.4	0.9	1.3	2.6	0.4	1.6	0.28	<0.01	0.18	2.9	5.1	9.7
Do..............	WB-15	3.18	28.3	8.9	0.7	1.1	3.5	0.2	1.5	0.80	<0.01	0.06	5.1	6.8	9.4
Do..............	WB-16	2.38	25.0	10.5	1.0	1.6	2.5	0.2	1.3	0.94	<0.01	0.21	6.5	8.4	12.5
Do..............	WB-17	2.91	27.9	9.6	1.1	1.5	2.9	0.4	1.5	0.23	0.02	0.10	2.2	5.5	9.7

LOI Loss on ignition.
NE Not evaluated.

contents of 20 and 9 pct, respectively. Poly-acrylate (PAC) and starch-graft copolymer (SGC), type J, also contained 20 pct sodium. The sodium content of the starch-graft copolymer, type A, was 11 pct while the other organic materials contained less than 3 pct sodium. The potassium and sulfur contents of all the organic samples were less than 0.3 and 0.05 pct, respectively. The pregelatinized starach (PS), No. 300, had the highest phosphorus content of 2.7 pct. The guar gumms (GG) phosphorus contents ranged from 0.06 to 0.2 pct, but all the other organics contained less than 0.06 pct phosphorus.

Iron Ore Concentrate. The iron ore used in this research was a magnetic taconite concentrate obtained from the eastern Mesabi Range. It was dried at 105°C overnight and screened to -28 mesh to remove any foreign material. The chemical analyses (pct) of the dried concentrate were Fe_T = 65.2, Fe^{2+} = 21.7, and Si = 2.6. The Blaine surface area and density were 0.17 m^2/g and 4.9 g/cc, respectively.

Experimental Methods

Raw Material Characterization Studies

Mineralogy. Mineral content of the augered glacial lake clays and some commercial samples were estimated using the results from x-ray diffraction data, chemical analyses, thermal gravimetric analyses (TGA), and differential thermal analyses (DTA).

X-Ray Analyses

In the initial x-ray diffraction studies, manual comparisons of integrated intensity peaks of the bulk samples were made. Mineral phases for the bulk-augered and selected commercial samples are shown in Tables III and IV, respectively; the relative peak intensity shows variations in the mineral quantities from sample to sample and does not indicate

TABLE III. - Relative Integrated Intensities of the Major X-Ray Diffraction Peaks for Minerals in Augered Glacial Lake Clay and a Shovel Samples[1]

Sample		Relative mineral peak heights							
Code	Footage	Quartz	Calcite	Dolomite	Plagioclase	Microcline	Mica	Chlorite	Kaolinite
I-07A....	15-20	1815	640	346	372	276	18	55	88
I-07B....	40-45	5300	1069	2714	666	ND	ND	ND	62
I-07D....	72-75	3684	342	428	471	231	58	85	146
II-32A...	5-10	6320	328	1069	1211	671	69	69	104
II-32C...	45-50	4186	1122	1376	548	286	21	64	ND
III-09A..	8-10	8724	502	1391	3014	835	34	28	72
III-09B..	25-30	4610	282	645	1289	420	71	144	213
IV-15A...	5-10	2500	1429	888	681	292	22	98	117
IV-15B...	20-25	2520	1303	1011	543	234	ND	121	102
IV-15C...	45-50	2570	1279	930	955	350	ND	106	149
V-23A....	10-15	6545	1190	2820	1142	906	46	ND	58
V-23C....	40-45	5670	1274	2450	1340	502	ND	ND	61
VI-36A...	15-20	2927	156	493	174	ND	12	100	94
VI-36B...	30-35	2098	328	506	296	ND	15	44	96
VI-36C...	70-75	3036	543	756	250	ND	49	ND	100
VI-37A...	25-30	2981	185	1050	328	182	31	69	128
VI-37C...	70-75	3102	576	980	361	234	64	77	98
VII-38A..	5-10	2181	740	640	635	282	ND	ND	76
VII-38C..	45-50	1648	942	595	441	172	ND	ND	74
VIII-39A.	15-20	2788	936	888	625	ND	ND	ND	85
VIII-39C.	45-50	1980	824	697	586	ND	ND	ND	58
IX-40A...	15-20	3091	1096	1823	412	388	ND	ND	ND
IX-40B...	45-50	3249	876	1656	480	228	ND	ND	ND
IX-41A...	15-20	3684	ND	2520	784	ND	ND	ND	ND
IX-41B...	45-50	12432	ND	1866	1030	299	ND	ND	ND
BRN-43...	NE	11990	ND	ND	ND	ND	ND	ND	ND
ITS-46[2]..	NE	49	ND	ND	ND	ND	ND	ND	21
WLH-52...	NE	2181	380	462	174	ND	ND	ND	ND
STC-53...	NE	11925	30	204	1056	ND	72	ND	86

ND None detected; NE Not evaluated.

[1]Relative integrated intensities are affected by several factors not corrected for here and therefore these results must be used with caution. Values given may be compared from sample to sample (vertically) as a clue to greater or lesser amounts of that phase, but cannot be used to determine relative abundance of the various phases within one sample (horizontally).

[2]Peak heights of goethite and hematite were 280 and 290, respectively.

TABLE IV. - Relative Integrated Intensities of the Major X-Ray Diffraction Peak for Minerals in Commercial Samples

Sample code	Relative mineral integrated peak heights[1]								
	Quartz	Calcite	Dolomite	Plagioclase	Microcline	Mica	Chlorite	Kaolinite	Cristobalite
KA-1...	1858	ND	ND	ND	ND	ND	ND	3457	ND
SB.....	1498	ND	ND	46	ND	40	ND	ND	ND
WB-3...	276	303	44	213	ND	ND	ND	ND	1253
WB-4...	276	108	169	615	237	59	ND	ND	1513
WB-5...	538	188	123	454	671	31	ND	ND	1452
WB-6...	955	ND	ND	296	ND	ND	ND	ND	96
WB-8...	475	620	ND	ND	ND	ND	ND	12	ND
WB-9...	2052	269	ND	ND	ND	174	ND	ND	ND
WB-10..	2830	339	ND	538	ND	66	ND	ND	ND
WB-11..	85	ND	36	462	44	56	ND	ND	1624
WB-12..	2916	111	ND	1136	462	77	ND	ND	ND
WB-13..	1274	135	ND	615	292	ND	ND	ND	ND
WB-14..	1116	ND	ND	778	376	262	ND	40	ND
WB-16..	289	ND	49	346	ND	ND	ND	ND	56
WB-17..	1475	117	72	918	286	ND	ND	ND	199

ND None detected.

[1]Relative integrated intensities are affected by several factors not corrected for here; therefore the data must be used with caution. Values given may be compared from sample to sample (vertically) as a clue to greater or lesser amounts of that phase, but cannot be used to determine relative abundance of the phases within one sample (horizontally).

quantitative composition of a given sample (horizontal comparison). Several augered samples were prepared with standard techniques to give preferred orientation, followed by glycolation to better identify the clay minerals (phyllosilicates).

The glacial lake clays were found to contain only small quantities of the clay minerals such as kaolinite, chlorite, dolomite, and mica derived from igneous and metamorphic rocks abraded by the glaciers. These nonclay minerals were found in sand-sizes, silt-sizes, and some in clay-sizes (-2 µm), as reported previously (5-6). Commercial western bentonite samples contained smectite, small quantities of quartz, cristobalite, feldspar, calcite, dolomite, mica, and kaolinite. X-ray data of the bulk sample were most useful for determining the impurities in the clay samples, but not very useful for determining the clay content.

To better quantify the smectite content of augered glacial lake clays, a few bulk samples were subjected to the sedimentation-glycolation method (12) used at the University of Minnesota to produce clay-size fractions. In this sedimentation test, the suspension (-2 µm) was centrifuged, and this dried sediment and the dried filtrate solids (-0.2 µm) were referred to as the coarse and fine clay fractions, respectively (Table V). The fine fractions contained more smectite than the coarse fractions. The combined fine and coarse fractions from sample VI-37A contained the most smectite (34-50 pct) of the Minnesota samples, but this smectite content was still less than the over 50 pct contained in the typical western bentonite (WB-17).

Chemistry

Chemical analysis helped in the delineation of the types of minerals present in clays. The aluminum and silicon elemental analyses of the commercial bentonites (Table II) were similar to the Midwest clays (Table I) except for sample ITS-46 which had much lower values. This material, paint rock, is a highly altered part of the slaty iron formation on the Mesabi Range. Alteration by weathering destroyed most of all the former quartz, iron silicates, and iron carbonate, leaving a clay-textured material high in aluminum silicates and iron oxides. Completely altered paint rock contains

TABLE V. - X-Ray Diffraction on Fractions of Selected Clay and Bentonite Samples

Sample	Clay[1]		Relative mineral[2] abundance[3]									
	Fraction	Pct	SM	VE	CH	I	KA	QU	FE	CA	D	Interstratification
I-7B....	Fine......	6	5	ND	ND	1	3	1	ND	ND	ND	I-VE = 4, I-SM = 4
	Coarse....	8	3	ND	2	2	2	2	2	ND	2	I-VE = 4, I-SM = 4
II-32C..	Fine......	15	4	ND	1	3	1	ND	2	1	ND	K-SM = 2, I-SM = 4
	Coarse....	16	3	ND	2	2	2	1	2	3	ND	I-VE = 4, I-SM = 4
III-9B..	Fine......	6	4	ND		2	2	1	2	ND	ND	K-SM = 3, I-SM = 5
	Coarse....	6	2	ND	2	3	3	2	4	ND	ND	I-VE = 2, I-SM = 2
IV-15B..	Fine......	2	2	ND	1	2	2	1	2	ND	ND	K-SM = 3, I-SM = 5
	Coarse....	14	2	ND	2	2	3	2	4	2	ND	I-VE = 3, I-SM = 3
V-23A...	Fine......	<1	4	ND	ND	2	3	1	2	1	ND	K-SM = 2, I-SM = 4
	Coarse....	27	2	1	1	2	2	2	2	3	3	I-VE = 3, I-SM = 3
VI-37A..	Combined..	74	5	1	1	3	3	2	3	ND	2	ND
VII-38A.	..do......	64	3	3	2	2	2	2	3	3	2	I-VE = 2
VIII-39A	..do......	46	2	4	1	2	2	2	3	4	2	I-VE = 1
IX-40A..	..do......	37	2	5	1	2	1	1	2	3	2	I-VE = 1
IX-41A..	..do......	21	4	4	2	3	2	2	3	ND	ND	ND
WB-17...	..do......	88	6	ND	ND	ND	ND	1	1	ND	ND	ND

ND None detected.

[1]The combined, coarse, and fine clay fractions refer to the >2-μm, 0.2 to 0.02-μm, and <0.02-μm particle size, respectively; the percentages were based on the initial sample weight. These tests were conducted by the University of MN Soils Survey Lab.

[2]The mineral abbreviations are SM = smectite, VE = vermiculite, CH = chlorite, I = illite, KA = kaolinite, QU = quartz, FE = feldspar CA = calcite, and D = dolomite.

[3]The abundance numbers refer to 1 = trace (<5 pct), 2 = very small (5-10 pct), 3 = small (11-20 pct), 4 = moderate (21-33 pct), 5 = abundant (34-50 pct), 6 = dominant (>50 pct).

mostly hematite, quartz, and kaolinite and small quantities of a smectite (the iron-bearing nontronite). Its claylike texture, high iron oxide and low silica content, and availability on the Iron Range prompted its testing as a binder for agglomerating iron ore concentrate.

Since bentonite contains mainly smectite (aluminum silicate mineral), the Si-to-Al ratio of the samples was evaluated to attempt to estimate the relative smectite content. Glacial lake clays, however, contained only a little smectite and various quantities of four or five other aluminum silicate minerals. Therefore, the Si-to-Al ratio was not a useful criterion for estimating the smectite content in glacial lake clays.

Thermal Analyses

Differential thermal and gravimetric analyses were also used for the determination of poorly crystallized complex minerals. With these methods, about 0.05-g sample was tested in a Perkin-Elmer thermoanalyzer at a heating rate of about 25°C/min. The thermogravimetric analysis (TGA) results (Figure 2) indicated that the greatest weight change started ot occur at about 400°C for illite (I-S) and the clay sample IV-15C and at about 700°C for a western bentonite (WB-17), dolomite (D), and limestone (L). The dolomite and limestone samples contained principally calcium magnesium carbonate and calcium carbonate, respectively.

The TGA data were more meaningful when they were used in conjunction with the differential thermal analysis (DTA) results shown in Figure 3. An exothermic peak obtained with the clay sample VI-37A between 350° and 450°C was probably due to the presence of organics. Dolomite, chlorite, and Minnesota clay samples had exothermic peaks at about 900°C. A strong endothermic peak, observed an all the samples between 100° and 300°C, was probably due to the removal of hygroscopic, interlayer and/or coordinated water. At 600°C, the kaolinite (KA-1) sample had a strong endothermic peak, which is in agreement with the results reported by Hoffman (13). Southern bentonite yielded only a small peak at 700°C, while western bentonite

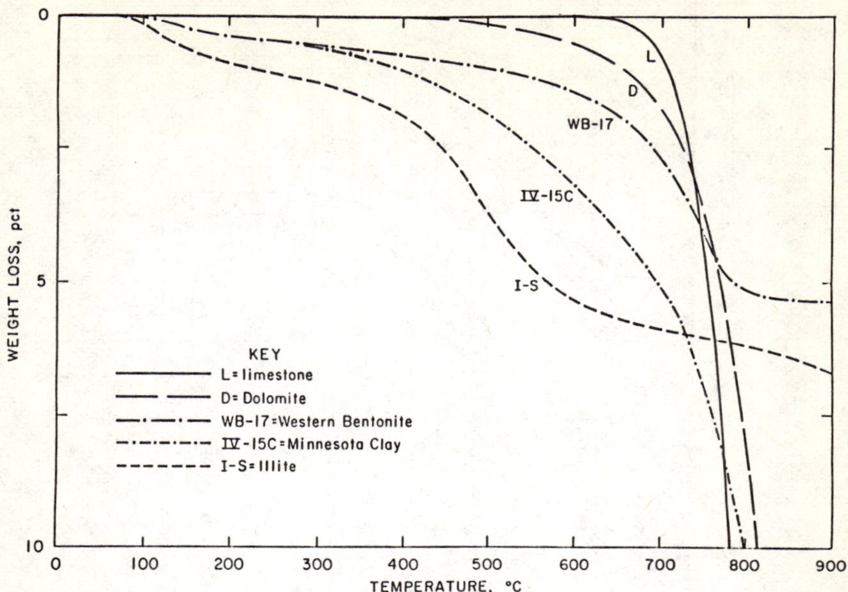

Fig. 2 - Thermogravimetric analysis of selected samples.

(WB-17) produced a large endothermic peak between 700° and 800°C. The Minnesota clay sample had a small endothermic peak at 600° and 800°C, which suggested that only small quantities of kaolinite, smectite, and limestone were present.

Shrinkage-Swelling and Clay Briquet Strengths. Three types of shrinkage-swelling procedures were conducted to characterize the inherent properties of the clay that could be ultimately related to the iron ore pellet strengths. One precedure involved the shrinkage and briquet strength, and two procedures involved swelling.

An attempt was made to develop an evaluation procedure to combine sample binding strength with its shrinkage properties. A simple briquet procedure was used for measuring the shrinkage of the samples by making briquets and determining the volume change on drying. Twenty-seven briquets were made by pressing a thick clay paste into 12-mm holes drilled through a 16-mm Teflon plate. About 20 pct water was added to the dried Minnesota clay powder and about 40 pct water to the western bentonites to make pastes of about the same consistency. The filled sample plate was dried overnight at 75°C. The briquet volumetric shrinkage was generally less than 20 pct with the Minnesota clays, but over 20 pct with the western bentonites. The dry clay briquet compressive strengths were less than 235 lb with the Minnesota clays and more than 275 lb with the western bentonites. This shrinkage and strength procedure did not appear to be a very precise clay evaluation method.

The first swelling procedure involved the dilatancy (DIL) method, in which dry clay powder was added slowly to water and the ratio of the final

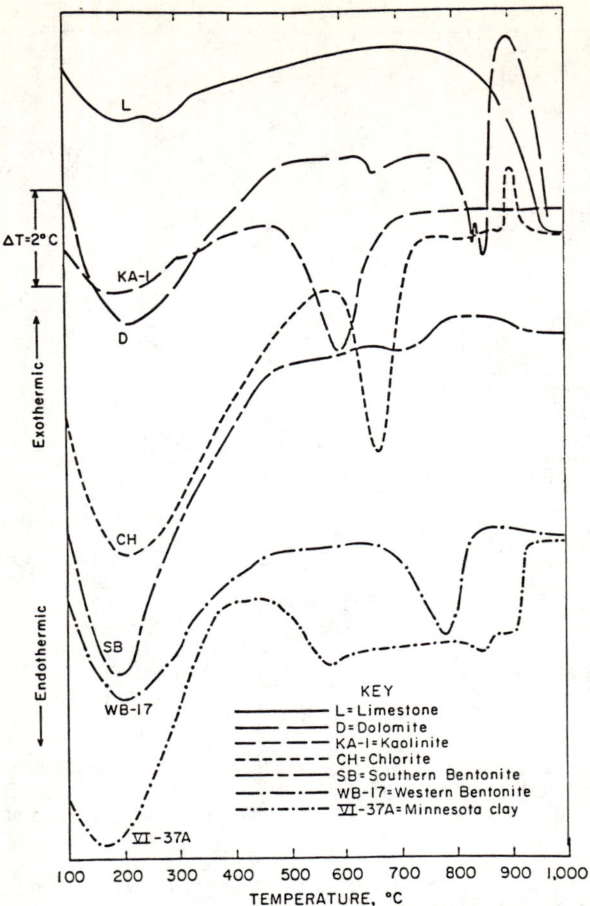

Fig. 3 - Influence of temperature on endothermic and exothermic behavior of different samples.

wet (^{V}f) and the initial dry (^{V}i) volume was determined. With the Minnesota clay samples, these ratios were less than 3 (Table VI) but with the western bentonites (WB's), ratios over 5 (Table VII) were obtained. With the augered Minnesota samples, the highest ratios were obtained with the Red River Valley samples.

A second swelling method (PWAT-V) was evaluated in conjunction with the PWAT number weight gain experiments. Similar swelling results were obtained as with the first method but the dilatancy method was more reproducible.

Particle Size. Particle size is usually an important variable in characterizing clay-bearing materials because one of the definitions of clay is a 2-μm-diam or smaller particle. However, nonclay minerals also occur in this size range and are referred to as "rock flour". Rock flour can be produced by the grinding action of a glacier. The particle size distributions of the clay samples were determined by optical laser and sedimentation methods. With the L&N Microtrac (laser method), the mass

TABLE VI. - Binder Characterization Data of Minnesota Samples

Sample		CEC, meq/100g		Electrode values[1]				Colloid, pct		Swell,[2] (V_f/V_i)		PWAT	η,[4] cp
Code	Footage	mb	BTAC-Na	Na^+	Ca^{2+}	Eh	pH	Total	Fine	DIL	PWAT-V	No. 3	
I-7A.....	15-20	15	NE	10.4	11.8	124	7.7	NE	NE	NE	0.61	100	NE
I-7B.....	40-45	15	0.4	9.2	23.4	135	7.8	13.8	5.6	1.16	0.33	30	9.5
I-7D.....	72-75	10	NE	1.7	12.8	235	8.5	14.1	4.6	NE	NE	106	NE
II-32A...	5-10	15	NE	0.7	21.2	245	8.1	23.1	14.8	NE	NE	69	NE
II-32C...	45-50	15	0.6	NE	NE	NE	NE	31.1	NE	1.97	0.83	95	6.5
III-9A...	8-10	5	NE	0.6	12.0	208	8.3	19.1	2.3	NE	NE	26	NE
III-9B...	25-30	10	0.5	1.1	11.4	206	8.3	11.7	5.9	0.60	0.47	41	6.5
IV-15A...	5-10	15	NE	1.1	14.8	304	8.2	4.7	1.4	NE	NE	78	NE
IV-15B...	20-25	20	NE	1.5	14.8	275	8.3	2.9	0.5	NE	NE	84	NE
IV-15C...	45-50	15	NE	1.9	17.1	245	8.2	5.6	0.1	1.10	0.78	101	6.0
V-23A....	10-15	10	0.10	0.9	16.8	260	7.9	26.9	0.1	1.72	1.25	72	6.75
V-23C....	40-45	10	NE	1.3	17.3	261	7.9	18.2	0.1	NE	0.82	60	NE
VI-36A...	15-20	20	0.90	16.0	79.0	112	7.7	2.3	NE	NE	1.38	106	NE
VI-36B...	30-35	25	1.10	6.2	22.3	119	7.7	8.2	NE	2.94	1.54	110	8.0
VI-36C...	70-75	20	0.70	3.9	23.1	111	7.9	29.9	NE	2.50	1.32	107	NE
VI-37A...	25-30	20	1.00	4.8	16.7	101	7.8	5.3	NE	2.41	1.08	92	7.0
VI-37C...	70-75	20	0.70	1.8	19.5	129	7.8	35.0	NE	2.12	1.23	102	NE
VII-38A..	5-10	15	0.0	8.0	7.2	143	8.2	36.9	2.5	NE	0.65	90	NE
VII-38C..	45-50	20	0.0	3.5	12.4	156	8.2	21.5	0.5	NE	0.76	109	NE
VIII-39A.	15-20	10	0.1	3.1	11.6	153	8.2	27.9	0.5	NE	0.64	73	NE
VIII-39C.	45-50	15	0.0	3.1	12.9	8	8.1	25.0	0.8	NE	0.74	98	NE
IX-40A...	15-20	10	0.0	1.9	15.0	8	8.2	23.2	1.7	NE	0.75	50	NE
IX-40B...	45-50	10	0.0	2.1	15.8	8	8.3	23.6	1.1	NE	0.78	56	NE
IX-41A...	15-20	10	0.0	1.7	9.5	8	8.2	13.9	1.8	NE	0.71	34	NE
IX-41B...	45-50	10	0.0	1.9	15.9	8	8.2	24.4	0.9	NE	0.72	50	NE
ITS-46...	NE	5	2.8	NE	NE	NE	NE	24.9	3.2	1.88	1.00	70	10.0

NE Not evaluated.

[1]The Na^+ and Ca^{2+} values (ppm) were determined by selective ion electrodes; the Eh (mv) was determined with a platinum electrode versus a standard normal calomel electrode.

[2]The terms V_f and V_i refer to the final wet and initial dry sample volume respectively; the PWAT-V values were determined from the same sample as was used on the PWAT No. test. DIL refers to dilatancy.

[3]The PWAT No. value was determined by multiplying the difference between the wet and dry weight (ΔW) by 100 and dividing by the initial weight (W_i).

[4]The viscosity (η) was determined at 600 rpm and expressed in centipoise (cp).

TABLE VII. - Binder Characterization Data for the Commercial Samples

Sample code	CEC-mb, meq/100g	CEC-BTAC, Meq/100g				Electrodes values[1]				Colloid, pct		Swell,[2] (V_f/V_i)		PWAT No.[3]	η,[4] cp
		Na^+	Ca^{2+}	K^+	Mg^{2+}	Na^+	Ca^{2+}	Eh	pH	Total	Fine	DIL	PWAT-V		
CH.......	5	0.3	0.2	0.00	0.6	2.5	1.2	96	7.3	3.7	0.0	1.25	0.65	33	9
M-B......	10	NE	NE	NE	NE	1.9	16.9	91	7.9	29.6	20.3	1.49	0.84	48	8
SG.......	5	3.4	0.3	0.10	0.0	32.9	1.7	127	6.6	12.3	1.6	0.76	0.59	67	7
Z........	NE	NE	NE	NE	NE	47.4	0.3	117	9.2	11.9	2.2	NE	0.72	50	9
KA-1.....	10	1.6	0.9	0.03	0.2	12.3	1.5	191	6.3	75.5	45.2	0.58	0.44	54	5
VG.......	5	0.2	2.1	0.10	2.0	2.3	10.5	113	7.4	11.9	0.4	1.67	0.90	21	8
KA-2.....	10	0.9	4.2	0.30	2.5	1.1	15.7	274	4.1	5.9	0.0	2.38	1.02	30	8
AT.......	15	0.5	8.2	0.30	4.1	1.5	11.0	94	7.8	83.6	76.6	1.16	0.75	259	26
I-S......	15	0.7	8.6	0.60	1.1	1.2	9.2	154	6.7	20.0	9.6	0.77	1.06	55	8
Al-C.....	5	NE	NE	NE	NE	NE	0.0	273	3.6	88.6	2.1	2.21	2.95	480	8
BE.......	25	19.1	11.6	0.80	2.1	31.7	4.9	103	7.6	35.6	5.6	1.92	2.05	109	7
SB.......	40	1.2	41.0	0.70	10.8	24.2	7.3	50	9.2	23.1	10.8	1.80	1.36	122	8
B-M......	20	3.0	0.5	0.00	2.0	28.5	0.8	107	6.3	21.4	3.5	1.70	1.75	264	8
WB-3.....	40	40.8	9.6	0.30	2.5	292.0	40.7	78	8.9	74.3	17.3	5.66	3.15	518	20
WB-4.....	40	24.6	7.2	0.50	3.6	136.0	26.6	137	8.3	73.0	26.4	5.90	3.75	542	20
WB-5.....	40	21.1	11.0	0.30	6.6	NE	NE	NE	NE	79.5	13.4	10.20	4.71	575	100
WB-6.....	45	41.7	11.6	1.10	3.1	279.0	22.2	89	8.9	78.4	10.2	15.00	6.21	811	43
WB-7.....	40	49.5	1.0	0.50	0.0	440.0	25.3	56	9.4	64.1	16.7	10.28	5.56	756	33
WB-8.....	55	41.7	10.6	1.30	4.4	192.0	20.7	62	8.7	85.4	30.7	16.17	7.73	811	97
WB-9.....	45	45.2	4.2	0.40	0.5	293.9	32.7	98	9.1	70.4	18.7	11.30	6.21	614	104
WB-10....	40	38.5	4.1	0.40	0.7	124.0	31.8	41	9.2	71.4	20.3	7.81	5.69	720	33
WB-11....	50	14.7	14.6	0.30	14.3	19.3	72.4	181	7.1	87.8	36.0	9.70	3.12	540	28
WB-12....	45	49.8	3.3	0.40	3.3	466.0	38.5	35	9.5	96.3	34.4	20.00	6.35	852	>150
WB-13....	50	36.8	8.4	1.30	2.3	297.0	40.9	40	9.3	79.6	31.0	15.59	7.50	1004	82
WB-14....	50	37.6	9.8	1.30	3.4	NE	NE	NE	NE	94.7	39.5	16.40	6.36	786	48
WB-15....	50	37.6	4.2	0.40	4.1	265.0	22.8	156	8.1	90.2	11.5	10.70	4.94	894	>150
WB-16....	55	34.7	17.6	0.70	12.1	289.0	81.8	78	8.7	93.6	14.7	15.17	7.61	918	94
WB-17....	55	32.4	12.9	1.00	5.9	247.0	46.0	37	8.2	82.0	29.5	19.60	6.90	930	88
WB-200[5]..	45	42.6	12.9	1.90	4.9	199.0	32.7	144	9.3	85.2	13.3	14.10	6.77	917	73
WB-300...	50	40.0	12.9	1.90	4.4	NE	NE	NE	NE	NE	NE	NE	NE	NE	65
WB-400...	45	42.6	12.9	2.00	4.4	161.0	28.8	143	9.2	75.9	11.1	12.20	6.73	864	49
WB-500...	40	39.2	13.9	2.20	3.8	NE	NE	NE	NE	NE	NE	NE	5.84	693	NE
WB-600...	35	40.9	14.9	2.40	4.1	133.0	12.8	158	8.5	48.9	7.0	8.75	5.00	612	9
WB-700...	20	33.2	12.7	NE	4.8	100.0	1.9	172	8.2	8.1	1.0	3.69	2.12	200	11
WB-800...	10	16.7	12.7	NE	4.4	68.1	NE	169	8.3	4.4	0.3	1.36	1.04	58	13
WB-PA[6]...	NE	NE	NE	NE	NE	NE	NE	NE	NE	NE	NE	NE	NE	NE	>150
ITS-GG-[7].	NE	NE	NE	NE	NE	NE	NE	NE	NE	NE	NE	NE	NE	242	>150

NE Not evaluated.

[1]The Na^+ and Ca^{2+} values (ppm) were determined by selective ion electrodes; the Eh (mv) was determined with a platinum electrode versus a standard normal calomel electrode.

[2]The terms V_f and V_i refer to the final wet and initial dry sample volume respectively; the PWAT-V values were determined from the same sample as was used on the PWAT No. test. DIL refers to dilatancy.

[3]The PWAT No. value was determined by multiplying the difference between the wet and dry weight (ΔW) by 100 and dividing by the initial weight (W_i).

[4]The viscosity (η) was determined at 600 rpm and expressed in centipoise (cp).

[5]These numbers refer to the drying temperatures (° C) of the WB-17 samples at 200° to 800° C. All other samples were dried at 75° C.

[6]This sample code refers to 9 pct polyacrylate (PA) added to WB-17.

[7]This sample consisted of 17 pct GG-211D added to ITS-46.

Unfired Pellet Properties

Commercial Samples. The commercial samples were investigated at the 1-pct addition level to obtain baseline unfired pellet data for comparison purposes. In general, the results in Table VIII indicated that the western bentonites had higher dry compressive strengths than the other materials. The lowest values of each of the unfired pellet properties of the western bentonites were used to define the minimum acceptable values that would have to be met by any potential binder. These target values were defined as

Dry compressive strength (DCS) = 14 lb (6.4 kg)
Wet compressive strength (WCS) = 3 lb (1.4 kg)
Drop number (No.) = 4

The average unfired pellet properties of the western bentonites were fairly well represented by WB-17, which was therefore used more extensively in this research (Figure 5). This sample was defined as the typical western bentonite. With the southern bentonite and bentonite-mica samples, these target values were almost reached.

The nonbentonite materials were selected for certain known properties and/or for being possible constituents in midwestern clays. The unfired pellet target values were not reached with any of the nonwestern bentonite samples.

The typical western bentonite (WB-17) was dried overnight at various temperatures and pellets were made at the 1-pct binder level. At this fixed addition level, more moles of aluminum and silicon oxide were added with the binders dried at the higher temperatures because drying removed moisture. However, the properties of unfired pellets made with the high temperature (600°C) dried binder were lower (Table VIII). The decreased strength values were possibly due to the collapse of the hydroxylated montmorillonite structure.

TABLE VIII. - Unfired Pellet Properties Made With 1 pct of Commercial Samples

Sample		Unfired pellet properties			
Type	Code	H_2O, pct	Drops, No.	WCS, kg	DCS, kg
Chlorite	CH	7.2	2.7	1.0	1.5
Meta bentonite	M-B	7.5	2.7	0.9	1.7
Silica gel	SG	7.6	3.4	1.2	1.8
Zeolite (4A)	Z	7.5	3.8	1.0	1.9
Kaolinite	KA-1	7.2	3.9	1.3	2.0
Vermiculite	VE	7.1	2.7	1.0	2.3
Fireclay	KA-2	7.2	2.9	1.3	2.4
Attapulgite	AT	7.1	2.9	1.1	2.7
Illite	I	6.9	3.0	1.5	2.7
Al_2O_3 - colloidal	Al-C	7.3	3.6	1.2	2.8
Beidellite	BE	7.1	3.6	1.3	4.3
Southern bentonite	SB	8.1	5.3	1.4	5.3
Bentonite-mica	B-M	8.6	8.7	1.2	6.3
Western bentonite	WB-3	7.6	4.0	1.6	6.5
Do	WB-4	7.2	4.7	1.7	7.5
Do	WB-5	8.7	10.0	1.5	7.7
Do	WB-6	8.1	6.0	1.3	8.2
Do	WB-7	7.0	4.5	2.0	8.3
Do	WB-8	7.8	6.3	1.6	8.4
Do	WB-9	7.4	5.0	1.6	8.4
Do	WB-10	7.0	5.9	2.0	8.6
Do	WB-11	8.7	8.5	1.5	8.8
Do	WB-12	7.0	8.2	2.4	9.0
Do	WB-13	7.4	4.6	1.7	9.6
Do	WB-14	8.5	6.3	1.9	9.8
Do	WB-15	8.2	9.9	1.6	10.3
Do	WB-16	7.2	7.6	1.9	11.0
Do	WB-17	7.6	9.2	1.9	9.1
Do	WB-600[1]	9.6	7.9	1.5	8.5
Do	WB-800[1]	7.2	3.8	1.4	3.2

WCS Wet compressive strength.
DCS Dry compressive strength.
[1]These WB-17 sampled were dried at 600° and 800° C, respectively. All other samples were dried at 75 ° C.

median particle sizes for some selected augered Minnesota samples ranged from 6 to 13 μm, while western bentonite particle sizes ranged from 4 to 6 μm. These results indicated very little difference in the mean particle size between the augered Minnesota samples and western bentonites.

The optical laser method only measured particles larger than 2 μm. Therefore, further tests were conducted with the USDA sedimentation method (14) which measured particle sizes down to 0.2 μm (Figure 4). One reason

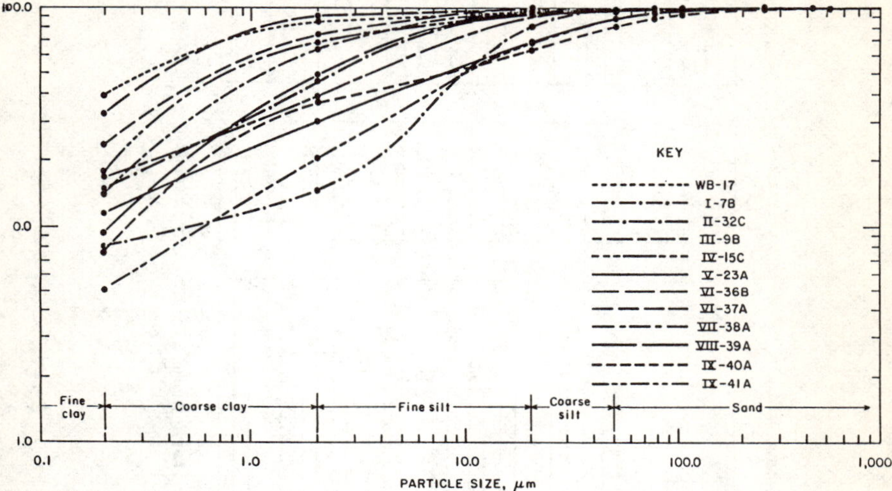

Fig. 4 - Particle size distribution of different samples with the sedimentation method.

that smaller apparent mean sizes were obtained with the sedimentation method was probably that of the low density of the expanded clay particles which resulted in slow settling rates. The typical western bentonites (WB-17) had the highest percentage of -0.2-μm particles and the next highest were the Red River Valley clay samples (VI-36B and 37A).

A modified sedimentation technique was used to determine the total and fine colloid content. The total colloid content was defined as the quantity of material that remained in suspension after 18 hr. The total colloid contents of the augered samples (Table VI) were less than 40 pct, compared with over 60 pct for the western bentonites dried at 75°C (Table VII). The colloid content decreased as the bentonite drying temperature was increased above 200°C. The 18-hr suspension was allowed to settle for an additional 6 days, and then this suspension was centrifuged. The centrifuged sediment was defined as the fine colloid fraction. In general, the fine colloid content was lower with the Minnesota samples than the with western bentonites.

It is well known that a colloid particle settling rate is dependent on the ions in solution. Therefore the sodium, calcium, pH, and Eh (oxidation or reduction) potential of the supernatant solutions from the colloid tests were determined. The solutions from the western bentonite samples (Table VII) contained over 100 ppm Na^+, while the augered Minnesota samples (Table VI) contained less tahn 20 ppm Na^+; the principal mineral in western bentonite is sodium montmorillonite, and the solubility of this mineral may be the reason for the high solution sodium analyses. No definite

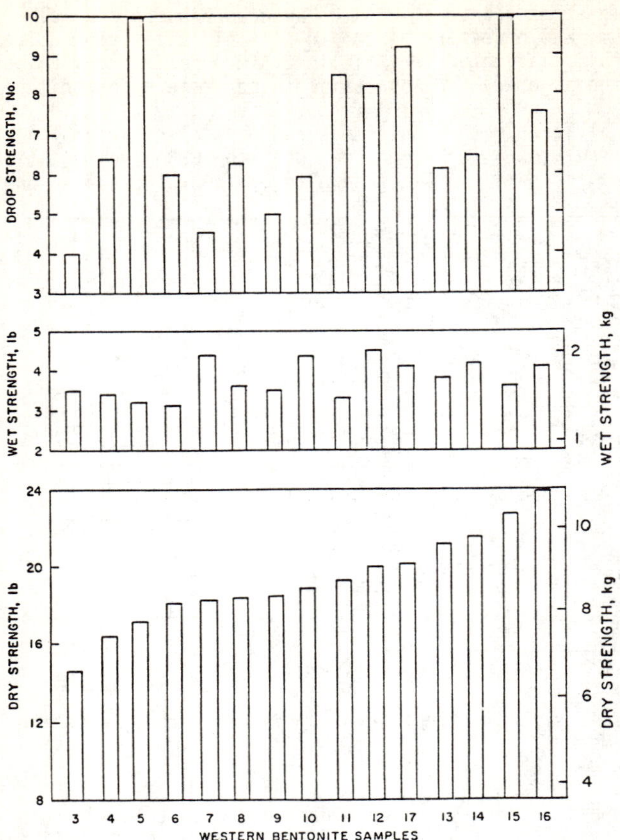

Fig. 5 - Influence of different western bentonites at 1 pct level on the unfired pellet.

Midwestern Samples. The initial pelletizing experiments were conducted with split-tube and shovel clay samples from various counties at the 1-pct level. The unfired pellet results in Figure 6 indicated that the largest dry compressive strengths from northwestern Minnesota in Clay, Kittson, Marshall, Norman, and Polk Counties, and from northeastern North Dakota in Walsh County. The wet compressive strength and drop numbers showed no particular pattern and were generally below the minimum target values obtained with the western bentonites.

To determine the binder percentage required to reach the minimum western bentonite target level, more extensive tests were conducted with the augered samples. The results in Figure 7 and Table IX indicated that the target unfired strength values were reached with 3 pct area IV-15C and VI-37A samples. However, this is an extremely high addition level, so attempts were made to improve the binding properties by the addition of soda ash.

The unfired pellet properties, made with the auger samples at the 1-pct binder level, generally improved with the addition of 0.1 pct soda ash, but

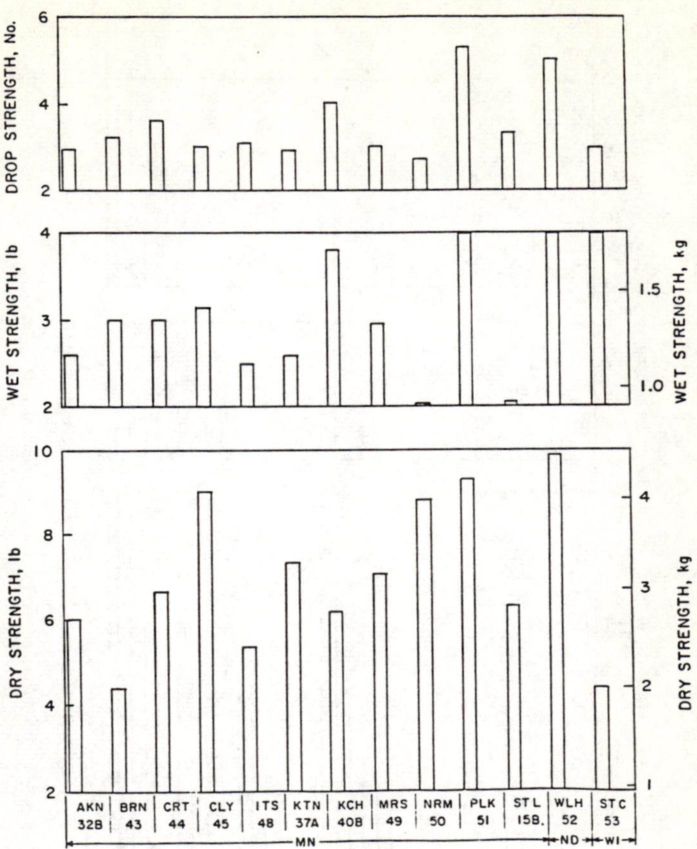

Fig. 6 - Unfired pellet properties made with 1-pct augered, split-tube, and shovel clay binders from various counties (Table I) in Minnesota, North Dakota, and Wisconsin.

the target level was still not obtained (Figure 8). The addition of soda ash to southern bentonite considerably improved the unfired pellet properties. These results indicated that the soda ash addition improved the unfired properties of samples that contained calcium montmorillonite, probably because the montmorillonite was converted to the more effective sodium form.

Higher binder levels with Minnesota clays using soda ash were investigated to determine what conditions were required to reach the target unfired strengths. Using 2 pct binder with area IV-15C clay and soda ash, the unfired pellet properties increased with increasing soda ash additions up to a maximum value. This maximum occurred with 32 parts clay to 1 part soda ash; at this addition level, all the unfired pellet target values were reached.

Different soda ash addition methods were investigated to further improve the unfired pellet properties. The first method involved mixing the

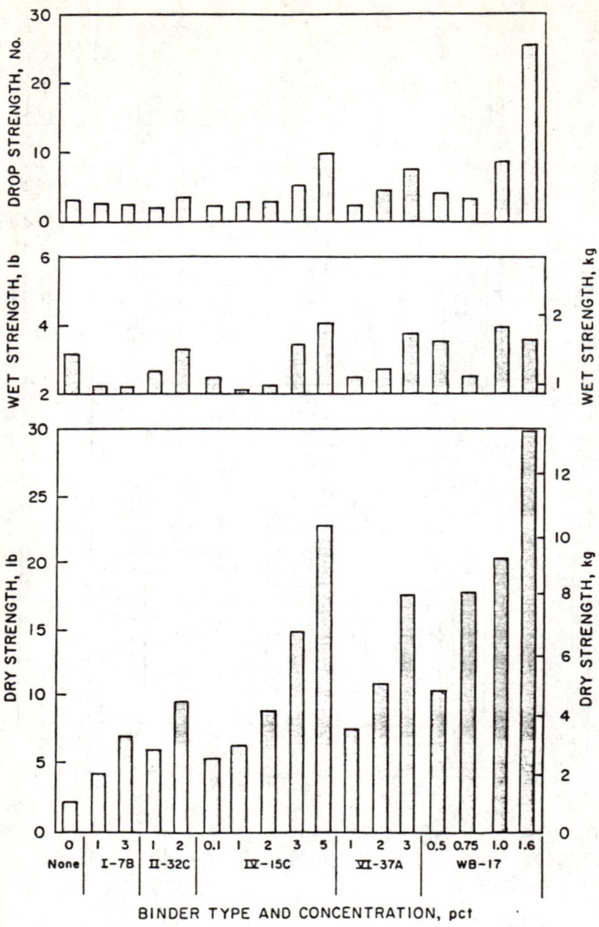

Fig. 7 - Influence of clay binder concentration on unfired pellet properties.

iron ore concentrate first with the clay and later with the soda ash. The second method involved changing the order of additions of the clay and soda ash and/or the water. The time before pelletizing the wet mixture was also varied from 1/2 hr to 3 days. Essentially the same unfired pellet strengths were obtained with all these soda ash treatment methods.

Binder Strength Versus Clay Characteristics. To predict the binder effectiveness, the pellet strength data were compared with the clay characterization data. The results in Figure 9 suggested that there was a definite correlation between dry pellet strength and some of the clay characterization variables. The highest correlation coefficient results indicated that the PWAT number and CEC-mb clay characterization variables were the best indicators of the dry pellet strengths.

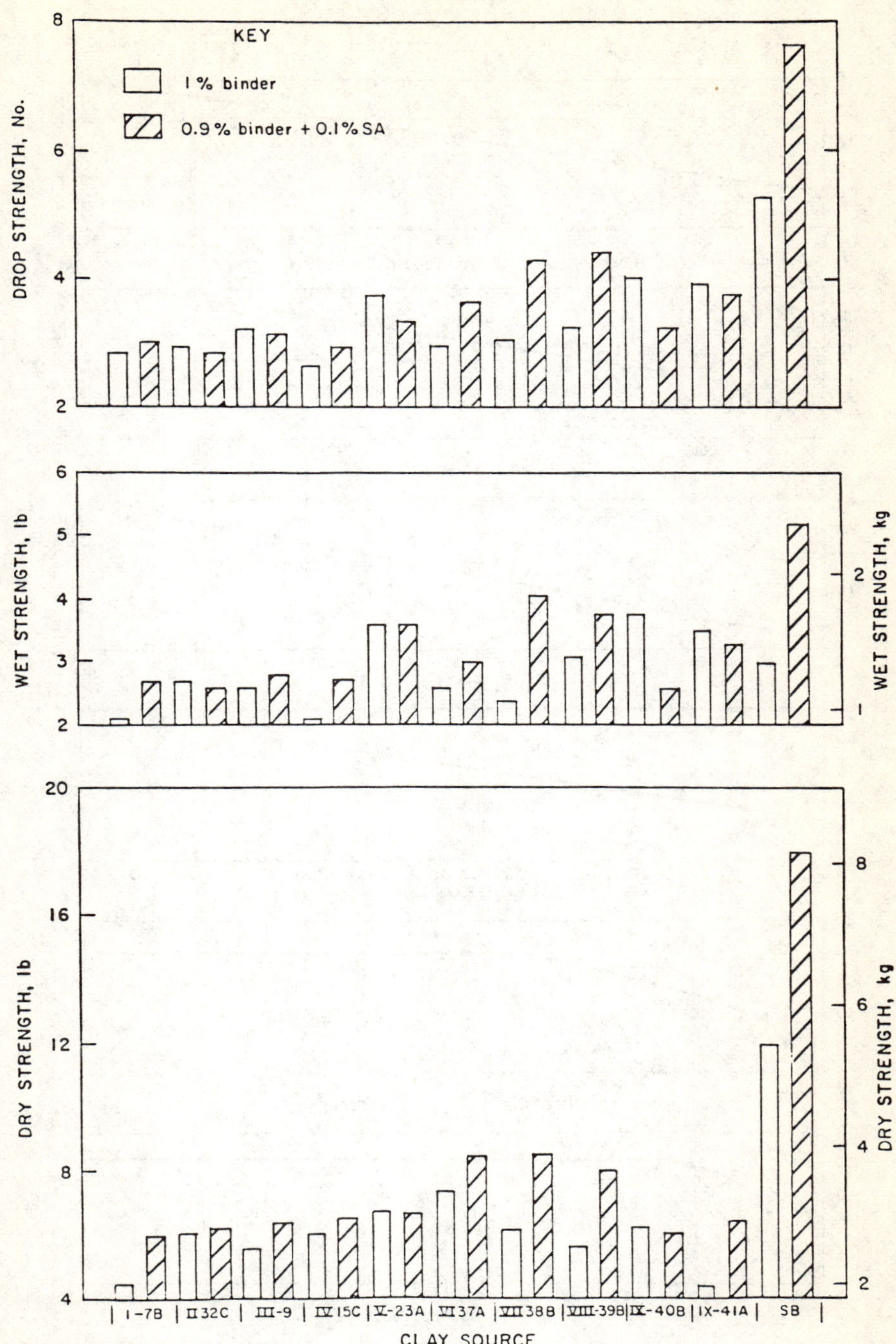

FIGURE 8. - Effect of the clay binder source with and without soda ash (SA) on the unfired pellet properties.

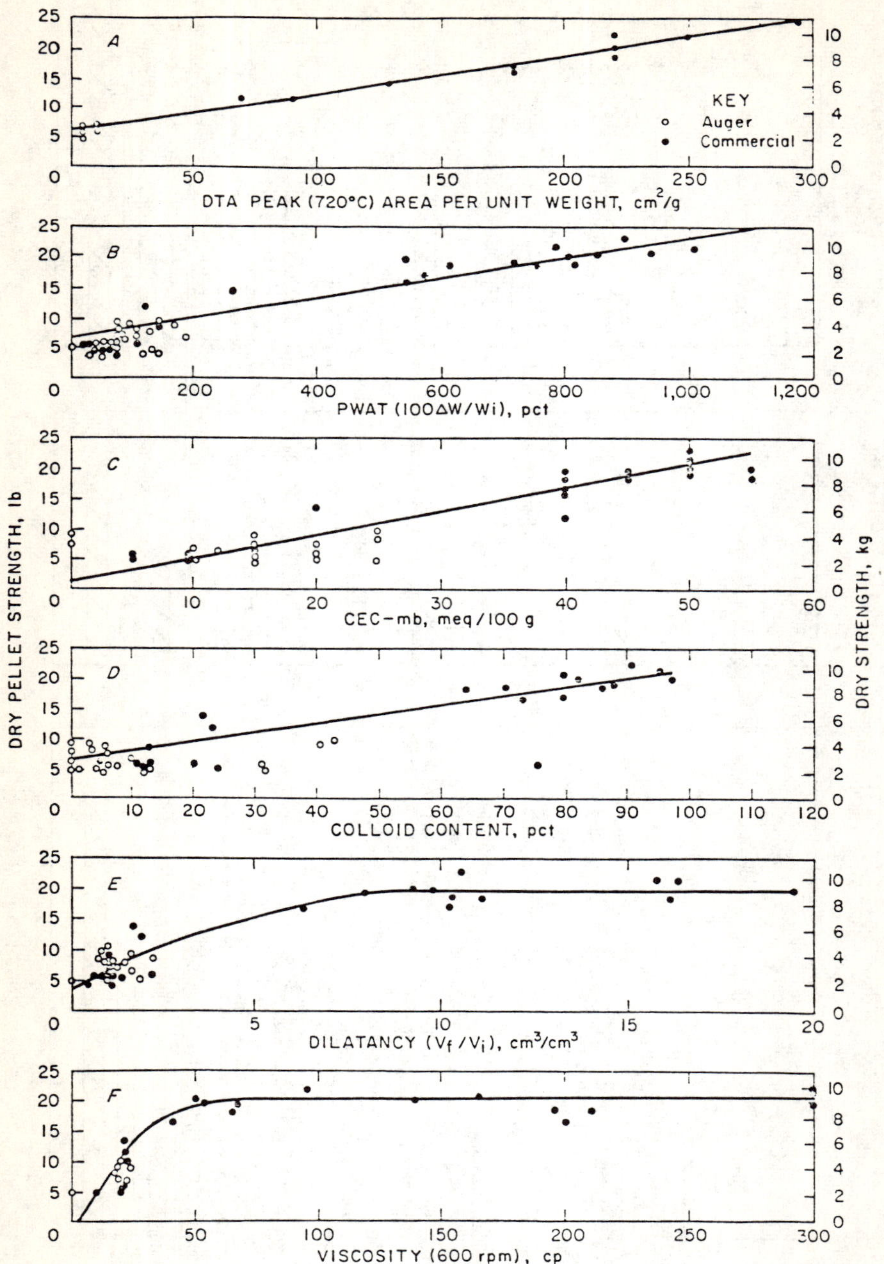

FIGURE 9. - Dependence of clay characterization variables on the dry unfired pellet strength.

TABLE IX. - Physical and Metallurgical Properties of Pellets Made From Selected Augered Minnesota Clay Samples With and Without Soda Ash

Code	Pct	Soda ash, pct	Unfired pellet properties[1]				Fired strength,[2] kg		R_{40}, pct/min[3]		t_{90}, min[4]	
			H_2O, pct	Drops, No	WCS, kg	DCS, kg	1200° C	1250° C	1200° C	1250° C	1200° C	1250° C
I-7B.....	1.00	0.00	7.4	3.0	1.0	2.0	209	317	NE	0.38	NE	282
	3.00	0	7.5	2.8	1.0	3.2	208	288	NE	NE	NE	NE
	0.99	0.01	7.3	2.9	1.1	2.9	155	286	0.49	NE	260	NE
	0.90	0.10	6.7	3.0	1.4	2.6	195	301	NE	NE	NE	NE
	0.90	0.10	7.3	3.0	1.2	2.7	173	NE	NE	NE	NE	NE
II-32C...	1.00	0	7.3	2.9	1.2	2.7	176	330	NE	0.35	NE	293
	2.00	0	6.7	5.7	1.8	4.4	231	NE	0.34	NE	333	NE
	0.90	0.10	7.2	2.8	1.2	2.8	187	283	0.49	NE	260	NE
III-9B...	1.00	0	7.3	3.2	1.2	2.5	244	303	NE	0.29	NE	358
	0.90	0.10	7.2	3.1	1.3	2.9	136	395	NE	NE	NE	NE
IV-15B...	0.10	0	7.1	2.9	1.1	2.5	173	NE	NE	NE	NE	NE
	1.00	0	7.4	2.6	0.9	2.7	191	335	0.39	0.33	285	314
	2.00	0	7.0	2.8	1.0	4.1	211	289	0.51	NE	261	NE
	3.00	0	7.4	5.7	1.8	6.7	260	301	NE	NE	NE	NE
	5.00	0	7.0	10.1	1.9	10.3	287	NE	NE	NE	NE	NE
	0.90	0.10	7.4	2.9	1.2	2.9	152	265	NE	NE	NE	NE
	1.98	0.02	7.0	3.1	1.3	4.0	195	295	NE	NE	NE	NE
	1.94	0.06	6.9	6.2	2.0	6.5	332	NE	0.45	NE	307	NE
	1.86	0.14	7.2	6.6	1.9	5.9	298	NE	NE	NE	NE	NE
	1.80	0.20	7.4	5.0	1.4	5.2	195	NE	0.46	NE	291	NE
V-23A....	1.00	0	7.1	3.7	1.6	3.0	237	298	NE	0.32	NE	302
	0.90	0.10	6.6	3.3	1.6	3.0	211	341	NE	NE	NE	NE
VI-37A...	1.00	0	7.2	2.9	1.2	3.3	216	255	0.54	0.47	232	258
	2.00	0	7.6	4.3	1.0	5.2	248	NE	0.54	NE	248	NE
	3.00	0	6.8	7.4	1.7	7.8	331	NE	NE	NE	NE	NE
	0.9	0.10	7.2	3.6	1.4	3.9	209	321	0.53	NE	267	NE
VII-38A..	1.00	0	7.2	3.0	1.1	2.8	191	NE	NE	NE	NE	NE
	0.90	0.10	6.5	4.3	1.8	3.9	276	NE	NE	NE	NE	NE
VIII-39A.	1.00	0	7.0	3.2	1.4	2.5	292	NE	NE	NE	NE	NE
	0.90	0.10	6.8	4.4	1.7	3.6	273	NE	NE	NE	NE	NE
IX-40A...	1.00	0	7.0	4.0	1.7	2.8	275	NE	NE	NE	NE	NE
	0.90	0.10	7.4	3.2	1.2	2.7	215	NE	NE	NE	NE	NE
IX-41A...	1.00	0	7.4	3.9	1.6	2.0	229	NE	NE	NE	NE	NE
	0.90	0.10	7.4	3.8	1.5	2.9	238	NE	NE	NE	NE	NE
WB-17[5]...	1.00	0	7.6	9.2	1.9	9.1	239	334	0.43	0.34	252	288
WB-600[6]..	1.00	0	8.2	7.9	1.5	8.5	253	NE	NE	NE	NE	NE
WB-800[7]..	1.00	0	7.2	3.8	1.4	3.2	148	NE	NE	NE	NE	NE
Na_2CO_3...	0	0.20	4.3	3.7	1.8	3.5	239	NE	NE	NE	NE	NE
No additive.	0	0	6.3	3.2	1.4	1.0	129	170	0.80	0.69	150	178

NE Not evaluated.

[1]The wet compressive strength and dry compressive strength were defined as WCS and DCS, respectively.

[2,3,4]These data were obtained with pellets fired at either 1200° or 1250° C; then the compressive strength, reduction rate at 40 pct (R_{40}), and time for 90 pct reduction (t_{90}) were determined. The reduction was conducted at 900° C with 30 pct CO and 0.2 pct H_2 in N_2 at 1.5 liters (STP) per min.

[5,6,7]The bentonite (WB-17) samples were dried at 75°, 600°, and 800° C, respectively.

Organic and/or Clay Binder Mixtures. The unfired pellet properties generally were improved by the use of organic binders, which is in agreement with results reported by other investigators (17-19). With 0.3 pct organic binder such as pregelatinized starches (PS-200 and PS-300), high dry pellet strengths of 41.8 and 19.6 lb were obtained, respectively (Table IX). When the organic additions were decreased to the 0.1-pct level, smaller improvements in the pellet properties were observed.

When the organic materials were added to Minnesota clays, the dry pellet strengths increased (Table X). For example, with a 0.5 pct IV-15C or ITS-46 and 0.1 pct PS-200 binder mixture, a dry compressive strength almost equivalent to that with 0.5 pct western bentonite (WB-17) was obtained. The paint rock sample (ITS-46) was preferred because it contained over 20 pct Si. The silica content of the binder is especially important in a direct reduced iron (DRI) process (20).

In general, the pellet drop numbers also were higher when the organic materials were added to clay. For example, when 0.08 pct polyacrylate (PAC) was added to 0.75 pct western bentonite binder, the drop numbers increased from 3.4 to 11.0 (Table X). The viscosities increased considerably when polyacrylate (PAC) or guar gum (GG) were added to the typical western bentonite (Table VII, last 2 rows). These results are consistent with the clay characterestics correlation data which indicated that the drop numbers were most dependent on the binder slurry viscosity.

The drop numbers were also dependent on the water content of the pellets. The results (Table IX) showed that the drop number increased considerably at the higher moisture levels, as indicated by the mixture of 0.5 pct IV-15C and 0.1 pct GG-416. Since dry usually present in the wet pellets; this moisture level was low compared with that used in commercial pelletizing plants, which usually operate at pellet moisture levels of about 9 pct.

Fired Pellet Properties

Clay Binders. The fired pellet properties were evaluated by the fired strengths, porosity, thermal shock test, reducibility, and reduction disintegration index (RDI). The greatest emphasis was placed on producing pellets with high reducibility and target strengths of over 330 lb (150 kg). The shock test was conducted at the same conditions (1/2 hr at 900°C) that were used in the first stage of the pellet induration procedure. The only pellets that broke apart in this test were those made with the typical western bentonite (WB-17) dried at 600°C or higher temperatures; more than 60 pct of the pellets broke apart in the shock test with pellets made with 1 pct of this bentonite dried at these high temperatures.

The reduction tests were conducted at 900°C with 30 pct CO and 0.2 pct H_2 according to the most recent ISO standard (21). The total gas flow rate per minute was 1.5 liters (calculated at standard temperature and pressure). The relative reduction was determined as described in reference 22. Some tests were conducted with the older ISO method (23) using 950°C and 40 pct CO. This resulted in a reduction rate per minute at the 40 pct reduction (R_{40}) of about twice as large as with 900°C and 30 pct CO. The strong dependence of the reduction rate on temperature and reductant concentration was consistent with previous research conducted in this laboratory (22).

The addition of 1 pct Minnesota clay to iron oxide concentrate lowered the reduction rate of the pellets considerably. When no binder was added to the pellet, the R_{40} value was high (last row in Table X), but

TABLE X. - PWAT Values and Pellet Properties Obtained With Different Clays and Organic Mixtures

Organic		Clay		Unfired pellet properties[1]				Fired strength, kg		R_{40},[2] pct/min		t_{90},[3] min		RDI,[4] pct		Porosity
Code[5]	Pct	Code	Pct	H_2O, pct	Drop No.	WCS, kg	DCS, kg	1200° C	1250° C	1200° C	1250° C	1200° C	1250° C	+6.3 mm	-0.5 mm	pct
ORGANIC BINDERS																
None.....	0.00	None..	0.00	6.6	3.2	1.4	1.0	129	170	0.80	0.69	150	178	NE	NE	25.6
CMC-P....	0.05	..do..	0.00	8.0	3.2	0.8	2.0	157	227	0.75	0.38	190	269	NE	NE	NE
CMC-P....	0.10	..do..	0.00	8.0	6.9	1.4	3.9	115	157	0.71	NE	178	NE	91.3	7.2	NE
CMC-P....	100.00	..do..	0.00	NE	NE	NE	NE	NE	NE	NE	NE	NE	NE	NE	NE	NE
CMC-H....	0.10	..do..	0.00	9.0	7.0	1.9	3.5	78	305	0.76	0.50	183	216	90.9	8.8	NE
CMC-H....	100.00	..do..	0.00	NE	NE	NE	NE	NE	NE	NE	NE	NE	NE	NE	NE	NE
SGC-A....	0.50	..do..	0.00	12.7	5.8	0.5	0.5	29	119	NE	NE	NE	NE	NE	NE	NE
SGC-A....	100.00	..do..	0.00	NE	NE	NE	NE	NE	NE	NE	NE	NE	NE	NE	NE	NE
SGC-J....	0.20	..do..	0.00	9.2	5.0	1.3	1.7	93	NE	NE	NE	NE	NE	NE	NE	NE
SGC-J....	100.00	..do..	0.00	NE	NE	NE	NE	NE	NE	NE	NE	NE	NE	NE	NE	NE
PS-200...	0.10	..do..	0.00	6.6	4.4	1.5	5.1	201	NE	0.57	NE	215	NE	89.1	3.5	23.6
PS-200...	0.30	..do..	0.00	7.7	10.4	2.4	19.0	124	293	0.75	NE	170	NE	90.9	3.6	NE
PS-200...	100.00	..do..	0.00	NE	NE	NE	NE	NE	NE	NE	NE	NE	NE	NE	NE	NE
PS-300...	0.30	..do..	0.00	7.3	4.4	1.2	8.9	131	349	NE	NE	NE	NE	86.4	5.6	NE
PS-300...	100.00	..do..	0.00	NE	NE	NE	NE	NE	NE	NE	NE	NE	NE	NE	NE	NE
GG-J.....	0.05	..do..	0.00	9.4	9.6	1.9	2.3	124	NE	NE	NE	NE	NE	NE	NE	NE
GG-J.....	0.10	..do..	0.00	12.5	29.2	1.8	3.4	83	NE	NE	NE	NE	NE	NE	NE	NE
GG-J.....	100.00	..do..	0.00	NE	NE	NE	NE	NE	NE	NE	NE	NE	NE	NE	NE	NE
GG-211D..	0.10	..do..	0.00	9.2	44.6	2.6	4.0	91	207	0.55	NE	205	NE	91.5	4.4	NE
GG-211D..	100.00	..do..	0.00	NE	NE	NE	NE	NE	NE	NE	NE	NE	NE	NE	NE	NE
CLAY BINDERS																
Do.....	0.00	IV-15C	1.00	7.4	2.6	0.9	2.7	225	398	0.39	0.33	285	314	95.9	2.2	23.5
Do.....	0.00	IV-15C	100.00	NE	NE	NE	NE	NE	NE	NE	NE	NE	NE	NE	NE	NE
Do.....	0.00	ITS-46	1.00	6.6	4.0	1.4	3.1	173	NE	0.48	NE	243	NE	93.5	3.9	NE
Do.....	0.00	ITS-46	100.00	NE	NE	NE	NE	NE	NE	NE	NE	NE	NE	NE	NE	NE
Do.....	0.00	WB-17.	1.00	7.5	9.2	1.9	9.1	239	334	0.43	0.34	252	288	93.1	2.6	21.9
Do.....	0.00	..do..	100.00	NE	NE	NE	NE	NE	NE	NE	NE	NE	NE	NE	NE	NE
Do.....	0.00	..do..	0.75	7.4	3.4	1.2	8.0	183	NE	NE	NE	NE	NE	NE	NE	NE
Do.....	0.00	..do..	0.50	7.0	4.2	1.6	4.9	170	420	0.56	NE	206	NE	90.8	3.4	NE
CLAY AND ORGANIC BINDERS																
PAC......	0.08	..do..	0.75	8.1	11.0	1.7	8.3	172	NE	NE	NE	NE	NE	NE	NE	NE
PS-200...	0.10	IV-15C	0.50	9.7	4.8	1.8	7.6	264	NE	0.56	NE	256	NE	94.9	2.5	23.3
GG-211D..	0.10	..do..	0.50	7.0	4.8	1.4	3.7	195	NE	0.59	NE	198	NE	94.9	3.6	NE
PS-200...	0.10	ITS-46	0.50	7.5	5.9	1.1	3.9	181	NE	0.56	NE	212	NE	NE	NE	25.6
PS-200...	0.30	..do..	0.50	7.1	4.3	1.2	10.2	123	321	0.74	NE	159	NE	91.6	4.4	NE
PS-300...	0.10	..do..	0.50	7.2	3.3	1.1	3.1	175	NE	0.53	NE	209	NE	92.1	4.1	25.4
PS-300...	0.30	..do..	0.50	7.0	4.6	1.1	8.8	131	360	0.76	NE	158	NE	90.5	4.0	NE
GG-211D..	0.10	..do..	0.50	6.8	3.3	1.5	3.2	151	363	0.62	NE	195	NE	92.4	2.7	25.6

NE Not evaluated.
[1]WCS refers to wet compressive strength and DCS refers to dry compressive strength.
[2]R_{40} refers to the rate at 40 pct reduction.
[3]t_{90} refers to the time required to obtain 90 pct reduction.
[4]RDI refers to percent +6.3 and -0.5 mm particles produced in the reduction disintegration index test.
[5]The abbreviations are CMC = carboxlyl methyl cellulose; SGC = starch graft copolymer; PS = pregelatinized starch; GG = guar gum; IV = area 4 clay; ITS = Itasca Co.; and WB = western bentonite.

R_{40} values decreased as the quantity of clay was increased. The highest reduction rates were obtained with an area VI sample and the lowest with an area III sample. The precise reasons for the lower reduction rates when different samples were added were not known, but one of the reasons was probably the presence of low-melting constituents in the samples. This observation was also supported by the decrease in R_{40} when a low-melting compound, soda ash, was added.

For pellets indurated for 15 minutes in a muffle furnace at 1,200°C, the target fired compressive strengths were obtained with all the Minnesota clay samples. With no binder, the target strength was reached only when the temperature was increased to 1,250°C, however, this decreased the R_{40} correspondingly.

Organic Binders. In general, when organic materials were used as a binder, the fired strengths were much lower than with the Minnesota clay or western bentonite samples and target strengths were not obtained at 1,200°C (Table X). The fired strengths decreased with increasing amounts of organic additives. Target strengths were obtained by increasing the firing temperature to 1,250°C, but unfortunately, the reduction rates also decreased.

Organic and Clay Binder Mixtures. The binder mixture studies were conducted with the organic materials that contained less than 3 pct sodium. The low-sodium binders were preferred because sodium is harmful to the blast furnace refractories.

When 0.1 pct organic materials were added to 0.5 pct Minnesota samples, fired (1,200°C) strengths of over 330 lb were obtained (Table X). Also the reduction rates were higher than with Minnesota clay binder alone, which is consistent with the generally higher porosities obtained with the mixtures.

The percentage of the +6.35-mm tumble fraction from the reduction disintegration index test generally was also higher with the mixtures of 0.1 pct organic material and 0.5 pct Minnesota clay samples than with straight organic binders. In general, the quantity of fines (-0.5 mm) also was lower with the mixtures than with the straight organic materials.

Summary and Conclusions

An extensive clay sampling and characterization investigation was conducted on samples obtained from former glacial lakes in the Midwest, mainly in Minnesota. Four glacial lakes were sampled in nine different areas. The sampling consisted of augering 15-cm-diam holes up to depths of 22 m. A total footage of 450 m was obtained from 40 holes. One hundred and fifty 1.5-m sample sections were selected for the laboratory studies. Most of the footage contained clay- and silt-sized particles, but most of the minerals were those from igneous and metamorphic rocks abraded by the glacier. Only a small quantity of the clay-sized particles were clay minerals (phyllosilicates), and those included chlorite, kaolinite, and some smectite. Considerably more smectite was found in conventional western bentonite binders which were studied for comparison purposes. With Midwest clays, the most smectite occurred in the Red River Valley samples, and these samples had the best bindering properties for agglomerating iron ore concentrates.

Fifteen clay characterization tests were conducted on the 30 midwestern clay samples and 15 western bentonites, and the results correlated with the pellet binding properties. The cation exchange capacity using the methylene blue procedure and the plate water absorption test were the best clay characterization techniques for predicting the efficiency of the clay sample for making pellets with high dry compressive strengths. This cation exchange capacity method was faster than the plate water absorption test, and several clay samples were evaluated in less than 1 hr. The clay slurry viscosity determination was the best test for evaluating the binder that produced the pellets with the highest drop numbers. The clay characterization tests were the most meaningful with clay samples that met the target strengths at the 1-pct addition level.

Adequate fired strengths were obtained with 1 pct Minnesota clay, and reduction rates of the pellets were similar to those made with 1 pct western bentonites. However, about 2 pct of this sample was required to produce similar unfired pellet properties as with 0.5 pct of a typical western bentonite. The addition of soda ash to Minnesota clay did not significantly enhance the binder properties.

The best combined physical and metallurgical pellet properties were obtained when organic materials were mixed with the Minnesota samples. With a mixture of 0.5 pct Minnesota clay (IV-15C) or paint rock (ITS-46) and 0.1 pct pregelatinized starch (PS-200), the unfired pellet properties were about the same as with 0.5 pct typical western bentonite. With pellets made using clay and organic mixtures, about the same reduction rates and percentages of +6.35-mm particles (RDI) were obtained as with a typical western bentonite. The Mesabi paint rock sample (ITS-46) was preferred because it contained only one-fifth of the silica as compared to western bentonite and several times more iron units.

Acknowledgements

The authors wish to express their gratitude to the following cooperators: Howard Hobbs from the Minnesota Geological Survey and Morris Eng from the Minnesota Department of Natural Resources for helping select the augering sites; Sam Dickinson from the Iron Range Resources and Rehabilitation Board for furnishing land ownership information, and the Board itself for covering part of the augering costs.

References

1. Mineral Commodity Summaries, U. S. Bureau of Mines, 1985, 76 pp.

2. R. J. Kipp, Minnesota Mining Directory, Mineral Resources Research Center, Univ. MN, 1984, 264 pp.

3. S. Schmickle, "Taconite Closing to Cost 450 Jobs," Minneapolis Star and Tribune, p. 1A, May 12, 1985.

4. S. C. Panigraby and M. Riguad, "Substitution of Bentonite With Peat Moss and Its Effect on Pellet Properties," presentation at 4th Int. Symp. on Agglomeration, Toronto, Canada, 7 pp., 1985.

5. J. H. Aase, and G. E. Leonhard, Field Investigation and Testing of a Minnesota Clay Resource for Iron Ore Pellet Bonding, BuMines, RI, 7206, 1968, 7 pp.

6. R. L. Bleifuss, Analysis of Clay I.R.R.R.C. Drilling Project Cook Area, St. Louis County, Univ. MN, Sch. Mines Exp. Sta., 1963, 7 pp.

7. J. S. Wakeman, E. C. Bogren, and D. J. Kawulok-Englund, " A Review of Bentonite Test Methods Being Developed by the Iron Ore Industry," pp. B120-132, in 3rd Int. Symp. on Agglomeration, 1981.

8. Standard Test Method for Water Absorption of Bentonite Porous Plate Method, American Society for Testing and Materials, ASTM E 946-83, 1983, 3 pp.

9. W. E. Parham, Clay Mineralogy and Geology of Minnesota's Kaolin Clays, MN Geol. Surv., SP-10, Univ. MN, 1970, 142 pp.

10. I. Auer, and R. L. Thayer, "Bentonite Update: Production, Reserves, Quality Control, and Testing," Min. Eng., 31, (1979), pp. 1467-1471.

11. L. A. Haas, J. A. Aldinger, R. L. Blake, S. A. Swan, "Characterization and Evaluation of Midwest Clays for Iron Ore Pellet Bonding," in Proc. of Minnesota AIME Sym., 1986.

12. D. J. Pluth, R. S. Adams, Jr., R. H. Rust, and J. R. Peterson, Characteristics of Selected Horizona From 16 Soil Series in Minnesota, Agriculture Experiment Station, Univ. MN, Tech. Bull. 272, 1970, 34 pp.

13. F. Hoffman, "Clay Minerals for Foundry Sands," The British Foundryman, 1959, 165 pp.

14. Procedures for Collecting Soil Samples and Methods of Analysis for Soil Survey, U. S. Soil Conservation Service, Soil Survey Investigation Report No. 1, 1972, 17-19 pp.

15. Standard Procedure for Field Testing Drilling Fluids, American Petroleum Institute, RP 13B, 10th ed., Washington, DC, 1984, 17-18 pp.

16. _____, "The Rheology of Oil-Well Drilling Fluids," American Petroleum Institute Bull. 13D, 1st ed., Washington, DC, 1980, 17-18 pp.

17. F. L. Shusterich, "Production of Peridur Pellets at Minorca," paper presented at 58th Annu. Meeting of Minnesota Sec. of AIME, Duluth, MN, 1985, 12 pp.

18. J. A. Clum, Substitutes for Western Bentonite in Magnetic Taconite Pellets, AIME/SME Preprint 76-B-11, 1976, 29 pp.

19. R. P. de Souza, C. F. Mendonca, and T. Kates, "Production of Acid Iron Ore Pellet for Direct Reduction, Using an Organic Binder," Min. Eng., 36, 1985, 1437-1431 pp.

20. International Organization for Standardization, Iron Ores--Determination of Relative Reductibility, ISO/DIS 7215, 1983, 8 pp.

21. L. A. Haas, J. C. Nigro, and R. K. Zahl, Utilization of Simulated Coal Gases for Reducing Iron Oxide Pellets, BuMines RI 8997, 1985, 14 pp.

22. International Organization for Standardization, Iron Ores--Determination of Reducibility, ISO/DIS 4695, 1982, 8 pp.

Process Mineralogy Applications to Metallurgical Products

VACUUM ELECTRON BEAM REFINING OF CLEAR COPPER

A. W. Fletcher and W. J. Mitchell

9422 East Placita Eunice
Tucson, Arizona 85715

Copper produced in Duval's CLEAR process by electrowinning from a strong chloride solution contains silver and certain other impurities that have necessitated treating it as blister grade, i.e. it must be melted, cast into anodes and electrolytically refined to cathode grade copper. An alternate method of refining is to remove certain impurities including silver, by selective volatilization at an elevated temperature in a high vacuum, conditions which are effectively provided by an electron beam furnace. The procedures by which the dendritic CLEAR copper product is melted and pre-refined in preparation for further treatment are outlined, and results are given for electron beam refining tests in laboratory and pilot scale equipment. Samples of refined metal were successfully drawn to rod and subsequently to fine wire. Detailed chemical analyses and the results of physical tests show the potential this process (patent pending), has to produce metal which meets the specifications of OFHC copper.

Introduction

Duval Corporation's successful operation of the CLEAR process for copper production from chalcopyrite concentrates for five years, reaching a scale of about 100 t/day, is clearly a landmark in chloride hydrometallurgy. The process decribed by Schweitzer and Livingston (1) involves dissolution of the chalcopyrite concentrate in cupric chloride solution, nearly saturated in chloride, to give a stable cuprous chloride solution from which copper is electrowon in cells of special design. Copper is deposited onto cathodes as dendritic crystals that fall onto a moving belt which continuously discharges the product. During electrowinning approximately half the copper in solution is reduced to metal at the cathode while at the same time the remaining cuprous ion is oxidized to cupric at the anode.

Cathode reaction: $Cu^{+} + e \rightarrow Cu^{\circ}$

Anode reaction: $Cu^{+} \rightarrow Cu^{2+} + e$

The anolyte containing cupric ion is recycled to leaching as shown schematically in Figure 1.

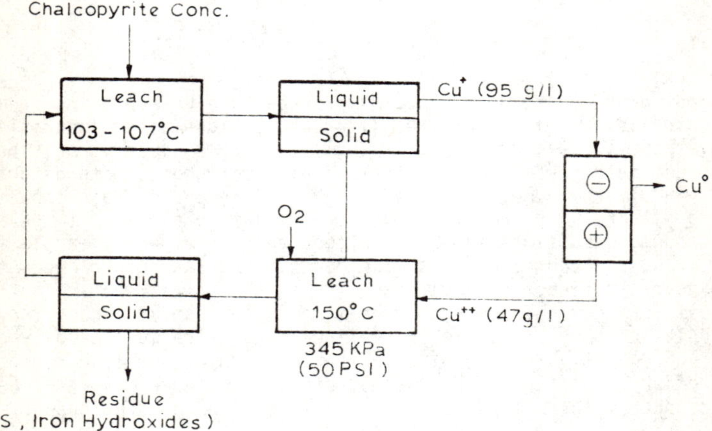

Fig. 1 - Simplified flowsheet of the CLEAR process

The 2-stage leaching effectively solubilizes over 99% of the copper with retention times of approximately 90 minutes per stage. The electrowinning step is also highly efficient in energy terms, (a power requirement of 1.8 Kwh/Kg Cu has been demonstrated on a commercial scale), and has many advantages compared with conventional electrowinning practice, e.g.:

In electrowinning from cuprous solution the current requirement is halved;

Use is made of the anode reaction to regenerate the lixiviant;

No handling of cathodes is required since the dendritic copper product is discharged from the cell continuously on a belt;

High current density operation (500 A/m^2) is practicable thereby reducing tankhouse size.

In evaluation the capital and operation costs of the process for comparison with pyrometallurgy and refining it is necessary to consider how the CLEAR product relates to conventionally produced cathode copper. Although CLEAR copper assays over 98% Cu, and with some modifications 99% Cu, it has to be treated as blister grade and therefore requires electrolytic refining to produce metal of conventional cathode quality. One of the main impurity elements which co-deposits with copper in the CLEAR tankhouse and which cannot be removed by fire refining is silver. In fact there is a high recovery of silver from concentrate to leach liquor and hence to the cathode deposit. Gold on the other hand does not go into solution during leaching and remains in the leach residue, from which it can be recovered if present in sufficient quantity.

Since CLEAR copper is of much higher grade than blister copper, it was of interest to consider alternatives to conventional electrolytic refining. Various fire refining procedures were considered including chlorination and it was at this point that we became aware of a suggestion made some years ago (2) that there is considerable potential in vacuum refining, i.e., in the removal of impurities by selective distillation, in an electron beam furnace. This report briefly outlines the tests carried out to prove the feasibility of such a refining route.

Previous Work

In a presentation to the Pacific Southwest Minerals Conference in March 1979 Hunt and McLean (2) stated that the process of metal purification by high vacuum distillation of impurities in electron beam heated hearths had been successfully applied commercially in the production of niobium and tantalum. In addition the process had been shown to be technically and economically feasible for the refining of a number of other metals such as Zr, Hf, Ti, and Fe, Ni and Co based alloys. With this background experience, supported by theoretical considerations, it was claimed that the process could have other applications, a good example being for silver removal from hydrometallurgically produced copper, and cost estimates were provided to justify the claim. In fact the feasibility of separating the more volatile silver from copper by electron beam distillation had been demonstrated some years earlier in the work carred out by Santala and Adams (3). Working with binary silver-copper alloys in wire form and containing from 8 to 42% Ag these workers studied the kinetics and thermodynamics of silver and copper evaporation using a lab scale 10 Kw Airco-Temescal constant current power source to operate a self-accelerated 180° bend electron beam gun. The operating pressure ranged from 1×10^{-5} to 10×10^{-5} torr (0.01 to 0.1, microns Hg). It was found that the classical Langmuir equation could be used to predict evaporation rates in the chosen binary system where activities and activity coefficients, and therefore the partial vapor pressure of each component, were known with sufficient accuracy. Santala and Adams simulated steady state conditions by continuously feeding the Ag-Cu wire to a graphite crucible exposed to electron beam radiation. The wire feed rate was controlled to maintain a fixed level of molten copper in the crucible while vapor was condensed as a film on a glass plate situated above the crucible. The crucible contents were sampled during evaporation by briefly dipping into the melt a cold gold surface which was later removed for analysis. The results given in Table 2 of the Santala and Adams paper have been re-arranged in Table I of the paper to show the effect of temperature and silver content of the feed on the selectivity of the distillation process. This is defined as a volatilization ratio, i.e. the ratio of silver in the condensate (which under steady state conditions equals the silver content of the feed) divided by the silver content of the crucible contents. It can be seen that the ratio increases, i.e. that the relative distillation rate of silver can be enhanced, by operationg at lower temperatures and lower feed silver contents. In the case of CLEAR copper the silver content would be in the region of 400-500 ppm which on theoretical grounds should give a relatively high volatilization ratio. This is important because the higher the ratio, the more selective the distillation process would be and the smaller the amount of copper distilled off with the silver. The amount of metal volatilized at a given pressure is a function of time and temperature, in other words electrical power, which largely controls process costs.

Table I. Variation of Silver Volatilization Ratio with Temperature and Silver Content of Feed. (After Santala and Adams (3).)

% Ag in feed (A)	Temp °K	% Ag in crucible (Calc) (B)	Volatilization Ratio Ag in condensate (i.e. feed)/ Ag in crucible, i.e. (A)/(B)
	1730	0.18	44.8
	1780	0.205	39.3
8.06	1845	0.243	33.2
	1890	0.274	29.4
	1940	0.305	26.4
15.7	1770	0.425	36.9
	1960	0.678	23.2
29.85	1695	0.796	37.5
	1955	1.550	19.3
42.24	1740	1.56	27.1
	1925	2.47	17.1

The volatilization ratio R may be expressed as the ratio:

$$R = \frac{(Ag)g/(Cu)g}{(Ag)\ell/(Cu)\ell}$$

Where g and ℓ refer to metal concentrations in the gas and liquid (source melt) phases:

$$\text{i.e. } R = \frac{(Ag)g}{(Ag)\ell} \times \frac{(Cu)\ell}{(Cu)g}$$

For a fairly pure copper like CLEAR copper where impurities are in the ppm range we can say that

which is the definition used above and in our studies.

To evaluate the feasibility of impurity removal by vacuum distillation in an electron beam furnace the following thermodynamic calculations provided an estimate of the minimum impurity levels that could be expected.

$$P = P°a$$
$$P = P°\gamma N$$
$$P/P° = \gamma N$$

where P = partial pressure of the impurity in copper.
P° = partial pressure of the pure impurity.
a = activity of the impurity in copper.
= activity coefficient of the impurity.
N = mole fraction of the impurity in copper.

For an ideal system we can equate the partial pressure of the impurity with the operational vacuum chamber pressure (Pch) at a maximum temperature of 2000°K.

i.e., $P = Pch$
and $Pch/P° = \gamma N$

At an operating pressure of 2×10^{-7} atmospheres (0.15 microns Hg) the calculated equilibrium values for various impurities is shown in Table II.

Table II

THERMODYNAMIC ESTIMATE OF FINAL IMPURITY LEVELS

Impurity	γ(a)	Impurity ppm Calculated	Actual
Ag	3	2	8
Pb	5	0.1	∼1
Bi	1	2	∼0.3
Fe	19	300 (b)	NA

(a) Approximate published values.
(b) Indicates that refining does not occur.

The calculated results, which were in reasonable agreement with later experimental results, show that Ag, Pb and Bi can be removed to low levels while iron will remain.

The question arises as to how much copper will volatilize during the vacuum refining step. Past observations (3) had shown that as an approximation one-half of the silver impurity was removed for each 2% of the copper volatilized.

Thus $x/y = 2^z$ (i)
Where x = initial silver concentration
y = final silver concentration
z = number of 2% copper volatilizations

Thus to reduce the silver content from 545 ppm to 2 ppm, eight, 2 percent copper volatilizations would be required and 15 percent of the copper would be volatilized. This was deemed acceptable from both a materials handling and energy viewpoint.

Iron Removal

Since iron cannot be removed by volatilization, it would have to be removed by oxidation in a preliminary fire refining step and the following criteria provide an estimate of the extent to which such removal would be feasible.

$$Fe + 1/2\ O_2 \rightarrow FeO$$

$$\underline{O} \rightarrow 1/2\ O_2$$

$$Fe + \underline{O} \rightarrow FeO$$

At equilibrium $\Delta G° = -RT\ln K$

$$\text{where } K = a_{FeO}/a_{Fe}\cdot a_{\underline{O}}$$

$$e^{-\Delta G°/RT} = a_{FeO}/a_{Fe}\cdot a_{\underline{O}}$$

$$a = \gamma N$$

$$N = (\text{ppm solute})(\text{MW solvent}) \times 10^{-6}/(\text{MW solute})$$

$$a_{Fe}.a_{\underline{O}} = a_{FeO}/e^{-\Delta G°/RT}$$

$$a_{Fe}\cdot a_{\underline{O}} = 1/e^{-\Delta G/RT} \text{ where } a_{FeO} \simeq 1$$

With this relationship it is possible to calculate the equilibrium iron content of copper at different oxygen contents. Thus at an O_2 content of 9500 ppm, which is close to the maximum solubility of oxygen in copper at operating temperature, the iron content would be 20 ppm. However if fluxes are added, a_{FeO} would be reduced to about 0.3. Thus $a_{Fe}\cdot a_{\underline{O}} = 0.3/e^{-\Delta G/RT}$ From this we calculate that the limiting iron content of the copper would be 5 ppm in equilibrium with a slag giving an FeO activity of 0.3.

Experimental

Laboratory Scale Tests

Preliminary experiments were carried out in a lab scale 5 kw electron beam furnace with a graphite crucible of approximately 57 mm diameter x 23 mm depth holding a charge of 300-500g. A general view of the furnace is shown in Figure 2. The transverse gun gave a beam which was deflected

Figure 2. Lab scale electron beam furnace

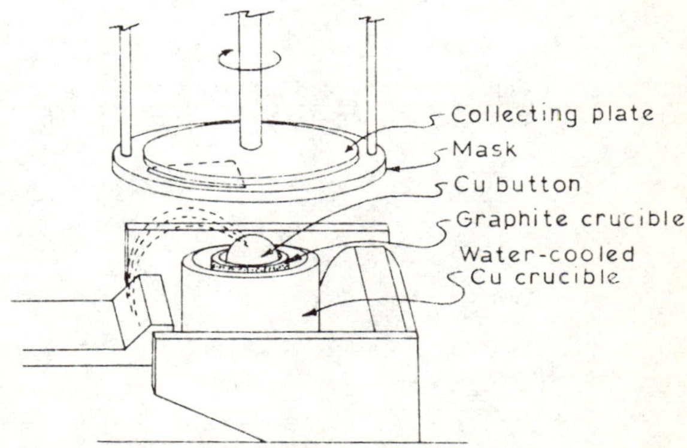

Figure 2A. Lab Electron beam Furnace, Orientation of beam and collecting plate.

through 180° so that a rotating collecting plate could be located directly above the crucible for sampling condensed vapor (Figure 2A). Early attempts to melt copper buttons prepared from CLEAR copper showed that it was essential to reduce the oxygen content of the copper to below about 200 ppm otherwise there was bubbling and excessive spitting, sometimes quite violent, in the molten copper and frequently this would cut out power to the gun. The bubbling was due to reaction between the oxygen and graphite producing carbon oxides.

It was found that satisfactory buttons could be obtained by melting CLEAR copper in a graphite crucible, in an electric furnace; however the graphite had to be of high purity otherwise the metal would readily pick up impurities from crucible walls. The weighed button was placed in the graphite crucible in the electron beam furnace, the door, which had a glass observation window, sealed, and the furnace evacuated by operationg first the backing pumps and then the oil diffusion pump till a vacuum of about 0.1 microns Hg (10^{-4} torr) was reached. The electron beam gun was then turned on at low power, with the beam directed onto the button and the power gradually increased until the button melted. At this point there was no visible sign of any glow above the melt indicating that vaporization was not occurring. The beam power was then increased to a pre-set value known to give measurable rates of distillation and the period of duration of the beam power at this value recorded. At this level a characteristic greenish colored glow was visible above the melt and copper could be seen condensing on the collecting plate. Although the control of beam power was not very accurate it was possible to demonstrate a relationship between the silver content of the final button and evaporation time or percent weight loss of the button at a fixed beam power, as shown in Figures 3 and 4. In Figure 4 the experimental result is compared with a theoretical curve assuming a volatilization ratio of 37.5. It can be seen that the shape of the Ag content versus weight loss (i.e., copper evaporated) curve is similar to that predicted from equation (i).

A more detailed result of these early tests is given in Table 3 where Stage I refers to the analysis of metal obtained by melting CLEAR copper in a gas fired furnace. After a 15 minute period of oxidation a coagulant was added and the surface slag was skimmed off to remove iron. The melt was found to have the composition shown in Stage II, Table III. Reduction of

TABLE III

Impurities in Copper after various Stages of Refining (Lab Scale)
(All concentrations in parts per million, ppm)

Stage	Se	Te	Bi	Sb	As	Pb	S	Ni	Fe	Ag	O_2
I	<2	<2	$\sim$4	5	7	$\sim$58	35	4	$\sim$200	$\sim$240	1554
II	<2	<2	$\sim$4	2	7	$\sim$58	28	4	26	$\sim$240	8400
III	<2	<2	0.1	5	5	<1	<2	5	6	$\sim$1	13

the melt by covering it with a 1 inch layer of graphite for about 15 minutes reduced the oxygen content to 204 ppm. The metal was poured into molds giving buttons weighing 300-500 g. These buttons were placed in a graphite crucible in the 5 kw electron beam furnace. At a vacuum of 0.1 microns Hg

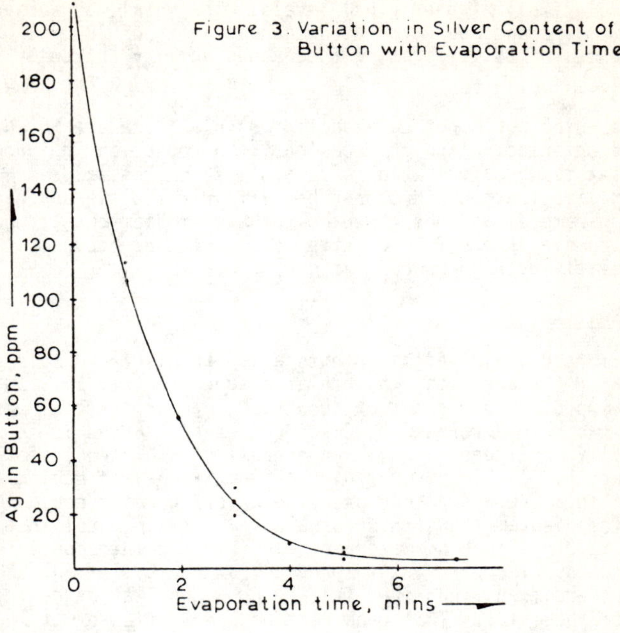

Figure 3. Variation in Silver Content of Button with Evaporation Time

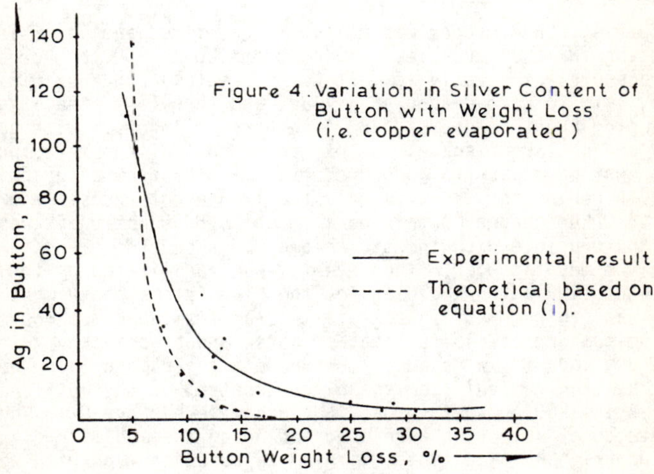

Figure 4. Variation in Silver Content of Button with Weight Loss (i.e. copper evaporated)

(10^{-4} torr), the buttons were first melted under low EB power (about 1.5 kw). The power was then increased to about 4.7 kw to give the characteristic green glow indicative of metal vaporization. The analysis of the metal remaining after 18% weight loss (i.e., 18% of the metal had been vaporized) is shown in Stage III, Table 3. The results demonstrate the marked reduction in the content of volatile impurities, Ag, Pb, Bi, S and O_2 despite the relatively high original levels. The metal was subjected to standard ASTM tests and found to have a conductivity of 101.2 TACS and a Rockwell hardness of 35, indicating good annealability characteristics for wire drawing.

Elements which would not be expected to volatilize are Sn, Ni and Fe and this was confirmed, although occasionally a reduction in iron content was noted but this was attributed to removal as an iron oxide scale rather than by volatilization. CLEAR copper has a fairly low tin and nickel content so non-removal was not expected to be a problem but it would be necessary to remove iron before electron beam refining because it could be present in relatively high concentrations.

Pilot Scale Tests

For the larger scale refining tests a 250 kw Electro-Metals electron beam cold hearth furnace with a hearth area approximately 23 x 10 x 1 inch depth and an expected throughput of about 180 lb/hour of metal was used. This furnace has been described by C. d'Hunt et al (4) and is shown diagrammatically in Figure 5, while an overall view of the installation is given in Figure 6. For copper refining a graphite hearth having the dimensions given above was placed over the existing water cooled copper hearth. As the diagram shows this furnace is fed with metal in the form of billet, (e.g., 3 or 4 inch diameter) or wire bar cut into convenient lengths of 6 to 12 inches. The plan was to produce refined billet for subsequent drawing to rod and then fine wire, this being a good way of demonstrating the quality of the metal. This required the preparation of at least 500 lb batches of metal in suitable form.

To do this, CLEAR copper was melted in gas fired and later induction furnaces using various crucibles holding from 100 to 800 lb charges. The general procedure was firstly to melt CLEAR crystals under oxidizing conditions, skim to remove iron as a slag, and then lower the oxygen content with a reducing agent before casting into bars or ingots. With gas firing, especially on the larger scale, the oxygen content of the melt increased repidly; it was necessary to add a reductant in the form of graphite or charcoal even during the iron removal step to prevent excessive oxidation. In fact with CLEAR copper containing as much as 8000 ppm O_2 it was necessary to charge the crucible with the stoichiometric amount of graphite powder mixed with the metal powder. Iron removal was facilitated by the addition of a siliceous flux which agglomerated the slag into a pasty mass which was relatively easy to remove. In some tests slag forming compounds such as silica, aluminum and calcium silicates, borax and proprietary materials used by the foundry industry were added to the melt to promote iron removal. Following the iron removal step the oxygen content of the metal had to be reduced to below 200 ppm before casting into wire bar or 20 lb ingots. Various methods of reduction were used including the addition of graphite, charcoal, proprietary reductants, lithium capsules and sparging with propane-argon or hydrogen-argon mixtures. With gas firing it was not possible to obtain satisfactory reduction of the melt by any of the methods even though the original silicon carbide crucibles were replaced with clay-graphite crucibles. During the course of this work it became apparent that quite a dramatic increase in iron content of the melt could occur at low metal oxygen concentrations and that there was a correlation between

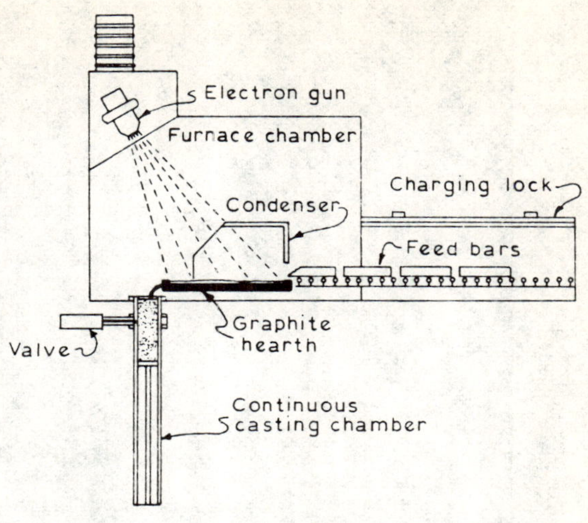

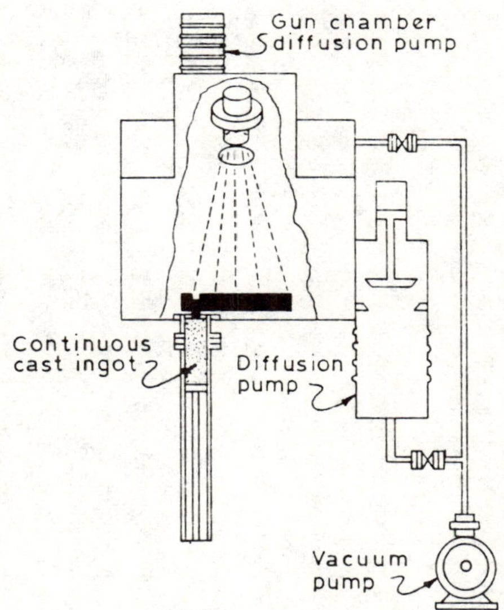

Figure 5. Electro-Metals electron beam furnace general layout

Figure 6. Electro-Metals Electron Beam Furnace Installation

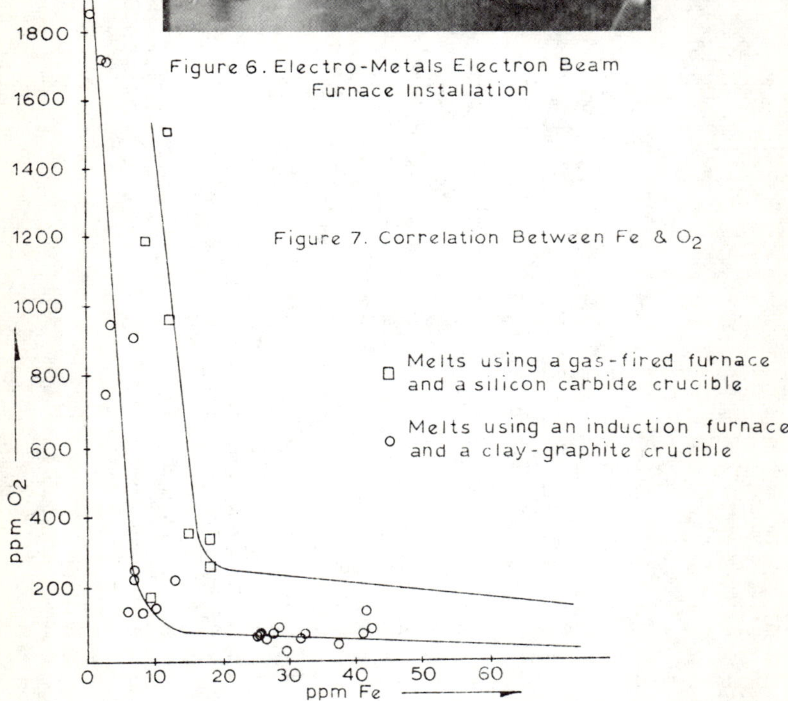

Figure 7. Correlation Between Fe & O_2

the two parameters as shown in Figure 7 and as predicted by theory. The source of the iron was in all probability the iron compounds in crucible walls and in solid reductants added to the melt. Very little work seems to have been done on the removal of iron in copper, down to the low levels (i.e., about 5 ppm) we hoped to achieve. Toshio Oishi et al (5) show that a relationship very similar to that shown in Figure 7 exists at iron contents from 200 to 1000 ppm and copper oxygen contents of 200 to 40 ppm.

The objective of removing iron to low levels and maintaing at this low level during the subsequent oxygen removal step was finally achieved by adopting a two-stage melting strategy using induction furnaces. The induction furnaces used were a 150 kw Inductotherm VIP, Powertrak 200-10 holding a 500 lb crucible and a 50 kw Ajax Magnethermic holding a 90 lb crucible. Firstly the metal was melted in a silicon carbide crucible and the melt allowed to stir in air until the oxygen content was raised to at least 1000-2000 ppm. The oxygen content was determined from pin samples using a LECO analyzer. The selected flux or slag coagulant was added and then skimmed off after allowing a contact time of about 15 minutes. The whole batch of metal of about 500 lbs was treated in this way and cast in ingots. The crucible in the induction frunace was then replaced with one machined from high purity graphite. The copper metal treated to remove iron was remelted in the graphite crucible and the melt was covered with a 1/2 inch layer of pure granular graphite. Power to the induction furnace was then adjusted to a level which gave gentle stirring of the metal for a period of about 10 to 15 minutes. The power was then turned down and the metal poured into graphite molds for casting into billets of 3 inch diameter x 12 inches length. This metal contained from 200-300 ppm O_2 and about 10 ppm Fe and was the feed to the electron beam refining furnace.

The treated metal now in the form of billet or ingots was placed in the charging lock of the electron beam furnace and the pump down commenced. When the desired pressure was reached, normally about 0.1 microns Hg, the electron beam was switched on at low power and the controls were adjusted to drip mist the metal onto the hearth and maintain the metal in molten state so that it flowed across the hearth into the continuous casting chamber, (Figure 5). With the controls it was possible to focus the beam and adjust its position so that it rapidly scanned the end of the billet and hearth area according to a prearranged programme. In the early tests it was convenient to first pass the metal through the furnace at low beam power to degas the metal, and cast it into a three inch diameter ingot. From the analysis given in Table IV it can be seen that the oxygen content of the metal was reduced to 39 ppm but that under these low beam power conditions there was little or no removal of volatile impurities. The metal was then passed through the furnace three times in three evaporation runs at a power rating of about 90 kw. Attempts were made to evaporate about 10% of the metal in each pass but this was difficult to control with small 200 lb. batches. Thereafter, a final casting run was made wherein the EB power was only sufficient to melt the metal and allow it to pass across the hearth into the casting chamber. During the evaporation runs, the evaporated metal was condensed onto a radiantly cooled graphite plate held about 1 ft. above the hearth surface. The analyses of the metal after the degassing run, the three evaporation runs and the final ingot casting run are given in Table IV. The results show that elements such as Bi, Pb, Te, and Se are rapidly distilled out of the copper in the EB furnace despite the relatively high starting amounts of impurities. After only the first evaporation run, the concentrations of these elements are reduced to very low levels. In the case of silver, the concentration was reduced from 450 ppm to 161, 48 and 13 ppm after the 1st, 2nd and 3rd evaporation stages respectively.

TABLE IV

Pilot Scale Refining of Copper in an Electron Beam Furnace
Analytical Results
(All concentrations in ppm)

Stage	Se	Te	Bi	Sb	As	Pb	S	Sn	Ni	Fe	Ag	Zn	O_2
Degas Run	0.1	2.0	1.5	4	5	13	6	12	10	11	450	3.5	39
1st Evap Run	<0.1	0.5	<0.1	4	4	0.4	10	12	11	12	161	2.2	47
2nd Evap Run	<0.1	<0.1	<0.1	3	5	<0.1	9	12	11	14	48	--	--
3rd Evap Run	<0.1	<0.1	<0.1	4	5	<0.1	6	11	--	15	13	0.5	45
Final Cast Ingot	<0.1	<0.1	<0.1	3	5	<0.1	--	12	10	14	16	<0.1	--

The final ingot produced in this run was drawn to a 5/8th inch bar and submitted for tensile testing according to ASTM spec E-8.

In the results given in Table V, the data for the electron beam refined copper are compared with those for a commercially produced rod for wire drawing. The results for the EB refined metal compare favorably with those reported for high grade OFHC copper which are: UTS - 32,000 psi, Elongation - 50% (max) and % Reduction in Area = 94 to 95%.

TABLE V. Summary of Tensile Test Results

Test Sample: length 1 in, diameter 0.25 in

Strain Rate: 0.1 in/min

Specimen		UTS (psi)	Yield Strength 0.2% OS (psi)	% Elongation in 1 inch	% Reduction In Area
Commercial	A	32,900	--	50	68
	B	32,575	14,400	50	72
EB Refined	A	32,790	16,560	50	94
	B	32,590	16,500	53	94

A and B designate two duplicate samples which were analyzed.

A larger scale test similar to that described above was carried out starting with the melting of 800 lbs of CLEAR crystals and refining for iron and oxygen removal using the 50 kw induction furnace. Five hundred lbs of the metal treated in this way in the form of three inch diameter billet were then refined in the electron beam furnace. The analyses of the starting metal and that after each of the five passes through the furnace are given in Table VI. Attempts were made to minimize the amount of metal volatilized by careful control fo beam power but this was not entirely successful. Thus

TABLE VI. RESULTS* OF LARGE SCALE TEST

Sample No.	Group 1			Group 2			Group 3	Group 4			Group 5	Oxy-gen	Anneal-ability RF	Conduc-tivity IACS %
	Se	Te	Bi	Sb	As	Pb	S	Sn	Ni	Fe	Ag			
CLEAR Crystals	0.6	4.0	1.4	3.4	5.0	1.3	32.0	0.5	<1	>19	573			
CLEAR Ingot	0.6	4.4	2.1	1.9	7.3	0.9	9.0	0.3	1.0	9.9	545	125		
E. Beam														
Pass 1	0.6	1.7	0.1		8.8	0.2	5.0	0.2		9.8	357			
Pass 2	0.5	1.3	0.1		7.3	0.3	7.0	0.3		9.1	255			
Pass 3	0.3	0.4	0.1		6.6	0.2	8.0	0.3	6.0	13.8	85			
Pass 4	0.8	0.2	0.1		8.1	0.2	8.0	0.4	11.0		35			
Pass 5	1.0	0.3	0.1		7.9	0.3	7.0	0.3	8.0	11.6	16			
Rod														
Lab 1	0.1	<0.1	<0.1		7.4	0.4					6.0			
Lab 2	<0.2	<2.0	<0.1	<1.0	8.0	<1.0	3.0	<2.0	>20.0	8.0	7.0	9.0	38	99.8
Lab 3		<2.0	<0.3	<2.6		2.5		<1.0	18.0	6.8	11.0	14.0	40	99.0
Lab 4	<0.1	<0.2	<0.1	1.0	>7.0	<0.1	3.5	<0.1	>8.0	7.0	6.3	12.0	37	100.0
Lab 5	<1.0	<1.0	<1.0	1.0	9.3	2.0	4.5	1.5	11.5	7.5	10.0	10.0	34	101.3

*Analyses in ppm.

in pass 2, for example, it seems that the beam power was too low judging by the high silver content of the metal.

The refined metal in the form of two, three inch diameter bars were hot rolled to 5/16 inch diameter rod, which in turn was drawn into magnet wire ranging in size from 44 to 36 gage. Samples of the rod were sent to 5 specialist commercial laboratories, and the results of their chemical analyses and physical tests are given in Table VI. The metal was contaminated with nickel during the electron beam refining, probably due to the fact that the Electro-Metals installation normally handles nickel alloys. The relatively high arsenic content in the original CLEAR copper crystals is rather untypical and the results show that it is not removed during melting or electron beam refining.

It can be seen from Table VI that there was reasonably good agreement in the chemical analyses reported by the various laboratories. Also there was general agreement that hot rolling resulted in a sound rod with a very good surface finish. An outstanding feature was the very low hardness values obtained, lower than that generally of the metal found in rod currently available. This is considered to be a desirable property for magnet wire applications. It was noted that the grain size was much larger than that found in commercially produced rod. The conductivity was not as good as that obtained in other tests and can be attributed to the rather high arsenic and nickle contents. The fine wire drawing tests were successful and several 5 to 10 lb reels of wire of gage 36 to 44 were producted. Breakages during drawing were largely attributed to the presence of slag inclusions, however electron microprobe examination showed the presence of some unusual features including particles of tungsten compounds at break points.

Discussion

The process for refining CLEAR copper by melting in induction furnaces under oxidizing and then reducing conditions followed by the removal of volatile impurities in an electron beam furnace and casting directly into billet or rod forms the basis of US Patent Application No. 503376 (6). In order to remove silver efficiently to the low levels required, a multistage method of refining with recycle of the condensed vapour from later stages is proposed and this forms the basis of US Patent Application No. 503377 (7). It is proposed that the molten copper of low oxygen content from the second induction furnace will be fed to the first hearth of a multihearth electron beam furnace via a conventional barometric seal leg. In the electron beam furnace the metal would flow from hearth to hearth by gravity and each hearth would be heated by appropriately positioned transverse electron beams which are deflected through 180° and programmed to scan the melt surface. The vapour from each hearth will be condensed as liquid on a surface placed above the hearths and this metal would flow by gravity to a preceding hearth. In this way it should be possible to concentrate the recycled volatile components by redistillation. Preliminary estimates indicate that it should be possible to reduce the silver content from 400 ppm to /20 ppm with 4 hearths. The silver rich fraction amounting to about only 1 or 2 percent of the feed would contain about 4% Ag and this could be discharged as such for further refining by conventional methods. Alternatively it could be passed to a forehearth where a further stage of electron beam refining would give a condensate containing about 39% Ag, leaving a residue containing 1200 ppm Ag which would join the main feed to the first hearth. The refined copper leaving the 4th hearth would be discharged from the furnace via a second barometric seal leg and could then be passed to a holding furnace for casting directly into billet or rod. It should be emphasized that at this stage these process design estimates are somewhat speculative, and that further pilot scale continuous refining testwork is required to obtain data on the effect of variables such as beam power and movement and hearth design on distillation rates of the important components of the feed. Also it will be necessary to design condensers and launders for handling the condensate. Preliminary estimates indicate that the technique would be highly competitive with conventional electrorefining of CLEAR copper and would give a product superior in certain respects to cathode grade copper.

REFERENCES

1. F. W. Schweitzer and R. W. Livingston, "Duval's CLEAR Hydrometallurgical Process", AIME Annual Meeting, Dallas, 1982.

2. C. d'A. Hunt and D. C. McLean, "Metal Refining in High Vacuum by Distillation on Electron Beam Hearths", Pacific Southwest Minerals Conf., San Francisco, March 25-28, 1979.

3. T. Santala and C. M. Adams, Jr., "Kinetics and Thermodynamics in Continuous Electron Beam Evaporation of Binary Alloys", J. of Vacuum Science and Technol., 7, No. 6, 1970.

4. C. d"A. Hunt, J. H. C. Lowe, and T. H. Harrington, "Electron Beam Coal-Hearth Refining Furnace for the Production of Nickel and Cobalt Base Super Alloys", in Electron Beam Melting and Refining, State of the Art 1983, R. Bakish (ed.) Bakish Materials Corp., P.O. Box 148, Englewood, NJ, 1983.

5. Toshio Oishi, Morinori Kamuo, Katsutoshi Ono, and Joichiro Morujama, "A Thermodynamic Study of Silica-Saturated Iron Silicate Slags in Equilibrium with Liquid Copper", Met. Trans. B, 14B (1983) pp. 101-104.

6. A. W. Fletcher and C. d'A. Hunt, "Electron Beam Refinement of Metals, Particularly Copper", US Patent Application No. 503376, June 10, 1983.

7. C. d'A. Hunt, "Horizontal Multistage Electron Beam Refinement of Metals with Recycle", US Patent Application No. 503377, June 10, 1983.

APPLICATIONS OF COPPER OXYCHLORIDE CHEMISTRY TO COPPER EXTRACTION FROM COMPLEX HYDROMETALLURGICAL SOLUTIONS

Daniel C. McLean

Davy McKee Corporation
2303 Camino Ramon
San Ramon, California

Copper oxychlorides (atacamite) are compounds which have received little recognition as viable process intermediates in copper and complex base metal hydrometallurgical circuits. They permit clean, simple separations of copper from base and precious metals and provide a means for making simple conversions from copper chloride extraction circuits to normal sulfate recovery processes.

Physical and chemical properties of the oxychlorides are described and several flowsheets are presented to show practical application of oxychloride technology.

Introduction

One of the few innovations in hydrometallurgy in recent years has been the utilization of the chemistry of basic iron sulfates called jarosites. These have been used to make iron-zinc separations in the processing of zinc concentrates and are a very important factor in both the Duval and Cymet copper chloride processes as a means of preventing iron build-up in the closed leaching circuits.

Copper oxychlorides are kindred compounds in that they are insoluble basic salts that form at low pH (1.5 to 3) and precipitate selectively from many base metal solutions. They have been reported in the chemical literature as far back as 1860, and a few of them are known to metallurgists in their mineralogical forms (particularly atacamite) usually as nuisance minerals because of their chloride content. Several patents describe their synthesis and use as fungicides and insecticides, and papers have discussed their formation as corrosion products on copper and brass piping (1).

The oxychlorides have received essentially no recognition or utilization in the hydrometallurgical field with the notable exception that the mineral atacamite is processed for copper recovery at the Mantos Blancos operation in the Atacama desert in Chile. This process is fully described in several publications (2).

The purpose of this paper is to review some of the chemistry of these very interesting compounds and to indicate several potential hydrometallurgical applications.

Chemistry of the Oxychlorides

Oxychlorides occur naturally as relatively rare mineral species. Dana describes five copper oxychlorides and six lead-copper varieties. The names and compositions of some of these are shown in Table I.

The actual chemical analyses reported vary significantly for each copper species. The overall range of copper content is from 51 to 60% Cu, and chloride values range from 12 to 28% Cl. All minerals are various shades of green.

TABLE I

NATURAL OXYCHLORIDES

Mineral	Formula	%Cu	%Cl
Atacamite (Paratacamite)	$CuCl_2 \cdot 3Cu(OH)_2$	59.2	16.6
Hydromelanothallite	$CuCl_2 \cdot Cu(OH)_2 \cdot H_2O$	50.8	28.4
Bottallachite	$CuCl_2 \cdot 3Cu(OH)_2 \cdot 2H_2O$	54.9	15.3
Tallingite	$CuCl_2 \cdot 4Cu(OH)_2 \cdot 4H_2O$	53.3	11.9
Cumingite	$4PbCl_2 \cdot 4CuO \cdot 5H_2O$		
Boleite	$9PbCl_2 \cdot 8CuO \cdot 3AgCl \cdot 9H_2O$		
Eglestonite	$2HgCl \cdot Hg_2O$		

A greater variety of copper oxychlorides have been made synthetically or have been observed to form as oxidation products on copper-base metals. Mellor mentions 18 compositions with ratios of $CuCl_2$:$Cu(OH)_2$:H_2O ranging from 1:1:0 to 1:8:12. In subsequent discussion, the copper oxychlorides will be described in terms of these ratios to simplify the written presentation.

Compositions of several synthetic oxychlorides are shown in Table II. The theoretical copper and chloride contents are about the same as the natural minerals (52 to 60% Cu: 14 to 33% Cl). Actual analyses of copper oxychloride prepared in the author's laboratory work have yielded copper assays of 50.5 to 60% and chloride values from 16 to 23%. Most of the assays, however, were near 54% Cu and 20% Cl with an approximate 5% moisture loss. This corresponds closely with the 1:2:1 formula, $CuCl_2 \cdot 2\ Cu\ (OH)_2 \cdot H_2O$.

The variety of oxychloride compositions which have been observed can be accounted for on the basis of the probable molecular structure as indicated below:

```
             H           H            H
             |           |            |
 Cl          O           O            O          H2O
   \       /   .       /   .        /   .       /
     Cu          Cu          Cu           Cu
   /       \   /    .      /    .       /    \
 Cl          O           O            O          H2O
             |           |            |
             H           H            H
```

This structure illustrates the bridging of copper ions by hydroxyl groups produced at high solution temperature by the process of "olation" (3). The length of the poly-metallic bridge would be determined by variables such as temperature, concentrations, and pH. This type of structure would also explain the various types of thermal decomposition which are described later. (See Table V).

TABLE II

SYNTHETIC COPPER OXYCHLORIDES

Compound	%Cu	%Cl
$CuCl_2 \cdot CuO$	59.3	33.1
$CuCl_2 \cdot Cu(OH)_2$	54.8	30.6
$CuCl_2 \cdot 2Cu(OH)_2$	57.9	21.5
$CuCl_2 \cdot 2Cu(OH)_2 \cdot H_2O$	54.9	20.4
$CuCl_2 \cdot 3Cu(OH)_2$	59.5	16.6
$CuCl_2 \cdot 3Cu(OH)_2 \cdot H_2O$	54.9	15.3
$5CuCl_2 \cdot 2CuO \cdot 4H_2O$	52.2	19.4
$2CuCl_2 \cdot 7Cu(OH)_2 \cdot 2H_2O$	57.9	14.4

Preparation of Oxychlorides

The preparation of oxychlorides has been described by many investigators (4). Basically, it is a simple hydrolysis type of reaction occurring over the pH range 1.5 to 4.0 starting with a copper chloride:

a. $4\ CuCl + O_2 + 4\ H_2O \longrightarrow CuCl_2 \cdot 3\ Cu(OH)_2 + 2\ HCl$

b. $6\ CuCl + 3/_2 O_2 + 3\ H_2O \longrightarrow CuCl_2 \cdot 3Cu(OH)_2 + 2\ CuCl_2$

c. $3\ CuCl_2 + 2\ H_2O + 4NaOH \longrightarrow CuCl_2 \cdot 2Cu(OH)_2 \cdot 2H_2O + 4NaCl$

These reactions are usually carried out in a temperature range of 75 to 85°C. The reaction is exothermic and will produce a spontaneous temperature rise of 10 to 15°C; hence, the starting temperature does not have to exceed 65°C. Reaction time at atmospheric pressure using oxygen is about 2 hours at 80°C for 80 to 100 G/L cuprous chloride conversion. The reaction is very slow at 25°C, and is reported to be "instantaneous" between 100° and 200°C.

Initially, a bright orange-red color may be observed due to the formation of Cu_2O. Reaction rate is a function of temperature, the use of air or oxygen, and oxygen pressure. A few pounds of pressure increases the rate significantly. Air can be used in place of oxygen but rates are usually slower. The presence of NaCl (up to 100 g/l) may also enhance reaction rates.

Referring to reversible reaction a, if the acid produced is not neutralized, the reaction will stop, and addition of an alkali is necessary to keep the reaction going. Notice in the case of reaction b, if an excess of CuCl is used, the acid is consumed by the oxidation of cuprous ion, and the reaction is self sustaining at essentially constant pH (about 2.5). An analysis of the reaction sequence involved is shown in Table III. This feature greatly simplifies reaction control and determination of completion of reaction.

Table III. Oxychloride-Cuprous Chloride Acid Balance

Reactants		Products
$6CuCl + 6H^+ + 3O$	$\longrightarrow$	$6Cu^{+2} + 3H_2O + 6Cl^-$
$6Cu^{+2} + 4Cl^- + 10H_2O$	$\longrightarrow$	$2CuCl_2 \cdot 2Cu(OH)_2 \cdot H_2O + 8H^+$
$2CuCl + 2H^+ + O$	$\longrightarrow$	$2Cu^{+2} + 2Cl^- + H_2O$
$4CuCl + O_2 + 3H_2O$	$\longrightarrow$	$CuCl_2 \cdot 2Cu(OH)_2 \cdot H_2O + CuCl_2$

The final product is a green salt with good settling and filtration characteristics. The degree of hydrolysis (or "olation") is a function of terminal pH, quantity of oxygen used, temperature, and the type of neutralizing agent used. If the final pH is between 2 and 3 (using excess CuCl), the 1:2:1 compound is the typical reaction product. This is fortunate, since this results in a minimum formation of $CuCl_2$ for re-cycle. The ratio of oxychloride to $CuCl_2$ produced as a function of the degree of hydrolysis obtained is shown in Table IV.

Table IV: CuCl Oxyhydrolysis Reactions

Reaction	I:S Ratio
$2CuCl + O + H_2O \longrightarrow CuCl_2 \cdot Cu(OH)_2$	2:0
$4CuCl + 2O + 2H_2O \longrightarrow CuCl_2 \cdot 2Cu(OH)_2 + CuCl_2$	3:1
$6CuCl + 3O + 3H_2O \longrightarrow CuCl_2 \cdot 3Cu(OH)_2 + 2CuCl_2$	2:1
$8CuCl + 4O + 4H_2O \longrightarrow CuCl_2 \cdot 4Cu(OH)_2 + 3CuCl_2$	1:67:1

Note: The greater the number of basic groups (OH) formed; the more acid formed; the more $CuCl_2$ formed.

The 1:2:1 oxychloride is the best situation with a ratio of 3 oxychloride to 1 cupric chloride. Fortunately, the 1:2:1 compound is the one usually produced under conditions indicated in Figure 2.

I:S RATIO = INSOLUBLE COPPER TO SOLUBLE COPPER RATIO

A flowsheet for the full scale production of oxychloride from CuCl is shown in Figure 1. Operating costs for this process (heat, chemicals, and labor) should be 2 to 3 cents per pound of copper using an equivalent copper starting solution of 100 grams per liter.

Product Evaluation

In the preparation of oxychlorides, it is important to note the similarity of analysis of the starting material and the final product on a copper basis. In experiments using CuCl, interpretation of results or completeness of reaction may be erroneous if based solely upon copper assays. With pure CuCl (Cu=64.2%), a change in copper content is fairly obvious. However, with technical grade CuCl, or wet-stored CuCl crystal, oxydation may present some difficulties. Wet cuprous chloride crystals as produced in the Cymet process,

FIGURE 1

COPPER OXYCHLORIDE SYNTHESIS FROM CuCl

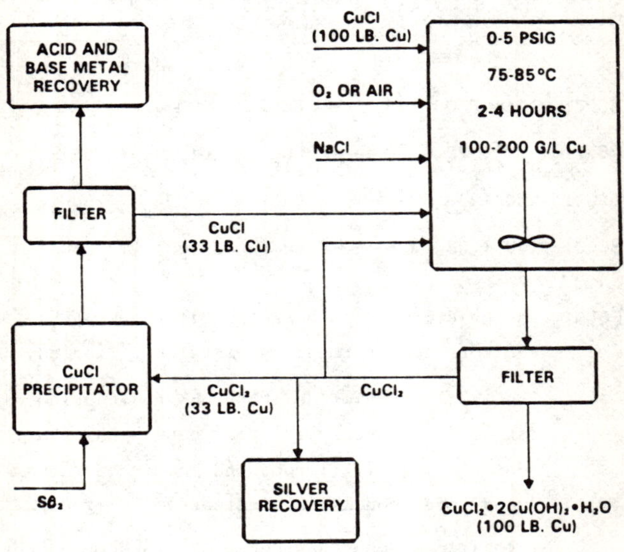

FIGURE 2

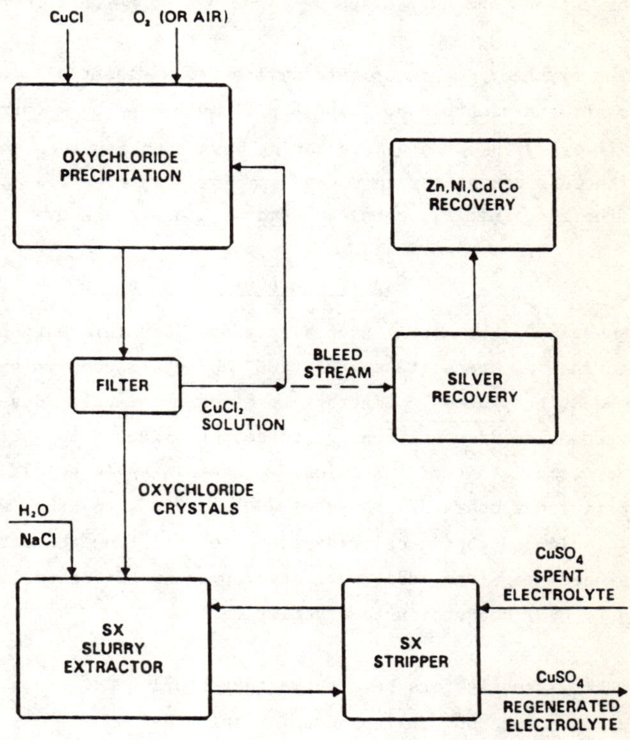
CHLORIDE TO SULFATE CONVERSION
CuCl
O_2 (OR AIR)
OXYCHLORIDE PRECIPITATION
Zn,Ni,Cd,Co RECOVERY
FILTER
BLEED STREAM
SILVER RECOVERY
$CuCl_2$ SOLUTION
OXYCHLORIDE CRYSTALS
H_2O
NaCl
$CuSO_4$ SPENT ELECTROLYTE
SX SLURRY EXTRACTOR
SX STRIPPER
$CuSO_4$ REGENERATED ELECTROLYTE

for example, if left exposed to the air for several days, oxidized appreciably with the copper assay dropping as low as 58% Cu and Cl to 32%. In this situation, one could be starting with the same (or lower) Cu content as that of the oxychloride being produced.

The best indication of product type and completeness of reaction is the chloride analysis (which should change from 35% to 15 to 20%). Careful analysis for cuprous and cupric species should also be performed.

The oxychloride precipitate will contain essentially all of any iron that is present and probably any lead and bismuth. A good separation can be made from silver, with oxychlorides having less than 5 ppm Ag being obtainable starting with CuCl containing up to 300 ppm Ag. Quantitative separations can be made from zinc, nickel, cobalt, cadmium, and manganese.

High Temperature Preparation

Woodall-Duckham (USA) have kindly provided information indicating that oxychlorides of atacamite composition can be produced in the Woodall-Duckham Spray Roasting process. Contrary to other metal chlorides (Fe, Mg, Al, etc.), cupric chloride does not convert to copper oxide in any of the spray oxidation roast systems. Instead, oxychlorides containing 14 to 22% Cl are obtained in the operating temperature range of 700° to 900°, together with some CuO. Hydrochloric acid and cupric chloride are recovered from the off gases for recycle. This process requires additional development work for optimization for application to CuCl conversion to oxychloride.

Other investigators have shown that heating CuCl in air to temperatures between 300° and 700° will produce oxychlorides.

Physical and Chemical Properties

The properties of atacamite are listed in Table V. The oxychlorides have a complex ionic structure as indicated previously, making them similar to polyvanadates, polytungstates and various heteropoly acid salts. They undergo several types of thermal decomposition as is shown in Table VI. The wide variety of products that can be obtained in synthesizing and decomposing oxychlorides (particularly the many reversible reactions that may occur) highlights the importance of maintaining close control of experimental conditions during laboratory and pilot plant test work.

TABLE V: PHYSICAL AND CHEMICAL PROPERTIES OF ATACAMITE

Crystalline Habit	- Orthorhombic
Hardness	- 3 - 3.5
Specific Gravity	- 3.76 - 3.78
Color	- Bright Green-Dark Emerald to Blackish Green

Blow Pipe Tests

Closed Tube - gives off acid and water and forms gray sublimate (oxy-chloride)

On charcoal brownish and greyish coating - metallic Cu

Fusion Temp (synthetic) 900-1000°C - decomposition products are HCl, $CuCl_2$, CuO

Solubility - water	@ 25°C	0.1 G/L
	@ 50°C	0.1 to 1.0 G/L

Soluble in acids, NH_4OH

Composition - 18 varieties with ratios of $CuCl_2:CuO:H_2O$ ranging from 1:1:1 to 1:8:12 (Mellor - See Tables I and II)

Heat of Formation (atacamite) = 23 KCal/g.mo/(Berthelot)

Uses:

1. Fungicidal pigmention paint - wood preservative
2. In concrete flooring, destroys fungi, molds, bacteria anti-cockroach in warm climates
3. Agricultural fungicide - non-toxic to humans

TABLE VI: THERMAL REACTIONS OF OXYCHLORIDES

1. $$4CuCl_2 + O_2 \longrightarrow 2(CuCl_2 \cdot CuO) + 2Cl_2$$

(Reference 5)

2. $$2(CuCl_2 \cdot CuO) \xrightarrow{400^\circ C} 4CuCl + O_2$$

(Reference 6)

3. $$CuCl_2 \cdot 3Cu(OH)_2 \xrightarrow{200\text{-}250^\circ C} \underbrace{CuCl_2 + 3CuO + 3H_2O}_{\text{Brown, Hygroscopic Mixture}}$$

$$3(CuCl_2 \cdot 3CuOH)_2) \xrightarrow{250^\circ C} 10CuO + 2CuCl_2 + 2HCl + 8H_2O$$

(Reference 7 and 8)

4. $$2Cu_2Cl_2 + O_2 \xrightarrow[\text{(Moist Air)}]{100^\circ C} 2(CuCl_2 \cdot CuO)$$

$$2(CuCl_2 \cdot CuO) \xrightarrow{400^\circ C} 2Cl_2Cl_2 + O_2$$

(Reference 1)

5. $$4CuCl_2 + 6H_2O \xrightarrow[(O_2)]{700^\circ\text{-}900^\circ C} CuCl_2 \cdot 3Cu(OH)_2 + 6HCl$$

(Reference 9)
Woodall-Duckham

6. $$CuCl_2 \cdot 2Cu(OH)_2 + 2C \xrightarrow{1200^\circ C} 2Cu + CuCl_2 + 2CO + 2H_2O$$

7. $$CuCl_2 \cdot 2Cu(OH)_2 + 3H_2 \xrightarrow{500^\circ C} 3Cu + 2HCl + 4H_2O$$

Characteristics Useful in Hydrometallurgical Applications

Several properties of the oxychlorides offer possibilities for separating copper from other base metals in both sulfate and chloride solutions, and provide simple means for converting from chloride to sulfate systems. Some of these properties are:

1. their low solubility in dilute acid (pH 1.2-3.0) which permits simple preparation from either cupric or cuprous chloride (slurries or solutions), or from many other copper salts.

2. they can be selectively and quantitatively precipitated from solutions containing high concentrations of Zn, Cd, Ag, Co, Ni, Sn, Mn and SO_4 at pH 1.5 to 2.5.

3. their basic nature makes them useful as neutralizing agents for acid copper solutions without introducing extraneous ions or requiring standard reagent alkalies.

4. they are unique <u>insoluble</u> chlorides which are inexpensive to synthesize and have interesting potential for chloride removal from process and waste streams at copper leaching and smelting operations.

5. they provide means for converting copper extracted in a chloride process to copper sulfate for direct use in conventional electrowinning circuits:
   a) they dissolve readily in sulfuric acid and yield high purity copper sulfate crystals with very low silver and chloride content. Reagent grade hydrochloric acid can be recovered from the mother liquor.
   b) they convert to sulfate via slurry solvent extraction. This is a variation of the Minimet Recherche process which converts CuCl to $CuSO_4$(10). Details are discussed in a latter section (see Figure 2).

6. they can be utilized in closed circuit loops to perform many hydrometallurgical duties utilizing cheap reagents such as air, SO_2 and lime. Sulfur dioxide can be produced from a variety of materials at most copper operations and lime is usually available locally. This decreases operating costs significantly.

7. most of the reactions of interest are exothermic, which greatly decreases energy costs.

Potential Process Applications

In the following section, several flowsheets will be presented primarily to indicate the hydrometallurgical potential for oxychloride technology. These flowsheets are not complete and are intended to stimulate further investigation.

1. Chloride to Sulfate Conversion

The flowsheet shown in Figure 2 provides a means for converting copper extracted in a chloride process to $CuSO_4$ for direct use in a conventional electrowinning circuit using SX-slurry extraction.

The chemicals reactions involved

$$CuCl_2 \cdot 2Cu(OH)_2 + 6HR \rightleftharpoons 3Cu\,R_2 + 4H_2O + 2HCl$$
$$3\,Cu\,R_2 + H_2\,SO_4 \longrightarrow 3\,Cu\,SO_4 + 6HR$$

By maintaining an excess of oxychloride as a slurry in the extraction mixer the pH equilibrates at about 2.1 and is self regulating. High copper solvent loadings (9 to 10 G/L) should be attainable using a 15% LIX-65N solvent at 30° to $40^{o}C$. The solvent disengages cleanly from the slurry with no emulsion formation. The solvent is stripped conventionally with spent electrolyte.

The oxychloride precipitation initially provides a means for separating the copper from silver and most base metals which may be present. Iron is co-precipitated.

2. Separation from Base Metals

The flowsheet in Figure 3 is based upon U.S. Patent 3,637,372 (1972) which actually utilizes the oxychloride precipitation step. This process was developed for treating lateritic ores but could be applied to copper-nickel scrap. The initial reduction roast converts the oxides to metallic copper and nickel with the iron converting to magnetite. The copper and nickel are dissolved in $CuCl_2$ leaving the magnetite intact. This is separated in a special apparatus described in U.S. Patent 1,273,465. The Cu-Ni chloride solution is then subjected to the simple hydrolysis step described in Figure 1, using calcium carbonate for pH control. It is interesting to note that this process reports the formation of the 1:1:0 oxychloride which is recovered as such for sale as an insecticide and fungicide. The nickel

FIGURE 3

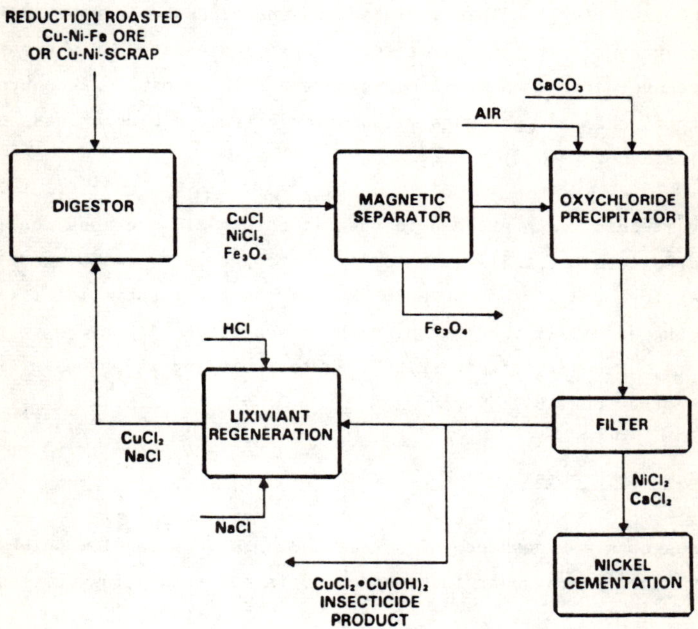
BASE METAL SEPARATION PROCESS
REDUCTION ROASTED
Cu-Ni-Fe ORE
OR Cu-Ni-SCRAP
DIGESTOR
CuCl
NiCl₂
Fe₃O₄
MAGNETIC
SEPARATOR
Fe₃O₄
AIR
CaCO₃
OXYCHLORIDE
PRECIPITATOR
HCl
LIXIVIANT
REGENERATION
CuCl₂
NaCl
NaCl
FILTER
NiCl₂
CaCl₂
NICKEL
CEMENTATION
CuCl₂•Cu(OH)₂
INSECTICIDE
PRODUCT

in the oxychloride filtrate is recovered by cementation. Part of the oxychloride is redissolved in HCl for recycle to the calcine digestor.

3. Complex Sulfide Separation Process

Figure 4 presents a flowsheet for the chloride leaching of a complex sulfide high in precious metals. The front end of the process is based upon the U.S. Bureau of Mines ferrous chloride-oxygen leaching technique disclosed in U.S. Patent 4,053,305 (1977). This variation of the process would be applicable where a cheap source of ferrous or ferric chloride existed or where iron could be recovered from the tailings by waste HCl leaching. A controlled level of total chloride prevents dissolution of the insoluble lead and silver chlorides formed.

The virtue of this leach process is that it throws all precious metals, Pb, Fe and As into the tailings, leaving the copper and remaining base metals in a form readily amenable to oxychloride treatment. This flowsheet indicates the possibility of reducing the oxychloride with hydrogen to anode grade copper. Recovery of copper by SX-electrowinning as shown in Figure 2 would be another possibility.

4. Chloride Removal Process

Figure 5 suggests a flowsheet which takes advantage of the low solubility of the oxychlorides to control chloride levels in circulating process streams:

$$MCl_2 + 2SO_2 + CuCl_2 \cdot 3Cu(OH)_2 \longrightarrow 4CuCl + MSO_4 + H_2SO_4 + 2H_2O$$

(M = Metal + 2)

$$4CuCl + O_2 + 2H_2O + 2NaOH \longrightarrow CuCl_2 \cdot 3Cu(OH)_2 + 2NaCl$$

This is a simple copper recycling system using cheap reagents which should be able to reduce chloride concentrations from 10 G/L and higher to less than 3 G/L. Any copper in the chloride stream would be recovered as oxychloride for further processing or sale as such. The filtrate from CuCl precipitation could be sent to copper leaching operations.

FIGURE 4

Cu-Pb-Zn-Ag-Au SULFIDE CONCENTRATE PROCESS

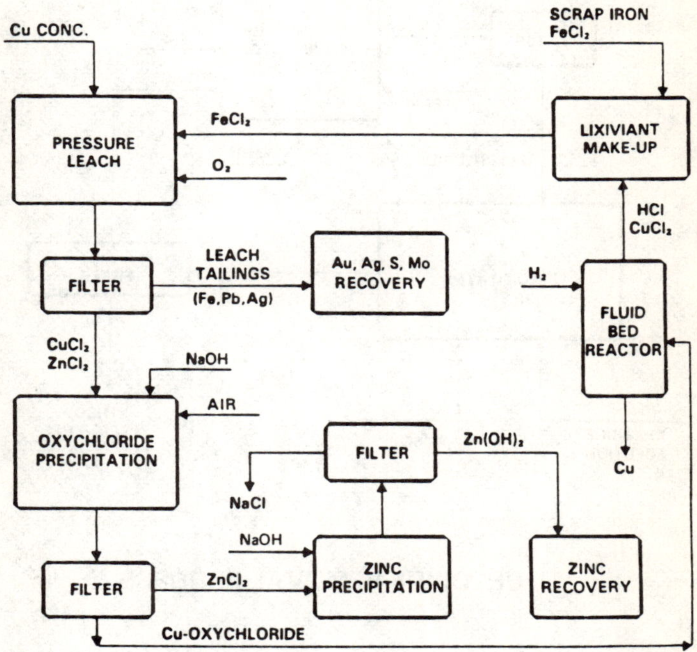

FIGURE 5

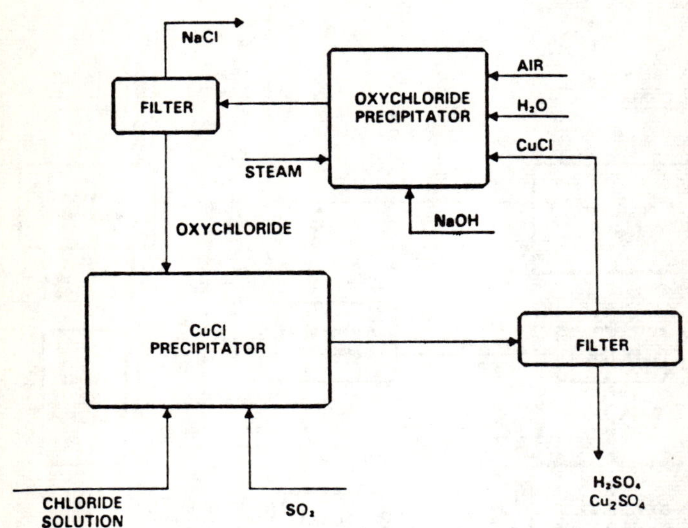

CHLORIDE REMOVAL PROCESS

Conclusions

Copper oxychlorides are compounds which have received little recognition as viable processing intermediates in copper hydrometallurgical circuits. They provide relatively simple means for making copper/silver and copper/silver/base metal separations in complex hydrometallurgical process streams.

References

1. G. Bengough and R. May, Can. Chem. and Met., (8)(1924), 268.

2. R. R. Knobler and W. Joseph, "The Mantos Blancos Operation," Mining Engineering, (1)(1962), 40-45.

3. L. Pokras, "On the Species Present in Aqueous Solutions," Part I, J. of Chem. Ed., (33)(4)(1956), 152-161.

4. C. A. Jacobson, Encyclopedia of Chemical Reactions (New York, N.Y.: Reinhold Publ. Co., 1949).

5. Gass. Chim, Intal 35, 324 (1905).

6. A. Mallet, Compt. Rend. 64, 226 (1867).

7. E. Ludwig, Annalen 169, 74 (1873).

8. E. Ludwig, Ber. 6, 553 (1873).

9. I. G. Farbenind A. G. et al, "Copper Oxychloride." Fr. Patent 822,048, Dec. 1937 (Chem. Abst. 32, P. 3561, 1938).

10. B. J. Scheiner et al, "Recovery of Copper and Silver from Sulfide Concentrates," U. S. Patent 4,053,305, Oct. 1977.

11. B. J. Scheiner, G. A. Smyres, P. R. Haskett, and R. E. Lindstrom, "Copper and Silver Recovery from a Sulfide Concentrate by Ferrous Chloride-Oxygen Leaching," U.S. Bureau of Mines, R.I. 8290, (1978).

J. M. DeMarthe, L. Gandon, and A. Gedrgeaux, "A New Hydro Metallurgical Process for Copper," Extractive Metallurgy of Copper (Yannopoulos and Agawal) Vol. II, (Warrendale, PA: The Metallurgical Socity, 1976), 825-848.

DeMarthe et al, "Method for Obtaining Copper from Copper-Bearing Ores." U.S. Patent 4,023,964, May 1977.

Y. Mayor and P. Tord, "Method for Recovering Nickel," U.S. Patent 3,637,372, Jan. 1972.

H. Huyvetter, "Copper Oxychlorides," Fr. Patent 819,860, Oct. 1937 (Chem Agst. Vol. 32, P. 3104).

J. Reduin and J. Mattern, "Reaction of 2-Vinylpyridine with Copper (II) Chloride," J. Chem. Soc. (78)(5)(1956), 2104-5.

THE PROCESS MINERALOGY STUDY OF REDUCTION OF TITANOMAGNETITE WITH CARBON

H. C. Hsiao and Z. L. Yang

Institute of Chemical Metallurgy, Academia Sinica, Beijing, China

The carbon reducing process for titanomagnetite, an intergrowth of magnetite and ulvospinel, has been studied using an optical microscope and a scanning electron microscope equipped with an energy dispersive X-ray spectrograph, at temperatures ranging from 800°C to 1200°C. The formation of metallic iron commences at about 1000°C. The mechanism of reduction varies because ulvospinel dissolves into magnetite at temperatures lower than 1000°C but coalesces at temperatures higher than 1000°C. At 1200°C, coalescence of ulvospinel and reduction of magnetite occur simultaneously. Ulvospinel first becomes reduced to ilmenite and then to anosovite solid solution. Ilmenite is a necessary intermediate reactant in the reduction of titanomagnetite, and it plays an important part in controlling the process of reduction.

Introduction

The main resources for titanium in the terrestrial crust are rutile, ilmenite and titanomagnetite. Because the reserves of rutile are limited, ilmenite is used as an alternative raw material for the manufacture of pigment-grade titanium dioxide and for the extraction of metallic titanium (1). Much research has been conducted on the oxidation and reduction of ilmenite and the separation of iron and titanium. The separation of the iron and titanium components in titanomagnetite by ore dressing is difficult due to the very fine intergrowths of magnetite and ulvospinel. The pyrometallurgical approach to their separation causes serious difficulties in blast furnace smelting as a result of the characteristics of high-titanium slag. Therefore low-titanium slag commonly is used and titanomagnetite serves mainly as a source for iron. In recent developments, high-titanium slag blast furnace smelting and direct reduction methods have gradually become available, and they are capable of extracting both iron and titanium from titanomagnetite. The study of the reduction mechanism of titanomagnetite at temperatures lower than that of slagging may provide information for prereduction processes prior to direct reduction or high temperature smelting. The purpose of this paper is to elucidate the reduction process of titanomagnetite at low temperatures by means of visual observation of the morphological changes in the microstructure of the ore.

Experiment

Raw Material

The raw material used for this study was high grade ore of titanomagnetite from deposits in southwestern China. Reflected light microscopic examination showed that the ore consisted principally of an intergrowth of magnetite and ulvospinel in which the latter formed a cloth texture that was exsolved from the former matrix. The width of ulvospinel laths are about 0.1-0.5 μm. Magnesia-alumina spinel also occur within the magnetite matrix, and they commonly are surrounded by ulvospinel that exhibits a box shape (Fig. 1). In addition to those minerals, a few ilmenite grains with polysynthetic twins are disseminated in the ore. Micro-area analyses with the energy dispersive spectrometer are given in Table I. These results can be summarized as follows: 1) magnetite includes a small amount of titanium, 2) the atomic ratio of Fe/Ti for the ulvospinel is stoichiometric, and 3) some iron is incorporated in magnesia-alumina spinel.

Table I. SEM-EDX Analysis of Titanomagnetite

minerals	wt%					mole					formula
	Fe	Ti	Al	Mg	O	Fe	Ti	Al	Mg	O	
magnetite	70.61	1.05	/	/	28.34	1.27	0.02	/	/	1.77	$Fe_{2.92}O_4$
ulvospinel	43.92	18.79	3.61	3.39	30.29	0.78	0.39	0.13	0.14	1.89	Fe_2TiO_4
spinel	9.33	0.53	26.78	18.29	45.06	0.17	0.01	0.99	0.76	2.82	$Mg_{1.09}Fe_{0.24}Al_{1.42}O_4$

Sample Treatment

The high grade titanomagnetite samples described above were cut and polished into 10 mm^3 cubes and embedded in carbon powder in a corundum crucible. The crucible was then placed into a vertical silica-carbon tube furnace where the samples were heated to 800°, 900°, 1000°, 1100° and 1200°C for 60 minutes separately, and then cooled rapidly to room temperature. Other samples were first heated to 900°C for 60 minutes, cooled to room temperature, then reheated to 1000°C and 1100°C for 60 minutes separately, and then finally cooled to room temperature. All the samples treated were subsequently prepared as polish sections and examined by reflected light microscopy and scanning electron microscopy.

Reducing Agent

Carbon was selected as the reducing agent in our experiments because it commonly is used as the reducing agent in some of the methods for direct reduction and because it has a relatively low rate of reduction. Thus, it is convenient to compare with the industrial products that were reduced by carbon, and it also promotes easy observation of the reaction process.

Optical Microscopic Observation

After the block samples were treated and prepared into polished sections, they were studied by reflected light microscopy. In order to examine the development of the reduction process, the surfaces of each sample were polished layer by layer, so that the phases formed from the outer to the inner layers could be identified and compared.

Energy Dispersive Spectrometer Analysis

The scanning electron microscope (JEOL 35-CF) equipped with an EDAX 9100 spectrometer was used to determine the phase composition of the samples before and after treatment. For each phase identified, point analyses were performed for at least 3 to 5 different positions and the average value was taken for further calculations. Quantitative determinations for Fe, Ti, Al and Mg were conducted using Fe_2O_3, TiO_2, Al_2O_3 and MgO standards provided by JEOL. The oxygen content (weight percentage) was calculated according to equation 1-(Fe+Ti+Al+Mg). For simplicity, the mole numbers of aluminum and magnesium and their combined oxygen were subtracted in the calculation of the formula for the iron-titanium oxide.

Experimental Results

Sample Reduced at 800°C

The microscopic examination of samples reduced at 800°C showed that most of the ulvospinels, which previously were present in the sample, had disappeared, and only a few which surrounded the Mg-Al spinel in a box shape still remained. Energy dispersive X-ray spectrometer (EDX) analysis of the matrix in which ulvospinel dissolved showed that its atomic ratio of Fe/Ti is 4.95. The granular ilmenites remained unchanged and their polysynthetic twins were clearly observed.

Sample Reduced at 900°C

The reflected light microscope and scanning electron microscope examination of the sample reduced at 900°C indicated that all of the ulvospinel grains had disappeared, but that the Mg-Al spinels remained unchanged (Fig. 2). EDX results showed that the atomic ratio of Fe/Ti for the matrix was 4.97 and that the calculated molecular formula was $Fe_{2.45}Ti_{0.5}O_4$ corresponding to a solid solution of ($Mt_{50}Usp_{50}$) (Table II).

In addition, no metallic iron was detected in the sample and ilmenite grains were unchanged.

Sample Reduced at 1000°C

In samples reduced at 1000°C, metallic iron was observed to be precipitated in the interstices of the ulvospinel cloth texture where phases magnetite was originally present. As a result of the very fine cloth texture and its oriented arrangement, the metallic iron precipitated in the interstices assumed a regularly graphic texture (Fig. 3). In addition, because of the previous existence of box-shaped ulvospinel surrounding the Mg-Al spinel, some of the metallic iron was formed inside or outside those box-shaped ulvospinels, and they roughly parallel the spindly Mg-Al spinels. Locally, the metallic iron that was distributed in the interstices is no longer present but the positions of its grains are marked by empty holes. These may have resulted from the coalescence of metallic iron or by plucking during polishing. In areas where no metallic iron was precipitated, a graphic texture composed of a light grey phase and a dark grey phase may be present. EDX analysis showed that the light phase is ferrous oxide and that the dark phase is ulvospinel (Table II).

Sample Reduced at 1100°C

Reflected light microscopic examination of the sample reduced at 1100°C showed that a large amount of metallic iron formed in vermicular shape surrounded by a grey phase (Fig. 4), and that ulvospinel, cloth texture, and spindly Mg-Al spinel were no longer present. EDX analysis of the grey phase indicated that it is ferrous oxide (Table II).

An examination of the microstructure of the sample after a thin layer of the sample surface was removed by polishing, showed that only small amounts of metallic iron was formed, fine grains coalesced together into dark grey colored areas and into a massive dark grey phase that apparently were formed by further coalescing, and a light grey matrix (Fig. 5). A few particles of metallic iron and Mg-Al spinel remain in the matrix. EDX analysis of the massive dark grey phase showed that its Fe/Ti atomic ratio was 2.09 and its formula was $Fe_{2.14}Ti_{1.00}O_4$. The light grey matrix proved to be ferrous oxide.

Sample Reduced at 1200°C

After reducing at 1200°C, a large amount of metallic iron was formed on the sample surface, but within the interstices of metallic iron. Two kinds of reduction processes formed the reaction zones. In one case, the microstructure consisted of a core portion with a grey color and a fringe portion with a violet grey color. EDX showed that the composition of the grey phase was close to that of Ti_2O_3 (i.e., $Fe_{0.24}Ti_{1.91}O_3$) and the violet grey phase is ilmenite ($Fe_{0.88}Ti_{0.94}O_3$). The border between them was not

Table II. SEM-EDX analysis of titanomagnetite reduced at different temperatures

sample treatment	description	WT%					mole					Fe/Ti	formula
		Fe	Ti	Al	Mg	O	Fe	Ti	Al	Mg	O		
900°C, 1 hr.	intergrowth disappears homogeneous matrix forms	56.08	9.67	2.80	1.65	29.80	1.00	0.20	0.10	0.07	1.87	4.97	$Fe_{2.45}Ti_{0.50}O_4$
1000°C, 1hr.	light grey phase	74.18	1.05	/	2.45	22.15	1.33	0.02	/	0.10	1.39	/	FeO
	dark grey phase	43.82	19.50	3.58	3.71	29.49	0.79	0.41	0.13	0.15	1.84	1.98	$Fe_{2.10}Ti_{1.06}O_4$
900°C, 1hr. cooled, then 1000°C, 1hr.	light grey vermicular phase	70.57	0.92	/	/	28.51	1.27	0.02	/	/	1.78	/	$Fe_{2.93}O_4$
	dark grey vermicular phase	44.19	19.08	3.25	3.84	29.64	0.79	0.40	0.12	0.16	1.85	1.98	$Fe_{2.05}Ti_{1.04}O_4$

continued

Table II. SEM-EDX analysis of titanomagnetite reduced at different temperatures

sample treatment	description	WT% Fe	WT% Ti	WT% Al	WT% Mg	WT% O	mole Fe	mole Ti	mole Al	mole Mg	mole O	Fe/Ti	formula
1100°C, 1hr.	light grey matrix	74.89	1.70	0.24	0.54	22.29	1.34	0.03	0.01	0.02	1.39	/	FeO
	grey coalescent grain	50.18	20.58	0.96	0.40	27.88	0.90	0.43	0.04	0.01	1.743	2.09	$Fe_{2.14}Ti_{1.06}O_4$
	grey fringe phase round metallic iron	77.60	/	/	/	22.40	1.39	/	/	/	1.40	/	FeO
1200°C, 1hr.	grey fringe phase round metallic iron	78.83	/	/	/	21.17	1.41	/	/	/	1.33	/	FeO
	violet fringe phase	44.00	19.26	3.65	3.69	29.41	0.79	0.40	0.14	0.16	1.85	1.98	$Fe_{2.13}Ti_{1.08}O_4$
	grey core phase	31.16	31.05	/	6.51	31.29	0.56	0.65	/	0.27	1.96	0.86	$Fe_{1.00}Ti_{1.16}O_3$
	violet grey fringe phase	31.13	28.76	0.11	5.50	34.50	0.56	0.60	/	0.23	2.15	0.93	$Fe_{0.88}Ti_{0.94}O_3$
	grey core phase	6.74	45.12	4.07	10.23	33.72	0.12	0.94	0.15	0.42	2.12	0.13	$Fe_{0.24}Ti_{1.91}O_3$
	grey tabular phase	9.13	43.55	3.51	5.35	38.46	0.16	0.91	0.13	0.22	2.40	0.18	$Fe_{0.45}Ti_{2.54}O_5$
	brown phase	37.07	20.92	4.16	5.22	32.63	0.66	0.44	0.15	0.22	2.04	1.49	$Fe_{1.24}Ti_{0.83}O_3$

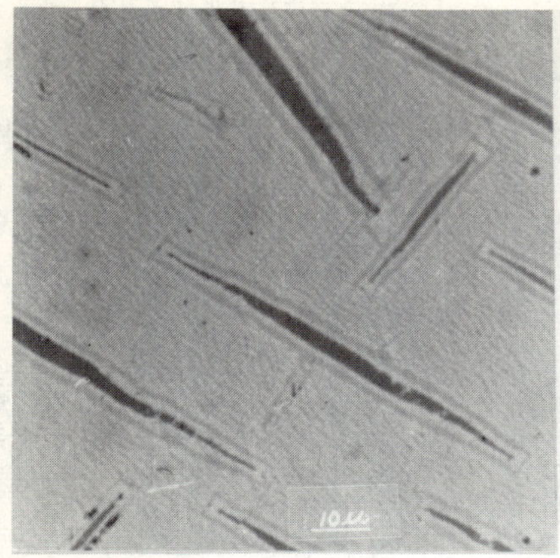

Fig.1 Optical micrograph of titanomagnetite showing magnetite-ulvospinel intergrowth and spinel

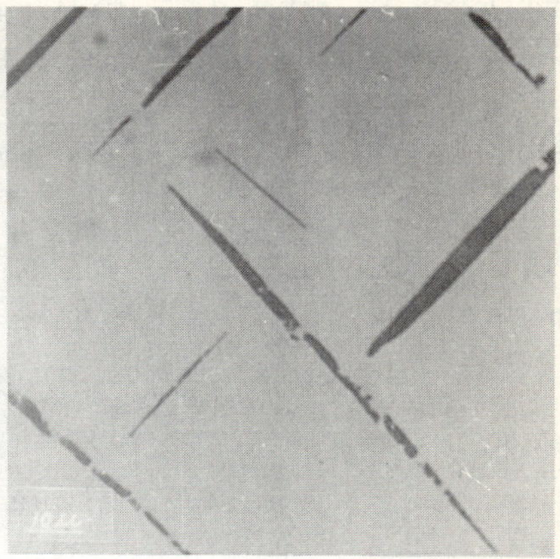

Fig.2 Optical micrograph of titanomagnetite reduced by carbon at 900°C, 1 hr., showing disappearance of cloth texture of ulvospinel

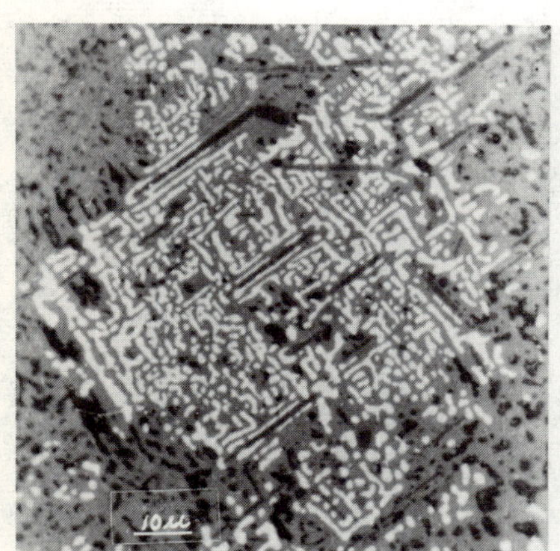

Fig.3 Optical micrograph of titanomagnetite reduced by carbon at 1000°C, 1 hr. showing regular graphic texture of metallic iron

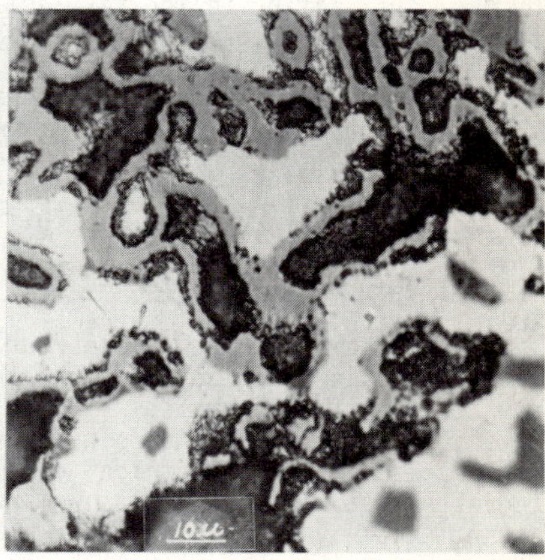

Fig.4 Optical micrograph of titanomagnetite reduced at 1100°C, 1 hr., showing metallic iron surrounded by wustite

clear (Fig. 6). In another case, the composition of core portion was equivalent to that of ilmenite ($Fe_{1.00}Ti_{1.16}O_3$) and the fringe portion was close to that of ulvospinel ($Fe_{2.13}Ti_{1.08}O_4$), and the border between the two phases was clear (Fig. 7). In both cases, the fringe phases were surrounded by metallic iron. A grey tabular phase also was observed to occur locally within the interstices of the metallic iron (Fig. 8). EDX analysis (Table II) showed that the grey tabular phase is a high titanium content phase with an Fe/Ti ratio of 0.18 and a calculated formula of $Fe_{0.45}Ti_{2.54}O_5$. This phase is equivalent to a M_3O_5 solid solution or anosovite solid solution.

In the inner layer of the sample, a brownish phase was commonly found to coexist with the vermicular metallic iron (Fig. 9). EDX analysis of this phase gave an atomic ratio of Fe/Ti of 1.49, a value between that of the ulvospinel and ilmenite. The calculated formula corresponds to $Fe_{1.24}Ti_{0.83}3$. Apparently, this was an intermediate product that formed during the process of ulvospinel reduction to ilmenite.

Sample Reduced at 900°C first, Cooled to Room Temperature, and Reduced again at 1000°C

As previously indicated, after reducing the sample at 900°C, the cloth texture of ulvospinel was no longer present. After the second reduction it was found microscopic examination (Fig. 10) that the metallic iron had precipitated among the spindly Mg-Al spinels that were in oriented arrangment. The metallic iron assumed a vermicular shape, was distributed freely without obvious orientation, and had a coarser grain size metallic iron that displayed graphic texture in the sample directly reduced at 1000°C. In addition to the vermicular metallic iron, two vermicular phases (Fig. 11), one dark grey and the other light grey, are present and they are mutually disseminated in some grains. Calculations from EDX analytical data indicated that the dark grey phase has a formula of $Fe_{2.05}Ti_{1.04}O_4$ and the light grey phase is $Fe_{2.93}O_4$.

Discussion

Homogenization of Magnetite and Ulvospinel Intergrowth

At low temperatures, whether or not a miscibility gap exists in titanomagnetite solid solution and the form of solvus are topics which have been studied by geologists. Vincent and Philip (3) indicated that the formation of magnetite-ulvospinel exsolution intergrowth in titanomagnetite was the result of a eutectoidal decomposition at 500°C. Homogenization studies of natural exsolution intergrowths of magnetite-ulvospinel by Vincent et al. (4) led him to draw an asymmetric solvus with a crest at 600°C. Price (5) showed that the solvus was relatively symmetrical and that the consolute temperature was lower than 490°C. Hydrothermal experiments by Lindsley (6) showed that a miscibitity gap is present, that the consolute temperature is 565 ± 15°C, and that the composition is close to that of $Mt_{55}Usp_{45}$. He considered that the consolute temperature found by Price (490°C) was due to the effect of MgO upon the sample.

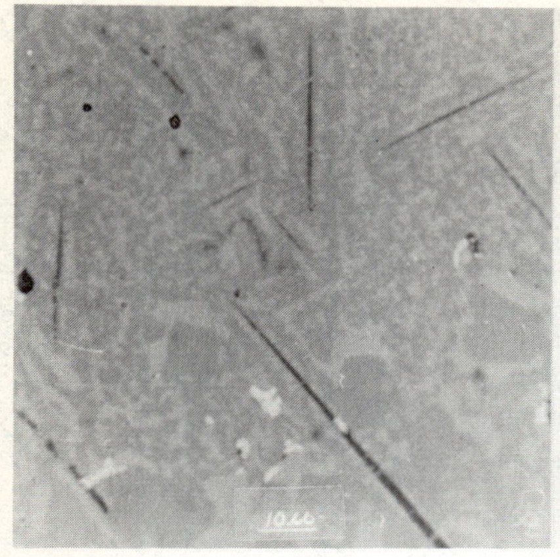

Fig.5 Optical micrograph of titanomagnetite reduced at 1100°C, 1 hr., showing internal microstructure consisting of coalesced ulvospinel (grey), spinel(black) and wustite (light grey)

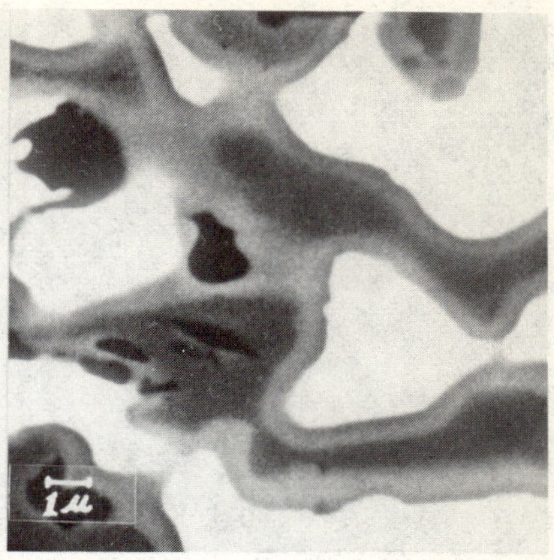

Fig.6 SEM micrograph of titanomagnetite reduced at 1200°C, 1 hr., showing reaction zone with dim border formed in interstices of metallic iron —— metallic iron (white), ilmenite (grey), Ti_2O_3(black)

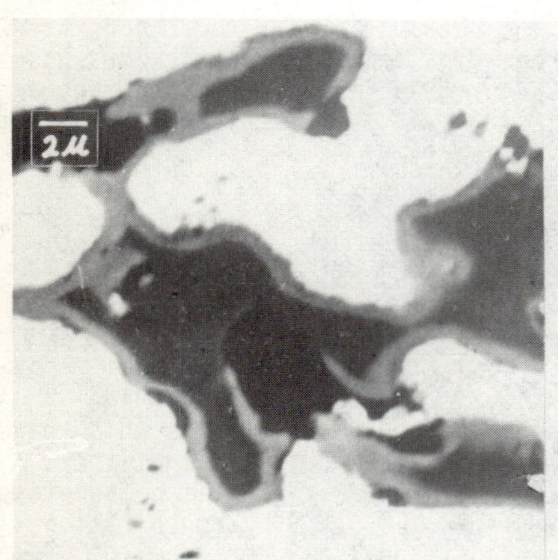

Fig.7 SEM micrograph of titanomagnetite reduced at 1200°C, 1 hr., showing reaction zone with clear border formed in interstices of metallic iron (white), ulvospinel (grey), ilmenite (black)

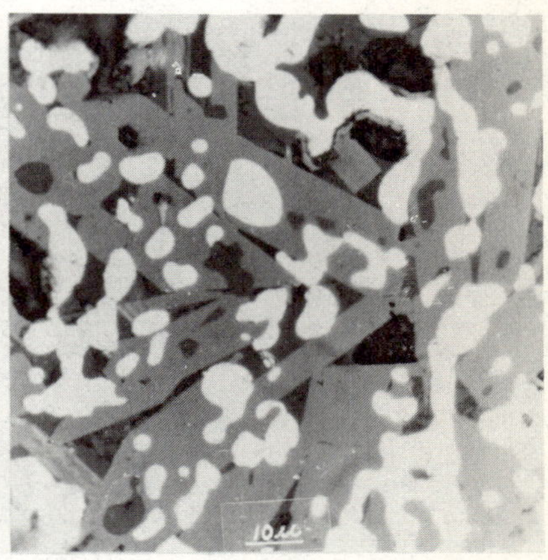

Fig.8 Optical micrograph of titanomagnetite reduced at 1200°C, 1 hr., showing tabular anosovite (grey) and metallic iron particles (white)

Fig.9 Optical micrograph of titanomagnetite reduced at 1200°C, 1 hr., showing vermicular metallic iron (white) and an intermediate phase $Fe_{1.24}Ti_{0.83}O_3$ (black)

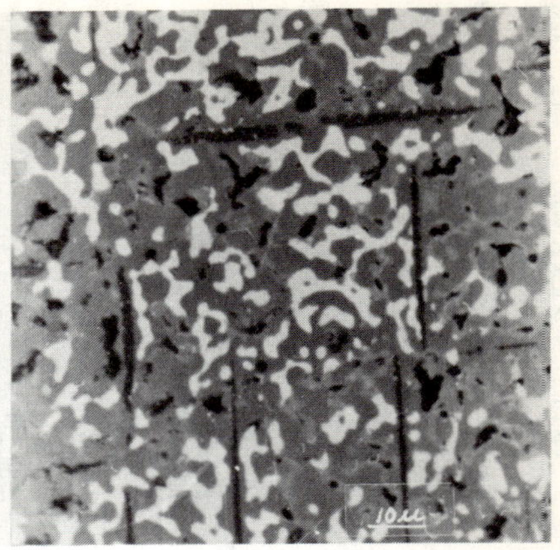

Fig.10 Optical micrograph of titanomagnetite, first reduced at 900°C, 1 hr., cooled then reduced at 1000°C, 1 hr., showing random arrangement of vermicular metallic iron (white) and spindly spinel (black)

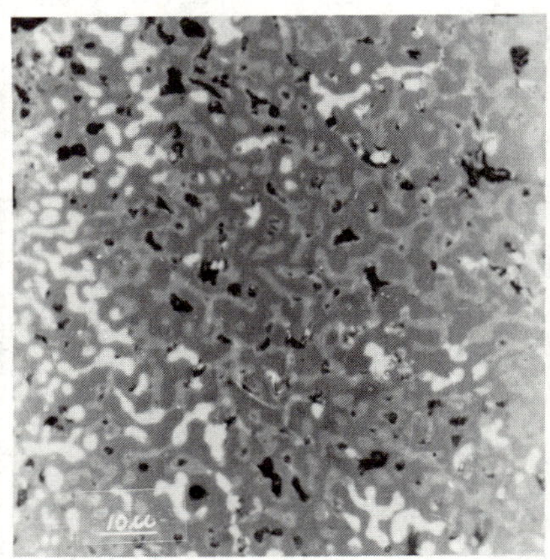

Fig.11 Optical micrograph of titanomagnetite, first reduced at 900°C, 1 hr., cooled, then reduced at 1000°C, 1 hr., showing vermicular phases of ulvospinel (grey) and magnetite (light grey)

In the research reported in this paper, no attempt has been made to study the miscibility gap of titanomagnetite, but rather on the effect of miscibility gap upon the reduction process of titanomagnetite is emphasized. The reduction experimental results of the present research show that: 1) after heating the sample at 800°C in carbon powder for 60 minutes, the cloth texture of ulvospinel typical of the original ore has partially disappeared and 2) that at 900°C for 60 minutes, the cloth texture and the box shape ulvospinel around the Mg-Al spinel both disappeared. These phenomena indicate that homogenization between magnetite and ulvospinel has occured. This is to be expected because the reduction temperatures are higher than those of the consolute temperatures found by Vincent (4), Lindsley (6) and Price (5) (600, 565±15, and 490°C, respectively), and the roast was conducted under weakly reducing atmosphere. These conditions permit homogenization to be brought about with comparative easy. It also would be expected that the very short heating time would prevent equilibrium to occur in the system. The calculations based on the average values of several point analyses by EDX showed that a solid solution of $Mt_{50}Usp_{50}$ had formed. This result coincides with the microstructural configuration of the intergrowth with an equal spacing arrangement of magnetite and ulvospinel prior to homogenization.

Coalescence of Ulvospinel

The experimental results have shown that when the sample of titanomagnetite undergoes reduction at 1100°C, a coalescence process of ulvospinel from fine particles to massive particles takes place in inner layers that are close to the surface. This coalescence process contrasts to the homogenization process at 800°-900°C. The question may be asked: Why does the coalescence process take place at the higher temperature? What is its effect on the reduction process? The EDX analytical results have shown that the ulvospinel coalesces within a matrix of wustite that was formed by the reduction of magnetite. It would be expected that at temperatures lower than 1000°C, both ulvospinel and unreduced magnetite in the intergrowth would have crystal structures identical to that of spinel. Consequently, they can form a solid solution, a condition necessary for homogenization to occur. If reduction proceeds at temperatures that are higher than 1000°C, e.g., 1100°C, , magnetite would continue to change to a lower valence iron oxide and finally to metallic iron as a result of suffering further reduction. Therefore, the crystal structure that originally was identical to that of ulvospinel would be changed, and the difference in crystal structure between the two phases would make them unable to form a solid solution. In contrast, high temperature diffusion will cause the fine ulvospinel crystals with identical crystal structures, as well as the wustite, to coalesce. Under further reduction, the metallic iron reduced from wustite will surround the massive ulvospinels and prevent them from further reduction. Thus, the coalescenced grains that occur within the inner portions of the samples are unfavorable to further reduction.

Effect of Microstructure on Reduction

Comparing the sample reduced at 1000°C with the sample first reduced at 900°C then re-reduced at 1000°C, an interesting phenomena becomes apparent. The differences in microstructural configuration samples prior to reduction at 1000°C has led to a difference in phase constitution and distribution after final reduction. These differences will also affect the results of further reduction.

If a sample is heated directly to 1000°C with no holding time at 900°C, the magnetite that is distributed within the interstices of the ulvospinel cloth texture will first be reduced to wustite or metallic iron, and the ulvospinel will not be reduced and will retain its cloth texture. EDX analyses have demonstrated this to be the case. The metallic iron that is formed develops an oriented graphic texture due to the presence of the cloth texture.

If a sample is first heated at 900°C for 60 minutes, and after cooling it is again reduced at 1000°C, the affect of the cloth texture configuration of ulvospinel on reduced phases is not important because of the homogenization of that intergrowth during reduction at 900°C. In this case, when the sample is further reduced at 1000°C, it is not the two phases that constitute the intergrowth but rather the solid solution formed by homogenization that undergoes the reducing reaction. The reflected light microscope indicated that in the sample reduced at 1000°C the solid solution had decomposed into two phases, ulvospinel and magnetite. It is interpreted that when the sample is reduced at 900°C for 60 minutes, the time is too short to achieve equilibrium in the system, and that during further reduction at 1000°C, the solid solution undergoes reactions to different extents within various microareas, probably as a result of compositional fluctuations of the solid solution. Thus, the light grey and dark grey phases that corresponded to magnetite and ulvospinel were formed. With further reduction, metallic iron, rather than magnetite, precipitated in those positions, and they assumed a vermicular texture lacking any special orientation.

To compare the cases discussed above, it is apparent that the very fine magnetite that was distributed in the interstices of the cloth texture of ulvospinel may be easily reduced not only because of the characteristics of magnetite itself, but also because of its increase in surface area. This leads to the formation of metallic iron with a graphic texture. In contrast, reduction at 1000°C of samples formerly homogenized at 900°C, causes the phase boundary or surface area to decrease because of the formation of solid solution. During reduction, a process of decomposition into two phases must proceed, and this delays the entire reaction process. As a result, large amounts of unreduced magnetite, together with metallic iron and ulvospinel, may persist. Also, the grain size of the vermicular texture in this sample is coarser than that of the graphic texture in the sample reduced directly at 1000°C. Thus, it can be concluded that no matter which reducing process is adopted for the pyrometallurgy of titanomagnetite, a long holding time, lower than that temperature at which the metallic iron commences to reduce, is also unfavorable for the reduction of titanomagnetite.

Reaction Sequence for Titanomagnetite During Reduction by Carbon

The titanomagnetite consists mainly of intergrowth of magnetite and ulvospinel. In the reduction process, these two minerals undergo reduction based upon their reducing characteristics. From the experimental results, it is clear that metallic iron commences to precipitate at 1000°C during the reducing process. Before the temperature exceeds 1000°C, the heating rate plays an important part in influencing the reduction that will proceed at and above 1000°C. Holding or heating slowly in the temperature range of 800°-900°C promotes the dissolution of the ulvospinel cloth texture, i.e., the homogenization of that intergrowth. In contrast, heating the sample to 1000°C rapidly, causes the magnetite to reduce before the dissolution of ulvospinel occurs. Therefore, the process of homogenization of that intergrowth is prevented from proceeding. Upon reducing the sample further at temperatures higher than 1000°C, the coalescence of the ulvospinel will gradually occur. The entire reduction process for titanomagnetite is not the sum of each single reducing process for magnetite and ulvospinel separately. Also, the mutual reactions between the intermediate reduction products inevitably occur at high temperatures. The experimental results demonstrate that the different heating methods used during reduction have led to two different reducing processes. Based upon the reaction patterns and the phases formed under the different temperatures, the experimental results are schematically summarized in Figure 12.

The formation of the reaction zone in samples reduced at 1200°C has been noted in the experimental results and EDX analysis has demonstrated that two kinds of reducing processes have occurred and that the levels of reduction for each is different. Combining both cases, the formation of the reaction zone and the development of the reduction process can be schematically shown in Figure 13 and explained as follows.

The magnetite in the sample is first reduced to form metallic iron according to the reaction (2)

$$Fe_3O_4 \longrightarrow 2FeO + Fe + O_2 \quad (2)$$

and then the reduction of ulvospinel proceeds according to the reaction (3)

$$Fe_2TiO_4 \longrightarrow FeTiO_3 + Fe + \tfrac{1}{2}O_2 \quad (3)$$

The ilmenite formed undergoes further reduction and its iron content decreases continuously. Finally, a core portion with a composition of $(FeTiO_3)_x(Ti_2O_3)_{1-x}$ is formed, leaving an outer layer of ilmenite. Due to the formation of a solid solution between the two phases, the small compositional difference of core and fringe portions leads to a dim border between them. When the sample is reduced at 1200°C, ilmenites with a series of different Fe/Ti atomic ratio will form. In fact, they are solid solutions between $FeTiO_3$ and Ti_2O_3 with Ti^{3+}. In a study of the Fe-Ti-O system, Grey et al. (7) have also considered that the $(FeTiO_3)_x(Ti_2O_3)_{1-x}$ was formed during the early stages of ilmenite reduction.

The ulvospinel reaction zone sometimes forms between ilmenite and metallic iron, as a result of the reaction (4)

$$FeTiO_3 + FeO \longrightarrow Fe_2TiO_4 \quad (4)$$

and it can be used to explain why there is no FeO reaction zone formed between them. As the reduction proceeds to the extent that not enough FeO exists, the ulvospinel will re-reduce to ilmenite that undergoes further reduction according to the reaction (5)

$$2FeTiO_3 \longrightarrow FeTi_2O_5 + Fe + \tfrac{1}{2}O_2 \quad (5)$$

to form a solid solution of M_3O_5 observed in our samples as a grey tabular phase with a composition of $Fe_{0.45}Ti_{2.54}O_5$ and that formed in the interstices between metallic iron grains.

Fig. 12 Two schemes for the reduction of titanomagnetite

	800°-900°C	1000°C	1100°C	1200°C
$(Mt - Usp)_{int}$ -------	----------------	Usp + Wus + iron -------	Usp + Wus + iron --------	Usp+Ilm+Ano+Wus+iron
		(reduction)	(coalescence) (reduction)	(reduction)
	$(Mt_{50}Usp_{50})_{sol}$ ---	Usp + Mt + iron -------	Usp + Wus + iron --------	Usp+Ilm+Ano+Wus+iron
	(homogenization)	(decomposition) (reduction)		

Mt: Magnetite; Usp: ulvospinel; Wus: Wustite; Ilm: Ilmenite; Ano: Anosovite; int: intergrowth

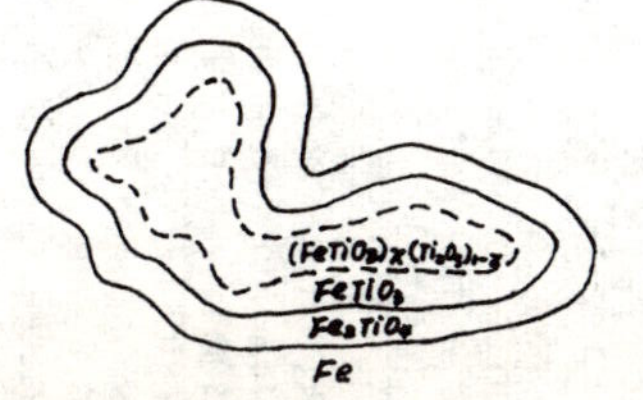

Fig.13 Reaction zones in the interstices of iron during reduction of titanomagnetite by carbon at 1200°C
______ sharp border
------ dim border

Conclusions

(a) If titanomagnetite is reduced in the temperature range of 800°-900°C and the holding time is longer than 60 minutes, homogenization of the original exsolution cloth texture intergrowth will occur and form a solid solution of ($Mt_{50}Usp_{50}$).

(b) Upon rereduction of the homogenized sample, the solid solution will again decompose as a result of not achieving equilibrium during the homogenization and this phenomena delays the development of the reduction process.

(c) Under reduction at 1100°C, the ulvospinel that is distributed in the inner portions of the sample will coalesce. Coexisting wustite has a different crystal structure but the high temperature of diffusion is favorable for its coalescence.

(d) The reducing process for titanomagnetite is more complex that that for ilmenite, because of the mutual reactions between intermediate reactants formed separately by reduction of magnetite and ulvospinel at high temperatures. The presence of wustite will affects the reduction of ulvospinel.

(e) At 1200°C, ilmenite is a necessary intermediate reactant that forms a series of solid solutions with Ti_2O_3 and plays an important role in controlling the reducing process.

(f) The anosovite solid solution that is finally formed in the reducing process contains a certain amount of iron that is difficult to reduce.

References

1. S. Thomas et al., "Alteration and Recovery of Ilmenite and Rutile", Australian Mining, 64 (11) (1972) pp. 18-44.

2. R. J. Fruehan "The Rate of Reduction of Iron Oxides by Carbon", Metall. Trans. B 8B (2) (1977) pp. 279-286.

3. E. A. Vincent and R. Phillips "Iron-Titanium Oxide Minerals in Layered Gabbros of the Skaergaard Intrusion, East Greenland", Geochimica et Cosmochimica Acta, 6 (1954) pp. 1-26.

4. R. A. Vincent et. al., "Heating Experiments in some Natural Titaniferous Magnetites", Mineral. Mag., 31 (1957) pp. 624-655.

5. G. D. Price, "Subsolidus Phase Relations in the Titanomagnetite Solid Solution Series", Am. Min., 66 (1981) pp. 751-758.

6. D. H. Lindsley, "Some Experiments Pertaining to the Magnetite-Ulvospinel Miscibility Gap", Am. Min., 66 (1981) pp. 759-762.

7. I. E. Grey et. al., "Reaction Sequences in the Reduction of Ilmenite: 4 - Interpretation in Terms of the Fe-Ti-O and Fe-Mn-Ti-O Phase Diagrams", Mineral Processing & Extractive Metallurgy, 83 (1974) pp. 105-111.

REFLECTED LIGHT MICROSCOPY AND ELECTRON MICROPROBE ANALYSIS OF FLASH SMELTING PRODUCTS OF A COPPER SULFIDES CONCENTRATE

Susanne Pignolet-Brandom and Richard D. Hagni

Department of Geology and Geophysics
and Generic Center for Pyrometallurgical Research
University of Missouri-Rolla, Rolla, MO 65401

and

Norman D. H. Munroe

Division of Metallurgy and Chemical Metallurgy
Henry Krumb School of Mines
Columbia University, New York, NY 10027

Experimental flash smelting products of a copper sulfide concentrate have been studied by reflected light microscopy, and electron microprobe analysis. Intermediate solid solution, bornite, chalcocite, metallic copper, magnetite, iron-copper spinel, pyrrhotite, hematite, delafossite, and cuprite have been identified in the product samples. The observed product phases agree closely with those predicted by thermodynamic considerations. The detailed reflected light microscopy study of the textural intergrowths of these phases has resulted in the identification of the process by which chalcopyrite is converted to copper matte and the conversion of pyrite to iron oxides.

Introduction

Two different approaches may be taken to the study of mineral reactions during flash smelting: 1) thermodynamic evaluations, and 2) microscopic observations. Although this paper is devoted primarily to the study of flash smelting reactions by direct observations under the reflecting microscope, thermodynamic and kinetic considerations can form a valuable method to predict reactions that would be expected.

Reflected light microscopy and electron microprobe analysis have not been extensively applied to the study of pyrometallurgical products. Reflected light microscopy provides precise information on the phases present, their abundance, and the nature of their intergrowths. The microscopic information is supplemented by electron microprobe analysis to obtain quantitative chemical data and confirm the optical identifications.

These methods have the potential for providing important information to metallurgists. As an example, the application of reflected light microscopy and electron microprobe analysis to experimental flash furnace products of a copper sulfides concentrate is discussed. In this study these techniques have been used to: 1) evaluate the sampling method, 2) identify intermediate reaction products and characterize their formation, 3) compare products formed under differing partial pressures of oxygen, 4) characterize the mechanism of separation of iron from copper in chalcopyrite, and 5) to compare the observed reactions with those predicted by thermodynamic considerations.

Thermodynamic and Kinetic Evaluation of Flash Smelting Reactions

The principal reactions of copper concentrate comprising of chalcopyrite and a minor amount of pyrite within the flash smelting reaction shaft can be written as follows:

$$5\ CuFeS_2 = Cu_5FeS_4 + 4\ Fe_{1-x}S + 4xFe = S_2 \qquad (1)$$

$$2FeS_2 = 2Fe_{1-x}S + S_2 \qquad (2)$$

$$S + O_2 = SO_2 \qquad (3)$$

$$3Fe + 2O_2 = Fe_3O_4 \qquad (4)$$

$$3Fe_{1-x}S + (5-2x)O_2 = (1-x)Fe_3O_4 + 3SO_2 \qquad (5)$$

$$2Cu_5FeS_4 + O_2 = 5Cu_2S + 2Fe_{1-x}S + 5Cu + 2xFe + SO_2 \qquad (6)$$

Thermodynamic considerations indicate that the sulfides of copper are less easily oxidized than those of iron and its sulfides (iron has a higher affinity for oxygen than copper), so that the principal oxidation reactions are the oxidation of dissociated sulfur and that of iron and its sulfide. The oxidation of copper and/or its sulfides become significant when parts of iron sulfide have reacted and the particle attains temperatures over about 1000°K.

In as much as smelting processes are thermally activated, classical kinetic theory indicates that the highest rates of reaction are obtained when pure oxygen is used at a high temperature. This is particularly true for the oxidation rate of pyrrhotite ($Fe_{1-x}S$), which has been theoretically calculated to be approximately 30 times the oxidation rate of chalcocite (Cu_2S) at 2000°K (1). Further details are available in Munroe (2).

Previous Investigations

Two types of previous investigation are pertinent to the mineralogical investigation of flash furnace products: 1) phase equilibrium and 2) direct phase observations. Phase equilibrium studies provide information on high temperature phases, their behavior during cooling, and are helpful in interpreting the textural intergrowths of the phases. Direct observations of polished sections of flash furnace products provide information on the phases that are formed.

Previous phase equilibrium studies of the Cu-Fe-S system have indicated that copper mattes can contain a wide range of phases with variable stoichiometry that results from the high temperature solid solutions. Three solid solutions occur at elevated temperatures: 1) intermediate solid solution (iss), 2) chalcocite-bornite, and 3) pyrrhotite (3,4,5). Phase equilibrium studies also provide data on transformation and exsolution of the phases during cooling. For example, iss undergoes a complex series of reactions such as inversions, Cu-Fe ordering, exsolutions (6) and transformations to mooihoekite ($Cu_9Fe_9S_{16}$), talnakhite ($Cu_9Fe_8S_{16}$), haycockite ($Cu_4Fe_5S_8$), cubanite (2) or quenchable iss and chalcopyrite (7).

Although previous studies of flash furnace products of copper concentrates lack detailed observations on the optical properties of the consituent phases, their intergrowths, and electron microprobe analyses of their composition, they provide basic information. Jorgensen (8) identified chalcopyrite, copper-ferrite solid solution, copper-bearing magnetite, chalcocite, and hematite. Ahlrichs (9) studied the change in oxidation products of copper sulfides with increasing time and found the first product to form was a bornite-like phase followed by a chalcocite-like phase. Increased oxidation resulted in the formation of metallic copper and cuprite (Cu_2O). Magnetite and fayalite were found to be the oxidation products of the iron sulfide.

Reflected light microscopy also has provided information on the prevalent textures found in flash furnace products. The formation of spherical particles and cenospheres, spheres with hollow cores, have been found to be especially common (8,9,10). Cenospheres of magnetite were observed in products from the combustion of pyrite in a laminar flow furnace (10) and in the products from copper sulfides fed to flash-smelting experiments (9). Cenosphere formation has been attributed to the sulfur-bearing gases of evolution during the molten stage (10).

Method of Study

Flash Smelting Reactor

A schematic flow sheet of the experimental flash smelter is shown in Figure 1. The reactor consists of a cylindrical chamber with an internal diameter of 12.7 cm and a height of 2 m. The reactor is heated electrically by four 7 kW sections, each regulated by a temperature controller activated by thermocouples that measure the temperature of the reactor wall. Copper sulfide concentrate is fed via a vibra screen feeder at a rate from 1 to 10 kg/h. The sulfide particles are dispersed into the oxidizing gas in co-gravity co-current flow through the vertical reactor shaft where smelting reactions occur.

As shown in Figure 1, the reactor shaft has three portholes (1, 2, and

3) for measuring the temperature of the reacting particles with a two-wavelength pyrometer. Three additional water-cooled portholes diametrically opposite to the portholes used for measuring temperature are used for sampling.

Samples 1 through 3 were taken simultaneously from each of these portholes using a pneumatic method of withdrawing the semi-molten particles. This technique involves applying a vacuum on the suspension for a short period of time through three probes placed within the sampling ports. The gas/solid stream leaving the reactor is directed through a filtering device equipped with a crucible with which the solid particles are removed. Additional samples are taken from a detachable crucible that collects the underflow of the cyclone separator (4) and the final product (5) collected from a detachable crucible at the bottom of the reactor (Figure 1). Samples were taken from four runs.

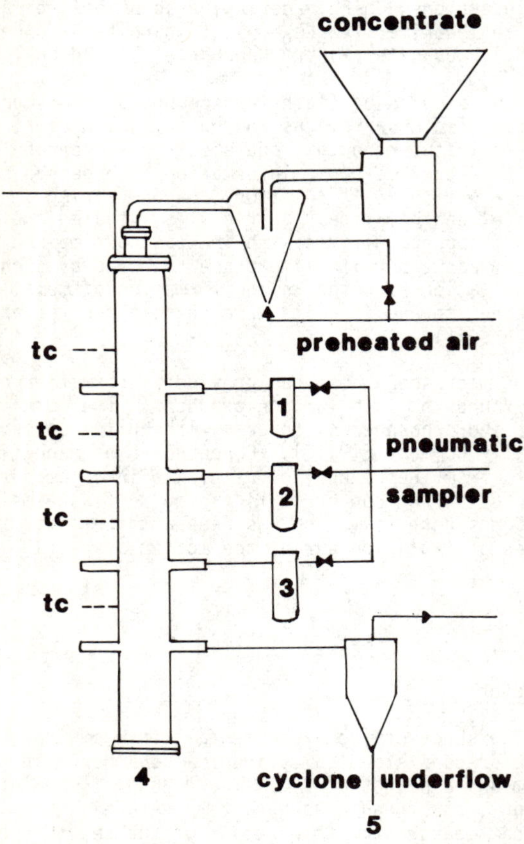

Fig. 1. - Schematic flow sheet of flash smelter reactor.

Experimental Conditions

Experimental runs were conducted at an average gas temperature of 1053°K. This temperature was measured during a preliminary experiment with only air flow through the reactor by using unsheathed chromel-alumel thermocouples that were inserted into the center of the gas stream through the sampling ports and through an opening at the top flange. The temperature during subsequent runs was 1400 - 1600°K due to exothermic mineral reactions.

The oxygen/concentrate ratio was changed by varying the feed rate of the concentrate with care being taken to maintain a constant flow of air. In experimental runs involving oxygen enrichment, calculated volumes of 95% pure commercial grade oxygen were mixed with air. The stoichiometric oxygen requirement for the oxidation of the copper concentrate to copper and magnetite corresponds to an oxygen/concentrate ratio of 0.46. Initial oxygen to concentrate ratios varied from 0.20 to 0.58. The oxygen content varies from the initial ratio within the reaction shaft as a result of the oxidation of sulfur and iron. Further information on the experimental conditions is available in Munroe (2).

Electron Microprobe Analysis

Sample grains for electron microprobe analyses were selected based upon the microscopic observations. The polished sections were carbon coated for the electron microprobe analyses. A JEOL JXA-733 electron probe X-ray microanalyzer, housed in the Earth and Planetary Science Department at Washington University in St. Louis, Missouri was used for the analyses. The unit is equipped with three wavelength spectrometers and an energy dispersive Si (Li) detector. The Tracor Northern automation system is controlled by a DEC PDP 11/34 computer. Count data are collected and converted to weight percent by the Bence-Albee correction technique. An accelerating potential of 15 KV, beam current of 30 nanoamps, and a one micrometer beam with 20 seconds count time was used. The background was measured for each analysis. Natural FeS and synthetic CuS and SiO_2 were used for standards.

The abundance of Cu, Fe, and S was determined by electron microprobe analyses in order to calculate the stoichiometry of the sulfides. Low totals for some of the analyses have resulted from the fact that the product grains are as small as 3 μm and are dispersed in a resin, inhibiting electrical conduction from the particles in the polished section. However, the consistency of the results on repeated analyses indicates that the compositions are useful for the present study.

Results

Evaluation of Feed

The copper concentrate was studied by reflected light microscopy. Chalcopyrite and pyrite are the most abundant sulfides in the feed material. Chalcopyrite grains range in size from 20 to 200 μm and are mostly free from coatings or inclusions of all other phases. Minor amounts of bornite and covellite compose less than 5% of the total sulfides, and magnetite is the most abundant oxide. The gangue phases are silicate minerals.

Evaluation of Sampling Method

Samples from the three portholes and the bottom crucible were examined by reflected light microscopy to determine if there was a significant variation in mineralogy with height in the reactor. However it was found that the majority of the samples collected from portholes 1, 2, and 3 consisted of extremely fine-grained particles that were less than 10 μm across. Fragmentation in known to occur during blast smelting in the reaction shaft (2,9), but the main cause for the fine particle size may have been that of the sampling method. Since particles collected from the bottom of the reactor have a mean particle size of 100-150 μm, it is apparent that the sampling method did not always obtain a representative sample. Therefore, only those samples that had a representative grain size distribution were used for further study.

Formation of Intermediate Products

The product samples contain sulfides, oxides, and silicates. It was found that all samples formed at pressures of oxygen less than that required for stoichiometric conversion of iron and sulfur contain the same phases in similar textural intergrowths.

Copper Sulfides. Phases with compositions in the field of intermediate solid solution (iss) are abundant in the flash furnace products. Since the quenchable iss phases and chalcopyrite have similar optical properties, electron microprobe analysis was used to distinguish iss from chalcopyrite (Table 1). The observed low temperature sulfide phases are referred to here as iss, the stable phase at furnace temperatures.

The morphology of the iss grains varies from irregular shapes (Figure 2) that are less angular than those of chalcopyrite in the feed material to spherical grains (Figure 3). The spherical grains resulted from melting, whereas the irregularly shaped grains were not completely molten. Intermediate morphologies include ovate grains. All of these grain types can contain hollow spherical inclusions that resulted from the formation of gaseous bubbles during smelting.

Irregular, ovate, and spherical grains as well as cenospheres of iss are present in all samples from the furnace. Spinel-type oxides, generally magnetite with minor amounts of copper and silica in solid solution, and less commonly copper-iron spinel, Fe_2CuO_4, are present as inclusions in the iss grains (Figures 2,3). These spinel-type oxides generally are less than 25 μm across and are subhedral to euhedral indicating crystallization from a melt. The inclusions tend to concentrate toward the margins of the particles where the concentration of the oxygen was the greatest.

The most abundant grains in all samples run at pressures of oxygen less than 0.50 atm. are those in which iss is intergrown with bornite (Figures 2,4,5). The intergrowths have formed by exsolution during cooling. Electron microprobe analysis of iss lamellae in bornite gave consistently low totals probably as a result of their narrow widths. Yet, multiple analyses from single lamellae (e.g., grains A, B, C, and F, Table I) have consistent compositions, and compositional variations from grain to grain can be detected. Four spherical grains that consisted of coarser grained iss show compositions that are similar to each other, but contain more iron and less copper than the exsolution lamellae (Table I). Bornite is present as lamellae in iss and vice versa (Figure 4). The length of the lamellae varies from 1 μm to 20 μm. Spinel-type oxides occur as inclusions in the

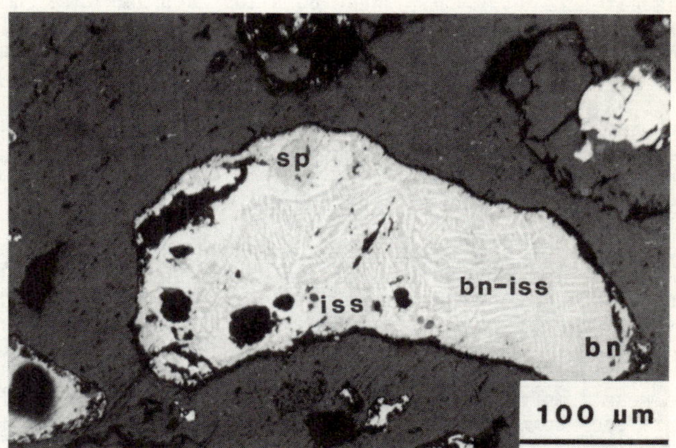

Fig. 2 - Photomicrograph showing an irregularly shaped iss-bornite (bn) grain with an exsolution intergrowth of iss-bornite and subhedral spinel-type oxide (sp) inclusions along its outer edge. Reflected light.

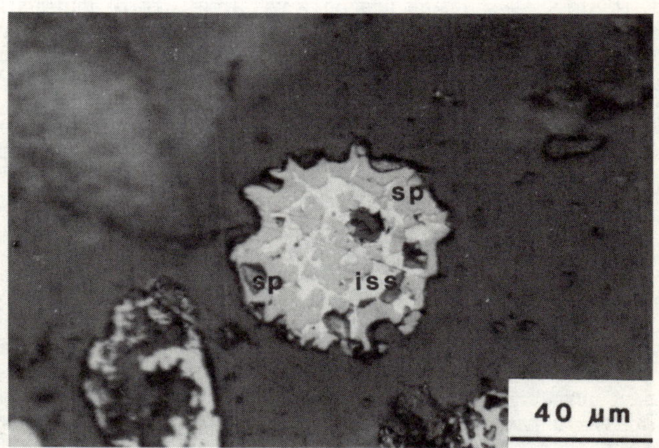

Fig. 3 - Photomicrograph showing spherical grain of iss with subhedral spinel at its edge and included within the iss. Reflected light.

iss-bornite grains. Most iss-bornite grains have irregular shapes, although ovate grains, spheres, and cenospheres are present. The relative amounts of the two phase in each grain are dependant upon the composition of the particle or melt droplets which varies from the composition of iss to that of bornite. Chalcocite also occurs as 10-15 μm exsolution lamellae in bornite that have formed from furnace products with compositions intermediate between chalcocite and bornite. These grains also contain spinel-type oxides. Bornite-spinel-type oxide grains, without lamellae of iss or chalcocite are uncommon. These bornites have variable compositions that have resulted from high temperature solid solution (Table I). However, there is not consistent variation in bornite composition with sampling port (Table 1).

Bornite also is present as rims on irregularly shaped iss grains. The bornite rims results from the reaction of oxygen with iss at the solid (or semi-molten)-gas boundary. This reaction progresses inward especially along favored crystallographic planes (Figure 5).

Chalcocite-spinel-type oxide grains are abundant in all furnace samples. These grains have irregular, ovate, and spherical (Figure 6) morphologies. Chalcocite-spinel-type oxides are the most abundant type of cenospheres and generally represent more than 50% of the total cenospheres present in a sample. Chalcocite-spinel-type oxide grains dominate the less than 50 μm particles indicating that the fine-grained portion of the feed concentrate reacts more rapidly than the coarser fractions. Chalcocite is present as rims on bornite-spinel-type oxide grains and it contains irregular patches of bornite (Figure 7). These textures indicate that bornite was converted to chalcocite during reaction with oxygen in the furnace.

Chalcocite grains also are rimmed or contain inclusions of hematite. The hematite generally is extremely fine-grained (less than 1 μm) and appears to replace the spinel-type oxides. Hematite replacement of spinel-type oxides in other copper sulfide product phases is rare and indicates that this reaction occurs only relatively late during the smelting of copper sulfides. Chalcocite shows variable amounts of iron substitution but no consistent variation in composition with sampling port (Table 1).

Minor amounts of covellite are present as product phases in a few samples. The covellite generally occurs as irregular fine-grained (less than 20 μm) particles associated with spinel-type oxides or hematite.

Metallic copper is present in trace amounts, less than 1%. It occurs as 2 to 10 μm spherical inclusions in copper sulfides (Figure 5). Only a single grain of metallic copper that is not included in sulfides has been observed. Larger grains of metallic copper, up to 300 μm, have only been observed in the final product (sample 4) from one of the runs.

Iron Sulfides. Pyrite (FeS_2) is the most abundant iron sulfide in the feed material. All of the pyrite that is present in the product samples is found there as unreacted pyrite; no pyrite has formed as a result of reactions in the furnace. The unreacted pyrite occurs as the cores of particles where they are surrounded by fibrous pyrrhotite that is, in turn, surrounded by magnetite (Fe_3O_4) and hematite (Fe_2O_3) (Figure 8). Such textures have resulted from the solid-gas reaction of pyrite and oxygen.

For those grains of iron sulfide that have been melted, the reaction with oxygen has resulted in an Fe-S-O melt. The resulting droplets are readily observed in polished sections as spherical grains and cenospheres (Figure 9). These spheres contain pyrrhotite and magnetite; they do not

Table I. Electron Microprobe Analyses of Flash Smelting Products

Sample	Phase	Grain	Type	Weight Percent Fe	Weight Percent Cu	Weight Percent S	Weight Percent Total	Molar Percent Fe	Molar Percent Cu	Molar Percent S
Run 1 #3	iss	A	lath	33.76	29.75	34.21	97.7	28.25	21.88	49.86
Run 1 #3	iss	A	lath	34.19	28.84	34.48	97.5	28.59	21.20	50.21
Run 1 #3	iss	A	lath	33.57	29.24	34.06	96.8	28.31	21.67	50.02
Run 1 #3	iss	A	lath	31.79	31.45	33.15	96.3	21.13	23.59	49.28
Run 1 #3	iss	A	lath	31.98	30.14	32.64	94.7	27.73	22.97	49.30
Run 1 #3	iss	B	lath	28.77	34.84	33.46	97.0	24.44	26.02	49.54
Run 1 #3	iss	B	lath	31.52	33.21	31.33	96.0	27.34	25.32	47.34
Run 1 #3	iss	B	lath	29.26	34.74	32.55	96.5	25.12	26.21	48.67
Run 1 #3	iss	C	lath	25.74	40.07	31.35	97.1	22.27	30.48	47.25
Run 1 #3	iss	C	lath	24.50	39.02	29.65	93.3	22.18	31.06	46.76
Run 2 #4	iss	F	lath	33.17	31.78	31.99	96.9	28.36	23.90	44.47
Run 2 #4	iss	F	lath	32.06	33.92	30.12	96.0	28.04	26.07	45.89
Run 2 #4	iss	F	lath	32.29	32.56	30.61	95.4	28.27	25.05	46.68
Run 2 #4	iss	F	lath	34.92	30.16	31.03	96.1	30.24	22.96	46.81
Run 2 #4	iss	C	sphere	39.26	25.49	30.20	94.9	34.35	19.90	46.58
Run 1 #4	iss	E	sphere	38.99	25.91	30.63	95.5	33.71	19.69	47.63
Run 1 #1	iss	A	sphere	37.97	25.87	32.22	96.0	32.49	19.45	48.05
Run 1 $1	iss	A	sphere	38.04	24.74	32.22	94.9	32.81	18.76	48.43
Run 1 #3	bn	A	host	19.37	50.37	29.93	99.7	16.73	38.24	45.03
Run 1 #3	bn	A	host	19.18	50.37	28.97	98.5	16.79	39.08	44.13
Run 1 #3	bn	B	host	16.30	54.76	28.71	99.8	14.24	42.05	43.71
Run 1 #3	bn	B	host	16.26	55.00	28.72	99.9	14.18	42.17	43.65
Run 1 #3	bn	C	host	16.68	54.04	28.39	99.1	14.68	41.79	43.53
Run 1 #3	bn	C	host	17.27	53.47	28.57	99.4	15.14	41.21	43.65
Run 1 #4	bn	B	host	15.28	57.82	26.41	99.5	13.63	49.81	41.04
Run 2 #4	bn	F	host	19.16	50.65	28.56	98.3	16.90	39.25	43.85
Run 1 #1	bn	B	sphere	17.79	54.99	27.28	100.0	15.65	42.53	41.82
Run 1 #3	cc	C	+Fe ox	3.59	75.30	22.03	100.9	3.32	61.18	35.50
Run 1 #3	cc	C	+Fe ox	2.97	74.10	22.58	99.7	2.77	60.62	36.61
Run 1 #3	cc	B	+Fe ox	4.20	73.40	22.50	100.1	3.93	59.72	36.35
Run 1 #3	cc	B	+Fe ox	3.08	75.3	22.48	100.8	2.84	61.05	36.11
Run 1 #1	cc	B	+Fe ox	1.47	76.88	21.60	99.9	1.38	63.35	35.27
Run 1 #1	cc	B	+Fe ox	6.91	71.01	23.44	101.4	6.27	56.66	37.06
Run 2 #4	cc	D	+Fe ox	4.01	73.08	21.99	99.0	3.76	60.30	35.94
Run 1 #4	cc	A	+Fe ox	7.60	66.44	24.37	98.4	7.01	53.85	39.14
Run 1 #4	cc	D	+Fe ox	1.77	75.60	22.40	99.7	1.65	61.97	36.38
Run 1 #4	cc	D	rim	0.55	75.27	22.65	98.4	0.52	62.32	37.16
Run 1 #4	cc	A	rim	3.52	74.72	22.07	100.3	3.27	61.01	35.72

iss = intermediate solid solution
bn = bornite
cc = chalcocite
Fe ox = iron oxide

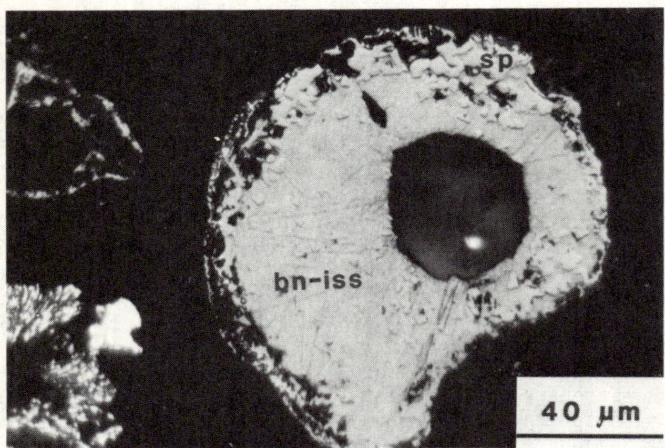

Fig. 4 - Photomicrograph showing an irregularly shaped cenosphere of bornite (bn) that contains fine-grained exsolution lamellae of iss. Spinel-type oxide forms a moderately thick rim and occurs as inclusions locally in the bornite and iss. Reflected light.

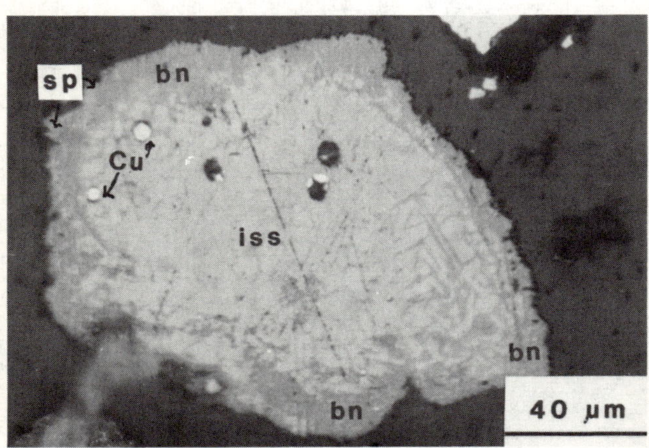

Fig. 5 - Photomicrograph showing progressive inward replacement of iss by bornite. Several spherical metallic copper (cu) inclusions, each less than 5 μm, are enclosed within iss. A very thin rim of spinel-type oxide is present. Reflected light.

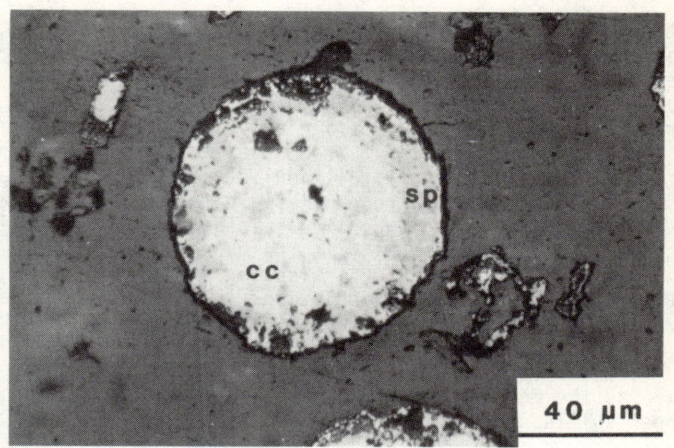

Fig. 6 - Photomicrograph showing a spherical grain of chalcocite (cc) that contains spinel-type oxide (sp) inclusions concentrated toward the outer margin. Reflected light.

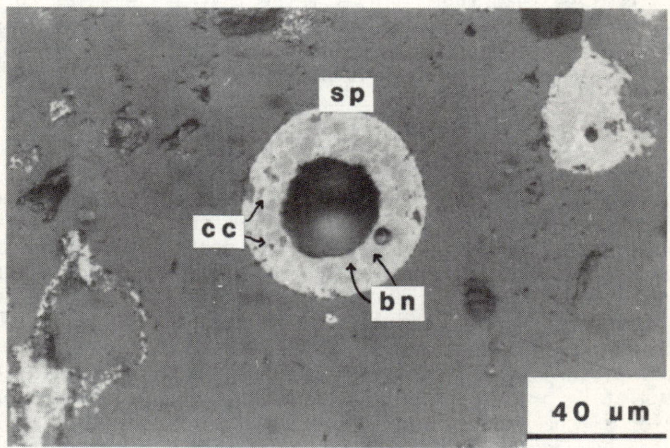

Fig. 7 - Photomicrograph showing a cenosphere that consists of bornite (bn) rimmed and partly replaced by chalcocite (cc). Spinel-type oxide (sp) forms a very thin rim. Reflected light.

contain any residual pyrite due to the complete conversion of pyrite to pyrrhotite and sulfur prior to melting. Pyrrhotite forms 3 to 5 μm blebs (Figure 9). Magnetite occurs as euhedral crystals that are especially concentrated toward the margins of the particles. Part of the magnetite forms skeletal crystals (Figure 10) suggesting that they formed by rapid crystallization during quenching. Magnetite and pyrrhotite are intergrown in a graphic pattern (Figure 11) that resulted either from cooling of a melt with a eutectic composition (11,12) or from rapid crystallization.

Separation of Iron from Copper. In addition to the spinel-type oxides that occur as inclusions in the sulfides (Figures 2,3,4,5,6,7) and the rims of magnetite and hematite on fibrous pyrrhotite with pyrite cores (Figure 8), oxides are present as irregular, ovate, and spherical grains. The irregular grains are composed of magnetite and hematite and vary in size up to 200 μm. These grains are distinct from unreacted magnetite in that they consist of magnetite that has recrystallized into 5-10 μm blebs of magnetite. The spherical grains are generally homogeneous and less than 50 um in diameter. These magnetite grains contain hematite that has developed along the octahedral planes and that rims the magnetite (Figure 11). Cuprite (Cu_2O) and delafossite ($CuFeO_2$) also are present as spherical grains. The spheres are present as inclusions in sulfide particles (Figure 12) and rarely are observed to be partially expelled from the sulfide (Figure 13). Based on this textural evidence, it is interpreted that the spheres have formed from an oxide melt that was immiscible in the sulfide liquid. This was the most important mechanism for the separation of iron from the copper sulfides.

Comparison of Product Samples Run at Differing Po_2

Samples run at pressure of oxygen greater than that required for stoichiometric conversion of the iron and sulfur contain only minor amounts of sulfide as product phases. The most abundant product phases are magnetite, hematite, cuprite, and silicates. The oxides occur largely as spherical grains less than 100 μm in diameter that are similar to the oxide spheres. Magnetite, hematite, and cuprite also are intimately intergrown with silicates in some spheres. The fine-grained nature of the intergrowths prevented electron microprobe analysis of these phases.

Discussion

The observed textural intergrowths and morphology of the product phases indicate that both gas-solid and gas-liquid reactions have occurred during the flash smelting of chalcopyrite. Spherical particles and cenospheres are products from melt. They have a wide range of bulk compositions from iss to chalcocite. During the quenching of these liquids, copper formed exsolution intergrowths of the stable sulfide phases, iss, bornite, and chalcocite. Iron formed spinel-type oxides that crystallized from the melt or formed immiscible liquids. This is in contrast to the predicted Fe_{1-x} phase. The relative proportions of the product phases was dependent upon the composition of the product liquid or semi-solid, which, in turn, was dependent upon the amount of dissolved oxygen and the rate of formation of sulfur-bearing gases. As more oxygen was dissolved or as more sulfur-bearing gases form, the composition of the liquid or semi-solid particle shifted from iss to bornite to chalcocite.

The dissolving of oxygen and the formation of sulfur-bearing gases are continuous processes that result in the formation of the final product phases, metallic copper and spinel-type oxide. The iron occurs in the

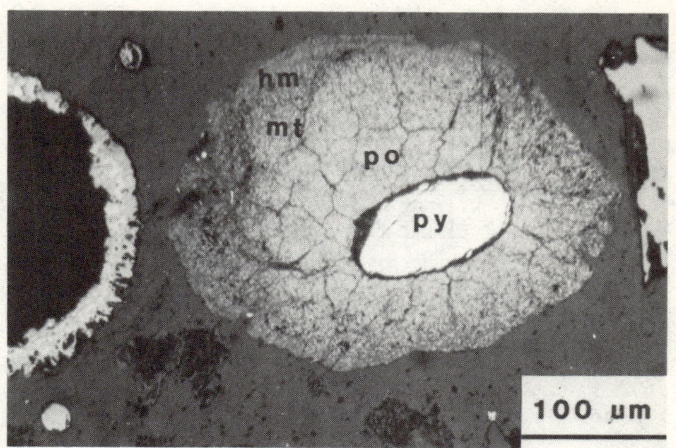

Fig. 8 - Photomicrograph showing a pyrite core that is surrounded successively by the phases, fine-grained pyrrhotite (po), magnetite (mt), and hematite (hm), which are formed from solid-gas reaction. Reflected light.

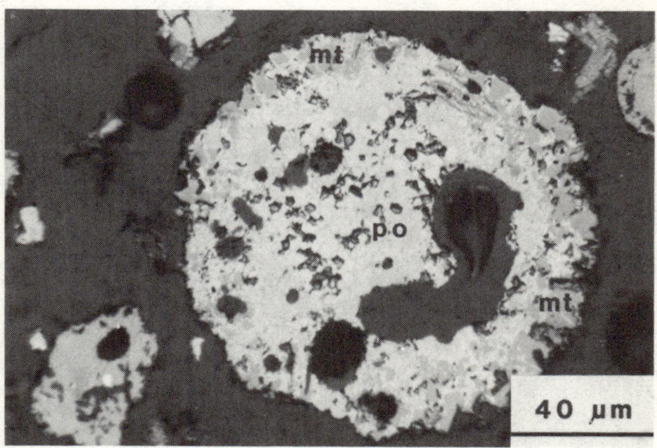

Fig. 9 - Photomicrograph showing a spherical grain that consists of pyrrhotite (po) and magnetite (mt) that formed from the cooling of an Fe-S-O liquid. The hollow inclusions (dark gray) were formed from gases. Reflected light.

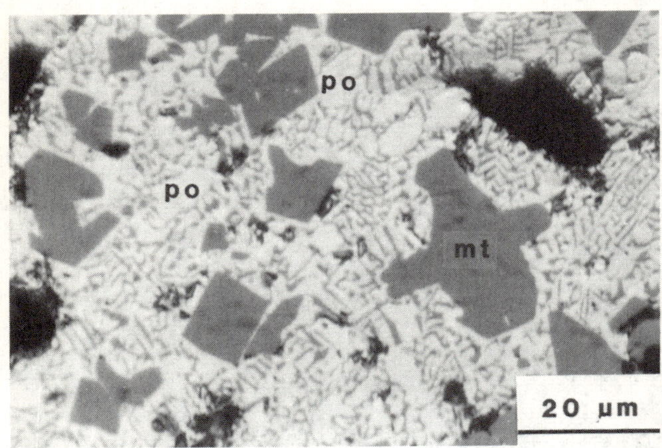

Fig. 10 - Photomicrograph showing skeletal magnetite (mt) and magnetite-pyrrhotite graphic intergrowth formed during rapid cooling of an Fe-S-O liquid. Reflected light.

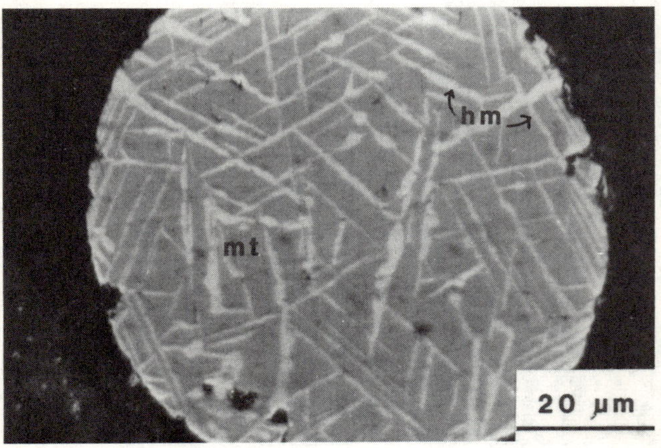

Fig. 11 - Photomicrograph showing hematite (hm) developed along octahedral crystal planes in a magnetite sphere. Reflected light.

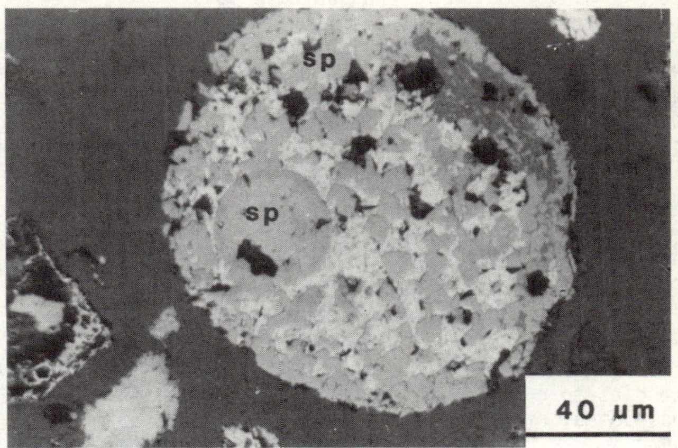

Fig. 12 - Photomicrograph showing a single spherical inclusion of spinel-type oxide (sp) that indicates an immiscible oxide liquid. Abundant subhedral crystalline inclusions of spinel-type oxide (sp) in iss. An unidentified silicate phase occurs in the upper right portion of the spherical grain. Reflected light.

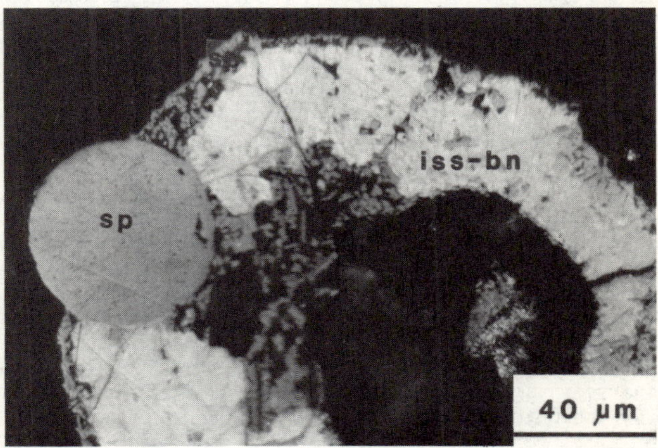

Fig. 13 - Photomicrograph showing spinel-type oxide sphere separating from a cenosphere of iss and bornite. Spinels also occur along the rim of the iss-bornite cenosphere. Reflected light.

intermediate products, iss and bornite, prior to conversion to the spinel-type oxide. Continued reaction of the particles with O_2 resulted in the conversion of the spinel-type oxide to hematite.

In contrast, the copper sulfides solid-gas reactions proceeded through a series of steps that are predicted by thermodynamic considerations. Chalcopyrite was converted to iss and spinel-type oxides, which are converted to bornite and spinel-type oxides. This is evidenced by the rimming of iss by bornite and bornite by chalcocite and by the presence of spinel-type oxide at the margins of irregular grains of the product phases. Fe_{1-x} was not produced even though predicted by thermodynamics.

Similar processes occur during the oxidation of pyrite and these reactions agree with the thermodynamic predictions. The series of solid-gas reactions proceeds as follows: pyrite to pyrrhotite to magnetite to hematite. These reactions are evidenced under the microscope by the consecutive rimming of these phases around residual pyrite and the fibrous morphology of the product phases. Molten iron sulfide forms Fe-S-O liquids from which pyrrhotite and magnetite crystallize during cooling.

Other techniques, such as transmitted light microscopy and X-ray diffraction, have not been utilized in this study, but could be incorporated in future studies to contribute additional information, especially on the silicate phases that are of interest in slag formation.

Conclusions

Reflected light microscopy of pyrometallurgical products and feed provides useful information on the nature of the product phases and their textural intergrowths and can be used to evasluate thermodynamically predicted reactions. This type of data can not be obtained from chemical data on product samples. Electron microprobe analysis provides quantitative chemical data for the individual product phases. The information microscopy and microprobe analyses has been used in this study to demonstrate its usefulness in determining the reaction mechanisms that have occurred during pyrometallurgical processes.

Acknowledgements

This research is supported by the U.S. Bureau of Mines, Generic Mineral Technology Center for Pyrometallurgical Research at the University of Missouri-Rolla. Arthur Morris is thanked for his suggestions and discussions on this project and review of the manuscript. Won C. Park, D. G. C. Robertson, and Thomas O'Keefe also reviewed the manuscript. The scanning electron microscope analyses were done by John Thompson at the University of Missouri-Rolla; the microprobe analyses by Daniel Kremser at Washington University, St. Louis, Missouri. Research on flash smelting at Columbia University is supported by the National Science Foundation (CPE-81-10526) and Outokumpu Oy Company of Finland.

References

1. G. Melcher, E. Muller, and H. Weigel, "The Kivcet Cyclone Smelting Process for Impure Copper Concentrates", Journal of Metals, July (1976) pp. 4-8.

2. Norman D. H. Munroe, "Rate Phenomena in the Outokumpu Flash Smelting Reaction Shaft, Unpub. Ph.D. dissertation, Henry Krumb School of Mines, Columbia University, New York, NY (in preparation).

3. H. E. Merwin and R. N. Lombard, "The System Cu-Fe-S", Econ. Geol., 32 (1937) pp. 203-284.

4. Louis J. Cabri, "New Data on Phase Relationships in the Cu-Fe-S System" Econ. Geol., 68 (1973) pp. 443-454.

5. Richard A. Yund and Gunnar Kullerud, Thermal Stability of Assemblages in the Cu-Fe-S System", Jour. Petrology, 7 (1966) pp. 354-388.

6. Paul B. Barton, Jr., "Solid Solution in the System Cu-Fe-S, Part I: The Cu-S and Cu-Fe-S Joins", Econ. Geol., 68 (1973) pp. 455-465.

7. J. E. Dutrizac, "Reactions in Cubanite and Chalcopyrite", Can. Mineral., 14 (1976) pp. 172-181.

8. John W. Ahlrichs, "Applications of Quantitative Mineralogy for Solving Metallurgical Problems", pp. 233-244 in Process Mineralogy: Extractive Metallurgy, Mineral Exploration, Energy Resources, Donald M. Hausen and Won C. Park, eds., AIME, New York, 1981.

9. F. R. A. Jorgensen, "Copper Flash Smelting Simulation Experiments", Proc. Australas Inst. Mining and Met., 261 (1977) pp. 39-46.

10. F. R. A. Jorgensen, "Cenosphere Formation during Combustion of Pyrite", Proc. Australas Inst. Mining and Met., 276 (1980) pp. 41-47.

11. Paul Ramdohr, The Ore Minerals and Their Intergrowths, 2nd English edition, Pergamon Press, Oxford 1980, 1205 pp.

12. Anthony J. Naldrett, "A Portion of the System Fe-S-O between 900 and 1080°C and its Application to Sulfide Ore Magmas, Jour. Petrology, 10 (1969) pp. 171-201.

Process Mineralogy Applications to Refractories

DEGRADATION OF DIRECT BONDED CHROME MAGNESIA BRICKS IN SECONDARY STEELMAKING UNDER VACUUM

D.H.H. Quon and K.E. Bell

Mineral Processing Laboratory, Energy, Mines and Resources Canada,
Ottawa, Canada

Abstract

Failure of direct bonded chrome magnesia bricks from steelmaking ladles was examined in terms of structural characteristics and chemical changes. Refractories degradation is in part caused by slag infiltration along the grain boundaries of the periclase and spinel. Other factors that cause the failure of refractories are the presence of intergranular phases such as Ca and Si and the loss of MgO under vacuum at high temperature. All these factors have a great influence in determining the corrosion/erosion resistance of the refractories under operating conditions.

Introduction

Various vacuum processes developed recently for different metallurgical objectives have created a need for low wear, high quality refractories for use in steelmaking ladles. Refractory wear can be of decisive importance for the economy of the process, and therefore it is important that the wear behaviour in practice and the laboratory be examined with the aim of reducing the wear rate. The most serious refractory failure has been ascribed to a combination of chemical attack and mechanical erosion where the refractories were in contact with slag and molten steel, as well as to thermal shock due to drastic temperature changes (1-11). Recently, attempts have been made to correlate refractory failure with the rate of loss of oxide at high temperature and low pressure (12-18).

The aim of this study is to examine the stability of direct bonded chrome magnesia bricks at high temperature and low pressure. This study forms part of a program to examine the failure of basic refractories in vacuum refining processes, which has been identified by the Canadian Steel and Iron Research Association (CSIRA) as a serious problem encountered in steel production.

Classification of Refractories

A wide variety of chrome magnesia refractories are available commercially for high temperature application. The five distinctive classes of chrome-magnesia refractories are:

a) Conventional chrome-magnesia brick,
b) Direct bonded chrome-magnesia brick,
c) Simultaneous sinter chrome-magnesia brick,
d) Fused grain or co-clinkered chrome-magnesia brick,
e) Fusion cast chrome-magnesia brick.

Mineralogically, all these bricks are made up of chrome iron spinel and periclase grains. However, the distribution of these phases differs in them and the degree of solid solution and exsolution of these phases also varies. There is also a distinct difference in bonding of the various phases in these bricks. In direct bonded brick, normally composed of relatively coarse chrome ore and fine magnesia, the nomenclature implies the direct bonding of the periclase and chrome spinel without an interrupting film or phase of silicate at the grain boundaries. The microstructural development of these bricks is controlled by the different processing techniques. In general, a progressive solid solution of chrome spinel in periclase occurs as the firing temperature increases. Traditionally, direct bonded bricks are used only in the low wear area in the ladle, such as above and below the slag line as dictated by the economical point of view.

Although chrome magnesia brick have been used extensively in the ladle in the past decade, still the nature of refractories degradation under different conditions is not fully resolved.

Experimental Procedure

Both used and unused bricks obtained from a Canadian steelmaker were first examined megascopically. Thin sections and/or polished sections of selected samples were made for microscopic examination. Scanning electron microscopy (SEM) was used to examine the microstructures that were not resolvable by optical microscopy. The energy dispersive X-ray (EDXR) technique was used to examine the distribution of various elements in the unused and used bricks.

The stability of the brick in contact with the slag was examined. In this test, unused bricks were machined into small cylindrical blocks of approximately 30 mm in diameter by 35 mm high. A hole of 14 mm diameter by 30 mm depth was drilled at the centre of the small block, and was subsequently packed with ground slag. The cup was then loaded into a molybdenum safety crucible and placed inside a graphite resistance furnace operated in vacuum at about 1700°C. The compositions of the slags used for the cup tests are given in Table I and had CaO/SiO_2 ratios of 0.92 and 1.44 respectively.

Table I. Compositions of slags used for the cup tests

	Slag 1	Slag 2
SiO_2	33.30	38.97
Cr_2O_3	0.63	0.67
Al_2O_3	10.50	10.32
CaO	44.80	35.88
MgO	9.81	12.69
Fe_2O_3	0.97	0.98
CaO/SiO_2	1.34	0.92

Results

Megascopically, the colouration of the used and unused bricks is identical, showing light brown grog grains of periclase in a dark brown groundmass of chrome spinel and periclase.

In the used bricks, fractures had developed, which tended to be arranged parallel to the slag and brick interface.

Typical microstructures of the unused brick are shown in Fig. 1. The dominant phases are periclase, as coarse grains, and chrome spinel. The latter occurs as euhedral to subhedral crystals distributed randomly in the periclase groundmass (Fig. 1a, 1b) and/or as variable amounts of an exsolved phase mainly within the periclase grains (Fig. 1c, 1d). Occasionally, chrome spinel tends to rim the periclase grains at the grain boundaries, but this is uncommon.

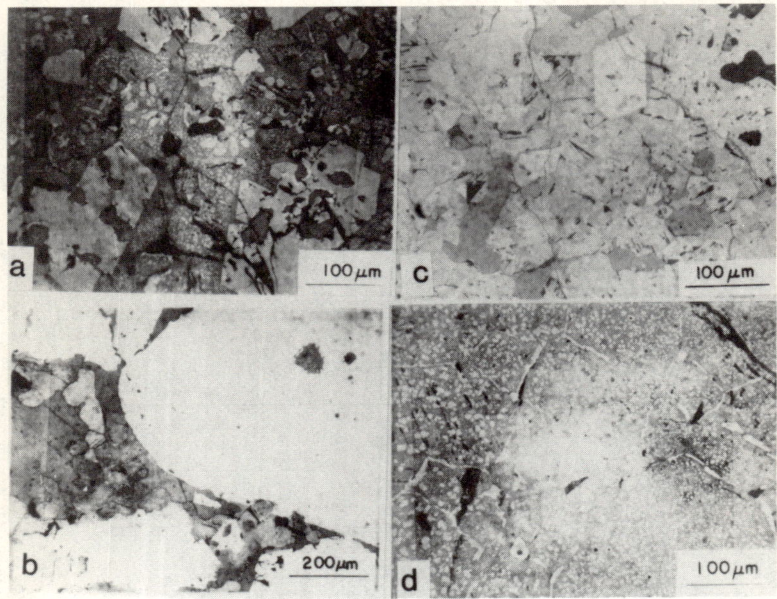

Fig. 1. Typical microstructures of an unused brick

A typical SEM micrograph and elemental distributions of the Ca, Si and Fe obtained by EDXR mapping of the unused brick are given in Fig. 2. EDXR maps revealed that the Ca and Si impurities occurred at the grain boundaries of the periclase whereas the iron appears to concentrate in both the spinel and periclase grains.

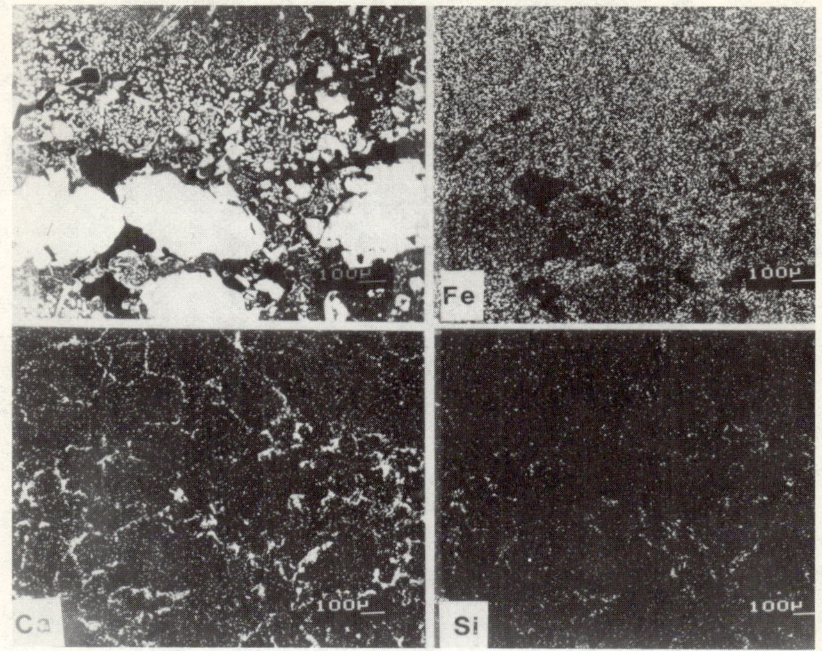

Fig. 2. SEM micrograph and EDXR maps of Ca, Fe, Si of an unused brick

The microstructure of a used brick at a zone of slag and brick interface is shown in Fig. 3 and 4. The polished section in Fig. 3 shows that the slag penetrated along a fracture of the brick, and lath-shaped monticellite crystals replaced the periclase grains. The rounded grains of the periclase and the islands of undissolved and exsolved blebs and euhedral crystals of the spinel phase indicate that the periclase dissolved and the chrome spinel remained intact during the slag infiltration.

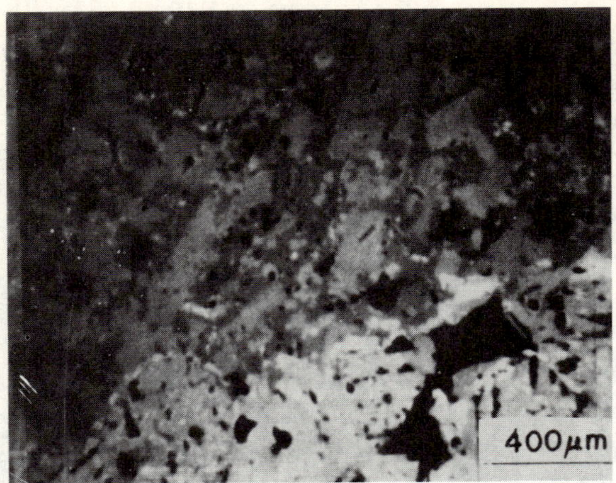

Fig. 3. Microstructure of a used brick at the slag brick interface

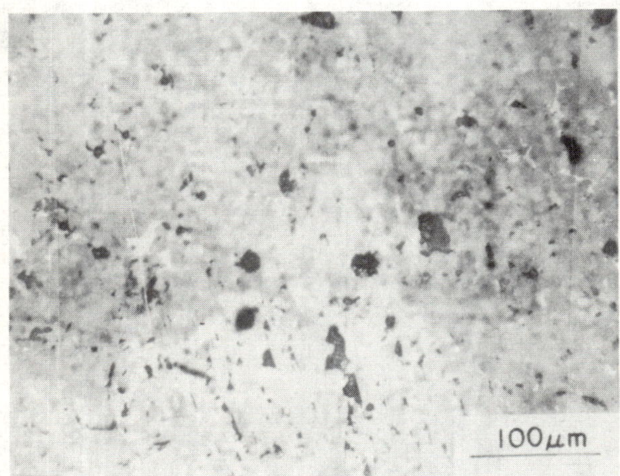

Fig. 4. Microstructure of a used brick showing highly slag resistance spinel phase in a groundmass of slag

Another typical feature observed in the used brick is the chrome spinel pseudomorph of the periclase outlining the ghost periclase grains, which were in part replaced by silicates (Fig. 4). An example of the disruption of the solid-solid bond of the periclase due to slag penetration is shown in Fig. 5. In this case the slag migrated along the grain boundaries of the large periclase grog grain.

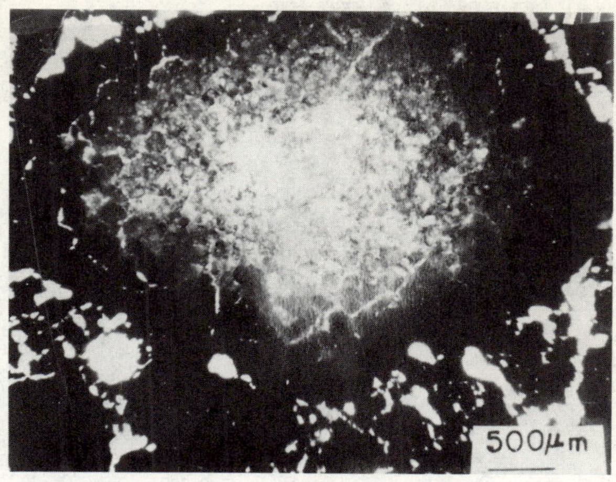

Fig. 5. Micrograph showing slag penetration in the grog grain of periclase and disruption of solid-solid bond of the grains

The SEM micrograph and EDXR maps for Mg, Ca, and Si of the used brick at a zone near the slag and brick interface are given in Fig. 6. The micrograph reveals euhedral chrome spinel crystals filled interstitially by lath-shaped silicate crystals. The EDXR maps show that the chrome spinel crystals are enriched in Mg and the silicate phase is outlined by the distribution of Ca and Si.

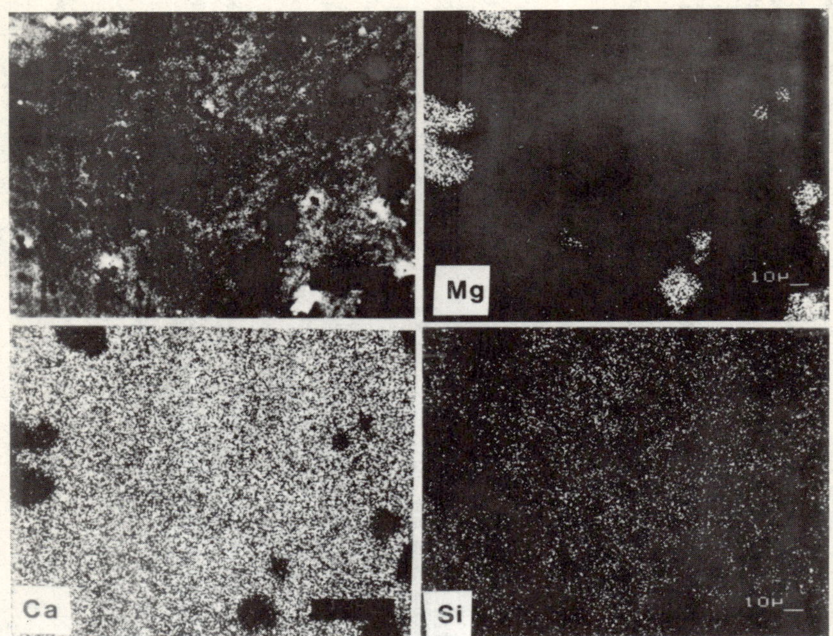

Fig. 6. SEM micrograph and EDXR maps of Ca, Mg, and Si of a slagged brick

The stability of the refractories in contact with slag was examined by the cup test technique, in which the materials were reacted for up to 5 hours under vacuum at 1700°C as opposed to the constantly changing dynamic conditions in the ladle. A photograph of the cross section of the cups after firing is shown in Fig. 7. The most evident feature is the formation of laminar fissures that have expanded to the point where the adjacent cup wall is slightly distorted. The laminar fissures are formed along planes of weakness resulting from the gases generated internally and trapped in the slag filled pores. These fissures represent structural weaknesses that would lead to increased wear by erosion/corrosion. A similar feature, but less pronounced, was also found in the altered zone of the used bricks. The microstructure of the cup in Fig. 8 shows that the slag has penetrated throughout the wall and bottom of the cup and silicates have replaced some of the periclase, leaving islands of chrome spinel in a silicate groundmass.

Fig. 7. Section through crucibles after slag infiltration

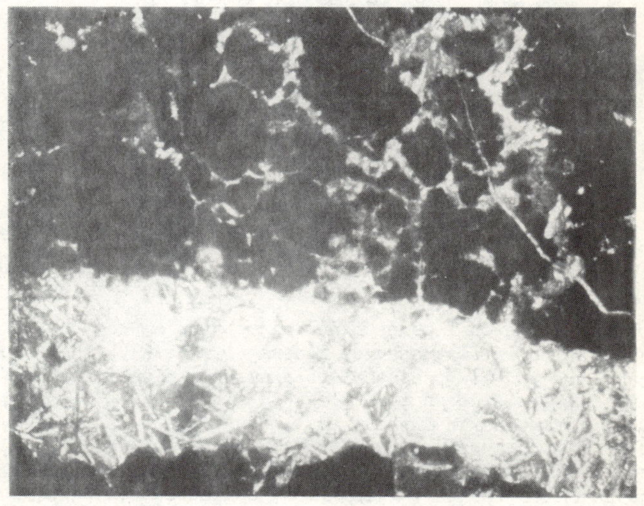

Fig. 8. Micrograph showing lath-shaped silicates formed in slagged brick

SEM micrograph and EDXR maps of the same area for Ca, Si, Al, and Fe of a slagged cup are given in Fig. 9. It can be seen that both Si and Ca are distributed at the grain boundaries of the refractories and outline the channels in which the slag migrated. Both Fe and Al tend to be found mainly in the chrome spinel grains and this suggests that these two oxides formed solid solution with the spinel phase.

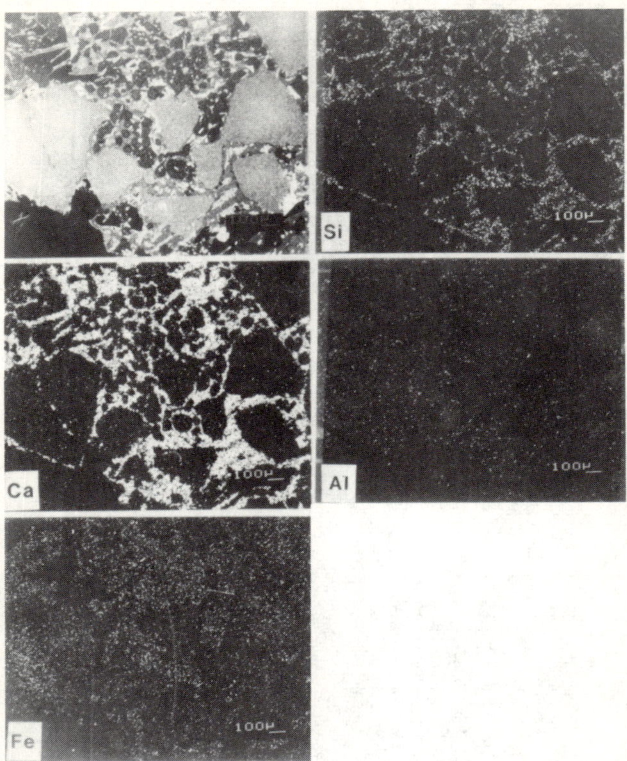

Fig. 9. SEM micrograph of EDXR map of Ca, Fe, Si and Al of a slag corroded crucible

The volatilization loss of oxide constituents from chrome magnesia refractories has been discussed by others (13,15,16). The loss of oxide at high temperature and low pressure was examined. Specimens of 1.4 x 1.4 x 0.6 mm were cut from the unused brick. A specimen was placed in the furnace in a similar manner as in the cup test, but without the presence of the slag. After the pressure of the furnace chamber was reduced to 0.13 Pa, the furnace temperature was raised to about 1700°C; samples were kept in the heated furnace for periods varying from 30 minutes to 5 hours. On completion of the heating cycle, the sample was removed from the furnace, and the weight change resulting from the heat treatment was obtained. The weight loss versus duration of heating of a series samples is graphically presented in Fig. 10. The results show that the weight loss increases almost linearly with increasing soak time. Megascopic and microscopic

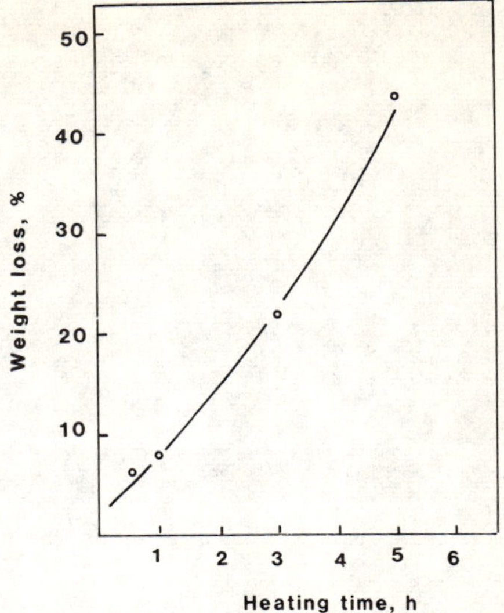

Fig. 10. Weight loss versus duration of heating at 1700°C of an unused brick

inspection of the heat treated samples all show varying degrees of surface degradation and loosening of the bonding (Fig. 11). SEM micrographs of the degraded surface are shown in Fig. 12 and 13. The characteristics

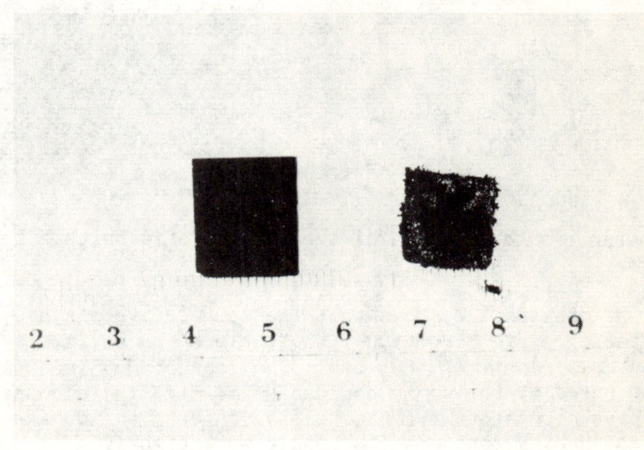

Fig. 11. Photograph showing brick prior to firing (left) and after firing at 1700°C for 5 h (right)

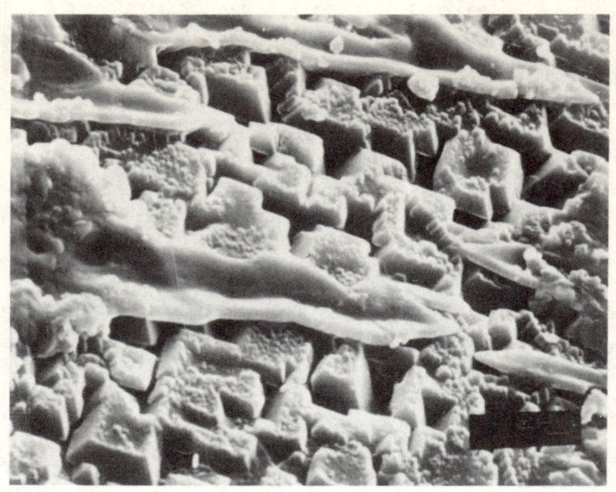

Fig. 12. SEM micrograph of the surface of a brick after firing

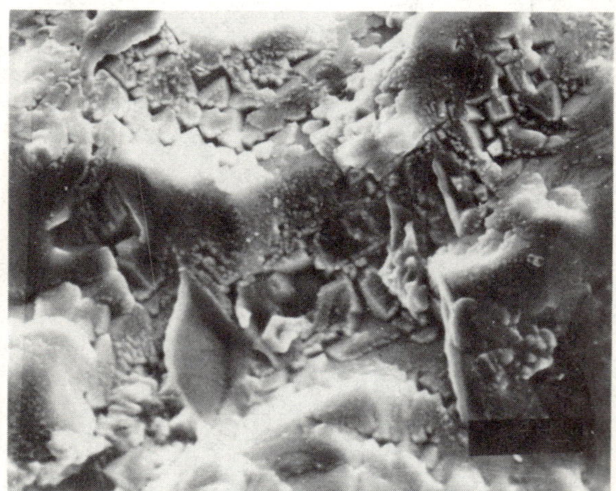

Fig. 13. SEM micrograph of the surface of a brick after firing

morphology of the surface shows that the spinel crystals are oriented in the (001) crystallographic direction with little or no periclase at the interstices as a result of the loss. The loss of MgO is apparent by the presence of pores at the grain boundaries of Fig. 13 in comparison with Fig. 14, which represents the fracture surface of an unused brick in which the periclase grains appear to be bonded more tightly.

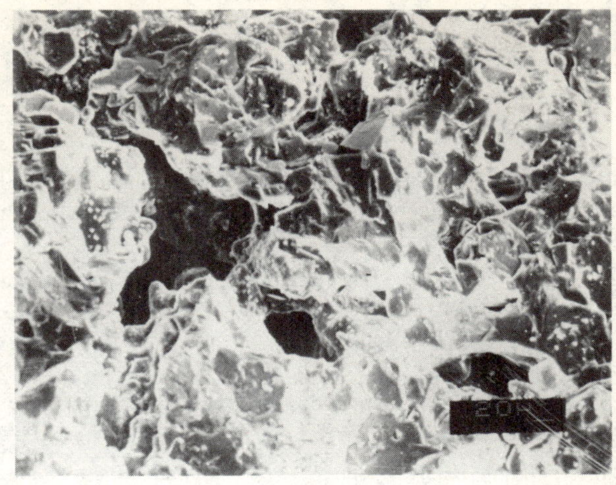

Fig. 14. SEM micrograph of a fracture surface of an unused brick

Concluding Remarks

Periclase bonded to periclase is a common texture in the direct bonded chrome magnesia brick. The lack of spinel rimming the periclase grains and spinel bonding in the brick makes it more susceptible to slag erosion and corrosion.

Optical and SEM microscopy examination of the used bricks indicate that slag penetration into the brick has caused partial dissolution of the refractories, and also confirm that the slag penetrates mainly along the grain boundaries of the spinel and periclase. The periclase grains are readily dissolved by contact with slag at high temperature. The chrome spinel is more slag resistant and tends to remain almost intact, forming islands in the groundmass of the slag.

The presence of intergranular Ca and Si phases in the direct bonded brick may also have some influence on the failure of the refractories as this phase tends to disrupt the bond of the grains, thus weakening the structure.

In vacuum at high temperature, a loss of MgO occurred and caused pores to form at the grain boundaries, opening up channels for slag to penetrate.

Acknowledgements

We wish to express gratitude to Atlas Steels Company for supplying the necessary materials for this study. Thanks are also due to D.R. Owen for performing the scanning electron microscopic examination, Y. Bourgoin for preparing thin and polished sections for this study, and to G. Lemieux for his assistance in the experimental work.

References

1. J.H. Chesters, Refractories for iron and steelmaking (The Metals Society; London; 1974), 501.

2. D.R.F. Spencer, "Development in LD refractories," Refract J, (1975), 8-26.

3. S. Alcock, and D.R.F. Spencer, "The applications and performance of basic refractories in secondary steelmaking ladles," Trans Br Ceram Soc, 77 (1978) 45-57.

4. D.R.F. Spencer, "Critical refractory properties and magnesia chrome clinker brick development for AOD application," Refract J, Jan-Feb, (1975) 10-16.

5. J.L. Evans, "Furnace design and operations - key to refractory performance. Part II. Basic oxygen steelmaking vessels and vacuum degassing," Refract J, Sept-Oct, (1980) 9-11.

6. J.R. Lakin, "Assessment techniques in refractories technology," Refract J, Nov-Dec, (1972) 7-10.

7. G.P. Carswell, M.P. Crosby, and R.F. Spencer, "Development of high performance ladle linings," Refract J, Jan-Feb, (1980) 9-20.

8. D.J. Gittins, D.R.F. Spencer, and N. Ball, "The application of basic materials in electric arc furnaces," Refract J, Sept-Oct, (1973) 12-30.

9. J. White, "An assessment of recent research on refractories (Part One)," Refract J, Nov-Dec, (1976) 10-19.

10. G.R. Rigby, F.R. Hutton, B.G. and Hamilton, "Reactions occurring in basic brick," J Am Ceram Soc, 33 (1963) 121-132.

11. T.F. Berry, W.C. Allen, and R.B. Snow, "Chemical change in basic brick during service," J Am Ceram Soc, 33 (1950) 121-132.

12. D. Binns, "The use of special ceramics in handling molten iron and steel: A review," Trans Br Ceram Soc, 77 (1978) 1-12.

13. D.H.H. Quon, and K.E. Bell, "Degradation of slagline chrome magnesia bricks in secondary steelmaking under vacuum environment: A progress report," Division Report MRP/MSL 81-83 (IR); CANMET, Energy Mines and Resources Canada; 1981.

14. B. Brezny, and R.L. Schultz, "Effect of service atmosphere in the structure changes of magnesia-chromite refractories," Open Hearth Proc, 58 (1975) 121-129.

15. V.A. Orlov, A.N. Sokolov, Yu.A. Polonskii, Ya.G. Caponov, E.P. Mezentsev, K.V. Simonov, V.S. Kavonov, S.A. Suvoror, G.I. Kuznetsov, A. Chermak, and V. Gavlik, "New refractories for plants handling the vacuum treatment of steel outside the furnace," Refractories (Translation from Russian OGNEUPORY), 21 (1980) 67-80.

16. S.A. Survorov, A.D. Mel'nikov, V.A. Orlov, K.V. Simonov, L.D. Bocharov, G.I. Kuznetsor, V.A. Pereplitsyn, and L.YA. Pvinik, "Wear of magnesia spinel refractories in production vacuum-processing steel plant," Refractories (Translated from Russian OGNEUPORY), 21 (1980) 67-80.

17. D.A. Whiteworth, F.D. Jackson, and R.F. Patrick, "Fused basic refractories in the argon-oxygen-decarburization process," Am Ceram Soc Bull, 53 (1974) 804-808.

18. K. Furumi, K. Semba, and N. Ono, "Wearing mechanism of basic brick at high-temperature in vacuum," Taikabutsu, 31 (1979) 133-137.

19. N. Naoyuki, T. Matsumura, and H. Adachi, "Analysis on direct bonded magnesite chromite brick used at slagline of steel ladle," Taikabutsu, 29 (1977) 8-12.

FRACTURE TOUGHNESS OF REINFORCED MgO-CHROMITE SYSTEMS

FOR CEMENT KILN HOT ZONES

R. E. Moore, J. L. Mendoza, P. Estrada and J. W. Ha

Department of Ceramic Engineering
University of Missouri-Rolla
Rolla, Missouri 65401

Abstract

The increase in maximum firing temperature to just above 1500°C has led to major economies in Portland Cement production costs. On the negative side has been the resultant inability to maintain hot zone lining materials presently fabricated from MgO-chrome ore materials systems. The failure mechanism of the materials is one in which altered surface regions (via reactions with the cement batch) tear away from the bulk of the liner shapes when thermally cycled. Research sponsored by an international refractories producer is designed to explore the specific effect of the reactions on relevant mechanical properties and to discover means to enhance these properties, specifically, fracture toughness. Measurements of bend strength and fracture toughness at 1500°C are essential to an understanding of the effects of cement batch reactions on liner materials. These measurements will be made after several heating and cooling cycles of liner parts which have been exposed to cement batch materials at temperature.

New liner development is to be based on a compositing approach; the incorporation of relatively long fibres oriented axially in the shapes (perpendicular to the hot face). Effects of fibre content on work of fracture and bend strength at 1500°C should demonstrate the feasibility of fibre compositing as a solution to "hot-face tearing" of these refractory systems.

Introduction

MgO-chrome refractory brick have been the material of choice for the reaction zone linings of cement kilns for several decades. They are currently being replaced by dolomite, MgO-dolomite and MgO brick due to the failure of MgO-chrome brick to provide dependable service life under more rigorous current practice. Early failure of these linings is due to incursion of cement batch chemicals which greatly weaken the brick in a shallow plane parallel to the hot face. The adherent accumulated batch forming a protective coating on the brick eventually exerts forces which cause the coating plus brick surface regions to separate from the bulk of the lining. The underlying brick are thus suddenly exposed to the maximum temperatures and atmospheres of the hot zone resulting in acceleration of the failure process.

The actual separation of brick matter is likened to a tearing away process. The most relevant high temperature resistance properties to this tearing are thought to be refractory creep and fracture toughness. The toughness may be more involved in resistance to thermal shock damage than to hot tearing. Thermal shock may occur at those times when the brick underneath the protective coating are suddenly exposed.

MgO-Chrome Refractories

Toughness of typical ceramics, e.g. polycrystalline aluminas or polyphasic porcelains is extremely low in comparison to metals. MgO-chrome refractories are polyphasic consisting of a chrome ore and/or MgO aggregate phase bonded by MgO and/or chrome ore fines. They are sintered at relatively high temperatures, above 1500°C, to densify them for service at 1550°C ± 25°C. They are compounded and fired to possess an affinity for the cement batch. Coatings up to 100 mm thickness are encouraged to develop throughout the hot zone; the coating serves to keep the cascading batch minerals away from the brick surface and to provide thermal insulation of the underlying brick. The refractories are in the form of long bricks which extend all the way back to the rotary kiln shell. The brick thus experience a rather severe thermal gradient. They are probably rigid to within 50 mm of the hot face (at the brick/coating interface). In this surface region the brick can experience significant high temperature creep and extension of cracks under the influence of the complex loading which occurs in a rotating kiln. It is essential then to know the refractory creep and fracture toughness properties of cement kiln lining brick.

Measurement of Fracture Toughness

It is possible to increase the typically low values of fracture toughness of oxide refractories by any one of four well developed methods:

a. Incorporation of fibres
b. Growing of fibres
c. Incorporation of phases which will induce internal stresses
d. Incorporation of a phase which undergoes a phase transformation, e.g. ZrO_2

Only the first two of these methods is of interest here because the third gives only moderate improvement in toughness and the fourth method is not effective at elevated temperatures. Both of the fibre methods are being attempted in this investigation but with an interesting departure from most fibre reinforcing strategies. Normally fibres are introduced in one or two-dimensional orientations parallel to the working face of a skin or structural part. In this study we plan to incorporate fibres parallel to the temperature gradient. This orientation will provide resistance to thermal shock induced crack extension due to sudden heating and should serve to counter the creep-tearing mechanism of failure.

Fracture surface energy is measured by a variety of methods. There are three aspects of this energy usually reported in $Jm.^{-2}$:

a. thermodynamic surface energy, γ_o
b. initiating fracture surface energy, γ_i
c. work of fracture, γ_f

γ_i represents the sum of all energy absorbing processes involved in crack extension. γ_f is simply defined as the work done on the specimens during the complete fracture process divided by twice the crossectional area remaining in a notched specimen. Refer to Figure 1. The shallow notched beam or the chevron notched beam are preferred for coarse-grained polycrystalline or polyphase ceramics as it is essential to have the crack extend through a sufficient 'sample' of the microstructure. Both of these beams are loaded in 3-point bending so as to slowly extend the crack to the opposite side of the beam. For such crack extension the so-called stress intensity factor K_{IC} relates to γ_i through the elastic constants, E and ν:[1]

$$K_{IC} = [\frac{2E\gamma_i}{1-\nu^2}]^{\frac{1}{2}} \quad (1)$$

The critical stress intensity factor, K_{IC}, can be viewed as a materials property integrating the elastic and surface energy properties, the latter often representing the sum of several energy absorbing mechanisms.

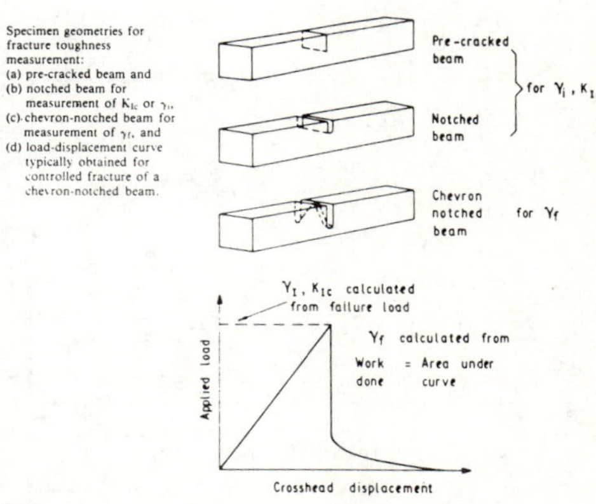

Figure 1. Specimen geometries for fracture toughness measurement.

More generally, K, the stress intensity factor is viewed by designers as a measure of the stress at the tip of a sharp crack:

$$\sigma(r,\theta) = K \cdot f(r,\theta) \tag{2}$$

where $f(r,\theta)$ is a spatial distribution function. For a sharp crack being stressed in a pure opening mode:

$$K = \sigma Y c^{\frac{1}{2}} \tag{3}$$

where Y is a constant depending on crack geometry and c measures the size of the crack.

Both relationships (1) and (3) assume homogeneity and isotropy of the medium containing the crack, conditions often not met in refractories and other ceramics. In the present study the crack extended from unaltered through altered refractory material.

Measurement of Creep

Figure 2 illustrates a typical creep curve for refractories. Region II on this figure is called the region of steady state creep. The rate of creep in this region is measured at fixed load or at fixed temperature and relates to other aspects of the creep process through:[2]

$$\dot{\varepsilon} = A\sigma^{b}e^{-\Delta H/RT} \tag{4}$$

where $\dot{\varepsilon}$ = strain rate, steady state
A = texture-related constant
σ = applied stress
n = stress exponent
ΔH = activation energy for the particular creep mechanism(s)

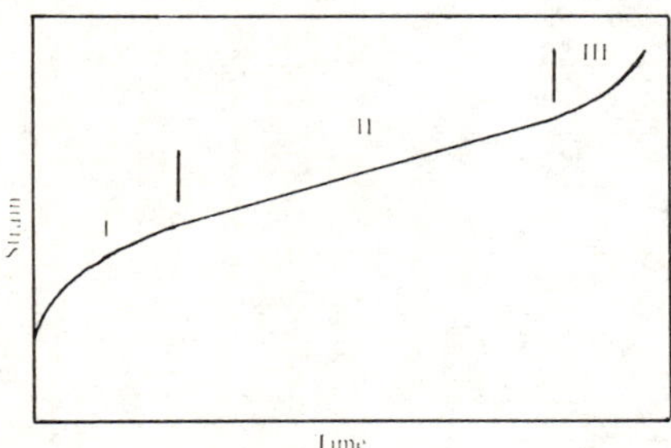

Figure 2. A schematic creep curve generally applicable to refractories.

In this study, both the stress exponent, n, and the creep activation energy, ΔH were measured. It must be emphasized that these quantities can only receive clear interpretation when they are measured on materials wherein a single phase is involved in creep. e.g. pure polycrystalline oxides or polyphase materials wherein only a pure matrix phase is creeping. In these cases the values of n and ΔH relate to specific creep mechanisms. The mechanisms most frequently deduced active in conventional refractories are viscous flow or diffusional responses. The stress exponent is equal to 1 for both of these mechanisms. Refractories with complex microstructures often yield values of $n < 1$, a result which suggests the structure was changing during the creep process.

Conduct of Investigation

1. Sample description and preparation.

The MgO-chrome refractory was obtained from a Mexican producer. Major oxide concentrations typical for this product are: MgO-63%, Cr_2O_3-15%, Al_2O_3-11%, Fe_2O_3-7%, CaO-2% and SiO_2-1.5%. It contained MgO and chrome ore grains as aggregate and MgO fines as matrix phase. The fired brick were approximately 17% porous. They were cut into 3" x 1" x 1" specimens for fracture toughness determinations and 1" x 1" x 1" specimens for measurements of steady state creep as a function of stress and temperature. To simulate service effects, half the specimens were impregnated with cement batch slurry and subsequently fired to 1550°C to react the enfused batch with the refractory. A simple weight change calculation shows that the 1 mm deep impregnated surface layer picked up about 4 weight percent of cement (fired basis). Figures 3 and 4 are photomicrographs of the impregnated specimens before and after firing. The reaction glass can be seen in most of the boundaries and triple-point regions between MgO matrix grains in Figure 4. The cement batch, unreacted, appears as a groundmass material in Figure 3.

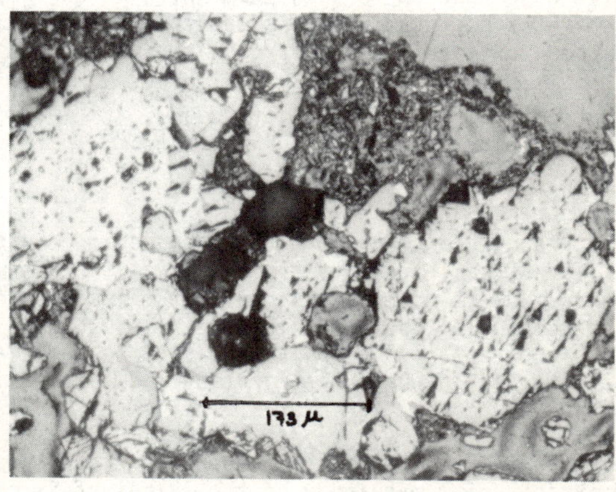

Figure 3. MgO-chrome refractory with impregnated cement batch materials before firing, 150X.

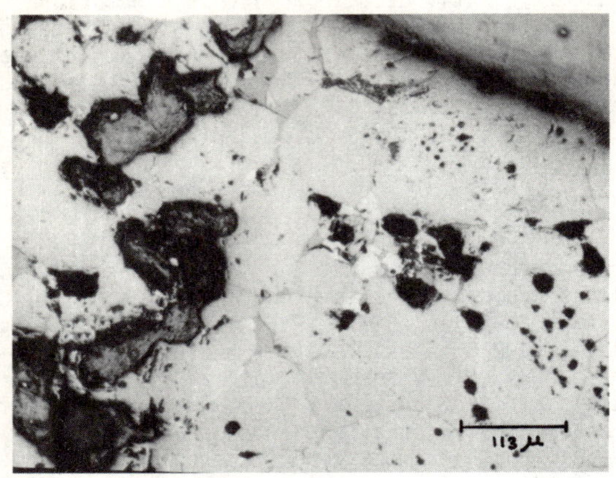

Figure 4. MgO-chrome refractory, impregnated with cement batch and fired to 1550°C, 150X.

The creep specimens were sawed to have the end faces flat and parallel; the fracture surface energy specimens were notched to various depths employing a diamond dicing saw. Creep was performed in a compressive mode. Specimens were tested with and without impregnation at various stresses and temperatures to obtain values of n and ΔH. Fracture surface energy specimens were notched so that the cracks terminated in the impregnated regions. The stress strain curves were similar to the schematic curve in Figure 1 and were used to calculate γ_f for the impregnated and unimpregnated refractories.

Experimental Results and Conclusions

Creep Experiments

Figure 5 shows results for the impregnated and unimpregnated specimens tested at 1360°C. The stress exponents are essentially the same. As the values are well below 1.0, it must be assumed that the microstructure is changing during the conduct of the experiment. This result is a very common one when creep tests are performed on polyphase ceramic materials. It has been suggested that changes in phase distribution and pore geometry are responsible.[3]

Figure 6 indicates that the activation energy for creep was not affected by impregnation. The magnitude of the ΔH values is consistent with a viscous flow or diffusion mechanism, as should be expected. Figure 7 reveals that the stress exponent is highly temperature dependent. The data were taken at 1450°C, right at the limit for obtaining steady-state creep. This fact suggests that the refractory under study may have an unusually low resistance to creep as the process service temperature approaches 1550°C.

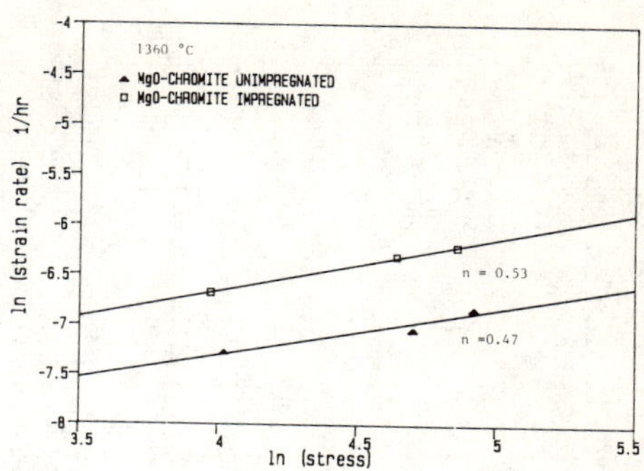

Figure 5. Stress exponent results at 1360°C at a stress of 56 psi.

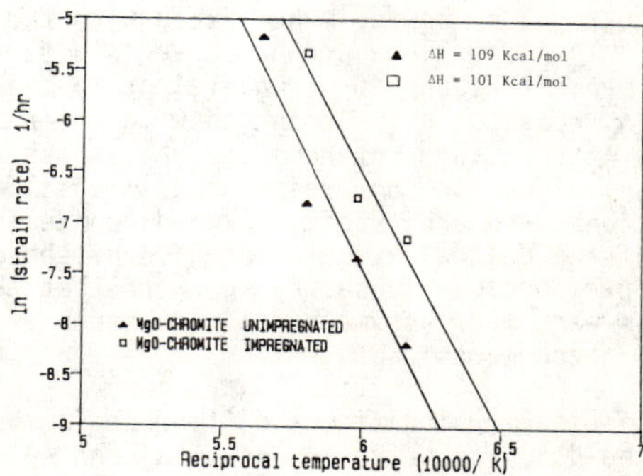

Figure 6. Activation energies for impregnated and unimpregnated refractories, measured at 56 psi.

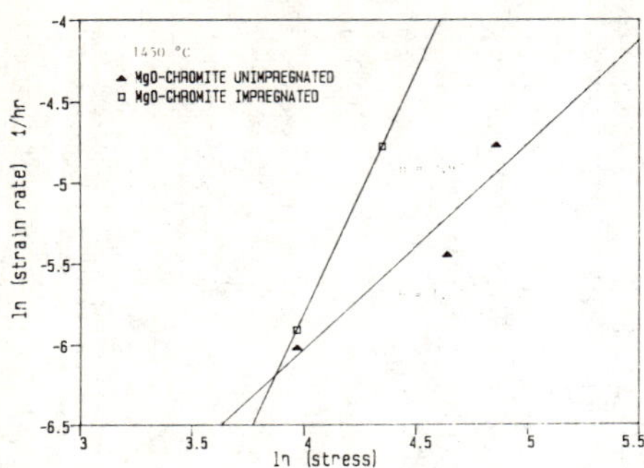

Figure 7. Stress exponent results at 1450°C at a stress of 56 psi.

Fracture Surface Energy Studies

Figure 8 is a schematic of a typical load-deflection curve for the impregnated and unimpregnated MgO-chrome ore specimens. Table I contains calculations of γ_f, the slopes of the load-deflection curves and C/A and X, measures of the degree of right-hand skewness of the curves. The values in the table are averages of 2 or 3 specimens listed with the respective standard deviations. The left slope of the curve is a direct measure of the stiffness of a specimen. As such, it is equivalent to the Young's modulus of elasticity. The γ_f results in in-lb/in^2 show the refractory to be very lacking in toughness at room temperature. The values are about an order of magnitude larger than values for commercial soda-lime-silica window glass. The skewness coefficients are arbitrarily defined as the fractional area lying to the right of the central symmetric region of the deflection curve. This measure was adopted because it seemed apparent that the curves for the impregnated specimens were more assymetrical than those of the unimpregnated specimens at the 20% notch depths.

There is a statistically significant difference in the stiffness (slope) values for the data for notch depths of 30 and 40%. The impregnated specimens are less stiff which could result from physical as well as chemical effects of the impregnation. Lowered stiffness could result from a higher level of microcracking due to the greater complexity of phases in the impregnated specimens. There is no statistically significant difference between the γ_f values. This may only reflect the small sample sizes and the rather large values of the standard deviations. Larger sample sizes need to be used in the future to discern difference in γ_f reflection of impregnation by cement batch.

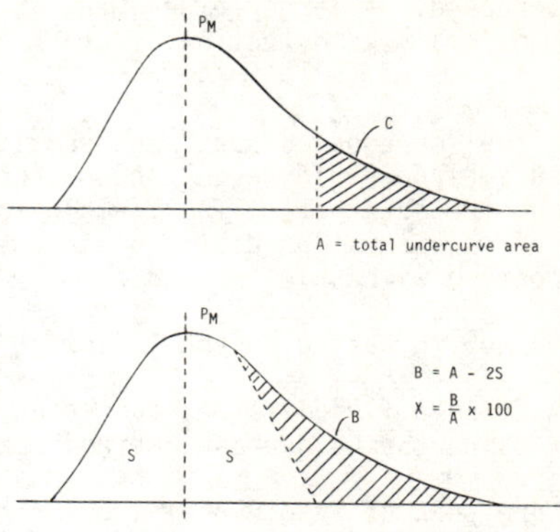

Figure 8. Schematics of the load-deflection curves illustrating the measurement of skewness coefficients, C/A and X.

Table I. Fracture Toughness Parameters for MgO-Chrome Specimens, Room Temperature.

Parameter	Notch Depth, %	Impregnated	Unimpregnated
slope (10^3 lb/in)	40	3.7 ± 0.2	4.6 ± 0.6
	30	3.8 ± 1.4	5.7 ± 3.3
	20	4.9 ± 0.7	4.7 ± 0.5
γ_f (10^{-3} in-lb/in^2)	40	46.8 ± 3.6	46.4 ± 0.2
	30	44.9 ± 3.6	47.9 ± 0.4
	20	54.8 ± 8.7	53.4 ± 11.9
C/A (%)	40	47.8 ± 3.0	49.6 ± 2.6*
	30	40.4 ± 9.1	46.2 ± 0.3
	20	37.3 ± 12.1	25.6 ± 7.5
X (%)	40	52.8 ± 3.7	56.5 ± 1.5*
	30	47.1 ± 11.5	52.5 ± 1.5
	20	42.0 ± 18.9	27.3 ± 9.9

*Refer to Figure 8 for calculation of this quantity.

Figure 8 illustrates the analysis method and Table I shows the skewness coefficient results for various notch depths. Note the systematic variation in the standard deviations. The 40% notch depth data are much lower in precision than the other two sets. The skewness coefficients were higher only for the 20% notch depths. Impregnation by the cement batch dramatically increases the skewness coefficients, c/a or x, relative to the coefficients for unimpregnated specimens. This effect is probably due to the slowing of the rate of crack extension with the longer crack path length extending into the batch impregnated regions. The stiffness data showed that the surface regions of impregnated specimens are less stiff and perhaps more extensively microcracked. A larger experiment is underway to verify whether the 20% depth notch allows detection of crack slowdown in altered surface regions.

The γ_f values for the impregnated specimens should be affected by the presence of the altered surface region even though there appeared to be no difference in γ_f between the data sets. The formula for K_{IC} is functional with E, ν and γ_f each of which must be different in the impregnated zone. The data must be interpreted with this in mind.

Investigations are now focussing on the following measurements:

1. Thermal shock using a fluidized bed quenching furnace.
2. Thermal shock using the Holderbank[4] method specific for cement kiln brick.
3. High temperature work of fracture of impregnated and non-impregnated specimens.
4. Fibre incorporation studies in MgO and MgO-chrome refractories.

Fibre reinforcement methods are to include attempts to grow in fibres and to mix in fibres. The incorporation of acetates as carbon fibre precursors will be attempted first employing MgO formulations to avoid reduction of the chrome ore constituents. The effects of chrome ore concentration on fibre development and the assessment of fiberizing of the refractories represent the ultimate goals of the study. It represents an effort to extend compositing of refractories to much higher temperatures.

References

1. R. Morrell, Handbook of Properties of Technical and Engineering Ceramics, Part 1: An Introduction for the Engineer and Designer, (National Physical Laboratory, Her Majesty's Stationery Office, London, 1985), 119-125.

2. R. W. Davidge, Mechanical Behavior of Ceramics, (Cambridge University Press, Cambridge, 1979), 71-74.

3. T. D. McGee, "Creep of Castable Refractories" (Proceedings of the 18th Annual Symposium on Refractories, St. Louis Section of the American Ceramic Society, 1982).

4. B. Oberfeuer and M. Koltermann, Untersuchung und Beurteilung von Tonerdesilikststeinen mit mehr Als 50% Al_2O_3 fur den Einsatz in der Eisenhuttenindustrie, (Radex-Rundschau), Vol. 3, October 1976, 749-750.

Subject Index

Mineral Index

Locality Index

$98=00